T0144603

REGULATION OF SLEEP AND CIRCADIAN RHYTHMS

LUNG BIOLOGY IN HEALTH AND DISEASE

Executive Editor

Claude Lenfant
Director, National Heart, Lung and Blood Institute
National Institutes of Health
Bethesda, Maryland

ADDITIONAL VOLUMES IN PREPARATION

The opinions expressed in these volumes do not necessarily represent the views of the National Institutes of Health.

REGULATION OF SLEEP AND CIRCADIAN RHYTHMS

Edited by

Fred W. Turek

Northwestern University
Evanston, Illinois

Phyllis C. Zee

Northwestern University
Chicago, Illinois

New York London

Informa Healthcare USA, Inc.
52 Vanderbilt Avenue
New York, NY 10017

© 2007 by Informa Healthcare USA, Inc.
Informa Healthcare is an Informa business

No claim to original U.S. Government works
Printed in the United States of America on acid-free paper
10 9 8 7 6 5 4

International Standard Book Number-10: 0-8247-0231-X (Hardcover)
International Standard Book Number-13: 978-0-8247-0231-1 (Hardcover)

Visit the Informa Web site at
www.informa.com

and the Informa Healthcare Web site at
www.informahealthcare.com

INTRODUCTION

The physiology of sleep has been of interest for millennia. In *Medicine: An Illustrated History*, Lyons and Petrucelli (1) reported that the philosopher–scientists of the pre-Hippocratic medicine period studied the mechanisms of sleep. In his book *Concerning Nature*, Alemeon, who lived in the fifteenth century, B.C., described the connection between the sense organs and the brain. Lyons and Petrucelli also wrote that Alemeon "speculated that sleep occurred when the blood vessels in the brain were filled; withdrawal of the blood from the brain caused wakening."

Of course, such a simplistic explanation of the origin of the sleep-wake cycle holds no more. Today, we know that sleepiness and wakefulness are determined by a higher organization—our biological clock—and that circulatory variations are a consequence, rather than the cause, of the sleep-wake cycle. Nonetheless, the interplay between all of these functions is very important, as disruption of the sleep pattern affects the circulatory, respiratory, and other functions. That is why this volume, with a title that may sound foreign to lung biologists, is included in the Lung Biology in Health and Disease series of monographs.

The series has already included two volumes on sleep and breathing. This particular one is the first of a subseries that will present many aspects of sleep biology. Although these future volumes will focus mostly on the physiology and pathology of sleep, I thought it was critical to provide the readership of this series of monographs with a comprehensive picture of the state of the art of the mechanism of sleep. That I was able to obtain the participation of Drs. Turek and Zee was an unbelievable coup. They sought and secured the contributions of the foremost experts in the area of regulation of sleep and circadian rhythms. Any clinical physician who sees patients with sleep disorders should read this volume.

As the Executive Editor of this series of monographs, I am grateful to the editors and authors for the opportunity to present this volume.

Claude Lenfant, M.D.
Bethesda, Maryland

Reference

1. Lyons AS, Petrucelli RJ. Medicine: An Illustrated History. New York: Harry H. Abrams, 1987.

PREFACE

One of the distinguishing characteristics of sleep throughout much of the animal kingdom is that the periods of sleep and wake occur at specific times of the day and/or night. In mammals, the central circadian (i.e., 24 hr) pacemaker that regulates the timing of sleep and wake as well as most, if not all, 24-hr rhythms, is located in a small region of the hypothalamus (the suprachiasmatic nucleus [SCN]), whereas the control of sleep and wake per se appears to involve many diverse regions of the brain. Although the expression of many 24-hr rhythms may be primarily under the control of the circadian clock in the SCN, many other rhythms are largely dependent on whether the organism is asleep or awake, regardless of the circadian clock time. Thus, the expression of most 24-hr rhythms at the behavioral, physiological, and biochemical levels depends on the integration of inputs from the circadian clock and sleep-wake state of the animal.

The search for the underlying mechanisms for the regulation of sleep and circadian rhythms has, in general, been carried out via two independent lines of research: one line focused on deep sleep mechanisms, the other on circadian mechanisms. A major impetus for producing the present volume is the working hypothesis that the circadian and sleep-wake regulatory systems are highly integrated with one another and that unraveling the linkage between them at various levels of organization will lead to a better understanding of the fundamental mechanisms underlying both systems. Thus, this monograph provides up-to-date reviews of what is known about the regulatory processes underlying sleep and circadian rhythms in anticipation that it will be an important reference source, not only for those who would like an overview of the sleep and circadian fields in one source, but also for individuals who seek to build on the present state of knowledge to elucidate the basic regulatory mechanisms underlying sleep and circadian rhythmicity.

Humans, unlike all other species, are often awake when their internal biological clock is telling them it is time to sleep, and humans are often trying to sleep when the circadian clock is sending signals throughout the brain and body that it is time to be awake. That is, humans routinely cognitively "override" their internal circadian clock so their timing of sleep and wake can be scheduled to meet the personal demands of their social and work schedules. In addition to this voluntary disruption of the normal temporal organization between the sleep and circadian clock systems, such disorganization can occur on an "involuntary" basis. Temporal disorganization can have severe consequences for human health, safety, performance, and productivity. In addition to providing an overview of the regulatory mechanisms that underlie sleep and circadian rhythmicity, this monograph provides an overview of the consequences of disruptive sleep and rhythmicity. These two overall objectives are presented within and between chapters in the framework that the sleep and circadian clock systems are highly integrated with one another. The publication of this volume comes at a very exciting time for the field of sleep and circadian rhythms. At a time when exciting discoveries of the circadian and sleep mechanisms are being made at a rapid pace, the medical profession and the public in general are becoming more and more aware of the importance of good sleep and overall temporal organization for human health and well-being.

Fred W. Turek
Phyllis C. Zee

CONTRIBUTORS

Donald L. Bliwise, Ph.D. Sleep Disorders Center, Wesley Woods Geriatric Hospital, and Associate Professor, Department of Neurology, Emory University School of Medicine, Atlanta, Georgia

Daniel J. Buysse, M.D. Associate Professor, Department of Psychiatry, University of Pittsburgh School of Medicine, Pittsburgh, Pennsylvania

Scott S. Campbell, Ph.D. Professor, Department of Psychiatry, Cornell University Medical College, and Institute for Circadian Physiology, White Plains, New York

Julie Carrier, Ph.D.* Postdoctoral Fellow, Department of Psychiatry, Western Psychiatric Institute and Clinic, University of Pittsburgh School of Medicine, Pittsburgh, Pennsylvania

---

Current affiliation: Assistant Professor in Research, Department of Psychiatry, University of Montreal, Montreal, Quebec, Canada.

Charles A. Czeisler, M.D., Ph.D. Professor of Medicine, Circadian, Neuro-endocrine, and Sleep Disorders Section, Department of Medicine, Brigham and Women's Hospital, Harvard Medical School, Boston, Massachusetts

Fred C. Davis, Ph.D. Associate Professor, Department of Biology, North-eastern University, Boston, Massachusetts

Derk-Jan Dijk, Ph.D. Assistant Professor of Medicine (Neuroscience), Circa-dian, Neuroendocrine, and Sleep Disorders Section, Department of Medicine, Brigham and Women's Hospital, Harvard Medical School, Boston, Massachusetts

Dale M. Edgar, Ph.D. Associate Professor, Department of Psychiatry and Behavioral Sciences, Stanford University School of Medicine, Stanford, California

Jidong Fang, M.D., Ph.D. Department of Veterinary and Comparative Anat-omy, Physiology, and Pharmacology, College of Veterinary Medicine, Washing-ton State University, Pullman, Washington

Rachael A. Floyd, Ph.D. Department of Physiology and Biophysics, University of Tennessee–Memphis, Memphis, Tennessee

Marcos G. Frank, Ph.D. Department of Physiology, University of California–San Francisco, San Francisco, California

Zoran M. Grujic, M.D. Clinical Instructor, Department of Neurology, North-western University Medical School, Chicago, Illinois

H. Craig Heller, Ph.D. Professor, Department of Biological Sciences, Stanford University, Stanford, California

J. Allan Hobson, M.D. Professor, Department of Psychiatry, Harvard Medical School, Boston, Massachusetts

Matcheri S. Keshavan, M.D. Professor, Department of Psychiatry, University of Pittsburgh School of Medicine, Pittsburgh, Pennsylvania

Thomas S. Kilduff, Ph.D. Center for Sleep and Circadian Neurobiology, De-partments of Biological Sciences and Psychiatry and Behavioral Sciences, Stan-ford University, Stanford, California

James M. Krueger, Ph.D. Professor and Chair, Department of Veterinary and Comparative Anatomy, Pharmacology, and Physiology, Washington State Uni-versity, Pullman, Washington

David J. Kupfer, M.D. Thomas Detre Professor and Chair, Department of Psychiatry, Western Psychiatric Institute and Clinic, University of Pittsburgh School of Medicine, Pittsburgh, Pennsylvania

Emmanuel Mignot, M.D., Ph.D.   Sleep Disorders Center, and Associate Professor, Departments of Psychiatry and Behavioral Sciences, Stanford University, Stanford, California

Timothy H. Monk, Ph.D., D.Sc.   Professor, Department of Psychiatry, Western Psychiatric Institute and Clinic, University of Pittsburgh School of Medicine, Pittsburgh, Pennsylvania

Eric A. Nofzinger, M.D.   Assistant Professor, Department of Psychiatry, University of Pittsburgh School of Medicine, Pittsburgh, Pennsylvania

Tarja Porkka-Heiskanen, M.D., Ph.D.*   Department of Psychiatry, Harvard University, Boston, and Neuroscience Laboratory, Brockton VA Medical Center, Brockton, Massachusetts

Charles F. Reynolds III, M.D.   Professor, Department of Psychiatry, University of Pittsburgh School of Medicine, Pittsburgh, Pennsylvania

William J. Schwartz, M.D.   Professor, Department of Neurology, University of Massachusetts Medical School, Worcester, Massachusetts

Karine Spiegel, Ph.D.   Research Associate, Department of Medicine, University of Chicago, Chicago, Illinois

Dag Stenberg, M.D., Ph.D.   Professor, Department of Physiology, Institute of Biomedicine, University of Helsinki, Helsinki, Finland

Joseph S. Takahashi, Ph.D.   Walter and Mary E. Glass Professor, Department of Neurobiology and Physiology, Howard Hughes Medical Institute, Northwestern University, Evanston, Illinois

Fred W. Turek, Ph.D.   Professor, Center for Circadian Biology and Medicine, Department of Neurobiology and Physiology, Northwestern University, Evanston, Illinois

Eve Van Cauter, Ph.D.   Professor, Department of Medicine, University of Chicago, Chicago, Illinois

David R. Weaver, Ph.D.   Associate Neurobiologist, Laboratory of Developmental Chronobiology, Pediatric Service, Massachusetts General Hospital, and Associate Professor of Pediatrics, Harvard Medical School, Boston, Massachusetts

Thomas A. Wehr, M.D.   Chief, Section of Biological Rhythms, National Institute of Mental Health, Bethesda, Maryland

---

Current affiliation: Senior Researcher, Department of Physiology, Institute of Biomedicine, University of Helsinki, Helsinki, Finland.

Jonathan P. Wisor, Ph.D. Sleep Research Center, Department of Psychiatry, Stanford University School of Medicine, Palo Alto, California

Kenneth P. Wright, Jr., Ph.D. Research Fellow in Medicine, Circadian, Neuro-endocrine, and Sleep Disorders Section, Department of Medicine, Brigham and Women's Hospital, Harvard Medical School, Boston, Massachusetts

Phyllis C. Zee, M.D., Ph.D. Associate Professor of Neurology, Departments of Neurology, Neurobiology, and Physiology, Northwestern University, Chicago, Illinois

Piotr Zlomanczuk, Ph.D. Department of Physiology, Rydygier Medical School, Bydgoszcz, Poland, and Department of Neurology, University of Massachusetts Medical School, Worcester, Massachusetts

CONTENTS

16. Sleep and Circadian Rhythm Disorders in Aging and Dementia   487

Donald L. Bliwise

17. Effects of Sleep and Circadian Rhythms on Performance   527

Julie Carrier and Timothy H. Monk

1

Introduction to Sleep and Circadian Rhythms

PHYLLIS C. ZEE

Northwestern University
Chicago, Illinois

FRED W. TUREK

Northwestern University
Evanston, Illinois

I. Introduction

One of the most obvious adaptive features of living organisms on earth is the ability of almost all species to change their behavior on a daily or 24-hr basis. In particular, many animals are active and awake only during certain times of the 24-hr day, and they are inactive (resting and/or sleeping) at other times of the day. Such changes in the rest-activity/sleep-wake cycle are highly regimented within a given species, whereas the time and duration of rest and activity times vary greatly between species. Daily changes in life-style are, of course, correlated with the dramatic changes that take place in the physical environment due to the rotation of the earth on its axis. Although not as readily apparent as the behavioral changes, just about every aspect of the internal environment of the organism also undergoes pronounced fluctuations over the course of the 24-hr day. Claude Bernard's concept of homeostasis, where the "milieu interieur" remains constant in living organisms, has been modified over the last 30–40 years to take into account the fact that regular and predictable 24-hr changes in the internal milieu are a hallmark of most living organisms.

A remarkable feature of the daily rhythms that are observed in organisms as

1

diverse as algae, fruit flies, and humans is that they are not simply a response to the 24-hr changes in the physical environment imposed by the principles of celestial mechanics, but instead arise from an internal time-keeping system (1). This time-keeping system, or biological clock(s), allows the organism to predict and prepare in advance for the changes in the physical environment that are associated with night and day. Thus, the organism adapts, both behaviorally and physiologically, to meet the challenges associated with the daily changes in the external environment, and there is temporal synchronization between the organism and the external environment. The most obvious example of such an adaptation to the physical environment is the finding that many animals are active only during the light period (diurnal species) or the dark period (nocturnal species) and are inactive during the other part of the day. Such "external synchronization" is of obvious importance for the survival of the species and insures that the organism does the "right thing" at the right time of the day. Of equal, but perhaps less appreciated, importance is the fact that this biological clock, like a conductor of a symphony orchestra, provides internal temporal organization and insures that internal changes take place in coordination with one another. Just as living organisms are organized spatially, they are also organized temporally to insure that there is "internal synchronization" between the myriad of biochemical and physiological systems in the body. While lack of synchrony between the organism and the external environment may lead to the immediate demise of the individual, e.g., as would be expected if a nocturnal rodent attempted to navigate the hazards of the diurnal world, lack of synchrony within the internal environment may lead to chronic difficulties with equally severe consequences for the health and well-being of the organism.

In mammals, the period of inactivity is often associated with the timing of sleep as defined by characteristic patterns in brain electroencephalographic (EEG) activity and behavior. Sleep and wakefulness are thought to be regulated by three basic processes: (1) a circadian process, which is driven by the central circadian clock regulating most, if not all, rhythms in mammals and is located in the suprachiasmatic nucleus of the hypothalamus, resulting in alternating cycles of high and low sleep propensity; (2) a homeostatic process, which is determined by prior sleep and wakefulness and, under normal conditions, modulates the circadian sleep propensity; and (3) an ultradian process within sleep, which defines the cycles of the two basic sleep states of non–rapid eye movement (NREM) and REM sleep (2–6). Although the sleep homeostatic and circadian processes are separate, it is their interaction that determines the temporal distribution and duration of sleep and wakefulness. Therefore, alterations in circadian rhythmicity accompany alterations in sleep. This chapter will discuss some of the general concepts that pertain to sleep and circadian rhythms and serves as an introduction for the remaining chapters in this volume.

II. Circadian Rhythms

A. Endogenous Self-Sustained Clocks Drive Circadian Rhythms

Under laboratory conditions devoid of any external time-giving cues from the physical environment, it has been found that just about all diurnal rhythms that are present under natural conditions continue to be expressed in the laboratory (7–9). However, under constant environmental conditions, the period of the rhythm rarely remains exactly 24 hr but instead in "about" 24 hr. Because the period of diurnal rhythms is close to, but not exactly, 24 hr, they are referred to as "circadian rhythms," from the latin *circa diem*, meaning "about a day." When a circadian rhythm is expressed in the absence of any 24-hr signals in the external environment, it is said to be "free-running"; i.e., the rhythm is not synchronized or entrained by any cyclic change in the physical environment. For a population of animals within a given species, the period of a given rhythm (e.g., drinking, body temperature, or locomotor activity) will be different between animals, but in general, all will lie in close proximity to 24 hr (e.g., from 23 to 25 hr). Strictly speaking, a diurnal rhythm should not be referred to as "circadian" until it has been demonstrated that such a rhythm persists under constant environmental conditions. The purpose of this distinction is to separate out those rhythms that are simply a response to 24-hr changes in the physical environment from those that are driven by some internal time-giving system. However, for practical purposes, there is little reason to make a distinction between "diurnal" and "circadian" rhythms as almost all diurnal rhythms expressed under natural conditions are found to persist under constant environmental conditions in the laboratory. Consequently, the term "circadian" is often used to refer to diurnal rhythms that are observed under either natural or laboratory conditions.

Genetic, physiological, and behavioral experiments have established that the timing system that underlies the generation of circadian rhythms is endogenous to the organism itself (see Chapter 13). That is, there is a circadian clock or clocks within the organism that somehow regulates the 24-hr fluctuations in diverse physiological and behavioral systems. Although literally thousands of rhythms have been monitored in plants and animals, it has not been possible to assay the state of a circadian clock directly in any experimental model to date. Thus, attempts to understand the properties of circadian clocks focus on the "hands" of the clock, i.e., the expression of overt rhythms regulated by the clock. While the list of biochemical and physiological processes that show circadian fluctuations is enormous, a few select behavioral rhythms (e.g., locomotor activity, drinking) are most often utilized to characterize the basic features of the clock system in mammals. Behavioral rhythms are utilized because of their ease of measurement for many cycles without disturbing the animal. Figure 1 provides

Time of Day

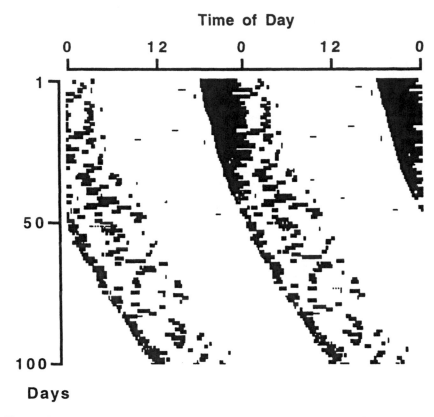

Figure 1 Continuous record of the circadian rhythm of wheel-running behavior in a golden hamster maintained in constant darkness for 100 days. Each revolution of the running wheel was recorded on-line via a personal computer. Black bars represent periods of activity. Successive days are plotted from top to bottom, and the activity record has been double-mounted on a 48-hr time interval to aid in the visualization of data. This record shows the remarkable precision of the activity rhythm under free-running conditions devoid of any timing cues. The onset of activity shows a period slightly greater than 24 hr throughout the 100 days, with deviations from the mean period being only a few minutes between any 2 days. The decrease in testicular size (and its associated decrease in serum testosterone levels), due to exposure to constant darkness, leads to a decrease in the total amount of activity per day during the latter half of this record.

an example of the circadian rhythm of locomotor activity in a male golden hamster held under free-running conditions for 100 days. Not only can this rhythm be monitored for essentially the lifetime of the animal without any interference of the sampling procedure on the rhythm itself, but automated sampling systems allow one to access the state of this rhythm continuously on essentially a minute-to-minute basis. For practical and economic reasons, such long-term and frequent sampling of biochemical and physiological rhythms is often not possible.

B. Entrainment of Circadian Clocks: Control of Period and Phase

The fact that the period of circadian rhythms is not exactly equal to that of the period of the rotation of the earth on its axis demands that 24-hr changes in the physical environment must somehow synchronize or entrain the internal clock system regulating circadian rhythms. Otherwise, even a clock with a period only a few minutes shorter or longer than 24 hr would soon be totally out of synchrony with the environmental day. An endogenous circadian clock that could not be reset by environmental signals would be of little use to organisms that need to time specific activities to particular times of the day. As discussed below, the light-dark (LD) cycle is clearly the major environmental entraining agent of circadian rhythms, although there is substantial evidence that other internal and external factors can influence how circadian rhythms are synchronized to the physical environment. It should be noted that while other stimuli might influence the expression of particular circadian rhythms by influencing a process between the endogenous clock and the effector systems (i.e., masking may occur), the focus in this section is on those agents that can control the phase of the circadian clock itself.

Entrainment by Light-Dark Cycles

Except in a few exotic species living in unusual environments (e.g., blind cave fish or eyeless mole rat), the light-dark cycle appears to be the primary environmental agent that synchronizes circadian rhythms to the 24-hr environmental cycle in the physical environment (7,9,10). Thus, in the presence of a 24-hr light-dark cycle the period of circadian rhythms exactly matches the period of the light-dark cycle (Fig. 2). From one day to the next, the time between successive recurrences of specific phase points of a rhythm (e.g., the onset of locomotor activity, the minimum in body temperature) is the same as the period or duration of the light-dark cycle. In addition to establishing "period control," an entraining light-dark cycle establishes "phase control" such that specific phases of the circadian rhythm occur at the same times in each cycle relative to the entraining agent. For example, in a hamster entrained to a 14:10 LD cycle (i.e., 14 hr of light followed by 10 hr of darkness every 24 hr), the onset of the main bout of daily activity always occurs within a few minutes after lights off, day after day (Fig. 2). Following a phase shift in the LD cycle, the rhythm reentrains (Fig. 2), although the development of a steady-state phase relationship between the circadian rhythm and the entraining LD cycle often takes many days owing to the limitation on the number of hours per day that the internal circadian clock can be phase-shifted by light (see below).

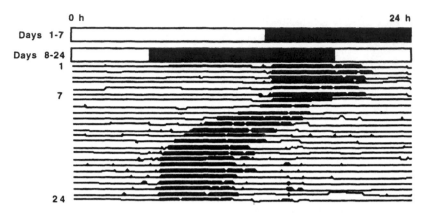

Figure 2 Daily rhythm of wheel-running behavior in a golden hamster exposed to a 14:10 light-dark cycle, which was phase-advanced by 8 hr on day 8 of this record. Timing of light-dark cycles before and after the advance in the light cycle is diagrammed at the top of the record. The animal was entrained to the initial cycle such that the onset of activity occurred within a few minutes of lights-off each day. Following the shift in the cycle, it took about 10 days for the animal to become reentrained to the new lighting schedule.

Although circadian rhythms can be entrained to LD cycles that are not exactly 24 hr in duration, entrainment is restricted to cycles with periods that are "close" to 24 hr in duration (7,9). The "range of entrainment" can vary from species to species, and is dependent on the experimental conditions (e.g., intensity of the LD cycle, whether the period of the LD cycle is changed gradually or rapidly), but in general animals do not entrain readily to LD cycles that are more than a few hours shorter or longer than the period of the endogenous free-running circadian rhythm. If the period of the LD cycle is too short or long for entrainment to occur, the circadian rhythm will free run with a period close to 24 hr.

For the past 35 years, one of the most widely used methods to examine how the LD cycle influences the circadian system has been to expose animals maintained in constant darkness (DD) to a brief pulse of light (e.g., 1–60 min in duration), and to return the animals to DD (1,11). The effects of the light pulse on a phase reference point of a circadian rhythm (e.g., onset of locomotor activity, minimum of body temperature) in subsequent cycles is then determined. This approach has demonstrated that light pulses can induce phase advances or phase delays or have no effect on free-running circadian rhythms. The direction and magnitude of the shifts are strongly dependent on the circadian time at which the light pulse occurs (Fig. 3). A plot of the phase shift induced by an environmental perturbation as a function of the circadian time at which the perturbation is given is called a "phase response curve" (PRC). Light pulse PRCs for all organisms

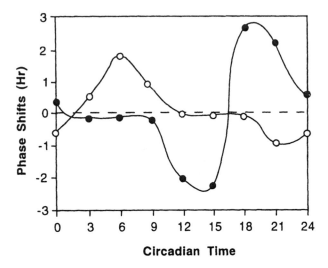

Figure 3 Schematic representation of two generalized phase response curves (PRC) for the phase-shifting effects on the circadian clock regulating the rhythm of locomotor activity following the presentation of either single pulses of light (5–60 min in duration) or stimuli that induce an acute increase in activity (e.g., exposure to a novel running wheel or injection of a short-acting benzodiazepine) in golden hamsters free-running in constant darkness. *Closed circles*: light pulse PRC; *open circles*: activity-induced PRC. Circadian time 12 refers to the time of activity onset in this nocturnal species. While exposure to light during the subjective daytime has little or no effect on the phase of the activity rhythm, exposure to light near the time of activity onset (i.e., near the time of sunset) induces a delay in the rhythm, while an equivalent light pulse given near the end of the subjective night (i.e., near the time of sunrise) induces an advance in the rhythm. While the amplitudes of the phase advances and phase delays that occur in response to the presentation of activity-inducing stimuli are similar to those observed in response to light pulses, the circadian times for the phase delay, phase advance, and unresponsive regions of the two PRCs are dramatically different.

share certain characteristics including the fact that light pulses presented near the onset of the subjective night (the subjective night and subjective day refer to those parts of the circadian cycle that would occur during the dark or light time, respectively, when the organism is exposed to a LD cycle) induce phase delays in the rhythm whereas light pulses presented in the late subjective night/early subjective day induce phase advances. In contrast, light pulses presented during most of the subjective day induce no phase shifts in the activity rhythm. Entrainment of the circadian clock to the LD cycle is thought to occur by light inducing phase advances and/or delays in the clock each day that equal the difference between the free-running period of the clock and the period of the daily LD cycle (7,9,12).

While the importance of the daily LD cycle for the entrainment of circadian rhythms in most plant and animal species has been recognized for many years, only recently has the importance of light for the entrainment of human circadian rhythms been appreciated. Early studies in humans held under conditions of temporal isolation indicated that the LD cycle was a very weak synchronizer of human circadian rhythms and that social environmental cues were more important for entrainment (13). Since the late 1970s, Czeisler and his colleagues (see Chapter 6) have carried out extensive studies demonstrating that the LD cycle could entrain human rhythms, and that light could be used to reset human rhythms under a variety of experimental conditions (14,15). Indeed, as in other animals, exposure to pulses of bright light can induce phase shifts in free-running human circadian rhythms. A major difference in the entrainment by light of circadian rhythms in humans and most other animal species is the need for apparently much brighter light to synchronize human rhythms, raising questions of the adequacy of normal indoor lighting for the entrainment of rhythms in humans who see very little natural light during the day (14,15).

Entrainment by Nonphotic Signals

A fundamental assumption in the early development of circadian rhythm research was that except for the LD cycle, endogenous circadian clocks were independent from most changes in the internal and external environment (16). Pittendrigh's early finding that circadian clocks are temperature-compensated (i.e., there is very little change in the period of the clock following an increase or decrease in temperature) led to the generalization that to keep accurate time, endogenous circadian clocks need to be buffered from most external and internal factors (17). Over the past decade, however, a number of internal and external stimuli have been found to influence circadian clocks in a variety of vertebrate species (18–20). Although the importance of nonphotic factors in the entrainment of circadian rhythms under natural conditions in mammals is not clear, under experimental laboratory conditions changes in ambient temperature, periodic presentation of food, and agents that alter the sleep-wake cycle can all alter the clock regulating overt circadian rhythms (18–21).

A great deal of attention has recently been focused on the possibility that changes in the activity-rest state of the animal can alter the circadian clock, which regulates the rhythm of activity as well as most other circadian rhythms in mammals. The acute presentation of a variety of pharmacological (e.g., injections of benzodiazepines) and nonpharmacological stimuli (e.g., by exposing hamsters to a novel running wheel or a pulse of darkness on a background of constant light), which can induce phase advances or phase delays in the free-running circadian rhythm of activity, also induce an acute increase in activity (18,19,22,23). These agents induce phase shifts in the circadian clock regulating the activity rhythm, as well as other behavioral and physiological rhythms (22,24,25), when they are presented at times when the hamsters are normally inactive. The PRCs generated to activity-inducing stimuli are about 180 degrees out of phase with the PRC to light pulses in the hamster (Fig. 3). The hypothesis that the increase in locomotor activity is itself somehow responsible for phase shifts in the circadian clock is supported by recent experiments in hamsters demonstrating that phase shifts induced by dark pulses or injections of short-acting benzodiazepines can be blocked by confining the animal to a small nest box or restraining tube during and for a period of time after the stimulus is presented (23,26). Rendering hamsters inactive by immobilizing them at a time when they are normally very active (i.e., during the early part of the subjective night) can also induce phase shifts in the circadian clock underlying the activity rhythm (27). Other experiments have demonstrated that induced activity can accelerate the rate of reentrainment of hamsters following a phase shift in the LD cycle (28) and that chronic exposure to a free or locked running wheel can influence the period of the circadian rhythm of locomotor activity in mice, hamsters, and rats (29–31). The overall implications of these findings for the normal entrainment and expression of circadian rhythms remain to be determined, but it is clear that changes in the behavioral state of the animal can influence the circadian time-keeping system.

The possible role of activity and/or social cues as synchronizing agents for the human circadian system has been recognized for more than three decades. Early studies indicated that human circadian rhythms can be entrained by scheduled bedtimes, mealtimes, and other social cues (7). Moreover, sleep schedules and social contacts have phase-resetting properties that can be separated from the effects of the LD cycle (32). Recent work by Van Reeth and colleagues (33) shows that appropriately timed acute physical exercise can produce phase delays in circadian rhythms, similar in magnitude to those in response to light. These observations indicate that social cues, sleep/wake state, and level of physical activity represent important synchronizing agents for human circadian rhythms.

From an historical perspective, it is interesting to note that while early studies in humans minimized the importance of the LD cycle, and focused on the role of social factors for the entrainment of human circadian rhythms, early studies in animals minimized the importance of behavioral changes, and focused on the almost exclusive role of the LD cycle in the regulation of circadian rhythms. The relative importance of photic and nonphotic signals for the entrainment of circadian rhythms may well vary between species, and is undoubtedly dependent on the evolutionary pressures faced by individual species adapting to the daily changes in the physical environment.

III. Sleep

The cyclic interchange of sleep and wakefulness is one of the most prominent and profound rhythms in life. Therefore, it is not surprising that interest in sleep has existed since the dawn of history. Yet, it was not until the twentieth century was well on its way that inquiry into the scientific basis of sleep and wakefulness began.

A. Sleep Architecture

Less than 70 years ago, sleep was thought to be a simple and passive state. This concept changed with the development of the EEG (34). In 1936, Loomis et al. (35) described different states of sleep characterized by distinct EEG patterns. The ability to analyze brain and physiological activity in a noninvasive manner during sleep (polysomnography), together with the discovery of REM sleep (36) and the basic sleep cycle (37), set the stage for modern sleep research.

Although sleep is perceived as rest, it is actually a period of substantial neurological and physiological activity. Electrophysiological recordings demonstrate two fundamentally distinct phases of sleep, NREM and REM sleep, each of which correlates with specific changes in brain activity, muscle tone, and autonomic activity. Based on the EEG, NREM sleep phase is subdivided into four very precisely, but relatively arbitrarily defined stages: stages 1, 2, 3, and 4 (38), with stages 3 and 4 often referred to as slow-wave sleep. Stage 1 sleep is a transitional stage between wakefulness and sleep and it is characterized by low-amplitude theta activity on the EEG. After a few minutes, stage 1 sleep transitions into stage 2 sleep (considered to be the onset of true sleep), defined by EEG spindles and K complexes. This is followed by slow-wave sleep, characterized by high-amplitude ($\geqslant 75$ μv) synchronized EEG waves of 0.5–2 Hz. The four NREM stages are associated with different "depth of sleep" and arousal thresholds, lowest in stage 1 and highest in stage 4. REM sleep is characterized by EEG activation, muscle atonia, and bursts of rapid eye movements. REM sleep is also referred to as "paradoxical sleep" because it is a state in which the brain is highly activated, but the body is paralyzed. One of its most fascinating aspects is that REM sleep is associated with the human experience of dreaming (39,40).

Under normal conditions sleep in humans begins with NREM sleep. After approximately 70–100 min of NREM sleep, the first period of REM sleep appears, and as the night progresses, REM sleep increases in duration with each successive cycle of approximately 90 min, whereas slow-wave sleep decreases. Therefore, normal sleep architecture in a young adult consists of approximately four alternating NREM and REM cycles, with REM sleep occupying about 25% of the total sleep time, and NREM sleep making up the remainder, as shown in Figure 4.

The strongest factor that influences normal sleep architecture and the temporal distribution of sleep is age. While the cyclic alternation between NREM and REM sleep is about 90 min in adults, it is more rapid (50–60 min) in infants. The fully developed EEG features of NREM sleep stages develop over the first 2–6 months of life. As cortical synaptic density develops, slow-wave sleep becomes maximal in children and decreases dramatically with age, whereas REM sleep as a percentage of total sleep remains fairly stable. At birth, sleep periods are distrib-

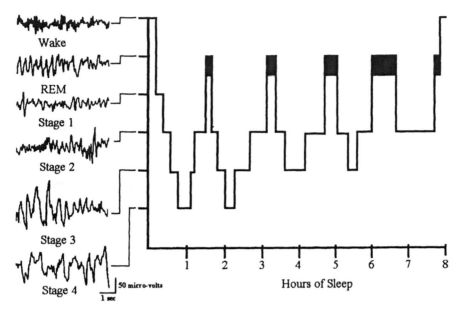

Figure 4 Typical sleep organization throughout the night with stages and representative EEG patterns shown on the left.

uted across the 24 hr, and children gradually develop a consolidated nocturnal sleep cycle (41).

B. Function of Sleep

While the need for sleep and the consequences of sleep deprivation are becoming better recognized by the public and health care professionals, the search for an understanding of the function of sleep continues today. Total sleep deprivation in rats results in the development of skin lesions, weight loss despite large increases in food intake (indicating increased energy expenditure and a role of sleep in metabolism), and death (42,43). Similar findings were obtained in studies of more selective REM and slow-wave sleep deprivation (43). Although a large amount of data is available on the physiological, psychological, and EEG effects of sleep deprivation in humans, the results are much less dramatic than in animals. What is evident from recent research is that inadequate quality and quantity of sleep have a profound negative impact on alertness, physical and neuropsychological performance, and the health of the individual (44–48).

Slow-Wave Sleep

The role of slow-wave sleep has been an area of great interest because following sleep deprivation, recovery sleep is characterized by an increase or rebound in the amount of slow-wave sleep, suggesting that it fulfills an essential need (49,50). The secretion of growth hormone, which stimulates growth and tissue repair, occurs during slow-wave sleep. The age-related decrease in growth hormone level (51–53) is correlated with the well-recognized decrease in slow-wave sleep with aging (53), and may contribute to the prevalence of sleep complaints in the elderly (see Chapter 14). In addition, slow-wave sleep may play a potentially significant role in enhancing immune system modulators, such as interleukin and tumor necrosis factors (54,55).

REM Sleep

The association of dreaming with REM sleep provides a unique opportunity to study the mind-body relationship and cognitive activity during sleep. Although philosophers and scientists alike have studied the psychophysiology of dreams (56–61), the function of dreams has remained elusive. The theory that dreaming is associated with emotional processing is supported by a recent study using positron emission tomography (PET) that found that the amygdala (among other areas) was activated during REM sleep (62).

Modern sleep research has provided important knowledge of the physiological, neurocellular, and neurochemical mechanisms involved in the generation of REM sleep (see Chapter 3) from which several theories of the function of REM sleep have emerged. These include the maintenance of catecholamine systems in the central nervous system (63); the regulation of central norepinephrine receptors, thus improving the organism's ability to sustain attention during wakefulness (64); the modulation of motivational behavior and mood (65,66); and the processing of information (67). Probably the most intriguing and best-studied function linked to REM sleep is its possible role in facilitating learning and memory. A number of animal studies have shown that REM sleep increases after learning and that REM sleep deprivation impairs retention (67–70). It has also been hypothesized that REM sleep may be a mechanism for "unlearning" certain irrelevant information (71). Support for a role of REM sleep in memory consolidation in humans comes from a recent report of improved performance on a perceptual learning task in humans following a normal night's sleep that was not seen in REM-sleep-deprived subjects (72). However, other studies of REM sleep deprivation in humans have found marginal effects on other types of memory (73).

Physiology in Sleep

Sleep has profound effects on many physiological and hormonal variables. For example, airway reflexes and bronchial smooth muscle reactivity are altered during sleep and time of the day, so that two-thirds of asthmatic persons have their lowest air flow rates between 10:00 P.M. and 8:00 A.M. (74). During NREM sleep, particularly slow-wave sleep, there is a reduction in physiological activity; metabolic rate, heart rate, blood pressure, and respiratory rate (75,76) are decreased. Whereas, in REM sleep, breathing becomes irregular, responsiveness to oxygen and carbon dioxide levels is altered (77), heart rate increases and becomes more variable, and blood pressure shows transient elevations (78–80). While the influence of sleep and circadian rhythms on gastrointestinal physiology is not well understood, there is a prominent circadian rhythm of acid secretion (81) and possible sleep-associated alterations in gastrointestinal motility (82,83).

IV. Importance of Maintaining Optimal Sleep and Circadian Temporal Organization

As described in several chapters (in particular, see Chapters 13–19), disturbed sleep and circadian rhythmicity, whether due to voluntary (e.g., jet lag and shift work) or involuntary (e.g., in illness and advanced age) circumstances, is associated with numerous physical and mental disorders and can severely impact the health, safety, performance, and productivity of humans. The importance of sleep and circadian rhythmicity for health and well-being has only recently been recognized and remains underappreciated by the both the general public and health care professionals.

References

1. Pittendrigh CS. Circadian rhythms and the circadian organization of living organisms. Cold Spring Harbor Symp Quant Biol 1960; 25:159–184.
2. Borbely AA. A two-process model of sleep regulation. Hum Neurobiol 1982; 1:195–204.
3. Borbely AA, Achermann P. Concepts and models of sleep regulation, an overview. J Sleep Res 1992; 1:63–79.
4. Dijk DJ, Duffy JF, Czeisler CA. Circadian and sleep-wake dependent aspects of subjective alertness and cognitive performance. J Sleep Res 1992; 1:112–117.
5. Dijk DJ, Czeisler CA. Contribution of the circadian pacemaker and the sleep homeostat to sleep propensity, sleep structure, electroencephalographic slow waves, and sleep spindle activity in humans. J Neurosci 1995; 15:3526–2528.
6. Czeisler CA, Zimmerman JC, Ronda JM, Moore-Ede MC, Weitzman ED. Timing of REM sleep is coupled to the circadian rhythm of body temperature in man. Sleep 1980; 2:329–346.
7. Aschoff J, Fatranska M, Giedke H, Doerr P, Stamm D, Wisser H. Human circadian rhythms in continuous darkness: entrainment by social cues. Science 1971; 171:213–215.
8. Moore-Ede MC, Sulzman FM, Fuller CA. The Clocks that Time Us. Cambridge: Harvard University Press, 1982.
9. Pittendrigh CS. Circadian systems: general perspective. In; Aschoff J, ed. Biological Rhythms. New York: Plenum Press, 1981:57–80.
10. Takahashi JS, Zatz M. Regulation of circadian rhythmicity. Science 1982; 217:1104–1111.

11. DeCoursey PJ. Phase control of activity in a rodent. Cold Spring Harbor Symp Quant Biol 1960; 25:49–55.

12. Takahashi JS, Murakami N, NIkaido SS, Pratt BL, Robertson LM. The avian pineal, a vertebrate model system of the Circadian oscillator: cellular regulation of circadian rhythms by light, second messengers, and macromolecular synthesis. Rec Prog Horm Res 1989; 45:279–352.

13. Wever R. Zur zeitgeber-staerke eines licht-dunkel-wechels fuer die circadiane periodik des menschen. Eur J Physiol 1970; 321:133–142.

14. Czeisler CA, Kronauer RE, Allan JS, Duffy JF, Jewett ME, Brown EN, Ronda JM. Bright light induction of strong (type 0) resetting of the human circadian pacemaker. Science 1989; 244:1328–1333.

15. Czeisler CA, Chiasera AJ, Duffy JF. Research on sleep, circadian rhythms and aging: applications to manned spaceflight. Exp Gerontol 1991; 26:217–232.

16. Turek FW, Van Reeth, O. Neural and pharmacological control of circadian rhythms. In: Nunez J, Dumont JE, Denton R, eds. Hormone and Cell Regulation. John Libey Eurotext, 1989:95–101.

17. Pittendrigh CS. On temperature independence in the clock system controlling emergence in *Drosophila*. Proc Natl Acad Sci USA 1954; 40:2697–2701.

18. Turek FW. Effects of stimulated activity on the circadian pacemaker of vertebrates. J Biol Rhythms 1989; 4:135–147.

19. Mrosovsky N. Phase response curves for social entrainment. J Comp Physiol A 1988; 162:35–46.

20. Mistlberger R, Rusak B. Mechanisms and models of the circadian timekeeping system. In: Kryger MH, Roth T, Dement WC, eds. Principles and Practice of Sleep Medicine. Philadelphia: WB Saunders, 1989:141–152.

21. Turek FW, Smith R, Reeth V, Wickland C. Disturbances of the activity rest cycle alter the circadian clock of mammals. In: Inouye S, Krieger JM, eds. Endogenous Sleep Factors. The Hague: SPB Academic Publishing, 1990:277–283.

22. Wickland C, Turek FW. Phase-shifting effects of acute increases in activity on circadian locomotor rhythms in hamsters. Am J Physiol 1991; 261:R1109–R1117.

23. Reebs S, Mrosovsky N. Running activity mediates the phase-advancing effects of dark pulses on hamster circadian rhythms. J Comp Physiol A 1989; 165:811–818.

24. Wickland C, Turek FW. Phase-shifting effect of triazolam on the hamster's circadian rhythm of activity is not mediated by a change in body temperature. Brain Res 1991; 560:12–16.

25. Turek FW, Losee-Olson S. The circadian rhythm of LH release can be shifted by injections of a benzodiazepine in female golden hamsters. Endocrinology 1988; 122: 756–758.

26. Van Reeth O, Turek FW. Stimulated activity mediates phase shifts in the hamster circadian clock induced by dark pulses or benzodiazepines. Nature 1989; 339:49–51.

27. Van Reeth O, Hinch D, Tecco JM, Turek FW. The effects of short periods of immobilization on the hamster circadian clock. Brain Res 1991; 545:208–214.

28. Mrosovsky N, Salmon PA. A behavioral method for accelerating re-entrainment of rhythms to new light-dark cycles. Nature 1987; 330:372–373.

29. Yamada N, Shimoda K, Ohi K, Takahashi K, Takahashi S. Free-access to a running wheel shortens the period of the free-running rhythm in blinded rats. Physiol Behav 1988; 42:87–91.

30. Edgar DM, Kilduff TS, Martin CE, Dement WC. Influence of running wheel activity on free-running sleep/wake and drinking circadian rhythms in mice. Physiol Behav 1991; 50:373–378.

31. Aschoff J, Figala J, Poppel E. Circadian rhythms of locomotor activity in the golden hamster (*Mesocricetus auratus*) measured with two different techniques. J Comp Physiol Psychol 1973; 85:20–28.

32. Honma K, Honma S, Nakamura K, Sasaki M, Endo T, Takahashi T. Differential effects of bright light and social cues on reentrainment of human circadian rhythms. Am J Physiol 1995; 268:R528–R535.

33. Van Reeth O, Sturis J, Bryne MM, Blackman JD, L'Hermite-Balériaux M, Leproult R, Oliner C, Refetoff S, Turek FW, Van Cauter E. Nocturnal exercise phase-delays the circadian rhythms of melatonin and thyrotropin secretion in normal men. Am J Physiol 1994; 266:E964–E974.

34. Berger H. Ueber das elektroekephalogramm des meschen. J Psychol Neurol 1930; 40:160–179.

35. Loomis AL, Harvey EN, Hobart GA. Electricial potentials of the human brain. J Exp Psychol 1936; 19:249–279.

36. Aserinsky E, Kleitman N. Regularly occurring periods of eye motility, and concomitant phenomena, during sleep. Science 1953; 118:273–274.

37. Dement WC, Kleitman N. Cyclic variations in EEG during sleep and their relation to eye movements, body motility, and dreaming. Electroencephalogr Clin Neurophysiol 1957; 9:673–690.

38. Rechtschaffen A, Kales A, eds. A Manual of Standardized Terminology, Techniques and Scoring System for Sleep Stages of Human Subjects. Los Angeles: UCLA Brain Information Service/Brain Research Institute, 1968.

39. Dement WC, Kleitman N. The relation of eye movements during sleep to dream activity: an objective method for the study of dreaming. J Exp Psychol 1957; 53:339–346.

40. Rechtschaffen A. Dream reports and dream experiences. Exp Neurol 1967; 4:4–15.

41. Carskadon MA, Dement WC. Normal human sleep: an overview. In: Kryger MH, Roth T, Dement WC, eds. Principles and Practices of Sleep Medicine. Philadelphia: WB Saunders, 1994:889–913.

42. Everson CA, Bergmann BM, Rechtschaffen A. Sleep deprivation in the rat. III. Total sleep deprivation. Sleep 1989; 12:13–21.

43. Kushida CA, Bergmann BM, Rechtschaffen A. Sleep deprivation in the rat. IV. Paradoxical sleep deprivation. Sleep 1989; 12:22–30.

44. Akerstedt T. Sleep/wake disturbances in working life. Lond Symp (EEG Suppl) 1987; 39:360–363.

45. Dinges DF, Pack F, Williams K, Gillen KA, Powell JW, Ott GE, Aptowicz C, Pack AI. Cumulative sleepiness, mood disturbance, and psychomotor vigilance performance decrements during a week of sleep restricted to 4–5 hours per night. Sleep 1997; 20:267.

46. Mitler MM, Cardadon MA, Czeisler CA, Dement WC, Dinges DF, Graeber RC. Catastrophes, sleep and public policy: consensus report. Sleep 1988; 11:100–109.

47. Pilcher JJ, Huffcutt AI. Effects of sleep deprivation on performance: a meta-analysis. Sleep 1996; 19:318–326.

48. Bonnet MH. Sleep deprivation. In: Kryger MH, Roth T, Dement WC, eds. Principles and Practice of Sleep Medicine. Philadelphia: WB Saunders, 1994:50–67.

49. Dement WC, Greenberg S. Changes in total amount of stage four sleep as a function of partial sleep deprivation. Electroencephalogr Clin Neurophysiol 1966; 20:523–526.

50. Webb JB, Agnew HW, Jr. Sleep: effects of a restricted regime. Science 1965; 150: 1745–1747.

51. Ho KY, Evans WS, Blizzard RM, Veldhuis JD, Merriam GR, Samojlik E, Furlanetto R, Rogol AD, Kaiser DL, Thorner MO. Effects of sex and age on the 24-hour profile of growth hormone secretion in man: importance of endogenous estradiol concentrations. J Clin Endocrinol Metab 1987; 64:51–58.

52. Iranmanesh A, Lizarralde GVJD. Age and relative adiposity are specific negative determinants of the frequency and amplitude of growth hormone (GH) secretory bursts and the half-life of endogenous GH in healthy men. J Clin Endocrinol Metab 1991; 73:1081–1088.

53. van Coevorden A, Mockel J, Laurent E, Kerkhofs M, N'Hermite-Balériaux M, Decoster C, Nève P, Van Cauter E. Neuroendocrine rhythms and sleep in aging. Am J Physiol 1991; 260:E651–E661.

54. Toth LA, Krueger JM. Alteration of sleep in rabbits by *Staphylococcus aureus* infection. Infect Immun 1988; 56:1785–1791.

55. Krueger JM, Majde JA. Sleep as a host defense: its regulation by microbial products and cytokines. Clin Immunol Immunopathol 1990; 57:188–199.

56. Hobson JA, McCarley RW. The brain as a dream state generator: an activation-synthesis hypothesis of the dream process. Am J Psych 1977; 134:1335–1348.

57. Hobson J, Hoffman S, Helfand R, Kostner D. Dream bizarreness and the activation-synthesis hypothesis. Hum Neurobiol 1987; 6:157–164.

58. Hobson JA. The Dreaming Brain. New York: Basic Books, 1988.

59. Kramer M, Kinney L, Scharf M. Dream incorporation and dream function. In: Koella WP, ed. Sleep 1982. Sixth European Congress on Sleep Research, Zurich, 1982. Basel: S Karger, 1983:369–371.

60. Freud S. The Interpretation of Dreams (originally published in 1900). New York: Basic Books, 1955.

61. Foulkes D. Dreaming: A Cognitive-Psychological Analysis. Hillsdale, NJ: Erlbaum, 1985.

62. Maquet P, Peters J, Aerts J, Delfiore G, Degueldre C, Luxen A, Franck G. Functional neuroanatomy of human rapid-eye movement sleep and dreaming. Nature 1996; 383: 163–166.

63. Stern WC, Morgane PJ. Theoretical view of REM sleep function: maintenance of catecholamine systems in the central nervous system. Behav Biol 1974; 11:1–32.

64. Siegel JM, Rogawski MA. A function of REM sleep: regulation of nonadrenergic receptor sensitivity. Brain Res 1988; 13:213–233.

65. Vogel GW, Thurmond A, Gibbons P, Sloan K, Boyd M, Walker M. REM sleep reduction effects on depression syndromes. Arch Gen Psychiatry 1975; 32:765–777.

66. Vogel GW, Vogel F, McAbee RS, Thurmond A. Improvement of depression by REM sleep deprivation. Arch Gen Psychiatry 1980; 37:247–253.

67. Pearlman CA. REM sleep and information processing: evidence from animal studies. Neurosci Biobehav Rev 1979; 3:57–68.

68. Pearlman CA. Sleep structure variation and performance. In: Webb WB, ed. Biological Rhythms, Sleep and Performance. New York: Wiley, 1982:143–173.

69. Smith C. Sleep states and learning: a review of the animal literature. Neurosci Biobehav Rev 1985; 9:157–168.
70. Tilley AJ, Empson JAC. REM sleep and memory consolidation. Biol Psychol 1978; 6:293–300.
71. Crick F, Mitchison G. The function of dream sleep. Nature 1983; 304:111–114.
72. Karni A, Tanne D, Rubenstein BS, Askenasy JM, Sagi D. Dependence of REM sleep of overnight improvement of a perceptual skill. Science 1994; 265:679–682.
73. McGrath MJ, Cohen DB. REM sleep facilitation of adaptive waking behavior: a review of the literature. Psychol Bull 1978; 85:24–57.
74. Connolly CK. Diurnal rhythms in airway obstruction. Br J Dis Chest 1979; 73: 357–366.
75. Mancia G, Zanchetti A. Cardiovascular regulation during sleep. In: Orem J, Barnes CD, eds. Physiology in Sleep. New York: Academic Press, 1981:2–56.
76. Sullivan CE, Issa FG, Berthon-Jones M, Eves L. Reversal of obstructive sleep apnea by continuous positive airway pressure applied through the nares. Lancet 1981; 1: 862–865.
77. Orem J. Control of the upper airways during sleep and the hypersomnia-sleep apnea syndrome. In: Orem J, Barnes CD, eds. Physiology in Sleep. New York: Academic Press, 1981:273–314.
78. Coccagna G, Mantovani M, Lugaresi E. Arterial pressure changes during spontaneous sleep in man. Electroencephalogr Clin Neurophysiol 1971; 31:277–281.
79. Jones JV, Sleight P, Smyth HS. Haemodynamic changes during sleep in man. In: Ganten D, Pfaff D, eds. Current Topics in Endocrinology. New York: Academic Press, 1982:213–272.
80. Parmeggiani PL. The autonomic nervous system in sleep. In: Kryger MH, Roth T, Dement WC, eds. Principles and Practice of Sleep Medicine. Philadelphia: WB Saunders, 1994:194–203.
81. Moore JC, Englert E. Circadian rhythm of gastric acid secretion in man. Nature 1970; 226:1261–1262.
82. Orr WC, Johnson LF, Robinson MG. The effect of sleep on swallowing, esophageal peristalsis, and acid clearance. Gastroenterology 1984; 86:814–819.
83. Orr WC. Gastrointestinal physiology. In: Kryger MH, Roth T, Dement WC, eds. Principles and Practice of Sleep Medicine. Philadelphia: WB Saunders, 1994:252–259.

2

Ontogeny of Sleep and Circadian Rhythms

FRED C. DAVIS

Northeastern University
Boston, Massachusetts

MARCOS G. FRANK

University of California–San Francisco
San Francisco, California

H. CRAIG HELLER

Stanford University
Stanford, California

I. Introduction

Sleep and waking are well-defined behavioral and physiological states that in many animals, including humans, normally occur at restricted times of day. The timing of sleep and waking is regulated by two processes: (1) a circadian pacemaker, which is entrained to the light-dark cycle and promotes wake during an active phase of the cycle and permits sleep during a rest phase of the cycle, and (2) a homeostatic process in which a need for sleep accumulates during waking and is dissipated during sleep (1–3). In addition, a third mechanism controls the ultradian alternation between the distinct states of sleep—rapid eye movement (REM) and non-REM sleep.

   The goal of this review is to examine the development of sleep and circadian rhythms in mammals. This includes the physiological characteristics of sleep, the homeostatic and ultradian regulation of sleep, and the mechanisms responsible for the circadian timing of sleep and waking. There are several reasons for trying to understand the development of any particular aspect of biological organization: (1) to identify general mechanisms of development; (2) to better understand the particular organ or system of interest as it exists and functions in the mature organism (using development as a "natural" experimental perturbation); (3) to learn whether the organ or system of interest has adaptive functions specific to stages of development; and (4) to specifically understand the etiology of normal and abnormal variations of the organ or system in humans. Each of these reasons requires, as a starting point, descriptive information about development.

II. Development of Sleep in Mammals Has Three Stages

Mammals are considered either altricial or precocial depending on their level of development at birth. Altricial species are less fully developed and more dependent on parental care, and precocial species are more fully developed and less dependent on parental care. Brain activity dramatically shows these different grades of development. In altricial species brain electrical activity is undifferentiated, reflecting the fact that most axons have not yet reached their targets and a great deal of synaptogenesis has yet to take place. In contrast, brain electrical activity in precocial species shows the characteristics of adult arousal states. Humans are unusual with respect to stage of development at birth. Humans are clearly altricial in most regards, but their brain activity displays differentiated arousal states.

Whether it occurs in utero in precocial species (and humans) or postnatally in altricial species, the emergence of adult sleep states in mammals from an undifferentiated condition consists of three general, but distinct, stages. The first stage has been called "indeterminate sleep" (4,5), "atypical sleep" (6), or "presleep" (7), and it is not associated with distinct cyclings of autonomic and brain activity typical of adult REM and non-REM sleep. The second stage of sleep development, called "concordance," encompasses the first appearance of classically defined REM and non-REM sleep states from the more undifferentiated presleep condition. The final stage of sleep development, called "maturation," includes the development of mature REM and non-REM sleep architecture, brain activity, and sleep regulatory mechanisms. These three stages of mammalian sleep development are discussed in the following sections.

A. Dissociation

In his study of premature human neonates, Hamburger used the term "presleep" to refer to the period of development when sleep, as it is classically defined, appears to be absent and the electroencephalogram (EEG) reflects background spontaneous activity of the immature nervous system (7). The distinguishing EEG characteristics of mammalian sleep are not present and other correlates of sleep states such as changes in respiratory pattern, heart rate, and phasic motor activity are dissociated from one another (8,9). While there may be periodicities in the spontaneous expression of individual variables, they are not coupled to each other into recognizable states (9) (Fig. 1). Similar findings are reported in fetal studies of

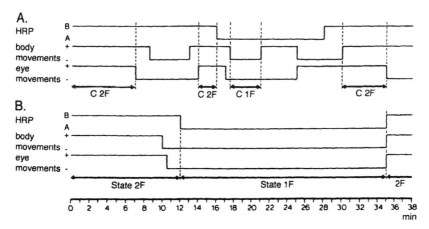

Figure 1 Dissociation of sleep parameters in the human fetus. Profiles of state variables obtained from ultrasound recordings at 32 (A) and 40 (B) weeks of gestation (HRP = heart rate pattern). Coincidence of 1F and 2F parameters at 32 gestational weeks is indicated by C1F and C2F, respectively. The state variables appear to cycle relatively independently of one another at 32 gestational weeks as shown by the periods of noncoincidence, whereas at 40 gestational weeks, the relative synchrony of the state transitions demonstrates linkage of the state variable. (Modified from Ref. 18, Copyright 1982, with permission from Elsevier Science.)

precocial species like the lamb. Prior to approximately 110–115 days gestational age (ga), precocial lamb fetuses do not appear to cycle between states of arousal and sleep (10,11). Their respiratory activity is nonepisodic, their motility is generally acyclic, and consistent patterns in EEG activity are absent (Fig. 2). Altricial species like the cat appear to also display a dissociated behavioral condition in the first 1–2 postnatal weeks typified by an absence of patterned EEG activity, and poor concordance between respiratory, electromyogram (EMG), and motor behaviors (12) (Fig. 3).

B. Concordance in Humans and Precocial Species

Precocial species complete a significant portion of their motor and neural development in utero. Many of these species display distinct sleep states near the end of gestation, and some, like the guinea pig, have adult forms of sleep at term (13). Most nonhuman primates [the chimpanzee appears to be an exception (14)] and humans (6,9,15–19) also develop distinct sleep states in utero, and for this reason are here grouped with precocial species. In precocial nonhuman species and humans, the transition from the dissociated presleep stage to REM and non-REM sleep states does not appear to include transitional precursor sleep states. In the

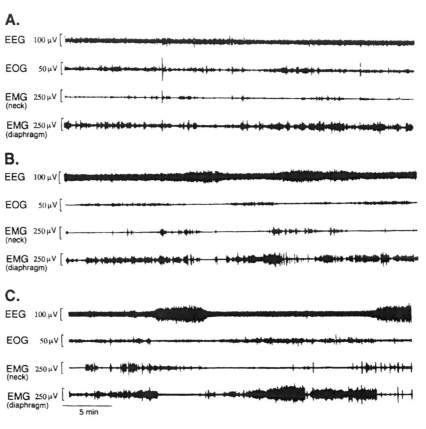

Figure 2 Dissociation of state variables in the precocial lamb fetus. (A) Polygraphic recording obtained from fetal lamb at 112 days of gestation. Note that discrete behavioral states cannot be identified at this age consistent with "presleep" in human fetuses. (B) Polygraphic recording obtained from the same fetal lamb as in (A) at 123 days of gestation. Order of the behavioral states shown as follows: REM sleep–arousal–non-REM sleep–REM sleep–non-REM sleep–REM sleep. (C) Polygraphic recording obtained from the same fetal lamb as in (A) and (B) at 137 days of gestation. Order of the behavioral states cycle shown is as follows: arousal–non-REM sleep–REM sleep–non-REM sleep. EEG, electroencephalogram; EOG, electrooculogram; EMG, electromyogram. (Modified from Ref. 11; Copyright 1985, American Sleep Disorders Association and Sleep Research Society, Rochester, MN. Reprinted by permission.)

fetal sheep (10,11,20) and baboon (21–23), sleep states comparable to REM and non-REM emerge simultaneously in the latter third of the gestational period. In fetal sheep, cyclic motor activity within the range of REM–non-REM sleep periodicities appears in conjunction with cycles of respiratory movements and EEG activity (11,20). Two distinct constellations of autonomic parameters and

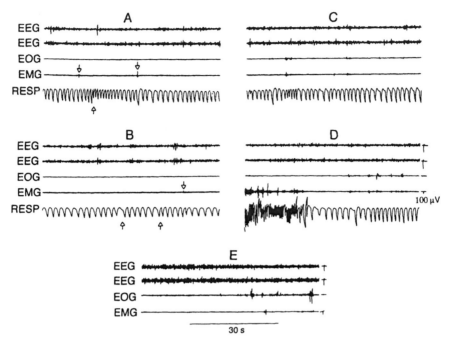

Figure 3 Dissociation of state variables in the altricial kitten. Polygraphic records from kittens in 10-day-old (A–D) and 20-day-old (E) kittens. State parameters are not consistently coupled with each other at these ages. Phasic motor activity and irregular respiration occur during quiet sleep (A and B) and during active sleep (defined by EOG activity) (C). A rapid transition from waking to AS is shown in (D). A transition from QS to AS in (E) involves only a slight change in EEG. (Modified from Ref. 59; Copyright 1977, reprinted by permission of John Wiley & Sons, Inc.)

brain activity comparable to REM and non-REM sleep are observed in sheep by 110–115 days ga (10,11,20). The REM-like state is characterized by increases in respiratory movements, phasic motor activity, and low-voltage EEG activity, and after 125+ days ga increases in cerebral blood flow (24–26) and brain metabolism (26–28). The non-REM-like state is characterized by decreases in respiratory movements and motor activity, high-voltage EEG activity, and, after 125+ days ga, decreases in cerebral blood flow (24–26) and brain metabolism (24–28).

Similarly in baboons between 143 and 153 days ga, two distinct behavioral states emerge that bear a strong resemblance to non-REM and REM sleep, respectively. State 1 is characterized by an EEG pattern of intermittent low-voltage slow activity (trace alternant EEG), low amounts of electrooculogram (EOG) activity, and low heart rate variability, and state 2 is characterized by increased high-frequency EEG activity, high EOG activity and greater heart rate variability. As is true of the fetal sheep, there are no reports of transient, precursor sleep states intermediate between presleep and states 1 and 2 in the fetal baboon (21,23,29).

In humans, concordance of individual sleep phenomena into distinct sleep states begins at approximately 28–32 weeks ga and is complete by 36–40 weeks ga (9,15–18,30). At about 28 weeks ga discrete periods characterized by REMs and respiratory movements begin alternating with periods of sustained motor quiescence with no or very slow EMs (19,31,32). Between 28 and 34 weeks, cycles of low and high variability in fetal heart rate are detected (33,34), and appear to be associated with periods of high and low REM frequency (35,36). Alternating periods of "continuous" and "discontinuous" EEG activity during sleep are first reported in studies of premature babies of approximately 28–31 weeks ga (8,37,38), and appear to associated with consistent changes in heart rate, REMs, and body movements (16,17).

The exact timing for the emergence of "true" sleep states is not clear, with some investigators reporting distinct REM and non-REM-like sleep states by 27–30 weeks ga (16,17,30) and others reporting state organization only after 32 weeks ga (9). In general, most studies report segregation of REMs, autonomic activity, body movements, and EEG activity into the fetal states 1F and 2F by at least 36–40 weeks ga (9,39). The fetal behavioral state 1F appears to be continuous with non-REM sleep in term infants (also referred to as "quiet sleep"). The fetal behavioral state 2F appears to be continuous with REM sleep in term infants (also referred to as "active sleep"). Therefore, states 1F and 2F most likely reflect the first appearance of these sleep states in human development (18,40).

As is the case for precocial nonhuman species, there do not appear to be precursor sleep states intermediate between presleep and the fetal states 1F and 2F in humans. Several investigators have reported periods of motor quiescence in fetuses younger than 30 weeks ga that bear some developmental resemblance to the fetal state 1F and non-REM sleep in adults (31,41–45). These quiescent periods are usually defined as brief episodes when gross body movements (in trunk or extremities) are undetectable. Sterman et al. reported a 20-min "rest-activity" cycle in gross body movements in utero in one fetus of 21 weeks ga (45,46). Similar findings are reported by Granat et al., who found rest-activity cycles of 40–80 min in fetuses of 23–37 weeks ga (43). To what extent these quiescent, or "rest" periods relate to non-REM sleep in adults is unclear. Approximately 45% of 20-week ga fetuses do not show quiescent periods (35,36). In addition, the Sterman et al. findings have not been replicated in more complete studies of fetal (47) behavior, and the Granat study provides no information on the frequency of rest-activity cycles in the youngest fetuses because it reports averages across 23–37 weeks ga. Some detailed analyses of fetal motor activity suggest that the apparent early appearance of activated and quiet states reflects chance coordination between independent patterns of fetal behavior, and that true sleep states are not observed until 32+ weeks ga (9,39,48).

C. Concordance in Altricial Species

In altricial species there are reports of precursor sleep states that appear 1–2 weeks before EEG differentiation. These presumptive precursor states are typically called active sleep (AS) and quiet sleep (QS) (13,49). It is important, however, to emphasize that though the terms AS and QS are used both in altricial species and in human premature and term fetuses, they do not refer to the same constellations of behavioral events. In human premature and term fetuses AS and QS refer to distinct sleep states comprised of well-defined EEG, autonomic, and motor patterns (16,17). In altricial species, the EEG is undifferentiated in the first few weeks ex utero and active sleep and quiet sleep are scored primarily with behavioral criteria (13,49). Behavioral AS (BAS), or "sleep with jerks," (50) refers to a state characterized by myoclonia, sometimes including REMs, and irregular respiration. Behavioral QS (BQS) is characterized by more regular respiration, and the absence of phasic activity and REMs (13,49,50). The behavioral criteria used to discriminate BAS and BQS, however, can vary considerably from study to study, and many studies of sleep in altricial species use only the presence or absence of myoclonia to assign states. There is also much weaker concordance between the behavioral criteria used to define BAS and BQS in altricial species compared to human fetuses (12).

Determining the nature of behavioral sleep is particularly important for any discussion of mammalian sleep ontogeny since the vast majority of studies on sleep ontogeny have been in altricial species. Behavioral sleep has traditionally been thought to represent intermediate "homolog" sleep states to REM and non-REM sleep in adult mammals (49). There are, however, important differences between behavioral sleep and sleep in adult mammals that argue against such an interpretation. On the other hand, there are strong similarities between behavioral sleep and the presleep stage of the human fetus and of fetuses of precocial species that suggest that behavioral sleep is simply an extension of the dissociated, presleep stage into the postnatal period, an interpretation entirely in keeping with the fact that altricial species complete much of their neural development ex utero.

D. Behaviorally Defined Sleep States of Altricial Species Do Not Appear to Be Homologous to Adult Sleep States

Other than very superficial behavioral similarities, BAS and BQS have few features in common with either the fetal states 2F and 1F or the adult sleep states of REM and non-REM sleep. BAS and REM sleep, for example are dissimilar with respect to brain activation. Studies of rat (51), kitten (52), and rabbit (53,54) cortical unit firing rates show no episodic firing patterns in individual cortical neurons prior to the appearance of EEG-defined sleep states. There are conflicting reports concerning the presence of subcortical activity during BAS. Increased firing during BAS of midbrain pontine reticular formation and forebrain units has been reported prior to EEG differentiation in rats (55–57). In the kitten, midline thalamic nuclei and brainstem tegmentum neurons also appear to increase their firing rates during BAS by P3–P4 (58), which is 1–2 weeks before a differentiated EEG is observed (12,59). Other studies of subcortical activity in neonatal rats and cats do not find state-specific firing in the pre-EEG period. In the rat, pontine tegmentum neurons selectively active during BAS are not detected until EEG differentiation (60). In the kitten, REM-sleep-like increases in pontine-geniculate-occipital (PGO) waves and associated lateral geniculate unit activity are not observed in BAS prior to EEG differentiation (61,62). Similarly, serotonergic raphe neurons are not significantly inhibited during AS (as they are during REM sleep) prior to the appearance of EEG-defined arousal states (63). In the rat (64) and rabbit (65) hippocampal theta activity is irregular, slow, and of low amplitude (<4.0 Hz) or absent prior to EEG-defined states.

BAS is not mediated by the same brain mechanisms responsible for REM sleep in adult mammals. In adult animals brainstem cholinergic nuclei, primarily via muscarinic receptors, mediate REM sleep phenomena, including cortical desynchronization, hippocampal theta, atonia, and PGO waves (66,67). At times when BAS is maximally expressed, the cholinergic system is largely immature. The amount of BAS expressed is inversely related to the development of muscarinic receptors (68,69), their coupled second-messenger systems (68), cholinergic efferents (70–73), cholinergic nuclei (74,75), neuronal responsiveness to acetylcholine (76), and cholinergically mediated behaviors (77,78). Cholinergic block has no effect on BAS in the rat, and cholinergic regulation of a REM-sleep-like state does not appear until the second postnatal week (79) (Fig. 4). Chemical or electrolytic lesions of pontine and midbrain REM sleep structures that suppress REM in adult and juvenile animals do not suppress behavioral BAS in neonatal rats or kittens (80). Transections of the brainstem from the spinal cord in P3–P8 rats does not abolish myoclonia (caudal to the section) during BAS (81). Similar transections in adult mammals abolish all signs of REM sleep caudal to the section (67).

Determining the precise relationship between BQS in altricial species and adult non-REM sleep has proven more difficult. The lack of narrowly defined behavioral criteria for BQS has led to conflicting accounts of BQS levels during development. BQS has been defined as periods of absolute behavioral quiescence (12,13,50), and also as periods of reduced activity relative to BAS (82,83). When BQS is defined as periods of regular respiration without phasic activity it makes up only 2% of total recording time (TRT) in P7–P8 rat pups (13). When small movements or "startles" are included in BQS scoring criteria, BQS levels increase to 30–35% TRT in P7–P8 rat pups (82,84). BQS coupled with regular respiration totals less than 3% TRT in P7 kittens. When respiratory patterns are excluded from consideration, the percentage of BQS increases to 20–25% TRT (12). Thus, the amounts of BQS reported during development are profoundly influenced by the behavioral criteria chosen by the investigator.

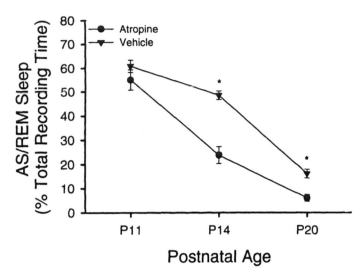

Figure 4 Active/REM (AS/REM) sleep in neonatal rats following muscarinic blockade. Six-hour mean (± SEMs) amounts of AS/REM sleep immediately following intra-peritoneal injections of atropine (6 mg/kg, *n* = 6) or vehicle (saline, *n* = 6) are shown. *Significant difference (*p* < 0.05) between atropine and vehicle groups for each age. Muscarinic blockade had no effect on AS/REM sleep amounts until P14. (Adapted from Ref. 79.)

The problem in classifying BQS is compounded by the wide range of ambient temperatures used in acute recording of sleep in altricial neonates. Brood animals like the rat pup quickly become hypothermic once removed from the dam or the litter huddle (85). Hypothermic rat pups become "torpid" and appear to be in a quiescent sleep state (86). Hypothermia in neonatal rats profoundly alters the amount of myoclonia observed during sleep (87), and since myoclonia is often used to discriminate BAS from BQS, hypothermia will also profoundly alter the amount of BQS scored in a given sleep record. Hypothermia-induced reductions in state criteria used to score BAS may also occur in larger animals like the kitten or rabbit, which have relatively immature thermoregulatory systems at birth (88,89).

Studies of neuronal activity during BQS have proven inconclusive. Unit activity studies in newborn animals report decreased firing rates in forebrain and brainstem neurons during BQS (55–57), but decreased firing rates may also accompany quiet wakefulness, attenuated BAS, and hypothermia. In other regions of the brain, like the thalamus (63) or pontine tegmentum (60), consistent changes in neuronal activity during BQS do not occur until EEG differentiation. No neurons selectively active during adult non-REM sleep (90) have been detected in neonatal forebrain neurons during BQS (57).

BQS does not appear to be continuous with non-REM sleep. Studies using the most conservative definitions of BQS (no phasic activity) indicate that behavioral QS is not a unique precursor to non-REM sleep. EEG slow-wave activity (SWA), the hallmark of non-REM sleep, can occur in periods of quiescence but it has also been reported during AS (12–14,59,91). This mixed condition (AS + SWA) gradually disappears concurrent with the development of EEG/EMG-defined non-REM sleep (12,13,91) (Fig. 5). These findings suggest that BQS is not a unique precursor to non-REM sleep.

E. BAS and BQS in Altricial Species Are Similar to Presleep in Precocial Species and Humans

The fetal presleep stages of humans and of precocial fetuses and behavioral sleep in altricial species are strikingly similar in terms of behavior and underlying neurophysiology. The behavioral components of BAS (e.g., irregular respiration, phasic motor activity, and REMs) are poorly coupled in the neonatal kitten prior to

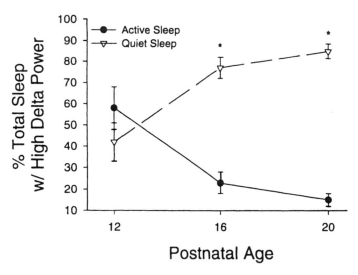

Figure 5 Concordance of EEG and behavioral state parameters in the altricial rat. Development of EEG delta power (0.5–4.0 Hz) in behaviorally identified active (myoclonia, REMs) and quiet (no myoclonia, no REMs) sleep (*n* = 5). Values are mean (± SEMs) amounts of epochs scored as either active sleep or quiet sleep that had identical levels of EEG delta power. The amount of active sleep associated with high levels of delta power significantly declined during postnatal development (*p < 0.05). (Adapted from Ref. 91.)

EEG differentiation (12). Respiratorion, motor activity, and REMs are likewise uncoupled in the fetal lamb and fetal human prior to EEG differentiation (9,11).

Neither the spontaneous activity typical of presleep nor the myoclonia of BAS requires intact sleep centers. The spontaneous motility of the presleep period is generated primarily by spinal and brainstem motor circuits, and does not depend on intact brainstem-forebrain sleep centers (7,92). The myoclonia of BAS also does not depend on intact brainstem-forebrain centers, since it persists following destruction of brainstem-midbrain REM sleep mechanisms (80) and after transection of the brainstem itself (81). Periodicities in spontaneous activity cycles in fetal humans and lambs prior to EEG differentiation are highly variable since regularly alternating periods of sustained quiescence and activity during the fetal period are rare (11,35,47).

In summary, BAS and BQS in altricial species appear to be unrelated to adult REM and non-REM sleep or the fetal states of 1F and 2F. BQS, for example, does not appear to be equivalent to nor continuous with non-REM sleep. The less conservative criteria for BQS make it essentially indistinguishable from descriptions of attenuated BAS or quiet waking prior to eye opening, and the amounts of BQS become vanishingly small when stringent behavioral criteria are used in conjunction with thermoneutral environments. Similarly, BAS and REM sleep are not comparable in terms of underlying physiology or behavior. Although BAS can be accompanied by increases in subcortical activity, it is unclear if this activity is generated by the same brain systems that generate REM sleep. Lesion studies and reports of cholinergic maturation suggest that such brain systems do not generate BAS. On the other hand, behavioral sleep and the presleep period of the human and nonhuman precocial fetus are comparable with respect to underlying neurophysiology and behavior. It would seem, therefore, that all mammals exhibit a period of spontaneous, dissociated activity (presleep) that progressively becomes organized into distinct sleep states. In humans and nonhuman precocial species, this concordance occurs in utero. In altricial species, which are born more immature, this concordance occurs ex utero. Furthermore, reports of intermediate, precursor states during the transition from presleep to REM and non-REM-like sleep states most likely reflect chance coordinations between dissociated state components (9), or misinterpretations of the cyclic nature of spontaneous motility in the immature nervous system (91).

F. Maturation

The emergence of distinct sleep states (concordance) is followed by dramatic maturational changes in the amounts, duration, and cyclicity of REM and non-REM sleep along with the brain activity associated with these states. Several general observations can be made for all mammals studied to date. The first is that the amount of REM sleep is initially much higher early in development than it is in later adult life. Conversely, the amount of non-REM sleep and wakefulness is lower early in development than it is in later life. The second general observation is that the stereotypical patterns of brain activity of REM and non-REM sleep become increasingly well defined in the postnatal period, although some precocial species (and humans) begin this process in utero (Fig. 2). The third general observation is that sleep regulatory mechanisms (ultradian, circadian, and homeostatic) develop comparatively late, and undergo important modifications in the neonatal period.

Sleep Architecture

One of the most striking observations in infant animals is the abundance of REM sleep early in development. This observation remains sound even when one takes into account the common misidentification of BAS in altricial species as being an immature form of REM sleep in adult mammals. REM and non-REM-sleep-like state appear at approximately the same time in development, but they do so in amounts that are quite distinct from what is observed in adult animals. In the human, a REM-sleep-like state (2F) appears between 30 and 32 weeks ga, and increases in amount until the first 1–2 postnatal weeks (9,16,17) at which time a decline toward adult values begins (93). Similar changes in REM sleep have been reported in the neonatal rhesus monkey (94) and chimpanzee (14,95). At the peak of REM sleep expression, the amount of REM sleep is much greater than that observed in adult animals. In the P14 rat, for example, the amounts of a REM-sleep-like state identified by EEG/EMG parameters and sensitive to cholinergic blockade are 2–3 times greater than at adulthood (3–6 months) (79). The duration of individual REM sleep periods also appears to display an initial lengthening (17,94), which, in the case of altricial species, is sometimes followed by a subsequent contraction to adult values in later development (13,83,91).

In contrast to REM sleep, the amount and duration of non-REM sleep periods rapidly reach adult values, and change comparatively little with subsequent development in humans (17,93,95), precocial (11), and altricial species (91). In the human fetus, the amount of non-REM sleep initially increases, and then stabilizes by 35–36 weeks ga, and only modest changes are noted in the neonatal period (93,97). In the fetal lamb, the amount of non-REM sleep is relatively constant in utero (11), and declines in absolute amounts only after parturition when wakefulness substantially increases (98). In the chimpanzee, non-REM sleep amounts also reach adult values very early in development (by 1 year postnatal), while the amount of REM sleep remains elevated for a considerably longer time (14,96). Similar findings have been reported in the altricial rat, with non-REM sleep amounts reaching adult values at approximately P16 (91).

As is the case with REM sleep, there is less consistency among reports of non-REM sleep bout durations during development. In humans, QS sleep bouts increase in length from about 30 weeks ga to 36 weeks ga. No further increases were noted at 41 weeks postconceptional age (17). In their study of the neonatal rhesus monkey, Meier and Berger (94) reported a prolonged development of adult-length non-REM sleep periods, which extended over several months. Similar findings have been reported in altricial kittens (59). In contrast, Szeto and Hinman (11) have shown no significant changes in non-REM bout durations in the fetal lamb in utero, and no significant changes in non-REM bout length are observed in the altricial rat after the third postnatal week (91).

Brain Activity

There are a number of excellent texts and reviews on the development of the human EEG (99,100); consequently the following sections will primarily cover animal studies. The general findings from these studies are that patterns of neural activity typical of REM and non-REM sleep become increasingly well defined during the latter half of gestation for precocial species like the lamb (10,11,20), and in the second to fourth postnatal weeks for altricial species like the rat or cat (12,13,91,101). There is some disagreement regarding the relative rates of maturation in REM and non-REM sleep brain activity. Several early studies reported high-frequency bursts in the neonatal rat and cat (13,50) and premature guinea pig (102) EEG during BAS, several days before EEG slow waves were detected in BQS. In other studies of the neonatal rat, however, EEG slow waves comparable to the trace alternant pattern of the human infant were reported before observations of REM-sleep-like EEG activation (71,101). Similar finding in the developing dog have been reported by Breazile, who noted that the development of EEG slow waves preceded the development of periods of desynchronized EEG (103). More quantitative studies of the developing EEG (91) and studies of cortical unit activity (51,104) show that patterns of non-REM and REM sleep brain activity generally appear at the same time. The main developmental changes in REM and non-REM brain activity are summarized below.

REM Sleep

The development of REM sleep brain activity is sequential, with changes in tonic cortical/subcortical activation preceding the maturation of phasic brain activity, and REM sleep atonia. The elevated neural activity typical of mature REM sleep is generally observed between the second and third postnatal weeks in altricial species like the rat or kitten (12,13,59,101), and in the latter third of gestation for precocial species like the lamb (10,11,20). In subsequent weeks, the EEG in REM sleep becomes increasingly dominated by faster frequencies and a slow theta rhythm (4.0–5.0 Hz) (91,101,104). The source of this theta rhythm appears to be from the hippocampus, which begins generating oscillations within the theta range (5.0–9.0 Hz) during sleep at approximately the time of EEG differentiation

(64,104,105). The increase in EEG activation during REM sleep is accompanied by the development of state-specific discharge in FTG cells of the pons (60) and serotonergic cells of the raphe nuclei (63). By the fourth postnatal week in the rat, the theta rhythm is a prominent constituent of cortical and hippocampal EEGs, and has shifted toward the faster frequencies typically observed in adult REM sleep (64,104,105). The maturation of EEG theta continues until the second postnatal month, at which time its frequency and amplitude become comparable to those of the adult animal (91,104,105). Similar developmental changes in hippocampal theta during REM sleep have been reported in the altricial rabbit (65). Other features of REM sleep develop more slowly. PGO waves are first detected in the kitten at approximately 21 postnatal days (5–10 days after EEG differentiation), but do not achieve completely adult properties until 1–2 weeks later (61,62). The motor inhibition typical of REM sleep atonia has an even more protracted development. Motor reflexes typically dampened in adult REM sleep are uninhibited or even augmented in the neonatal kitten prior to the fourth postnatal week [masseteric reflex (106)] or as late as 60 postnatal days [monosynaptic reflex (107, 108)]. A delayed maturation of REM sleep atonic mechanisms has also been observed in humans (109,110) and nonhuman precocial species (111).

Non-REM Sleep

The EEG of non-REM sleep in adult mammals is characterized by slow waves (delta bands: 0.5–4.0 Hz) and sleep spindles (7–14 Hz). These two EEG features of non-REM sleep appear to develop at slightly different rates. EEG slow waves are generally reported to develop first, appearing as isolated slow waves in a burst-suppression EEG pattern (14,23,101). This immature EEG pattern, called trace alternant in the human infant (30) and associated with state 1 in the fetal baboon (23), is replaced with a more continuous slow-wave pattern during the course of development. Sleep spindles are also noted in the EEG at later points of development. In the rat, for example, EEG slow waves first appear between P9 and P10 in a burst-suppression pattern, and increase in number and amplitude over the next 2 postnatal weeks. Sleep spindles appear at approximately P14–P16 (91,101).

In contrast to REM sleep, where the development of electrographic features of REM sleep is quite prolonged, the maturation of non-REM brain activity is completed in a relatively short time. The prominence of EEG slow waves measured as delta power (DP) in the rat increases until approximately the second to third postnatal weeks, after which time adult values are attained (91,101,104). Similar findings have been reported in the fetal lamb (11) and human neonate (112). The level of EEG delta activity during non-REM sleep does not appreciably change during the latter third of gestation, reaching stable values by 120 days ga in the lamb fetus (11). There may also be transient overexpressions of EEG delta activity during development, followed by subsequent declines to adult values. In the rat, for example, non-REM delta power peaks in the fourth postnatal week, exceeding non-REM delta power obtained in adult rats (91,101,104). Similar overexpression of non-REM delta activity has been reported in humans (70).

III. Development of Sleep Regulatory Mechanisms

Mammalian sleep is regulated by three distinct mechanisms. An ultradian mechanism determines the alternation of REM and non-REM sleep states within a given rest period. A circadian mechanism determines the distribution of sleep and wakefulness across the 24-hr day. A homeostatic mechanism determines the intensity and, to a lesser extent, the amount of sleep expressed as a function of prior sleep-wake history. These three mechanisms in concert produce the organized sleep-wake patterns typical of normal adult sleep (2).

A. Ultradian Regulation

In adult mammals, REM and non-REM sleep alternate during the course of the rest period. This "sleep cycle" has been postulated to be generated by an endogenous oscillator (113), but recent work has produced evidence that a sleep homeostat governs the timing of REM and non-REM sleep without the need for an oscillator (114). In humans, the REM–non-REM sleep cycle is approximately 90 min long, and in the rat about 12 min long, although in both cases a large amount of variability exists between individuals (114). In the developing mammal, sleep cyclicity develops relatively slowly and tends to be considerably shorter than in adults. In the developing kitten, for example, significant ultradian periodicities in REM and non-REM cycles are not consistently observed until the third to fourth postnatal week (12). The few kittens that do display faint periodicity in their sleep cycles prior to the third postnatal week generally display sleep cycles of much shorter duration compared to adult cats (12). Human infants also appear to exhibit shorter and more variable sleep cycles as compared to adults (46,47,115). Stern et al. (47) and Scher et al. (115) examined sleep cyclicity in human premature infants ($\leqslant$32 weeks ga) and found highly variable periodicities, which, on average, were significantly shorter (10–12 min) than those reported in term infants, toddlers (30–50 min) (46,116,117), or adults.

B. Sleep Homeostasis

Several findings in adult mammals suggest that sleep is homeostatically regulated. Periods of enforced waking lead to increased sleep drive, or sleepiness, which is relieved by subsequent sleep (1,2). Sleep deprivation produces compensatory increases in REM and non-REM sleep time (118) and non-REM delta activity during recovery (1,2). Non-REM delta activity has also been shown to accumulate during normal periods of consolidated wakefulness and discharge (decline) during subsequent non-REM sleep (1,2). These changes in sleep drive and sleep expression are thought to reflect the accumulation of sleep need during enforced waking and the homeostatic discharge of sleep need during recovery (1,2). It has been further suggested that the discharge of sleep need is necessary for normal waking function (119,120).

Sleep homeostatic mechanisms undergo several important modifications during the neonatal period. One important developmental change in sleep homeostasis is in the tolerance of sleep pressure. The amount of waking is very low during the neonatal period in human and altricial species, and neonates are unable to maintain consolidated bouts of waking comparable to those typically observed in adult animals (13,91,93). Short periods of sleep deprivation that have negligible effects in adult animals lead to a rapid rise in sleep pressure and produce compensatory increases in sleep time and/or intensity during recovery (121–124). These findings suggest that sleep pressure accumulates at a greater rate in infant animals compared to adult animals. The discharge of sleep pressure also differs between neonatal and adult animals. Total sleep deprivation in neonatal rats increases non-REM sleep time, but does not increase non-REM sleep delta activity until the beginning of the fourth postnatal week (125) (Fig. 6). The development of adult-like responses to sleep deprivation is paralleled by the appearance of a declining trend in EEG delta activity across the rest period (126) (Fig. 7).

Similarly in humans, declining trends in EEG delta activity are not observed until approximately the second postnatal month (127). Human infants also appear to respond to sleep deprivation in a manner comparable to the neonatal rat. Selective (122) or total sleep deprivation (123,124) in human neonates leads to compensatory increases only in non-REM sleep time. Exactly when sleep deprivation produces increases in EEG slow wave activity in human neonates is at present unknown. An interesting observation in both altricial and human neonates is the absence of a REM sleep "rebound" following selective REM or total sleep deprivation at ages when non-REM sleep increases are reported (122,125). In neonatal rats, REM sleep time rebounds are first detected near the end of the third postnatal week following 3 hr of total sleep deprivation (125). Non-REM sleep time rebounds, however, are observed as early as P12 in the rat (125). It would seem, as is true for other REM sleep phenomena, that REM sleep homeostatic mechanisms have a more protracted development compared to non-REM sleep.

The reason for the dramatic changes in the response to sleep deprivation in neonatal animals is unknown. The ability to increase non-REM sleep DP after sleep loss could be related to the development of thalamocortical networks necessary for increased synchronized firing in non-REM sleep (128). In the neonatal rat, synaptogenesis and myelination extend into the fourth postnatal week (129). Prior to P24, neonatal rats may not be able to increase synchronized activity in thalamocortical networks due to the relative immaturity of these systems in the second and third postnatal weeks (129). The increase in non-REM sleep time

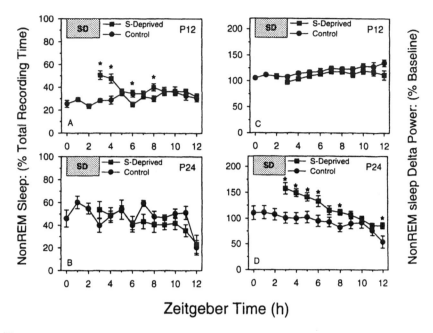

Zeitgeber Time (h)

Figure 6 Effects of 3-hr sleep deprivation by forced locomotion in neonatal (P12) and juvenile (P24) rats ($n = 7$). Mean ($\pm$ SEMs) hourly amounts of non-REM sleep (A, B) and non-REM sleep EEG delta power (0.5–4.0 Hz) in sleep-deprived ($n = 7$) and control rats ($n = 7$) are shown. Sleep deprivation in P12 rats increased NREM sleep time (A) but had no effect on NREM sleep delta power (C). In contrast, sleep deprivation in P24 rats had no effect on NREM sleep time (B), but increased NREM sleep delta power. SD = interval of sleep deprivation. *Significant difference between sleep-deprived and control groups ($p < 0.05$). (Adapted from Ref. 125.)

following sleep deprivation may represent a compensatory response to sleep deprivation in an animal that, due to developmental constraints, cannot intensify non-REM sleep DP. Of course, this hypothesis assumes a direct relationship between non-REM sleep DP and the restorative function of sleep. We do not know what that relationship or function is.

One suggestion is that the function of non-REM sleep could be the restoration of brain energy reserves. In this model it has been hypothesized that activation of the A1 receptor by adenosine, which is released extracellularly as a function of metabolic rate, mediates the increases in non-REM sleep DP following sleep deprivation (120). The number of mammalian brain A1 receptors (130,131) and concentrations of the synthetic enzyme for adenosine (5′ nucleotidase) (132,133) increase during the first 3 postnatal weeks, reaching adult levels between P20 and P30. Brain concentrations of the degradative enzyme for adenosine (adenosine

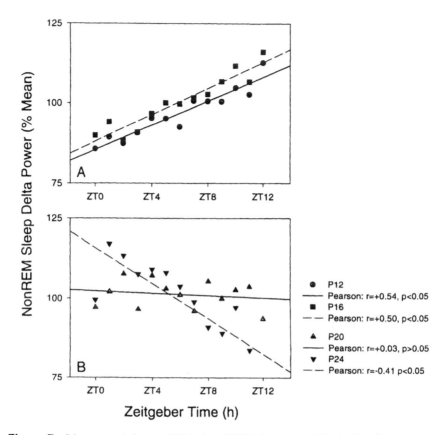

Figure 7 Linear trends in non-REM sleep EEG delta power (0.5–4.0 Hz) during the rest phase (ZT0–ZT12) in neonatal rats ($n = 8$). Mean hourly delta power values are expressed as a percentage of the recording period average. Best-fit regression lines and corresponding Pearson correlation values are shown for P12 and P16 (A), and P20 and P24 (B) rats. A declining trend in non-REM EEG delta power across the rest phase was not observed until P24. (Adapted from Ref. 126.)

deaminase) decrease during the same period (134). Therefore, the developmental increase in non-REM sleep DP following sleep deprivation between P20 and P24 may represent the maturation of adenosinergic modulation of non-REM sleep DP.

C. Circadian Regulation

In adult mammals the distribution of sleep and wakefulness across the circadian day is regulated by an endogenous oscillator located within the suprachiasmatic nucleus of the hypothalamus (SCN) (see Chapter 4). Light information conveyed to the SCN via the retinohypothalamic tract (RHT) entrains this endogenous oscillator to external time cues, which in turn allows for synchrony between endogenous sleep-wake cycles and temporal cycles in the external environment. Other, lesser, zeitgebers like food availability and social cues can also entrain the endogenous oscillator.

The development of circadian regulation is reviewed in detail in Section IV. Briefly, circadian regulation of sleep and waking is robust by 2–3 months after birth in humans and may begin sooner; rhythms in activity and rest have been reported in preterm infants and in fetuses. In rats circadian control over sleep and waking begins to appear by about postnatal day 15 and is robust by the fourth postnatal week. In all mammals it is likely that the circadian pacemaker within the SCN is functional during fetal development and is initially entrained by rhythms of the mother. Entrainment by light occurs by the end of the first postnatal week in rats and is probably possible in humans by the time of birth. As described above, several important aspects of sleep structure and regulation undergo developmental changes coincident with the development of the circadian regulation of sleep. An important, but as yet unanswered, question is whether the regulation of sleep by the circadian pacemaker plays a role in the development of sleep itself.

D. Is Sleep Important for Developing Animals?

The large amount of sleep in infants suggests an important role for sleep in developing animals (93,135,136). REM sleep, for example, has received considerable attention as a state that may facilitate brain development. Infants seem to need more REM sleep than adults since REM sleep amounts progressively decrease as animals mature (13). Brain activation is important for neural development (137), and REM sleep provides endogenous brain stimulation at times when wake amounts are very low (93).

Two classes of experiments provide evidence for a role for REM sleep in neural development. Correlative studies have shown associations between the amount of REM sleep, or REM sleep phasic activity, and brain development. When juvenile rats (30+ days) are placed in enriched environments, the weight and synaptic densities of their brains increase, and these changes are associated with increased REM sleep (138). In rats, phasic REMs increase in duration and intensity during REM sleep near the time of eye opening. This elevation in phasic activity is also observed in dark-reared rat pups and is believed to represent preparatory activations of the visual system prior to waking experience (139).

The function of REM sleep has also been studied by depriving infants of REM sleep, or REM sleep phenomena. Neonatal sleep deprivation is particularly difficult to do since sleep pressure is very high in infants (121), and the majority of these studies have relied on pharmacological blocks of REM sleep. Pharmacological suppression of REM sleep leads to chronic deficits in sexual, aggressive, and sleep behaviors in adulthood (135,136). These findings should be interpreted cautiously because of the multiple effects these drugs have on the developing brain, and in light of recent experiments that suggest that neonatal REM sleep suppression is not the causal factor in the adult deficits in sleep behaviors (140). In the earlier work, a pharmacological block was used that elevates both serotonin and norepinephrine (NE), which had the effect of suppressing REM sleep (135,136). The use of more selective agents has made it possible to suppress REM sleep by elevating either 5-hydroxytryptamine (5HT) or NE. The result was that the selective elevation of NE did not produce the long-term effects but the elevation of 5HT did. Since both treatments suppressed REM sleep, the long-term effects must be associated with the 5HT modifications rather than with the loss of REM sleep per se (140).

Davenne and Adrien (141), Oksenberg et al. (142), and Pompeiano et al. (143) have independently demonstrated a possible role for REM sleep in the developing visual system. In the former study, PGO activity, which is maximally expressed during REM sleep, was abolished in neonatal kittens by bilateral lesions of brainstem PGO centers. The loss of PGO activity during development resulted in smaller lateral geniculate (LGN) neurons and lowered LGN responses to optic stimulation (141). These findings suggest that the phasic activity of REM sleep may be important for the development of LGN neurons. Using the large-and-small-pedestal technique of REM sleep deprivation, Oksenberg et al. have shown that REM sleep deprivation in kittens enhances the effects of monocular deprivation in LGN neurons (142). Similar findings have been reported by Pompeiano et al. (1993), who used gentle handling to deprive kittens of both REM and non-REM sleep (117). These results suggest that brain activity during REM sleep may play a role in the establishment of particular patterns of connectivity that occur during "critical periods" in the developing brain.

The presence of non-REM sleep regulation in both neonatal rats (125) and humans (122–124) suggests that non-REM sleep may also be important for developing animals. The maturation of non-REM sleep coincides with the formation of thalamocortical and intracortical patterns of innervation (144,145) and periods of heightened synaptogenesis (70,129). Non-REM sleep is associated with two processes important in synaptic remodeling: elevations of intracellular Ca^{2+} and synchronized firing in neuronal networks (146,147). During non-REM sleep, waking patterns of neuronal activity are reactivated, suggesting that information acquired during wakefulness is further processed during sleep (148,149). It is therefore possible that non-REM sleep contributes to synaptic remodeling by providing an endogenous source of repetitive, synchronized activity within specific neuronal paths (150). Alternatively, non-REM sleep may replenish neural substrates reduced during the increased neural activity characteristic of the maturing brain (119,120). The maturation of non-REM sleep also parallels developmental increases in waking activity (13,91) and brain metabolism (151). Consequently non-REM sleep may be important in the restoration of neuronal substrates depleted during brain development.

IV. Development of Circadian Timing

Circadian pacemaker, used in reference to mammals, is almost synonymous with the suprachiasmatic nucleus (SCN). "Circadian pacemaker" has, however, a formal definition encompassing properties that may or may not be attributable to the activity of the SCN alone. Formally, a circadian pacemaker regulates circadian timing, in particular the phase and period of a circadian rhythm. While one study may address the development of a circadian pacemaker as defined by formal criteria, another might examine the development of the SCN. It is likely that the results of one study are directly relevant to the other, but this should not be assumed.

A. Development of the Suprachiasmatic Nucleus

Much is known about the anatomy of the SCN in mammals, including that of humans, but the relationships between its anatomical features and its function as a circadian pacemaker are poorly understood (see Chapter 4). The SCN is not a homogeneous population of cells. In addition to neurons and glia, the neurons themselves can be classified into several types on the basis of neurotransmitter or peptide content, morphology, or electrical activity. A given type of neuron is generally not uniformly distributed within the SCN, and the SCN of some mammals, including rats and humans, is described as comprising distinct subdivisions. How the diversity of cells within the SCN is generated and how its distinct subdivisions, as well as specific patterns of afferents and efferents, arise are not understood. These are fundamental questions of developmental neurobiology, and require as a starting point for their answers basic information about SCN development such as the site and timing of SCN neuron production and the timing of phenotypic differentiation.

Formation and Differentiation of the SCN

The SCN arises from proliferating cells of the neuroepithelium in the anterior, ventral diencephalon just caudal to the preoptic recess and optic fissures (152–156). Neurogenesis of the SCN occurs between E14 and E17 in rats and E10 and E13 in Syrian hamsters, 5 and 3 days before birth, respectively (152,154,156). In rats, the SCN has been generalized by Altman and Bayer (154) to arise as part of a third wave of hypothalamic development. There are gradients in neurogenesis within the SCN itself (152,154,156), raising the possibility that the time when a cell becomes postmitotic is related to the specification of its fate and to the formation of subdivisions. In contrast to many other central nervous system structures, the overproduction of cells followed by cell death does not appear to have a significant role in SCN development. Instead, cell number appears to be regulated primarily by the rate and duration of proliferation and by the coalescence of cells into nuclei (157–162).

Within 2 or 3 days of SCN neurogenesis in rodents, mRNAs (VP, VIP), proteins (VP, PHI, VIP), and specific ligand binding (VIP, melatonin), characteristic of SCN cells, can be detected (163–172). In humans, the SCN is formed before 20 weeks' gestation (173) and both VP and VIP immunoreactive neurons have been observed at approximately 30 weeks' gestation (174,175). These neurons appear to increase in number during the first postnatal year and may continue to change in number and pattern for several years. Two genes (*clock* and *per*) have recently been cloned in mammals that may code for molecular components of the circadian oscillator mechanism (176–178). However, the expression of these genes during development has not yet been examined.

Although there is rapid differentiation of SCN neurons during the first few days after the cells are produced, the differentiation of the rodent SCN as a whole continues over several weeks. For example, in rats, VIP-expressing cells continue to increase for more than 3 weeks after birth (172). Synaptogenesis within SCN, probably involving both intrinsic and afferent connections, occurs primarily between E21 and P10 in rats and E15 and P4 in Syrian hamsters (157,179). The SCN receives three principal afferent projections: from the retina (retinohypothalamic tract, RHT), from the intergeniculate leaflet (IGL) of the lateral geniculate nucleus (GHT), and a serotonergic projection from the midbrain raphe nuclei (180). RHT innervation of the rat and Syrian hamster SCN occurs most intensively between P1 and P10 and P4 and P15, respectively (179). Based on immunocytochemistry for serotonin, the serotonergic innervation of the SCN develops between E22 and P21 in rats (181) and between P3 and P21 in Syrian hamsters (182). In hamsters, the GHT, based on neuropeptide Y immunoreactivity, develops between P4 and P11 (183). The IGL itself is formed in the hamster around embryonic day 14 (183).

In rats, the RHT initially projects to a larger field that includes areas outside of the SCN. These ectopic projections disappear between P4 and P10 (179) and it is possible that, as elsewhere in the nervous system, the survival of afferents depends on competition for a limited resource provided by functionally appropriate targets. If one eye is removed early in development (P2) in rats, projections of the remaining eye increase (184), possibly in response to the availability of additional synaptic sites and an increased availability of trophic substances. While interactions with target cells may influence development of the RHT, there is little evidence that the RHT or other SCN afferents influence the development of the SCN. For example, ablation of the SCN before RHT development does not result in the respecification of another tissue as the pacemaker (185,186), and to the extent it has been studied, deafferentation of the SCN early in development has little effect on its morphology or properties (161,187).

When an SCN-controlled rhythm appears during development, this means that not only is the pacemaker functioning at that time, but the output pathway to that function has also developed. The best-characterized efferent pathway of the SCN is that for the control of the pineal melatonin rhythm. In rats, a rhythm in activity of *N*-acetyltranseferase (NAT, the rate-limiting enzyme for the synthesis of melatonin) begins on postnatal day 2–4. This is after the onset of pacemaker function (see below), but coincident with the development of sympathetic inner-vation of the pineal (188). By the start of the third postnatal week circadian organization of sleep and wakefulness has been seen in rats (H. C. Heller and M. G. Frank, unpublished data) and in both rats and Syrian hamsters an activity rhythm may be present (187,189,190). Following transplantation of fetal tissue in hamsters, rhythmicity may be restored within 10 days, suggesting that SCN output is sufficient to drive a rhythm in activity and rest by at least the first postnatal week (191). Developmental change in pacemaker output mechanisms is suggested by studies of temperature rhythms in rats. In artificially raised rat pups kept in constant light at ambient temperatures below thermoneutrality, a temperature rhythm can be measured between postnatal days 6 and 15. This rhythm disappears by about day 17 and is subsequently replaced by a new temperature rhythm with different characteristics (192). A similar result has been reported in lambs (193).

Specification of Fate

Little is known about the mechanisms that specify the fate of pacemaker cells. It is not even known to what extent the ability to generate circadian oscillations is a differentiated property of specialized cells rather than a relatively common prop-erty of many cells and tissues. The ability to generate circadian oscillations may be a property that is more commonly lost by cells during differentiation rather than gained. What is known about the specification of pacemaker cells indicates that the SCN, at least to the extent it has been examined, is relatively unaffected by the environment in which it develops. In rats, fetal anterior hypothalamic tissue was transplanted to the anterior chamber of the eye in adult hosts before SCN neuro-genesis (E13). Clusters of cells characteristic of the SCN were observed in the tissues despite development in an abnormal environment (194). Thus before SCN cells have become postmitotic and before SCN afferents develop, the fate of anterior hypothalamic tissue to produce SCN-like neurons has been determined. On the other hand, the time course of cellular differentiation may be influenced by the site of transplantation (195,196). In some studies, fetal tissue has been minced or cells dispersed prior to transplantation (196,197), yet SCN phenotypes devel-oped, indicating that specific spatial relationships among SCN cells during differ-entiation are not required for phenotypic expression. Similarly, cultures of post-natally obtained SCN cells or tissue are also capable of expressing phenotypes characteristic of the in situ SCN, including the generation of circadian oscillations (198–200).

B. Functional Development of the Circadian Pacemaker

The functional development of the pacemaker refers to the onset of circadian oscillations, the development, by functional criteria, of input to and output from the SCN, and developmental changes in properties such as the free-running period. An increase in amplitude is a common feature of developing rhythms (190), but it is less clear to what extent a change in amplitude reflects changes in the pacemaker itself rather than a change in the maturation of the systems regulated by the pacemaker.

Initiation of Oscillations

Circadian oscillations appear to be generated by the SCN at least as early as a day or two after SCN neurons become postmitotic. The first evidence for this was that entrainment begins before birth in rats. In 1975, Deguchi (201) reported that the postnatal phase of the rhythm in rat pineal NAT activity was determined in part by the phase of maternal rhythmicity experienced before birth. This was substantiated and extended by Reppert et al. (202), who demonstrated that not only was phase initially established before birth but also maternal rhythmicity was responsible rather than the external light/dark cycle. Similar evidence for prenatal maternal entrainment was also obtain for the plasma corticosterone (203) and temperature (192) rhythms in rats, and for the wheel-running activity rhythms in Syrian hamsters (204) and the spiny mouse (205). Because these rhythms are regulated by the SCN in adults, the prenatal establishment of phase in them is likely to be accomplished by the prenatal entrainment of oscillations generated by the SCN, thus providing evidence that the SCN is generating oscillations before birth.

Rhythmicity in the fetal SCN itself was first demonstrated by Reppert and Schwartz in rats (206,207) and later in a fetal primate (Fig. 8) (208). They measured 2-deoxy-D-glucose (2DG) uptake in the fetal SCN at two times of day. A clear difference in 2DG uptake was observed between day and night in both the fetuses and their dams with uptake being higher during the subjective day. In rats, the day/night difference in 2DG uptake was first seen at about E19.5. It is possible that oscillations had started before E19.5 but were not yet expressed in 2DG uptake. Importantly, Shibata and Moore (209,210) measured 2DG uptake and electrophysiological activity in slices from E22 rats at two times of day 1–4 hr after taking the fetuses from the dam, indicating that rhythms in the SCN did not represent passive responses to maternal rhythms. Reppert and Uhl (164) also demonstrated a rhythm in vasopressin mRNA levels in vivo on E21.5. A rhythm in VIP mRNA has also been reported in rats at least as early as postnatal day 10 (172). Interestingly the endogenous component of this rhythm seemed to disappear by day 20 when mRNA levels were more directly regulated by light as they are in adults.

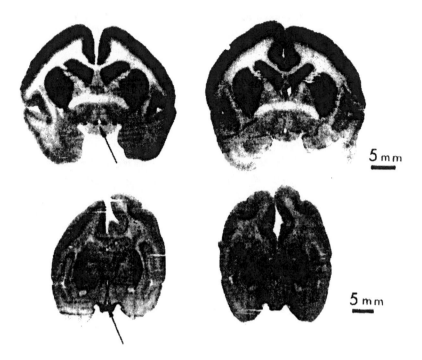

Figure 8 A rhythm in deoxyglucose (DG) uptake in the SCN of a fetal primate and its mother (squirrel monkey, *Saimiri sciureus*). Autoradiographic images of maternal and fetal sections are shown above histological sections from the fetus (approximately 90% of term). (Left) Images from animals given 2DG during the subjective day; (right) images from animals given DG during subjective night. The metabolically active SCN are indicated on the subjective day autoradiographs by arrows. (From Ref. 208; reproduced by permission of Elsevier Science Ireland Ltd.)

Although no fetal rhythm has been directly shown to free-run in the absence of maternal rhythmicity, the hypothesis that these rhythms represent oscillations generated by the fetal SCN is consistent with the evidence for prenatal entrainment. In addition, 2DG uptake has been measured in fetuses in vivo or in fetal brain slices in vitro after the dams received SCN lesions early in gestation (210,211). While the fetuses from such mothers, as a group, did not express a rhythm, they showed a range of values suggesting that individual fetuses were expressing rhythms that were out of phase with each other. Similarly, when pregnant hamsters received SCN lesions early in gestation (day 7) the pups expressed normal wheel-running activity rhythms at weaning, but the synchrony among the pups within a litter was disrupted (212). In rats, maternal rhythms have also been disrupted by exposing the dams to constant light (210,213). In one study, temperature rhythms in individual pups were measured in constant light after isolation from the dam on P9. In some litters, synchrony among the pups within a litter was absent but in others it was not. Although it is possible that maternal rhythms were not completely disrupted by constant light, it is also possible that synchrony can exist in the absence of maternal rhythmicity. Synchrony could have been established by direct interactions among pups prior to isolation even though the available evidence does not support this (202,214). Alternatively, such "residual" synchrony could result from developmental synchrony; instances of such synchrony have also been reported following SCN lesions of the dam early in gestation (211,215a). This possibility would require that the initiation of oscillations occurs at a specific time in development and at a specific phase of the oscillations.

Development of Entrainment

As already discussed in the context of pacemaker initiation, entrainment in rodents begins before birth and is initially mediated by maternal rhythms. Even when the animals are raised under constant lighting conditions, the rhythms of the dam and her offspring normally show a consistent phase relationship with each other (Fig. 9). In addition, at the first appearance of rhythmicity within the fetal SCN the rhythms are in phase with those of the dam. As also discussed, SCN lesions of the dam disrupt the initial entrainment. In hamsters, lesions on day 14 of gestation are less disruptive than those on day 7, suggesting that entrainment occurred before day 14 (212). If circadian oscillations do not begin until sometime after SCN neurogenesis, then these results indicate that entrainment occurs between about E12 and E14 in the hamster. Entrainment appears be accomplished rapidly, suggesting that oscillations, when they are first generated, are especially sensitive to resetting stimuli.

With the discovery that entrainment by the mother occurs before birth, effort has been given to identifying specific signals. Reppert and Schwartz (211) removed, in separate experimental groups, the adrenal, pituitary, or pineal glands, or the ovaries and measured the phase of the pups' rhythms. They saw no effect on entrainment of the fetal SCN 2DG rhythm or the postnatal pineal NAT rhythm, which was equivalent to the effect of ablating the dam's SCN. The only possible effect of these ablations was an effect of pinealectomy on the 2DG rhythm, where some abnormally low values were observed (215b).

A second approach to identifying candidate signals was to experimentally administer signals to pregnant rats or hamsters. Using this approach, three treatments have been shown to entrain the rhythms of the pups measured postnatally: restricted feeding schedules in rats (216), and in Syrian hamsters, melatonin injections or injections of a D1-dopamine receptor agonist, SKF 38393 (217–219). In no case has it been established that natural rhythms mimicked by these treatments are physiological entraining signals, but at the same time it is possible that all three are in some way involved.

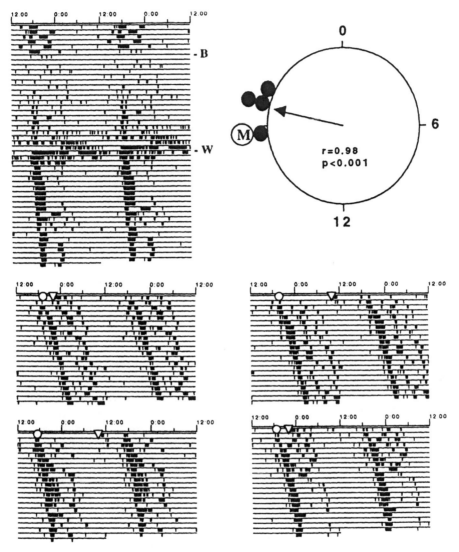

Figure 9 Maternal entrainment of Syrian hamster behavioral rhythms. (Top left) The double-plotted wheel-running activity record of a pregnant hamster in dim constant light. Her litter was born (B) and weaned (W) on the days indicated. The records of four of her pups, also kept in constant light, beginning on the day of weaning are shown below. On each of the pups' records the time of weaning is indicated with a triangle and the phase of activity onset, on that day, is indicated with a circle. The activity onset phases are also plotted on a circle representing the 24 hr of the day of weaning (filled circles). The pups' phases were clustered among themselves and around the dam's activity onset (circled M). The arrow within the circle indicates the pups' average phase, and its length, r, indicates the degree of synchrony among them. The p value is based on the Rayleigh test and indicates that the distribution of pups' phases is significantly different from uniform. (Unpublished data courtesy of N. Viswanathan and F. C. Davis.)

There is strong precedence for the pineal gland hormone, melatonin, to function as a circadian signal. Throughout vertebrates, including mammals, a rhythm in plasma melatonin is likely to function as an internal signal for nighttime (elevated melatonin levels) and for daytime (low melatonin levels) (220). Maternal melatonin passes the placenta producing within the fetus a rhythm in melatonin (221–224). Exogenous melatonin can entrain the circadian rhythms of adult rats (225) and possibly those of humans as well (226,227). Furthermore, in Siberian hamsters, the maternal pineal and melatonin rhythm convey photoperiodic information to the fetus (228,229).

To test the hypothesis that melatonin is a maternal entraining signal, Davis and Mannion injected SCN-lesioned, pregnant hamsters with melatonin on days 7–15 of gestation (217). All of the dams actually received two injections per day, one with melatonin and one with vehicle only. Two groups received the melatonin at different times of day 12 hr apart. A third group received vehicle only at both times of day. The phases of the pups' wheel-running activity rhythms at weaning were measured. The melatonin injections restored synchrony among the pups within individual litters, and set the average phases of the two melatonin groups 180° apart (Fig. 10). Because the dams were handled and given injections twice a day, the cause of the different average phases must have been the timing of melatonin exposure. The fetal SCN is a likely target of melatonin since melatonin receptors have been localized to the fetal SCN in several species including Syrian hamsters, rats, and humans (169,173,230–233). In addition, melatonin injections given directly to pups on postnatal days 1–5 were also shown to cause entrainment, suggesting that fetuses are also directly affected by melatonin (234).

The case for melatonin as a signal mediating the initial entrainment of the circadian pacemaker in mammals is strong, yet, as previously mentioned, pinealectomy in rats does not disrupt maternal entrainment. The effects of pinealectomy may be more subtle, and/or there may be redundant signals as indicated by the entraining effects of treatments other than melatonin. Another such treatment shown to cause entrainment in Syrian hamsters is injections of the D1-dopamine receptor agonist SKF 38393 (218). SKF 38393 also induces expression of c-*fos* in the fetal rat, hamster, and mouse SCN and cocaine induces both c-*fos* and *jun*-B in the fetal rat SCN (218,235,236). In both rats and hamsters, D1 receptor mRNA is expressed in the fetal SCN (218,236). Nicotine injections to pregnant rats also induce c-*fos* expression in the fetal SCN (Fig. 11) suggesting that a cholinergic system is also involved in entrainment of the fetal circadian pacemaker (237). Exactly what maternal signals might be mimicked by dopaminergic or nicotinic activation within the fetal SCN is unknown. Ligands of maternal origin could themselves be signals or maternal signals could stimulate the release of ligands within the fetus (238).

An interesting aspect of prenatal entrainment by SKF 38393 in Syrian hamsters is that the average phase of the pups' rhythms is approximately 180° different

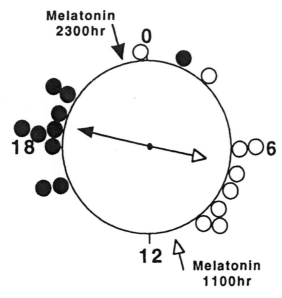

Figure 10 Entrainment of Syrian hamster pups by prenatal melatonin injections to SCN-lesioned dams during the last week of gestation. The large circle represents the 24 hr of the day of weaning and the small symbols are the average phases of entire litters. The dams of the litters represented by closed symbols received melatonin at night and vehicle alone in the morning while those of the litters represented by open symbols received melatonin in the morning and vehicle alone at night. The arrows indicate the average phases of the two groups, which differed by 180°. (Redrawn from Ref. 217.)

from that produced by melatonin (219) (Fig. 12). Melatonin sets the oscillations in such a way that mid-subjective night becomes coincident with the time of the injection while SKF sets mid-subjective day to the time of the injection. Also, unlike prenatal SKF and light pulses in adults, it appears that prenatal melatonin does not induce *c-fos* (219). Together these observations suggest the hypothesis that elevated maternal melatonin is a signal representing nighttime while dopaminergic activation is a signal representing daytime (Fig. 13).

As already noted, the fetal SCN appears to be highly sensitive to entraining signals; a single prenatal injection of either melatonin or SKF 38393 is sufficient to entrain the pups. Since these injections were given to SCN-lesioned dams it is likely that the phases of the fetal pacemakers were widely distributed prior to the injections. The production of synchrony from such distributions by single injections would require phase shifts larger than those produced by light in adult hamsters (219). The sensitivity to these drugs rapidly disappears after birth. Both melatonin and SKF injections given directly to hamster pups on postnatal days 1–

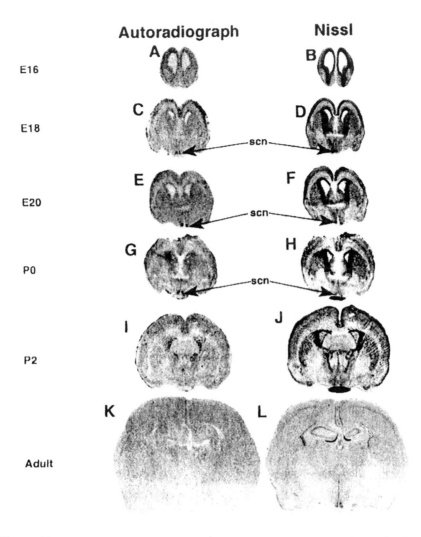

Figure 11 Induction of c-*fos* expression in the fetal and newborn rat SCN by nicotine. In situ hybridization autoradiographs (left) and corresponding Nissl-stained sections (right) hybridized with [^{35}S] c-*fos* cRNA probes after treatment with nicotine are shown. Hybridization to c-*fos* cRNA was seen at fetal ages E18 and E20 and on the first day after birth (P0) but not 2 days later or in adults. (From Ref. 237; reproduced by permission of Elsevier Science B.V.)

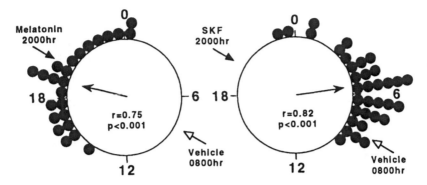

Figure 12 Prenatal entrainment of Syrian hamster circadian rhythms by single prenatal injections of melatonin or the D1-dopamine receptor agonist SKF 38393 (SKF). Injections were given to SCN-lesioned dams on E15, and the filled symbols represent the activity-onset phases of individual pups. The single injections were able to establish synchrony within the groups, and although the two substances were given at the same time of day, they established average phases differing by approximately 180°. (Redrawn from Ref. 219.)

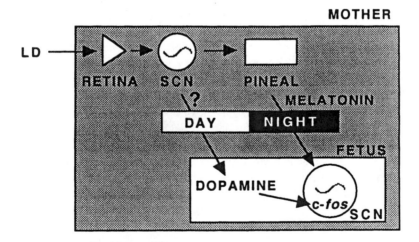

Figure 13 Conceptual scheme for maternal entrainment of the fetal pacemaker. Maternal melatonin secreted at night directly affects the fetal pacemaker through membrane receptors but by as-yet-unidentified transduction mechanisms. Other maternal signals act through a dopaminergic (and possible cholinergic) mechanism to induce c-*fos* expression in the SCN. These two signaling systems have opposite effects on the phase of the fetal pacemaker where artificially activated at the same time of day. Thus, if active at opposite times of day, as depicted here, systems would act together to entrain the pacemaker to one and the same phase.

5 caused entrainment, but injections in days 6–10 did not (234). The induction of c-*fos* expression in the fetal rat SCN by SKF 38393 or by nicotine also ends within 2–5 days after birth (237,239). Interestingly, at about the same time, the induction of c-*fos* by light begins (see below). The loss of c-*fos* induction by SKF is not, however, caused by the development of retinal afferents since the loss occurs in rats enucleated on the day of birth (239).

Although maternal entrainment begins before birth, it continues to varying degrees after birth (202,240,241). The signals mediating postnatal entrainment could be the same as or different from those mediating prenatal entrainment. For example, melatonin has been found in human milk and has been shown to pass from the mother to her pups via this route in rats (242,243). In the field mouse (*Mus budooga*), postnatal maternal entrainment can be mediated by artificial cycles in the presence and absence of the dam and the presence of the dam is always interpreted as day (241).

In rats, the ability of the circadian pacemaker within the SCN to be entrained by light develops during the first postnatal week, coincident with the declining strength of maternal entrainment (234,244). The RHT begins to innervate the SCN on postnatal day 1 and the SCN is well innervated by day 7, about the time when entrainment of the rat pineal NAT rhythm by light is first observed (244). The possibility that entrainment by light can occur earlier is suggested by the responsiveness of the SCN to light on day 1 or 2 in rats (using 2DG uptake or expression of c-*fos*), or day 4 in Syrian hamsters and day 3 in Siberian hamsters using Fos protein expression (239,245–248). The slightly later development of a response to light in Syrian hamsters relative to rats could be related to differences in the development of the RHT (179). A summary of both the anatomical and functional aspects of SCN development in rodents is shown in Figure 14.

It is clear from the early development of responsiveness to light that the opening of the eyes in these altricial rodents is not required for the perception of light. Weaver and Reppert demonstrated that light can penetrate the intrauterine environment as well; the SCN is already innervated by the RHT on the day of birth in the precocious rodent, the spiny mouse, and just prior to birth, light caused increased 2DG uptake in the fetal SCN (249). It is unclear, however, whether light would have caused entrainment. In sheep, the SCN is innervated by the RHT before birth (250) and there is suggestive evidence for this in humans (251), raising the possibility that entrainment by light could occur in utero.

Specification of Functional Properties

Circadian pacemakers express specific properties of functional significance. For example, the value of the free-running period and the magnitude of light-induced phase shifts affect the phase relationship of rhythms to environmental cycles. The average values of such parameters as well as the variation in these values vary

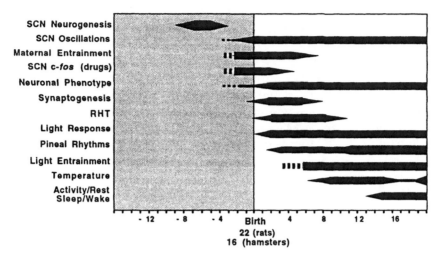

Figure 14 Structural and functional development of the rat and Syrian hamster circadian system (SCN). The time scale shows days relative to the day of birth. The black lines indicate the times in development after which structures or functions exist, or during which developmental processes are occurring. Dashed lines indicated uncertainty about the beginning or ending of a feature, and variation in line thickness roughly indicates the magnitude of the feature. Maternal entrainment includes entrainment by exogenous melatonin. Drug effects on c-*fos* refer to the effects of dopamine agonists and nicotine. Light responses refer to changes in c-*fos* expression or 2DG uptake in the SCN.

among species and are probably adaptive products of natural selection (252). Little is known, however, about what determines the specific value of, for example, the free-running period for an individual or species. It is clear that genetic constitution influences the properties of circadian pacemakers; single gene mutations with large effects on free-running period have been found (253,254) and strain differences in average period suggest polygenic influences (255,256). Genes could exert their effects at several levels, from molecular components of the oscillatory mechanism itself to less direct effects on the regulation of pacemaker development. For example, a mutation of a particular isoform of the neural cell adhesion molecule (NCAM) affects both free-running period and the numbers of VIP-expressing neurons in the mouse SCN (257). It is uncertain, however, whether, the effect on period is the consequence of abnormal development since disruption of NCAM in adults also affects period.

Studies early in the field of circadian rhythms in which organisms were raised under constant conditions demonstrated that it is unnecessary for an organism to experience periodicity in the environment to generate and express circadian rhythms when mature (258,259). More recently, it was found that hamsters or rats born to SCN-lesioned dams in constant conditions develop normal circadian rhythms (211,212) and SCN grafts transplanted as early as embryonic day 11 in hamsters restore rhythmicity (191). Because activity-dependent processes are widespread during the development of the nervous system, it would not be surprising if some modification of neural activity during pacemaker development were able to modify pacemaker properties. Thus far, however, evidence for this is lacking. In the most rigorous test of this possibility, the free-running period of mice was uninfluenced by being raised on non-24-hr maternal and light/dark cycles (260). Similarly, deafferentation is known to affect the development of neural structures, especially within the visual system, yet effects of deafferentation on circadian rhythmicity have not been seen. Specifically, neonatal enucleation had no effect on the free-running period in hamsters (187) and the periods of the rhythmicity restored by grafts (which are abnormal in many ways including the absence of retinal afferents) are on average normal (261,262).

Circadian oscillations are generated by individual SCN neurons (198). The periods expressed by individual neurons in culture are not, however, identical; Welsh et al. (198) reported a range of 21.25–26.25 hr in rat SCN cells. Although it is not known to what extent this variation exists in vivo, an interesting possibility is that the variation arises during differentiation and is important for overall pacemaker function.

An aspect of development that would influence pacemaker properties is the proliferation of SCN cells. Variation in the rate or duration of proliferation during SCN neurogenesis could affect the relative or total numbers of SCN cells and thereby affect properties of the population. Lesion and transplantation studies suggest that cell number can affect free-running period (197,261,263). Hormones might also affect pacemaker development; sex differences have been reported in circadian rhythms and in SCN anatomy, in both humans and other animals (175, 264–267).

V. Functional Aspects of Circadian Rhythmicity During Development

The mammalian circadian pacemaker is functional and entrainable early in a mammal's life at a time when it might seem as though coordination with 24-hr periodicity in the environment is irrelevant. Even in utero, however, the embryo and fetus are exposed to 24-hr periodicity in parameters that may be critical for normal growth and development, including rhythms in nutrients, hormones, temperature, and possibly toxic substances (268). It may be adaptive for aspects of fetal physiology or development to occur only at certain times within the mother's circadian cycle. Daily rhythms in fetal growth, cell proliferation, and susceptibility to toxic substances have been reported in rodents (269–273). The possible necessity for coordination between maternal and fetal physiology does not, however, offer a satisfactory explanation for why the fetal pacemaker should function in utero and be entrained by the mother. In rodents, where the evidence for a fetal pacemaker is strongest, there is no evidence of a rhythm, outside of the SCN itself, that is regulated by the pacemaker. Thus there is no known output through which the pacemaker might play a role in coordinating fetal physiology with that of the mother. In other animals, including humans, there is evidence for rhythms in fetal physiology, but to what extent these rhythms are controlled by an endogenous fetal pacemaker is unclear (see below).

A likely function for prenatal entrainment of the pacemaker is that it is important for an animal's overt rhythms to be in phase with the environment when the rhythms appear some time after birth. Prenatal entrainment of the pacemaker may facilitate this. In this case the environment would continue to include maternal rhythmicity, but also rhythmicity such as the light/dark cycle. In seasonally breeding mammals postnatal entrainment to the light/dark cycle is important for obtaining seasonal information.

In Siberian hamsters, as well as some other species, the photoperiod experienced by the dam during gestation influences how the offspring will respond to the photoperiod they experience after birth, and reproductive development occurs with different timing depending on the time of year (274,275). This prenatal effect of photoperiod is mediated by the maternal melatonin rhythm (228,229), and may involve entrainment of the fetal pacemaker (276). Thus one function for the early development of the pacemaker may be to mediate the transfer of photoperiodic information from the dam to the fetus.

The best example of coordination between the rhythms of the mother and her offspring is that of rabbits in which the doe feeds her pups only once a day. The doe avoids the nest most of the day returning for only a few minutes to feed the pups. The pups anticipate the doe's arrival, probably as a result of entrainment to the doe's rhythm (277–279). A pup that does not anticipate a feeding may miss it, with eventual consequences for its growth and survival.

Another suggestion for the function of early pacemaker development is involvement in the timing of birth. In many mammals, including humans, birth occurs at restricted times of day (280). Because the fetus is known to play a role in the initiation of parturition, and through maternal entrainment has time of day information, the fetus could also influence the time of day when birth occurs. While there is some evidence for this (281), other evidence points to a predominant role for maternal circadian rhythms (282).

The significance of prenatal entrainment could be that it plays a role in the normal development of the pacemaker itself. For example, without entrainment at some critical time in development, the pacemaker may not function normally or there may be a delay in the development of the pacemaker's control over certain functions. There is as yet little evidence for this and additional research is needed, especially in humans where disruption of rhythmicity early in development may become increasingly common and small differences in the development of rhythms such as the sleep/wake cycle could have a large impact on the general health and well-being of both infants and parents.

VI. Development of Human Circadian and Sleep/Wake Cycles

In humans, the pattern of sleep and wakefulness changes from one of multiple, relatively short sleep and wake bouts per day with little distinction between day and night, to one of consolidated sleep at night and continuous waking during the day (Figs. 10 and 11). Much of this change occurs during the first 3 months after birth, so that by 3 months of age clear differences between day and night sleep are established (283–287) (Fig. 15). An increase in the average duration of sleep episodes is more pronounced than an increase in the average duration of waking (285,286,288), and the average duration of sleep episodes may be the first aspect of sleep/wake patterns to be influenced by circadian timing (284–286,288).

The consolidation of sleep that occurs during the first months after birth is sometimes referred to as entrainment (286–288), since as sleep is becoming

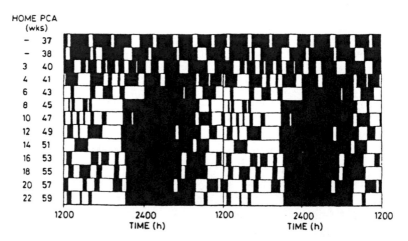

Figure 15 Double-plotted sleep/wake activity record of a female infant born at gestational age 34 weeks and discharged home at postconceptional age 38 weeks. Behavioral state was determined by observation for 24 hr every 1–2 weeks. The dark bars indicate times asleep. The sleep/wake activity rhythm of this infant was considered to have been entrained after 8 weeks in the home environment. (From Ref. 287; reproduced by permission of International Pediatric Research Foundation, Inc.)

consolidated it is coincident with nighttime. The emergence of a rhythm in sleep and waking may, however, indicate little about the entrainment of an underlying circadian pacemaker. The pacemaker may be entrained well before the sleep/wake rhythm appears. The appearance of the sleep/wake rhythm would instead reflect the maturation of pacemaker control over sleep and arousal mechanisms. Consistent with this, Kleitman and Engelmann (284) reported that most infants are in phase with the environment when the sleep/wake rhythm develops. However, they highlighted one exceptional case in which the sleep/wake record of an infant (recorded from 2 to 26 weeks of age) showed a free-running rhythm with an average period of about 24.8 hr between weeks 6 and 15 (Fig. 16). It wasn't until 24 weeks that the rhythm showed a stable 24-hr period with the normal phase relationship to the light/dark cycle. The light and dark exposure of this infant was

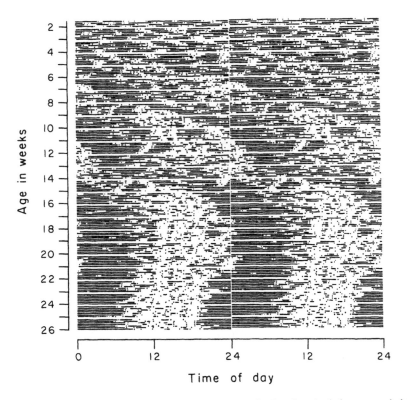

Figure 16 Double-plotted sleep/wakefulness record of a female infant recorded by observation from 2 to 26 weeks of age. The lines indicate time asleep and dots are feedings. (From Ref. 190; reproduced by permission of Plenum Publishing Corp.; original data from Ref. 284.)

not described but it was maintained on a self-demand feeding schedule. At least two interpretations of this record are possible. One is that the circadian pacemaker began to express itself with a period of about 24.8 hr before it became entrained. As it free-ran the phase relationship between the pacemaker and the environment changed causing a modulation in the consolidation of sleep (284,288). Eventually the pacemaker became entrained with a further increase in the consolidation of sleep at night. Alternatively, the pacemaker could have been entrained all along to the 24-hr environment, and the 24.8-hr rhythm was derived from the homeostatic component of sleep regulation. Sleep/wake rhythms with circadian periods (usually longer than 24 hr) can be expressed in adults even though it is clear in those cases that the circadian pacemaker is free-running with a different period (289). Thus this infant's record suggests that the consolidation of sleep and wake occurs to some extent independent of regulation by the circadian pacemaker. The circadian pacemaker is still likely, however, to have a strong influence on the consolidation of sleep, eventually dictating the time of day when consolidated sleep occurred; the largest increases in the duration of sleep episodes in the infant's record occurred when sleep was coincident with what would have been the subjective night phase of the pacemaker (288).

Differences in sleep between day and night are clearly established by the third postnatal month; small, but significant, differences in the activity of infants or in sleep episodes have also been seen within the first postnatal month (283–285,290,291) (Fig. 17) and in some preterm infants (292,293). Rhythms in rectal or skin temperature or in heart rate have also been observed in preterm infants but in many cases rhythmicity is not seen or is only transiently expressed (293–295). Studies of preterm infants have been conducted in hospitals under relatively constant conditions, especially with respect to light/dark cycles. It is difficult, however, to exclude effects of day/night differences in the care of infants (292), and clear instances of free-running rhythms with non-24-hr periods in which a phase reference consistently drifts with respect to clock time have not been demonstrated.

Although the evidence for free-running, endogenously generated circadian rhythms in newborn full-term or preterm infants is not yet strong, it is likely that the fetus and newborn have functional circadian pacemakers. The presence of melatonin-binding sites within the human fetal SCN (18–19 weeks of gestation) (173) and the observation of a 2DG rhythm in the fetal SCN of squirrel monkeys (Fig. 1) (208) indicate that, like rodents, the human fetus has a functional and entrainable pacemaker. There is also suggestive evidence for a fetal pacemaker in sheep; a rhythm in the number of cells showing Fos-like immunoreactivity has been described in fetal sheep (296) and a rhythm in cerebrospinal fluid vasopression is likely to originate from the SCN (297). Thus, the pacemaker may be functioning well before birth in humans, but the output mechanisms that mediate its influence over a measurable rhythm might not mature until after birth.

Circadian rhythms have been directly measured in the fetuses of some mammals while in utero, including humans. In humans, the clearest such rhythms are in heart rate and activity, with activity highest in the evening and heart rate lowest in the morning (298–302). Because these rhythms are generally clearer in fetuses than they are in preterm or newborn infants, it is likely that they are largely passive responses to maternal rhythms. This is supported by one case in which rhythms in fetal heart rate and movement were absent in the fetus of a woman lacking adrenal glands (303). On the other hand, in a study of fetal baboons, fetal heart rate and fetal breathing movements were both rhythmic. The fetal heart rate rhythm showed a consistent phase relationship to the mother, but the rhythm in breathing movement did not, suggesting that it was endogenously generated by the fetus (304).

If there is a functional fetal pacemaker in humans it is likely to be entrained by maternal rhythms. Based on evidence in rodents and the presence of melatonin-binding sites in the human fetal SCN (173), it is also likely that melatonin is a signal for this entrainment. A number of other substances may also affect the fetal SCN, including dopamine agonists, cocaine, and nicotine (218,236,237). Regardless of what the normal maternal signal for entrainment is, exposing the fetus to such substances may phase-shift the fetal pacemaker with long-term consequences for the eventual normal entrainment of the pacemaker and for the development of circadian rhythms.

At some point in development, as in rodents, the human circadian pacemaker is likely to make a transition from entrainment by maternal rhythms to entrainment by other environmental cycles, in particular the light/dark cycle. Although it is not known when the human circadian pacemaker becomes entrainable by light, it is likely that it is entrainable soon after birth; the retinal hypothalamic projection appears to be present before birth (251) and plasma melatonin, although not yet rhythmic, is influenced by light within the first 2 days after birth (305). In newborn baboons, light entrains circadian rhythms and induces c-*fos* expression of the SCN (306). In several studies describing the development of human circadian rhythms, the rhythms are in phase with the environment when they appear, suggesting prior entrainment of the underlying pacemaker or the influence of an uncontrolled environmental factor (284–287,292,302,307–310). In one study, the phases of body temperature rhythms, although not random on day 2, became more synchronized by the fourth week (290).

In a study of term and preterm infants, McMillen et al. (287) found that the development of the sleep/wake rhythm was influenced by the length of time infants had been home from the hospital. An obvious difference between the hospital and home was the strength of the environmental light/dark cycle. Their results, therefore, suggested an influence of the light/dark cycle as early as 35 weeks postconception. It is also possible, however, that differences in the care of the infants had an effect. Even if the light/dark cycle was important, it is unclear

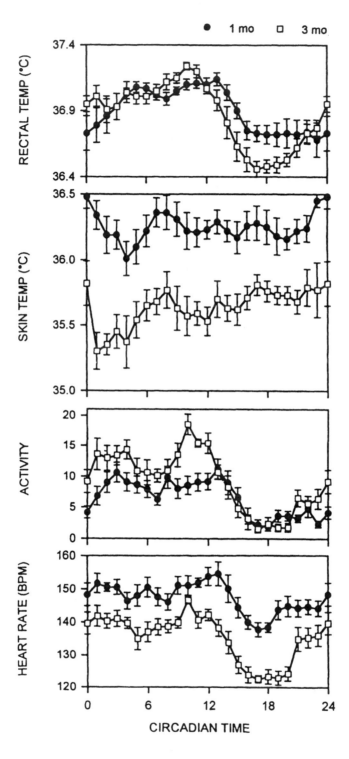

whether its effects were due to earlier entrainment of the circadian pacemaker or to the promotion of sleep at an appropriate circadian phase for consolidated sleep. Mann et al. (311) studied two groups of premature infants kept in nurseries differing in the extent of environmental periodicity, especially the light/dark cycle. Infants from the more rhythmic environment spent more time asleep and less time feeding and gained more weight after discharge than those from the relatively arrhythmic environment. These results indicate that the effects of a rhythmic environment can be long-lasting. Kennaway et al. (312) found that the delayed development of a melatonin rhythm in some premature infants could be accelerated by exposing the infants to dark at night, also suggesting an environmental influence on the maturation of circadian rhythms. It remains unclear to what extent the effects in these studies resulted from influences on pacemaker development or on the regulated functions, or on both.

Sudden infant death syndrome (SIDS) is a major cause of death within the first year of life, occurring primarily between the ages of 4 and 24 weeks (313), an age when many developmental changes in physiology are occurring including the emergence of a clear 24-hr sleep/wake rhythm as well as rhythms in melatonin (Fig. 18) and cortisol secretion (307,314). Research on SIDS has led to numerous hypotheses concerning its etiology.

One hypothesis that has received considerable attention is that SIDS is due to a generalized arousal deficit early in development (315–319). This hypothesis proposes that SIDS infants die because they are unable to arouse from deep sleep. Attenuation of arousal mechanisms is evident from 1 week to 6 months of age in humans, but appears to be most severe at 3 months of age (320) and corresponds with the time of peak SIDS incidence (321,322). SIDS infants that have sleep patterns that put them at risk for SIDS ("near miss" infants) have fewer awakenings than control infants at 2–3 months of age (323–325). Because high arousal thresholds might be an important cause of SIDS, aspects of sleep physiology and sleep homeostasis that influence depth of sleep and arousability warrant considerable attention. Factors that influence sleep structure, sleep continuity, and sleep depth include prior sleep deprivation, internal and external thermal conditions, circadian rhythms, recency of feeding, and of course, interactions between these factors.

It is recognized that there are a number of risk factors for SIDS. They include low birth weight, mothers who smoked or used substances such as cocaine

---

Figure 17 Averaged daily rhythms of 10 infants at 1 and 3 months of age. Recordings were made at home for 3 consecutive days for each infant and values are averaged within 1-hr bins. The rhythms are plotted according to clock time with circadian time 0 being equal to 8:00 A.M. Significant rhythms were observed at both ages in all variables except skin temperature. Between 1 and 3 months of age the amplitudes but not the phases of the rhythms changed. (From Ref. 283; reproduced by permission of Pediatrics, 94:482–488, 1994.)

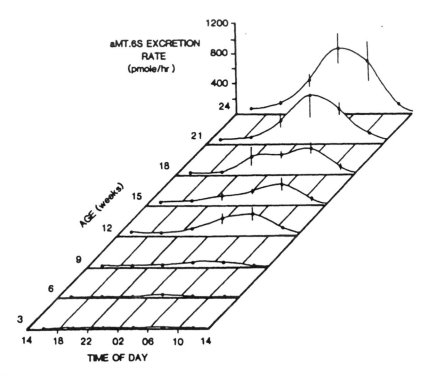

Figure 18 Development of the daily rhythm in 6-sulfatoxymelatonin excretion in full-term human infants. (From Ref. 314; reproduced by permission of the Endocrine Society.)

during pregnancy, and overbundling. Whereas some factors such as overbundling could have a direct influence on enhancing depth of sleep and arousability, others such as low birth weight and nicotine or cocaine exposure could alter developmental patterns of sleep control/homeostasis and circadian control (312,322,326).

The impact of sleep/wake cycle development on the general health and well-being of a family and its possible linkage to less common but more devastating outcomes such as SIDS, together with the fact that the fetus is normally exposed to and affected by 24-hr rhythmicity, points to the need for understanding the consequences of abnormal rhythmicity during development. Although anecdotal, disruption of rhythmicity in fetal sheep reduces their survival; a rhythm in cerebrospinal fluid vasopressin in sheep was disrupted when the ewes were maintained in constant light and their fetuses did not survive. When ewes were kept on a light/dark cycle, most fetuses survived (297). As already discussed, there is evidence that the relatively arrhythmic conditions of hospital nurseries may be detrimental to human infants. It is possible that other aspects of modern life contribute to the disruption of both maternal and fetal rhythmicity. Drugs may directly affect the fetal pacemaker and shift work performed by pregnant women may contribute to low birth weight (327–329). Other potential, but poorly documented, disruptions of rhythmicity during the neonatal period may include excessive use of bright light during the night at home and the feeding of breast milk collected at one time of day but given at another; human milk is known to contain melatonin and may differ in other ways between day and night as well (242).

VII. Summary

The ontogeny of mammalian REM and non-REM sleep occurs in three general stages. The earliest stage of sleep ontogeny ("dissociation") is characterized by the absence of distinct sleep states. Individual electrophysiological and behavioral components of non-REM and REM sleep are either undetectable or dissociated from each other. These dissociated sleep components begin to cluster into distinct REM and non-REM sleep states in utero for precocial species and ex utero for altricial species. This second stage of sleep ontogeny ("concordance") does not appear to involve intermediate, precursor sleep states. The emergence of distinct REM and non-REM sleep states is followed by the continued development of electrophysiological markers of REM and non-REM sleep, and by the maturation of ultradian and homeostatic sleep regulatory mechanisms ("maturation"). Sleep homeostatic mechanisms undergo dramatic changes in the postnatal period. Neonatal rats show compensatory increases in sleep time, but do not intensify non-REM EEG slow-wave activity following sleep deprivation until the postweaning period. The presence of sleep homeostasis in neonates suggests that sleep may serve important functions during early development. Neonatal sleep has been hypothesized to contribute to brain maturation and periods of heightened plasticity, but at present, little empirical evidence exists to support these hypotheses.

Evidence primarily from rodents indicates that the fetal circadian pacemaker located in the SCN begins to generate oscillations soon after the nucleus is formed. This would be within the first trimester in humans. The circadian regulation of sleep states does not begin until much later. Oscillations with the SCN are initially entrained by maternal rhythms beginning just after the nucleus is formed. More than one maternal rhythm may be involved in entrainment but it is likely that the pineal melatonin rhythm is one of these. In rodents and probably in humans as well, entrainment by light is possible within the first week after birth. Soon after birth, but possibly earlier in humans, overt functions, including sleep and waking, begin to be regulated by the pacemaker. In humans, episodes of sleep and waking increase in duration during the first few postnatal weeks. This may involve the development of homeostatic mechanisms as well as increasing regulation by the pacemaker. Because the circadian phase at which sleep occurs influences the duration of sleep episodes, consolidated sleep predominates during subjective night and wakefulness predominates during subjective day. As the pacemaker becomes entrained to the light/dark cycle, subjective night and consolidated sleep occur during the dark phase of the environmental light/dark cycle. It is likely that the development of a normal, adult-like pattern of sleep and waking requires a circadian pacemaker that is both entrained to the environment and able to regulate sleep/wake mechanisms. If, for whatever reason, pacemaker development is abnormal, affecting either its output or entrainment, then the development of consolidated sleep at night may be delayed. Although still poorly documented, the work, sleep, and eating habits as well as drug use, including smoking, by pregnant women may be among the environmental variables that affect pacemaker output and entrainment.

Acknowledgments

The preparation of this review was supported in part by the National Institute of Child Health and Human Development Grant HD-18686 to FCD and National Institute of Child Health and Human Development Grant P50-HD-29732 to HCH.

References

1. Borbely AA. Sleep regulation: circadian rhythm and homeostasis. Current topics in neuroendocrinology. In: Ganten D, Pfaff D, eds. Sleep: Clinical and Experimental Aspects. Berlin: Springer-Verlag, 1982:83–103.
2. Borbely AA. Sleep homeostasis and models of sleep regulation. In: Kryger MH, Roth T, Dement WC, eds. Principles and Practice of Sleep Medicine. Philadelphia: WB Saunders, 1994:309–320.
3. Edgar DM, Dement WC, Fuller CA. Effect of SCN lesions on sleep in squirrel monkeys: evidence for opponent processes in sleep-wake regulation. J Neurosci 1992; 13:1065–1079.
4. Hoppenbrouwers T. Polysomnography in newborns and young infants: sleep architecture. J Clin Neurophysiol 1992; 9:32–47.
5. Korner AF. REM organization in neonates. Arch Gen Psychiatry 1968; 19:330–340.
6. Dreyfus-Brisac C. Ontogenesis of sleep in human prematures after 32 weeks of conceptional age. Dev Psychobiol 1970; 3:91–121.
7. Hamburger V. Fetal behavior. In: Hafez ES, ed. The Mammalian Fetus: Comparative Biology and methodology. Springfield, IL: Charles C Thomas, 1975.
8. Dreyfus-Brisac C. The electroencephalogram of the premature infant. World Neurol 1968; 3:5–15.
9. Prechtl HFR. Ultrasound studies of human fetal behavior. Early Hum Dev 1985; 12:91–98.
10. Dawes GS, Gardner WM, Johnston BM, Walker DW. Activity of intercostal muscles in relation to breathing movements, electrocortical activity and gestational age in fetal lambs. J Physiol (Lond) 1980; 307:47–48.

11. Szeto H, Hinman DJ. Prenatal development of sleep-wake patterns in sheep. Sleep 1985; 8:347–355.
12. Hoppenbrouwers T, Sterman MB. Development of sleep state patterns in the kitten. Exp Neurol 1975; 49:822–838.
13. Jouvet-Mounier D, Astic L, Lacote D. Ontogenesis of the states of sleep in rat, cat, and guinea pig during the first postnatal month. Dev Psychobiol 1970; 2:216–239.
14. Balzamo E. Sleep ontogeny in the chimpanzee: from birth to two months. Electroenceph Clin Neurophysiol 1972; 33:41–46.
15. Arabin B, Riedewald S. An attempt to quantify characteristics of developmental states. Am J Perinatol 1992; 9:115–119.
16. Curzi-Dascalova L, Figueroa JM, Eiselt M, Christova E, Virassamy A, D'Allest AM, Guimares H, Gaultier C, Dehan M. Sleep state organization in premature infants of less than 35 weeks gestational age. Pediatr Res 1993; 34:624–628.
17. Curzi-Dascalova L, Peirano P, Morel-Kahn F. Development of sleep states in normal premature and full-term newborns. Dev Psychobiol 1988; 21:431–444.
18. Nijhuis JG, Prechtl HFR, Martin CBJ, Bots RSGM. Are there behavioral states in the human fetus? Early Hum Dev 1982; 6: 177–195.
19. Okai T, Kozuma S, Shinozuka N, Kuwabara Y, Mizuno MA. Study on the development of sleep-wakefulness cycle in the human fetus. Early Hum Dev 1992; 29:391–396.
20. Clewlow F, Dawes GS, Johnston BM, Walker DW. Changes in breathing, electrocortical and muscle activity in unanesthetized fetal lambs with age. J Physiol 1983; 341:463–476.
21. Grieve PG, Myers MM, Stark RI. Behavioral states in the fetal baboon. Early Hum Dev 1994; 39:159–175.
22. Stark RI, Daniel SS, Kim Y-I, Leung K, Rey HR, Tropper PJ. Patterns of development in fetal breathing activity in the latter third of gestation of the baboon. Early Hum Dev 1993; 32:31–47.
23. Stark RI, Haiken J, Nordli D, Myers MM. Characterization of electroencephalographic state in fetal baboons. Am J Physiol 1991; 261:R496–R500.
24. Abrams RM, Gerhardt KJ, Burchfield DJ. Behavioral state transition and local cerebral blood flow in fetal sheep. J Dev Physiol 1991; 15:283–288.
25. Rankin JHG, Landauer M, Tian Q, Phernetton TM. Ovine fetal electrocortical activity and regional blood flow. J Dev Physiol 1987; 9:537–542.
26. Richardson BS, Caetano H, Homan J, Carmichael L. Regional brain blood flow in the ovine fetus during transition to the low-voltage electrocortical state. Dev Brain Res 1994; 81:10–16.
27. Abrams RM, Hutchison AA, Jay TM, Sokoloff L, Kennedy C. Local cerebral glucose utilization non-selectively elevated in rapid eye movement sleep of the fetus. Dev Brain Res 1988; 40:65–70.
28. Richardson BS, Carmichael L, Homan J, Gagnon R. Cerebral oxidative metabolism in lambs during perinatal period: relationship to electrocortical state. Am J Physiol 1989; 257:R1251–R1257.
29. Stark RI, Daniel SS, Kim Y-I, Leung K, Myers MM, Tropper PJ. Patterns of fetal breathing in the baboon vary with EEG sleep state. Early Hum Dev 1994; 38:11–26.
30. Dreyfus-Brisac C. Neurophysiological studies in human premature and full-term neonates. Biol Psychiatry 1975; 10:485–496.

31. Hayes MJ, Plante LS, Fielding BA, Kumar SP, Delivoria-Papadopoulos M. Functional analyses of spontaneous movements in preterm infants. Dev Psychobiol 1994; 27:271–287.

32. Prechtl HFR, Nijhuis JG. Eye movements in the human fetus and newborn. Behav Brain Res 1983; 10:119–124.

33. Dawes GS, Houghton CRS, Redman CWG, Visser GHA. Pattern of the normal human fetal heart rate. Br J Obstet Gynecol 1982; 89:276–284.

34. Wheeler T, Murrills A. Patterns of fetal heart rate during normal pregnency. Br J Obstet Gynecol 1978; 85:18–27.

35. Swartjes JM, van Geijn HP, Mantel R, van Woerden EE, Schoemaker HC. Coincidence of behavioral state parameters in the human fetus at three gestational ages. Early Hum Dev 1990; 23:75–83.

36. Swartjes JM, van Geijn HP, Meinardi H, van Alphen M, Schoemaker HC. Fetal rest-activity cycles and chronic exposure to antiepileptic drugs. Epilepsia 1991; 32: 722–728.

37. Lombroso CT. Quantified electrographic scales on 10 pre-term healthy newborns followed up to 40–43 weeks of conceptional age by serial polygraphic recordings. Electroenceph Clin Neurophysiol 1979; 46:460–474.

38. Parmelee AHJ, Schulte FJ, Akiyama Y, Waldemar HW, Wenner W, Schultz MA, Stern E. Maturation of EEG activity during sleep in premature infants. Electroenceph Clin Neurophysiol 1968; 24:319–329.

39. Faienza C, Capone C, Galgano MC, Sani F. The emergence of the sleep-wake cycle in infancy. Int J Neurol Sci 1986; 5(Suppl):37–42.

40. Pillai M, James D. Are the behavioral states of the newborn comparable to those of the fetus? Early Hum Dev 1990; 22:39–49.

41. de Vries JIP, Visse GHA, Prechtl HFR. The emergence of fetal behavior. II. Quantitative aspects. Early Hum Dev 1985; 12:99–120.

42. Dierker LJ, Pillay SK, Sorokin Y, Rosen M. Active and quiet periods in the preterm and term fetus. Obstet Gynecol 1982; 60:65–70.

43. Granat M, Lavie P, Adar D, Sharf M. Short-term cycles in human fetal activity. Am J Obstet Gynecol 1979; 134:696–701.

44. Hayes MJ, Kumar SP, Delivoria-Papadopoulos M. Spontaneous motility in premature infants: features of behavioral activity and rhythmic organization. Dev Psychobiol 1993; 26:279–291.

45. Sterman MB, Hoppenbrouwers T. The development of sleep-waking and rest-activity patterns from fetus to man adult in man. In: Sterman MB, McGinty DJ, Adinolfi AM, eds. Brain Development and Behavior. New York: Academic Press, 1971:203–227.

46. Sterman MB. The basic rest-activity cycle and sleep: developmental considerations in man and cat. In: Clemente CD, Purpura DP Mayer FE, eds. Sleep and the Maturing Nervous System. New York: Academic Press, 1972:175–197.

47. Stern E, Parmelee AH, Harris MA. Sleep state periodicity in prematures and young infants. Dev Psychobiol 1972; 6:357–365.

48. Shinozuka N, Okai T, Kuwabara Y, Mizuno M. The development of sleep-wakefulness cycle and its correlation to other behavior in the human fetus. Asia-Oceania J Obstet Gynecol 1989; 15:395–402.

49. Ellingson RJ. Development of wakefulness-sleep cycles and associated EEG patterns in mammals. In: Clemente CD, Purpura DP, Mayer FE, eds. Sleep and the Maturing Nervous System. New York: Academic Press, 1972:165–173.

50. Shimizu A, Himwich H. The ontogeny of sleep in kittens and young rabbits. Electroenceph Clin Neurophysiol 1968; 24:307–318.

51. Mirmiran M, Corner M. Neuronal discharge patterns in the occipital cortex of developing rats during active and quiet sleep. Dev Brain Res 1982; 3:37–48.

52. Huttenlocher PR. Development of cortical neuronal activity in the neonatal cat. Exp Neurol 1967; 17:247–262.

53. Garma L, Verley R. Activities cellulaires corticales etudiees par electrodes implantess chez le lapin nouveau-ne. J Physiol Paris 1967; 59:357–376.

54. Garma L, Verley R. Ontogenese des etats de veille et de sommeil chez les mammiferes. Rev Neuropsychiatrie Infantile 1965; 17:487–504.

55. Tamasy V, Koranyi L. Multiunits in the mesencephalic reticular formation: ontogenetic development of wakefulness and sleep cycle in the rat. Neurosci Lett 1980; 17:143–147.

56. Tamasy V, Koranyi L, LIssak K. The developing mesecephalic reticular formation: changes in responsiveness during ontogeny of the rat. Acta Physiol Acad Sci Hung 1980; 56:187–201.

57. Tamasy V, Koranyi L, Lissak K. Early postnatal development of wakefulness-sleep cycle and neuronal responsiveness: a multiunit activity study on freely moving newborn rat. Electroenceph Clin Neurophysiol 1980; 49:102–111.

58. Davies TL, Lindsay RD, Scheibel ME, Scheiber AB. Ontogenetic development of somatosensory thalamus II. Electrogenesis. Exp Neurol 1976; 52:13–29.

59. McGinty RJ, Stevenson M, Hoppenbrouwers T, Harper RM, Sterman MB, Hodgman J. Polygraphic studies of kitten development: sleep state patterns. Dev Psychobiol 1977; 10:455–469.

60. Corner MA, Bour HL. Postnatal development of spontaneous neuronal discharges in the pontine reticular formation of free-moving rats during sleep and wakefulness. Exp Brain Res 1984; 54:66–72.

61. Adrien J, Roffwarg HP. The development of unit activity on the lateral geniculate nucleus of the kitten. Exp Neurol 1974; 43:261–275.

62. Bowe-Anders C, Adrien J, Roffwarg HP. Ontogenesis of ponto-geniculo-occipital activity in the lateral genicualte nucleus of the kitten. Exp Neurol 1974; 43:242–260.

63. Adrien J, Lanfumey L. Ontogenesis of unit activity in the raphe dorsalis of the behaving kitten: its relationship with the states of vigilance. Brain Res 1986; 366:10–21.

64. Leblanc MO, Bland BH. Developmental aspects of hippocampal electrical acitvity and motor behavior in the rat. Exp Neurol 1979; 66:220–237.

65. Creery BL, Bland BH. Ontogeny of fascia dentata electrical activity and motor behavior in the dutch belted rabbit. Exp Neurol 1980; 67:554–572.

66. Jones B. Paradoxical sleep and its chemical/structural substrates in the brain. Neuroscience 1991; 40:637–656.

67. Siegel JM. Brainstem mechanisms generating REM sleep. In: Kryger MH, Roth T, Dement WC, eds. Principles and Practice of Sleep Medicine. Philadelphia: WB Saunders, 1994.

68. Lee W, Nicklaus KJ, Manning DR, Wolfe BB. Ontogeny of cortical muscarinic receptor subtypes and muscarinic receptor-mediated responses in rat. J Pharmacol Exp Ther 1990; 252:484–490.

69. Miyoshi R, Kito S, Shimizu M, Matsubayashi H. Ontogeny of muscarinic receptors in the rat brain with emphasis on the differentiation of m1 and m2 subtypes—semi-quantitative in vitro autoradiography. Brain Res 1987; 420:302–312.

70. Feinberg I, Thode Jr HC, Chugani HT, March JD. Gamma distribution model describes maturational curves for delta wave amplitude, cortical metbaolic rate and synaptic density. J Theoret Biol 1990; 142:149–161.

71. Gramsbergen APS, Prechtl HFR. The postnatal development of behavioral states in the rat. Dev Psychobiol 1970; 3:267–280.

72. Kiss J, Patel AJ. Development of the cholinergic fibres innervating the cerebral cortex of the rat. Int J Dev Neurosci 1992; 10:153–170.

73. Nyakas C, Buwalda B, Kramers RJK, Trabert J, Luiten PGM. Postnatal development of hippocampal and neocortical cholinergic and serotonergic innveration in rat: effects of nitrite-induced prenatal hypoxia and nimodipine treatment. Neuroscience 1994; 59:541–559.

74. Brady DR, Phelps PE, Vaughn JE. Neurogenesis of basal forebrain cholinergic neurons in rat. Dev Brain Res 1989; 47:81–92.

75. Gould E, Woolf NJ, Buthcher LL. Postnatal development of cholinergic neurons in the rat. I. Forebrain. Brain Res 1988; 27:767–789.

76. Strejckova A, Mares P, Raevsky K. Changes of activity of cortical neurons induced by acetylcholine in adult and young rats. Exp Neurol 1987; 98:555–562.

77. Burt DK, Hungerford SM, Crowner ML, Baez LA. Postnatal development of a cholinergic influence on neuroleptic-induced catalepsy. Pharamacol Biochem Behav 1982; 16:535–540.

78. Fibiger HC, Lytle LD, Campbell BA. Cholinergic modulation of adrenergic arousal in the developing rat. J Comp Physiol Psychol 1970; 72:384–389.

79. Frank MG, Page J, Heller HC. The effects of REM-sleep inhibiting drugs in neonatal rats: evidence for a distinction between neonatal active sleep and REM sleep. Brain Res 1997; 778:64–72.

80. Adrien J, Davenne D, Lanfumey L, Robain O. Effects of ponto-mesencephalic lesions on the development of the brain and of the sleep patterns. In: Koella WP, Ruther E, eds. Sleep. Bochum: Pontenagel Press, 1984.

81. Blumberg MS, Lucas DE. Dual mechanisms of twitching during sleep in neonatal rats. Behav Neurosci 1994; 108:1196–1202.

82. Hilakivi LA, Hilakivi IT. Sleep-wake behavior of newborn rats recorded with movement sensitive method. Behav Brain Res 1986; 19:241–248.

83. Thoman EB, Waite SP, Desantis DT, Denenberg VH. Ontogeny of sleep and wake states in the rabbit. Anim Behav 1979; 27:95–106.

84. Hilakivi LA, Hilakivi I, Ahtee H, Haikala H, Attila M. Effect of neonatal nomifensine exposure on adult behavior and brain monoamines in rats. J Neural Transm 1987; 70:99–116.

85. Alperts JR. Huddling by rat pups: group behavioral mechanisms of temperature regulation and energy conservation. J Comp Physiol Psychol 1978; 92:231–245.

86. Swanson HH, Bolwerk E, Brenner E. Effects of cooling in infant rats on growth, maturation, sleep patterns and responses to food deprivation. Br J Nutr 1984; 52: 139–148.

87. Blumberg MS, Stolba MA. Thermogenesis, myoclonic twitching, and ultrasonic vocalization in neonatal rats during moderate and extreme cold exposure. Behav Neurosci 1996; 110:305–314.

88. Hull D. Oxygen consumption and body temperature of newborn rabbits and kittens exposed to cold. J Physiol 1965; 177:192–202.

89. Hull D, Hull J, Vintner J. Preferred environmental temperature of newborn rabbits. Biol Neonate 1986; 50:323–330.

90. Szymusiak R. Magnocellular nuclei of the basal forebrain: substrates of sleep and arousal regulation. Sleep 1995; 18:478–500.

91. Frank MG, Heller HC. Development of REM and slow wave sleep in the rat. Am J Physiol 1997; 272:R1792–R1799.

92. Robertson SS. Oscillation and complexity in early infant behavior. Child Dev 1993; 64:1022–1035.

93. Roffwarg HF, Muzio J, Dement WC. Ontogenetic development of the human sleep-wakefulness cycle. Science 1966; 152:604–619.

94. Meier G, Berger RJ. Development of sleep and wakefulness patterns in the infant rhesus monkey. Exp Neurol 1965; 12:257–277.

95. Balzamo E, Bradley RJ, Rhodes JM. Sleep ontogeny in the chimpanzee: from two months to forty-one months. Electroenceph Clin Neurophysiol 1972; 33:47–60.

96. Louis J, Cannard C, Bastuji H, Challamel M.-J. Sleep ontogenesis revisited; a longitudinal 24-hour home polygraphic study on 15 normal infants during the first two years of life. Sleep 1997; 20:323–333.

97. Groome LJ, Swiber MJ, Atterbury JL, Bentz LS, Holland SB. Similarities and differences in behavioral state organization during sleep periods in the perinatal infant before and after birth. Child Dev 1997; 68:1–11.

98. Ruckebusch Y. Development of sleep and wakefulness in the foetal lamb. Electroeneceph Clin Neurophysiol 1972; 32:119–128.

99. Havlicek V, Childiaeva R, Chernick V. EEG frequency spectrum characteristics of sleep states in full-term and preterm infants. Neuropadiatrie 1974; 6:24–40.

100. Tharp B. Electrophysiological brain maturation in premature infants: an historical perspective. J Clin Neurophysiol 1990; 7:302–314.

101. Gramsbergen A. The development of the EEG in the rat. Dev Psychobiol 1976; 9:501–515.

102. Astic L, Sastre J-P, Brando A-M, Etude polygraphique des etats de vigilance chez le foetus de cobaye. Physiol Behav 1973; 11:647–654.

103. Breazile JE. Neurologic and behavioral development in the puppy. Vet Clin North Am 1978; 8:31–45.

104. Bronzino JD, Siok CK, Austin K, Austin-LaFrance RJ, Morgane P. Spectral analyses of the electroencephalogram in the developing rat. Dev Brain Res 1987; 35: 257–267.

105. Cavoy A, Delacour TE. Le rhythme theta de SP comme indice de maturation cerebrale: etude chez le rat et le cobaye. Physiol Behav 1981; 26:233–240.

106. Chase MH. Brain stem somatic reflex activity in neonatal kittens during sleep and wakefulness. Physiol Behav 1971; 7:165–172.

107. Iwamura Y, Tsuda K, Kud N, Kohama K. Monosynaptic reflex during natural sleep in the kitten. Brain Res 1968; 11:456–459.

108. Iwamura Y. Development of supraspinal modulation of motor activity during sleep and wakefulness. In: Sterman MB, McGinty DJ, Adinolfi AM, eds. Brain Development and Behavior. New York: Academic Press, 1971:129–143.

109. Kohyama J. A quantitative assesment of the maturation of phasic motor inhibition during REM sleep. J Neurol Sci 1996; 143:150–155.

110. Kohyama J, Shimohira M, Yoshihide I. Maturation of motility and motor inhibition in rapid-eye-movement sleep. J Pediatr 1997; 130:117–122.

111. Ioffe S, Jansen AH, Russel BJ, Chernick V. Sleep, wakefulness and the monosynaptic reflex in fetal and newborn lambs. Pflügers Arch 1980; 388:149–157.

112. Harper RM, Frostig, Taube D. Infant sleep development. In: Mayes A, ed. Sleep Mechanisms and Functions in Humans and Animals—An Evolutionary Perspective. Cambridge, UK: Van Nostrand Reinhold (UK), 1983:106–124.

113. Dement WC, Kleitman N. Cyclic variations in EEG during sleep and their relation to eye movements, body motility, and dreaming. Electroenceph Clin Neurophysiol 1957; 9:673–690.

114. Benington JH, Heller HC. Does the function of REM sleep concern non-REM sleep or waking? Prog Neurobiol 1994; 44:433–449.

115. Scher MS, Steppe DA, Dokianakis SG, Guthri RD. Maturation of phasic and continuity measures during sleep in preterm neonates. Pediatr Res 1994; 36:732–737.

116. Emde RN, Swedberg J, Suzuji B. Human wakefulness and biological rhythms after birth. Arch Gen Psychiatry 1975; 32:780–783.

117. Stern E, Parmelee AH, Akiyama Y, Schultz MA, Wenner WH. Sleep cycle characteristics in infants. Pediatrics 1969; 43:65–70.

118. Ellman SJ, Spielman AJ, Luck D, Steiner SS, Halperin R. REM sleep deprivation: a review. In: Ellman S, Antrobus J, eds. The Mind in Sleep, Psychology, and Psychophysiology. New York: Wiley, 1991:329–369.

119. Horne JA. Sleep function with particular reference to sleep deprivation. Ann Clin Res 1985; 17:199–208.

120. Benington JH, Heller HC. Restoration of brain energy as the function of sleep. Prog Neurobiol 1995; 45:347–360.

121. Alfoldi P, Tobler I, Borbely AA. Sleep regulation in rats during early development. Am J Physiol 1990; 258:R634–R644.

122. Anders TF, Roffwarg HP. The effects of selective interruption and deprivation of sleep in the human newborn. Dev Psychobiol 1973; 6:79–91.

123. Canat E, Gaultier CL, D'Allest AM, Dehan M. Effects of sleep deprivation on respiratory events during sleep in healthy infants. J Appl Physiol 1989; 66:1158–1163.

124. Thomas DA, Poole K, McArdle EK, Goodenough PC, Thompson J, Beardsmore CS Simpson H. The effect of sleep deprivation on sleep states, breathing events, peripheral chemoresponsiveness and arousal propensity in healthy 3 month old infants. Eur Respir J 1996; 9:932–938.

125. Frank MG, Morrissette R, Heller HC. Effects of sleep deprivation in neonatal rats. Am J Physiol 1998; 275:R148–R157.

126. Frank MG, Heller HC. Development of diurnal organization of EEG slow activity and slow wave sleep in the rat. Am J Physiol 1997; 273:R472–R478.

127. Schectman VL, Harper RJ, Harper RM. Distribution of slow-wave EEG activity across the night in developing infants. Sleep 1994; 17:316–322.

128. Steriade M, McCormick DA, Sejnowski TJ. Thalamocortical oscillations on the sleeping and aroused brain. Science 1993; 262:679–685.

129. Jacobson M. Developmental Neurobiology. 1. New York: Plenum Press, 1991.

130. Geiger JD, LaBella FS, Nagy JI. Ontogenesis of adenosine receptors in the central nervous system of the rat. Brain Res 1984; 315:97–104.

131. Marangos PJ, Patel J, Stivers J. Ontogeny of adenosine binding sites in rat forebrain and cerebellum. J Neurochem 1982; 39:267–270.

132. Schoen SW, Leutenecker GW, Kruetzberg GW, Singer, W. Developmental changes of 5+ nucleotidase distributions in the kitten visual cortex. Neursci Lett 1986; 26(Suppl):S57.

133. Schoen SW, Graeber MB, Toth L, Kreutzberg GW. 5' Nucleotidase in postnatal ontogeny of rat cerebellum: a marker for migrating nerve cells? Dev Brain Res 1988; 39:125–136.

134. Geiger JD, Nagy JI. Ontogenesis of adenosine deaminase activity in rat brain. J Neurochem 1987; 48:147–153.

135. Mirmiran M. The function of fetal/neonatal rapid eye movement sleep. Behav Brain Res 1995; 69:13–22.

136. Mirmiran M, Van Someran E. The importance of REM sleep for brain maturation. J Sleep Res 1993; 2:188–192.

137. Zheng D, Purves D. Effects of increased neural activity on brain growth. Proc Natl Acad Sci USA 1995; 92:1802–1806.

138. Mirmiran M, van den Dungen H, Uylings HBM. Sleep patterns during rearing under different environmental conditions in juvenile rats. Brain Res 1982; 233:287–298.

139. van Someren EJW, Mirmiran M, Bos NPA, Lamur A, Kumar A, Molenaar PCM. Quantitative analyses of eye movements during REM-sleep in developing rats. Dev Psychobiol 1990; 23:55–61.

140. Frank MG, Heller HC. Neonatal treatments with the serotonin uptake inhibitors clomipramine and zimelidine, but not the noradrenaline uptake inhibitor desipramine, disrupt sleep patterns in adult rats. Brain Res 1997; 768:287–293.

141. Davenne D, Adrien J. Suppression of PGO waves in the kitten: anatomical effects on the lateral geniculate nucleus. Neurosci Lett 1984; 45:33–38.

142. Oksenberg A, Shaffery JP, Marks GA, Speciale SG, Mihailoff G, Roffwarg H. Rapid eye movement sleep deprivation in kittens amplifies LGN cell-size disparity induced by monocular deprivation. Dev Brain Res 1996; 97:51–61.

143. Pompeiano O, Pompeiano M, Corvaja N. Effects of sleep deprivation on the postnatal development of visual-deprived cells in the cat's lateral geniculate nucleus. Arch Ital Biol 1995; 134:121–140.

144. Ivy GO, Killackey HP. The ontogeny of the distribution of callosal projection neurons in the rat parietal cortex. J Comp Neurol 1981; 195:367–389.

145. Ivy GOP, Killackey HP. Ontogenetic changes in the projections of neocortical neurons. J Neurosci 1982; 2:735–743.

146. Bear MF, Malenka RC. Synaptic plasticity: LTP and LTD. Curr Opin Neurobiol 1994; 4:389–399.

147. Cramer KS, Sur M. Activity dependent remodeling of connections in the mammalian visual system. Curr Opin Neurobiol 1995; 5:106–111.

148. Wilson MA, McNaughton BL. Reactivation of hippocampal ensemble memories during sleep. Science 1994; 265:676–679.

149. Skaggs WE, McNaughton BL. Replay of neuronal firing sequences in rat hippocampus during sleep following spatial experience. Science 1996; 271:1870–1873.

150. Kavanau JL. Sleep and dynamic stabilization of neural circuitry: a review and synthesis. Behav Brain Res 1994; 63:111–126.

151. Nehlig A, de Vasconcelos AP, Boyet S. Quantitative audoradiographic measurement of local cerebral glucose utilization in freely moving rats during postnatal development. J Neurosci 1988; 8:2231–2333.

152. Altman J, Bayer SA. Development of the diencephalon in the rat. I. Autoradiographic study of the time of origin and settling patterns of neurons of the hypothalamus. J Comp Neurol 1978; 182:945–972.

153. Altman J, Bayer SA. Development of the diencephalon in the rat. II. Correlation of the embryonic development of the hypothalamus with the time of origin of its neurons. J Comp Neurol 1978; 182:973–994.

154. Altman J, Bayer SA. The development of the rat hypothalamus. Adv Anat Embryol Cell Biol 1986; 100:1–178.

155. Altman J, Bayer SA. Atlas of Prenatal Rat Brain Development. Boca Raton, FL: CRC Press, 1995.

156. Davis FC, Boada R, LeDeaux J. Neurogenesis of the hamster suprachiasmatic nucleus. Brain Res 1990; 519:192–199.

157. Moore RY, Bernstein ME. Synaptogenesis in the rat suprachiasmatic nucleus demonstrated by electron microscopy and synapsin I immunoreactivity. J Neurosci 1989; 9:2151–2162.

158. Noguchi T, Sugisaki T, Kudo M, Satoh I. Retarded growth of the suprachiasmatic nucleus and pineal body in dw and lit dwarf mice. Dev Brain Res 1986; 26:161–172.

159. Silver J. Abnormal development of the suprachiasmatic nuclei of the hypothalamus in a strain of genetically anophthalmic mice. J Comp Neurol 1977; 176:589–606.

160. Scheuch GC, Silver J. Ontogeny of the suprachiasmatic nuclei in genetically anopthalmic mice: anatomical and behavioral studies. In: Klein DC, ed. Melatonin Rhythm Generating System: Developmental Aspects. Basel: S. Karger AG, 1982:20–41.

161. Lenn NJ, Beebe B, Moore RY. Postnatal development of the suprachiasmatic hypothalamic nucleus of the rat. Cell Tissue Res 1977; 178:463–475.

162. Ibuka N. Circadian rhythms in sleep-wakefulness and wheel-running activity in a congenitally anophthalmic rat mutant. Physiol Behav 1987; 39:321–326.

163. Whitnall MH, Key S, Ben-Barak Y, Ozato K, Gainer H. Neurophysin in the hypothalamo-neurohypophysial system. J Neurosci 1985; 5:98–109.

164. Reppert SM, Uhl GR. Vasopressin messenger ribonucleic acid in supraoptic and suprachiasmatic nuclei: appearance and circadian regulation during development. Endocrinology 1987; 120:2483–2487.

165. Botchkina GI, Morin LP. Ontogeny of radial glia, astrocytes and vasoactive intestinal peptide immunoreactive neurons in hamster suprachiasmatic nucleus. Dev Brain Res 1995; 86:48–56.

166. Kuhlman S, Watts AG, Sanchez-Watts G, Davis FC. Developmental expression of preprovasoactive intestinal polypeptide (VIP) mRNA in the Syrian hamster suprachiasmatic nucleus. Soc Neurosci Abstr 1995; 21:452 (abstract).

167. Romero M-T, Silver R. Time course of peptidergic expression in fetal suprachiasmatic nucleus transplanted into adult hamster. Dev Brain Res 1990; 57:1–6.

168. Williams LM, Martinoli MG, Titchener LT, Pelletier G. The ontogeny of central melatonin binding sites in the rat. Endocrinology 1991; 128:2083–2090.

169. Duncan MJ, Davis FC. Developmental appearance and age related changes in specific 2-[^{125}I]iodomelatonin binding sites in the suprachiasmatic nuclei of female Syrian hamsters. Dev Brain Res 1993; 73:205–212.

170. Robinson ML, Fuchs JL. [^{125}I]Vasoactive intestinal peptide binding in rodent suprachiasmatic nucleus: developmental and circadian studies. Brain Res 1993; 605:271–279.

171. Terman JS, Remé CE, Terman M. Rod outer segment disk shedding in rats with lesions of the suprachiasmatic nucleus. Brain Res 1993; 605:256–264.

172. Ban Y, Shigeyoshi Y, Okamura H. Development of vasoactive intestinal peptide mRNA rhythm in the rat suprachiasmatic nucleus. J Neurosci 1997; 17:3920–3931.

173. Reppert SM, Weaver DR, Rivkees SA, Stopa EG. Putative melatonin receptors in a human biological clock. Science 1988; 242:78–84.

174. Swaab DF, Hofman MA, Honnebier MBOM. Development of vasopressin neurons in the human suprachiasmatic nucleus in relation to birth. Dev Brain Res 1990; 52:289–293.

175. Swaab DF, Zhou JN, Ehlhart T, Hofman MA. Development of vasoactive intestinal polypeptide neurons in the human suprachiasmatic nucleus in relation to birth and sex. Dev Brain Res 1994; 79:249–259.

176. King DP, Zhao YL, Sangoram AM, et al. Positional cloning of the mouse circadian clock gene. Cell 1997; 89:641–653.

177. Tel H, Okamura H, Shigeyoshi Y, et al. Circadian oscillation of a mammalian homologue of the *Drosophila* period gene. Nature 1997; 389:512–516.

178. Sun ZS, Albrecht U, Zhuchenko O, Baily J, Eichele G, Lee CC. RIGUI, a putative mammalian ortholog of the *Drosophila* period gene. Cell 1997; 90:1003–1011.

179. Speh JC, Moore RY. Retinohypothalamic tract development in the hamster and rat. Dev Brain Res 1993; 76:171–181.

180. Morin LP. The circadian visual system. Brain Res Rev 1994; 19:102–127.

181. Ugrumov MV, Popov AP, Vladimirov SV, Kasmambetova S, Novodjilova AP, Tramu G. Development of the suprachiasmatic nucleus in rats during ontogenesis: serotonin-immunopositive fibers. Neuroscience 1994; 58:161–165.

182. Botchkina GI, Morin LP. Development of the hamster serotoninergic system: cell groups and diencephalic projections. J Comp Neurol 1993; 338:405–431.

183. Botchkina GI, Morin LP. Organization of permanent and transient neuropeptide Y–immunoreactive neuron groups and fiber systems in the developing hamster diencephalon. J Comp Neurol 1995; 357:573–602.

184. Stanfield B, Cowan WM. Evidence for a change in the retino-hypothalamic projection in the rat following early removal of one eye. Brain Res 1976; 104:129–136.

185. Mosko S, Moore RY. Retinohypothalamic tract development: alteration if supra-chiasmatic lesions in the neonatal rat. Brain Res 1979; 164:1–15.
186. Mosko S, Moore RY. Neonatal suprachiasmatic nucleus ablation: absence of functional and morphological plasticity. Proc Natl Acad Sci USA 1978; 75:6243–6246.
187. Davis FC. Development of the suprachiasmatic nuclei and other circadian pace-makers. In: Klein DC, ed. Melatonin Rhythm Generating System: Developmental Aspects. Basel: Karger, 1982:1–19.
188. Ellison N, Weller JL, Klein DC. Development of a circadian rhythm in the activity of pineal serotonin *N*-acetyltransferase. J Neurochem 1972; 19:1335–1341.
189. Teicher MH, Flaum LE. Ontogeny of ultradian and nocturnal activity rhythms in the isolated albino rat. Dev Psychobiol 1979; 12:441–454.
190. Davis FC. Ontogeny of circadian rhythms. In: Aschoff J, ed. Handbook of Behavioral Neurobiology, Vol 4. New York: Plenum Press, 1981:257–274.
191. Kaufman CM, Menaker M. Effect of transplanting suprachiasmatic nuclei from donors of different ages into completely SCN lesioned hamsters. J Neural Transplant Plast 1993; 4:257–265.
192. Nuesslein B, Schmidt I. Development of circadian cycle of core temperature in juvenile rats. Am J Physiol 1990; 259:R270–R276.
193. Davidson TL, Fewell JE. Ontogeny of a circadian rhythm in body temperature in newborn lambs reared independently of maternal time cues. J Dev Physiol 1993; 19:51–56.
194. Roberts MH, Bernstein MF, Moore RY. Differentiation of the suprachiasmatic nucleus in fetal rat anterior hypothalamic transplants in oculo. Dev Brain Res 1987; 32:59–66.
195. Griffioen HA, Duindam H, Van der Woude TP, Rietveld WJ, Boer GJ. Functional development of fetal suprachiasmatic nucleus grafts in suprachiasmatic nucleus-lesioned rats. Brain Res Bull 1993; 31:145–160.
196. Wiegand SJ, Gash DM. Organization and efferent connections of transplanted suprachiasmatic nuclei. J Comp Neurol 1988; 267:562–579.
197. Silver R, Lehman MN, Gibson M, Gladstone WR, Bittman EL. Dispersed cell suspensions of fetal SCN restore circadian rhythmicity in SCN-lesioned adult hamsters. Brain Res 1990; 525:45–58.
198. Welsh DJ, Logothetis DE, Meister M, Reppert SM. Individual neurons dissociated from rat suprachiasmatic nucleus express independently phased circadian firing rhythms. Neuron 1995; 14:697–706.
199. Wray S, Castel M, Gainer H. Characterization of the suprachiasmatic nucleus in organotypic slice explant cultures. Microsc Res Tech 1993; 25:46–60.
200. Tominaga K, Inouye S-IT, Okamura H. Organotypic slice culture of the rat supra-chiasmatic nucleus: sustenance of cellular architecture and circadian rhythm. Neuroscience 1994; 59:1025–1042.
201. Deguchi T. Ontogenesis of a biological clock for serotonin: acetyl coenzyme A *N*-acetyltransferase in pineal gland of rat. Proc Natl Acad Sci USA 1975; 72:2814–2818.
202. Reppert SM, Coleman RJ, Heath HW, Swedlow JR. Pineal *N*-acetyltransferase activity in 10-day-old rats: a paradigm for studying the developing circadian system. Endocrinology 1984; 115:918–925.

203. Hiroshige T, Honma K, Watanabe K. Prenatal onset and maternal modifications of the circadian rhythm of plasma corticosterone in blind infantile rats. J Physiol 1982; 325:521–532.

204. Davis FC, Gorski RA. Development of hamster circadian rhythms: prenatal entrainment of the pacemaker. J Biol Rhythms 1985; 1:77–89.

205. Weaver DR, Reppert SM. Maternal-fetal communication of circadian phase in a precocious rodent, the spiny mouse. Am J Physiol 1987; 253:E401–E409.

206. Reppert SM, Schwartz WJ. Maternal coordination of the fetal biological clock in utero. Science 1983; 220:969–971.

207. Reppert SM, Schwartz WJ. The suprachiasmatic nuclei of the fetal rat: characterization of a functional circadian clock using ^{14}C-labeled deoxyglucose. J Neurosci 1984; 4:1677–1682.

208. Reppert SM, Schwartz WJ. Functional activity of the suprachiasmatic nuclei in the fetal primate. Neurosci Lett 1984; 46:145–149.

209. Shibata S, Moore RY. Development of neuronal activity in the rat suprachiasmatic nucleus. Dev Brain Res 1987; 34:311–315.

210. Shibata S, Moore RY. Development of a fetal circadian rhythm after disruption of the maternal circadian system. Dev Brain Res 1988; 41:313–317.

211. Reppert SM, Schwartz WJ. Maternal suprachiasmatic nuclei are necessary for maternal coordination of the developing circadian system. J Neurosci 1986; 6:2724–2729.

212. Davis FC, Gorski RA. Development of hamster circadian rhythms: role of the maternal suprachiasmatic nucleus. J Comp Physiol A 1988; 601:610.

213. Nuesslein-Hildesheim B, Schmidt I. Manipulation of potential perinatal zeitgebers for the juvenile circadian temperature rhythm in rats. Am J Physiol Regul Integr Comp Physiol 1996; 271:R1388–P1395.

214. Hiroshige T, Honma K, Watanabe K. Possible zeitgebers for external entrainment of the circadian rhythm of plasma corticosterone in blind infantile rats. J Physiol 1982; 325:507–519.

215a. Honma S, Honma K, Shirakawa T, Hiroshige T. Effects of elimination of maternal ceircadian rhythms during pregnancy on the postnatal development of circadian corticosterone rhythm in blinded infantile rats. Endocrinology 1984; 114:44–50.

215b. Reppert SM, Schwartz WJ. Maternal endocrine exterpations do not abolish maternal coordination of the fetal circadian clock. Endocrinology 1986; 119:1763–1767.

215c. Weaver DR, Reppert SM. Periodic feeding of SCN-lesioned pregnant rats entrains the fetal biological clock. Dev Brain Res 1989; 46:291–296.

215. Davis FC, Mannion J. Entrainment of hamster pup circadian rhythms by prenatal melatonin injections to the mother. Am J Physiol 1988; 255:R439–R448.

216. Viswanathan N, Weaver DR, Reppert SM, Davis FC. Entrainment of the fetal hamster circadian pacemaker by prenatal injections of the dopamine agonist, SKF 38393. J Neurosci 1994; 14:5393–5398.

217. Viswanathan N, Davis FC. Single prenatal injections of melatonin or the D1-dopamine receptor agonist SKF 38393 to pregnant hamsters sets the offsprings' circadian rhythms to phases 180° apart. J Comp Physiol A 1997; 180:339–346.

218. Armstrong SM. Melatonin and circadian control in mammals. Experientia 1989; 45:933–938.

219. McMillen IC, Nowak R. Maternal pinealectomy abolishes the diurnal rhythm in plasma melatonin concentrations in the fetal sheep and pregnant ewe during late gestation. J Endocrinol 1989; 120:459–464.

220. Yellon SM, Longo LD. Effect of maternal pinealectomy and reverse photoperiod on the circadian melatonin rhythm in the sheep and fetus during the last trimester of pregnancy. Biol Reprod 1988; 39:1093–1099.

221. Reppert SM, Shea RA, Anderson A, Klein DC. Maternal-fetal transfer of melatonin in a non-human primate. Pediatr Res 1979; 13:788–791.

222. Klein DC. Evidence for placental transfer of ^{3}H-acetyl-melatonin. Nature 1972; 237:117–119.

223. Redman J, Armstrong S, Ng KT. Free-Running activity rhythms in the rat: entrainment by melatonin. Science 1983; 210:1089–1091.

224. Deacon S, Arendt J. Melatonin shifts human circadian rhythms according to a phase-response curve. Chronobiol Int 1992; 9:380–392.

225. Lewy AJ. Melatonin-induced temperature suppression and its acute phase-shifting effects correlate in a dose-dependent manner in humans. Brain Res 1995; 688:77–85.

226. Weaver DR, Reppert SM. Maternal melatonin communicates datlength to the fetus in djungarian hamsters. Endocrinology 1986; 119:2861–2863.

227. Elliott JA, Goldman BD. Reception of photoperiodic information by fetal Siberian hamsters: role of the mother's pineal gland. J Exp Zool 1989; 252:237–244.

228. Weaver DR, Rivkees SA, Reppert SM. Localization and characterization of melatonin receptors in rodent brain by in vitro autoradiography. J Neurosci 1989; 9: 2581–2590.

229. Roca AL, Godson C, Weaver DR, Reppert SM. Structure, characterization, and expression of the gene encoding the mouse Mel1a melatonin receptor. Endocrinology 1996; 137:3469–3477.

230. Carlson LL, Weaver DR, Reppert SM. Melatonin receptors and signal transduction during development in Siberian hamsters (*Phodopus sungorus*). Dev Brain Res 1991; 59:83–88.

231. Rivkees SA, Reppert SM. Appearance of melatonin receptors during embryonic life in Siberian hamsters (*Phodopus sungorus*). Brain Res 1991; 568:345–349.

232. Grosse J, Velickovic A, Davis FC. Entrainment of Syrian hamster circadian activity rhythms by neonatal melatonin injections. Am J Physiol Regul Integr Comp Physiol 1996; 270:R533–R540.

233. Weaver DR, Roca AL, Reppert SM. c-*fos* and *jun*-B mRNAs are transiently expressed in fetal rodent suprachiasmatic nucleus following dopaminergic stimulation. Dev Brain Res 1995; 85:293–297.

234. Weaver DR, Rivkees SA, Reppert SM. D1-dopamine receptors activate c-*fos* expression in the fetal suprachiasmatic nuclei. Proc Nat Acad Sci USA 1992; 89:9201–9204.

235. Clegg DA, O'Hara BF, Heller HC, Kilduff TS. Nicotine administration differentially affects gene expression in the maternal and fetal circadian clock. Dev Brain Res 1995; 84:46–54.

236. Ugrumov MV, Popov AP, Vladimirov SV, Kasmambetova S, Thibault J. Development of the suprachiasmatic nucleus in rats during ontogenesis: tyrosine hydroxylase immunopositive cell bodies and fibers. Neuroscience 1994; 58:151–160.

237. Weaver DR, Reppert SM. Definition of the developmental transition from dopaminergic to photic regulation of c-*fos* gene expression in the rat suprachiasmatic nucleus. Mountain Brain Res 1995; 33:136–148.

238. Sasaki Y, Murakami N, Takahashi K. Critical period for the entrainment of the circadian rhythm in blinded pups by dams. Physiol Behav 1984; 33:105–109.

239. Viswanathan N, Chandrashekaran MJ. Cycles of presence and absence of mother mouse entrain the circadian clock of pups. Nature 1985; 317:530–531.

240. Illnerová H, Buresová M, Presl J. Melatonin rhythm in human milk. J Clin Endocrinol Metab 1993; 77:838–841.

241. Reppert SM, Klein DC. Transport of maternal [³H]melatonin to suckling rats and the fate of [³H]melatonin in the neonatal rat. Endocrinology 1978; 102:582–588.

242. Duncan MJ, Banister MJ, Reppert SM. Developmental appearance of light-dark entrainment in the rat. Brain Res 1986; 369:326–330.

243. Fuchs JL, Moore RY. Development of circadian rhythmicity and light responsiveness in the rat suprachiasmatic nucleus: a study using the 2-deoxy[1-¹⁴C]glucose method. Proc Nat Acad Sci USA 1980; 77:1204–1208.

244. Leard LE, Macdonald ES, Heller HC, Kilduff TS. Ontogeny of photic-induced c-*fos* mRNA expression in rat suprachiasmatic nuclei. Neuroreport 1994; 5:2683–2687.

245. Kaufman CM, Menaker M. Ontogeny of light-induced Fos-like immunoreactivity in the hamster suprachiasmatic nucleus. Brain Res 1994; 633:162–166.

246. Duffield GE, Dickerson JM, Alexander IHM, Ebling FJP. Ontogeny of a photic response in the suprachiasmatic nucleus in the Siberian hamster (*Phodopus sungorus*). Eur J Neurosci 1995; 7:1089–1096.

247. Weaver DR, Reppert SM. Direct in utero perception of light by the mammalian fetus. Dev Brain Res 1989; 47:151–155.

248. Torrealba F, Parraguez VH, Reyes T, Valenzuela G, Serón-Ferré M. Prenatal development of the retinohypothalamic pathway and the suprachiasmatic nucleus in the sheep. J Comp Neurol 1993; 338:304–316.

249. Glotzbach ST, Sollars P, Ariagno RL, Pickard GE. Development of the human retinohypoithalamic tract. Soc Neurosci Abstr 1992; 18:875 (abstract).

250. Pittendrigh CS, Daan S. A functional analysis of circadian pacemakers in nocturnal rodents. IV. Entrainment: pacemaker as clock. J Comp Physiol A 1976; 106:291–331.

251. Ralph MR, Menaker M. A mutation of the circadian system in golden hamsters. Science 1988; 241:1225–1227.

252. Vitaterna MH, King DP, Chang A-M, et al. Mutagenesis and mapping of a mouse gene, clock, essential for circadian behavior. Science 1994; 264:719–725.

253. Schwartz WJ, Zimmerman P. Circadian timekeeping in BALB/c and C57BL/6 inbred mouse strains. J Neurosci 1990; 10:3685–3694.

254. Ebihara S, Tsuji K, Kondo K. Strain differences of the mouse's free-running circadian rhythm in continuous darkness. Physiol Behav 1978; 20:795–799.

255. Shen HM, Watanabe M, Tomasiewicz H, Rutishauser U, Magnuson T, Glass JD. Role of neural cell adhesion molecule and polysialic acid in mouse circadian clock function. J Neurosci 1997; 17:5221–5229.

256. Pittendrigh CS. On temperature independence in the clock system controlling emergence time in *Drosophila*. Proc Nat Acad Sci USA 1954; 40:1018–1029.

257. Aschoff J. Exogenous and endogenous components in circadian rhythms. Cold Spring Harbor Symp Quant Biol 1960; 25:11–28.
258. Davis FC, Menaker M. Development of the mouse circadian pacemaker: independence from environmental cycles. J Comp Physiol A 1981; 143:527–539.
259. Davis FC, Viswanathan N. The effect of transplanting one or two suprachiasmatic nuclei on the period of the restored rhythm. J Biol Rhythms 1996; 11:291–301.
260. Viswanathan N, Davis FC. Suprachiasmatic nucleus grafts restore circadian function in aged hamsters. Brain Res 1995; 686:10–16.
261. Davis FC, Gorski RA. Unilateral lesions of the hamster suprachiasmatic nuclei: evidence for redundant control of circadian rhythms. J Comp Physiol A 1984; 154: 221–232.
262. De Vries GJ, Buijs RM, Swaab DF. Ontogeny of the vasopressinergic neurons of the suprachiasmatic nucleus and their extrahypothalamic projections in the rat brain— presence of a sex difference in the lateral septum. Brain Res 1981; 218:67–78.
263. Davis FC, Darrow JM, Menaker M. Sex differences in the circadian control of hamster wheel-running activity. Am J Physiol 1983; 244:R93–R105.
264. Wever RA. Sex differences in human circadian rhythms: intrinsic periods and sleep fractions. Experientia 1984; 40:1226–1234.
265. Zucker I, Fitzgerald KM, Morin LP. Sex differentiation of the circadian system in the golden hamster. Am J Physiol 1980; 238:R97–R101.
266. Serón-Ferré M, Ducsay CA, Valenzuela GJ. Circadian rhythms during pregnancy. Endocr Rev 1993; 14:594–609.
267. Barr MJ. Prenatal growth of wistar rats: circadian periodicity of fetal growth late in gestation. Teratology 1973; 7:283–287.
268. Davis FC. Daily Variation in maternal and fetal weight gain in mice and hamsters. J Exp Zool 1989; 250:273–282.
269. Miller MW. Circadian rhythm of cell proliferation in the telencephalic ventricular zone: effect of in utero exposure to ethanol. Brain Res 1992; 595:17–24.
270. Sauerbier I. Circadian variation in teratogenic response to dexamethasone in mice. Drug Chem Toxicol 1986; 9:25–31.
271. Sauerbier I. Circadian modification of ethanol damage in utero to mice. Am J Anat 1987; 178:170–174.
272. Horton TH. Growth and maturation in microtus montanus: effects of photoperiods before and after weaning. Can J Zool 1983; 62:1741–1746.
273. Stetson MH, Elliott JA, Goldman BD. Maternal transfer of photoperiodic information influences the photoperiodic response of prepubertal djungarian hamsters. Biol Reprod 1986; 34:664–669.
274. Shaw D, Goldman BD. Gender differences in influence of prenatal photoperiods on postnatal pineal melatonin rhythms and serum prolactin and follicle-stimulating hormone in the Siberian hamster (*Phodopus sungorus*). Endocrinology 1995; 136: 4237–4246.
275. Hudson R, Distel H. Temporal pattern of suckling in rabbit pups: a model of circadian synchrony between mother and young. In: Reppert SM, ed. Development of Rhythmicity and Photoperiodism in Mammals. Ithaca, NY: Perinatology Press, 1989:83–102.

276. Jilge B. The ontogeny of circadian rhythms in the rabbit. J Biol Rhythms 1993; 8:247–260.

277. Jilge B. Ontogeny of the rabbit's circadian rhythms without an external zeitgeber. Physiol Behav 1995; 58:131–140.

278. Honnebier MBOM, Swaab DF, Mirmiran M. Diurnal rhythmicity during early human development. In: Reppert SM, ed. Development of Circadian Rhythmicity and Photoperiodism in Mammals. Ithaca, NY: Perinatology Press, 1989:221–244.

279. Reppert SM, Henshaw D, Schwartz WJ, Weaver DR. The circadian-gated timing of birth in rats: disruption by maternal SCN lesions or by removal of the fetal brain. Brain Res 1987; 403:398–402.

280. Viswanathan N, Davis FC. The fetal circadian pacemaker is not involved in the timing of birth in hamsters. Biol Reprod 1993; 48:530–537.

281. Glotzbach SF, Edgar DM, Boeddiker M, Ariagno RL. Biological rhythmicity in normal infants during the first 3 months of life. Pediatrics 1994; 94:482–488.

282. Kleitman N, Engelmann TG. Sleep characteristics of infants. J Appl Physiol 1953; 6:269–282.

283. Parmelee AH, Wenner WH, Schulz HR. Infant sleep patterns: from birth to 16 weeks of age. J Pediatr 1964; 65:576–582.

284. Coons S, Guilleminault C. Development of consolidated sleep and wakeful periods in relation to the day/night cycle in infancy. Dev Med Child Neurol 1984; 26: 169–176.

285. McMillen IC, Kok JSM, Adamson TM, Deayton JM, Nowak R. Development of circadian sleep-wake rhythms in preterm and full-term infants. Pediatr Res 1991; 29:381–384.

286. Pollak CP. Regulation of sleep rate and circadian consolidation of sleep and wakefulness in an infant. Sleep 1994; 17:567–575.

287. Aschoff J, Gerecke U, Wever R. Desynchronization of human circadian rhythms. Jpn J Physiol 1967; 17:450–457.

288. Sitka U, Weinert D, Berle K, Rumler W, Schuh J. Investigations of the rhythmic function of heart rate, blood pressure and temperature in neonates. Eur J Pediatr 1994; 153:117–122.

289. Sadeh A, Dark I, Vohr BR. Newborns' sleep-wake patterns: the role of maternal, delivery and infant factors. Early Hum Dev 1996; 44:113–126.

290. Borghese IF, Minard KL, Thoman EB. Sleep rhythmicity in premature infants: implications for developmental status. Sleep 1995; 18:523–530.

291. Mirmiran M, Kok JH. Circadian rhythms in early human development. Early Hum Dev 1991; 26:121–128.

292. Tenreiro S, Dowse HB, D'Souza S, et al. The development of ultradian and circadian rhythms in premature babies maintained in constant conditions. Early Hum Dev 1991; 27:33–52.

293. Glotzbach SF, Edgar DM, Ariagno RL. Biological rhythmicity in preterm infants prior to discharge from neonatal intensive care. Pediatrics 1995; 95:231–237.

294. Constandil L, Parraguez VH, Torrealba F, Valenzuela G, Serón-Ferré M. Day-night changes in c-*fos* expression in the fetal sheep suprachiasmatic nucleus at late gestation. Reprod Fertil Dev 1995; 7:411–413.

295. Stark RI, Daniel SS. Circadian rhythm of vasopressin levels in cerebrospinal fluid of the fetus: effect of continuous light. Endocrinology 1989; 124:3095–3101.

296. Ehrstrom C. Circadian rhythm of fetal movements. Acta Obstet Gynecol Scand 1984; 63:539–541.

297. Patrick J, Campbell K, Carmichael L, Natale R, Richardson B. Patterns of gross fetal body movements over 24-hour observation intervals during the last 10 weeks of pregnancy. Am J Obstet Gynecol 1982; 142:363–371.

298. Patrick J, Campbell K, Carmichael L, Probert C. Influence of maternal heart rate and gross fetal body movements on the daily pattern of fetal heart rate near term. Am J Obstet Gynecol 1982; 144:533–538.

299. Visser GHA, Goodman JDS, Levine DH, Dawes GS. Diurnal and other cyclic variations in human fetal heart rate near term. Am J Obstet Gynecol 1982; 142: 535–544.

300. Hellbrugge T, Lange JE, Rutenfranz J, Stehr K. Circadian periodicity of physiological functions in different stages of infancy and childhood. Ann NY Acad Sci 1964; 117:361–373.

301. Arduini D, Rizzo G, Parlati E, Dell'Acqua S, Romanini C, Mancuso S. Loss of circadian rhythms of fetal behaviour in a totally adrenalectomized pregnant women. Gynecol Obstet Invest 1987; 23:226–229.

302. Fletcher KL, Leung K, Myers MM, Stark RI. Diurnal rhythms in cardiorespiratory function of the fetal baboon. Early Hum Dev 1996; 46:27–42.

303. Jaldo-Alba F, Munoz-Hoyos A, Molina-Carballo A, Molina-Font MA, Acuna-Castroviejo D. Light deprivation increases plasma levels of melatonin during the first 72 h of life in human infants. Acta Endocrinol (Copenh) 1993; 129:442–445.

304. Rivkees SA, Hofman PL, Fortman J. Newborn primate infants are entrained by low intensity lighting. Proc Natl Acad Sci USA 1997; 94:292–297.

305. Spangler G. The emergence of adrenocortical circadian function in newborns and infants and its relationship to sleep, feeding and material adrenocortical activity. Early Hum Dev 1991; 25:197–208.

306. Weinert D, Sitka U, Minors DS, Waterhouse JM. The development of circadian rhythmicity in neonates. Early Hum Dev 1994; 36:117–126.

307. Updike PA, Accurso FJ, Jones RH. Physiologic circadian rhythmicity in preterm infants. Nurs Res 1985; 34:160–163.

308. Terman M, Lewy AJ, Dijk DJ, Boulos Z, Eastman CI, Campbell SS. Light treatment for sleep disorders: consensus report. 4. Sleep phase and duration disturbances. J Biol Rhythms 1995; 10:135–147.

309. Mann NP, Haddow R, Stokes L, Goodley S, Rutter N. Effect of night and day on preterm infants in a newborn nursery: randomised trial. Br Med J 1986; 293:1265–1267.

310. Kennaway DJ, Goble FC, Stamp GE. Factors influencing the development of melatonin rhythmicity in humans. J Clin Endocrinol Metab 1996; 81:1525–1532.

311. Gibson AAM. Current epidemiology of SIDS. J Clin Pathol 1992; 45(Suppl):7–10.

312. Kennaway DJ, Stamp GE, Goble FC. Development of melatonin production in infants and the impact of prematurity. J Clin Endocrinol Metab 1992; 76:367–369.

313. Brady JP, Chir B, McCann EM. Control of ventilation in subsequent siblings of victims of sudden infant death syndrome. J Paediatr 1980; 106:212–217.

314. McCulloch L, Brouilette RT, Guzzetta AJ, Hunt CE. Arousal responses in near-miss sudden infant death syndrome and in normal infants. J Paediatr 1982; 101:911–917.

315. Van der Hal AL, Rodriguez AM, Sargent CW, Platzker ACG, Keens TG. Hypoxic and hypercapnic arousal responses and predictions of subsequent apneas in apnea of infancy. Pediatrics 1985; 75:848–854.

316. McGinty DJ, Hoppenbrouwer T. The reticular formation, breathing disorders during sleep and SIDS. In: Tildon JT, Roeder L, Steinschneider A, eds. Sudden Infant Death Syndrom. New York: Academic Press, 1983:375–399.

317. Harper RM. State-related physiological changes and risk for the sudden infant death syndrome. Aust Paediatr J 1986; 22(Suppl):55–58.

318. Newman NM, Trinder JA, Phillips KA, Jordan K, Cruickshank J. Arousal deficit: mechanism of the sudden infant death syndrome? Austr Pediatr J 1989; 25:196–201.

319. Goldberg J, Hormung R, Yamashita T, Wehrmacher W. Age a death and risk factors in sudden infant death syndrome. Austr Pediatr J 1986; 22(Suppl):29–32.

320. Newman NM. Sudden infant death syndrome in Tasmania: 1975–1981. Aust Pediatr J 1986; 22(Suppl):17–19.

321. Harper RM, Leake B, Hoffman H, et al. Periodicity of sleep states is altered in infants at risk for the sudden infant death syndrome. Science 1981; 213:1030–1032.

322. Harper RM, Frostig Z, Taube D, Hoppenbrouwers T, Hodgman J. Development of sleep-waking temporal sequencing in infants at risk for the sudden infant death syndrome. Exp Neurol 1983; 79:821–829.

323. Navelet Y, Payan C, Guilhaume A, Benoit O. Nocturnal sleep organization in infants "at risk" for sudden infant death syndrome. Pediatr Res 1984; 18:654–657.

324. Becker LE. Links in the chain of events leading to sudden infant death syndrome. Dev Brain Dysfunct 1996; 9:232–242.

325. Axelsson G, Rylander R, Molin I. Outcome of pregnancy in relation to irregular and inconvenient work schedules. Br J Indust Med 1989; 46:393–398.

326. Nurminen T. Shift work, fetal development and course of pregnancy. Scand J Work Environ Health 1989; 15:395–403.

327. Armstrong BG, Nolin AD, McDonald AD. Work in pregnancy and birth weight for gestational age. Br J Indust Med 1989; 46:196–199.

3

Neural Control of Sleep

J. ALLAN HOBSON

Harvard Medical School
Boston, Massachusetts

I. Conceptual Overview

Modern neurobiology is founded upon the reflex paradigm of Charles Sherrington (1) and undergirded by the neuron doctrine and connectivity paradigm of Ramon Y. Cajal. Unfortunately many robust physiological and behavioral phenomena are not easily reducible to analysis within this powerful conceptual framework. The phenomena include (1) the spontaneous activity of the CNS, which precedes and outlasts the effects of stimulation that cause reflexive response; (2) the organized fluctuations of spontaneous activity whose rhythmicity persists even when synchronizing stimuli are masked; and (3) the organization of phases of the rhythmic fluctuations of physiological activity into states whose properties determine behaviors including conscious experience. Perhaps the most cogent argument for the complementarily of the reflex and state concepts is that the reflexes are themselves state dependent.

In this chapter I focus on the states of waking, sleeping, and dreaming as a way of making clear the continuous, but regularly varying, spontaneous activity of the brain. To explain these properties I invoke the notion of neuronal oscillators or clocks and show how the spontaneous activity of neurons can interact via synaptic connectivities whose excitability can be periodically altered via changes in the neuromodulatory microclimate of the brain.

I will begin by briefly defining the behavioral states and then describe their control by the best-known spontaneous oscillation in the brain, the circadian clock in the hypothalamus. I then turn to a more detailed analysis of the nature of sleep and its neuronal underpinnings. To illustrate the concept of state dependency, emphasis will be placed on the cellular and molecular mechanisms by which consciousness fluctuates in level and in quality over the sleep wake cycle.

II. Sleep

Sleep is a behavioral state that alternates with waking. It is characterized by a recumbent posture, a raised threshold to sensory stimulation, low levels of motor output, and a unique behavior, dreaming. The complex neurobiology of these behavioral features of sleep has been explored at the system, cellular, and molecular levels for more than 35 years. While these studies have provided substantial insight into the physiology and pathology of sleep, we have yet to obtain definitive answers to the adaptive significance of sleep even though this behavioral state takes up a third of our lives. (See Fig. 1.)

In contrast to sleep, the conscious behavior of waking is characterized by an active and deliberate sensorimotor discourse with the environment. To maintain waking behavior, animals must keep open the neural gates for sensory input and motor output, keep their brains in a condition of highly tuned activation, and provide the chemical microclimate appropriate to the processing and recording of information.

Wakefulness is accompanied by conscious experience that reaches its highest level of complexity in adult humans. Waking consciousness includes a number of components such as sensation, perception, attention, memory, instinct, emotion, volition, cognition, and language, which comprise the integrated awareness of the world, body, and self and form the basis for an adaptive interaction with our environment. Here we focus on the mechanisms of behavioral state control that make such interactions possible with a special emphasis on sleep as a component of adaptive behavior. Because the evidence so strongly favors an integration of psychological and physiological state features we will refer to the substrate of conscious experience as the brain-mind.

A. Sleep and the Circadian System

Sleep is one of many bodily functions under the timing control of the circadian clock located in the suprachiasmatic nucleus of the hypothalamus. Thus while sleep and waking continue to alternate regularly in subjects isolated from time

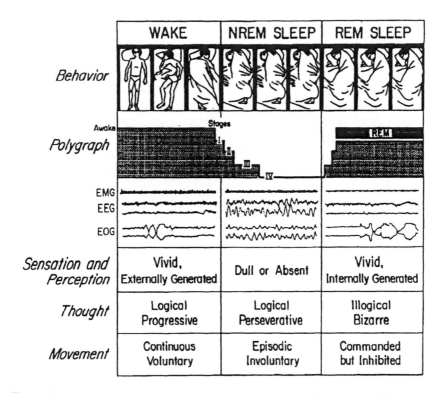

	WAKE	NREM SLEEP	REM SLEEP
Behavior			
Polygraph			
EMG			
EEG			
EOG			
Sensation and Perception	Vivid, Externally Generated	Dull or Absent	Vivid, Internally Generated
Thought	Logical Progressive	Logical Perseverative	Illogical Bizarre
Movement	Continuous Voluntary	Episodic Involuntary	Commanded but Inhibited

Figure 1 Behavioral states in humans. States of waking, NREM sleep, and REM sleep have behavioral, polygraphic, and psychological manifestations. In behavior channel, posture shifts (detectable by time-lapse photography or video) can occur during waking and in concert with phase changes of sleep cycle. Two different mechanisms account for sleep immobility: disfacilitation (during stages I–IV of NREM sleep) and inhibition (during REM sleep). In dreams we imagine that we move but we do not. Sequence of these stages represented in polygraph channel. Sample tracings of three variables used to distinguish state are also shown: electromyogram (EMG), which is highest in waking, intermediate in NREM sleep, and lowest in REM sleep; and electroencephalogram (EEG) and electro-oculogram (EOG), which are both activated in waking and REM sleep and inactivated in NREM sleep. Each sample record is 20 sec. Three lower channels describe other subjective and objective state variables.

cues, the period length of that alteration is usually longer than 24 hr (2). This free-running sleep-wake oscillation is usually in phase with the free-running rhythm of body temperature but occasionally the two rhythms become dissociated (3). This process, called internal desynchronization, has raised questions about the unity (vs. duality) of the circadian clock in the brain (4).

B. Sleep and Temperature Regulation

Sleep onset normally occurs on the descending limb of the curve describing the circadian body temperature rhythm, and a further drop in body temperature occurs with the first episode of non-rapid eye movement (NREM) sleep. Furthermore, two distinct thermal adaptations, shallow torpor (a temperature drop occurring daily in small mammals) and hibernation (a more profound and prolonged seasonal drop in body temperature), occur during NREM sleep (5). These facts combine to favor the view that sleep is part of a continuum of diverse energy conservation strategies used by mammals to cope with varying levels and sources of heat and light. Whatever benefits sleep may accrue for tomorrow, one clear function is to conserve calories for today.

In addition to a drop in body temperature, reduced responsiveness to changes in ambient temperature is also observed in NREM sleep; shivering in response to cold and sweating in response to heat are both diminished. Once NREM sleep is established and the temperature nadir is passed, REM sleep supervenes, during which thermoregulatory reflexes disappear altogether. In recent studies of the brain basis of these phenomena, Heller and co-workers at Stanford (5) and Parmeggiani in Bologna (6) have shown that the responsiveness of temperature sensor neurons in the preoptic hypothalamus dips to a lower level in NREM sleep than in waking, and bottoms out in REM sleep. In sleep, the animal apparently substitutes behavioral control for its neuronal thermostat. It seems unlikely that an animal would develop such a high-cost maneuver to gain only a small and short-term calorie saving.

Daily variations in human body temperature of 1.5° were observed in the first phase of sleep research and led to the discovery of the circadian rhythm (3), which is now widely believed to be regulated by the suprachiasmatic hypothalamus (7). The circadian rhythm of body temperature has been shown to be tightly coupled to the circadian rhythm of sleep and waking, and the degree to which these can be dissociated continues to be debated. In recent studies, Zulley and Campbell (8) showed that subjects isolated from time cues were likely to enter long sleep episodes only at or near minimal body temperature. Sleep was virtually impossible at other points on the body temperature curve. Naps could occur at other times, but sleep was not maintained.

These studies questioned the previously reported phenomenon of "internal desynchronization" of the two rhythms by showing that the observed desynchronization was in part an experimental artifact arising from injunctions not to nap, which led to the underreporting of sleep by subjects taking part in the original bunker experiments (3). The dual oscillator theories resulting from these studies—one oscillator for body temperature and the other for sleep—have stimulated elaborate modeling efforts (4).

An important problem to be solved is the neural basis of the usually strong coupling between the hypothalamic circadian oscillator and the pontine infradian sleep oscillator. Complementing earlier reports of surgical dissociation of the oscillators, Jouvet and colleagues (9) have recently shown that the two oscillators can also be chemically uncoupled. However, the cellular and molecular basis of the circadian oscillator and the mechanism of its coupling to the other oscillators remain to be fathomed.

III. Brain Stem Regulation of Sleep

A. The NREM-REM Sleep Cycle

All of the defining behavioral signs of sleep vary systematically over any given sleep bout as posture, threshold to arousal, and motor output all change in a stereotyped and cyclic manner. While these changes can be directly observed, their study is facilitated by electrographic recording. Two distinct substates of sleep can be discerned, both of which contrast with waking in an interesting and informative manner.

At sleep onset, awareness of the outside world is lost as the brain deactivates. A well-described sequence of thalamocortical events is responsible for the progressive electroencephalographic (EEG) slowing of this sleep phase. Arousal threshold rises in proportion to the degree of EEG slowing, and at the greatest depth of sleep, awakenings are often difficult, incomplete, and short-lived. This sleep state is designated slow-wave sleep, or NREM, sleep. Subjects have little or no recall of conscious experience in deep NREM sleep. Many autonomic and regulatory functions such as heart rate, blood pressure, and respiration diminish in NREM sleep, but there is significant neuroendocrine activity with the pulsatile release of pituitary growth and sexual maturation hormones at maximal levels. Over 95% of the daily output occurs in NREM sleep.

At regular intervals through sleep, the brain reactivates into a state characterized by low-voltage, fast activity, muscle atonia, and rapid eye movements. This sleep state, termed REM sleep because of the prominent eye movements, differs from waking because of an inhibition of sensory input and motor output. In REM sleep there is loss of temperature control and cardiopulmonary activation, which may occur with marked irregularity of these functions. Postural shifts precede and follow REM while eye movements, intermittent small muscle twitching, and penile erection occur within it. The presence (or absence) of REM sleep erection is used clinically to distinguish between psychological and physiological male impotence. In REM sleep many autonomic functions change; there is loss of temperature control and variability of cardiorespiratory functions.

Subjects aroused from NREM sleep, especially early in the night, are confused, have difficulty reporting conscious experience and return to sleep rapidly. Subjects aroused from REM sleep, especially from epochs with frequent eye

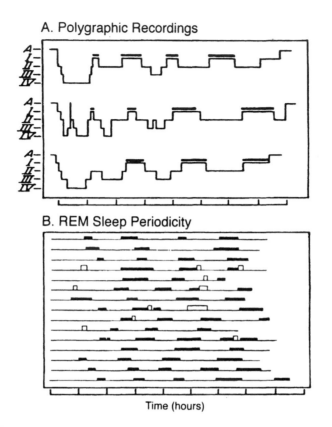

A. Polygraphic Recordings

B. REM Sleep Periodicity

Time (hours)

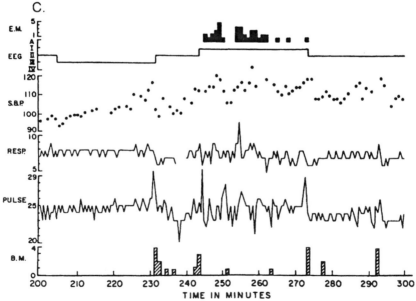

C.

movements, give detailed narrative dream reports characterized by vivid hallucinatory percepts, bizarre cognition, and intense emotion. These observations indicate that dreaming is the behavioral concomitant of the brain activation in REM sleep.

The NREM and REM phases alternate throughout sleep at intervals of 90 min in adult humans. The NREM phase is deeper and longer early in the night. Together with the regular period length, this property suggests that a damped oscillator is the underlying neural mechanism for NREM-REM cycles. (See Fig. 2.)

B. The Anatomy and Physiology of Brainstem Regulatory Systems

The brainstem reticular formation contains specific neuronal groups involved in behavioral state regulation. Moruzzi and Magoun's original concept of a nonspecific reticular activating system (10) has been greatly elaborated and modified by subsequent anatomical and physiological studies. Two general principles have emerged. One is that most of the classic reticular core neurons are neither diffusely projecting nor nonspecific in their connections. Rather, they have very specific afferent inputs and highly organized outputs. The other is that the reticular formation contains small groups of neurons that perform a chemically specific neuromodulatory function by sending their widely ramifying axons to distant parts of the brain.

The norepinephrine neurons [designated A1–A7 by Dahlstrom and Fuxe (11)] are located in the pons and medulla in two major groups. One group consists of scattered neurons largely located in the ventral and lateral reticular formation. The most caudal neurons (A1–A3 of Dahlstrom and Fuxe) project rostrally to the brainstem, hypothalamus, and basal forebrain whereas the rostral groups (Dahlstrom and Fuxe's A5 and A7) project caudally to the brainstem and spinal cord. These

---

Figure 2   (A and B) Ultradian sleep cycle of NREM and REM sleep shown in detailed sleep-stage graphs of three human subjects (A) and REM sleep periodograms of 15 human subjects (B). In polysomnograms of (A), note typical preponderance of deepest stages (III and IV) of NREM sleep in the first two or three cycles of night; REM sleep is correspondingly brief (subjects 1 and 2) or even aborted (subject 3). During the last two cycles of night, NREM sleep is restricted to lighter stage (II), and REM periods occupy proportionally more of the time with individual episodes often exceeding 60 min (all three subjects). Same tendency to increase REM sleep duration is seen in (B). In these records, all of which begin at sleep onset, not clock time, note variable latency to onset of first (usually short) REM sleep epoch. Thereafter inter-REM period length is relatively constant. For both (A) and (B) time is in hours. (F. Snyder and J.A. Hobson, unpublished observations.) (C) Eye movements (E.M., EEG), systolic blood pressure (S.B.P.), respiration (RESP.), pulse, and boyd movements (B.M.) in a 100-min sample of uninterrupted sleep over successive minutes of typical sleep cycle. Entire interval from minute 242 to 273 is considered to be REM period, even though eye movements (heavy bars) are not continuous.

*neuronal groups, designated together as the lateral tegmental neuron group (12), appear to be involved in hypothalamic regulation and motor control.

The major norepinephrine cell group is the locus coeruleus (A4 and A6 of Dahlstrom and Fuxe). This compact cell group is located in the rostral pontine reticular formation and central gray. The neurons of the locus coeruleus project widely but in a highly specific pattern. One group appears to project largely caudally to sensory regions of the brainstem and spinal cord. Other locus neurons project widely to cerebella cortex, dorsal thalamus, and cerebral cortex (12). Thus, the projection patterns of the locus coeruleus appear to be primarily to sensory structures and to cortical structures involved in integrative activities. From this we could expect the locus coeruleus to be involved in the regulation of sensory input and cortical activation and, as we shall see, this is in accord with the available functional information.

Two additional subsets of chemically identified neurons appear to be involved critically in behavioral state regulation. The first to be discovered were the serotonin neurons of the brainstem raphe (11) (B1–B9 in the nomenclature of Dahlstrom and Fuxe). These neurons extend from caudal medulla to midbrain and are located predominantly in the raphe nuclei, a set of neuronal groups located in the midline of the brainstem reticular formation. The largest number of serotonin neurons is found in the midbrain nuclei, the dorsal raphe, and median raphe nuclei (B8 and B9). These groups project largely rostrally, innervating nearly the entire forebrain (13) in a pattern that also suggests a role in behavioral state regulation.

The other important set of reticular formation nuclei that are involved in behavioral state control are ones that produce acetylcholine as their neuro-transmitter. It is well known that acetylcholine is the transmitter of motor neurons and early work indicated that acetylcholine is found in nonmotor brain areas. Two sets of cholinergic neurons are involved in behavioral state control. The first consists of two pontine nuclei, the laterodorsal tegmental nucleus and the pedunculo-pontine nucleus. Cholinergic neurons in these nuclei project to the brainstem reticular formation, hypothalamus, thalamus, and basal forebrain. The latter projection involves the second set of cholinergic neurons, those in the medial septum, the nucleus of the diagonal band, and the substantia innominata–nucleus basalis complex. These project to limbic forebrain, including the hippocampus, and to the neocortex.

In a further contrast to the modulatory neurons that are characterized by their production of norepinephrine, serotonin, and acetylcholine, the neurons in reticular formation nuclei that are involved in sensorimotor integration typically produce either the excitatory transmitter, glutamate, or the inhibitory transmitter, GABA.

IV.   NREM Sleep and Thalamocortical Function

Philosophical speculation regarding the nature of consciousness is as old as recorded history and many materialist philosophers including the Ionian Greeks interestingly anticipated the physicalistic models that have only recently assumed the specific articulation of modern neuroscience (14). The signal event that demarcates the modern scientific era is the discovery of the electrical nature of nervous activity and, more specifically, the 1928 discovery of the human EEG by the German psychiatrist Adolf Berger.

It was the state-dependent nature of the EEG that helped Berger convince his skeptical critics that the rhythmic oscillations he recorded across the human scalp and skull with his galvanometer were of brain origin and not artifacts of movement or of scalp muscle activity. When his subjects relaxed, closed their eyes, or dozed off into drowsiness, the low-voltage front brain wave activity associated with alertness gave way to higher-voltage, lower-frequency patterns. And these in turn were rapidly blocked when the subjects were aroused.

A. The Activation Concept

Following Berger's discovery there ensued a flurry of descriptive and experimental studies aimed at understanding the EEG itself, the full range of its state-dependent variability, and the control of that variability by the brain. Loomis and associates (15) were the first to describe the tendency of the EEG to show a level of alertness to fall as subjects fell asleep at night.

Because other mammals shared the same cross-correlation between arousal level and the EEG, the Belgian physiologist Frederick Bremer (16) made experimental transections of the brain to determine the nature and source of EEG activation (the low-voltage front pattern of waking) and deactivation (the high-voltage slow pattern of sleep).

Thinking it probably depended upon sensory input and in keeping with the regnant reflex doctrine, Bremer transected the brain at the level of the first cervical spinal cord segment producing a preparation called the *encephale isolé*. He was surprised to find that the isolated forebrain was activated and alert despite such major deafferentation of its sensory input. But when he then produced the *cerveau isolé* by transecting the midbrain at the intercollicular level, he observed persistent EEG slowing and unresponsiveness. Interpreting this observation Bremer incorrectly inferred that it was deafferentation of the trigeminal nerve inputs (which entered the brainstem between the level of the two cuts) that accounted for the sleep-like state of the *cerveau isolé*.

B. The Electrical Induction of Sleep

That sleep might be an active intrinsic brain process—and not simply the absence of waking as Bremer supposed—was first experimentally suggested by the work of the Swiss Nobel Laureate W. R. Hess (17). As part of a broad program of investigation of the effects of electrical stimulation of the subcortical regions mediating autonomic control (especially the hypothalamus), Hess discovered that by driving the thalamocortical system at the frequencies of intrinsic EEG spindles and slow waves, he could induce the behavioral and electrographic signs of sleep in unanesthetized cats (17). This discovery opened the door to the idea that sleep and waking might both be active processes each with its own specific cellular and metabolic mechanisms and functional consequences, this paradigm has since borne abundant scientific fruit as the precise details of spindle and slow-wave elaboration have been worked out (18, 19). In his concept of arousal as ergotrophic (or energy consuming) and sleep as trophotropic (or energy conserving) Hess also correctly anticipated the recent finding that central sympathetic neurons mediate waking and central cholinergic neurons mediate REM sleep.

The intrinsic nature of brain activation was clearly demonstrated in 1949 when Giuseppe Moruzzi and Horace Magoun (10) discovered that EEG desynchronization and behavioral arousal could be produced by high-frequency electrical stimulation of the midbrain. To explain their observation Moruzzi and Magoun advanced the concept of the nonspecific (i.e., nonsensory) reticular activating system operating in series and in parallel with the ascending sensory pathways. This concept not only allowed for the translation of afferent stimuli into central activation (Bremer's idea) but opened the door to the more radical idea of autoactivation of the brain-mind sui generis. Now that we accept as a given the spontaneous activity of neurons and the continuity and elaboration of such activity in determining the several substages of sleep, it is difficult for us to appreciate how strong and persistent was its antecedent concept. To make this point clear, we have only to consult such scientific giants as Ivan Pavlov (20) and Charles Sherrington (1) both of whom were so inbred with the reflex doctrine that they were convinced that brain activity simply ceased in the absence of sensory input.

C. Steriade's Model of Thalamocortical Regulation

About 75% of sleep time is spent in the NREM phase and the neurobiology of this important state has been greatly advanced by Steriade's detailed studies of cellular activity in thalamocortical circuits (see Ref. 21 for review). One important and useful unifying concept is that the forebrain, and particularly the cortex, tends toward spontaneous oscillation—with resulting unconsciousness—unless it is activated by the brainstem.

The initiation of NREM sleep is gradual and is characterized by slowing of EEG frequency. This initial phase is termed stage 1 and is succeeded by stage 2, which is characterized by a further decrease in EEG frequency with intermittent, high-frequency spike clusters called sleep spindles. Sleep spindles decrease in stage 3 as the amplitude of slow waves increases; this mixed pattern gives way to very high-amplitude, delta waves in the deepest sleep, stage 4.

The critical circuitry consists of reciprocally interconnected thalamic and cortical neurons that oscillate to produce the NREM sleep spindle waves of stages 2 and 3 of NREM sleep. The thalamocortical and thalamic reticular neurons that regulate this state are shifted into the burst firing mode by deactivation and demodulation from the hypothalamus and brainstem. During waking, these circuits are modulated by the activity of neurons that produce ACh, NE, 5HT, and glutamate whereas ACh is the principal modulator released during REM sleep. This indicates that the activated brain sustains consciousness according to the net brainstem and hypothalamic drive on the thalamocortical system and produces the different kinds of consciousness seen in waking and dreaming according to the differential modulatory inputs to the thalamocortical circuitry.

As sleep becomes deeper and spindling diminishes (in NREM stages 3 and 4), the thalamocortical cells become progressively hyperpolarized. Cortical neurons then generate their own spontaneous delta (1–4 Hz) and slow (1 Hz) oscillations. When these high-voltage slow waves are present, subjects are difficult to arouse, evince confusion, may confabulate, and cannot perform intellectual tests following awakening. Steriade has speculated that the thalamocortical and cortical neurons may use the oscillatory modes of NREM sleep to balance ionic currents and intracellular regulatory mechanisms in such a way as to incorporate experience from previous waking episodes and prepare them for subsequent ones. NREM sleep is a quiescent state for the brain in which both blood flow and glucose utilization are decreased by more than 40%. A recent PET study has shown that decrease in blood flow is particularly marked in the brainstem and diencephalon (22).

V. REM Sleep

A. The Discovery of REM Sleep

In 1953, Eugene Aserinsky and Nathaniel Kleitman (23), working in Chicago, discovered that the brain-mind did indeed self-activate, especially during sleep, when at regular intervals they observed the spontaneous emergence of EEG desynchronization, accompanied by clusters of rapid saccadic eye movements (or REMs) together with acute accelerations of heart and respiration rates. Working with Kleitman, William Dement was then able to show that these periods of spontaneous autoactivation of the brain-mind were associated with dreaming (24,25) and that this autoactivation process was also found in cats (26).

In adult humans, the intrinsic cycle of inactivation (NREM sleep) and activation (REM sleep) recurred with a period length of 90–100 min with REM sleep occupying 20–25% of the recording time and NREM the rest (75–80%). The NREM phases of the first two cycles were deep and long while REM occupied more of the last two or three cycles and sleep lightened.

In the early days of EEG recording, brain waves with frequencies in excess of 25 c/sec were either ignored or filtered out because of the problem of interference by 50 or 60 c/sec artifacts from electrical power sources. Recently research aimed at solving what has been called the binding problem has focused its attention on the 35–40 c/sec activity that is found in adjacent neuronal ensembles when animals attend to external stimuli (27). It has been proposed that by synchronizing multiple and distant brain units, the spatiotemporal unity of conscious experience may be achieved. Llinas has further proposed that as the cortex is scanned by the thalamus, 40 c/sec waves are propagated from the frontal to the occipital poles. Noting that 40 c/sec activity is observed in REM sleep as well as waking, Llinas has emphasized the importance of intrinsic neuronal oscillations in the genesis of all states of consciousness and suggested that the main difference between waking and dreaming arises from input-output gating (28).

B. Input-Output Gating

The paradoxical preservation of sleep in the face of brain-mind activation during sleep began to be explained when Michel and Jouvet, working in Lyon in 1959, demonstrated that active muscle inhibition was a regular component of REM sleep in cats. Using transection, lesion, and stimulation techniques, the Jouvet team also discovered that the control system for REM sleep was localized to the pontine brainstem and that the pons was the source of the EEG activation and the REMs themselves while the muscle inhibition was mediated by pontine signals relayed via the bulbar inhibitory reticular formation to the spinal cord (29). Synchronous with each flurry of REMs, phasic activation signals or ponto-geniculo-occipital (PGO) waves were sent from the pons up into the forebrain (and down into the spinal cord). Jouvet also discovered that cellular level studies later revealed that the PGO waves triggered bursts of firing by geniculate and cortical neurons and that other signals of brainstem origin damped both sensory input (via presynaptic inhibition) and motor output (via postsynaptic inhibition) (30). It thus became clear that in REM sleep the autoactivated brain-mind was effectively off-line with respect to both external inputs and outputs and that it stimulates itself.

The cellular and molecular basis of these dramatic changes in input-output gating mechanisms have been detailed using Sherrington's reflex paradigm together with extracellular and intracellular recording techniques (31). For example, it became clear that during REM sleep, each motor neuron was subjected to 10 mV of tonic hyperpolarization, which blocked all but a few of the phasic activation signals generated by the REM-PGO system and that this inhibition was mediated by glycine (32,33). When this motor inhibition was experimentally disrupted by lesioning the pons, the cats—still in a REM-like sleep state but without the atonia—evinced stereotyped behaviors (such as fear-defense and aggressive-attack postures) that expressed the activation, in REM, of the specific motor-pattern generators for these instinctual fixed acts (34,35).

The clear implication was that in normal REM sleep the motor inhibition served to contain the motor commands of the instinctual behavior pattern genera-

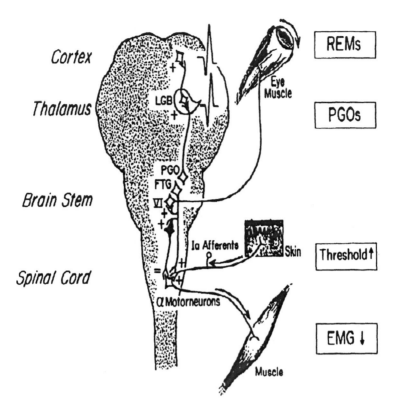

Figure 3 Three mechanisms underlying state-dependent changes in AIM model factor I: *Efferent copy*: During REM sleep, neurons of the pontine reticular formation (FTG) are activated. When they fire in bursts, ipsiversive REMs are generated and ipsilateral PGO waves are triggered in the LGB and posterolateral cortex. *Presynaptic inhibition*: Via axons descending from the FTG, the primary afferent terminals of the group Ia cutaneous afferents are depolarized, making them less responsive to incoming volleys from the skin. In this way sensory thresholds are raised. *Postsynaptic inhibition*: Cells of the medullary reticular formation are also excited by volleys from the pons, but they convey inhibitory signals to the anterior horn motor neurons and muscle tone is suppressed (EMG). In this way the threshold to motor activation is raised.

tors rendering them virtual as far as the outside world was concerned and fictive for the internal world of the brain-mind itself. The strong significance of these findings for a theory of dream consciousness lies in their ability to explain the ubiquity of imagined movement in dreams (36). From a functional point of view, it may also be important to recognize that the autoactivation process of REM sleep is highly patterned (and hence nonrandom) from a sensorimotor perspective.

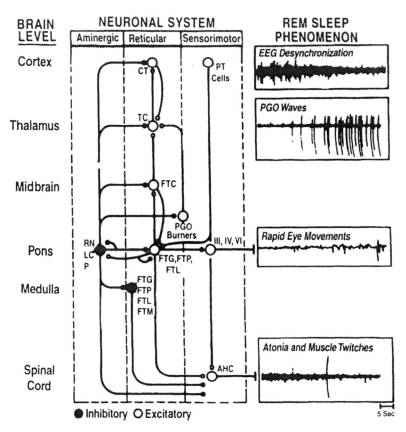

BRAIN LEVEL — **NEURONAL SYSTEM** (Aminergic | Reticular | Sensorimotor) — **REM SLEEP PHENOMENON**

Cortex — CT, PT Cells — *EEG Desynchronization*

Thalamus — TC — *PGO Waves*

Midbrain — FTC

Pons — RN LC P, PGO Burners, FTG,FTP, FTL, III, IV, VI — *Rapid Eye Movements*

Medulla — FTG FTP FTL FTM

Spinal Cord — AHC — *Atonia and Muscle Twitches*

● Inhibitory ○ Excitatory   5 Sec

Figure 4 Schematic representation of the REM sleep generation process. A distributed network involves cells at many brain levels (left). The network is represented as comprising three neuronal systems (center) that mediate REM sleep electrographic phenomena (right). Postulated inhibitory connections are shown as solid circles; postulated excitatory connections, as open circles. In this diagram no distinction is made between neurotransmission and neuromodulatory functions of the depicted neurons. The actual synaptic signs of many of the aminergic and reticular pathways remain to be demonstrated, and, in many cases (e.g., the thalamus and cortex), the neuronal architecture is known to be far more complex than indicated here. Two additive effects of the marked reduction in firing rate by aminergic neurons at REM sleep onset are postulated: disinhibition (through removal of negative restraint) and facilitation (through positive feedback). The net result is strong tonic and phasic activation of reticular and sensorimotor neurons in REM sleep. REM sleep phenomena are postulated to be mediated as follows: EEG desynchronization results from a net tonic increase in reticular, thalamocortical, and cortical neuronal firing rates. PGO waves are the result of tonic disinhibiton and phasic excitation of burst cells in the lateral pontomesencephalic tegmentum. Rapid eye movements are the consequence of phasic firing by reticular and vestibular cells; the latter (not shown) directly excite oculomotor

The recording of individual neurons in behaving animals was pioneered by Evarts (37) using the movable microelectrode system of Hubel (38). Evarts was able to show that both the tonic EEG activation and the phasic PGO wave activation signals of REM reflected the excitation of neurons throughout the forebrain including the visual motor and association cortices (39) as well as the thalamic nuclei reciprocally connected to them (40,41). It may come as a surprise to realize that we now know more about the neurophysiology of REM sleep and dreaming than we know about waking. This paradox is explained by the fact that in REM the animal is paralyzed and partially anesthetized by intrinsic modulation of its motor and sensory systems! REM sleep is therefore a natural state favoring deep neurophysiological analysis.

The picture that emerges is of global, but specific, alterations of neuronal activation and information flow throughout the brain. This picture is highly relevant to our understanding of the differences that distinguish the conscious states of waking from that of NREM sleep and REM sleep with dreaming. In waking the activated brain-mind gives priority to processing data from the outside world and acting upon that world accordingly. (In NREM sleep the system is taken off-line passively via deactivation.) In REM sleep, by extreme contrast with waking, it is the internal representations of that world that become "inputs" and the action that is summoned (but not executed) is one of those inputs (28).

C. Modulation

The chemical means by which these dramatic changes in brain-mind state are affected are mediated by the modulatory neuronal systems of the brainstem (42). In this regard the noradrenergic locus coeruleus and the serotonergic raphe neurons of the pons are particularly critical. Both of these aminergic populations contain pacemaker elements that fire spontaneously and tonically throughout waking. They also phasically increase their output in response to salient stimuli and decrease their output during interstimulus lulls and at sleep onset (43–45). In their phasic response to stimuli, they are joined by the cholinergic neurons of the far lateral pedunculopontine nucleus but these cells are not pacemakers and tend to be otherwise quiescent in waking. Thus the waking brain is bathed in constant

---

neurons. Muscular atonia is the consequence of tonic postsynaptic inhibition of spinal anterior horn cells by the pontomedullary reticular formation. Muscle twitches occur when excitation by reticular and pyramidal tract motor neurons phasically overcomes the tonic inhibition of the anterior horn cells. RN, raphe nuclei; LC, locus ceruleus; P, peribrachial region; FTG, gigantocellular tegmental field; FTC, central tegmental field; FTP, parvocellular tegmental field; FTM, magnocellular tegmental field; TC, thalamocortical; CT, cortical; PT cell, pyramidal cee; III, oculomotor; IV, trochlear; V, trigmenial motor nuclei; AHC, anterior horn cell. (Modified from Ref. 18.)

levels of norepinephrine and serotonin and receives pulsatile boosts of these two chemicals and of acetylcholine when novel input data call for them. These observations suggest that the chemistry of attentive, mnemonic waking is an aminergic-cholinergic collaboration.

All three modulatory systems abate at sleep onset. But as NREM sleep deepens the activity of the two aminergic systems gradually and spontaneously declines (45–47) while the activity of the cholinergic neurons gradually and spontaneously increases (46). At REM sleep onset the aminergic system is completely arrested while the cholinergic system is unabatedly autoactive. The net effect, which is confirmed by measurements of transmitter release (48), is a shift from an aminergic cholinergic microclimate in waking to an exclusively cholinergic microclimate in REM.

These physiological findings have inspired—and are strengthened by—the results of systemic and local pharmacological experiments showing that both antiaminergic and procholinergic drugs tend to increase REM while both pro-aminergic and anticholinergic agents suppress it. As predicted by the physiological model to be presented in the next section, these drugs have reciprocal effects upon waking and NREM sleep. Among the wealth of pharmacological studies yielding these conclusions (18), two are particularly impressive and both reveal dramatic REM sleep enhancement. When the cholinergic agonist carbachol or the anticholinesterase neostigmine is injected into the paramedian pontine brainstem, it causes immediate, intense, and prolonged REM sleep episodes (18,49,50). This effect, which wanes in about 6 hr, is called short-term REM sleep enhancement. When injected into the far lateral peribrachial pons, these drugs cause immediate but only unilateral PGO wave enhancement and delayed (24–48 hr) enhancement of REM sleep episodes of normal lengths for 6–10 days; this is called long-term REM sleep enhancement (51,52). See Figure 6.

VI. Cognitive Correlates of Behavioral States

The physiological differences between waking, NREM sleep, and REM sleep that have been described can now be linked in a tentative way to the changes in conscious state that are correlated with them (Table 1). In the domain of sensation and perception, responsiveness to the external world is progressively lost with the cortical deactivation of sleep onset; responsiveness declines further as NREM sleep deepens. During REM sleep the brain reactivates but presynaptic inhibition blocks the exteroception of sensory signals and the phasic sensorimotor activation embodied by the REMs, and their associated PGO waves come to constitute internal stimuli. These endogenous signals send specific information about the eye movements from the brainstem to the thalamocortical visual system possibly accounting for the intense visual hallucination of dreams. Furthermore, these PGO

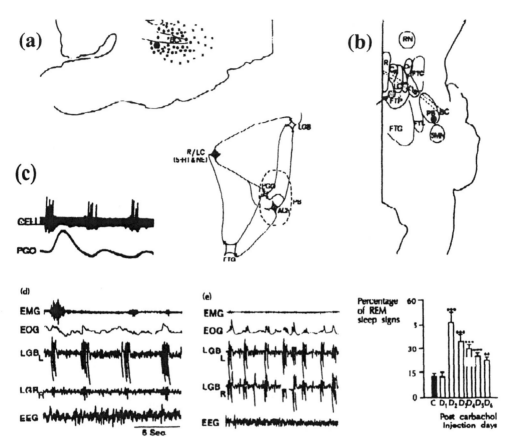

Figure 5 Relationship of PGO burst cells to cholinergic REM induction sites. (a) Filled circle indicates site of injection of carbachol into peribrachial pons. Small dots indicate location of cholinergic cells and crosses indicate location of PGO burst cells. (b) Location of peribrachial long-term (filled circle) and paramedian short-term (open circle with dots) REM induction sites. (c) On the left is an extracellular unit recording of a PGO burst cell and PGO waves in the ipsilateral LGB. On the right is a hypothetical wiring diagram of the PGO trigger zone (PB) elements (PGO and ACh) with modulatory aminergic raphe (R) and locus coeruleus (LC), thalamic (LGB), and paramedian pontine reticular cells (FTC). (d) Injections of carbachol at the PB site shown in (a) and (b) produce immediate PGO waves in the ipsilateral LGB. (e) After 24 hr REM triples and remains elevated for 6 days; this is the long-term REM effect. Reticular tegmental nuclei (FTC, FTP, FTL, FTC). Aminergic nuclei (R,LC). Cholinergic nuclei (C5,C6). Reference structures shown are red nucleus (RN), trigeminal motor nucleus (MN), and brachium conjuctivum (BC). (a, c, and e reproduced with permission from Ref. 52.)

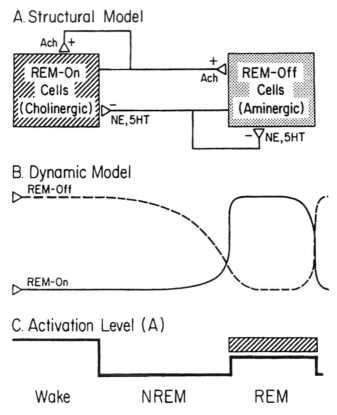

Figure 6 Physiological mechanisms determining alterations in activation level. (A) Structural model of reciprocal interaction. REM-on cells of the pontine reticular formation are cholinoceptively excited and/or cholinergically excitatory (ACH+) at their synaptic endings (open boxes). Pontine REM-off cells are noradrenergically (NE) or serotonergically (5HT) inhibitory (−) at their synapses (filled boxes). (B) Dynamic model. During waking the pontine aminergic (filled box) system is tonically activated and inhibits the pontine cholinergic (open box) system. During NREM sleep aminergic inhibition gradually wanes and cholinergic excitation reciprocally waxes. At REM sleep onset aminergic inhibition is shut off and cholinergic excitation reaches its high point. (C) Activation level (A). As a consequence of the interplay of the neuronal systems shown in (A) and (B), the net activation level of the brain (A) is at equally high levels in waking and REM sleep and at about half this peak level in NREM sleep. (From Antrobus JS, Bertini M, eds. The Neuropsychology of Sleep and Dreaming. A New Model of Brain-Mind State: Activation Level, Input Source, and Mode of Processing (AIM). Hillsdale, NJ: Lawrence Erlbaum, 1992:227–245.)

Table 1 Physiological Basis of Differences Between Waking and Dreaming

Function	Nature of difference	Causal hypothesis
Sensory input	Blocked	Presynaptic inhibition
Perception (external)	Diminished	Blockade of sensory input
Perception (internal)	Enhanced	Disinhibition of networks storing sensory representations
Attention	Lost	Decreased aminergic modulation causes a decrease in signal-to-noise ratio
Memory (recent)	Diminished	Because of aminergic demodulation, activated representations are not restored in memory
Memory (remote)	Enhanced	Disinhibition of networks storing mnemonic representations increases access to consciousness
Orientation	Unstable	Internally inconsistent orienting signals are generated by cholinergic system
Thought	Reasoning ad hoc; logical rigor weak, processing hyperassociative	Loss of attention memory and volition leads to failure of sequencing and rule inconstancy, analogy replaces analysis
Insight	Self-reflection lost	Failures of attention, logic, and memory weaken second (and third) order representations
Language (internal)	Confabulatory	Aminergic demodulation frees narrative synthesis from logical restraints
Emotion	Episodically strong	Cholinergic hyperstimulation of amygdala and related temporal lobe structures triggers emotional storms, which are unmodulated by aminergic restraint
Instinct	Episodically strong	Cholinergic hyperstimulation of hypothalamus and limbic forebrain triggers fixed action motor programs, which are experienced fictively but not enacted
Volition	Weak	Top-down motor control and frontal executive power cannot compete with disinhibited subcortical network activation
Output	Blocked	Postsynaptic inhibition

signals cholinergically drive not only the visual system but also the amygdala (perhaps accounting for such dream emotions as anxiety and surprise).

Cognition also undergoes a reliable shift from the wake state as attention, orientation, and logically directed thought give way first to the nonprogressive ruminations of NREM sleep (18,53,54) and then to the disoriented, undirected, and illogical cognition of dreaming. In addition to the loss of the temporal and spatial constancies conferred by the sleep-dependent sensorimotor deafferentation, the brain in REM sleep is aminergically demodulated, which may contribute to the failures of attention, orientation, memory, and logic that characterize dreaming itself and make dreams so hard to remember (46,55–58).

VII. Dysfunctional Sleep

It is significant for the pathophysiology of sleep that the damping of the NREM-REM oscillator is incomplete. During the activity phase, numerous studies have revealed a weak, but significant, periodicity of behavior in the 90–100-min range, suggesting that the pontine clock may be timing at low amplitude throughout the day, thus causing the waxing and waning of attention and of motor activity that normally punctuate waking. A better understanding of this circadian ultradian clock interaction is an important problem for future research.

A. Central Autonomic Control Systems Are State Dependent

Cognitive states are not the only functions linked to the extreme changes in central modulatory systems that have been discovered through cellular- and molecular-level sleep research. Numerous vital vegetative functions are also altered in reliable and occasionally problematic ways. For example, active hypothalamic temperature control is diminished or abandoned in REM sleep (5,6) and the sensitivity of the modulatory respiratory control system is likewise diminished. The upshot is

---

Figure 7 The activation-synthesis model. (A) Systems model. As a result of disinhibition caused by cessation of aminergic neuronal firing, brainstem reticular systems autoactivate. Their outputs have effects including depolarization of afferent terminals causing phasic presynaptic inhibition and blockade of external stimuli, especially during the bursts of REM, and postsynaptic hyperpolarization causing tonic inhibition of motor neurons that effectively counteract concomitant motor commands so that somatic movement is blocked. Only the oculomotor commands are read out as eye movements because these motor neurons are not inhibited. The forebrain, activated by the reticular formation and also aminergically disinhibited, receives efferent copy or corollary discharge information about somatic motor and oculomotor commands from which it may synthesize such internally generated perceptions as visual imagery and the sensation of movement, both of which typify dream mentation. The forebrain may, in turn, generate its own motor commands that help to perpetuate the process via positive feedback to the reticular formation. (B) Synaptic

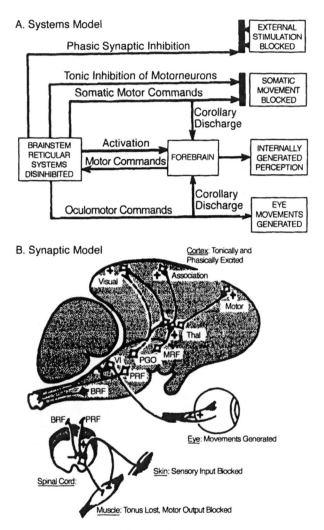

A. Systems Model

Phasic Synaptic Inhibition → EXTERNAL STIMULATION BLOCKED

Tonic Inhibition of Motorneurons
Somatic Motor Commands → SOMATIC MOVEMENT BLOCKED

Corollary Discharge

BRAINSTEM RETICULAR SYSTEMS DISINHIBITED → Activation / Motor Commands → FOREBRAIN → INTERNALLY GENERATED PERCEPTION

Corollary Discharge

Oculomotor Commands → EYE MOVEMENTS GENERATED

B. Synaptic Model

Cortex: Tonically and Phasically Excited
Association
Visual
Motor
Thal
MRF
VI PGO
PRF
BRF
BRF PRF
Eye: Movements Generated
Skin: Sensory Input Blocked
Spinal Cord:
Muscle: Tonus Lost, Motor Output Blocked

model. Some of the directly and indirectly disinhibited neuronal systems are schematized together with their supposed contributions to REM sleep phenomena. At the level of the brainstem, five neuronal types are illustrated: MRF, midbrain reticular neurons projecting to thalamus convey tonic and phasic activating signals rostrally; PGO, burst cells in the peribrachial region convey phasic activation and specific eye movement information to the geniculate body and cortex (pathway dashed line indicates uncertainty of direct projection); PRF, pontine reticular formation neurons transmit phasic activation signals to oculomotor neurons (VI) and spinal cord which generate eye movements, twitches of extremities, and presynaptic inhibition; BRF, bulbar reticular formation neurons send tonic hyperpolarizing signals to motorneurons in spinal cord. As a consequence of these descending influences, sensory input and motor output are blocked at the level of the spinal cord. At the level of the forebrain, visual association and motor cortex neurons all receive tonic and phasic activation signals from both nonspecific and specific thalamic relays.

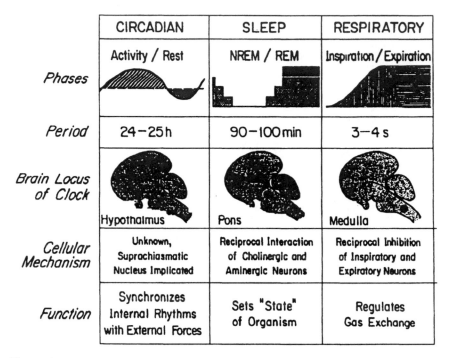

	CIRCADIAN	SLEEP	RESPIRATORY
Phases	Activity / Rest	NREM / REM	Inspiration / Expiration
Period	24–25 h	90–100 min	3–4 s
Brain Locus of Clock	Hypothalmus	Pons	Medulla
Cellular Mechanism	Unknown, Suprachiasmatic Nucleus Implicated	Reciprocal Interaction of Cholinergic and Aminergic Neurons	Reciprocal Inhibition of Inspiratory and Expiratory Neurons
Function	Synchronizes Internal Rhythms with External Forces	Sets "State" of Organism	Regulates Gas Exchange

Figure 8   Biological rhythms and brainstem clocks. Three rhythms interact to determine cyclic order of sleep and waking states. Circadian rhythms are endogenous fluctuations of many bodily functions, including rest and activity, with periods of approximately 24 hr. As seen in schematic sagittal brain sections, the suprachiasmatic nucleus of hypothalamus is a key part of this control system that serves to synchronize internal processes with external forces. The ultradian sleep cycle, with its 90–100-min period of NREM and REM sleep, is one of the physiological functions whose expression is circadian. It is controlled by reciprocal interaction of cholinergic and aminergic pontine reticular neurons, which oscillate out of phase with one another. This clock determines behavioral state (wake, NREM sleep, and REM sleep) of the organism. The mechanism by which the circadian clock sets the threshold of the sleep-cycle clock is unknown. Many homeostatic regulatory functions, including respiration, are influenced by circadian rhythm and sleep-waking cycle. The respiratory oscillator is similar in neuronal design to sleep-cycle clock but has a shorter period (3 sec) determined by reciprocal inhibition of expiratory and inspiratory neurons in the medulla.

that a temperature- or PO_2-sensitive neuron (in waking) loses that responsiveness in REM (59). Sleep, then, involves a significant reorganization of reflex pattern.

These changes in the CNS controllers of vegetative function together with those already described in the sensorimotor domain can be combined to create a set of principles for the understanding of such sleep-related dysfunctions as the insomnias, the hypersomnias, and the parasomnias where those three general categories reflect, respectively, too little sleep, too much sleep, and unwanted motor activity in sleep (60,61).

B. Neurobiology and Pathophysiology

When the level of CNS aminergic activity is raised, the brain is shifted in the direction of hyperarousal with resulting insomnia and even stress. If both stress and insomnia are prolonged, the organism remains in a state of chronic sympathetic overdrive with potentially deleterious cardiovascular, cognitive, and behavioral consequences. The need for, and the efficacy of, those behavioral and pharmacological interventions that tune down the overdriven sympathetic system may be understood as specific reversals of the peripheral and central excitatory stimuli that keep the locus coeruleus and raphe systems from obtaining the periodic respite of sleep. It goes without saying that insomnia is also a form of sleep deprivation that prevents the cholinergic system from exerting its own restorative effects.

Conversely, the lowering of the level of CNS aminergic activity (as occurs in depression or narcolepsy) will have exactly converse effects on cardiovascular, cognitive, and behavioral systems. With the attendant sleepiness, attentiveness and intellectual capabilities decline and productive action is impaired. Under these conditions the cholinergic system is disinhibited and the REM generator is triggered abnormally easily with unwanted sleepiness or even frank REM sleep attacks as its symptomatic manifestations. In these conditions the sympathetic system needs the kind of pharmacological boost that is afforded by the aminergic agonists, the aminergic reuptake blockers, and the anticholinergics. Indeed, the best drugs for the treatment of the hypersomnias have both proaminergic and anticholinergic actions.

C. Respiratory Dysfunction in Sleep

One of the most dramatic and life-threatening problems associated with sleep is an exaggeration of the natural tendency for respiratory drive to decline at sleep onset and for respiratory irregularity to be provoked by the chaotic brainstem neuronal stem of REM. This sleep-dependent disruption of a CNS autonomic controller may begin as central sleep apnea, the simple diminution or cessation of breathing effort that characterizes NREM sleep in males. But it may become both aggravated and complicated, especially in obese subjects, by collapse or compression of the airway when arousals and desperate breathing efforts are triggered by the resulting anoxia. It is then called peripheral sleep apnea because the locus of the pathology has shifted from the brainstem to the oropharynx.

Sleep apnea sufferers may be hypersomnolent (because they are chronically sleep deprived) and complain of insomnia (because truly they cannot sleep). They need to be treated (for example, by continuous positive airway pressure, CPAP) to

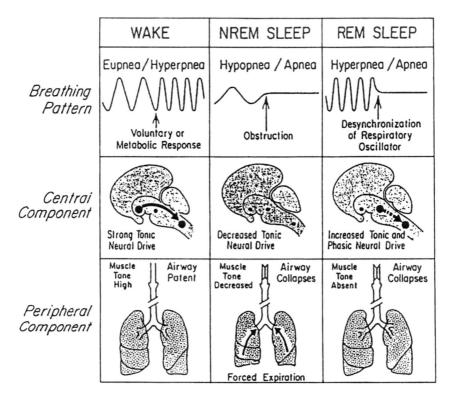

Figure 9 During waking, the respiratory oscillator of the medulla receives tonic drive from other neural structures and can respond to voluntary and metabolic signals to change the breathing pattern. Ventilation is assured by active maintenance of the airway pathway via tonus of the oropharyngeal musculature. In NREM sleep, central drive on both respiratory oscillator and peripheral muscles declines due to disfaciliation. Respiratory rate and amplitude thus diminish and airway is subject to collapse. If obstruction occurs, forced expiratory effort may actually aggravate airway construction and prolonged apneas with marked hypoventilation and hypoxia may occur. During REM sleep, activation of pontine generator neurons produces tonic and phasic driving of respiratory oscillator, which may desynchronize leading to hyperpnea and/or apnea. In addition, the medullary oscillator becomes unresponsive to metabolic signals. In patients with a tendency to airway collapse, these processes may multiply deleterious effects of ventilation.

promote normal sleep and normal daytime function and to prevent such serious secondary cardiovascular complications as cor pulmonale.

D. Motor Disturbances of Sleep

The parasomnias that affect the skeletal motor system (like sleep walking, sleep talking, tooth grinding, and night terrors) may be less disabling than sleep apnea but are nonetheless unwelcome and embarrassing to their youthful victims. All are the result of the activation, in NREM sleep, of the central motor pattern generators that must be first disinhibited (at sleep onset) and later inhibited (to prevent enactment of their commands in REM sleep). And indeed, some elderly male subjects suffer severe sleep disruption because of their failure to inhibit the expression of motor acts in REM. These men may literally enact their dream scenarios in their bedrooms with tragicomic results. This is not sleep walking (a NREM sleep phenomenon of the young) but rather the REM sleep behavior disorder, a syndrome that may herald degenerative brain disease.

A strikingly similar syndrome can be experimentally induced by making discrete bilateral lesions of the pontine tegmentum in cats. These animals then evince what Jouvet (9) has called "hallucinatory behaviors" and what Morrison (62) has called "REM sleep without atonia." In other words, when their capacity to inhibit REM sleep motor pattern commands is impaired, they "read out" these commands as stereotyped behavior sequences such as hissing, piloerection, pouncing, and jumping. The strikingly instinctual or "fixed act" aspect of these released behaviors is relevant to the psychophysiology of dreaming and has important implications for the hypothesis that REM sleep serves an active maintenance function.

VIII. Functional Significance of Sleep

While it is not at all difficult to conceptualize the function of waking, and even to understand the adaptive value of consciousness, it has been difficult to move beyond the subjectively compelling, but scientifically unsatisfactory, folk psychology notion of sleep as rest. Recent progress strongly suggests that sleep has a more positive anabolic and actively conservative function related to the increased complexity of the brain in mammals.

The most dramatic evidence of this positive homeostatic function comes from sleep deprivation studies in rats, which imply that metabolic caloric balance, thermal equilibrium, and immune competence are all preserved by sleep (63,64). These inferences arise from the discovery that sleep deprivation is universally fatal when it persists for 4–6 weeks. Early in the deprivation period the rats begin to eat more but cannot maintain their body weight; they are in a negative metabolic caloric drain. Later, they develop thermal dyscontrol, cannot maintain body temperature, and develop strong heat-seeking behavior: they are in a negative heat caloric drain. Finally, they die of overwhelming sepsis because of immunodeficiency.

The link between immune function and sleep has been studied in detail by Krueger (65,66) following up on Pappenheimer's (67) finding that the NREM sleep of rabbits was enhanced by dimuramyl peptides of bacterial cell wall origin (factor S) extracted from the spinal fluid and urine of sleep-deprived goats and

rabbits. NREM sleep is also enhanced by the cytokines interleukin-1 and inter-
leukin-2 and these substances are, in turn, released during NREM sleep.

Contributing further to the notion of an active anabolic function is the
superabundance of sleep in early life. REM sleep predominates, in utero, where its
stereotypic motor pattern activation and specific chemical microenvironment
could promote CNS development (68), a function that would be expected to
decline, just as REM does in childhood. Growth and development could also be
enhanced by the release of growth hormone and gonadotropins in NREM sleep, a
function that also declines, along with NREM sleep after growth and sexual
maturation are complete at the end of the third decade of life.

The commonsense rest theory of sleep is enriched by the recognition that
sleep behavior is an energy- and heat-saving behavior that tends to occur at night
when ambient temperatures are low (14). Safety from predators, as well as
warmth, may also be conferred by sleeping in nests with conspecifics. This is
especially true for the most vulnerable young members of a species, which are, for
other reasons, also sleeping more.

Not only energy but also information may be conserved in sleep. Numerous
studies show that animals have more REM when learning a normal task and that
REM deprivation interferes with learning (69–73). In this regard it may be
significant that sleep and especially REM is associated with such dramatic
changes in neuromodulation. The very chemicals needed to form recent memories
(e.g., norepinephrine and serotonin) could be conserved (and their receptors
regulated or sensitized) by the arrest of firing of their neurons in REM. Then the
hypercholinergic state of the brain could serve to enhance consolidation of those
memories already in the system. Finally, cognitive and emotional homeostasis,
which are the clear subjective beneficiaries of sleep, could be mediated by the
reversal of wake-state neuromodulatory balance in REM (14).

References

1. Sherrington C. Man on His Nature. New York: Doubleday, 1955.
2. Aschoff J. Circadian Clocks. Amsterdam: North-Holland, 1976.
3. Aschoff J. Circadian rhythms in man. Science 1965; 148:1427–1432.
4. Kronauer RE, Czeisler CA, Pilato SF, Moore-Ede MC, Weitzman ED. Mathematical
   Model of the Human Circadian System with Two Interacting Oscillators. Bethesda:
   American Physiological Society, 1982:R3–R24.
5. Heller CS, Glotzbach S, Grahn D, Radehe C. Sleep dependent changes in the thermoregulatory system. In: Lydic R, Biebuyck JF, eds. Clinical Physiology of Sleep.
   Bethesda: American Physiological Society. 1988:145–158.
6. Parmeggiani PL. Thermoregulation during sleep from the viewpoint of homeostasis.
   In: Lydic R, Biebuyck JF, eds. Clinical Physiology of Sleep. Bethesda: American
   Physiological Society. 1988:159–170.

7. Moore RY. The anatomy of central neural mechanisms regulating endocrine rhythms. In: Krieger DT, ed. Endocrine Rhythms. New York: Raven Press, 1979:63–87.

8. Zulley J, Campbell SS. Naping behavior during "spontaneous internal desynchronization": sleep remains in synchrony with body temperature. Hum Neurobiol 1985; 4: 123–126.

9. Jouvet M, Buda L, Denges M, Kitahama K, Sallanon M, Sastre J. Hypothalamic regulation of paradoxical sleep. In: Onian T, ed., Neurobiology of Sleep-Wakefulness Cycle. Metsniereba, Tbilsi: Georgian Academy of Sciences, 1988:1–17.

10. Moruzzi G, Magoun HW. Brainstem reticular formation and activation of the EEG. Electroenceph Clin Neurophysiol 1949; 1:455–473.

11. Dahlstrom A, Fuxe K. Evidence for the existence of monoamine neurons in the central nervous system. I. Demonstration of monoamines in the cell bodies of brain stem neurons. Acta Physiol Scand 1964; 62(S232):1–55.

12. Moore RY, Card JP. Noradrenaline-Containing Neuron Systems. Amsterdam: Elsevier, 1984.

13. Jacobs BL, Azmitia EC. Structure and function of the brain serotonin system. Physiol Rev 1992; 72:165–229.

14. Hobson JA. The Dreaming Brain. New York: Basic Books, 1988.

15. Loomis AL, Harvey EN, Hobart GA. Cerebral states during sleep as studied by human brain potentials. J Exp Psychol 1937; 21:127.

16. Bremer F. L'activite cerebrale au cours du sommeil et de la narcose. Contribution a l'etude du mecanisme du sommeil. Bull Acad R Med Belg 1937; 4:68–86.

17. Hess WR. Diencephalon: Autonomic and Extrapyramidal Functions. New York: Grune & Stratton, 1954.

18. Hobson JA, Steriade M. The neuronal basis of behavioral state control. In: Bloom FE ed. Handbook of Physiology. Section I: The Nervous System. Vol IV. Intrinsic Regulatory Systems of the Brain. Bethesda: American Physiological Society, 1986: 701–823.

19. Steriade M, McCarley RW. Brainstem Control of Wakefulness and Sleep. New York: Plenum Press, 1990.

20. Pavlov IP. Conditioned Reflexes: An Investigation of the Physiological Activity of the Cerebral Cortex, transl. New York: G. V. Anrep, Dover, 1960.

21. Steriade et al. Synchronized sleep oscillations and their proxysmal developments. 1994; TINS 17:199–208.

22. Maquet et al. Functional neuroanatomy of human slow wave sleep. J Neurosci 1997; 17:2807–2812.

23. Aserinsky E, Kleitman N. Regularly occurring periods of ocular mobility and concomitant phenomena during sleep. Science 1963; 118:361–375.

24. Dement W, Kleitman N. Cyclic variations in EEG during sleep and their relation to eye movements, body mobility and dreaming. Electroenceph Clin Neurophysiol 1955; 9:673–690.

25. Dement W, Kleitman N. The relation of eye movements during sleep to dream activity: an objective method for the study of dreaming. J Exp Psychol 1957; 53: 89–97.

26. Dement W. The occurrence of low voltage, fast, electroencephalogram patterns during behavioral sleep in the cat. EEG Clin Neurophysiol 1958; 10:291–296.

27. Singer W. Central-core control of visual cortex functions. In: Schmitt RFO, Worden FG, eds. The Neurosciences: Fourth Study Program. Cambridge, MA: MIT Press, 1979.

28. Llinas RR, Pare D. Of dreaming and wakefulness. Neuroscience 1991; 44:521–535.

29. Jouvet M. Recherche sur les structures nerveuses et les mecanismes responsables des differentes phases du sommeil physiologique. Arch Ital Biol 1962; 100:125–206.

30. Callaway CW, Lydic R, Baghdoyan HA, Hobson JA. Ponto-geniculo-occipital waves: Spontaneous visual system activation occurring in REM sleep. Cell Mol Neurobiol 1987; 7:105–149.

31. Pompeiano O. The neurophysiological mechanisms of the postural and motor events during desynchronized sleep. Proc Assoc Res Nerv Ment Dis 1967; 45:351–423.

32. Morales FR, Chase MH. Postsynaptic control of lumbar motoneuron excitability during active sleep in the chronic cat. Brain Res 1981; 225:279–295.

33. Soja PJ, Finch DM, Chase MH. Effect of inhibitory amino acid antagonists on masseteric reflex suppression during active sleep. Exp Neurol 1987; 96(1):178–193.

34. Jouvet M, Delorme F. Locus coeruleus et sommeil paradoxal. Soc Biol 1965; 159:895.

35. Henley K, Morrison AR. A re-evaluation of the effects of lesions of the pontine tegmentum and locus coeruleus on phenomena of paradoxical sleep in the cat. Acta Neurobiol Exp 1974; 34:215–232.

36. Porte HS, Hobson JA. Physical motion in dreams: one measure of three theories. J Abnorm Psychol 1993; 105:329–335.

37. Evarts EV. Effects of sleep and waking on spontaneous and evoked discharge of single units in visual cortex. Fed Proc 1960; 4(Suppl):828–837.

38. Hubel DH. Single unit activity in striate cortex of unrestrained cats. J Physiol (Lond) 1959; 147:226–238.

39. Adey WR, Kado RT, Rhodes JM. Sleep: cortical and subcortical recordings in the chimpanzee. Science 1963; 141:932.

40. Steriade M, Pare D, Parent A, Smith Y. Projections of cholinergic and non-cholinergic neurons of the brainstem core to relay and associational thalamic nuclei in the cat and macaque monkey. Neuroscience 1988; 25:47–67.

41. Bizzi E, Brooks DC. Functional connections between pontine reticular formation and lateral geniculate nucleus during deep sleep. Arch Ital Biol 1963; 101:666–680.

42. Jouvet M. The role of monoamines and acetylcholine-containing neurons in the regulation of the sleep-waking cycle. Ergeb Physiol Biol Chem Exp Pharmakol 1972; 64:166–307.

43. Aston Jones G, Bloom FE. Activity of norepinephrine containing locus coeruleus neurons in behaving rats anticipates fluctuations in the sleep-waking cycle. J Neurosci 1981; 1:876–886.

44. Aston Jones G, Bloom FE. Norepinephrine-containing locus coeruleus neurons in behaving rats exhibit pronounced responses to nonnoxious environmental stimuli. J Neurosci 1981; 1:877–900.

45. Chu NS, Bloom FE. Activity patterns of catecholamine-containing pontine neurons in the dorsolateral tegmentum of unrestrained cats. J Neurobiol 1974; 5:527–544.

46. Hobson JA, McCarley RW, Qyzinki PW. Sleep cycle oscillation: reciprocal discharge by two brainstem neuronal groups. Science 1975; 189:55–58.

47. McGinty DJ, Harper RM. Dorsal raphe neurons: Depression of firing during sleep in cats. Brain Res 1976; 101:569–575.

48. Lydic R, Baghdoyan HA, Lorinc Z. Microdialysis of cat pons reveals enhanced acetylcholine release during state-dependent respiratory depression. Am J Physiol 1991; 261:766–770.

49. Baghdoyan HA, Monaco AP, Rodrigo-Angulo ML, Assens F, McCarley RW, Hobson JA. Microinjection of neostigmine into the pontine reticular formation of cats enhances desynchronized sleep signs. J Pharmacol Exp Ther 1984; 231:173–180.

50. Baghdoyan HA, Rodrigo-Angulo ML, McCarley RW, Hobson JA. Site-specific enhancement and suppression of desynchronized sleep signs following cholinergic stimulation of three brainstem regions. Brain Res 1984; 306:39–52.

51. Calvo J, Datta S, Quattrochi JJ, Hobson JA. Cholinergic microstimulation of the peribrachial nucleus in the cat. II. Delayed and prolonged increases in REM sleep. Arch Ital Biol 1992; 130:285–301.

52. Datta S, Calvo J, Quattrochi JJ, Hobson JA. Cholinergic microstimulation of the peribrachial nucleus in the cat. I. Immediate and prolonged increases in ponto-geniculo-occipital waves. Arch Ital Biol 1992; 130:263–284.

53. Foulkes D. Dreaming: A Cognitive-Psychological Analysis. Hillsdale, NJ: Erlbaum, 1985.

54. Arkin A, Antrobus J, Ellman S, The Mind in Sleep, 2nd ed. Hillsdale, NJ: Erlbaum, 1991.

55. Hobson JA. Neuropsychology of sleep and cognition. Schacter D, ed. The Conscious State Paradigm: A Neurocognitive Approach to Waking, Sleeping and Dreaming. In: Gazzaniga M, ed. The Cognitive Neurosciences. Cambridge, MA: MIT Press, 1993.

56. Hobson JA. Consciousness as a State-Dependent Phenomenon. In: Cohen J, Schooler J, eds. Scientific Approaches to the Question of Consciousness. Hillsdale, NJ: Lawrence Erlbaum, 1994.

57. Hobson JA. Consciousness: lessons for anesthesia from sleep research. In: Biebuyck JF, ed. Anaesthesia: Biologic Foundations. New York: Raven Press, 1994:423–431.

58. Hobson JA, Stickgold R. The conscious state paradigm: a neurocognitive approach to waking, sleeping and dreaming. In: Gazzaniga M, ed. The Cognitive Neurosciences. 1994:1373–1389.

59. Phillipson EA. Control of breathing during sleep. Am Rev Respir Dis 1978; 118:909–939.

60. Hobson JA. Sleep and its disorders. In: Cecil Textbook of Medicine. Philadelphia: WB Saunders, 1982:1930–1935.

61. Kryger MH, Roth T, Dement WC. The Principles and Practice of Sleep Medicine. Philadelphia: WB Saunders, 1994.

62. Henley K, Morrison AR. A re-evaluation of the effects of lesions of the pontine tegmentum and locus coeruleus on phenomena of paradoxical sleep in the cat. Acta Neurobiol Exp 1974; 34:251–232.

63. Rechtschaffen A, Bergmann BM, Everson CA, Kushida CA, Gilliland MA. Sleep deprivation in the rat. I. Conceptual issues. Sleep 1989; 12:1–4.

64. Rechtschaffen A, Bergmann BM, Everson CA, Kushida CA, Gilliland MA. Sleep deprivation in the rat. X. Integration and discussion of the findings. Sleep 1989; 12:68–87.

65. Krueger JM, Walter J, Dinarello CA, Chedid L. Induction of slow-wave sleep by interleukin-1. In: Kloger MJ, Oppenheim JJ, Powanda MC, eds. The Physiologic, Metabolic, and Immunologic Actions of Interleukin-1. New York: Alan R Liss, 1985: 161–170.

66. Krueger JM, Walter J, Levin C. Factor S and related somnogens: an immune theory for slow-wave sleep. In McGinty DJ, et al, eds. Brain Mechanisms of Sleep. New York: Raven Press, 1985:253–275.

67. Pappenheimer JR, Koski G, Fenci V, Karnovsky ML, Krueger J. Extraction of sleep-promoting factor S from cerebrospinal fluid and from brain of sleep-deprived animals. J Neurophysiol 1975; 38:1299–1311.

68. Roffwarg JP, Muzio JM, Dement WC. Ontogenetic development of the human sleep-dream cycle. Science 1966; 152:604–619.

69. Fishbein WC, Kastaniotis C, Chattman D. Paradoxical sleep: prolonged augmentation following learning. Brain Res 1974; 79:61–77.

70. Smith C, Kitahama K, Valatx JL, Jouvet M. Increased paradoxical sleep in mice during acquisition of a shock avoidance task. Brain Res 1974; 77:221–230.

71. Smith C, Young J, Young W. Prolonged increases in paradoxical sleep during and after avoidance-task acquision. Sleep 1980; 3(1):67–81.

72. Karni A, Tanne D, Rubenstein BS, Askenasy JM, Sagi D. Dependence on REM sleep of overnight improvement of a perceptual skill. Science 1994; 265:679–682.

73. Pavlides C, Winson J. Influences of hippocampal place cell firing in the awake state on the activity of these cells during subsequent sleep episodes. J Neurosci 1989. 9:2907–2918.

4

Circadian and Homeostatic Control of Wakefulness and Sleep

DERK-JAN DIJK

Brigham and Women's Hospital
Harvard Medical School
Boston, Massachusetts

DALE M. EDGAR

Stanford University School of Medicine
Stanford, California

I. Introduction

With the invention of the first commercially practical incandescent lamp in 1879, Thomas Alva Edison revolutionized the way human beings live and work. The addition of light in factories has made shift work possible, allowing unprecedented levels of industrial productivity and economic competition. But working the night shift—times of day in which human sleep is normally concentrated—is very difficult for most individuals. Impaired alertness and performance during the night-shift contribute to accidents on the job and automobile accidents on the way home. Similar risks undermine the health and safety of persons suffering from jet lag, in which the circadian rhythm of alertness is temporarily out of phase with local time. Shift work and transmeridian travel are now central to the fabric of the industrialized world, and the risk to public health continues to mount. Unfortunately, treatments for impaired alertness and functioning are limited. This is because the most fundamental mechanisms of sleep-wake regulation—the mechanisms responsible for circadian rhythms in alertness and the increased sleep tendency that results from extended wakefulness—remain poorly understood.

Although a comprehensive review of sleep physiology is beyond the scope of this chapter, we provide here a review of some important changes in the way sleep-wake regulation is now conceptualized. There have, indeed, been important developments over the last decade that converge to form a rational consensus regarding the circadian and homeostatic control of sleep-wakefulness in mammals. We must, as a matter of prudence, strongly emphasize that sleep regulation is highly integrated with systemic physiology at many levels (immune, neuroendocrine, thermoregulatory, etc.), each of which has complex oscillating feedback mechanisms that are also modulated by the circadian timekeeping system. For the sake of clarity, we will focus our discourse on the functional role of the circadian pacemaker located in the suprachiasmatic nucleus of the hypothalamus (SCN) in mammalian sleep-wake regulation. In addition, we will highlight the interaction of the circadian process with the other main regulatory process involved in sleep regulation—sleep homeostasis.

II. Sleep and Wakefulness in Biological Time

The circadian timekeeping system imparts distinct temporal niches that are thought to be genetically defined (1), and which ultimately serve to lower competition between species sharing resources in overlapping physical niches. Species concentrate most of their vital behavioral activities during the day (diurnal species), during the night (nocturnal species), or during dawn and dusk (crepuscular species), but very few species consolidate all of their wakefulness and sleep into separate halves of the circadian cycle the way humans and many nonhuman primates do. Humans, although decidedly diurnal in their circadian organization, are also one of only a few species capable of exerting volitional control over their temporal niche (e.g., in the form of self-imposed sleep deprivation, night work, and high-speed travel across multiple time zones). Analogous behaviors are limited primarily to higher carnivores (e.g., felines) which are known to hunt continuously (day and night). After the hunt comes to a successful completion, wakefulness gives way to compensatory sleep that can last for many hours or even days.

   Domesticated cats have been used extensively in sleep research to understand brainstem mechanisms mediating sleep-wakefulness in general, and rapid eye movement (REM) sleep mechanisms in particular (2–5). The effects of sleep deprivation have also been well documented in cats (6). Unfortunately, the lack of robust circadian rhythms in domesticated cats (7) has rendered this species a poor model for understanding the circadian control of sleep-wakefulness or the interaction of the circadian timing system with the mechanisms responsible for compensatory sleep responses to sleep deprivation. Thus, the bulk of neurobiological work in this area has come from numerous studies of inbred nocturnal rats and mice, and fewer, but equally important, studies of diurnal primates. Rodent and monkey data, when taken together with the innovative approaches developed to study sleep-wake regulation in humans, comprise the research foundation on which the contemporary notions of sleep-wake control by the circadian system are currently based.

III.   Manifest Rhythms of Sleep and Wakefulness

A.   Exogenous Factors

The daily rhythm of sleep and wakefulness exhibited by animals in their natural environment is the result of endogenous modulation by the circadian time keeping system and exogenous components derived from the environment and from the animal's volitional behavior. These exogenous influences are sometimes referred to as masking effects because they obscure the underlying endogenous circadian rhythm of interest (Fig. 1). Examples of masking effects include light-dark cycle modulation of body temperature, stress-induced increase in plasma cortisol, light-induced suppression of melatonin secretion, activity-dependent increase in body temperature, and both light and activity influences on sleep-wake tendency, to name a few.

   In diurnal species, locomotor activity behavior is inhibited in darkness, even during the usual activity phase of the circadian cycle (8). During the usual rest phase of the circadian cycle, light increases activity in diurnal species. Although light facilitation of activity during the rest phase necessarily involves wakeful-

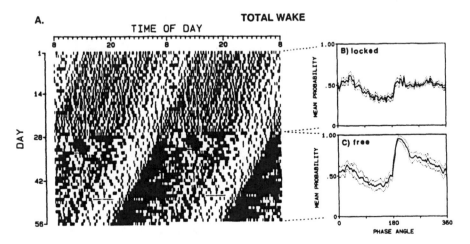

Figure 1   Activity-dependent masking of sleep-wakefulness in the mouse. Mice housed in a cage with running wheels exhibit more robust sleep-wake circadian rhythms, primarily due to activity-dependent consolidation of wakefulness. Activity-dependent masking in the mouse also decreases total sleep time per circadian day. (From Ref. 18.)

phase of the circadian cycle, light increases activity in diurnal species. Although light facilitation of activity during the rest phase necessarily involves wakefulness, the converse is not necessarily true during the dark phase. For example, if lights are turned off during the subjective day, the squirrel monkey, a diurnal new world primate, will show very little locomotor activity, yet vigilance remains high as measured by electroencephalographic techniques (9). Indeed, wakefulness remains robust throughout the subjective day in this species, suggesting that some aspect of circadian control facilitates the maintenance of wakefulness at particular times in the circadian cycle (9,10).

In humans, objective measures of sleep latency suggest that alertness varies across the subjective day. Latency to sleep onset decreases with sleep loss, reflecting increased physiological sleep tendency (11–13). Multiple sleep latency tests have revealed a biphasic pattern of daytime alertness in humans, with a "midday dip" in alertness occurring in the midafternoon (14). The latter is consonant with manifest patterns of sleep-wakefulness in some parts of the world where afternoon naps are considered not only normal but healthful. But in most industrialized countries, where daytime sleeping is labeled as socially unacceptable, individuals successfully overcome the midday dip in physiological alertness by volitionally employing behavioral or pharmacological countermeasures. These include the use of stimulants (caffeine), social interaction, engaging intellectual pursuits, and exercise. All of these techniques serve to enhance wakefulness and therefore mask the underlying rhythm in physiological sleepiness/alertness. To accurately assay the endogenous circadian variation in physiological sleep tendency in humans, it is therefore essential to isolate the subject from or otherwise control all environmental and behavioral masking effects. To these ends, human subjects are now routinely studied in a "constant routine" protocol where the masking influences of ambient light, temperature, noise, food consumption, and activity level are carefully controlled. The utility of this technique for understanding the circadian control of sleep is detailed later in this chapter.

In nocturnal mammalian species, including rats, mice, and hamsters, light-dark cycle masking is an important determinant of manifest daily rhythms in activity behavior. Light inhibits activity behavior in nocturnal rodents and facilitates sleep (15,16). The converse is true in darkness. Unlike primates, however, where the masking effects of light and dark have little effect on arousal state during the usual activity phase of the circadian cycle, rodents show light-dark modulations of arousal state across the entire circadian cycle.

In some species, the masking effects of light (or other environmental factors) on activity behavior may be a major determinant of the manifest daily rhythm of physiological sleepiness and alertness. In mice, for example, the opportunity to run in a running wheel dramatically increases the consolidation of wakefulness during the active phase of the circadian cycle (17,18). In all mammals, including humans, exercise increases brain monoamine levels (dopamine, norepinephrine, serotonin [19–22]) that have well-documented action on arousal state and increases core body temperature. The opportunity to express vigorous motivated behavior in mice not only increases the duration of wakefulness episodes, but results in a net decrease in total sleep time per circadian day (18). Thus, from a regulatory perspective, sleep homeostasis may have a dynamic, or plastic, component that can be affected by the volitional exercise behavior(s) of a given species. Furthermore, the spontaneous placement of running activity within the circadian cycle can feed back to and phase-shift the circadian pacemaker on a daily basis (23,24), thereby altering the manifest free-running period of sleep-wakefulness (25) and perhaps affecting the phase position of the sleep-wake rhythm with respect to photic zeitgebers (26). The latter may have important relevance in aging, where sedentary behavior may negatively affect sleep-wake cycle consolidation and timing.

Masking effects have special importance in the modern pursuit of sleep-wake regulatory mechanisms. In recent years there has been a rush to identify genes (and ultimately gene products) that comprise the fundamental mechanisms of sleep-wake regulation. In light of the profound effect spontaneous activity behavior and other highly integrated physiological factors (thermoregulation, endocrine, etc.) can have on manifest sleep-wakefulness (especially in mice, the species of choice for molecular genetics studies), it becomes clear that simply screening manifest sleep-wake patterns in genetically engineered animals (e.g., selected through mutagenesis, specific gene knockouts, or transgenic approaches) will inevitably lead to the discovery of numerous genes that are indirectly associated with sleep-wake regulation. Indeed, great care must be exercised to differentiate genomic constructs that mediate sleep-wake regulation per se from those that alter some easily overlooked exogenous element that only masks the expression of sleep-wakefulness.

B. Endogenous Circadian Rhythm of Wakefulness and Sleep

In mammals, the suprachiasmatic nuclei (SCN) are now known to be both necessary and sufficient for the generation of circadian rhythms in systemic physiological function and organismal behavior (27). Shortly after the discovery of the SCN in 1972 (28,29), many studies investigated whether these small, bilaterally paired hypothalamic nuclei were necessary for the generation of sleep-wake circadian rhythms in animals (10,30–39). The rationale for such studies was quite sound, considering the important observations of Nathanial Kleitman in his classic book *Sleep and Wakefulness* (40). In the process of reviewing virtually the entire history of sleep research, Kleitman noted many published case reports of fragmented sleep-wake patterns and increased sleep time in patients suffering from tumors or cerebral vascular accidents involving the base of the hypothalamus.

Early SCN lesion studies in animals, however, produced variable findings. Some reports claimed that only certain aspects of sleep-wake circadian rhythms were abolished after SCN lesions. For example, in a few of these studies it was reported that circadian oscillations in REM sleep (also known as paradoxical sleep) could persist, but non-REM (NREM) sleep circadian oscillations were eliminated after SCN lesions (35,37,39) (Fig. 2). Unfortunately, these early conclusions were confounded by the fact that SCN-lesioned animals were often studied in light-dark cycles. It is now well known that light-dark cycle masking of sleep-wakefulness can differentially influence arousal states and is mediated independently of the SCN (9,41).

By the early to mid-1980s there was heated debate regarding how many central neural pacemakers were responsible for the generation of circadian rhythms in general, and how the circadian timing system controlled sleep-wakefulness and sleep stages in particular. Several studies had demonstrated the elimination of sleep-wake and sleep stage circadian rhythms in rats (30,34,38) and Siberian chipmunks (36) monitored in constant conditions, and in blinded rats housed in light-dark cycles (32). Eastman (30) offered one of the more compelling early studies, demonstrating that both sleep-wakefulness and body temperature

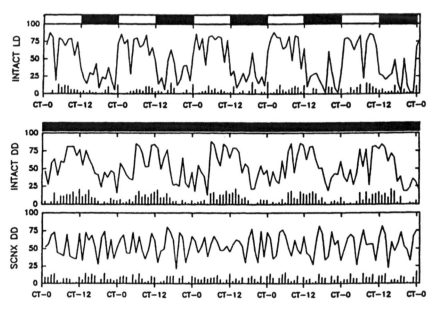

Figure 2 Sleep rhythms in control rats housed in an LD 12:12 light:dark cycle (top), and in constant darkness (middle), and in SCN-lesioned rats housed in constant darkness (bottom). Data are plotted as percent time in NREM sleep (solid line) and percent time in REM sleep (vertical bars) per hour. NREM and REM sleep circadian rhythms remain robust in constant conditions but are abolished after SCN lesions. Ultradian rhythms seen in SCN-lesioned rats exhibit period lengths of 3 to 4 hours. (From Ref. 92.)

circadian rhythms were abolished in SCN-lesioned rats. At the time, the generality of these findings in rodents seemed doubtful in light of what seemed like overwhelming evidence for multiple circadian pacemakers in mammals. As will be detailed further in our discussion of the human literature (below), a series of classical studies conducted first by Aschoff (42,43) and later by Weitzman and Czeisler (44-46) showed that humans maintained in long-term isolation exhibited rest-activity and body temperature circadian rhythms with distinct circadian periods, a phenomenon known as spontaneous internal desynchronization. These findings stimulated great interest in sleep-wakefulness modeling that still goes on today.

Based largely on the observation of internal desynchronization, Kronauer and colleagues (47) developed a mathematical model in which two interacting Van der Pol oscillators could predict internal desynchronization of human circadian rhythms with remarkable accuracy. They posited that a vegetative class of circadian variables (body temperature, REM sleep) was controlled by one pacemaker, whereas overt behavioral rhythms (feeding, drinking, sleep-wakefulness) was controlled by the other pacemaker. In rodents, evidence for at least two circadian oscillators also supported the two-pacemaker hypothesis. For example, the phase relationship of two rhythmic components of the activity circadian rhythm, a morning component and an evening component, could "split apart" from each other when housed in constant conditions for prolonged intervals (48–50). Under these conditions, the "split" rhythmic components exhibited distinct circadian periods that, even today, are best explained by two self-sustained circadian oscillators. The notion of multiple circadian oscillators was also supported by SCN lesion studies in monkeys. Fuller and colleagues (51) reported that radio frequency thermal lesions focused in the SCN of squirrel monkeys eliminated behavioral circadian rhythms but did not eliminate circadian oscillations in body temperature. Reppert and colleagues (52) also reported that cortisol circadian rhythms persisted after SCN lesions in rhesus monkeys.

As the multiple pacemaker model of sleep-wake regulation gained favor, Borbély and colleagues (53,54) began positing a regulatory scheme that took into account not only the circadian timekeeping system but also the well-known compensatory sleep responses to sleep deprivation seen in virtually all mammals. Borbély's "two-process" model differed most notably, however, in that only one circadian pacemaker was necessary for the circadian control of sleep-wakefulness. Daan and Borbély (54) posited that the circadian timing system gated sleep expression and that a separate "homeostatic process" was the source of physiological sleep drive (e.g., that which causes sleepiness). In this model, the circadian system is thought to be unnecessary for sleep homeostatic function, a notion supported by sleep deprivation studies in SCN-lesioned laboratory rats (34,38). Remarkably, however, this critically important generality has not been demonstrated in other species.

IV. Sleep Homeostasis

The notion that sleep is homeostatically conserved stems from sleep deprivation studies in animals and humans dating back more than a century (55). In animals and humans, the depth of sleep (as indexed by EEG delta power during NREM sleep) is proportional to the duration of prior wakefulness (56–64). Deep sleep is also characterized by an increase in arousal threshold, which explains why it is more difficult to awaken a highly sleep-deprived individual. The increased depth of sleep may assure the restorative benefits of sleep continuity (65) by rendering the individual less susceptible to arousing stimuli in the environment. In laboratory rats, which normally exhibit polyphasic sleep-wake patterns, the depth of sleep (EEG delta power in NREM, sleep episode duration or "sleep bout length," and NREM sleep per unit time are highly interrelated and increase systematically in response to sleep deprivation (66). Thus, these variables can be useful quantitative indices of the compensatory sleep response to sleep deprivation. Maximum compensatory sleep responses are thought, for a particular species, to occur after a finite amount of sleep deprivation. In rats, maximum sleep drive is achieved after 10 to 12 hours of sleep deprivation (60). Maximum sleepiness in humans is reportedly reached after only 30 hours of continuous wakefulness (67,68). Interestingly, repeated partial sleep loss in humans appears to be cumulative (13), evidenced in part by the tendency to make up for lost sleep on weekends and holidays.

Although the neurobiological substrates underlying sleep homeostasis are uncertain, cumulative sleepiness in humans suggests that one or more of the neurobiological components responsible for physiological sleepiness can remain stable for extended periods of time until acted upon by the sleeping process. Many neurotransmitters have been implicated in the regulation of sleep; these include acetylcholine, the biogenic amines (norepinephrine, serotonin, dopamine, and histamine), the amino acids glutamate and GABA, various neuropeptides, prostaglandins, and immune factors, and more recently, the nucleoside adenosine (see 69–78 for reviews). Psychostimulants such as amphetamine, pemoline, and methylphenidate are presumed to exert their vigilance and behavioral activation-promoting effects through increased cholinergic and monoaminergic neurotransmission (79–81). However, due to the multiple effects of psychostimulants on various neurotransmitter systems, it has been difficult to isolate the precise neurochemical mechanisms governing wakefulness.

Adenosine, a ubiquitous inhibitory neurotransmitter in the central nervous system, has long been thought to play a role in physiological sleepiness (82–84). Methylxanthines such as caffeine and theophylline, which act primarily as antagonists on adenosine receptors, profoundly enhance wakefulness (85–87). It was recently reported that adenosine accumulates in the region of the cat NBM during prolonged wakefulness and may be responsible for the sleep-promoting effects of sleep deprivation (6). Adenosine is a potent inhibitor of mesopontine cholinergic neurons (88) and may also act upon basal forebrain cholinergic neurons to influence wakefulness (89).

V. Functional Role of the Suprachiasmatic Nucleus

Although the precise mechanisms through which the SCN controls wakefulness and sleep are not known, it is clear that the circadian timing system functions as one of at least two major processes controlling sleep-wakefulness. We generally think of sleepiness as the behavioral expression of homeostatic sleep drive (which increases with wakefulness duration). However, physiological sleep tendency is also determined by circadian control mechanisms. In both monkeys and humans, the circadian timing system promote and maintain wakefulness across the subjective day and opposes the accumulating "homeostatic sleep drive" resulting from that wakefulness (10,90,91). Edgar and colleagues coined the antagonistic interaction between sleep homeostasis and SCN-dependent alerting "opponent processes" based on a larger neurobiological framework from which explicit hypotheses could be tested (10) (Fig. 3). Using the forced desynchrony protocol,

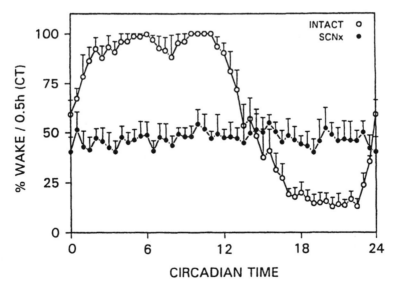

Figure 3 Temporal distribution of wakefulness in intact and SCN-lesioned squirrel monkeys. Intact monkeys exhibit a ratio of wake:rest of approximately 16:8. SCN-lesioned monkeys sleep approximately 50% of the time at all times of day. Thus, SCN-lesions result in a 4-hour increase in total sleep time. (From Ref. 10.)

Dijk and Czeisler have provided compelling evidence that the circadian contribution to alertness in humans increases over the subjective day and is greatest just before habitual bedtime (90,91). The circadian contribution to alertness is lowest in the hours preceding habitual wake-up time (approximately 3 to 6 AM, when the circadian body temperature rhythm is also lowest).

In monkeys with experimentally induced lesions of the SCN (10) and in humans who have experienced injury to the basal hypothalamus (40), sleep becomes highly fragmented, circadian rhythmicity is lost, and there is a marked increase in total sleep time (10). Thus, the circadian timing system not only influences when sleep can occur, but plays a role in determining how much sleep a primate can get over the 24-hour day. Indeed, diurnal monkeys, like humans, have asymmetric sleep-wake cycles with about 16 hours of wakefulness to 8 hours of sleep (10). In the absence of circadian control, sleep probability becomes essentially random in the monkey. The resulting loss of the asymmetric circadian waveform results in a 4-hour net increase in sleep time after SCN lesions that can not be explained by collateral hypothalamic injury. In monkeys (10) and rats (92), the relatively long spontaneous bouts of wakefulness during the circadian activity phase decrease dramatically after SCN lesions, approximating the durations normally observed during the usual rest phase of the circadian cycle. Sleep latency, which is also normally long during the subjective day in the diurnal squirrel monkey, decreases to subjective night durations after LCN lesions as well (92). In nocturnal rodents, SCN lesions leave the sleep homeostatic process intact (e.g., animals exhibit recovery sleep in response to sleep deprivation), but baseline ratios of sleep and wakefulness (approximately 1:1 over 24 hours) remain essentially unchanged (34). Some may argue that the rodent SCN lesion studies, by virtue of the fact that they have been replicated several times, "prove" that the circadian system has nothing to do with the regulation of daily total sleep time. But the polyphasic sleep patterns of nocturnal rodents are so profoundly different from the consolidated sleep-wake patterns of diurnal primates that one must critically question sweeping generalizations based only on laboratory rat data. It is, in fact, the remarkable similarities between monkey and human sleep that make the SCN lesion studies in squirrel monkeys important (Fig. 4).

Although there is now a general consensus that SCN-dependent alerting mechanisms promotes and maintains wakefulness during the day in diurnal primates, there is less agreement as to why humans remain asleep throughout the night. By 3 to 4 AM, the sleeping process has presumably satisfied most of the underlying sleep deficit in humans, yet most of us do not awaken spontaneously until 7 AM or later. Dijk and Czeisler (90,91) have postulated that the SCN may not only promote wakefulness during the day, but actively promote sleep at night, thereby explaining nocturnal sleep maintenance. This postulate is attractive because the propensity toward REM sleep is greatest during the circadian body temperature minimum (even during forced desynchrony protocols where the

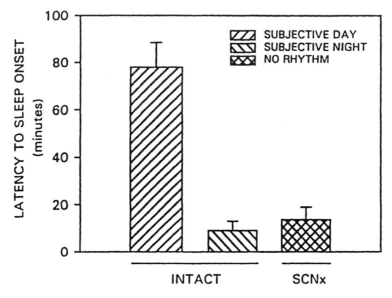

Figure 4 Latency to sleep onset in intact and SCN-lesioned squirrel monkeys. When housed in an L:D 2:2 cycle, latency to sleep onset on transition to darkness provides an estimate of physiological sleepiness akin to the multiple sleep latency test used in humans. SCN-lesioned moneys show sleep latencies at all times of day that approximate that seen only during the subjective night (normal rest phase) in intact animals. (From Ref. 92.)

sleep-wakefulness cycle is experimentally dissociated from the underlying body temperature rhythm). Experimentally induced SCN lesions eliminate circadian rhythms in REM sleep (and all other sleep stages) in rodents and monkeys (10,30), also arguing for circadian mechanisms controlling REM sleep. Putting REM sleep aside for a moment, sleep maintenance late at night need not be due to an SCN-dependent sleep-promoting mechanism (Fig. 5). For example, continual withdrawal of SCN-dependent alerting to lowest levels at 3 to 6 AM, corresponding with the body temperature minimum, could account for maximum sleep propensity (shortest sleep latency and latency to return to sleep after awakening) at this time of day. In this view, humans would not spontaneously awaken and remain awake at the body temperature minimum because of the very low levels of SCN-dependent alerting at that time of day. At a neurobiological level, the withdrawal of SCN-dependent mechanisms initiating cortical and behavioral activation may be essential for gating the brainstem initiation and maintenance of REM sleep. Thus, the SCN need only function to promote cortical and behavioral activation to give the appearance of active control over non-REM and REM sleep. Operationally, this may be a semantic argument, but from a neurobiological perspective,

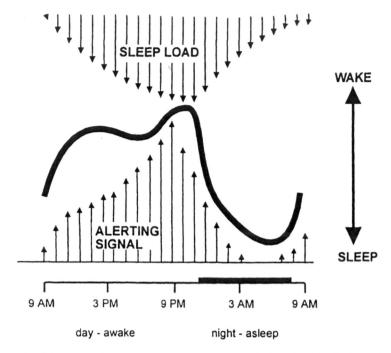

Figure 5 Schematic of the "opponent processes" mediating physiological sleepiness as a function of time of day. Sleep drive increases in response to wakefulness imposed and/or maintained by the suprachiasmatic pacemaker. Increasing levels of SCN-dependent alerting over the subjective day opposes homeostatic sleep drive, both of which peak shortly before the habitual sleep phase. (From Ref. 66.)

these distinctions may be critically important as new studies are designed to try to resolve this important question.

VI. Neuroanatomy of Circadian Control

The neural pathways and transmitter systems responsible for SCN-dependent alerting and sleep homeostasis constitute the frontier of sleep neuroscience. Although much work has been done, the pathways responsible for SCN-dependent alerting and the target tissues modulated by these projections are not known. The SCN and secondary projections from peri-SCN regions terminate in many brain regions that could be important for cortical and behavioral arousal (93–95). The neurons in the SCN are thought to be largely GABAergic and are colocalized with various neuropeptides including vasoactive intestinal polypeptide (VIP), vasopressin, and somatostatin (96). When VIP is injected into the CSF, rodents exhibit increased REM sleep (97), but there is no evidence that VIP in the SCN has any functional relevance to the circadian control of REM sleep. In light of the many lesions, knife cuts, and microinjection studies conducted over the last three decades, it is remarkable that no one brain region (other than the SCN) has been found vital for the circadian modulation of arousal states. It seems likely that several brain regions can mediate circadian rhythms of wakefulness and sleep; however, midbrain transection studies do not eliminate circadian variation in sleep-wakefulness (98), suggesting that circadian control is mediated by SCN efferent projections to diencephalic and forebrain regions. There is now evidence that a small collection of neurons in the ventrolateral pre optic area (VLPO) could be important in sleep regulation and timing (99). Posited as a "sleep gate," there is much interest in this site because of its proximity to the SCN and its projections to the histaminergic magnocellular neurons of the tuberomammalary/posterior hypothalamic areas. These histaminergic neurons are known to have diffuse activating projections throughout the brain and are thought to be one of several transmitter systems that mediate cortical arousal (100–102). At this juncture, however, one cannot rule out the importance of any monoaminergic transmitter system, particularly those projecting to cholinergic neurons in the basal forebrain and perhaps the midbrain as well.

At the neurobiological level, opponent process regulation is envisioned to involve one or more brain regions (e.g., cholinergic basal forebrain and midbrain, histaminergic tuberomammilary posterior hypothalamus) responsible for cortical and behavioral activation that are stimulated (directly or indirectly) by projections from the SCN, but inhibited by one or more sleep factors that accumulate during waking (e.g., adenosine, neurosteroids). The results of the two opposing processes would determine physiological sleep tendency at any given time of day.

VII. Circadian Control of Sleep-Wakefulness in Humans

Contemporary studies of the circadian control of sleep and wakefulness in humans are characterized by long-term recordings of a variety of physiologic variables such as hormones and body temperature to monitor aspects of the circadian process. In addition, neurobehavioral and electrophysiologic variables, such as reaction times and the electroencephalogram, are recorded frequently or even continuously, to characterize wakefulness and sleep. This intensive physiologic monitoring is often conducted in specialized facilities in which humans can be shielded from external 24-hour synchronizers and in which environmental parameters such as light intensity and temperature can be controlled.

In the early days of human circadian rhythm, research scientists descended into natural caves, underground bunkers or hospital basements, but long-term polysomnographic monitoring of sleep and wakefulness was not yet feasible (40,103,104). Those pioneering studies established that the timing of rest and activity and core body temperature were modulated by a process with a near 24-hour period. These studies have also shown that in the absence of knowledge of clock-time and the 24-hour light-dark cycle, the alternation between rest and activity does not occur with clockwise precision. On the contrary, during long-term experiments, the alternation between sleep and wakefulness can become so variable that mathematical techniques may need to be applied to elucidate the contribution of the circadian process to the timing of rest-activity cycles (105).

A. Sleep During Synchronized and Desynchronized Free-Running Conditions

The circadian regulation of rest-activity cycles and its variability are well documented in Aschoff's two classic papers (42,104). Aschoff and Wever described the timing of rest and activity in nine human subjects who were isolated from the 24-hour cycles of light, temperature, etc., and had no clocks available for the duration (on average 14 days) of the experiment (106). The main findings reported in this and a subsequent paper entitled "Circadian Rhythms in Man" (104) were that humans maintain a monophasic sleep-wake cycle and that the sleep-wake ratio remains approximately 1:2. The sleep-wake cycle which remained synchronized with the core body temperature rhythm exhibited a free-running period close to 25 hours. These and other data (40,107,108) demonstrated the endogenous origin of the monophasic sleep-wake cycle in the adult human. The serial correlation between the duration of subsequent sleep-wake cycles and the correlation between the duration of activity and subsequent rest both turned negative. This implies that a self-sustaining oscillator, not a stochastic process or renewal process, underlies the observed rhythmicity (109). These early data were in accordance with animal studies and supported the hypothesis that, also in humans, a single circadian pacemaker underlies circadian rhythmicity in a variety of variables including the human rest-activity cycle and the rhythms of body temperature, urine production, and the variation in urinary electrolytes. This observation furthermore suggested that in humans, as in animals, activity onset was a reliable marker of the circadian process.

Two years later the paper entitled "Desynchronization of Human Circadian Rhythms" (42) highlighted the variability of the sleep-wake cycle especially in long-term experiments, i.e., 2 weeks or longer. During these long recordings the period of the sleep-wake cycle was different from the period of the temperature cycle and this phenomenon was called internal desynchronization. In extreme cases of internal desynchronization, which has been reported by several investigators, the period of the sleep-wake cycle would be as long as close to 48 hours or as short as approximately 13.5, whereas the period of the core body temperature cycle remained close to 25.0 hours. On the basis of this phenomenon it was suggested that the sleep-wake cycle was governed by an oscillator separate from the oscillator controlling core body temperature.

The phenomenon of internal desynchronization has been the source of considerable controversy and has played an important role in the development of models of the regulation of the human sleep-wake cycle. Part of the confusion and controversy is related to the different ways in which animal data and human data are analyzed. In most rodent studies, the alternation between sleep and wakefulness is not analyzed. Activity onset, defined as that part of the circadian cycle in which a sudden increase in motor activity is observed, serves as the primary phase marker. The frequent alternation between sleep and wakefulness, which occur during both the activity phase and the rest phase, are not considered. In humans, the alternation between sleep and wakefulness has been used as the primary phase marker of the circadian system. Even though in humans sleep and wakefulness, as recorded polysomnographically or by self-report, exhibit far less frequent alternations than in rodents, more than one sleep episode may occur per circadian cycle. Subjects may experience these sleep episodes either as naps or major sleep episodes. Zully and Campbell have demonstrated that the results of analyses of the periodicity of the sleep-wake cycle during spontaneous desynchronization depend to a considerable extent on whether or not naps are included (110). A major finding from their analyses was that even though the major sleep-wake episode and associated behaviors such as breakfast, lunch, and dinner, and the subjective estimate of the passage of time, may desynchronize from the body temperature rhythm, sleep propensity, as indexed by the timing and duration of both naps and major sleep episodes, remains in many cases in synchrony with the body temperature rhythm.

Experiments by Zulley and Campbell have shown that the frequency of the alternation between sleep and wakefulness in humans can be increased dramatically by reducing the number of behavioral options available to the subjects. For instance, during continuous bed rest subjects take many naps per circadian cycle, but the instruction to a subject prior to entering a temporal isolation study not to nap may reduce the frequency of the alternation between wakefulness and sleep (110). Although the analyses by Zulley and Campbell do not explain the phenomenon of internal desynchronization, they highlight that many factors besides circadian phase contribute to the timing of a specific behavior, and that the timing of sleep and wakefulness may become highly variable under certain conditions. The primary implications of these observations are that the coupling between the circadian process and the sleep-wake cycle is only weak, that activity onset and sleep onset are not reliable markers of the circadian process, and that the characteristics of the oscillatory process involved in the timing of sleep and wakefulness are very different from the characteristics of the oscillator driving the rhythm of body temperature. The nature of the oscillatory process involved in the regulation of the sleep-wake cycle will be described below, after we have discussed the contribution of the circadian process to sleep timing and duration.

B. Sleep Duration and the Circadian Phase of Sleep Onset and Offset

Analyses of sleep timing, sleep duration, and sleep structure during spontaneous desynchrony have provided important insights in the interaction of the circadian process and the sleep-wake cycle. Even though sleep duration is highly variable, it exhibits a relation with circadian phase of sleep onset. During entrainment, sleep onset occurs approximately 6 hours before and awakening occurs approximately 2 hours after the temperature minimum, resulting in the typical 8-hour sleep episode (111–113). In healthy young adults the temperature minimum will be located at approximately 6 AM, and sleep onset will thus occur several hours after dusk and wake onset will occur several hours after dawn for most months of the year (114). When subjects are released into free-running conditions, and the sleep-wake cycle remains synchronized with the body temperature cycle, subjects typically select to go to sleep close to the temperature minimum and awaken on the rising limb of the body temperature rhythm; i.e., the sleep-wake cycle delays approximately 4 hours relative to the temperature cycle.

It is currently not known whether this phase delay is related to the absence of the external 24-hour light-dark cycle or some other aspect of these protocols. During spontaneous desynchrony, sleep onset and sleep termination cover an even wider range of phase relationships with the temperature cycle than during synchronized free run. In addition, the cycle-to-cycle variations in the duration of both sleep and wakefulness increases dramatically (115). Despite this variability the circadian phases of sleep onset and sleep termination still exhibit some clustering, such that most sleep initiations occur on the falling limb of the temperature cycle and most sleep terminations occur on the rising limb of the temperature cycle. The circadian phase at which sleep is initiated is a major determinant of sleep duration in that longest sleep episodes are initiated 10 to 15 hours before the temperature minimum and shortest sleep episodes are initiated 0 to 5 hours after temperature minimum (44,111). Most sleep terminations occur between 4 and 7 hours after the temperature minimum, which is later in the circadian cycle than when we wake up under entrained conditions (111,116).

During spontaneous desynchrony, sleep initiations are not distributed uniformly across the circadian cycle. Surprisingly, sleep is rarely initiated at approximately 8 hours before the temperature minimum, which would correspond to 9 to 10 PM under entrained conditions. In addition to this evening wake maintenance zone, Strogatz et al. identified a morning wake maintenance zone approximately 5 hours after the temperature minimum, which would correspond to 10 to 11 AM under entrained conditions (117). This wake maintenance zone is, however, far less pronounced than the evening wake maintenance zone. Both zones refer to circadian phases at which subjects select to go to sleep less frequently than other circadian phases, and these zones do not constitute forbidden zones for sleep since subjects can sleep through these phases and actually can initiate sleep at these phases albeit with a somewhat longer sleep latency.

Although common sense and the concept of sleep homeostasis may lead to the suggestion that the duration of the preceding wake episode should be associated with the duration of the subsequent sleep episodes, this has not been easy to demonstrate. During synchronized and desynchronized free run the correlations between the duration of activity and rest turn negative. This is related to the dependency of sleep duration on the circadian phase of sleep onset. Strogatz (116) attempted to identify a contribution of the duration of the previous wake episode after he had accounted for the contribution of the circadian process but failed. Subsequently, Chandrashekaran and co-workers described a number of subjects in whom during spontaneous desynchrony a positive correlation was observed between the duration of the preceding wake episodes and subsequent sleep episodes (118). In a laboratory study in which subjects were entrained, Åkerstedt and co-workers demonstrated that the duration of daytime sleep increased progressively as the duration of the preceding night sleep was reduced progressively, providing additional evidence for the contribution of sleep homeostasis to sleep duration (119).

C. Sleep Structure During Synchronized and Desynchronized Free-Running Conditions

Polysomnographic recordings of sleep during synchronized and desynchronized free run have revealed the role of the circadian process in the timing of REM sleep. Concomitant with the change in the phase relationship between the sleep-wake cycle and the endogenous circadian component of the body temperature rhythm is a change in the redistribution of REM sleep across the sleep episode. Whereas under entrained conditions, REM sleep exhibits its polarity, i.e., the duration of REM sleep episodes increases in the course of the sleep episode, during synchronized free run. REM episodes in the beginning of the sleep episode are already long, and no further increase in their duration is observed. During spontaneous desynchrony, REM sleep propensity exhibits its crest at or shortly after the trough of the body temperature rhythm, and it is well established that REM sleep is under strong circadian control (45,120,121).

In contrast to the strong influence of the circadian process on REM sleep, SWS and its time course is remarkably insensitive to changes in the circadian system. Thus SWS is always most abundant in the initial part of sleep episodes and declines throughout the sleep episode. At the very end of some of the very long (18 to 22 hour) sleep episodes during spontaneous desynchrony a small recurrence of

SWS has been reported (122). These data could reflect a sleep-dependent rhythm in SWS or a circadian modulation of SWS, or this recurrence of SWS could be related to intermittent wakefulness in these very long sleep episodes. Studies of the time course of SWS and computerized analyses of the EEG during long sleep episodes during entrainment have suggested that the sporadic recurrence of SWS reflects a circadian modulation of SWS (123–125). In particular, it seems that SWS is somewhat suppressed close to the temperature nadir, when REM sleep episodes are long and non-REM sleep episodes are short.

D. Circadian Regulation of Sleep Propensity and Sleep Structure

In the synchronized and desynchronized free run the circadian regulation of sleep is studied under conditions in which the subjects self-select when to go to sleep and when to get out of bed. Although the advantage of this approach is that the spontaneous timing of sleep initiation and sleep termination can be studied, disadvantages of this approach include:

1. The large variability in the duration of wakefulness preceding sleep and its associated changes in homeostatic sleep pressure which will affect subsequent sleep structure
2. Nonuniform distribution across the circadian cycle
3. The poorly understood contribution of behavioral options available to the subject to the timing of sleep and wakefulness
4. Social isolation
5. Interindividual differences in the interpretation of instructions to the subject on how to schedule his or her sleep-wake cycle, i.e., include or avoid naps, etc.

The circadian regulation of sleep propensity and sleep structure can also be studied by scheduling the subjects' sleep-wake cycle, i.e., instruct the subject to go to sleep at particular circadian phases and then study the characteristics of polysomnographically recorded sleep. These protocols differ primarily in the period length of the sleep-wake cycle which may vary from as short as 20 min to as long as 28 or even 42.85 hours (90,126–128). Another difference between these protocols concerns the number of circadian cycles studied. This may vary from less than one, in most multiple sleep latency tests, to more than 20 circadian cycles in some forced desynchrony protocols. When the period length of the scheduled sleep-wake cycle is short, the information obtained will primarily be related to the circadian regulation of sleep initiation, i.e., sleep latency. When the period length of the scheduled sleep-wake cycle is long, as in the 28-hour forced desynchrony, information on the circadian regulation of sleep initiation, sleep consolidation, sleep termination, sleep structure, and waking neurobehavioral function as well as the interaction between sleep-wake dependent (i.e., homeostatic) regulation and the circadian regulation of these variables can be obtained. In all of these protocols, with the exception of the multiple sleep latency test, the ratio of the duration of scheduled sleep episode to scheduled wake episodes has been 1:2. The main rationale for this is that homeostatic pressure should remain stable in the course of the experiment.

There is a remarkable convergence of the results obtained in all these different protocols and the results obtained in the synchronized and desynchronized free run (Figs. 6 and 9). Sleep propensity is highest at or shortly after the temperature minimum. Sleep propensity does not exhibit a sudden or marked decline in the morning hours, i.e., on the initial part of the rising limb of the temperature curve. After its trough in the morning hours a gradual decline, i.e., increase in sleep latencies, has been observed in all protocols (with the exception of the MSLT [14]). The observation that sleep latencies are long in the morning hours in the MSLT protocol is most likely related to the very short durations of prior wakefulness at this time and the fact that later in the day wakefulness preceding the MSLT increases (this is a consequence of the non 1:2 ratio of sleep to wakefulness in this protocol). In all protocols, sleep latencies are longest in the evening hours, i.e., approximately 14 to 17 hours after the temperature minimum and close to habitual bedtime under entrained condition. In protocols in which sleep propensity was assessed on the basis of the propensity to terminate sleep, highest sleep propensity was again located close to the nadir of the temperature rhythm and lowest sleep propensity was again located close to the wake maintenance zone, which was followed by a sudden increase in sleep propensity (Fig. 10) (90,91). The circadian drive for wakefulness at the end of the habitual day is so strong that in protocols in which sleep initiation trials were as short as 7 min, sleep was rarely initiated (129). This resulted in this zone being called the "forbidden zone for sleep," even though subjects will eventually fall asleep at this circadian phase. The use of the term zone is somewhat inaccurate because the increase in sleep latencies, i.e., the increase of the circadian alerting signal, starts shortly after the temperature nadir, continues throughout the waking day, and reaches its crest in the evening hours. The term zone also refers to the clock concept, i.e., a mechanism which produces particular signals, like an alarm, at certain times. The analyses of sleep propensity in all of these protocols indicate that it may be better to consider the circadian pacemaker's role in sleep-wake regulation as a process that produces a continuous signal which waxes and wanes across the circadian cycle. This convergence indicates that the propensity to initiate and terminate sleep is most likely dependent on one circadian signal. However, this does not imply that the mechanisms of awakening and sleep initiation are similar.

The MSLT data indicate that under conditions in which the duration of wakefulness preceding sleep initiation and circadian phase change simultaneously, there is a transient increase in sleep propensity in the afternoon (14). This increase

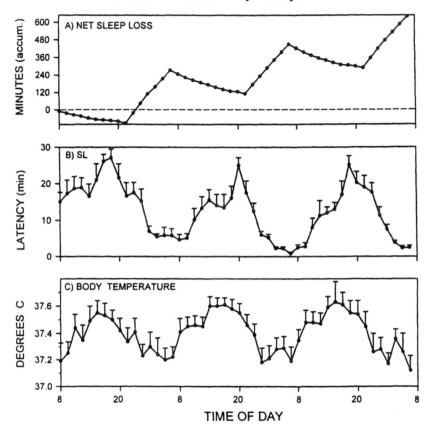

Figure 6   Circadian rhythms in sleep latency and body temperature as a function of sleep loss during a human 90-minute day protocol. Circadian rhythms in physiological alertness persist over 3 days, with maximum alertness occurring late in the subjective day despite accumulating sleep loss inherent to the study design. Note that sleep latency is most sensitive to the effects of sleep loss early in the subjective day (when SCN-dependent alerting is hypothesized to be relatively weak). (From Ref. 149.)

in sleep propensity is minor compared to the increase in sleep propensity during the nocturnal hours and may be absent in many subjects, especially when sleep propensity is assessed on the basis of wakefulness within scheduled sleep episodes. Such interindividual variability in the waveform of the sleep propensity function may reflect interindividual differences in the waveform of the output signal from one pacemaker or alternatively reflect interindividual differences in the phase relationship between the morning and evening oscillator, each of which may promote wakefulness at a particular phase of its endogenous cycle. Theoretically, changes in the photoperiod should change the phase relationship between the two oscillators; indeed, Wehr obtained some evidence that during exposure to long nights and long sleep episodes, the sleep propensity function (and the pattern of melatonin secretion) during the nocturnal hours becomes bimodal (130,131). It remains to be established whether this bimodality is related to the long sleep episodes and associated changes in the homeostatic sleep pressure, or a change in the circadian process.

E. Interaction of Circadian and Homeostatic Processes During Forced Desynchrony

The contribution of sleep-dependent processes to the propensity to wake up has been demonstrated unequivocally in forced desynchrony protocols (90,91) (Figs. 7, 8). The interaction between the circadian and sleep-dependent or homeostatic process is such that in the initial part of sleep, sleep efficiency is high independently of circadian phase. This is because after 18 hours of wakefulness homeostatic sleep pressure is sufficiently high to overcome the circadian drive for wakefulness at any circadian phase. In the sleep episodes during the forced desynchrony protocol, sleep efficiency remains high throughout the sleep episode when the end of the sleep episode coincides with the nadir if the temperature cycle, or shortly thereafter (Fig. 7). In contrast, when the end of the scheduled sleep episode coincides with the second half of the rising limb of the temperature cycle, sleep efficiency drops to very low levels. This implies that sleep episodes will be long and have a high sleep efficiency when they are initiated close to the crest of the circadian drive for wakefulness and end at the peak of the circadian sleep propensity rhythm. This result is consistent with the data of the spontaneous desynchrony protocols. Analyses of neurobehavioral function during wakefulness in the forced desynchrony protocol have provided additional evidence for the importance of the interaction of circadian and wake-dependent, (i.e., homeostatic) process in the regulation of vigilance (132).

The trough of the circadian rhythm of subjective alertness and cognitive throughput is located close to the temperature nadir, i.e., close to habitual wake time, and the crest is located 12 to 16 hours later (Fig. 9). High levels of alertness and throughput can be observed at all circadian phases when the duration of preceding wakefulness is short. However, when the duration of wakefulness exceeds 8 hours, high levels of alertness and throughput can be maintained only when the homeostatic drive is opposed by the circadian alerting signal and alertness drops to very low levels when long durations of prior wakefulness coincide with the temperature trough. In other words. the amplitude of the circadian variation in alertness increases with increasing duration of prior wakefulness.

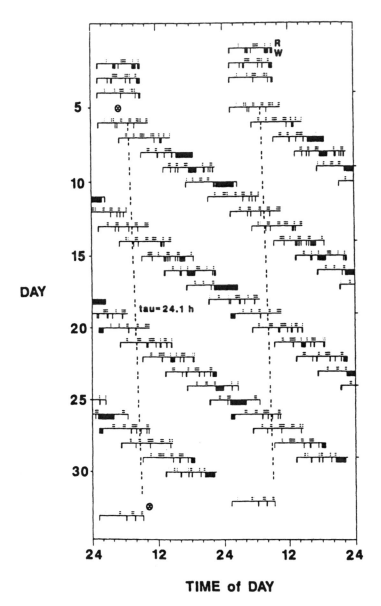

Figure 7 Double plot of the forced desynchrony protocol. Successive days are plotted both next to and beneath each other. In each sleep episode (horizontal solid line) wakefulness (solid black bars below sleep episode) and REM sleep (dotted lines/bars above sleep episode) are indicated for subject 1136. An intrinsic circadian temperature cycle with a period of 24.1 hours is estimated by a nonparametric spectral analysis of the core body temperature data during the forced desynchrony part of the protocol. The time of the

This indicates that the circadian and homeostatic contributions to alertness interact; i.e., alertness is not just the sum of the circadian and homeostatic process.

VIII. Sleep Structure and Human EEG

The analysis of sleep structure in forced desynchrony protocols has confirmed the conclusions from the spontaneous desynchrony experiments, i.e., that SWS is virtually independent of circadian phase (Fig. 8) and the crest of REM sleep is located shortly after the temperature minimum (Fig. 9). The forced desynchrony protocol has demonstrated that in young subjects REM sleep exhibits, in addition to the circadian modulation, a sleep-dependent inhibition (91). Thus, under entrained conditions the sleep-dependent and circadian modulation of REM sleep will result in very long REM sleep episodes close to habitual wake time. The application of quantitative EEG analyses in these forced desynchrony experiments has shown that slow components in the EEG are subject to a small circadian modulation, in accordance with sleep displacement experiments (58,133). Somewhat surprisingly, the circadian rhythm of slow EEG components during non-REM sleep exhibits its nadir close to the crest of the sleep propensity rhythm, and the crest of this EEG rhythm is located close to the nadir of the circadian sleep propensity rhythm. Quantitative EEG analyses have also demonstrated a circadian modulation of sleep spindle activity (91,134). The crest of this circadian rhythm is located in the nighttime. Since sleep spindles are major inhibitory events, which block the transfer of sensory stimuli, this circadian rhythm may represent a mechanism by which the circadian pacemaker reduces arousability from sleep. EEG components during REM sleep exhibit little circadian variation with the exception of alpha activity—in particular, low-frequency alpha activity. The circadian rhythm of this EEG activity exhibits its nadir at the crest of the REM sleep propensity rhythm, and its crest coincides with the crest of the circadian variation in cognitive throughput.

A. Characteristics of the Oscillatory Sleep Homeostatic Process

The phenomenon of spontaneous desynchronization and the initial observation that SCN lesions do not abolish the temperature rhythm but do abolish the activity-rest rhythm, have led to the suggestion that the sleep-wake cycle is

---

minimum of the circadian temperature rhythm as estimated by the nonparametric spectral analysis is indicated by the dashed line. The minimum of the endogenous circadian rhythm of core body temperature as assessed in the initial and final constant routine is indicated by an encircled X. The data are plotted with respect to clock time. (From Ref. 91.)

Slow Wave Activity

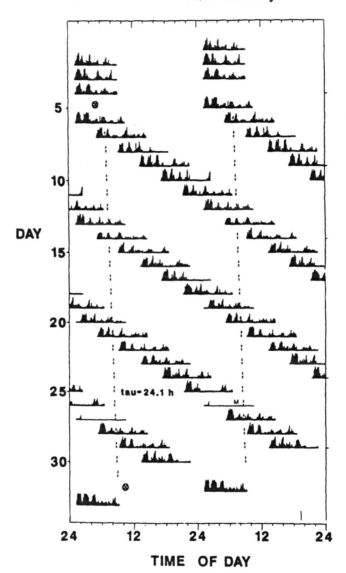

Figure 8 Double plot of the time course of slow-wake activity during sleep episodes in the forced desynchrony protocol (subject 1136). Artifacts during sleep and wakefulness within the scheduled sleep episodes were removed. One-minute averaged values are plotted for slow-wave activity. The time of the minimum of the circadian temperature rhythm, estimated by nonparametric spectral analysis, is indicated by the dashed line. Each mini-

governed by an oscillator separate from the body temperature oscillator. More recent evidence suggests that SCN lesions abolish both the circadian rhythm of body temperature and the circadian distribution of sleep and wakefulness in the squirrel monkey. Nevertheless, data from both spontaneous and forced de-synchrony protocols indicate that the temporal variation of sleep propensity and sleep consolidation as well as the variation in alertness and cognitive throughput cannot be explained with only one oscillatory process, but requires the interaction of two processes. One of these oscillatory processes may be referred to as the strong oscillator in view of its stability, and is generally assumed to be located in the SCN. The other process is intimately linked with the sleep-wake cycle. Actually, manipulations of the sleep-wake cycle such as sleep displacement or sleep deprivation appear to immediately shift this oscillatory process, i.e., affect sleep propensity. Although in several species multiple localized oscillators have been identified, it is unclear whether this sleep-wake oscillatory process in mammals, including humans, is localized in a specific brain region. In addition, it is unclear whether this oscillatory process continues to oscillate in the absence of a sleep-wake cycle. It therefore seems reasonable to assume that this process is generated by the sleep-wake cycle. A multitude of sleep deprivation, sleep displacement, and nap studies have indicated that slow waves in the EEG and other sleep parameters exhibit predictable changes in response to these manipulations (reviewed in 95). In particular it has been demonstrated that slow waves in the EEG during sleep and slow waves in the EEG during wakefulness increase in an exponential saturating way as a function of the duration of wakefulness over a broad range of durations of wakefulness.

B. Physiology of the Circadian Process

The animal and human data are consistent with the notion that the circadian pacemaker provides signals to the rest of the CNS, thereby modulating the probability of the occurrence of the three vigilance states—wakefulness, REM sleep, and non-REM sleep. The nature of these signals and their neuroanatomical targets remains obscure. In research on the circadian regulation of human sleep, a number of correlates of the circadian variation of sleep propensity and sleep structure have been identified. The two best-known correlates are plasma melatonin and core body temperature. The circadian rhythm of plasma melatonin is one of the most robust markers of the circadian pacemaker, and the pathways by which the

---

mum of the endogenous circadian rhythm of core body temperature as assessed in the initial and final constant routine is indicated by an encircled X. M indicates missing data. The data were plotted with respect to clock time. The vertical line in the right-hand lower corner correspond to 200 μ V^2/Hz. (From Ref. 91.)

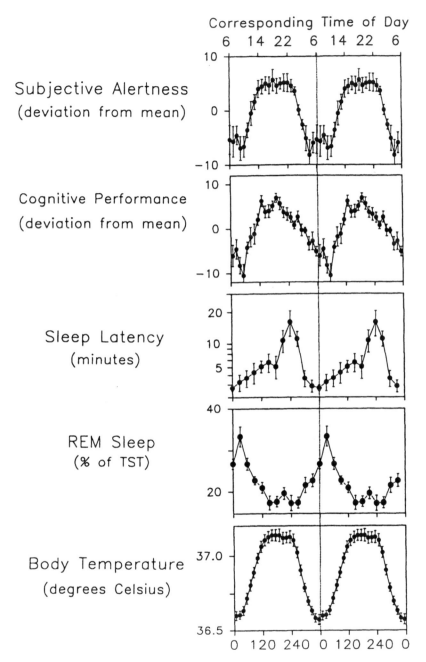

Figure 9 Circadian waveform of body temperature, REM sleep, sleep latency, cognitive performance (throughput), and subjective alertness as assessed in a forced desynchrony protocol. Circadian phase 0 corresponds to the fitted minimum of the circadian rhythm of body temperature. (From Refs. 90 and 132.)

suprachiasmatic nuclei innervate the pineal have been identified. In humans, the onset and offset of the phase during which melatonin is present in plasma demarcate reliably the phase during which sleep propensity is high, even during forced desynchrony of the scheduled sleep-wake cycle and the endogenous circadian timing system (134–138) (Fig. 10).

Evidence for a causal role for melatonin in the modulation of sleep propensity has been obtained in experiments in which melatonin was administered during the daytime hours, i.e., when endogenous levels of melatonin were low. In these experiments an increase of sleep propensity has been observed (137). The mechanisms by which melatonin exerts these sleep-inducing effects remain unclear. Pharmacological doses of melatonin were used in most experiments and it remains to be demonstrated conclusively that physiologic doses exert similar sleep inducing effects although some positive effects have been reported. Melatonin could exert its effects through melatonin receptors located outside the SCN or through one of the two melatonin receptors found in the SCN. It has been suggested that melatonin exerts its effects by inhibiting the alerting signal which is thought to be generated by the SCN and which peaks shortly before the onset of melatonin secretion (135). Such a mechanism would suggest that the SCN modulates the distribution of sleep and wakefulness primarily by providing an alerting signal. Evidence that the SCN also actively promotes sleep may be derived from the observation that REM sleep and sleep propensity as indexed by sleep latency, peak approximately 9 hours after the onset of endogenous melatonin secretion.

It seems reasonable to assume that melatonin is only one of more signals that modulate sleep propensity, some of which may inhibit the alerting signal whereas others may actively promote sleep. Identification of these signals hold great promise for the development of new sleep inducing drugs.

IX. Sleep-Wake Behavior and Light Exposure

The circadian rhythm of sleep propensity is driven by the light-sensitive pacemaker (139–142). The circadian rhythm of light sensitivity—i.e., the magnitude of the phase shifts, or drive onto the circadian pacemaker elicited by a light pulse of a given strength—has its crest close to the temperature nadir (143,144). Even though the human circadian pacemaker is sensitive throughout the circadian cycle, largest phase delays are induced by light pulses given shortly before the temperature minimum, and largest phase advances are induced by light pulses given shortly after the temperature minimum. Young adults typically initiate sleep approximately 6 hours before the temperature minimum, i.e., shortly after the crest of the circadian drive for wakefulness, and end sleep approximately 2 hours after the temperature minimum, i.e., shortly after the crest of the circadian drive for sleep. The putative functional significance of this paradoxical phase relationship is the consolidation of sleep and wakefulness. The mechanism by which this phase relationship is maintained, i.e., how the sleep-wake cycle is entrained to the

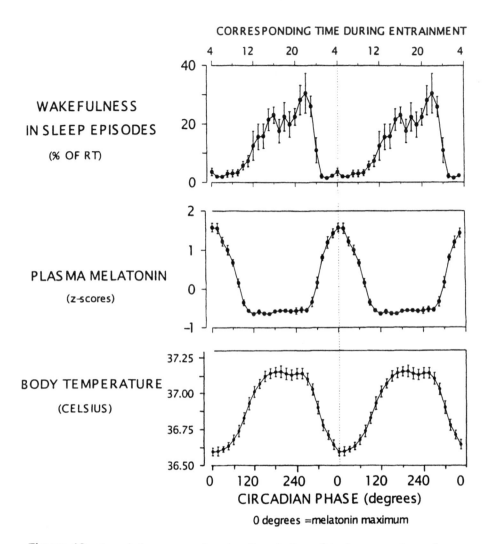

Figure 10 Association among the circadian rhythm of body temperature, plasma melatonin, and wakefulness within scheduled sleep episodes as assessed in a forced desynchrony protocol. Circadian phase 0 corresponds to the fitted maximum of the plasma melatonin rhythm. (From Ref. 134.)

endogenous circadian rhythm of sleep propensity, is unknown. In humans, sleep is associated with darkness (humans close the curtains, turn off the light, and close their eyes), and wakefulness is associated with light (humans open their eyes, turn on the artificial light, and open the curtains). This implies that the sleep-wake cycle and its phase relation with the circadian rhythm of light sensitivity become major determinants of light exposure of the circadian pacemaker. Because the circadian pacemaker is sensitive to light intensities that are present in our living rooms (145), the sleep-wake cycle also becomes a major determinant of entrained phase, i.e., the phase of the circadian cycle of light sensitivity and the associated rhythms of melatonin, cortisol, sleep propensity, etc., relative to clock time and the natural light-dark cycle. Some evidence for this notion comes from the analyses of the effects of the seasonal changes in photoperiod, and the change from standard time to daylight savings time on entrained phase of the melatonin rhythm (146). Under these conditions the phase of the melatonin rhythm remains associated with clock time. This is because our behavior while living in society is dictated by clock time.

As mentioned above, the sleep-wake cycle delays approximately 4 hours when subjects are released into free run, i.e. in the absence of external synchronizers. This implies that external synchronizers contribute to the phase relationship between the sleep-wake cycle and the endogenous circadian timing system. Further evidence for the contribution of external light-dark cycle to this phase relationship may be derived from the observed dependency of the phase angle difference between the light-dark cycle and the sleep-wake cycle on the period of the artificial light-dark cycle (147,148).

X. Conclusion

Our current understanding of the regulation of the human sleep-wake cycle highlights that sleep-wake behavior is not simply controlled by a clock located in the brain. Sleep-wake behavior is generated by a complex interaction of environmental factors, and endogenous circadian and homeostatic processes. Whereas this complexity may make it more challenging to predict sleep-wake behavior, this multitude of factors contributing to sleep propensity also offers a multitude of potential points of control of this behavior.

References

1. Ralph MR, Menaker M. A mutation of the circadian system in golden hamsters. Science 1988; 241:1225–1227.
2. Jouvet M. Recent data on experimental insomnia in the cat. Rev Neurol 1966; 115: 454–456.
3. Jouvet M. Biogenic amines and the states of sleep. Science 1969; 163:31–41.
4. Jouvet M, Buda C, Debilly G, Dittmar A, Sastre JP. Central temperature is the principal factor of regulation of paradoxical sleep in pontile cats. C R Acad Sci Serie III, Sciences de la Vie 1988; 306:69–73.
5. Morrison AR. Paradoxical sleep without atonia. Arch Ital Biol 1988; 126:275–289.
6. Porkka-Heiskanen T, Strecker RE, Thakkar M, Bjorkum AA, Greene RW, McCarley

RW. Adenosine: a mediator of the sleep-inducing effects of prolonged wakefulness. Science 1997; 276:1265–1268.

7. Sterman MB, Knauss T, Lehman D. Circadian sleep and waking patterns in the laboratory cat. Electroencephalogr Clin Neurophysiol 1965; 19:509–517.

8. Gander PH, Moore-Ede MC. Light-dark masking of circadian temperature and activity rhythms in squirrel monkeys. Am J Physiol 1983; 245:R927–R934.

9. Edgar DM. Circadian Timekeeping in the Squirrel Monkey: Neural and Photic Control of Sleep, Brain Temperature, and Drinking. Riverside, CA: Dept. Biological Science, University of California, 1986:360.

10. Edgar DM, Dement WC, Fuller CA. Effect of SCN lesions on sleep in squirrel monkeys: evidence for opponent processes in sleep-wake regulation. J Neurosci 1993; 13:1065–1079.

11. Carskadon MA, Dement WC. Effects of total sleep loss on sleep tendency. Percept Mot Skills 1979; 48:495–506.

12. Carskadon MA, Harvey K, Dement WC. Sleep loss in young adolescents. Sleep 1981; 4:299–312.

13. Carskadon MA, Dement WC. Cumulative effects of sleep restriction on daytime sleepiness. Psychophysiology 1981; 18:107–113.

14. Richardson GS, Carskadon MA, Orav EJ, Dement WC. Circadian variation of sleep tendency in elderly and young adult subjects. Sleep 1982; 5(suppl 2):S82–S94.

15. Borbély AA, Huston JP, Waser PG. Control of sleep states in the rat by short light-dark cycles. Brain Res 1975; 95:89–101.

16. Borbély AA. Effects of light on sleep and activity rhythms. Prog Neurobiol 1978; 10:1–31.

17. Welsh D, Richardson GS, Dement WC. Effect of running wheel availability on circadian patterns of sleep and wakefulness in mice. Physiol Behav 1988; 43: 771–777.

18. Edgar DM, Kilduff TS, Martin CE, Dement WC. Influence of running wheel activity on free-running sleep/wake and drinking circadian rhythms in mice. Physiol Behav 1991; 50:373–378.

19. Jacobs BL, Fornal CA. Activity of brain serotonergic neurons in the behaving animal. Pharmacol Rev 1991; 43:563–578.

20. Jacobs BL, Azmitia EC. Structure and function of the brain serotonin system. Physiol Rev 1992; 72:165–229.

21. Jacobs BL, Fornal CA. 5-HT and motor control: a hypothesis. Trends Neurosci 1993; 16:346–352.

22. Miller JD, Farber J, Gatz P, Roffwarg H, German DC. Activity of mesencephalic dopamine and non-dopamine neurons across stages of sleep and walking in the rat. Brain Res 1983; 273:133–141.

23. Edgar DM, Dement WC. Regularly scheduled voluntary exercise synchronizes the mouse circadian clock. Am J Physiol 1991; 261:R928–R933.

24. Mrosovsky N. Locomotor activity and non-photic influences on circadian clocks. Biol Rev Camb Philosoph Soc 1996; 71:343–372.

25. Edgar DM, Martin CE, Dement WC. Activity feedback to the mammalian circadian pacemaker: influence on observed measures of rhythm period length. J Biol Rhythms 1991; 6:185–199.

26. Bradbury MZ, Dement WC, Edgar DM. 5HT-containing fibers in the hypothalamus attenuate light-induced phase delay in mice. Brain Res 1997; 768:125–134.

27. Ralph MR, Foster RG, Davis FC, Menaker M. Transplanted suprachiasmatic nucleus determines circadian period. Science 1990; 247:975–978.

28. Moore RY, Lenn NJ. A retinohypothalamic projection in the rat. J Comp Neurol 1972; 146:1–14.

29. Stephan FK, Zucker I. Circadian rhythms in drinking behavior and locomotor activity of rats are eliminated by hypothalamic lesions. Proc Natl Acad Sci USA 1972; 69:1583–1586.

30. Eastman CI, Mistlberger RE, Rechtschaffen A. Suprachiasmatic nuclei lesions eliminate circadian temperature and sleep rhythms in the rat. Physiol Behav 1984; 32:357–368.

31. Ibuka NS, Kawamura H. Loss of circadian rhythm in sleep-wakefulness cycle in the rat by suprachiasmatic nucleus lesions. Brain Res 1975; 96:76–81.

32. Ibuka N, Inouye ST, Kawamura H. Analysis of sleep-wakefulness rhythms in male rats after suprachiasmatic nucleus lesions and ocular enucleation. Brain Res 1977; 122:33–47.

33. Ibuka N, Nihonmatsu I, Sekiguchi S. Sleep-wakefulness rhythms in mice after suprachiasmatic nucleus lesions. Waking Sleeping 1980; 4:167–173.

34. Mistlberger RE, Bergmann BM, Waldenar W, Rechtschaffen A. Recovery sleep following sleep deprivation in intact and suprachiasmatic nuclei-lesioned rats. Sleep 1983; 6:217–233.

35. Mouret J, Coindet J, Debilly G, Chouvet G. Suprachiasmatic lesions in the rat: alterations in sleep circadian rhythms. Electroencephal Clin Neurophysiol 1978; 45: 402–408.

36. Sato T, Kawamura H. Effects of bilateral suprachiasmatic nucleus lesions on circadian rhythms in a diurnal rodent, the Siberian chipmunk (*Eutamias sibiricus*). J Comp Physiol 1984; 155:745–752.

37. Stephan FJ, Nunez AA. Elimination of circadian rhythms in drinking, activity, sleep, and temperature by isolation of the suprachiasmatic nuclei. Behav Biol 1977; 20: 1–16.

38. Tobler I, Borbély AA, Groos G. The effect of sleep deprivation on sleep in rats with suprachiasmatic lesions. Neurosci Lett 1983; 42:49–54.

39. Yamaoka S. Participation of limbic-hypothalamic structures in circadian rhythm of slow wave sleep and paradoxical sleep in the rat. Brain Res 1978; 151:255–268.

40. Kleitman N. Sleep and Wakefulness. Chicago: University of Chicago Press, 1963:552.

41. Sisk CL, Stephan FJ. Central visual pathways and the distribution of sleep in 24-hr and 1-hr light-dark cycles. Physiol Behav 1982; 29:231–239.

42. Aschoff J, Gerecke U, Wever R. Desynchronization of human circadian rhythms. Jpn J Physiol 1967; 17:450–457.

43. Aschoff J. Desynchronization and resynchronization of human circadian rhythms. Aerospace Med 1969; 40:844–849.

44. Czeisler CA, Weitzman E, Moore-Ede MC, Zimmerman JC, Knauer RS. Human sleep: its duration and organization depend on its circadian phase. Science 1980; 210:1264–1267.

45. Czeisler CA, Zimmerman JC, Ronda JM, Moore-Ede MC, Weitzman ED. Timing of REM sleep is coupled to the circadian rhythm of body temperature in man. Sleep 1980; 2:329–346.

46. Weitzman ED, Czeisler CA, Zimmerman JC, Ronda JM. Timing of REM and stages 3 + 4 sleep during temporal isolation in man. Sleep 1980; 2:391–407.

47. Kronauer RE, Czeisler CA, Pilato SF, Moore-Ede MC, Weitzman ED. Mathematical model of the human circadian system with two interacting oscillators. Am J Physiol 1982; 242:R3–R17.

48. Pickard GE, Turek FW. The suprachiasmatic nuclei: two circadian clocks? Brain Res 1983; 268:201–210.

49. Pittendrigh CS. Circadian rhythms and the circadian organization of living systems. Cold Spring Harbor Symp Quant Biol 1960; 25:159–184.

50. Boulos Z, Terman M. Splitting of circadian rhythms in the rat. J Comp Physiol 1979; 134:75–83.

51. Fuller CA, Lydic R, Sulzman FM, Albers HE, Tepper B, Moore-Ede MC. Circadian rhythm of body temperature persists after suprachiasmatic lesions in the squirrel monkey. Am J Physiol 1981; 241:R385–R391.

52. Reppert SM, Perlow MJ, Underleider LG, et al. Effects of damage to the suprachiasmatic area of the anterior hypothalamus on the daily melatonin and cortisol rhythms in the rhesus monkey. J Neurosci 1981; 1:1414–1425.

53. Borbély AA. A two process model of sleep regulation. Hum Neurobiol 1982; 1: 195–204.

54. Daan S, Beersma DG, Borbély AA. Timing of human sleep: recovery process gated by a circadian pacemaker. Am J Physiol 1984; 246:R161–R183.

55. Patrick GTW, Gilbert JA. Studies from the psychological laboratory of the University of Iowa. On the effects of sleep loss. Psychol Rev 1896; 3:469–483.

56. Dijk DJ, Beersma DG, Daan S. EEG power density during nap sleep: reflection of an hourglass measuring the duration of prior wakefulness. J Biol Rhythms 1987; 2: 207–219.

57. Dijk DJ, Beersma DG. Effects of SWS deprivation on subsequent EEG power density and spontaneous sleep duration. Electroencephalog Clin Neurophysiol 1989; 72:312–320.

58. Dijk DJ, Brunner DP, Beersma DG, Borbély AA. Electrogram power density and slow wave sleep as a function of prior waking and circadian phase. Sleep 1990; 13: 430–440.

59. Dijk DJ. EEG slow waves and sleep spindles: windows on the sleeping brain. Behav Brain Res 1995; 69:109–116.

60. Tobler I, Borbély AA. Sleep EEG in the rat as a function of prior waking. Electroencephalog Clin Neurophysiol 1986; 64:74–76.

61. Tobler I, Scherschlicht R. Sleep and EEG slow-wave activity in the domestic cat: effect of sleep deprivation. Behav Brain Res 1990; 37:109–118.

62. Tobler I, Borbély AA. The effect of 3-h and 6-h sleep deprivation on sleep and EEG spectra of the rat. Behav Brain Res 1990; 36:73–78.

63. Trachsel L, Tobler I, Borbély AA. Sleep regulation in rats: effects of sleep deprivation, light, and circadian phase. Am J Physiol 1986; 251:R1037–R1044.

64. Trachsel L, Tobler I, Borbély AA. Effect of sleep deprivation on EEG slow wave

activity within non-REM sleep episodes in the rat. Electroencephalogr Clin Neurophysiol 1989; 73:167–171.

65. Roehrs T, Merlotti L, Petrucelli N, Stepanski E, Roth T. Experimental sleep fragmentation. Sleep 1994; 17:438–443.

66. Edgar DM. Circadian control of sleep/wakefulness: implications in shiftwork and therapeutic strategies. In: Shiraki K, Sagawa S, Yousef MK, eds. Physiological Basis of Occupational Health: Stressful Environments. Amsterdam: Academic Publishing, 1996:253–265.

67. Webb WB, Agnew HW. Stage 4 sleep: influence of time course variables. Science 1971; 174:1354–1356.

68. Webb WB, Agnew HW. Stage 4 sleep: influence of time course variables. Science 1971; 174:1354–1356.

69. Benington JH, Kodali SK, Heller HC. Stimulation of A_1 adenosine receptors mimics the electroencephalographic effects of sleep deprivation. Brain Res 1995; 692:79–85.

70. Gaillard J-M, Nicholson AN, Pascoe PA. Neurotransmitter systems. In: Kryger MH, Roth T, Dement WC, eds. Principles and Practice of Sleep Medicine. Philadelphia: W.B. Saunders, 1994:338–348.

71. Gaillard JM. Neurochemical regulation of the states of alertness. Ann Clin Res 1985; 17:175–184.

72. Hayaishi O. Sleep-wake regulation by PGD2 and E2. J Biol Chem 1988; 263:14593–14596.

73. Inoue S. Sleep substances. Clin Neurosci 1987; 5:22–30.

74. Jones B. Basic mechanisms of sleep-wake states. In: Kryger MH, Roth T, Dement WC, eds. Principles and Practice of Sleep Medicine. Philadelphia: W.B. Saunders, 1994:145–162.

75. Krueger JM. Muramyl peptides and interleukin-1 as promotors of slow wake sleep. In: Inoue S, Borbély AA, eds. Endogenous Sleep Substances and Sleep Regulation. Toyko: Japan Scientific Societies Press, 1985:181–195.

76. Ongini E, Longo VG. Dopamine receptor subtypes and arousal. Int Rev Neurobiol 1989; 31:239–255.

77. Opp MR, Kapas L, Toth LA. Cytokine involvement in the regulation of sleep. Proc Soc Exp Biol Med 1992; 201:16–27.

78. Toth LA, Krueger MJ. Infectious disease, cytokines and sleep. In: Mancia M, Marini G, eds. The Diencephalon and Sleep. New York: Raven Press, 1990:331–341.

79. Cho AK. Ice: d-methamphetamine hydrochloride. In: Korenman SG, Barchas JD, eds. Biological Basis of Substance Abuse. New York: Oxford University Press, 1993:299–307.

80. Holman RB. Biological effects of central nervous system stimulants. Addiction 1994; 89:1435–1441.

81. Nishino S, Sampathkumaran R, Tafti M, et al. Is presynaptic activation of dopaminergic transmission important for the EEG arousal effect of stimulant compounds? Sleep Res 1995; 24:310.

82. Radulovacki M, Virus RM, Djuricic-Nedelson M, Green RD. Adenosine analogs and sleep-like behavior in rats. J Pharmacol Exp Ther 1983; 228:268–274.

83. Radulovacki M, Virus RM, Yanik G, Green RD. Adenosine and sleep in rats. In:

Koella WP, Ruther E, Schulz H, eds. Sleep '84. New York: Gustav Fischer Verlag, 1985:17–22.

84. Radulovacki M. Role of adenosine in sleep in rats. Rev Clin Basic Pharmacol 1985; 5:327–339.

85. Nehlig A, Daval J-L, Debry G. Caffeine and the central nervous system: mechanisms of action, biochemical, metabolic and psychostimulant effects. Brain Res Rev 1992; 17:139–170.

86. Virus RM, Ticho S, Pilditch M, Radulovacki M. A comparison of the effects of caffeine, 8-cyclopentyltheophylline, and alloxazine on sleep in rats. Neuropsychopharmacology 1990; 3:243–249.

87. Williams M. Purine receptors in mammalian tissues: pharmacology and functional significance. Annu Rev Pharmacol Toxicol 1987; 27:315–345.

88. Rainnie DG, Grunze HC, McCarley RW, Greene RW. Adenosine inhibition of mesopontine cholinergic neurons: implications for EEG arousal. Science 1994; 263: 689–692.

89. Portas CM, Thakkar M, Rainnie DG, Greene RW, McCarley RW. Role of adenosine in behavioral state modulation: a microdialysis study in the freely moving cat. Neuroscience 1997; 79:225–235.

90. Dijk DJ, Czeisler CA. Paradoxical timing of the circadian rhythm of sleep propensity serves to consolidate sleep and wakefulness in humans. Neurosci Lett 1994; 166: 63–68.

91. Dijk DJ, Czeisler CA. Contribution of the circadian pacemaker and the sleep homeostat to sleep propensity, sleep structure, electroencephalographic slow waves, and sleep spindle activity. J Neurosci 1995; 15:3526–3538.

92. Edgar DM. Functional role of the suprachiasmatic nuclei in the regulation of sleep and wakefulness. In: Guilleminault C, Lugaresi E, Montagna P, Gambetti P, eds. Fatal Familial Insomnia: Inherited Prion Diseases, Sleep, and the Thalamus. New York: Raven Press, 1994:203–213.

93. Watts AG, Swanson LW, Sanchez-Watts G. Efferent projections of the suprachiasmatic nucleus. I. Studies using anterograde transport of *Phaseolus vulgaris* leucoagglutinin in the rat. J Comp Neurol 1987; 258:204–229.

94. Watts AG. The efferent projections of the suprachiasmatic nucleus: anatomical insights into the control of circadian rhythms. In: Klein DC, Moore RY, Reppert SM, eds. Suprachiasmatic Nucleus: The Mind's Clock. New York: Oxford University Press, 1991:77–106.

95. Watts AG, Swanson LW. Efferent projections of the suprachiasmatic nucleus. II. Studies using retrograde transport of fluorescent dyes and simultaneous peptide immunohistochemistry in the rat. J Comp Neurol 1987; 258:230–252.

96. Van den Pol AN, Tsujimoto JL. Neurotransmitters of the hypothalamic suprachiasmatic nucleus: immunocytochemical analysis of 25 neuronal antigens. Neuroscience 1985; 15:1049–1086.

97. Riou F, Cespuglio R, Jouvet M. Endogenous peptides and sleep in the rat. III. The hypnogenic properties of vasoactive intestinal polypeptide. Neuropeptides 1982; 2: 265–277.

98. Hanada Y, Kawamura H. Sleep-waking electrocorticographic rhythms in chronic cerveau isole rats. Physiol Behav 1981; 26:725–728.

99. Sherin JE, Shiromani PJ, McCarley RW, Saper CB. Activation of ventrolateral preoptic neurons during sleep. Science 1996; 271:216–219.

100. Lin JS, Sakai K, Jouvet M. Evidence for histaminergic arousal mechanisms in the hypothalamus of cat. Neuropharmacology 1988; 27:111–122.

101. Lin JS, Sakai K, Vanni-Mercier G, Jouvet M. A critical role of the posterior hypothalamus in the mechanisms of wakefulness determined by microinjection of muscimol in freely moving cats. Brain Res 1989; 479:225–240.

102. Lin HJ, Sakai K, Vanni-Mercier G, et al. Involvement of histaminergic neurons in arousal mechanisms demonstrated with H^3-receptor ligands in the cat. Brain Res 1990; 523:325–330.

103. Siffre M. Six months alone in a cave. Natl Geogr 1975; 147:426–435.

104. Aschoff J. Circadian rhythms in man: a self-sustained oscillator with an inherent frequency underlies human 24-hour periodicity. Science 1965; 148:1427–1432.

105. Winfree AT. The Geometry of Biological Time. New York: Springer-Verlag, 1980.

106. Aschoff J, Wever R. Spontanperiodik des Menschen bei Ausschluss aller Zeitgeber. Naturwissenschaften 1962; 49:337–342.

107. Webb WB, Agnew HW Jr. Sleep and waking in an environment free of cues to time. Psychophysiology 1972; 9:133–133.

108. Mills JN. Circadian rhythms during and after three months in solitude underground. J Physiol (Lond) 1964; 174:217–231.

109. Wever RA. The Circadian System of Man: Results of Experiments Under Temporal Isolation. New York: Springer-Verlag, 1979.

110. Zulley J, Campbell SS. Napping behavior during "spontaneous internal desynchronization": sleep remains in synchrony with body temperature. Hum Neurobiol 1985; 4:123–126.

111. Zulley J, Wever R, Aschoff J. The dependence of onset and duration of sleep on the circadian rhythm of rectal temperature. Pflugers Arch 1981; 391:314–318.

112. Zulley J, Wever RA. Interaction between the sleep-wake cycle and the rhythm of rectal temperature. In: Aschoff J, Daan S, Groos G, editors. Vertebrate Circadian Systems. Berlin: Springer-Verlag, 1982:253–261.

113. Czeisler CA. Human circadian physiology: internal organization of temperature, sleep-wake, and neuroendocrine rhythms monitored in an environment free of time cues. PhD dissertation, Stanford University, Stanford, CA, 1978.

114. Czeisler CA, Dumont M, Duffy JF, et al. Association of sleep-wake habits in older people with changes in output of circadian pacemaker. Lancet 1992; 340:933–936.

115. Strogatz SH. The Mathematical Structure of the Human Sleep Wake Cycle. New York: Springer-Verlag, 1986.

116. Strogatz SH, Kronauer RE, Czeisler CA. Circadian regulation dominates homeostatic control of sleep length and prior wake length in humans. Sleep 1986; 9:353–364.

117. Strogatz SH, Kronauer RE, Czeisler CA. Circadian pacemaker interferes with sleep onset at specific times each day: role in insomnia. Am J Physiol 1987; 253:R172–R178.

118. Chandrashekaran MK, Marimuthu G, Geetha L. Correlations between sleep and wake in internally synchronized and desynchronized circadian rhythms in humans under prolonged isolation. J Biol Rhythms 1997; 12:26–33.

119. Åkerstedt T, Gillberg M. A dose-response study of sleep loss and spontaneous sleep termination. Psychophysiology 1986; 23:293–297.
120. Hume KI, Mills JN. The circadian rhythm of REM sleep. J Physiol 1977; 270:32.
121. Hume K, Mills J. Rhythms of REM and slow-wave sleep in subjects living on abnormal time schedules. Waking Sleeping 1977; 1:291–296.
122. Weitzman ED, Czeisler CA, Zimmerman JC, Ronda JM. Timing of REM and stages 3 and 4 sleep during temporal isolation in man. Sleep 1980; 2:391–407.
123. Dijk D-J, Brunner DP, Borbély AA. Time course of EEG power density during long sleep in humans. Am J Physiol 1990; 258:R650–R661.
124. Dijk D-J, Cajochen C, Tobler I, Borbély AA. Sleep extension in humans: sleep stages, EEG power spectra and body temperature. Sleep 1991; 14:294–306.
125. Achermann P, Dijk D-J, Brunner DP, Borbély AA. A model of human sleep homeostasis based on EEG slow-wave activity: quantitative comparison of data and simulations. Brain Res Bull 1993; 31:97–113.
126. Lavie P, Scherson A. Ultrashort sleep-waking schedule. I. evidence of ultradian rhythmicity in 'sleep ability'. Electroencephogr Clin Neurophysiol 1981; 52: 163–174.
127. Weitzman ED, Nogeire C, Perlow M, et al. Effects of a prolonged 3-hour sleep-wake cycle on sleep stages, plasma cortisol, growth hormone, and body temperature in man. J Clin Endocrinol Metab 1974; 38:1018–1030.
128. Wyatt JK, Dijk D-J, Ronda JM, et al. Interaction of circadian and sleep-wake homeostatic-processes modulate psychomotor vigilance test (PVT) performance. Sleep Res 1997; 26:759.
129. Lavie P. Ultrashort sleep-waking schedule III. "Gates" and "forbidden zones" for sleep. Electroencephogr Clin Neurophysiol 1986; 63:414–425.
130. Wehr TA. The durations of human melatonin secretion and sleep respond to changes in daylength (photoperiod). J Clin Endocrinol Metab 1991; 73:1276–1280.
131. Wehr TA, Schwartz PJ, Turner EH, Feldman-Naim S, Drake CL, Rosenthal NE. Bimodal patterns of human melatonin secretion consistent with a two-oscillator model of regulation. Neurosci Lett 1995; 194:105–108.
132. Dijk D-J, Duffy JF, Czeisler CA. Circadian and sleep/wake dependent aspects of subjective alertness and cognitive performance. J Sleep Res 1992; 1:112–117.
133. Dijk D-J, Brunner DP, Borbély AA. EEG power density during recovery sleep in the morning. Electroencephogr Clin Neurophysiol 1991; 78:203–214.
134. Dijk D-J, Shanahan TL, Duffy JF, Ronda JM, Czeisler CA. Variation of electroencephalographic activity during nonREM and REM sleep with phase circadian melatonin rhythm in humans. J Physiol (Lond) 1997; 505:851–858.
135. Lavie P. Melatonin: role in gating nocturnal rise in sleep propensity. J Biol Rhythms 1997; 12:657–665.
136. Shochat T, Luboshitzky R, Lavie P. Nocturnal melatonin onset is phase locked to the primary sleep gate. Am J Physiol 1997; 273:R364–R370.
137. Tzischinsky O, Lavie P. Melatonin possess time-sdependent hypnotic effects. Sleep 1994; 17:638–645.
138. Dijk D-J, Cajochen C. Melatonin and the circadian regulation of sleep intiation, consolodation, structure, and the sleep EEG. J Biol Rhythms 1997; 12:627–635.
139. Dijk D-J, Visscher CA, Bloem GM, Beersma DGM, Daan S. Reduction of human

sleep duration after bright light exposure in the morning. Neurosci Lett 1987; 73: 181–186.

140. Campbell SS, Dawson D. Aging young sleep: a test of the phase advance hypothesis of sleep disturbance in the elderly. Sleep Res 1991; 20:447.

141. Campbell SS, Dawson D, Anderson MW. Alleviation of sleep maintenance insomnia with timed exposure to bright light. J Am Geriatr Soc 1993; 41:829–836.

142. Carrier J, Dumont M. Sleep propensity and sleep architecture after bright light exposure at three different times of day. J Sleep Res 1995; 4:202–211.

143. Czeisler CA, Kronauer RE, Allan JS, et al. Bright light induction of strong (type 0) resetting of the human circadian pacemaker. Science 1989; 244:1328–1333.

144. Jewett ME, Rimmer DW, Duffy JF, Klerman EB, Kronauer RE, Czeisler CA. The human circadian pacemaker is sensitive to light throughout subjective day without evidence of transients. Am J Physiol 1997; 273:R1800–R1809.

145. Boivin DB, Duffy JF, Kronauer RE, Czeisler CA. Dose-response relationships for resetting of human circadian clock by light. Nature 1996; 379:540–542.

146. Wehr TA, Giesen HA, Moul DE, Turner EH, Schwartz PJ. Suppression of men's responses to seasonal changes in day length by modern artificial lighting. Am J Physiol 1995; 269:R173–R178.

147. Aschoff J. Desynchronization and resynchronization of human circadian rhythms. Aerospace Med 1969; 40:844–849.

148. Aschoff J, Pöppel E, Wever R. Circadiane Periodik des Menschen unter dem Einfluss von Licht-Dunkel-Wechseln unterschiedlicher Periode. Pflugers Arch 1969; 306:58–70.

149. Dantz B, Edgar DM, Dement WC. Circadian rhythms in narcolepsy: studies on a 90-minute day. Electroencephal Clin Neurophysiol 1994; 90:24–35.

5

Influence of Light on Circadian Rhythmicity in Humans

CHARLES A. CZEISLER and KENNETH P. WRIGHT, JR.

Brigham and Women's Hospital
Harvard Medical School
Boston, Massachusetts

I. Introduction

Terrestrial life evolved with a sidereal period of Earth's revolution about the sun close to 365 days and an axial rotation near 24 hr. It is thus not surprising that nearly all organisms studied, including prokaryotes and eukaryotes, possess a(n) internal biological clock(s) with a near-24-hr periodicity. These internal clock mechanisms or endogenous pacemakers determine the appropriate environmental timing for species specific biological and behavioral events (e.g., sleep, endocrine function, temperature regulation). It is generally accepted that the periodic light-dark cycle, associated with the earth's daily axial rotation, is the dominant environmental synchronizer used by nearly all organisms to entrain to the 24-hr geophysical day. Entrainment is a fundamental property of endogenous oscillators by which the period of the pacemaker is synchronized to the period of the entraining stimuli. In the absence of periodic synchronizing agents, endogenous pacemakers oscillate at their own intrinsic period. Controlled experimental studies have shown the endogenous pacemaker in humans to cycle with an average period that is somewhat longer than 24 hr (1–4). Stable entrainment of the endogenous pacemaker to the light-dark cycle is achieved when the resetting response to light exposure offsets the drift due to the longer-than-24-hr intrinsic period of the human circadian pacemaker.

II. Neuroanatomy of the Mammalian Circadian Pacemaker

Studies of the neuroanatomy of the circadian pacemaker in mammals have identified the structural elements necessary to mediate circadian phase resetting by light. In mammals, the primary endogenous circadian pacemaker is located in the suprachiasmatic nucleus (SCN) of the hypothalamus, which is itself composed of individual neurons capable of self-sustained oscillation (5). Ablation of the SCN leads to a loss of endogenous rhythmicity. The activity rhythm can subsequently be restored by transplantation of a fetal SCN from another animal. When circadian rhythmicity is restored by SCN transplants, the intrinsic period is determined by the genotype of the donor animal (6). Photic information is communicated to the SCN from specialized retinal ganglion cells via a monosynaptic retinohypothalamic tract (RHT) and a multisynaptic pathway involving the geniculohypothalamic tract (GHT) (7). Comparable neuroanatomical structures to those demonstrated to drive circadian rhythmicity (the SCN) and mediate photic entrainment (the RHT) in other mammals have also been identified in human subjects (8–11), although the GHT may not be as well developed in humans (12). Less is known about the efferent pathways emanating from the SCN, especially in humans, although a number of efferent projections from the human SCN to nearby structures (such as the paraventricular nucleus and the dorsomedial nucleus, both of the hypothalamus, and the SCN itself) were recently shown to be similar to those observed in other mammals (13). A key efferent neural pathway by which photic information affects melatonin secretion involves structures as diverse as the intermediolateral cell column of the spinal cord and the superior cervical ganglion, linking the SCN to the pineal via sympathetic enervation. In addition to these direct neural connections, evidence from transplant studies indicates that the SCN may communicate an output signal sufficient to drive rhythmicity, in some but not all rhythmic variables, via a diffusible substance (14).

Results from several lines of research indicate that multiple independent pacemakers may exist in the circadian time-keeping system in many species, from single cellular algae (15–17) to mammals. Tosini and Menaker (18) reported a rhythmic pacemaker to exist in the mammalian retina. They showed a circadian rhythm in retinal melatonin synthesis in vitro that could be entrained by exposure to light. In addition, *tau* mutant hamsters, which have an intrinsic activity period that is shorter than wild-type hamsters, also showed a shorter period of retinal melatonin synthesis. Exciting progress has also been made in the genetic and molecular regulation of circadian oscillators, which has even resulted in a molecular model for the phase-resetting effects of light (19). It has also been reported that gene expression controls circadian oscillators and that expression of these genes can be observed in many brain and body areas (20–22). The contribution of these pacemakers outside of the SCN to mammalian circadian organization remains to be determined.

III. Assessment of Endogenous Rhythmicity in Humans

Accurate estimation of the phase and amplitude of the output of endogenous circadian pacemaker(s) is crucial for making inferences about the properties of human biological clock(s). Unfortunately, physiological responses evoked by environmental and behavioral stimuli can mask the endogenous component of many commonly observed cyclic processes, such as body temperature, melatonin, and cortisol levels (23–25). Over the past 15 years, Czeisler and colleagues developed and validated a method to assess in 1–2 days the phase and amplitude of the endogenous circadian pacemaker by monitoring endogenous circadian rhythms under constant environmental and behavioral conditions, using a modification of the constant routine (CR) technique first proposed by Mills (26–28). Figure 1 shows the body temperature cycle of a 66-year-old woman studied first under routine ambulatory conditions in the laboratory (scheduled sleep episode

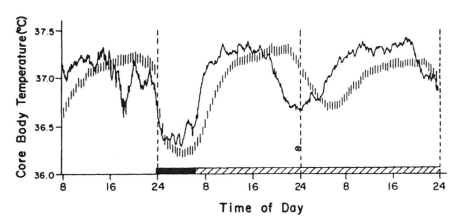

Figure 1 The core body temperature (solid line) of a healthy, 66-year-old woman under baseline (first 24 hr) and constant routine (remaining 40 hr) conditions. These data are superimposed upon average (±SEM) temperature data collected from 29 healthy young men on the same protocol (vertical hatch marks). Data from the controls are averaged with respect to their habitual bedtime, normalized to her bedtime of 24:00 hr. Black bar represents her bed rest episode. Hatched bar represents the duration of the constant routine. The encircled cross marks the nadir of the endogenous component of her core body temperature rhythm. (Reproduced from Ref. 27, with permission.)

between midnight and 6 A.M.) and then on a CR, during which time she remained continuously awake in a semirecumbent posture in artificial indoor room light, with meals taken at hourly intervals. During routine ambulatory conditions, her body temperature cycle appears to reach its nadir midway between bedtime and wake time, as would be expected for a normally entrained individual. However, during the CR, the nadir of the endogenous component of her body temperature cycle is revealed to be markedly phase-advanced as compared to that of a group of normal young men. This abnormality was obscured or masked under routine ambulatory conditions by both activity (which elicited an increase in her body temperature during the late evening) and sleep in a supine posture (which elicited a decrease in her body temperature in the latter half of the night).

Another technique that has been widely used to assess circadian phase (but not amplitude) has been estimation of the time at which melatonin levels first rise in the evening. Since light suppresses the nocturnal elevation of melatonin, this procedure must be carried out in dim light. Hence, it is called the dim light melatonin onset (DLMO) procedure (29). Given recent evidence that posture (30–32) and physical activity (33,34) may affect melatonin levels, and that the duration of melatonin secretion may reflect pacemaker amplitude (32,35), assessment of the entire melatonin profile under constant routine conditions (32,35,36) may yield the most precise estimates of the endogenous circadian phase and amplitude of the pacemaker driving that rhythm (presumed to be the SCN).

IV. Resetting Responses to Light

A. Resetting Effects of Light on Endogenous Rhythms

The circadian resetting response to light was first characterized 40 years ago by Hastings and Sweeney in a single-cell dinoflagellate, *Gonyaulax polyedra* (37). They demonstrated the relationship between the induced phase shifts and the phase of administration of the light stimulus (known as a phase-response curve, PRC) and the strength of the light stimulus (known as a dose-response curve, DRC). They found light stimuli induced a phase-delay shift when administered early in the subjective night, whereas that same stimulus administered late in the subjective night induced a phase-advance shift. Smaller phase shifts were observed when the light stimuli were administered during the subjective day. PRCs to light have since been observed to share similar properties across a wide array of species, from unicellular algae to primates (38,39), and more recently, humans (40–44). DRCs to light also share similar properties across species (45–50). The relationship between the strength of light stimuli and the magnitude of induced phase shifts generally follows a nonlinear function that is dependent upon the illuminance level (45,46,51) and the duration (47,51,52) of the light stimuli. The spectral composition of light stimuli also affects the induced phase shifts (53).

In humans, a number of studies have demonstrated that light is a powerful circadian synchronizer. In a case study, Czeisler et al. found that exposure to bright light could reset the circadian pacemaker independent of the timing of the sleep/wake cycle (27). Dijk et al. (54,55) reported that three consecutive mornings of exposure to bright light advanced the body temperature, melatonin, and sleep propensity rhythms by an hour, also when the timing of the sleep-wake cycle was held constant; Broadway et al. reported a similar ~2-hr phase advance of the melatonin rhythm following 6 weeks of exposure to bright light in the morning and evening during the Antarctic winter (56). Foret et al. (57) and Clodoré et al. (58) reported that the rhythms of cortisol, alertness, and performance were phase-advanced by repeated morning exposure to bright light. Honma et al. (59) reported phase advances of the melatonin and temperature rhythms after repeated exposures to bright light. Drennan et al. (60) and Lewy et al. (61) reported that the body temperature cycle and the onset of nocturnal melatonin secretion, respectively, were delayed by 2–3 hr after 3–7 consecutive evenings of bright light, respectively, even when the timing of sleep was held constant. Lack et al. (62) reported two nights of bright light exposure to produce a phase delay in the temperature rhythm and delay the time of awakening in early-morning-awakening insomniacs. Campbell et al. (63) reported that elderly individuals with sleep-maintenance insomnia, who were exposed to evening bright light across multiple days, exhibited phase delays in their temperature nadir and latency to REM sleep and improvement in their sleep quality (reduced wake after sleep onset, increased sleep efficiency and stage 2 sleep, reduced number of awakenings). Murphy and Campbell (64) also reported 3 months of bright-light exposure to delay the body temperature rhythm and improve sleep efficiency in sleep-maintenance insomniacs. Anderson et al. (65) and Brunner et al. (66) reported that 5 days of bright-light treatment improved sleep efficiency in people suffering from seasonal affective disorder. Carrier and Dumont (67) and Dumont and Carrier (68) reported 3 days of exposure to evening bright light phase-delayed sleep-onset latency and the temperature rhythm compared to similar exposures to morning or afternoon light. Tzischinsky and Lavie (69) reported evening exposure to 2 hr of bright light for 5 days to delay the acrophase in oral temperature and the rhythm in sleep propensity. Wever reported that bright light (~4000 lux) could expand the physiological range of entrainment, although he concluded that light below 1500 lux exerts no direct physiological effects (70) other than through an influence on the timing of behavior (71). Furthermore, Honma et al. reported entrainment of human circadian rhythms by the light/dark cycle (72) and phase-dependent shifts of human circadian rhythms in response to single bright-light pulses (43), although the responses they observed were largely limited to phase-advance shifts, leading them to question whether the human circadian pacemaker could achieve phase-delay shifts in normal subjects (43). Burešová et al. (74), Kennaway et al. (75), Minors et al. (42), Dawson et al. (76), van Cauter et al. (77) and Jewett et al. (44) have

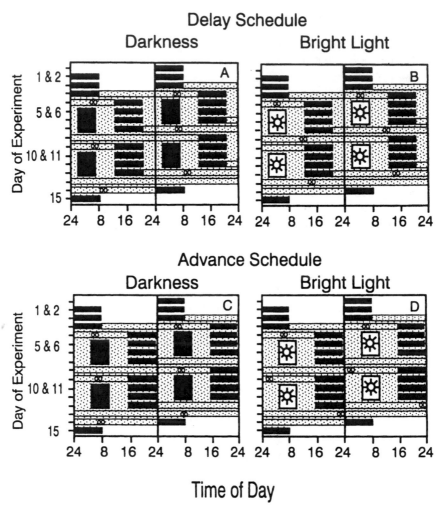

Figure 2 Double raster plot of average phase shifts observed after two series of exposures to three-cycle stimuli consisting of 5 hr of exposure to bright light (7000–13,000 lux) or darkness (<0.03 lux) per day. Time of day is plotted on the x-axis and successive days of the experiment are plotted both beside and underneath each other. Filled bars represent scheduled bed rest episodes, stippled bars represent constant routine (CR) procedures, stippled area represents ambient lighting of 10–15 lux, open boxes with sun symbol represent 5 hr/day of bright light, shaded boxes represent 5 hr/day of darkness, ⊗ represent the estimated circadian phase of the temperature minimum (ECP_{min}) during each constant routine. SEM are smaller than the symbols used. Delay schedule: The center of both the wake episode and the 5-hr stimulus was scheduled 22.5 hr after the ECP_{min} as measured on CR1. The center of the 8-hr bed rest episode was scheduled 12 hr opposite the center of the wake episode. This protocol resulted in the rest-activity schedule being shifted by an average of 12.2 hr later than subjects' habitual bedtimes. (A) Results for darkness group: CR1-CR2, 0.26 ± 0.48 h phase delay; CR2-CR3, 2.55 ± 0.74 h phase delay. (B) Results for the bright-light group: CR1-CR2, 6.91 ± 0.49 h phase delay; CR2-CR3, 2.41 ± 0.35 h phase delay. Advance schedule: The center of both the wake episode and the 5-hr stimulus

reported modest phase-shifting responses to single bright-light exposures. For a recent review of studies that employed melatonin as a phase marker, see Shanahan et al. (35).

Daily exposure to bright light for 3 consecutive days can reset the endogenous circadian temperature, cortisol, melatonin, alertness, performance, and TSH rhythms by up to 12 hr (32,36,40,78–83). Figure 2 shows phase resetting in humans in response to two series of three-cycle stimuli consisting of exposure to bright light or darkness. Circadian phase assessments were obtained using the constant routine procedure before, between, and after the series of light exposures. In groups A (darkness) and B (bright light) the timing of the light pulse was centered at a phase of the human PRC that was predicted to produce a phase delay, whereas the timing of the light pulse in groups C (darkness) and D (bright light) was centered at a phase predicted to produce a phase advance. Both groups who were exposed to bright light exhibited large shifts in phase whereas the groups exposed to dim light and darkness showed modest phase delays consistent with the longer-than-24-hr intrinsic period of the human circadian pacemaker.

In humans, blindness is often associated with loss of entrainment (84–89) of endogenous circadian rhythms and this loss of entrainment is associated with cyclic recurring sleep disruption. Blind subjects in whom light input to the SCN is preserved, as assayed by light-induced suppression of plasma melatonin (see Section V), remain synchronized to the environmental light-dark cycle (90,91). There is a nonlinear relationship between the light intensity and the resetting response, such that ordinary indoor room light can exert a significant resetting effect on the human circadian pacemaker (50). Furthermore, humans are sensitive to light throughout the subjective day (92). Taken together, these data demonstrate that the human circadian pacemaker is exquisitely sensitive to light as a circadian synchronizer (81).

B. Phase-Response Curve to Light

Two general PRC morphologies have been reported: a low-amplitude PRC (with maximal phase shifts of only a few hours), and a high-amplitude PRC (with phase shifts as great as $\pm$ 12 hr). Winfree (93) has designated these two classes of PRCs as reflecting weak type 1 and strong type 0 phase resetting, respectively, based on the topology of their resetting contours (see below). Both type 1 and type 0

---

was scheduled 1.5 hr after the ECP_{min} as measured on CR1. The center of the 8-hr bed rest episode was scheduled 12 hr opposite the center of the wake episode. This protocol resulted in the rest-activity schedule being shifted by an average to 8.4 hr earlier than subjects' habitual bedtimes. (C) Results for darkness group: CR1-CR2, 1.05 $\pm$ 0.38 h phase delay; CR2-CR3, 0.69 $\pm$ 0.44 h phase delay. (D) Results for the bright-light group: CR1-CR2, 4.75 $\pm$ 0.41 h phase advance; CR2-CR3, 2.85 $\pm$ 0.31 h phase advance. (Adapted from Ref. 82, with permission.)

resetting have been observed in many organisms from unicellular algae to mammals. Species differences in the resetting response to light have been found in the magnitude or strength of stimulus necessary to produce type 1 or type 0 resetting. For example, 12 hr of exposure to extremely bright light (~80,000 lux) is necessary to produce type 0 resetting in the cockroach *Leukophae maderae* (94), whereas 55 sec of exposure to dim blue light (1 W/m^2) can produce type 0 resetting in the pupal eclosion rhythm of *Drosophila pseudoobscura* (95).

Strong Type 0 Resetting

In strong type 0 resetting, suppression of circadian amplitude occurs en route to achieving the largest phase shifts in response to light. Type 0 resetting has most often been reported in lower organisms. Until recently, it was believed that higher organisms, such as mammals, were limited to weak type 1 resetting (93). The magnitude and shape of the resetting contour observed after bright-light exposure indicates that human subjects are capable of strong type 0 resetting (40,78,79), and can be reset to any desired phase by scheduled exposure to light for 2–3 days. This type 0 PRC for human subjects is qualitatively indistinguishable from the type 0 PRCs found in insects and plants and indicates that the responsiveness of the human circadian pacemaker to light is within the range of sensitivity observed in lower organisms (81). Figure 3 (right panels) shows the type 0 human PRC to bright light. Phase shifts in response to a three-cycle stimulus of bright light (~7000–10,000 lux) are very large when the light stimulus is centered near the initial circadian minimum. The farther away from the minimum the light stimulus is placed, the smaller the phase shift. Figure 4 (right panel) shows the similarity in the type 0 resetting contour for the mosquito *Culex pipiens quinquefasciatus* from the work of Peterson and Saunders (96) to that observed in the human (Fig 3, upper right panel).

Weak Type 1 Resetting

In every organism for which type 0 resetting has been found, a stimulus of reduced strength has been shown to lead to type 1 resetting (79,93). Kronauer's mathematical model of the effect of light on the human circadian pacemaker predicts that reduction of the strength of the light stimulus will result in the induction of weak (type 1) resetting (78,79,97). Characterization of the type (1 or 0) of resetting response to a light stimulus of given strength is important since it affects the topology of a PRC and hence any therapeutic strategies designed to reset phase. Minors et al. (42) and Honma and Honma (43) have reported results consistent with type 1 resetting in human subjects after exposure to a single bright-light stimulus during free-running conditions. Unfortunately, the results of their studies were confounded by masking effects (79,81) and the strength and duration of the light exposures were uncertain (43). Burešová et al. (74), Kennaway et al. (75), Dawson et al. (76), and van Cauter et al. (77) have also reported modest phase

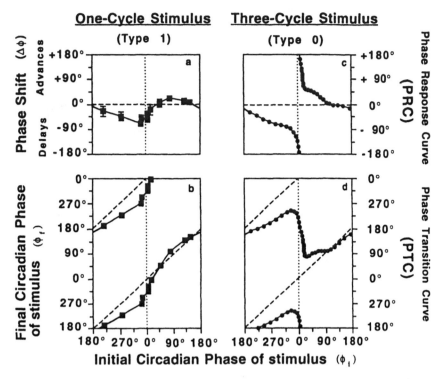

Figure 3 Human PRCs to light. (a) Smoothed PRC for a one-cycle bright-light stimulus (light intensity: 7000–12,000 lux; stimulus duration: 5.24 ± 0.22 hr). The magnitude and direction of the phase shifts achieved are plotted against the phase of the stimulus. The nadir of the fitted endogenous temperature cycle is assigned a phase of 0°. Phase advances are represented as positive numbers and phase delays as negative numbers. Error bars show SEMs for the points in which the error is larger than the symbol used. (b) Smoothed phase-transition curve (PTC) to the one-cycle data, in which the final phase of the stimulus is plotted against the initial phase. (c) Smoothed PRC to a three-cycle bright-light stimulus (light intensity: 7000–12,000 lux; stimulus duration: 5 hr). SEMs are generally the same size as the symbols used, and are thus not included. (d) Smoothed PTC to the three-cycle data. (Adapted from Ref. 44, with permission.)

shifts to a single light exposure, consistent with the type 1 resetting predicted by Kronauer's model after exposure to a single bright-light stimulus (78,79). Jewett et al. (44) have reported the results of a preliminary series of experiments exploring the response of the human circadian pacemaker to a single episode of bright-light exposure across circadian phases. Results for these 11 trials reveal a type 1 resetting contour that is consistent with the predictions of the Kronauer model (78,98). Figure 3 (left panels) shows the type 1 human PRC to bright light. A single exposure to bright light (~7000–10,000 lux) produced small delay shifts

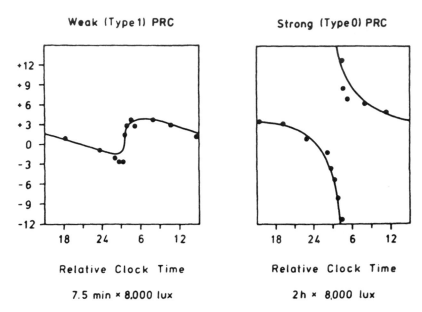

Figure 4 Light PRCs for the mosquito *Culex pipiens quinquefasciatus*. (a) Smoothed PRC for a one-cycle bright-light stimulus (light intensity: 8000 lux; stimulus duration: 7.5 min). The magnitude and direction of the phase shifts achieved are plotted against the phase of the stimulus. The evening activity peak (flight activity) is assigned a phase of 0°. (b) Smoothed PRC to a one-cycle bright-light stimulus (light intensity: 8000 lux; stimulus duration: 2 hr). (Adapted from Ref. 96, with permission of the publisher, Academic Press Limited, London.)

when centered before the temperature minimum and small advance-phase shifts after the temperature minimum. Figure 4 (left panel) reveals the similarity in the type 1 resetting contour of the mosquito to that observed in the human (Fig. 3, left upper panel). However, additional experiments will be required to quantify precisely the response of the human circadian pacemaker to a single light pulse at all circadian phases. In human subjects, the critical zone of stimulus application that is essential for an experimental discrimination between type 1 or type 0 resetting is only a few hours wide, centered near the endogenous temperature minimum, about 2 hr before habitual wake time.

Sensitivity to Light During the Subjective Day

In some animals, such as the nocturnal deermouse *Peromyscus leucopus*, a zone of insensitivity to the resetting effects of bright light has been reported to occur during the subjective day (99). In these animals, a light stimulus administered during the so-called "dead zone" does not appear to phase-shift the circadian pacemaker. Other animals, such as the diurnal red squirrel *Tamiasciurus hudsonicus*, do not exhibit dead zones in their PRCs and thus are sensitive to light at all circadian phases (100). As described above, the human circadian pacemaker appears to be sensitive to light throughout the subjective day and night with no apparent dead zone of insensitivity (92). Hashimoto et al. (101) reported exposure to midday bright light for 3 days phase-advanced the nocturnal rise in the melatonin rhythm and increased the area under the curve of melatonin secretion compared to dim light exposure. These data suggest that light exposure across the waking day contributes to circadian entrainment in humans.

Amplitude Suppression

Experiments in single-cell organisms (102,103), plants (104), insects (95,105,106), and humans (97) have verified that in organisms that can achieve type 0 resetting, a light stimulus of specific timing, duration, and intensity can attenuate circadian amplitude such that the organism appears to be near the theoretical "singularity" region (97). An analysis of resetting trials with one, two, or three consecutive cycles of light exposure has revealed that the human circadian timing system behaves as a nonsimple oscillator with at least two state variables (e.g., phase and amplitude) (44,78,79). This implies that light exposure can affect both phase and amplitude of endogenous circadian rhythms (44,97). Indeed, critically timed exposure to light centered on the nadir of the endogenous circadian rhythm of core body temperature rhythm can markedly reduce the amplitude of core body temperature, plasma cortisol, and melatonin rhythms (32,97). Figure 5 shows the temperature and cortisol rhythms of an 18-year-old man before and after exposure to a two-cycle stimulus of bright light. The prestimulus CR shows the commonly observed nocturnal nadir in temperature and peak in cortisol secretion. After exposure to the bright-light stimulus, the fitted temperature and cortisol amplitudes were markedly reduced.

C. Dose-Response Curve to Light and Wavelength Sensitivity

In 1990, Kronauer (98) hypothesized that the relationship between the phase-shifting effect of light and its illuminance levels in human subjects is nonlinear, with resetting preserved despite marked reduction in light intensity. In fact, Boivin et al. recently confirmed this hypothesis (50,107). They showed that ordinary room light (~180 lux) can significantly reset the human circadian pacemaker and found that the strength of the resetting response to bright light is dependent on the illuminance level of the light stimulus. These data are consistent with the earlier report (40) that the timing of room light (~150 lux) can substantially modulate the magnitude and direction of phase shifts produced by 5-hr bright-light exposures (~7000–13,000 lux) . Additional data are needed to evaluate the phase-resetting effects of light with an intensity below that of ordinary room light on the circadian pacemaker in humans.

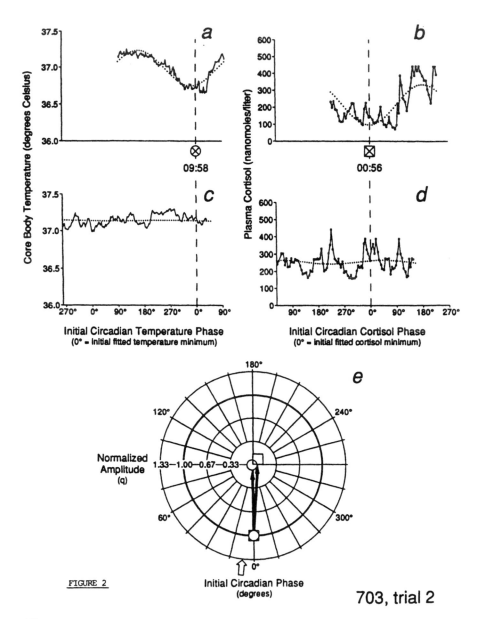

FIGURE 2

703, trial 2

Figure 5 Endogenous circadian rhythms of core body temperature and plasma cortisol levels measured in an 18-year-old man. (a) A dual harmonic regression model (dotted line) was fitted using nonlinear least squares to the temperature data (solid line) collected during the initial prestimulus constant routine to identify the fitted temperature minimum (encircled cross). That fitted minimum has been assigned a reference value of 0 degrees on

Zeitzer et al. (108) examined the wavelength sensitivity of the resetting response to light of the human circadian pacemaker. They found that stimulation of the cone photoreceptive system that mediates photopic responses to light is sufficient to induce phase shifts in both melatonin and core body temperature rhythms (108). Figure 6 shows similar phase shifts in both temperature (A) and melatonin (B) rhythms after exposure to white ordinary room light and moderately bright red light. Since the red light stimulus that was used is outside the spectral sensitivity of rhodopsin-like photoreceptors, it was concluded that a cone-based or a novel photoreceptor in the mammalian eye is sufficient to mediate human circadian photoreception. In fact, recent data suggest that a novel opsin may mediate circadian photoreception via an unknown photoreceptor (189).

V. Acute Effects of Light

In 1978, Wetterberg (109) reported that exposure to light while remaining awake at night could markedly diminish the nocturnal secretion of melatonin in a 25-year-old man. Lewy et al. (110) demonstrated in a group of subjects that elevated nocturnal plasma melatonin levels could be acutely suppressed during exposure to bright light (1500–2500 lux), and that melatonin secretion would resume upon return to darkness. It has since been demonstrated that a significant reduction of melatonin secretion at night can be achieved with even lower light intensities (<200 lux) (111). The relationship between the melatonin-suppressing effect of light at night and its illuminance levels in humans has been described by a dose-response curve (112–115). Wavelength-specific light-induced melatonin suppression in humans has been described by Brainard et al. (116). They reported that melatonin suppression in humans was greater in response to middle-wavelength-light exposure (i.e., green light) than to longer-wavelength light (i.e., red light) or

---

the abscissa. The initial fitted amplitude of the single harmonic component of the model was 0.24°C. (b) The same model was used to estimate the amplitude and phase of the circadian component of the cortisol values collected during the original CR. The amplitude of the fitted curve for cortisol was 119 nmol/L. During the poststimulus constant routine the fitted temperature amplitude (c) was reduced to 0.005°C, and the fitted amplitude of the cortisol rhythm (d) was reduced to 11 nmol/L. (e) A polar representation of the above data in which circadian phase is expressed in degrees (of both temperature and cortisol cycles) and the fractional reduction in amplitude (q) indicated on the radius, with the initial amplitude of both the temperature (open circle) and cortisol rhythm (open box) normalized to a value of 1.00. Polar vectors (filled arrows) represent changes in amplitude and phase of the oscillations. The open arrow represents the circadian phase of the weighted midpoint of the total daily light exposure referenced to the temperature cycle. (Reproduced from Ref. 97, with permission.)

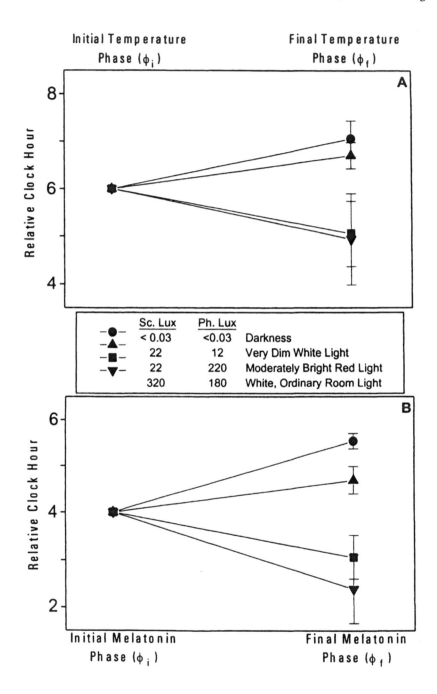

possibly shorter-wavelength light (i.e., blue light). Humans with color blindness, however, exhibit melatonin suppression similar to that of normal-sighted individuals (117). The latter data and data from studies of the blind showing that some blind individuals, without conscious light perception, maintain input into the circadian pacemaker (Fig. 7) suggest that the photoreceptors necessary for conscious light perception, although perhaps sufficient for circadian photoreception, may not be necessary for circadian photoreception. Figure 7 shows melatonin suppression in response to a 90–100-min pulse of 10,000 lux of light in a blind individual with no conscious light perception (right panel). When the same individual wears a blindfold over his eyes (left panel), there is no suppression of melatonin in response to the light pulse.

Badia et al. (118) were among the first to demonstrate the effect of bright light on nighttime body temperature levels. By exposing individuals to alternating blocks of bright and dim light every 90 min, they demonstrated that temperature levels increased during exposure to bright light and decreased during exposure to dim light. Unlike melatonin, a dose-response curve to light for effects on nighttime temperature levels has not been demonstrated (119). However, attempts to document such a dose-response relationship have been limited to the evening hours, before melatonin levels remain consistently elevated.

Exposure to bright light can also modestly improve performance during the subjective night. When melatonin is suppressed and temperature increased, it has been reported that cognitive performance is enhanced by exposure to bright light during the nighttime work hours and across sleep deprivation (119–126). Figure 8 shows the effects of nighttime exposure to ~2000 lux of bright light versus dim room light <100 lux on melatonin, body temperature, and performance across two nights of sleep deprivation. During continuous exposure to all-night bright light, melatonin levels are lower and temperature and performance levels are higher than during dim light. The effects of bright light on temperature and performance appear to be limited to the nighttime hours; exposure to bright light outside of the nocturnal melatonin secretory interval has little effect on temperature and performance (118,124). Some studies have also shown nocturnal bright-light exposure to produce signs of EEG arousal (118,124) while others report no effect of bright light on EEG signs of alertness (119,121,127).

Figure 6 Light-induced phase shifts of the temperature (A) and melatonin (B) rhythms. Initial phase is defined as the fitted temperature nadir or the melatonin midpoint assessed during the first constant routine. Final phase was assessed during the second constant routine after exposure to 3 consecutive days of 5 hr of very dim white light, moderately bright red light, normal room light, and darkness. Phase change ($\pm$SEM) was calculated by subtracting the initial from the final phase estimate. (Reproduced from Ref. 108, with permission.)

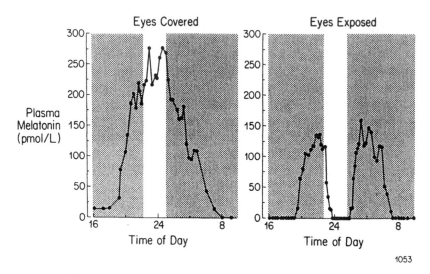

Figure 7 Results of melatonin-suppression test with and without a blindfold in a blind subject. Plasma melatonin concentrations did not fall during the bright-light exposure when the blindfold was in place (left panel), but they fell abruptly during the exposure when the subject's eyes were not blindfolded (right panel). (Reproduced from Ref. 90, with permission.)

Acute effects of bright light on sleep have also been described. Dijk et al. (128) and Cajochen et al. (127) reported that exposure to bright light for 4 hr immediately prior to bedtime increased sleep-onset latency and temperature levels during sleep with no effect on sleep staging. However, some changes in electroencephalographic (EEG) power density during sleep were observed after exposure to bright light. Bunnell et al. (129) reported that a single 2-hr pulse of bright light prior to bedtime increased temperature levels during sleep, delayed rapid-eye-movement sleep, and changed EEG power.

At present, its is unclear whether the acute effects of bright light on melatonin, temperature, performance, and sleep reflect a direct action of light on these variables or are related to changes in the amplitude and phase of the endogenous circadian pacemaker. In fact, Lakin-Thomas has proposed that with respect to melatonin, both may reflect the same process (130).

VI. Photoperiodic Effects of Light

Seasonal changes in the daily light exposure pattern in the temperate climates of earth play an important role in regulating reproductive cycles in seasonal breeding animals as well as other seasonal physiological and behavioral events (131–135).

Broadway et al. (56) and Vondrasová et al. (136) reported seasonal changes in the melatonin rhythm in humans. Wehr et al. (131,132,137) reported that the duration of melatonin secretion in humans is longer after chronic exposure to long nights and shorter after exposure to short nights. These studies suggest that the capacity to respond to seasonal alterations in light exposure may be conserved in humans. Whether humans utilize information from seasonal changes in light exposure remains to be determined. The use of artificial lighting, however, separates humans from other life forms on earth in that we exercise voluntary control over our daily light exposure, deciding when the lights go on or off regardless of the time of day. The biological impact of this self-selected light exposure, which has only been available to us as a species for an instant on the evolutionary time scale, is not known.

VII. Future Directions

A. Applications to Shift Work

Bright-light exposure has been used in various clinical settings to induce physiological adaptation in patients suffering from circadian rhythm disorders (138). These syndromes form a distinct class of sleep and arousal disorders and are characterized by a misalignment between the sleep/wake schedule and the endogenous circadian pacemaker, which is a major determinant of the timing of sleep, sleep stages, subjective alertness, and cognitive performance (139–146). For instance, misalignment of circadian phase and work-sleep schedules can overpower the night-shift worker's ability to remain awake and attentive while at work and to sleep during available daytime hours (147–151), which can lead to impaired job performance and higher rates of accidents and injuries (152–154). Exposure to bright light and darkness can induce rapid physiological adaptation of the circadian pacemaker to a single week of night work (83) and can facilitate rapid entrainment to a rotating work schedule (155). Figure 9 shows the plasma melatonin rhythm in two subjects on a rotating shift schedule. The individual who received bright-light treatment during the night work shift showed adaptation of his melatonin rhythm to the rotating shift schedule. In contrast, the melatonin rhythm of the control subject who was exposed to normal room light did not adapt to the night shift, although other control subjects did show substantial adaptation.

Czeisler et al. also implemented a bright-light treatment program for crew members of NASA's Space Shuttle (beginning with STS-35), which produced successful realignment of the endogenous circadian pacemaker driving the melatonin rhythm with the new sleep/wake cycle required for the dual-shift mission (156), a procedure that is now routinely used on nearly all shuttle flights requiring a ≥3-hr schedule displacement. Finally, data from clinical studies suggest that evening exposure to bright light and early-morning exposure to bright

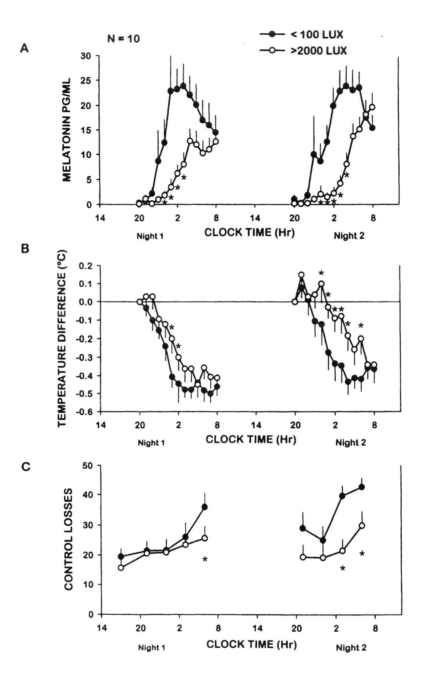

light are successful in the treatment of advanced sleep phase syndrome (63,157–159) and delayed sleep phase syndrome (27,157,158,160–163), respectively. This appears to be well tolerated by patients, providing a more rapid and practical treatment than chronotherapy (164).

B. Nonphotic Synchronizers

While light is the most powerful synchronizer of circadian rhythms in mammals, several other stimuli such as food availability and activity have been reported to act as circadian synchronizers in nonhumans (165–170). Whether nutrient intake can entrain human circadian rhythms is unknown. Several studies have reported that exercise may shift circadian rhythms in humans (77,171–173), whereas exercise distributed evenly across all circadian phases does not appear to affect the period of the human pacemaker (174). Other nonphotic cues such as the timing of the sleep-wake cycle, knowledge of clock time, and social cues are reported to be weak synchronizers of human circadian rhythms, especially compared to exposure to light (2,82). However, there is substantial evidence that nonphotic cues may be able to maintain circadian entrainment to the 24-hr day in many totally blind (complete loss of both sight and circadian photoreception) persons (175).

Amir and Stewart (176) recently reported that the response to light of the mammalian circadian timekeeping system can be classically conditioned to occur in response to a neutral nonphotic stimulus. They showed that exposure to a 20-min gentle air stimulus is, by itself, capable of activating Fos expression in the SCN and also phase-shifted the activity and temperature rhythms of rats after the animal learned that the air stimulus predicted subsequent exposure to light.

Administration of exogenous melatonin has been reported to phase-shift human circadian rhythms (177–181). However, the strength of the evidence supporting the potential use of exogenous melatonin alone as a synchronizing agent in humans has been questioned, since melatonin is unable to entrain free-running circadian rhythms in blind individuals and since none of the studies claiming to demonstrate phase-shifting effects of exogenous melatonin administration carefully controlled for exposure to light (182). Dawson et al. (183)

Figure 8 Salivary melatonin (A), tympanic temperature (B), and dual task control loss performance (C) during two nights of sleep deprivation. Melatonin and temperature data were collected in hourly intervals from 20:00 to 08:00 hr each night. Temperature data are presented as a difference score from 20:00 hr baseline on night 1. Cognitive performance was assessed every 3 hr each night. Higher scores on the dual task control loss performance task represent worse performance. During the daytime hours all subjects were exposed to <100 lux. Between 20:00 and 08:00 hr, subjects in the bright-light condition were exposed to >2000 lux. Error bars represent standard error of the mean. *Significant difference between dim and bright light exposure. (Adapted from Ref. 120, with permission.)

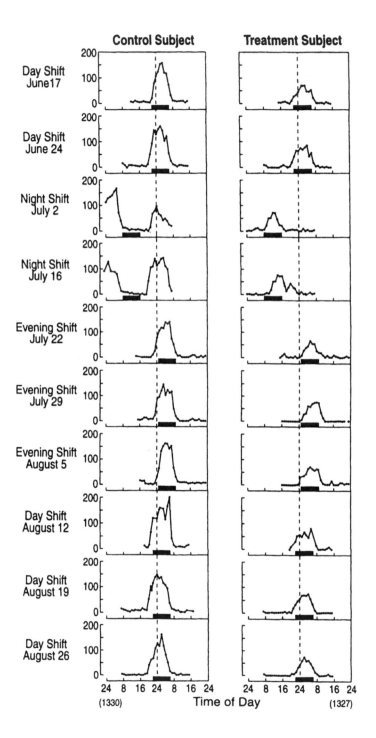

Time of Day

controlled subjects' light exposure and reported no difference between exogenous melatonin and placebo in their ability to produce a phase-delay shift during a simulated night shift protocol. Melatonin is also unable to prevent the phase-shifting effects of bright light in humans (184). These results suggest that if melatonin alone does synchronize human circadian rhythms, the effect is likely to be much weaker than that of bright light.

C. Limits of Current Knowledge

While much is known about the effects of light exposure on circadian phase resetting, very little is known about the minimum light intensity needed to maintain synchronization of the circadian pacemaker to the 24-hr day. Little research has been conducted on photic entrainment in humans, in part because these studies require subjects to remain in the laboratory under controlled conditions for months at a time. Early research examining circadian entrainment in humans were compromised by lack of control over light exposure (reviewed in Ref. 81). Czeisler et al. have conducted two preliminary experiments to evaluate whether human circadian rhythms could be entrained by exposure to ordinary room light (100–200 lux) and darkness (185,186). A strict light-dark cycle that was comparable to that used in entrainment studies of plants and animals was used. Results show that 16 hr of exposure to ordinary room light alternating with 8 hr of exposure to total darkness could entrain both the rest-activity and body temperature cycles of human subjects to a 24-hr period, as reported in animal studies (185,186). However, while it was found that exposure to a cycle of ordinary indoor room light alternating with darkness was able to entrain the human circadian system, in the absence of exposure to brighter outdoor light the

---

Figure 9 Rotating shift work simulation. After 2 weeks of baseline evaluation (sleep episode 22:00–6:00; work 07:00–15:00), subjects worked 3 weeks of night shift (sleep 08:00–16:00; work 23:00–07:00), 3 weeks of the evening shift (sleep 01:00–09:00; work 15:00–23:00), and 3 weeks of the day shift (same as baseline). The treatment subject (right) received bright light (up to 5000 lux) during the work shift; the control subject (left) remained in room light (~150 lux) during the work shift. Treatment subjects were provided blackout shades for their bedrooms at home to ensure complete darkness during scheduled sleep episodes. Vertical dashed line represents local midnight; solid horizontal bars represent sleep episodes. The melatonin rhythm in the bright light-treated subject peaked during the sleep episode on all work rotations, and this adaptation to rotating shifts was evident during the first week of a new work schedule. The melatonin rhythm in the control subject continued to peak during the hours of 22:00–10:00, despite multiple shifts of his sleep-wake cycle. Therefore, the rhythm in this control subject failed to adapt to the rotating shift work schedule and often peaked during the work/wake episode, although other control subjects showed substantial adaptation during their night shift rotations. (Reproduced from Ref. 35, with permission.)

phase relationship between the underlying rhythm and the imposed light-dark cycle appears to have been delayed, although constant routine phase assessments were not conducted at that time. Under these conditions, sleep efficiency appears to have been adversely affected by entrainment to this weakened synchronizer. Additional research is needed to determine the minimum light intensity necessary to entrain human circadian rhythms. In addition, the relationship between light intensity and the phase angle of entrainment has not been derived from humans as it has for other species (187,188). Finally, the effect of individual differences in intrinsic period on entrainment and phase resetting by light remains to be explored.

Acknowledgments

We thank Dr. Derk-Jan Dijk for helpful discussions regarding the manuscript and Dr. Christian Cajochen and Jeanne Duffy for their assistance in preparing the figures. The work described in this article was supported in part by Grants NIMH-1-R01-MH45130 from the National Institute of Mental Health; NAG9-524 from the National Aeronautics and Space Administration; NIA-1-R01-AG06072 and NIA-P01-AG09975 from the National Institutes on Aging; the National Space Biomedical Research Institute; by General Clinical Research Center Grant NCRR-GCRC-M01-RR02635 from the National Center for Research Resources; by the National Geographic Society and the Brigham and Women's Hospital; K.P.W. has been supported by a a Non-Service Fellowship Award of the Bowling Green State University and an Institutional National Research Service Award by Army Research Institute MDA 903-93-K-002, U.S. Army Medical Research Acquisition and Activity DAMD17-95-2-5015; and the Ohio Board of Regents Selective Excellence Program.

References

1. Czeisler CA, Duffy JF, Shanahan TL, et al. Reassessment of the intrinsic period (τ) of the human circadian pacemaker in young and older subjects. Sleep Res 1995; 24A:505.
2. Middleton B, Arendt J, Stone BM. Human circadian rhythms in constant dim light (8 lux) with knowledge of clock time. J Sleep Res 1996; 5:69–76.
3. Campbell SS, Dawson D, Zulley J. When the human circadian system is caught napping: evidence for endogenous rhythms close to 24 hours. Sleep 1993; 16(7):638–640.
4. Hiddinga AE, Beersma DG, van den Hoofdakker RH. Endogenous and exogenous components in the circadian variation of core body temperature in humans. J Sleep Res 1997; 6:156–163.
5. Welsh DK, Logothetis DE, Meister M, Reppert SM. Individual neurons dissociated from rat suprachiasmatic nucleus express independently phased circadian firing rhythms. Neuron 1995; 14:697–706.

6. Ralph MR, Foster RG, Davis FC, Menaker M. Transplanted suprachiasmatic nucleus determines circadian period. Science 1990; 247:975–978.
7. Moore RY. Organization of the mammalian circadian system, In: Waterhouse JM, ed. Circadian Clocks and Their Adjustment. Chichester (Ciba Foundation Symp 183): Wiley, 1994:88–99.
8. Lydic R, Schoene WC, Czeisler CA, Moore-Ede MC. Suprachiasmatic region of the human hypothalamus: homologue to the primate circadian pacemaker? Sleep 1980; 2:355–361.
9. Sadun AA, Schaechter JD, Smith LEH. A retinohypothalamic pathway in man: light mediation of circadian rhythms. Brain Res 1984; 302:371–377.
10. Stopa EG, King JC, Lydic R, Schoene WC. Human brain contains vasopressin and vasoactive intestinal polypeptide neuronal subpopulations in the suprachiasmatic region. Brain Res 1984; 297:159–163.
11. Moore RY, Speh JC. A putative retinohypothalamic projection containing substance P in the human. Brain Res 1994; 659:249–253.
12. Moore RY. The organization of the human circadian timing system. Prog Brain Res 1992; 93:101–117.
13. Dai J, Swaab DF, Buijs RM. Distribution of vasopressin and vasoactive intestinal polypeptide (VIP) fibers in human hypothalamus with special emphasis on suprachiasmatic nucleus effect projections. J Comp Neurol 1997; 383:397–414.
14. Silver R, Lesauter J, Tresco PA, Lehman MN. A diffusible coupling signal from transplanted suprachiasmatic nucleus controlling circadian locomotor rhythms. Nature 1996; 382:810–813.
15. Roenneberg T, Morse D. Two circadian oscillators in one cell. Nature 1993; 362:362–364.
16. Morse D, Hastings JW, Roenneberg T. Differential phase responses of two circadian oscillators in *Gonyaulax*. J Biol Rhythms 1994; 9:263–274.
17. Johnson CH, Hastings JW. The elusive mechanism of the circadian clock. Am Sci 1986; 74:29–36.
18. Tosini G, Menaker M. Circadian rhythms in cultured mammalian retina. Science 1996; 272:419–421.
19. Myers MP, Wager-Smith K, Rothenfluh-Hilfiker A, Young MW. Light-induced degradation of TIMELESS and entrainment of the *Drosophila* circadian clock. Science 1996; 271:1736–1740.
20. Tei H, Okamura H, Shigeyoshi Y, et al. Circadian oscillation of a mammalian homologue of the *Drosophila period* gene. Nature 1997; 389:512–516.
21. Sun ZS, Albrecht U, Zhuchenko O, Bailey J, Eichele G, Lee CC. RIGUI, a putative mammalian ortholog of the *Drosophila period* gene. Cell 1997; 90:1003–1011.
22. Plautz JD, Kaneko M, Hall JC, Kay SA. Independent photoreceptive circadian clocks throughout *Drosophila*. Science 1997; 278:1632–1635.
23. Waterhouse J, Minors D, Redfern P. Some comments on the measurement of circadian rhythms after time-zone transitions and during night work. Chronobiol Int 1997; 14:125–132.
24. Waterhouse J, Minors D, Åkerstedt T, Hume K, Kerkhof G. Circadian rhythm adjustment: difficulties in assessment caused by masking. Path Biol (Paris) 1996; 44:205–207.

25. Czeisler CA, Klerman EB: Circadian and sleep-dependent regulation of hormone release in humans. Recent Prog Horm Res 1998; (in press).

26. Czeisler CA, Khalsa SBS. The human circadian timing system and sleep-wake regulation. Principles Pract Sleep Med 1998; (in press).

27. Czeisler CA, Allan JS, Strogatz SH, et al. Bright light resets the human circadian pacemaker independent of the timing of the sleep-wake cycle. Science 1986; 233:667–671.

28. Mills JN, Minors DS, Waterhouse JM. Adaptation to abrupt time shifts of the oscillator(s) controlling human circadian rhythms. J Physiol (Lond) 1978; 285:455–470.

29. Lewy AJ, Sack RL. The dim light melatonin onset as a marker for circadian phase position. Chronobiol Int 1989; 6:93–102.

30. Deacon S, Arendt J. Posture influences melatonin concentrations in plasma and saliva in humans. Neurosci Lett 1994; 167:191–194.

31. Deacon SJ, Arendt J, English J. Posture: a possible masking factor of the melatonin circadian rhythm, In: Touitou Y, Arendt J, Pevet P, eds. Melatonin and the Pineal Gland—From Basic Science to Clinical Application. New York: Elsevier Science Publishers, 1993:387–390.

32. Shanahan TL. Circadian physiology and the plasma melatonin rhythm in humans. M.D. dissertation, Harvard Medical School, Boston, MA, 1995:1–221.

33. Monteleone P, Maj M, Fusco M, Orazzo C, Kemali D. Physical exercise at night blunts the nocturnal increase of plasma melatonin levels in healthy humans. Life Sci 1990; 47:1989–1995.

34. Carr DB, Reppert SM, Bullen B, et al. Plasma melatonin increases during exercise in women. J Clin Endocrinol Metab 1981; 53:224–225.

35. Shanahan TL, Zeitzer JM, Czeisler CA. Resetting the melatonin rhythm with light in humans. J Biol Rhythms 1997; 12:556–567.

36. Shanahan TL, Czeisler CA. Light exposure induces equivalent phase shifts of the endogenous circadian rhythms of circulating plasma melatonin and core body temperature in men. J Clin Endocrinol Metab 1991; 73:227–235.

37. Hastings JW, Sweeney BM. A persistent diurnal rhythm of luminescence in *Gonyaulax polyedra*. Biol Bull 1958; 115:440–458.

38. Pittendrigh CS. Circadian systems: entrainment. In: Aschoff J, ed. Handbook of Behavioral Biology: Biological Rhythms. New York: Plenum Press, 1981:95–124.

39. Hoban TM, Sulzman FM. Light effects on circadian timing system of a diurnal primate, the squirrel monkey. Am J Physiol 1985; 249:R274–R280.

40. Czeisler CA, Kronauer RE, Allan JS, et al. Bright light induction of strong (type 0) resetting of the human circadian pacemaker. Science 1989; 244:1328–1333.

41. Strogatz SH. Interpreting the human phase response curve to multiple bright-light exposures. J Biol Rhythms 1990; 5(2):169–174.

42. Minors DS, Waterhouse JM, Wirz-Justice A. A human phase-response curve to light. Neurosci Lett 1991; 133:36–40.

43. Honma K, Honma S. A human phase response curve for bright light pulses. Jpn J Psychiatry Neurol 1988; 42(1):167–168.

44. Jewett ME, Kronauer RE, Czeisler CA. Phase/amplitude resetting of the human circadian pacemaker via bright light: a further analysis. J Biol Rhythms 1994; 9: 295–314.

45. Nelson DE, Takahashi JS. Sensitivity and integration in a visual pathway for ciradian entrainment in the hamster (*Mesocricetus auratus*). J Physiol 1991; 439:115–145.
46. Chandrashekaran MK, Loher W. The relationship between the intensity of light pulses and the extent of phase shifts of the circadian rhythm in the eclosion rate of *Drosophila pseudoobscura*. J Exp Zool 1969; 172:147–152.
47. Takahashi JS, DeCoursey PJ, Bauman L, Menaker M. Spectral sensitivity of a novel photoreceptive system mediating entrainment of mammalian circadian rhythms. Nature 1984; 308:186–188.
48. McGuire RA, Rand WM, Wurtman RJ. Entrainment of the body temperature rhythm in rats: effect of color and intensity of environmental light. Science 1973; 181: 956–957.
49. Allan JS, Czeisler CA, Duffy JF, Kronauer RE. Non-linear dose response of the human circadian pacemaker to light. 154th Annual AAAS Meeting, AAAS Publication No 897-30, 1988:101.
50. Boivin DB, Duffy JF, Kronauer RE, Czeisler CA. Dose-response relationships for resetting of human circadian clock by light. Nature 1996; 379:540–542.
51. Johnson CH, Hastings JW. Circadian phototransduction: phase resetting and frequency of the circadian clock of *Gonyaulax* cells in red light. J Biol Rhythms 1989; 4(4):417–437.
52. Colepicolo P, Roenneberg T, Morse D, Taylor WR, Hastings JW. Circadian regulation of bioluminescence in the dinoflagellate *Pyrocystis lunula*. J Psychol 1993; 29:173–179.
53. DeCoursey PJ. Monochromatic phase response curves for a nocturnal rodent. 1st Mtg, Soc Res Biol Rhythms 1988; 1:118.
54. Dijk D-J, Visscher CA, Bloem GM, Beersma DGM, Daan S. Reduction of human sleep duration after bright light exposure in the morning. Neurosci Lett 1987; 73:181–186.
55. Dijk D-J, Beersma DGM, Daan S, Lewy AJ. Bright morning light advances the human circadian system without affecting NREM sleep homeostasis. Am J Physiol 1989; 256:R106–R111.
56. Broadway J, Arendt J, Folkard S. Bright light phase shifts the human melatonin rhythm during the Antarctic winter. Neurosci Lett 1987; 79:185–189.
57. Foret J, Aguirre A, Touitou Y, Clodoré M, Benoit O. Effect of morning bright light on body temperature, plasma cortisol and wrist motility measured during 24 hour of constant conditions. Neurosci Lett 1993; 155:155–158.
58. Clodoré M, Foret J, Benoit O, et al. Psychophysiological effects of early morning bright light exposure in young adults. Psychoneuroendocrinology 1990; 15(3):193–205.
59. Honma K-I, Honma S, Nakamura K, Sasaki M, Endo T, Takahashi T. Differential effects of bright light and social cues on reentrainment of human circadian rhythms. Am J Physiol 1995; 268:R528–R535.
60. Drennan M, Kripke DF, Gillin JC. Bright light can delay human temperature rhythm independent of sleep. Am J Physiol 1989; 257:R136–R141.
61. Lewy AJ, Sack RL, Miller LS, Hoban TM. Antidepressant and circadian phase-shifting effects of light. Science 1987; 235:352–354.
62. Lack LC, Mercer JD, Wright H. Circadian rhythms of early morning awakening insomniacs. J Sleep Res 1996; 5:211–219.

63. Campbell SS, Dawson D, Anderson MW. Alleviation of sleep maintenance insomnia with timed exposure to bright light. J Am Geriatr Soc 1993; 41:829–836.

64. Murphy PJ, Campbell SS. Enhanced performance in elderly subjects following bright light treatment of sleep maintenance insomnia. J Sleep Res 1996; 5:165–172.

65. Anderson JL, Rosen LN, Mendelson WB, et al. Sleep in fall/winter seasonal affective disorder: effects of light and changing seasons. J Psychosom Res 1995; 38:323–337.

66. Brunner DP, Krauchi K, Dijk D-J, Leonhardt G, Huag HJ, Wirz-Justice A. Sleep electroencephalogram in seasonal affective disorder and in control women: effects of midday light treatment and sleep deprivation. Biol Psychiatry 1997; 40:485–496.

67. Carrier J, Dumont M. Sleep propensity and sleep architecture after bright light exposure at three different times of day. J Sleep Res 1995; 4:202–211.

68. Dumont M, Carrier J. Daytime sleep propensity after moderate circadian phase shifts induced with bright light exposure. Sleep 1997; 20:11–17.

69. Tzischinsky O, Lavie P. The effects of evening bright light on next-day sleep propensity. J Biol Rhythms 1997; 12:259–265.

70. Wever RA. Light effects on human circadian rhythms: a review of recent Andechs experiments. J Biol Rhythms 1989; 4:161–185.

71. Wever RA, Polasek J, Wildgruber CM. Bright light affects human circadian rhythms. Pflügers Arch 1983; 396:85–87.

72. Honma K, Honma S, Wada T. Entrainment of human circadian rhythms by artificial bright light cycles. Experientia 1987; 43:572–574.

73. Honma K, Honma S, Wada T. Phase-dependent shift of free-running human circadian rhythms in response to a single bright light pulse. Experientia 1987; 43:1205–1207.

74. Burešová M, Dvoráková M, Zvolsky P, Illnerová H. Early morning bright light phase advances the human circadian pacemaker within one day. Neurosci Lett 1991; 121:47–50.

75. Kennaway DJ, Earl CR, Shaw PF, Royles P, Carbone F, Webb H. Phase delay of the rhythm of 6-sulphatoxy melatonin excretion by artificial light. J Pineal Res 1987; 4:315–320.

76. Dawson D, Lack L, Morris M. Phase resetting of the human circadian pacemaker with use of a single pulse of bright light. Chronobiol Int 1993; 10(2):94–102.

77. Van Cauter E, Sturis J, Byrne MM, et al. Preliminary studies on the immediate phase-shifting effects of light and exercise on the human circadian clock. J Biol Rhythms 1993; 8:S99–S108.

78. Kronauer RE, Czeisler CA. Understanding the use of light to control the circadian pacemaker in humans, In: Wetterberg L, ed. Light and Biological Rhythms in Man. Oxford: Pergamon Press, 1993; 217–236.

79. Kronauer RE, Jewett ME, Czeisler CA. Commentary: The human circadian response to light: strong *and* weak resetting. J Biol Rhythms 1993; 8:351–360.

80. Allan JS, Czeisler CA. Persistence of the circadian thyrotropin rhythm under constant conditions and after light-induced shifts of circadian phase. J Clin Endocrinol Metab 1994; 79:508–512.

81. Czeisler CA. The effect of light on the human circadian pacemaker. In: Waterhouse JM, ed. Circadian Clocks and Their Adjustment. Chichester (Ciba Found Symp 183): Wiley, 1995:254–302.

82. Duffy JF, Kronauer RE, Czeisler CA. Phase-shifting human circadian rhythms: Influence of sleep timing, social contact and light exposure. J Physiol (Lond) 1996; 495:289–297.

83. Czeisler CA, Johnson MP, Duffy JF, Brown EN, Ronda JM, Kronauer RE. Exposure to bright light and darkness to treat physiologic maladaptation to night work. N Engl J Med 1990; 322:1253–1259.

84. Miles LEM, Raynal DM, Wilson MA, Blind man living in normal society has circadian rhythms of 24.9 hours. Science 1977; 198:421–423.

85. Orth DN, Besser GM, King PH, Nicholson WE. Free-running circadian plasma cortisol rhythm in a blind human subject. Clin Endocrinol 1979; 10:603–617.

86. Martens H, Endlich H, Hildebrandt G, Moog R. Sleep/wake distribution in blind subjects with and without sleep complaints. Sleep Res 1990; 19:398.

87. Sack RL, Lewy AJ, Blood ML, Keith LD, Nakagawa H. Circadian rhythm abnormalities in totally blind people: incidence and clinical significance. J Clin Endocrinol Metab 1992; 75:127–134.

88. Nakagawa H, Sack RL, Lewy AJ. Sleep propensity free-runs with the temperature, melatonin, and cortisol rhythms in a totally blind person. Sleep 1992; 15:330–336.

89. Klein T, Martens H, Dijk D-J, Kronauer RE, Seely EW, Czeisler CA. Chronic non-24-hour circadian rhythm sleep disorder in a blind man with a regular 24-hour sleep-wake schedule. Sleep 1993; 16(4):333–343.

90. Czeisler CA, Shanahan TL, Klerman EB, et al. Suppression of melatonin secretion in some blind patients by exposure to bright light. N Engl J Med 1995; 332: 6–11.

91. Martens H, Klein T, Rizzo, III, Shanahan TL, Czeisler CA. Light-induced melatonin suppression in a blind man. 3rd Mtg, Soc for Res on Biol Rhythms 1992; 58.

92. Jewett ME, Rimmer DW, Duffy JF, Klerman EB, Kronauer RE, Czeisler CA. Human circadian pacemaker is sensitive to light throughout the subjective day without evidence of transients. Am J Physiol 1997; 273:R1800–R1809.

93. Winfree AT. The Geometry of Biological Time. New York: Springer-Verlag, 1980.

94. Wiedenmann G. Weak and strong phase shifting in the activity rhythm of *Leucophaea maderae* (*blaberidae*) after light pulses of high intensity. Z Naturforsch 1977; 32:464–465.

95. Winfree AT. Integrated view of resetting a circadian clock. J Theor Biol 1970; 28: 327–374.

96. Peterson EL. A limit cycle interpretation of a mosquito circadian oscillator. J Theor Biol 1980; 84:281–310.

97. Jewett ME, Kronauer RE, Czeisler CA. Light-induced suppression of endogenous circadian amplitude in humans. Nature 1991; 350:59–62.

98. Kronauer RE. A quantitative model for the effects of light on the amplitude and phase of the deep circadian pacemaker, based on human data, In: Horne J, ed. Sleep '90, Proceedings of the Tenth European Congress on Sleep Research, Dusseldorf: Pontenagel Press, 1990:306–309.

99. Daan S, Pittendrigh CS. A functional analysis of circadian pacemakers in nocturnal rodents. II. the variability of phase response curves. J Comp Physiol A 1976; 106: 253–266.

100. Pohl H. Characteristics and variability in entrainment of circadian rhythms to light in

diurnal rodents, In: Aschoff J, Daan S, Groos GA, eds. Vertebrate Circadian Systems: Structure and Physiology. Berlin: Springer-Verlag, 1982; 339–346.

101. Hashimoto S, Kohsaka M, Nakamura K, Honma H, Honma S, Honma K-I. Midday exposure to bright light changes the circadian organization of plasma melatonin rhythm in humans. Neurosci Lett 1997; 221:89–92.

102. Malinowski JR, Laval-Martin DL, Edmunds LN, Jr. Circadian oscillators, cell cycles, and singularities: light perturbations of the free-running rhythm of cell division in *Euglena*. J Comp Physiol B 1985; 155:257–267.

103. Walz B, Sweeney BM. Kinetics of the cycloheximide-induced phase changes in the biological clock in *Gonyaulax*. Proc Natl Acad Sci USA 1979; 76(12):6443–6447.

104. Engelmann W, Johnson A. Attenuation of the petal movement rhythm in kalanchoë with light pulses. Physiol Plant 1978; 43:68–76.

105. Saunders DS. An experimental and theoretical analysis of photoperiodic induction in the flesh-fly, *Sarcophaga argyrostoma*. J Comp Physiol 1978; 124:75–95.

106. Peterson EL. Phase-resetting a mosquito circadian oscillator. I. Phase-resetting surface. J Comp Physiol 1980; 138:201–211.

107. Boivin DB, Czeisler CA. Resetting the circadian melatonin and cortisol rhythms in humans by ordinary room light. Neuroreport 1998; 9:779–782.

108. Zeitzer JM, Kronauer RE, Czeisler CA. Photopic transduction implicated in human circadian entrainment. Neurosci Lett 1997; 232:135–138.

109. Wetterberg L. Melatonin in humans—physiological and clinical studies. J Neural Transm 1978; 13:289–310.

110. Lewy AJ, Wehr TA, Goodwin FK, Newsome DA, Markey SP. Light suppresses melatonin secretion in humans. Science 1980; 210:1267–1269.

111. Brainard GC, Rollag MD, Hanifin JP. Photic regulation of melatonin in humans: ocular and neural signal transduction. J Biol Rhythms 1997; 12:537–546.

112. Bojkowski CJ, Aldhous ME, English J, et al. Suppression of nocturnal plasma melatonin and 6-sulphatoxymelatonin by bright and dim light in man. Horm Metab Res 1987; 19:437–440.

113. McIntyre IM, Norman TR, Burrows GD, Armstrong SM. Human melatonin suppression by light is intensity dependent. J Pineal Res 1989; 6:149–156.

114. Brainard GC, Lewy AJ, Menaker M, et al. Dose-response relationship between light irradiance and the suppression of plasma melatonin in human volunteers. Brain Res 1988; 454:212–218.

115. Trinder J, Armstrong SM, O'Brien C, Luke D, Martin MJ. Inhibition of melatonin secretion onset by low levels of illumination. J Sleep Res 1996; 5:77–82.

116. Brainard GC, Lewy AJ, Menaker M, et al. Effect of light wavelength on the suppression of nocturnal plasma melatonin in normal volunteers. Ann NY Acad Sci 1985; 453:376–378.

117. Ruberg FL, Skene DJ, Hanifin JP, et al. Melatonin regulation in humans with color vision deficiencies. J Clin Endocrinol Metab 1996; 81:2980–2985.

118. Badia P, Myers B, Boecker M, Culpepper J, Harsch JR. Bright light effects on body temperature, alertness, EEG and behavior. Physiol Behav 1991; 50:583–588.

119. Myers BL, Badia P. Immediate effects of different light intensities on body temperature and alertness. Physiol Behav 1993; 54:199–202.

120. Wright Jr. KP, Badia P, Myers BL, Plenzler SC, Hakel M. Caffeine and light effects

on nighttime melatonin and temperature levels in sleep-deprived humans. Brain Res 1997; 747:78–84.

121. Wright Jr. KP, Badia P, Myers BL, Plenzler SC. Combination of bright light and caffeine as a countermeasure for impaired alertness and performance during extended sleep deprivation. J Sleep Res 1997; 6:26–35.

122. Campbell SS, Dawson D. Enhancement of nighttime alertness and performance with bright ambient light. Physiol Behav 1990; 48:317–320.

123. Dawson D, Campbell SS. Timed exposure to bright light improves sleep and alertness during simulated night shifts. Sleep 1991; 14(6):511–516.

124. Daurat A, Aguirre A, Foret J, Gonnet P, Keromes A, Benoit O. Bright light affects alertness and performance rhythms during a 24-h constant routine. Physiol Behav 1993; 53:929–936.

125. Leproult R, Van Reeth O, Byrne MM, Sturis J, Van Cauter E. Sleepiness, performance, and neuroendocrine function during sleep deprivation: effects of exposure to bright light or exercise. J Biol Rhythms 1997; 12:245–258.

126. Murphy PJ, Badia P, Wright Jr KP. Boecker M, Hakel M. Bright light and nonsteroidal anti-inflammatory drug effects on performance and alertness during extended sleep deprivation. Sleep Res 1995; 24:532.

127. Cajochen C, Dijk D-J, Borbély AA. Dynamics of EEG slow-wave activity and core body temperature in human sleep after exposure to bright light. Sleep 1992; 15:337–343.

128. Dijk D-J, Cajochen C, Borbély AA. Effect of a single 3-hour exposure to bright light on core body temperature and sleep in humans. Neurosci Lett 1991; 121:59–62.

129. Bunnell DE, Treiber SP, Phillips NH, Berger RJ. Effects of evening bright light exposure on melatonin, body temperature and sleep. J Sleep Res 1992; 1:17–23.

130. Lakin-Thomas PL. Commentary: Effects of photic and nonphotic stimuli on melatonin secretion. J Biol Rhythms 1997; 12:575–578.

131. Wehr TA. Melatonin and seasonal rhythms. J Biol Rhythms 1997; 12:517–526.

132. Wehr TA, Moul DE, Barbato G, et al. Conservation of photoperiod-responsive mechanisms in humans. Am J Physiol 1993; 265:R846–R857.

133. Eskes GA, Zucker I. Photoperiodic regulation of the hamster testis: dependence on circadian rhythms. Proc Natl Acad Sci USA 1978; 75:1034–1038.

134. Karsch FJ, Woodfill CJI, Malpaux B, Robinson JE, Wayne NL. Melatonin and mammalian photoperiodism: synchronization of annual reproductive cycles, In: Klein DC, Moore RY, Reppert SM, eds. Suprachiasmatic Nucleus: The Mind's Clock. New York: Oxford University Press, 1991:217–232.

135. Pittendrigh CS. The Photoperiodic phenomena:seasonal modulation of the "day within." J Biol Rhythms 1988; 3:173–188.

136. Vondrasová D, Hájek I, Illnerová H. Exposure to long summer days affects the human melatonin and cortisol rhythms. Brain Res 1997; 759:166–170.

137. Wehr TA. The durations of human melatonin secretion and sleep respond to changes in daylength (photoperiod). J Clin Endocrinol Metab 1991; 73:1276–1280.

138. Terman M, Lewy AJ, Dijk D-J, Boulos Z, Eastman CI, Campbell SS. Light treatment for sleep disorders: Consensus Report. IV. Sleep phase and duration disturbances. J Biol Rhythms 1995; 10:135–147.

139. Carskadon MA, Dement WC. Sleep studies on a 90-minute day. Electroenceph Clin Neurophysiol 1975; 39:145–155.

140. Carskadon MA, Dement WC. Multiple sleep latency tests during the constant routine. Sleep 1992; 15:396–399.

141. ASDC. Association of Sleep Disorders Centers: Diagnostic classification of sleep and arousal disorders (H. Roffwarg, Chairman, Nosology Committee). Sleep 1979; 2:1–137.

142. Johnson MP, Duffy JF, Dijk D-J, Ronda JM, Dyal CM, Czeisler CA. Short-term memory, alertness and performance: a reappraisal of their relationship to body temperature. J Sleep Res 1992; 1:24–29.

143. Zulley J, Wever R, Aschoff J. The dependence of onset and duration of sleep on the circadian rhythm of rectal temperature. Pflügers Arch 1981; 391:314–318.

144. Dijk D-J, Czeisler CA. Paradoxical timing of the circadian rhythm of sleep propensity serves to consolidate sleep and wakefulness in humans. Neurosci Lett 1994; 166:63–68.

145. Czeisler CA, Dijk D-J. Human circadian physiology and sleep-wake regulation. Handbook of Behavioral Neurobiology: Circadian Clocks 1998; (in press).

146. Dijk D-J, Duffy JF, Czeisler CA. Circadian and sleep/wake dependent aspects of subjective alertness and cognitive performance. J Sleep Res 1992; 1:112–117.

147. Strogatz SH, Kronauer RE, Czeisler CA. Circadian regulation dominates homeostatic control of sleep length and prior wake length in humans. Sleep 1986; 9:353–364.

148. Rosekind MR, Gander PH, Dinges DF. Alertness management in flight operations: strategic napping. Aerospace Technology Conference and Exposition, Long Beach CA, 1991:1–12.

149. Rutenfranz J, Aschoff J, Mann H. The effects of a cumulative sleep deficit, duration of preceding sleep period and body-temperature on multiple choice reaction time, In: Colquhoun WP, ed. Aspects of Human Efficiency: Diurnal Rhythm and Loss of Sleep. London: English Universities Press, 1972:217–229.

150. Tilley AJ, Wilkinson RT, Warren PSG, Watson B, Drud M. The sleep and performance of shift workers. Hum Factors 1982; 24:629–641.

151. Czeisler CA, Dijk D-J. Use of bright light to treat maladaption to night shift work and circadian rhythm sleep disorders. J Sleep Res 1995; 4:70–73.

152. Smith MJ, Colligan MJ, Tasto DL. Health and safety consequences of shift work in the food processing industry. Ergonomics 1982; 25:133–144.

153. Leger D. The cost of sleep-related accidents: a report for the national commission on sleep disorders research. Sleep 1994; 17:84–93.

154. Mitler MM, Carskadon MA, Czeisler CA, Dement WC, Dinges DF, Graeber RC. Catastrophes, sleep, and public policy: Consensus report. Sleep 1988; 11:100–109.

155. Czeisler CA, Allan JS. Acute circadian phase reversal in man via bright light exposure: application to jet-lag. Sleep Res 1987; 16:605.

156. Czeisler CA, Chiasera AJ, Duffy JF. Research on sleep, circadian rhythms and aging: applications to manned spaceflight. Exp Gerontol 1991; 26:217–232.

157. Czeisler CA, Kronauer RE, Johnson MP, Allan JS, Johnson TS, Dumont M. Action of light on the human circadian pacemaker: treatment of patients with circadian rhythm sleep disorders. In: Horne J, ed. Sleep '88. Stuttgart: Gustav Fischer Verlag, 1989:42–47.

158. Terman M, Lewy AJ, Dijk DJ, Boulos Z, Eastman CI, Campbell SS. Light treatment for sleep disorders: consensus report. IV. Sleep phase and duration disturbances. J Biol Rhythms 1995; 10:135–147.

159. Campbell S, Satlin A, Volicer L, Ross V, Herz L. Management of behavioral and sleep disturbance in Alzheimer's patients using timed exposure to bright light. Sleep Res 1991; 20:446.

160. Campbell SS, Terman M, Lewy AJ, Dijk DJ, Eastman CI, Boulos Z. Light treatment for sleep disorders: consensus report. V. Age-related disturbances. J Biol Rhythms 1995; 10:151–154.

161. Lewy AJ, Sack RL, Singer CM. Treating phase typed chronobiologic sleep and mood disorders using appropriately timed bright artificial light. Psychopharmacol Bull 1985; 21:368–372.

162. Joseph-Vanderpool JR, Kelly KG, Schulz PM, Allen R, Souetre E, Rosenthal NE. Delayed sleep phase syndrome revisited: preliminary effects of light and triazolam. Sleep Res 1988; 17:381.

163. Rosenthal NE, Joseph-Vanderpool JR, Levendosky AA, et al. Phase-shifting effects of bright morning light as treatment for delayed sleep phase syndrome. Sleep 1990; 13(4):354–361.

164. Czeisler CA, Richardson GS, Coleman RM, et al. Chronotherapy: resetting the circadian clocks of patients with delayed sleep phase insomnia. Sleep 1981; 4:1–21.

165. Moore-Ede MC, Sulzman FM, Fuller CA. The Clocks that Time Us: Physiology of the Circadian Timing System. Cambridge, MA: Harvard University Press, 1982.

166. Mrosovsky N, Salmon PA. A behavioral method for accelerating re-entrainment of rhythms to new light-dark cycles. Nature 1987; 330:372–373.

167. Van Reeth O, Turek FW. Stimulated activity mediates phase shifts in the hamster circadian clock induced by dark pulses or benzodiazepines. Nature 1989; 339(6219):49–51.

168. Turek FW. Effects of stimulated physical activity on the circadian pacemaker of vertebrates. J Biol Rhythms 1989; 4(2):135–147.

169. Mistlberger RE. Scheduled daily exercise or feeding alters the phase of photic entrainment in Syrian hamsters. Physiol Behav 1991; 50:1257–1260.

170. Edgar DM, Dement WC. Regularly scheduled voluntary exercise synchronizes the mouse circadian clock. Am J Physiol 1991; 261:R928–R933.

171. Van Reeth O, Sturis J, Byrne MM, et al. Nocturnal exercise phase delays circadian rhythms of melatonin and thyrotropin secretion in normal men. Am J Physiol 1994; 266:E964–E974.

172. Eastman CI, Hoese EK, Youngstedt SD, Liu L. Phase-shifting human circadian rhythms with exercise during the night shift. Physiol Behav 1995; 58:1287–1291.

173. Piercy J, Lack L. Daily exercise can shift the endogenous circadian phase. Sleep Res 1988; 17:393.

174. Rimmer DW, Czeisler CA. Exercise of moderate intensity does not affect the period of the human circadian pacemaker. Sleep Res 1997; 26:749.

175. Klerman EB, Rimmer DW, Dijk D-J, et al. Nonphotic entrainment of the human circadian pacemaker. Am J Physiol 1998; 43:R991–R996.

176. Amir S, Stewart J. Resetting of the circadian clock by a conditioned stimulus. Nature 1996; 379:542–545.

177. Lewy AJ, Ahmed S, Sack RL. Phase shifting the human circadian clock using melatonin. Behav Brain Res 1996; 73:131–134.

178. Redman JR. Circadian entrainment and phase shifting in mammals with melatonin. J Biol Rhythms 1997; 12:581–587.

179. Lewy AJ, Sack RL. Exogenous melatonin's phase-shifting effects on the endogenous melatonin profile in sighted humans: a brief review and critique of the literature. J Biol Rhythms 1997; 12:588–594.

180. Sack RL, Lewy AJ. Melatonin as a chronobiotic: treatment of circadian desynchrony in night workers and the blind. J Biol Rhythms 1997; 12:595–603.

181. Arendt J, Skene DJ, Middleton B, Lockley SW, Deacon S. Efficacy of melatonin treatment in jet lag, shift work, and blindness. J Biol Rhythms 1997; 12:604–617.

182. Czeisler CA. Commentary: Evidence for melatonin as a circadian phase-shifting agent. J Biol Rhythms 1997; 12:618–623.

183. Dawson D, Encel N, Lushington K. Improving adaptation to simulated night shift: timed exposure to bright light versus daytime melatonin administration. Sleep 1995; 18:11–21.

184. Kräuchi K, Cajochen C, Danilenko KV, Wirz-Justice A. The hypothermic effect of late evening melatonin does not block the phase delay induced by current bright light in human subjects. Neurosci Lett 1997; 232:57–61.

185. Czeisler CA. Human circadian physiology: internal organization of temperature, sleep-wake, and neuroendocrine rhythms monitored in an environment free of time cues. Ph.D. thesis, Stanford University, Stanford, CA, 1978:1–346.

186. Czeisler CA, Richardson GS, Zimmerman JC, Moore-Ede MC, Weitzman ED. Entrainment of human circadian rhythms by light-dark cycles: a reassessment. Photochem Photobiol 1981; 34:239–247.

187. Pittendrigh CS, Daan S. A functional analysis of circadian pacemakers in nocturnal rodents. IV. Entrainment: pacemaker as clock. J Comp Physiol A 1976; 106:291–331.

188. Aschoff J, Pohl J. Phase relations between a circadian rhythm and its zeitgeber within the range of entrainment. Naturwissenschaften 1978; 65:80–84.

189. Soni BG, Philip AR, Foster RG, et al. Novel retinal photoreceptors. Nature 1998; 394:27–28.

6

Role of Melatonin in the Regulation of Sleep

FRED W. TUREK

Northwestern University
Evanston, Illinois

CHARLES A. CZEISLER

Brigham and Women's Hospital
Harvard Medical School
Boston, Massachusetts

I. Introduction

Melatonin is a hormone produced in the pineal gland of vertebrates. While early medical textbooks often referred to the pineal gland as a vestigial organ in mammals, the discovery of melatonin by Aaron Lerner in the late 1950s (1) ushered in an exciting era of research on the function of melatonin and the pineal gland, which has become very intense over the past decade. Of primary importance for understanding the function of melatonin is the signature feature of its production and release into the general circulation: it is secreted primarily during the darkness in all mammalian species (2), with there being very low levels of melatonin in the circulation during the daytime. Thus, in humans high levels of melatonin are present in the blood stream during the normal time of sleep (3) (Fig. 1). This correlation with the sleep-wake cycle coupled with two responses to exogenous melatonin have led to a great deal of interest in recent years that melatonin may be involved in the regulation of sleep, and that exogenous melatonin and/or analogs of melatonin may represent a new class of hypnotics for the treatment of insomnia. First, although controversial and contradictory in nature, a fairly extensive literature now exists on the hypnotic effects of melatonin. Second, as discussed more fully in Chapter 7 (4), melatonin can have chronobiotic properties, meaning it can

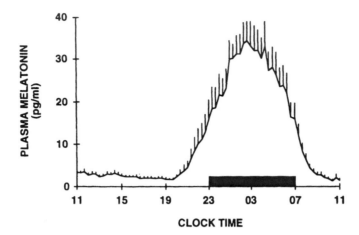

Figure 1 Mean melatonin profiles (±SEM) over a 24-hr time period from eight normal adults (four men, four women) studied in dim light conditions (<200 lux). The dark bar represents the sleep period in darkness. (Van Cauter, unpublished results)

influence the phase and/or the period of the circadian clock regulating most if not all circadian (i.e. 24 hr) rhythms in mammals. Since the circadian clock plays a central role in the timing of sleep and wakefulness, any effects of melatonin on the clock would influence the timing of sleep under normal conditions as well as when sleep is occurring at abnormal times, as occurs during jet lag or in shift workers.

A recent issue of the *Journal of Biological Rhythms* (JBR) (5) was devoted entirely to the subject of the control and action of melatonin, and the reader is directed to that issue as well as a recent book by Arendtt (2) and a "white paper" published by the National Sleep Foundation (6) for comprehensive reviews on melatonin. In particular, five papers in the JBR (3,7–10) and a commentary on these papers (11) provide a thorough evaluation of the pros and cons for the evidence that melatonin is or is not involved in the regulation of sleep.

At this time, the one clear statement that can be made about the role of melatonin in the regulation of sleep in humans is that no clear consensus has emerged. Indeed, while one can read statements in the recent literature like "These data demonstrate that melatonin exerts effects on the main characteristics of human sleep, i.e. latency to sleep onset, sleep consolidation, slow waves, sleep spindles and REM sleep" (3), one can also read, "It was concluded that there is not a convincing body of evidence, using generally accepted measures, that melatonin administration improves sleep in insomniacs with non-circadian sleep disturbance" (8). Both statements are fully supported by the literature.

The controversy over the role of melatonin in the regulation of sleep is due to differences in the way experiments were carried out and the questions about sleep that were addressed. Different studies have administered melatonin at different times of the day, have used different routes of administration, have used different doses of melatonin from the physiological to the pharmacological, have examined sleep in normal controls and in subjects with a wide variety and often unclearly defined sleep problems, have used a wide variety of outcome measures of treatment ranging from subjective alertness to subjective appraisal of sleep to electroencephalographic (EEG) measurements (8). Even when sleep has been monitored by activity, the extent to which EEG data were collected and evaluated varies tremendously from one study to another (8).

This chapter will attempt to provide a balanced review of the data indicating that treatment with melatonin can influence both the homeostatic drive for sleep (Section II), and thus act as a hypnotic, and the circadian control of the timing of sleep (Section III), and thus act as a chronobiotic. In addition, the possible physiological mechanisms by which melatonin could influence the sleep-wake cycle are discussed in Section IV from an evolutionary perspective of the function of the pineal gland in other vertebrate species.

II. Evidence for the Hypnotic Effects of Melatonin

Much of the evidence that treatment with melatonin can have hypnotic properties has come from studies on normal volunteers in which melatonin was administered either just prior to scheduled naps during the normal wake time, or just prior to the normal sleep time. There is a general consensus in the literature that treatment with melatonin during the day, at times when circulating endogenous levels of melatonin are low, induces subjective sensations of fatigue and sleepiness (9,10, 12–14), and has effects on EEG activity during both daytime wakefulness and daytime naps (15,16). While early studies usually involved administering melatonin at doses that led to supraphysiological levels of circulating melatonin, more recent studies suggest that much lower doses of melatonin, generating peak serum melatonin levels during the day that are closer to the range of normal nighttime levels, can also induce subjective feelings of sleepiness (17,18).

In contrast, studies on the hypnotic effects of melatonin administered near the time of normal sleep (and thus near the time of rising endogenous melatonin levels) have not yielded consistent results, and in general, increased sleepiness is only achieved after administration of high doses of melatonin during the night. While in one study there was no effect of melatonin (1 or 5 mg oral dosage) on the time to sleep onset, on the duration of sleep, or on measures of mood and alertness the next day (19), another study reported essentially the opposite results, finding an enhancement in all of these measures after treatment with melatonin (80 mg oral dosage) when compared to placebo control nights (20). While these contradictory results could be due to the different doses of melatonin used, they may also reflect the difficulty of demonstrating an improvement in sleep efficiency using melatonin in normally entrained subjects sleeping at night. As noted by Lavie (9), the effect of nighttime administration of melatonin is bounded by the endogenous increase in nocturnal sleepiness, and no further increase in sleep may be possible in healthy individuals without sleep complaints treated with melatonin during the rising phase of endogenous melatonin. Thus, the lack of sleep-inducing effects of melatonin delivered during the evening to good sleepers may be telling us very little about its hypnotic properties. Indeed, this is a general problem in examining the hypnotic effects of melatonin delivered at night in subjects with no sleep problems.

Lavie (9) has found a precise coupling between the endogenous increase in melatonin secretion and the opening of what he calls a "sleep gate" (Fig. 2). He argues that melatonin participates in the regulation of the sleep-wake cycle by inhibiting a CNS wakefulness-generating system. Sack and Lewy take this one step further and hypothesize that melatonin may conteract the daily wake-promoting signal emanating from the suprachiasmatic nuclei (SCN) (21). This hypothesis is intriguing in view of the finding that the SCN contains a high concentration of melatonin receptors and that melatonin may quiet the activity of the SCN (22). SCN lesions in monkeys result in an increased need for sleep (23), indicating that the SCN is a site for a wakefulness-regulating mechanism in primates. Whether SCN output in mammals promotes wakefulness, sleep, or both, depending on time of day, has yet to be resolved.

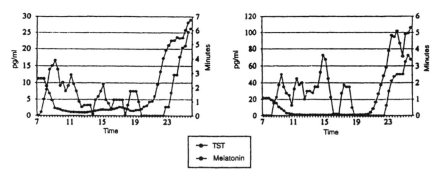

Figure 2   Sleep propensity measured by the ultrashort sleep-wake 7/13 paradigm (7 min sleep, 13 min wake) from 07:00 to 02:00 hr and plasma melatonin levels measured every 20 min in two representative subjects. A "sleep gate" is tightly correlated with the rising phase of the melatonin rhythm. Each point represents total sleep time (TST) for a given 7-min sleep attempt in each trial. [Courtesy of Lavie (9).]

It may well be that melatonin has very little effect on the sleep of normal individuals who are not experiencing insomnia. Surprisingly, few attempts have been made to determine the effects of melatonin as a hypnotic in people suffering from insomnia, particularly in individuals under the age of 65. In one study involving treatment with supraphysiological doses of melatonin (75 mg oral dosage) for 14 consecutive days, there was a significant increase in the subjective assessment of total sleep time and daytime alertness (24); however, there were no objective polysomnographic sleep recordings to verify these self-reports. A second study in which sleep was polysomnographically recorded (25) failed to find any effects on the onset or duration of sleep or on mood or alertness the following day after treatment with a single low dose of melatonin (1 or 5 mg oral dosage). The differences in dose, duration of treatment, and outcome variable make it difficult to compare the results of such studies. In a few children with low melatonin levels due to a rare genetic disorder (Angelman syndrome), or to a pineal tumor, who have severe sleep problems, treatment with exogenous melatonin is reported to increase sleep time, but again no polysomnographic data are reported in these studies (7).

Evidence exists that sleep-onset insomnia secondary to misalignment of circadian phase can be effectively treated with melatonin, even in normal young subjects. A recent study reported that even very low doses of melatonin (0.3 or 1.0 mg oral dosage) when given in the evening before opening of the "sleepgate" had sleep-inducing effects compared to a control group (26). The college students in the control group for that study had long sleep latencies, typical of subjects attempting to sleep during the wake maintenance zone (27). Treatment with melatonin reduced these latencies to within the range for normal subjects sleeping at their habitual bedtime. This is consistent with the hypothesis, noted above, that melatonin may counteract a wake-promoting signal emanating from the SCN. These hypotheses are also consistent with data reported from studies of disrupted sleep in blind patients who have lost circadian entrainment and are no longer synchronized to the 24-hr day. In such blind individuals, daily melatonin administration has been found to stabilize and promote nocturnal sleep, while reducing the need for daytime naps, even while endogenous rhythms of circadian temperature, cortisol, and melatonin itself continue to free-run (28).

While there are little data in the literature to indicate that melatonin can improve sleep in insomniacs under the age of 65, there have been a number of recent reports indicating beneficial effects of melatonin for elderly insomniacs. Recent studies in which motor activity of the wrist was used as an outcome variable have reported that elderly subjects complaining of insomnia have reduced motor activity at night following treatment with both high and low doses of melatonin just prior to bedtime (29–31). This has raised the hope that melatonin administration may prove to be effective in the treatment of disrupted sleep in the elderly, although again objective polysomnographic sleep recording is necessary before this finding can be accepted. Until such recordings are made in a randomized, double-blind, placebo-controlled trial with an adequate sample size, the positive effects of melatonin on sleep in elderly insomniacs will remain an attractive but unproven hypothesis. The potential use of melatonin replacement for treating sleep-wake disorders in the elderly is in fact a particularly attractive hypothesis since disturbed sleep becomes more prevalent (32–34) and melatonin levels have been reported to decline with advancing age (35). Furthermore, it has been reported that melatonin levels are significantly lower in elderly insomniac patients than in age-matched controls (9,36).

Most of the studies that have examined the effects of melatonin for the treatment of jet lag (following real or simulated phase shifts in the time structure of the external environment) have assessed the effects of treatment on subjective well-being or on the phase of circadian rhythms (see next section). Even in those studies that have examined the effects of melatonin on sleep in a real or simulated jet lag paradigm, the measures of sleepiness have only been subjectively evaluated (37–42). Indeed, although it has been suggested that the major effects of melatonin for alleviating the adverse effects of jet lag are due to its improvement of sleep, objective measures of the effects of melatonin on sleep under jet lag conditions have yet to be made. There is clearly a need for randomized, double-blind, placebo-controlled trials of adequate sample size to examine the possible sleep-inducing effects of melatonin in humans after traveling rapidly across time zones. Only then might any firm conclusions be made about the effects of melatonin on sleep under such conditions.

As suggested by Zhdanova and Wurtman (7), exogenous treatment with melatonin may not lead to sleep per se, but instead may promote a condition of general relaxation and sedation, which under favorable conditions might facilitate sleep onset. Recent findings that both subjective and objective measures of sleepiness induced by melatonin are suppressed when subjects change posture to standing (10) are consistent with the hypothesis that melatonin is not inducing a sleep-like state. The effects of posture on melatonin's effects on sleep (Fig. 3) are intriguing, and if verified, have important implications for the use of melatonin as a hypnotic. While there is now a good deal of literature to indicate that melatonin may induce some sort of "quiet wakefulness" state, it is still too early to conclude that melatonin is effective for any particular disorder of the sleep-wake cycle (8,11). Furthermore, indiscriminate use of melatonin by self-mediation may in fact lead to disruption of normal sleep in some individuals (43), the mechanism for which is unknown.

As noted by Roth and Richardson (11), today's standard for determining sleep efficacy is polysomnographic evidence of decreased sleep latency, decreased wake time after sleep onset, and increased total sleep time or sleep efficiency. The effectiveness of the use of melatonin in individuals suffering from

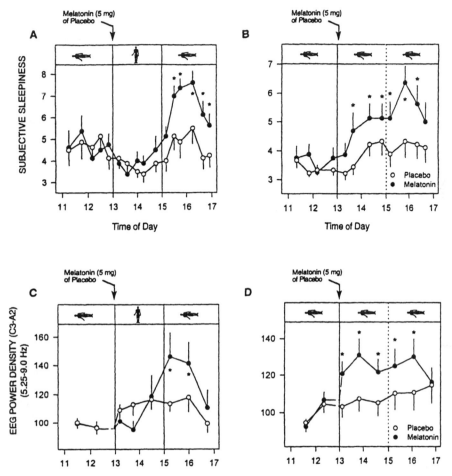

Figure 3 Time course of subjective sleepiness ratings (A and B) (Karolinska Sleepiness Scale, KSS) and EEG power density in the range of 5.25–9.0 Hz (C and D) in subjects receiving a single administration of placebo (○) or 5 mg of melatonin (●) at 13 hr. While all subjects were initially in the supine position, the subjects depicted in A and C had a postural change from the supine to the upright at 13 hr and from the upright to the supine at 15 hr as depicted by the human symbols at the top of the figure. No postural change took place for the subjects whose data are depicted in B and D. All values are mean hourly values ± SEM. *Significant differences between the placebo and melatonin groups. [Courtesy of Cajochen et al. (10).]

any particular form of acute or chronic insomnia has not been proven using "today's standard." Until this occurs, we will not know (1) if melatonin will be a useful aid for the medical management of insomnia, (2) in which patients it might be beneficial, or (3) how melatonin compares to other pharmacological or behavioral therapies for insomnia (11). In a recent survey by the Consumers Union in the United States (44), about one-fourth of the respondents found melatonin to be "very helpful" as a sleeping pill—about half the response to the use of prescription sleeping pills. More importantly, melatonin had the highest rate of any treatment for a response of "not at all helpful."

III. Evidence for the Chronobiotic Effects of Melatonin

Since the role of melatonin in the circadian clock system is discussed in detail in Chapter 7, this section will be limited to the possible use of melatonin as a potential phase-shifting stimulus for the sleep-wake cycle.

For the chronobiotic properties of melatonin to be useful in the treatment of insomnia, the underlying cause of the insomnia would need to be due to an abnormal timing of the daily sleep propensity rhythm. Such abnormal timing occurs in what is known as the "delayed sleep phase syndrome" (DSPS), in which sleep occurs at a delayed clock time relative to the light-dark cycle and to social, economic, and family demands (45). DSPS, while originally described in adults (46), often begins in childhood and is relatively common among adolescents (47,48). In the first use of melatonin in patients with DSPS, it was found that when given 5 hr before sleep onset for a period of 4 weeks, melatonin (5 mg oral dosage) advanced sleep onset and wake times when compared to placebo controls (49). Although the sample size in this study was small ($N = 8$), a similar study, also on a small sample size ($N = 7$), has produced similar results (50). While more studies are clearly needed, these initial findings indicate that melatonin may be an effective treatment for this specific sleep disorder. However, this tells us little about the efficacy of melatonin to shift directly the circadian pacemaker driving the sleep propensity rhythm. In fact, evidence from studies in the blind cited above indicate that daily administration of melatonin in free-running blind humans fails to entrain the circadian pacemaker to the 24-hr day, even when the daily phase advance required is only 0.2–0.3 hr (51). In DSPS patients, melatonin administration might well facilitate sleep during the evening wake maintenance zone, just as it did in the young normal subjects studied by Zhadanova et al. (26). This might enable the DSPS patients to awaken at an earlier hour, as desired. Subsequent morning light exposure would then be expected to advance their circadian pacemaker (52,53).

In contrast to the timing of sleep in DSPS, one of the most common sleep complaints in the elderly is that the timing of sleep onset or offset is advanced

(35). Therefore, in principle, if melatonin could be used to phase-delay the circadian clock timing the sleep-wake cycle, it could perhaps normalize the timing of sleep in the elderly who have an advanced endogenous circadian phase (54). However, to date no such studies have been reported. Furthermore, the evidence that melatonin has phase-delaying effects in humans is based on a very small number of data points (55), and these few delays were observed in response to melatonin administered in the morning hours. Thus, treating elderly humans with a phase-advanced sleep-wake cycle with melatonin in the early morning hours might be counterproductive if melatonin also acted as a hypnotic at wake time. There are clearly many issues that relate to the appropriate timing of melatonin treatment for insomnia, particularly for the elderly. These issues need to be addressed in large clinical trials before any recommendations can be made for melatonin's use in the elderly.

A number of studies have attempted to use melatonin to alleviate the perceived effects of jet lag (primarily sleep disturbances) under both placebo-controlled and uncontrolled conditions. Indeed, one laboratory has reported that in a population of 474 subjects taking melatonin versus 112 subjects taking placebo, there was an overall 50% reduction in self-*related* jet lag (38). A number of authors have hypothesized that at least part of the beneficial effects of melatonin for jet lag may be due to a phase-shifting effect (38,56,57). However, the largest randomized, double-blind, placebo-controlled clinical trial of melatonin to treat jet lag has failed to detect an improvement in: (1) self-reported jet lag symptoms; (2) reported sleep onset; (3) reported time of awakenings; or (4) reported hours slept or nap sleep (58). Furthermore, the mechanism by which any positive subjective effect might occur in such studies cannot be determined at this time. In fact, no objective assessment of any markers of circadian phase or of sleep efficiency were made in these studies. Nonetheless, based on the reported phase-response curve to melatonin, elaborate timetables have been developed to advise travelers when they should take melatonin following eastward and westward travels over different numbers of time zones (57). One difficulty in using melatonin for jet lag is that its use as a chronobiotic may require administration at times when it will have undesired hypnotic properties (38,57,59).

It has become routine for authors to assume that melatonin can reduce the number of days it takes endogenous circadian rhythms, including the sleep-wake cycle, to reentrain following rapid travel across time zones (38). However, there is very little evidence that under jet-lag conditions, treatment with melatonin does indeed have phase-shifting effects on objectively measured phase markers of human circadian rhythms under control of the central circadian clock. In fact, the jury is still out on the efficacy of melatonin as a phase-shifting agent in humans (51). In one study that involved a simulated 9-hr time shift, melatonin treatment was found to enhance the resynchronization speed of some, but not all, monitored hormone and electrolyte secretion rhythms (60). However, the authors of this study concluded that the enhanced synchronization of rhythms was not great enough to warrant melatonin's use as a phase-shifting agent for alleviating the effects of jet lag.

IV. Possible Physiological Mechanisms by Which Melatonin Could Influence the Sleep-Wake Cycle

As noted by Lavie in a recent review (9), "If one had to guess which one of the brain hormones is involved in sleep regulation just from its 24 hr pattern of secretion, melatonin would no doubt be a prime candidate. What would be a better agent to regulate sleep than a brain substance exclusively produced and secreted during the dark hours?" As with just about every statement concerning melatonin and sleep, one can argue from the opposite point of view. Thus, one could argue that since melatonin levels are high at night in *all* mammals, including nocturnal mammals who are awake at night when melatonin levels are elevated, it is unlikely to be involved in regulating the timing of sleep. These two seemingly opposite points of view can in fact be reconciled if one takes a broad evolutionary view of the function of melatonin in vertebrates in general, and in mammals in particular. In lower vertebrates, particularly well studied in reptiles and birds, removal of the pineal gland or treatment with exogenous melatonin can have severely disruptive effects on the 24-hr rhythm of rest and activity (61). Melatonin is playing an important role in controlling the overall temporal organization of at least some lower vertebrates—it is conveying information about the "time of day."

Pinealectomy has little, if any, effect on the ability of mammals to tell the "time of day"; 24-hr rhythmicity is normal in pinealectomized rodents (62). However, pinealectomy severely disrupts the ability of mammals to determine "time of year" (63–65). For many mammalian species, there is a seasonal rhythm to reproductive activity, and for many species it is the seasonal change in day length that controls the timing of this annual cycle (66). It has been firmly established that it is the seasonal change in the duration of elevated nighttime melatonin levels (e.g., long-duration elevated levels during long nights of winter) that mediates the effects of day length on reproductive function. A particular elevated melatonin duration signal can have opposite effects on different species. Thus, in sheep, for example, the short-duration nighttime signal associated with long days is inhibitory to reproduction while the same signal is stimulatory to reproduction in hamsters (63,65). In other words, the duration of high circulating nighttime melatonin levels is in itself not necessarily an inhibitory or stimulatory signal to reproductive function: instead it signals the brain how long the night (or day) is, and in the context of the evolutionary history of the organism the signal induces a particular species-specific response. Similarly, in both nocturnal and diurnal species, circulating melatonin levels may represent a signal that informs the brain and the rest of the body of the time of day, and the response to that signal will depend again on the evolutionary history of the organism.

As noted above, treatment with melatonin during the daytime has been shown to induce sleep, reduce alertness, and induce changes in EEG activity that are similar to those observed during sleep (3,9). Such results have led to the hypothesis that melatonin can have direct hypnotic effects on the brain. Other reports that melatonin can phase-shift circadian rhythms have led to the hypothesis that by shifting the circadian clock regulating the sleep-wake cycle, melatonin affects the sleep-wake cycle. Both hypotheses are compatible with the suggestion that melatonin is a signal informing the brain that it is day or night—and various areas of the brain respond accordingly.

Thus, there is a clear precedence to support the hypothesis that melatonin's main action is as a signal informing the brain of the time of day, since it has already been demonstrated to be a signal that informs the brain (and the rest of the body) of the season of the year. Such information could have a variety of effects. Such a broad interpretation of how melatonin affects the brain is useful for explaining why melatonin appears to have so many different effects. When melatonin is present, at normal or abnormal times, the brain and body may respond as if it were night. For example, high circulating melatonin levels are associated with a decrease in body temperature (Fig. 2), and one effect of melatonin when given during the daytime is to lower body temperature in humans (10). Indeed, the hypnotic effects of melatonin during the daytime have been attributed to its lowering of the body temperature (10). Such a decrease would be a natural response of a neural center regulating body temperature in a diurnal mammal if it received a signal that it was nighttime.

Acknowledgments

Some portions of this paper were included in a White Paper entitled, "Is Melatonin a Treatment for Insomnia and Jet Lag?," published by the National Sleep Foundation.

References

1. Lerner AB, Case JD. Melatonin. Fed Proc 1960; 19:590–592.
2. Arendt J. Melatonin and the Mammalian Pineal Gland. London: Chapman & Hall, 1995.
3. Dijk D-J, Cajochen C. Melatonin and the circadian regulation of sleep initiation, consolidation, structure and the sleep EEG. J Biol Rhythms 1997; 12:627–635.
4. Weaver DR. Role of melatonin in the regulation of sleep. In: Turek FW, Zee P, eds. Neurobiology of Sleep and Circadian Rhythms. New York: Marcel Dekker, 1998.

5. Czeisler C, Turek FW, eds. Melatonin, sleep and circadian rhythms: current progress and controversies. J Biol Rhythms 1997; 12:1–710.

6. Czeisler C, Turek FW. Is Melatonin a Treatment for Insomnia and Jet Lag? Washington DC: National Sleep Foundation, 1997.

7. Zhdanova IV, Wurtman RJ. Efficacy of melatonin as a sleep promoting agent. J Biol Rhythms 1997; 12:644–650.

8. Mendelson WB. Efficacy of melatonin as a hypnotic agent. J Biol Rhythms 1997; 12: 651–656.

9. Lavie P. Melatonin: role in gating nocturnal rise in sleep propensity. J Biol Rhythms 1997; 12:657–665.

10. Cajochen C, Kräuchi K, Wirz-Justice A. The acute soporific action of daytime melatonin administration: effects on the EEG during wakefulness and subjective alertness. J Biol Rhythms 1997; 12:636–643.

11. Roth T, Richardson G. Is melatonin administration an effective hypnotic? Commentary. J Biol Rhythms 1997; 12:666–672.

12. Nave R, Peled R, Lavie P. Melatonin improves evening napping. Eur J Pharmacol 1995; 275:213–216.

13. Dawson D, Encel N. Melatonin and sleep in humans. J Pineal Res 1993; 15:1–12.

14. Hughes RJ, Badia P. Sleep-promoting and hypothermic effects of daytime melatonin administration in humans. Sleep 1997; 20:124–131.

15. Dijk D-J, Roth C, Landolt H-P, Werth E, Aeppli M, Achermann P, Borbély A. Melatonin reduces low-frequency EEG activity during daytime sleep. Sleep Res 1995; 24A:117 (abstract).

16. Dijk D-J, Roth C, Landolt H-P, Werth E, Aeppli M, Achermann P, Borbély A. Melatonin effect on daytime sleep in men: suppression of EEG low frequency activity and enhancement of spindle frequency activity. Neurosci Lett 1995; 201:13–16.

17. Tzischinsky O, Lavie P. Melatonin possesses a hypnotic effect which is time-dependent. Sleep 1994; 17:638.

18. Dollins AB, Zhdanova IV, Wurtman RJ, Lynch HJ, Deng MH. Effect of inducing nocturnal serum melatonin concentrations in daytime on sleep, mood, body temperature, and performance. Proc Natl Acad Sci USA 1994; 91:1824–1828.

19. James SP, Mendelson WB, Sack DA, Rosenthal NE, Wehr TA. The effect of melatonin on normal sleep. Neuropsychopharmacology 1988; 1:41–44.

20. Waldhauser F, Saletu B, Trinchard-Lugan I. Sleep laboratory investigations on hypnotic properties of melatonin. Psychopharmacology 1990; 100:222–226.

21. Sack RL, Lewy AJ. Melatonin as a chronobiotic treatment of circadian desynchrony in night workers and the blind. J Biol Rhythms 1997; 12:595–603.

22. Reppert SM, Weaver DR, Rivkees SA, Stopa EG. Putative melatonin receptors in a human biological clock. Science 1988; 242:78–81.

23. Edgar DM, Miller JD, Prosser RA, Dean RR, Dement WC. Serotonin and the mammalian circadian system. II. Phase-shifting rat behavioral rhythms with serotonergic agonists. J Biol Rhythms 1993; 8:17–31.

24. MacFarlane JG, Cleghorn JM, Brown GM, Streiner DL. The effects of exogenous melatonin on the total sleep time and daytime alertness of chronic insomniacs: a preliminary study. Biol Psychiatry 1991; 30:371–376.

25. James SP, Sack DA, Rosenthal NE, Mendelson WB. Melatonin administration in insomnia. Neuropsychopharmacology 1989; 3:19–23.

26. Zhdanova IV, Wurtman RJ, Lynch HJ, Ives JR, Dollins AB, Morabito C, Matheson JK, Schomer DL. Sleep-inducing effects of low doses of melatonin ingested in the evening. Clin Pharmacol Ther 1995; 57:552–558.

27. Strogatz SH, Kronauer RE, Czeisler CA. Circadian regulation dominates homeostatic control of sleep length and prior wake length in humans. Sleep 1986; 9:353–364.

28. Arendt, J, Skene DJ, Middleton B, Lockley SW, Deacon S. Efficacy of melatonin treatment in jet lag, shift work, and blindness. J Biol Rhythms 1997; 12:604–617.

29. Haimov I, Lavie P. Potential of melatonin replacement therapy in older patients with sleep disorders. Drugs Aging 1995; 7:75–78.

30. Haimov I, Lavie P, Laudon M, Herer P, Zisapel N. Melatonin replacement therapy of elderly insomniacs. Sleep 1995; 8:598–603.

31. Garfinkel D, Laudon M, Nof D, Zisapel N. Improvement of sleep quality in elderly people by controlled-release melatonin. Lancet 1995; 346:541–544.

32. Prinz PN. Sleep and sleep disorders in older adults. J Clin Neurophysiol 1995; 12: 139–146.

33. Dement WC, Miles L, Carskadon M. "White paper" on sleep and the elderly. J Am Geriatr Soc 1982; 30:25–50.

34. Bliwise DL. Normal aging. In: Kryger MH, Roth T, Dement WC, eds. Principles and Practices of Sleep Medicine. Philadelphia: WB Saunders, 1994:26–39.

35. Van Coevorden A, Mockel J, Laurent E, Kerkhofs M, L-Hermite-Baleriaux M, Decoster C, Neve P, Van Cauter E. Neuroendocrine rhythms and sleep in aging men. Am J Physiol 1991; 260:E651–E661.

36. Haimov I, Laudon M, Zisapel N, Souroujon M, Nof D, Shlitner A, Herer P, Tzischinsky L, Lavie P. Sleep disorders and melatonin rhythms in elderly people. Br Med J 1994; 309:167.

37. Arendt J, Aldhous M, English J, Marks V, Arendt JH. Some effects of jet-lag and their alleviation by melatonin. Ergonomics 1987; 30:1379–1393.

38. Arendt J, Deacon S, English J, Hampton S, Morgan L. Melatonin and adjustment to phase shift. J Sleep Res 1995; 4:74–79.

39. Claustrat B, Brun J, David M, Sassolas G, Chazot G. Melatonin and jet lag: confirmatory result using a simplified protocol. Biol Psychiatry 1992; 32:705–711.

40. Deacon S, Arendt J. Adapting to phase shifts. II. Effects of melatonin and conflicting light treatment. Physiol Behav 1996; 59:415.

41. Petrie K, Congaglen JV, Thompson L, Chamberlain K. Effect of melatonin on jet lag after long haul flights. Br Med J 1989; 298:705–707.

42. Petrie K, Dawson AG, Thompson L, Brook R. A double-blind trial of melatonin as a treatment for jet lag in international cabin crew. Biol Psychiatry 1993; 33:526–530.

43. Middleton BA, Stone BM, Arendt J. Melatonin and fragmented sleep patterns. Lancet 1996; 348:551–552.

44. Consumers Union. Overcoming insomnia. Consum Rep 1997; 62:10–13.

45. Parkes JD. Melatonin and sleep. In: Fraschini F, Reuter RH, Stanov B, eds. The Pineal Gland and Its Hormones. New York: Plenum Press, 1995:183–197.

46. Weitzman ED, Czeisler CA, Coleman RM, Spielman AJ, Zimmerman JC, Dement WC, Richardson GS, Pollak CP. Delayed sleep phase insomnia: a chronobiologic disorder associated with sleep onset insomnia. Arch Gen Psychiatry 1981; 38:737–746.

47. Pelayo RP, Thorpy MJ, Glovinsky P. Prevalence of delayed sleep phase syndrome among adolescents. Sleep Res 1988; 17:391.

48. Thorpy MJ, Korman E, Spielman AJ, Glovinsky PB. Delayed sleep phase syndrome in adolescents. J Adolesc Health Care 1988; 9:22–27.
49. Dahlitz M, Alvarez B, Vignau J, English J, Arendt J, Parkes JD. Delayed sleep phase syndrome response to melatonin. Lancet 1991; 337:1121–1124.
50. Oldani A, Ferini-Strambi L, Zucconi M, Stanskov B, Fraschini F, Smirne S. Melatonin and delayed sleep phase syndrome: ambulatory polygraphic evaluation. NeuroReport 1994; 6:132–134.
51. Czeisler CA. Is melatonin an effective phase-shifting agent? Commentary. J Biol Rhythms 1997; 12:618–626.
52. Jewett ME, Rimmer DW, Duffy JD, Klerman EB, Kronauer RE, Czeisler CA. Human circadian pacemaker is sensitive to light throughout subjective day without evidence of transients. Am J Physiol 1997; 273:R1800–R1809.
53. Boivin DB, Duffy JF, Kronauer RE, Czeisler, CA. Dose-response relationships for resetting of human circadian clock by light. Nature 1996; 379:540–542.
54. Czeisler CA, Dumont M, Duffy JF, Steinberg JD, Richardson GS, Brown EN, Sanchez R, Ríos CD, Ronda JM. Association of sleep-wake habits in older people with changes in output of circadian pacemaker. Lancet 1992; 340:933–936.
55. Lewy AJ, Ahmed S, Jackson JML, Sack RL. Melatonin shifts human circadian rhythms according to a phase-response curve. Chronobiol Int 1992; 9:380–392.
56. Lewy AJ, Sack RL, Blood ML, Bauer VK, Cutler NL, Thomas KH. Melatonin marks circadian phase position and resets the endogenous circadian pacemaker in humans. In: Ciba Foundation Symposium 183: Circadian Clocks and Their Adjustments. Chichester: Wiley, 1995.
57. Arendt J, Aldhous M, Marks V. Alleviation of jet lag by melatonin: preliminary results of controlled double blind trial. Br Med J 1986; 292:1170.
58. Spitzer RL, Terman M, Malt U, Singer F, Terman JS, Williams JBW, Lewy AJ. Failure of melatonin to affect jet lag in a randomized double-blind trial. Soc Light Treatment Biol Rhythms Abstr 1997; 9:1.
59. Lino A, Silvy S, Condorelli L, Rusconi AC. Melatonin and jet lag: treatment schedule. Biol Psychiatry 1993; 34:587–588.
60. Samel A, Wegmann H-M, Vejvoda M, Maab H, Gundel A, Schütz M. Influence of melatonin treatment on human circadian rhythmicity before and after a simulated 9-hr time shift. J Biol Rhythms 1991; 6:235–248.
61. Underwood H, Goldman BD. Vertebrate circadian and photoperiodic systems: role of the pineal gland and melatonin. J Biol Rhythms 1987; 2:279–315.
62. Aschoff J, Gerecke U, Von Goetz C, Groos GA, Turek FW. Phase responses and characteristics of free-running activity rhythms in the golden hamster: independence of the pineal gland. In: Aschoff J, Daan S, Groos G, eds. Vertebrate Circadian Systems: Structure and Physiology. Berlin: Springer-Verlag, 1982:129–140.
63. Karsch FJ, Woodfill CJI, Malpaux B, Robinson JE, Wayne NL. Melatonin and mammalian photoperiodism: synchronization of annual reproductive cycles. In: Klein DC, Moore RY, Reppert SM, eds. Suprachiasmatic Nucleus: The Mind's Clock. New York: Oxford University Press, 1991:217–232.
64. Reiter RJ. Neuroendocrine effects of the pineal gland and of melatonin. In: Ganong WF, Martini L, eds. Frontiers in Neuroendocrinology. New York: Raven Press, 1982: 287–316.

65. Goldman BD, Elliott JA. Photoperiodism and seasonality in hamsters: role of the pineal gland. In: Stetson MH, ed. Processing of Environmental Information in Vertebrates. New York: Springer-Verlag, 1988:203–218.
66. Turek FW, Van Reeth O. Circadian Rhythms. In: Fregly MJ, Blatteis CM, eds. Handbook of Physiology: Chapter 4, Environmental Physiology. Oxford: Oxford University Press, 1996:1329–1360.

7

Melatonin and Circadian Rhythmicity in Vertebrates
Physiological Roles and Pharmacological Effects

DAVID R. WEAVER

Massachusetts General Hospital
Harvard Medical School
Boston, Massachusetts

I. Introduction

Two major physiological functions have been established for the pineal hormone melatonin (*N*-acetyl-5-methoxy tryptamine) in vertebrate species. First, melatonin is important for maintaining circadian organization in some species, while more subtle effects of melatonin on circadian organization have been demonstrated in other species (for reviews, see 1–12). Second, melatonin is critical for the regulation of reproduction in many seasonally breeding mammals (12). Other roles have been proposed for circulating melatonin, but these (e.g., effects on the immune system; local actions in gut and retina; 13,14) are less firmly established.

   The objective of this chapter is to review the literature describing the effects of melatonin on circadian rhythmicity. There is an amazing degree of diversity in the contribution of melatonin to circadian organization when one compares across vertebrate species. In view of these major species differences, an attempt will be made to describe the results from each species separately without broad cross-species generalizations.

II. Organization of Vertebrate Circadian Timing Systems

Circadian rhythms are rhythms in behavior or physiology with a cycle length of approximately 1 day, and which are generated by an endogenous timekeeping system (15,16). A defining characteristic of a circadian rhythm is that the rhythm persists with a cycle length (period) close to 24 hr in the absence of periodic environmental input.

A. Components of Circadian Timing Systems

Conceptually, a circadian timing system can be viewed as consisting of three components (Fig. 1). The central component is a circadian clock, or oscillator, which has an intrinsic mechanism for circadian time measurement. The oscillator can be reset by environmental or internal stimuli that are perceived and relayed to the oscillator via one or more input pathways. Light is the most prevalent periodic stimulus for daily resetting of biological clocks. The overt expression of the circadian oscillator is through the circadian regulation of hormones, behavior (including sleep), and physiology (15,16).

B. Tissues Involved in Vertebrate Circadian Timing Systems

There is remarkable diversity among vertebrate species in the roles played by the tissues that comprise the circadian timing system. The least redundant situation exists in mammals, where the photoreceptive, oscillatory, and output roles appear to occur separately in distinct tissues (Fig. 2A, but see next paragraph). Lighting information perceived by the retinae is relayed to the suprachiasmatic nuclei (SCN), which function as the master circadian oscillator in mammals (16). The SCN regulate rhythmic outputs via a multitude of neuronal pathways, the most well defined of which is the multisynaptic pathway regulating rhythmic melatonin production from the pineal gland. Light affects the circadian pacemaker only via the retina. Light perceived by the retina can also affect the expression of rhythmicity via noncircadian mechanisms, e.g., by directly affecting the expression of behavior ("masking").

Recent studies demonstrate that the hamster retina contains a circadian oscillator capable of regulating rhythmic melatonin production (17). It is possible that the location of potential oscillators is equally redundant in mammals and nonmammalian vertebrates; it may be only the technical ease with which oscillatory function can be detected that differs. Nevertheless, the SCN contain the master functional pacemaker in mammals, and the retinal oscillator is neither necessary nor sufficient for the maintenance of rhythmicity at the whole-animal level.

The circadian timing system in nonmammalian vertebrates is more complicated due to redundancy in the anatomical location of system components (Fig 2B).

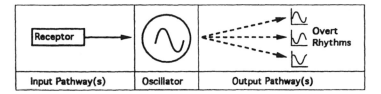

Figure 1 Conceptual components of a circadian timing system.

A.

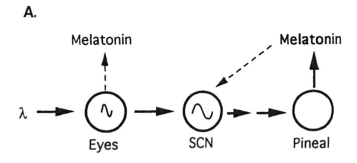

B.

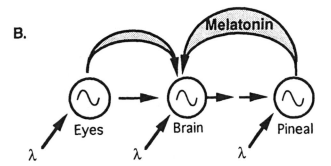

Figure 2 Schematic illustration of components of the circadian timing system in (A) mammals and (B) nonmammalian vertebrates. Note that the mammalian system appears less redundant, in that oscillatory and photoreceptive functions are present in fewer tissues. λ indicates light. The sinusoidal curve denotes tissues that contain a circadian oscillator.

Functional photoreceptors impinging upon the circadian system exist in the retinae of the lateral eyes, as in mammals, but are also found in the pineal gland and in the brain (so-called extraretinal photoreceptors; 1,2). Circadian oscillators are contained within the brain, likely in the visual SCN (vSCN, the apparent homolog of the mammalian SCN), and also within the retinae and pineal gland of several species (18–20). The pineal glands of most (but not all) nonmammalian vertebrates will secrete melatonin rhythmically in vitro for several cycles, and the melatonin rhythms can be entrained by light exposure in vitro. These studies demonstrate the presence of input, oscillator, and output functions within a single tissue. Indeed, studies with dispersed cells indicate that the entire circadian system can reside within single photoreceptor cells from retina and pineal (21–24).

As indicated in Figure 2, melatonin is involved at several points within the circadian timing system of vertebrates. First, as indicated above, the rhythmic secretion of melatonin from the pineal gland is a well-characterized output of the circadian timing system (25). In addition, the ability of periodic melatonin treatment to reset the circadian clock in some species indicates that melatonin can also act as an input to the circadian oscillator (1–11). Finally, the very maintenance of rhythmicity is dependent upon rhythmic melatonin exposure in some species (4,5); melatonin is necessary for normal circadian function due to its role as a coupling agent between oscillatory tissues within a multioscillatory circadian system. The relative importance of melatonin as an input and as a potential clock component varies greatly by species, as discussed in greater detail below.

III. Melatonin Production

A. Regulation of Pineal Melatonin Secretion

The major source of circulating melatonin is the pineal gland (25,26). Melatonin secretion is rhythmic, with peak melatonin levels occurring at night. Melatonin is produced from serotonin by the sequential action of the enzymes serotonin N-acetyltransferase (NAT) and hydroxyindole-O-methyltransferase (HIOMT) (Fig. 3). The rate of melatonin production is controlled primarily by regulation of NAT activity. In rats, transcriptional regulation of NAT mRNA is the major regulator of NAT activity, although significant posttranslational influences on NAT enzymatic activity are also apparent (27,28). In sheep, the rhythms of NAT activity and NAT mRNA levels are considerably more shallow, indicating a more important role for posttranscriptional mechanisms (28). NAT and HIOMT are primarily localized to the pineal gland and retina, recognized sources of melatonin (18).

The regulation of pineal melatonin production can be categorized as being either one of two distinct types. In the first type, melatonin production is regulated by neural input to the pineal gland. This pattern is characteristic of all mammals studied to date. In rats, for example, melatonin production is stimulated by nor-

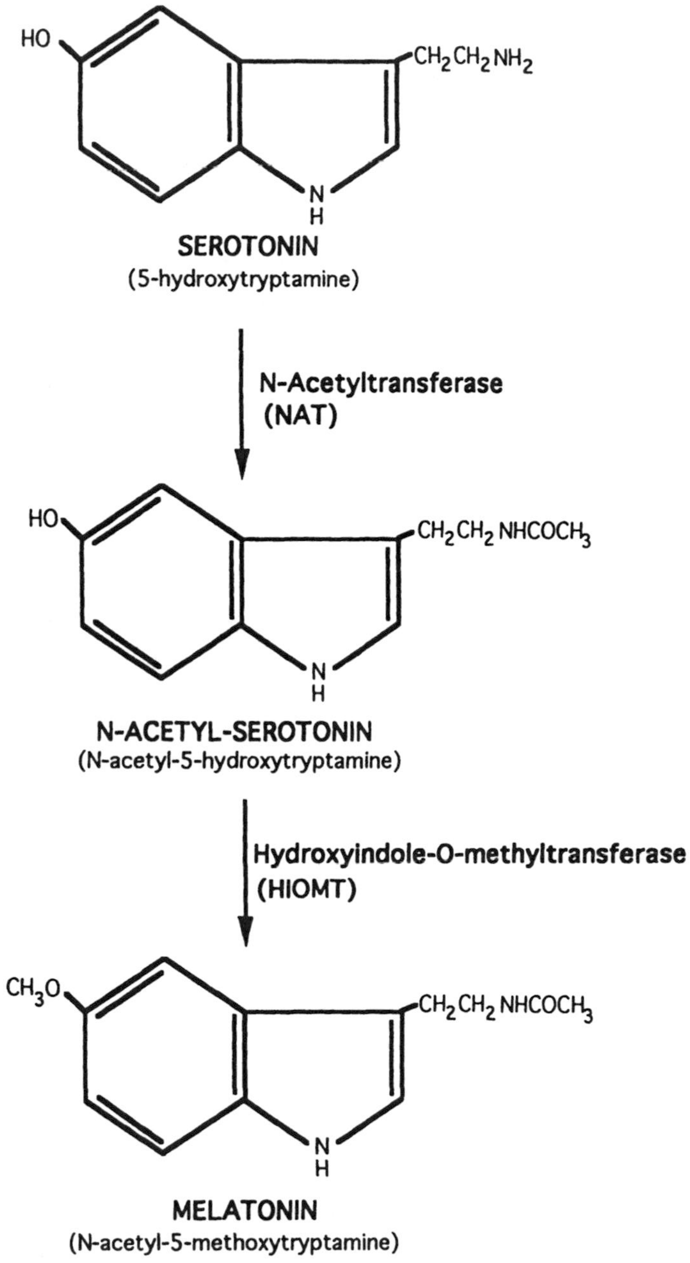

Figure 3 Biosynthesis of melatonin from serotonin.

adrenergic input transmitted to the pineal gland by a multisynaptic pathway originating in the SCN and culminating in the projection from the superior cervical ganglion to the pineal. Stimulation of β-adrenergic receptors stimulates cAMP accumulation, and activation of other signal transduction pathways, leading to an increase in NAT activity (for review see 25). The second type of melatonin regulatory pattern is present in many nonmammalian vertebrates, and is characterized by both intrinsic and extrinsic control of melatonin production. In these species, the pineal itself is photoreceptive and oscillatory, and the oscillations can be observed by the rhythmic secretion of melatonin in vitro (19,20,22–24,29,30). The chicken pineal has been studied extensively owing to its ability to maintain rhythmic melatonin secretion for several days and to entrain to lighting cycles in vitro. Neuronal input does play a role in regulation of pineal melatonin production in the intact animal, however. In chickens, norepinephrine (NE) turnover in the pineal is highest during the day (30). NE inhibits cAMP accumulation via alpha-2 adrenergic receptors, which inhibit NAT activity and melatonin synthesis during the daytime (30). Thus, the same transmitter used by the mammalian sympathetic nervous system to activate melatonin production at night is responsible for inhibition of melatonin production in the bird during the daytime.

Despite the species diversity in the mechanisms regulating pineal melatonin production, one aspect of the melatonin rhythm-generating system is invariant: *melatonin is the hormone of darkness*. The hormone is characterized by rhythmic secretion, and in all species examined, the peak in melatonin levels occurs at night (25).

B. Other Sources of Melatonin

Melatonin is often characterized as a hormone of the vertebrate pineal gland. While this statement is true, it may suggest that other sources of melatonin are physiologically unimportant. On the contrary, the retinae also produce melatonin, as noted above, and in some nonmammalian species, melatonin produced by the retinae is detectable in the systemic circulation and contributes to physiological responses (31–35). Species differences in the relative contribution of pineal, ocular, and other sources to circulating melatonin levels mean that removal of the pineal gland can have markedly different effects on the blood melatonin rhythm. Rhythmic ocular melatonin production in vitro indicates the presence of circadian oscillators within the eyes of hamsters (17), *Xenopus* (19,21), and zebrafish (36). Ocular melatonin is rhythmic in vivo in pigeons, quail, and chickens (31–35,37). Locally produced melatonin appears to influence retinal physiology (14,38). Other sources of melatonin include the gastrointestinal tract and the Harderian gland (39,40). In rats, gut melatonin content exceeds the melatonin content of the pineal (39). Other studies indicate that the pineal gland is the sole source of *circulating* melatonin in rats (26), however.

C. Melatonin Catabolism

In mammalian species, melatonin disappears from the circulation as a result of 6-hydroxylation in the liver, followed by excretion in a sulfatoxy conjugated form. The half-life of melatonin in plasma is under 1 hr (41). Due to the apparent absence of a secretory or storage mechanism, melatonin levels in blood are regulated by the rate of melatonin production. Recently, however, another mechanism for regulating melatonin levels has been demonstrated in nonmammalian vertebrates. Enzymatic deacetylation of melatonin occurs in the retinae, pineal, and brain of a variety of vertebrates (42,43). This finding, along with the detection of low amounts of NAT mRNA in tissues other than the pineal and retina (28), suggests that melatonin may be produced and degraded locally, having a paracrine or autocrine (rather than endocrine) role in some tissues.

IV. Sites and Mechanisms of Melatonin Action

A. Introduction

For an agent to affect circadian rhythmicity, the presence of the agent must be detected and transmitted to the circadian oscillator. The suprachiasmatic nuclei are the major circadian pacemaker in mammals (reviewed in 16), and structural homologs of the SCN in nonmammalian species appear to be functional homologs as well (6,44–48). Destruction of the rat SCN prevents the influence of melatonin on circadian rhythmicity in rats (49). Several lines of evidence suggest that melatonin elicits effects on circadian rhythmicity by interacting with melatonin receptors located within the SCN.

B. High-Affinity Melatonin Receptors Are Located in the SCN

Dubocovich and Takahashi (50) provided the first convincing characterization of a high-affinity melatonin receptor, by demonstrating the very close correlation between binding characteristics of the melatonin analog, 2-^{125}I-iodomelatonin (^{125}I-MEL) in chicken retina and the effects of melatonin on dopamine release from rabbit retina. Shortly thereafter, autoradiographic localization of ^{125}I-MEL binding in rat suprachiasmatic nucleus was reported by Vanecek et al. (51). Many studies describing characteristics and localization of high-affinity melatonin receptor binding have followed, demonstrating discrete localization of pharmacologically specific, high-affinity ($K_d < 200$ pM) melatonin receptors. These studies provide a consistent picture of high-affinity melatonin receptors as being G-protein-coupled receptors with similar pharmacological characteristics across a variety of tissues and species. The consensus rank order of drug potency in inhibiting ^{125}I-MEL binding to high-affinity melatonin receptors (across diverse species and tissues) is generally: 2-iodomelatonin > 6-chloromelatonin ≥ melatonin > 6-hydroxymelatonin > *N*-acetylserotonin ≫ 5-hydroxytryptamine.

High-affinity melatonin receptors are present in the SCN of many species (11,51–73, Table 1, Fig. 4). ^{125}I-MEL binding is present in the vSCN of all avian species examined (48) and in the SCN of all mammalian species in which a circa-

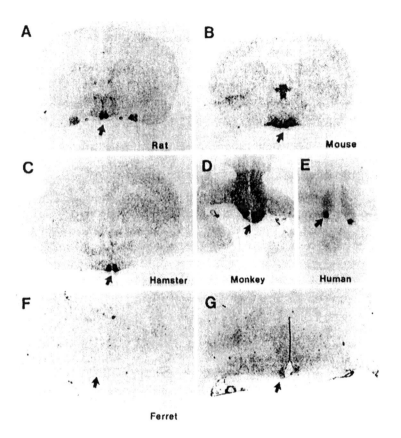

Figure 4 Autoradiographic localization of ^{125}I-MEL binding in the SCN. Sections through the SCN were processed for detection of ^{125}I-MEL binding. Nonspecific binding was defined by incubation of adjacent sections in ^{125}I-MEL (30–100 pM) plus 1 μM melatonin. Dark areas above the section background represent specific ^{125}I-MEL binding, except in the human and monkey specimens, where blood vessels contain nonspecific binding. The location of the SCN is indicated by an arrow in each panel. (A) Rat; (B) mouse; (C) Siberian hamster; (D) rhesus monkey; (E) human adult; (F,G) ferret. (F) and (G) represent the same section; (F) is the autoradiogram and (G) is a photomicrograph of the section after staining with cresyl violet. Note the complete absence of detectable ^{125}I-MEL binding in ferret SCN. For experimental details, see Refs. 52,55,64,71,81,87.

Table 1 Autoradiographic Localization of ^{125}I-MEL Binding in the Mammalian SCN

Species	^{125}I-MEL binding	Ref.[a]
Rodents		
Rat	Yes	51,52
White-footed mice	Yes	53
House mouse (C3H and C57Bl6 strains)	Yes	54,55
Edible dormouse	Yes	56
Garden dormouse	Yes	56
Siberian hamster	Yes	52,57
Syrian hamster	Yes	52,58
Jerboa	Yes	56
European hamster	Yes	56
13-lined ground squirrel	Yes	Thomas, Bittman, and Zucker, unpublished, cited in 59.
Golden-mantled ground squirrel	Yes	60 (but see 61)
Other herbivores		
Guinea pig	Yes	60
Rabbits	Yes	60,62
Insectivores		
Musk shrew	Yes	59
Little brown bat	Yes	59; Weaver unpublished
Hedgehog	Yes	56
Carnivores		
Mink	No	63
Ferret	No	64
Skunk	No	65
Dog	Yes	66
Farm animals/ungulates		
Sheep	No	67
Horse	No	68
Cow	No	68
Donkey	No	68
Goat	No	69
Primates		
Vervet monkey	Yes	70
Baboon	Yes	70
Rhesus monkey	Yes	71
Human	Yes	71–73

[a]References were selected for inclusion based on priority and with a bias toward papers describing the distrubution of binding throughout the brain. Numerous additional papers exist examining SCN binding in several of the rodent species listed.

dian effect of melatonin has been reported. ^{125}I-MEL binding is absent from the mink SCN (63), as in other members of the weasel family (64,65), and melatonin does not appear to influence circadian rhythmicity in mink (63).

While the presence of melatonin receptors in the SCN is often taken as strong evidence that melatonin acts directly within the SCN, this is not a logical necessity. Melatonin could act at sites distant from the SCN, and influence the SCN via afferents. Conversely, the absence of detectable ^{125}I-MEL binding in the SCN does not preclude an effect of the hormone: melatonin could act at distant sites, or it is possible that an undetectable density of receptors present within the SCN is effective in eliciting physiological responses.

Other lines of evidence suggest that the SCN are a site of melatonin action. Melatonin treatment in vivo inhibits SCN metabolic activity (74) only at circadian times when melatonin in capable of influencing circadian rhythmicity. Melatonin treatment has also been reported to inhibit the photic induction of c-*fos* gene expression in the mouse SCN at night (11). Melatonin induces acute changes in rodent SCN neuronal firing rate and other acute responses in vitro (75–80), suggesting the presence of functional circadian receptors on SCN cells. SCN multiunit activity is inhibited by melatonin, and this effect of melatonin is absent in mice with targeted deletion of the Mel_{1a} melatonin receptor subtype (81). Pinealectomy or exposure to constant light increases the number of SCN cells responding acutely to melatonin, and alters the rhythm of responsiveness to melatonin observed in vitro (79,80), suggesting that endogenous melatonin influences SCN neuronal responsiveness to exogenous melatonin. The most direct evidence for the SCN being the direct site of melatonin action in eliciting its effects on circadian phase comes from in vitro studies of the rodent SCN (81–86; see Section V.C).

C. Molecular Identification of Melatonin Receptor Subtypes

Recent molecular studies have led to the identification of three melatonin receptor subtypes, which comprise a distinct subfamily within the superfamily of G-protein-coupled receptors (Table 2; 55,86–92; for review see 93). The pharmacological characteristics of the three subtypes are functionally indistinguishable with currently available drugs; all possess the general rank order of drug potency defined for high-affinity melatonin receptors from tissues (above).

Two melatonin receptor subtypes have been identified in mammals, the Mel_{1a} and the Mel_{1b} receptor subtypes (88,89). Both bind ^{125}I-MEL with high affinity. The close correlation between the distribution of ^{125}I-MEL binding (detected by autoradiography) and the distribution of Mel_{1a} receptor mRNA (detected by in situ hybridization) suggested that the Mel_{1a} receptor is responsible for most (if not all) ^{125}I-MEL binding observed by autoradiography (55,73,87); this has been confirmed by the observation that ^{125}I-MEL binding is not detectable in the

Table 2 Characteristics of Recombinant Melatonin Receptors

	Mel_{1a}	Mel_{1b}	Mel_{1c}	MelRR-1
Class/species				
Mammals				
Human	Yes[a]	Yes[a]	—	Yes
Sheep	Yes[a]	—	—	Yes[b]
Siberian hamster	Yes[a]	Yes[b,c]	—	Yes[b]
Syrian hamster	Yes[b]	Yes[b,c]	—	—
Mouse	Yes[a]	Yes	—	Yes
Rat	Yes[b]	Yes[b]	—	Yes[b]
Chicken	Yes[a]	Yes[b]	Yes[a]	—
Zebrafish	Yes[b]	Yes[b]	Yes[b]	—
Xenopus	Yes[b]	Yes[b]	Yes[a]	—
Receptor properties[d]				
K_d (^{125}I-MEL, pM)	20–60	160	30–60	N.A.[e]
Effect on cAMP	Decrease	Decrease	Decrease	N.A.[e]
Structure				
Exons	2	2	2	2
Amino acid identity[f]	80–94%	80–90%	80%	80–90%

[a]cDNAs encoding functional receptor proteins have been isolated and expressed.
[b]Only fragments of genomic DNA and/or cDNA have been isolated.
[c]Receptor cDNA contains one or more stop codons in exon 2; gene product cannot encode a functional receptor.
[d]Receptor properties were determined by expression of cloned receptor cDNAs in mammalian cell lines. The rank order of potency for inhibiting ^{125}I-MEL binding for each receptor subtype is 2-iodomelatonin > 6-chloromelatonin ≥ melatonin > N-acetylserotonin > 6-hydroxymelatonin ≫ serotonin.
[e]The human MelRR1 does not bind ^{125}I-MEL or ^{3}H-melatonin, and no signal transduction responses to melatonin have been identified to date.
[f]Amino acid identity among members within each receptor subfamily.
Overlapping regions generally represent the entire sequence between transmembrane domains 3 and 6 (inclusive) of the receptor protein. Comparisons across receptor subtype are lower; e.g., for the full-length clones identities are: human Mel_{1a} vs. human Mel_{1b} 60% , human Me_{1a} vs. human MelRR1 45%, human Mel_{1b} vs. human MelRR1 45%, chick Mel_{1a} vs. chick Mel_{1c} 68%.
Source: Data from Refs. 55, 86–92 and S. M. Reppert and/or D. R. Weaver, unpublished data. Table modified from Ref. 93.

brains from mice with targeted deletion of the Mel_{1a} melatonin receptor (81). In these Mel_{1a} melatonin receptor knockout mice, the acute effects of melatonin on SCN multiunit activity are completely absent, while the phase-shifting response to melatonin appears normal when physiological concentrations (1 nM) are used. Similarly, a 1-nM concentration of 2-iodomelatonin causes comparable phase shifts in Mel$_{1a}$ knockout and littermate control mice. Only at low doses of 2-iodo-melatonin (10 pM) does an effect of the absence of Mel$_{1a}$ receptors become detectable; even this low concentration causes detectable phase shifts in slices from both wild-type and knockout mice, but the shifts are smaller in the knockout mice (81). These data indicate that another receptor subtype is capable of mediating phase shifts to melatonin, in the absence of the Mel$_{1a}$ receptor. The ability of melatonin to phase-shift the electrical activity rhythm in the SCN of Mel$_{1a}$-receptor-deficient mice is blocked by a pertussis-toxin-sensitive mechanism, indicating that a G-protein-coupled receptor likely mediates the response. This focuses attention on the Mel$_{1b}$ receptor, and indeed, the Mel$_{1b}$ receptor mRNA is detectable in the SCN by RT-PCR (although not by in situ hybridization). Furthermore, the low concentrations of 2-iodomelatonin that cause phase shifts in the SCN of Mel$_{1a}$-receptor-deficient mice indicates that the relevant receptor has high affinity for ^{125}I-MEL (81). That ^{125}I-MEL binding is not detectable in the SCN of Mel$_{1a}$-receptor-deficient mice strongly suggests that a very low density of melatonin receptors in capable of producing the full response. Collectively, these data suggest functional redundancy between the Mel$_{1a}$ and Mel$_{1b}$ receptor subtypes. Generation of mice lacking both the Mel$_{1a}$ and Mel$_{1b}$ receptor subtypes is necessary to test this hypothesis. However, the Mel$_{1b}$ receptor is not necessary for in vitro circadian responses to melatonin in Siberian hamsters (86).

The relative contribution of the Mel$_{1a}$, Mel$_{1b}$, and Mel$_{1c}$ receptor subtypes in mediating physiological responses in nonmammalian vertebrates is unclear. In several species, genes encoding all three receptor subtypes are present (e.g., chicken, zebrafish; *Xenopus*; 89, 92; see Table 2), although the entire coding region has not been cloned for all three subtypes in any species. In chickens, both Mel$_{1a}$ and Mel$_{1c}$ melatonin receptor subtypes appear to contribute to the pattern of ^{125}I-MEL binding observed (89); the distribution of the Mel$_{1b}$ receptor subtype has not been described. Each of these receptor subtypes may contribute to physiological responses to melatonin.

D. Other Melatonin-Binding Sites

Other ^{125}I-MEL-binding sites have been described. By one schema, the high-affinity ^{125}I-MEL-binding sites described above are termed ML-1 sites (94). It is now apparent that the sites designated "ML-1 sites" in various tissues and species represent expression of one or more of the three subtypes whose cDNAs have been isolated. A second site, termed the ML-2 site (57,94), initially detected in hamster brain, is characterized by lower affinity (K_d ca. 1 nM), detection only at reduced temperatures (0–4°C), and a unique pharmacological profile (2-iodo-melatonin $\geq$ 6 chloromelatonin $\geq$ prazosin $>$ *N*-acetylserotonin $\geq$ melatonin $\geq$ 6-hydroxymelatonin $\gg$ serotonin). This site does not appear to be G-protein-coupled, and its physiological significance remains to be established.

Nuclear melatonin-binding sites have also been described (95). Reiter and co-workers have reported accumulation of melatonin in cell nuclei, and binding to nuclear proteins, suggestive of a nuclear uptake mechanism or nuclear binding site. Several studies by Carlberg et al., have described melatonin as the ligand for RZRβ, an orphan receptor of the nuclear receptor family (for review, see 96). This finding has not been replicated by other groups, however (97,98), and its relationship to the nuclear melatonin-binding sites described by Reiter's group is unclear. The present consensus, to the extent there is one, indicates that RZRβ is not a nuclear melatonin receptor (97,98).

The massive doses of melatonin employed in most studies of circadian rhythms are likely to produce levels of the hormone far above the physiological levels, and far above the levels that are likely to retain specificity for high-affinity melatonin receptors (40,41,99). Injection of large doses of melatonin could influence levels of other neurotransmitters (e.g., 5HT), providing a mechanism by which melatonin can have indirect effects on the SCN via other receptor subtypes (but see 100). In addition, the possibility that additional melatonin receptor subtypes exist cannot be excluded at this time.

E. Melatonin Receptor Signal Transduction

Signal transduction pathways influenced by melatonin have generally been identified by examining responses to melatonin receptor occupation in tissues containing ^{125}I-MEL binding. The sheep pars tuberalis, hamster pars tuberalis, neonatal rat pituitary, rat tail and cerebral vessels, and brain tissue from various regions have been studied most extensively (see 101 for review). In these tissues, melatonin inhibits cAMP accumulation via a pertussis toxin (PTX)-sensitive G-protein. Morgan and colleagues have also demonstrated coupling of melatonin receptors by PTX-insensitive, but cholera-toxin-sensitive G-proteins (102). Multiple melatonin receptor subtypes could be expressed in each tissue. Only with the recent development of cell lines stably expressing individual recombinant melatonin receptor cDNAs has it been possible to definitively link occupation of specific melatonin receptor subtypes with transduction responses (103).

Other effects of melatonin on cellular signal transduction pathways demonstrated in neonatal pituicytes by Vanecek and colleagues include inhibition of cGMP accumulation, diacylglycerol, and arachadonic acid, and inhibition of membrane depolarization and intracellular calcium levels (see 101 for review). Melatonin also inhibits GABA release from rabbit cortex (104) and can potentiate cAMP accumulation under some circumstances (105,106). Melatonin potentiates prostaglandin-stimulated arachidonate release in cells stably expressing the recombinant Mel$_{1a}$ melatonin receptor (103). Melatonin has also been reported to cause a transient activation of protein kinase C activity in the rat SCN (83; see below). Effects of melatonin to stimulate inositol phosphate metabolism in chicken brain and melanoma cells appear to be mediated by a receptor with different pharmacological characteristics (107,108).

A variety of downstream responses to melatonin receptor occupation have been described using in vitro systems, e.g., inhibition of dopamine release in retina (38) and brain (50,109), inhibition of GnRH-stimulated LH release from explants of the anterior pituitary (see 101), and inhibition of phosphorylation of cAMP response element binding protein (110). Whether these responses occur in SCN in response to melatonin is unclear.

Two studies have examined signal transduction pathways activated by melatonin in rat SCN that lead to phase shifts. Using similar methods, one group has implicated the inositol phosphate pathway (83), while another indicated that the phase-shifting effects of melatonin are mediated via activation of nitric oxide synthase (85).

V. Experimental Approaches

Several approaches have been used to examine whether melatonin influences entrainment and/or circadian rhythmicity in vertebrates. The general types of experiments will be outlined here, and their relative advantages and disadvantages noted.

A. Pinealectomy

The most obvious means to assess the role of the pineal gland in circadian organization is to remove the pineal gland and examine circadian rhythmicity in constant conditions. Studies of this type can detect whether the pineal gland is needed for maintenance of rhythmicity. More subtle effects, including alterations in free-running period, activity duration, or activity consolidation, can be detected.

By examining the effects of pinealectomy in circadian behavior in light-dark (LD) cycles, it is possible to detect changes in phase angle of entrainment (the timing of activity onset relative to the LD transition), or rate of reentrainment following a shift of the LD cycle. Interpretation of studies conducted in LD cycles is complicated by the possibility that pinealectomy alters the stimulus properties of light (7,111,112). Cyclic lighting normally has "masking" influences on circadian rhythms, meaning that the presence or absence of light affects the rhythm measured, independently of the circadian system. For example, locomotor activity in most nocturnal rodents is acutely suppressed by light exposure at any circadian time. Conversely, an increase in activity frequently is induced by a dark pulse. These responses to light are influenced by both ambient and stimulus light intensity. Altered perception of light intensity can have a profound impact on the expression of circadian rhythmicity, including free-running period, even without affecting the underlying circadian clock. Behavioral patterns recorded in a LD cycle are likely not an accurate reflection of the neuronal output of the circadian clock, due to the "masking" effects of light on behavior (see 7,112 for a more complete discussion).

A major disadvantage of pinealectomy as an experimental approach is that the pineal is not the sole source of rhythmic melatonin in many nonmammalian vertebrates. In quail and pigeons, for example, retinal melatonin contributes to the rhythmic levels of melatonin in blood (31–33). The recent demonstration that rhythmic melatonin production can be driven by a circadian clock in the hamster retina (17) raises a similar concern in mammals, although the pineal is generally regarded as the sole source of *circulating* melatonin in mammals under normal conditions (26). Another disadvantage of pinealectomy as an experimental approach is that other products of the pineal gland could contribute to circadian organization, whether secreted rhythmically or not.

Among studies using pinealectomy as an experimental approach, the most readily interpreted are those that demonstrate alterations in circadian period or circadian organization, in species in which the pineal is the sole source of circulating melatonin.

B. Treatment with Melatonin

Another class of studies involves the use of exogenous melatonin. Several paradigms have evolved. The simplest version involves injecting a single dose of melatonin into subjects in constant lighting conditions, followed by a period of assessment of rhythmicity to determine whether the phase of the rhythm has been altered. By determining the response to treatments administered at varying circadian times, it is possible to define the "phase-response curve" (PRC, a plot of the phase-shift amplitude in response to treatment at various phases of the circadian cycle; Figs. 5 and 6). In many studies, however, the effects of melatonin are so subtle as to require repeated treatment (e.g., several days of injection) to detect an effect. The extreme of this line of studies involves repeated injection of melatonin at the same time of day each day for weeks. The repeated daily injection results in small daily phase advances, or small changes in period length, such that the period of rhythmicity equals the period of the injection ($T = 24$), resulting in entrainment.

In some cases, melatonin treatment has been combined with other treatments (e.g., light pulse or benzodiazepine treatment) to assess interactions. A common approach involves determining the effect of repeated melatonin treatment on the rate of reentrainment to a phase shift of the LD cycle. Again, issues of altered stimulus properties of light can confound the interpretation of these studies.

Finally, continuous treatment with melatonin (e.g., by implantation of melatonin-filled silastic capsules) has been used to produce chronic, high levels of melatonin, thus "swamping out" any rhythmicity in melatonin produced from endogenous sources. This method is often a complementary approach to removal of sources of melatonin, as one would predict similar results from these two types

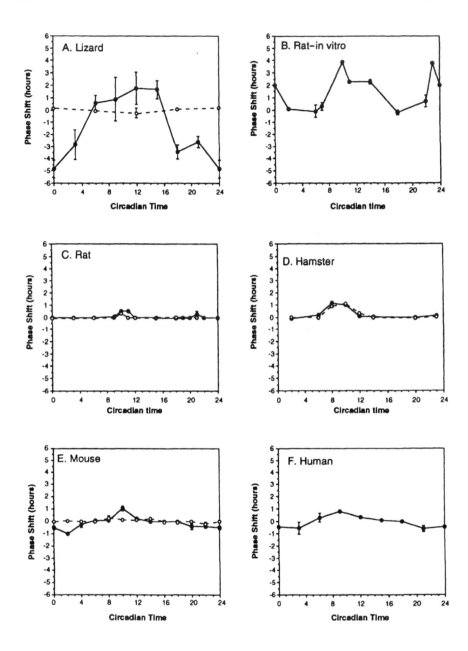

of studies if rhythmicity of melatonin were important for the outcome being measured.

Generally speaking, studies involving administration of melatonin use a gross excess of melatonin, leading to supraphysiological levels (41,99,113,114), and the possibility that other receptors are being influenced by the melatonin treatment. Furthermore, handling animals for injection may lead to arousal, thus producing activity-related phase shifts that are independent of the effects of melatonin (115). Examination of vehicle-treated controls is an obvious and critical control.

Many studies examining the effects of exogenous melatonin on circadian rhythms use pineal-intact animals. Failure to observe effects of melatonin administered at night could be due to the occupation of melatonin receptors at the time of treatment by endogenous melatonin.

C. In Vitro Preparations

As indicated above, the most compelling evidence that melatonin acts within the SCN to influence circadian rhythmicity in mammals comes from an in vitro preparation. Hypothalamic explants containing the SCN maintain rhythms in

---

Figure 5 Phase-response curves to melatonin. Data are plotted as the mean ± SEM phase shift induced by the treatment. Positive values represent phase advances; negative values represent phase delays. All data have been plotted as circadian time (CT), where CT 0 represents the beginning of subjective day, and CT 12 represents the beginning of subjective night. Phase shifts were detected by monitoring the locomotor activity rhythm, except as noted. Data at the end of each panel are double-plotted. To emphasize the species difference in the magnitude of melatonin-induced phase shifts, the same scale has been used for all panels. (A) Lizards (*Sceloporus occidentalis*) free-running in constant dim illumination were injected with melatonin (10 μg; approximately 10 mg/kg; filled circles) or vehicle (open circles). (Data from Ref. 170.) (B) SCN slices from rats were treated with melatonin (1 nM for 1 hr) in vitro. Phase shifts were assessed by determining the time of peak electrical activity (firing rate) in individual slices. (Data from Refs. 82 and 83.) (C) Rats free-running in DD were injected with melatonin (50 mg/kg, filled circles) or vehicle (open circles). Three of five rats injected with vehicle at CT 10 had significant phase shifts. (Data from Ref. 8.) (D) Syrian hamsters free-running in DD were injected with melatonin (1 mg/ kg, filled circles) or vehicle (open circles). No specific response to melatonin was detected. (Data from Ref. 115.) (E) Mice free-running in constant darkness were injected with melatonin (90 μg; approximately 3 mg/kg; filled circles) or vehicle (open circles) on three consecutive days. (Data from Ref. 21) (F) Human subjects maintained in dim light were treated with melatonin (0.5 mg; approximately 0.007 mg/kg) and circadian phase was assessed by determining the time of subsequent endogenous melatonin secretion. The dim light melatonin onset of each subject prior to treatment was defined as CT 14. Data from Ref. 232 have been averaged into 3-hr bins; values are plotted at the midpoint of each bin.

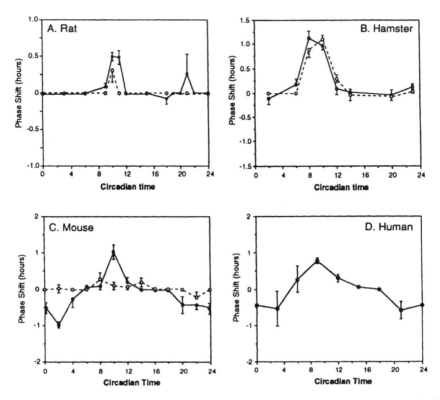

Figure 6 Phase-response curves to melatonin. The data presented in Figure 5 (panels C, D, E, and F) have been plotted to maximize the ability to see small phase shifts by adjusting the y-axis for each panel.

single-unit activity (firing rate) for up to 3 days in vitro. Treatment of the rat SCN with melatonin in vitro induces a phase shift of 2–4 hr (82–85). The response to melatonin is dependent upon the time of treatment, and the PRC of the SCN in vitro maintains a reasonable resemblance to the PRC to melatonin administered in vivo, except that the amplitude of shifts observed in vitro is much larger (Fig. 5B, 5C). Melatonin is extremely potent in shifting the electrical activity rhythm of the SCN, with EC_{50} values less than 10 pM (83,84). The agonist 2-iodomelatonin mimics the effects of melatonin (81–83). As noted above, the phase-shifting effects of melatonin and 2-iodomelatonin are preserved in mice lacking the Mel_{1a} receptor, despite the absence of detectable melatonin receptor binding in these animals (81). The SCN appear exquisitely sensitive to melatonin.

This in vitro system allows examination of signal transduction pathways not amenable to manipulation in vivo. The effects of melatonin are prevented by pretreatment with PTX, indicating involvement of a guanine nucleotide binding protein, likely Gi (81–83). The inhibition of cAMP appears unlikely to mediate the response to melatonin, however. One laboratory has reported that melatonin induces phase shifts via effects on protein kinase C (PKC) activity (83). The effects on PKC activity are shockingly brief. Using a very similar paradigm, another investigator reports that melatonin and serotonin causes phase shifts in the rat SCN via effects on nitric oxide (NO) (85). NO mediates the phase-shifting effects of glutamate, the principal retinohypothalamic transmitter. Low levels of nitric oxide synthase (NOS) activity and NOS immunoreactivity have been demonstrated in the SCN, but there is little evidence from other approaches to indicate how melatonin might influence NO levels in the SCN. In other systems, NO is thought to elicit its cellular effects by stimulation of guanylyl cyclase activity in target cells, but melatonin inhibits cGMP accumulation in other systems (101). Furthermore, treatment of SCN slices with cGMP causes phase-dependent phase shifts in SCN electrical activity (116). These effects are limited to phase advances in response to cGMP treatment at night; treatment during the subjective day is without effect (116). It is difficult to envision how melatonin could elicit a phase shift via stimulation of NOS activity and cGMP at a time when direct application of cGMP is without effect. Clearly, additional study is needed to reconcile these studies with respect to the specific mechanisms of melatonin action.

VI. The Role of Melatonin in Regulation of Vertebrate Circadian Rhythms

A. Nonmammalian Vertebrates

In many avian and reptilian species, endogenous melatonin is an important component of the circadian timing system. In these species, there is evidence for a central circadian oscillator, which is likely to reside (at least in part) in the homolog of the SCN. The pineal gland is directly photoreceptive and oscillatory (as are the retinae in some cases), and these structures interact with the central oscillator via secreted melatonin. The most extensively studied species is the house sparrow (*Passer domesticus*). These studies will be reviewed in detail to illustrate the importance of melatonin for circadian organization in sparrows; this is in contrast to the subtle effects of melatonin in mammals.

The Pineal Is the Driving Pacemaker Within the Sparrow Circadian Timing System

In an extensive and elegant series of studies, Menaker and colleagues have demonstrated that the pineal hormone plays a major role in the circadian organization of house sparrows (see 4–6, for reviews). In fact, the pineal gland is the driving oscillator in a multioscillatory system. Melatonin secreted from the pineal

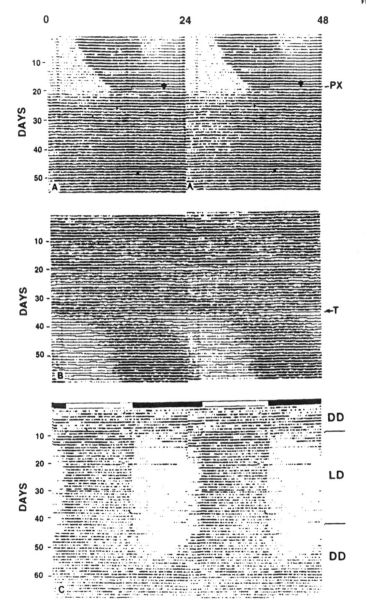

gland interacts with other components of the system to regulate behavioral rhythmicity (see below).

Removal of the pineal gland disrupts circadian rhythmicity in perch-hopping behavior (117) (Fig. 7). Pinealectomized sparrows nevertheless have clear evidence of a (damped) circadian oscillator: activity rhythms can entrain to LD cycles, the phase position of pinealectomized animals is altered in LD, rhythmicity is lost gradually (rather than immediately) following transfer of pinealectomized birds from LD to constant darkness (DD), a light pulse will induce several days of rhythmicity in otherwise arrhythmic sparrows, and there are limits of entrainment to melatonin treatments (117–119). These studies indicate that the pineal gland interacts with another oscillator in the regulation of locomotor rhythmicity. The SCN appears to be that secondary oscillator; lesion studies show that the anterior hypothalamus (including SCN) is necessary for maintenance of rhythmicity in house sparrows (44).

Initial evidence for the importance of a humoral output from the pineal gland came from the observation that neither disruption of the sympathetic input to the gland (with 6-hydroxydopamine lesions) nor disruption of neural efferents from the pineal gland disrupts free-running rhythmicity (120). Furthermore, treatment with a constant-release form of melatonin shortens the free-running period in DD or induces continuous activity in pineal-intact sparrows (121). Finally, resto-

---

Figure 7   The pineal is a driving pacemaker in the sparrow circadian timing system. Each panel represents the perching activity rhythm of a sparrow double-plotted in actogram format. Each horizontal line represents a 48-hr period. The data from successive 24-hr periods are reproduced both to the right and below the first 24-hr period, resulting in a double-plotted record for ease of visualization. Movement of the bird to the perch causes a pen deflection from the baseline; period of intense perching activity appear as thick bars. (A) Pinealectomy disrupts rhythmicity in a sparrow housed in constant darkness. The recurrence of activity at regular intervals despite the absence of environmental time cues is evidence for endogenous rhythmicity. The activity occurs later on each day of the record, indicating that the free-running cycle length (period) is greater than 24 hr. Removal of the pineal gland (PX, arrow) causes an immediate loss of rhythmicity in DD. (Adapted from Ref. 117) (B) Rhythmicity is restored in a pinealectomized sparrow following transplantation (T, arrow) of a donor pineal into the anterior chamber of the eye. The animal was housed in DD except for a brief light pulse at the time of surgery. (Adapted from Ref. 4.) (C) Locomotor activity of a blinded, pinealectomized sparrow. In constant darkness (DD), the animal is arrhythmic. Exposure to a light-dark cycle (12L:12D; LD) entrains rhythmicity via photoreceptors located in the brain. Upon transfer to DD, rhythmicity decays gradually (in marked contrast to the immediate cessation of rhythmicity upon pinealectomy of animals in DD, as in A). This transient rhythmicity indicates the presence of a circadian oscillator outside the eye and pineal, which is not robust enough to be self-sustaining. (Adapted from Ref. 122.)

ration of free-running rhythmicity by pineal transplantation into the anterior chamber of the eye indicated a humoral output from the pineal (120,123). Rhythmicity was restored within 24 hr after pineal transplantation surgery in several cases. Furthermore, pinealectomized sparrows bearing intraocular pineal transplants entrain to LD cycles, free-run when placed in DD, and phase-shift to light pulses as do pineal-intact birds (123). The possibility that pineal melatonin played a permissive role in synchronizing the rhythmicity of other oscillators was entertained until additional transplantation studies were performed. In these studies, pinealectomized sparrows received pineal transplants from donor birds maintained on opposite lighting schedules. The phase of restored rhythmicity was determined by the phase of entrainment of the donor pineal gland (123). Thus, the pineal gland appears to be a self-sustaining, driving pacemaker in the sparrow circadian system. Recent studies using timed administration of melatonin have shown that rhythmic melatonin restores behavioral rhythmicity in pinealectomized house sparrows (124). The sparrow pineal thus appears to act as a pacemaker through its rhythmic production of melatonin.

Other recent studies have addressed the sites and mechanisms by which melatonin influences circadian organization in sparrows. High-affinity ^{125}I-MEL binding sites are present within the sparrow SCN (125,126). Metabolic activity in the sparrow vSCN is rhythmic, with glucose utilization being higher during the day than at night (126). Acute injection of melatonin reduces vSCN metabolic activity (126a). Pinealectomy leads to disruption of metabolic activity rhythms and to disruption of rhythmicity in melatonin receptor density (125) within several days.

To summarize these findings: melatonin is a critical component of the sparrow circadian timing system. The rhythmic secretion of melatonin from the pineal gland imposes rhythmicity on subservient oscillators within central structures. These central structures, which include the SCN, directly regulate the expression of overt rhythms. Neural input to the pineal gland does not appear necessary for the maintenance of rhythmicity.

The Neuroendocrine Loop Model of the Avian Circadian Timing System

Studies conducted in other avian species indicate that a more typical avian pattern involves a less self-sustaining pineal oscillator than is the case in house sparrows. In some other avian species, relatively weak oscillators are coupled to each other in a manner that results in their mutual inhibition (see 4–7,127). It is this mutually inhibitory loop, a neuroendocrine loop, that leads to a self-sustaining oscillatory system. The locus of one oscillator is the pineal gland, while central structures comprise another oscillator. The extent of rhythm disruption following removal of sources of rhythmic melatonin will depend upon the autonomy of other oscillators within the circadian timing system and their ability to interact in the absence of melatonin. Thus, removal of melatonin may have little effect, or it may alter phase of entrainment in LD, alter circadian period, lead to rhythm instability, or lead to clear arrhythmicity.

The components of the proposed neuroendocrine loop are indicated in Figure 2B. In this scheme, melatonin is produced rhythmically by the pineal gland (and eyes, in some cases). Circulating melatonin inhibits the activity of central oscillators, presumed to include the homolog of the SCN. The SCN are in turn responsible for inhibiting the activity of the pineal gland during the day via the sympathetic nervous system. The components of the system thus provide a loop with alternating periods of activity, each approximately 12 hr in length.

The data from other avian species fit with the neuroendocrine loop hypothesis with varying degrees of success. In some cases, it is necessary simply to recognize that the eyes are also a potential source of rhythmic melatonin secretion to explain discrepancies between the data and this model. In other cases, however, it is clear that the basic circadian organization differs.

One example of a difference from the "neuroendocrine loop" model occurs in sparrows, in that sympathetic innervation of the pineal does not appear necessary for maintenance of circadian rhythmicity. As discussed above, a sparrow pineal gland transplanted to the eye restore rhythmicity in locomotor activity of pinealectomized birds, in the apparent absence of sympathetic innervation of the gland (4,120,123). Nevertheless, the sparrow pineal gland is apparently not capable of sustaining rhythmic melatonin production when placed under constant conditions in vitro (128). It is possible that the restoration of rhythmicity in vivo is dependent upon innervation of the pineal gland by sympathetic fibers that normally innervate the iris, but this has not been demonstrated. (If this were the case, the choice of an ocular transplantation site was fortuitous in leading to success in this line of studies.) Rhythmic melatonin secretion from pineals of most other avian species also decays to arrhythmicity over the course of several days in constant conditions in vitro (20,30). This "damping" of rhythmicity has been taken to indicate that periodic neural input from the superior cervical ganglion (or photic input) is necessary to maintain the pinealocytes in a coordinated condition (30). In contrast, the sparrow pineal gland appears capable of self-sustaining rhythmicity in melatonin production in vivo, but not in vitro. (In contrast, pineal glands from anoles and zebrafish appear capable of maintaining long-term rhythmicity in vitro (36,129). This raises the possibility that "coupling,"—in this case meaning coupling of pineal cells to form a coherent oscillator in the absence of adrenergic or photic input—can be disrupted by in vitro conditions. A similar phenomenon may underlie the failure of rat SCN circadian clock cells to synchronize under in vitro conditions (130).

A second example for a departure from the neuroendocrine loop model is in quail, where the evidence is direct. Superior cervical ganglionectomy does not disrupt the pineal melatonin rhythm in Japanese quail (131,132). In this case, the eyes appear to be self-sustaining circadian oscillators that communicate with the pineal by another, likely neural, route. The observation that the anatomical pathway by which information from the brain reaches the pineal gland does not involve the superior cervical ganglion does not contradict the gist of the neuroendocrine loop model, however.

More fundamental divergence from the neuroendocrine loop model exists in species where rhythmic melatonin secretion does not appear to be necessary for maintenance of rhythmicity. In other words, avian species exist in which the central circadian oscillator is sufficiently stable/robust to be able to maintain oscillatory function in the absence of periodic melatonin (see below). An intermediate situation appears to be more common, in which removal of periodic melatonin secretion results in altered rhythmicity, such as instability or alterations of circadian period (cycle length). This likely indicates that melatonin plays a modulatory, not necessary, role in circadian organization.

There are differences among avian species in the importance of melatonin in circadian organization, in the extent to which the pineal gland contributes to circulating melatonin levels in blood, in the role of the retina as a photoreceptor and as a source of circulating melatonin, and in the degree to which the SCN contain a self-sustaining circadian oscillator. In view of this variability, it is difficult to characterize "the avian circadian system" as a single entity. This dictates a species-by-species description of the effects of removal of the pineal gland and of exogenous melatonin treatment on circadian organization (see Tables 3 and 4 and below). Furthermore, observation of significant differences among avian species makes generalization, e.g., from avian species to mammals, difficult at best. Despite this caveat, the neuroendocrine loop model has been a useful framework in which to view the effects of melatonin on circadian organization in a variety of species.

The Pineal Gland, Melatonin, and Rhythmicity in Other Avian Species

Other Sparrow Species

Several species display a house-sparrow-like dependence on pineal melatonin. Pinealectomy causes arrhythmicity or period changes in house finches, white-crowned sparrows, and white-throated sparrows (133–135). Of particular interest among these studies is the finding that pinealectomy disrupts not only the timing of normal (daytime) activity, but also disrupts *nocturnal* migratory restlessness in a day-active species (133). Studies with melatonin administration have not been performed in these species.

Pinealectomy also disrupts circadian rhythmicity in Java sparrows (*Padda oryzivora*). Continuous melatonin treatment of pineal-intact birds results in arrhythmicity (45). Hypothalamic lesions disrupt locomotor rhythmicity, indicating that the hypothalamus contains a component of the circadian timing system (44), as also shown for house sparrows (44). The homolog of the SCN is a potential site for the circadian effects of melatonin, as this brain region (the vSCN) contains putative melatonin receptors in Java sparrows (Rivkees unpublished data, cited in 136), as in house sparrows (125), chickens (136,137), and other avian species (48).

European Starlings (*Sturnus vulgaris*)

In European starlings, activity rhythms are generally altered by pinealectomy. Decreased free-running period and arrhythmicity have been reported after pinealectomy (127,138), while one report indicates no detectable effect of pinealectomy on free-running rhythms (139). Pinealectomy prevents rhythms of melatonin in the blood stream (140). Rhythmicity can be entrained in starlings by repeated daily injections of melatonin (141).

Pigeons (*Columba livia*)

In pigeons, the pineal and retinae each contribute to rhythmic blood levels of the hormone (33,142). Removal of the eyes or of the pineal alone does not disrupt rhythmicity, but rhythmicity is abolished if both procedures are performed (143). Conversely, constant-release melatonin implants disrupt rhythmicity in intact birds (143,144). In pinealectomized pigeons, free-running rhythms of feeding behavior are entrained by infusion of physiological doses of melatonin (142). Furthermore, rhythmic melatonin infusions into arrhythmic, pinealectomized-plus-enucleated pigeons restores rhythmicity (142,145). The presence in variability in the phase angle of entrainment and of damping cycles after cessation of melatonin treatment supports the existence of a residual circadian oscillator (142) outside of the pineal gland and eyes. These data suggest that melatonin secreted by the pineal and retinae are components of the circadian timing system, and are consistent with a neuroendocrine loop hypothesis, which recognizes both the eyes and pineal as sources of melatonin. The importance of innervation of the pineal in forming the outward limb of the loop is not clear; e.g., it is not clear whether it is more appropriate to refer to the pineal as a driving pacemaker or as a coupled oscillator in this species.

Chicken (*Gallus gallus*)

In the chicken, free-running circadian rhythms in locomotor activity persist following pinealectomy, but some disruption of circadian behavior has been reported (146,147).

There is controversy over the extent to which pinealectomy disrupts the rhythm of melatonin in the blood stream. Several studies demonstrate that the pineal gland is the major source of plasma melatonin, with at most a small contribution from the retina under normal circumstances (34,148,149). Pinealectomy greatly reduces nighttime melatonin levels and leads to loss of melatonin rhythmicity in plasma (34,148,149) and brain (150). Bilateral enucleation (in combination with pinealectomy) has little additional effect on the melatonin

Table 3 Effects of Pinealectomy on Circadian Organization in Vertebrates

	MEL rhythm?[a]	Effect of pinealectomy on free-running rhythms	Ref.
Amphibians			
Newts	?	Arrhythmicity	156
Fishes			
Burbot	?	Period change by 2 hr, direction depends on season	161
Lake chub	?	Period change	160
White suckers	?	Period change (direction variable), splitting	3,162
Killifish	?	Arrhythmicity in the coloration rhythm	163
Catfish	?	Arrhythmicity	164
Reptiles			
Texas spiny lizard	?	Period increase, splitting, arrhythmicity	165,166
Western fence lizard	?	Period increase, splitting, arrhythmicity, instability	166–170
Anole	Absent	Arrhythmicity	174
Ruin lizard	Absent	Period change, splitting	175,176
Desert iguana	Absent	No effect	180

	MEL rhythm	Effect	Refs.
Birds			
House sparrow	Absent	Arrhythmicity	117
White-crowned sparrow	?	Arrhythmicity	119
White-throated sparrow	?	Arrhythmicity	133
House finch	?	Arrhythmicity	134,135
Java sparrow	?	Arrhythmicity	45
European starling	Absent	Period decrease or arrhythmicity	127,141
		No effect	138
Pigeon	Present	PX: period change	143
	Absent	PX + EX: arrhythmic	142,143,145
Chicken	Absent	Little effect	146
		Period change or arrhythmicity	147
Japanese quail	Present	PX: no effect	31,47
	Absent	PX + EX: arrhythmic	31,32
Mammals			
Rat	Absent	Altered period in LL	111
		Increased disruption of rhythms in LL	187
Syrian hamster	Absent	No effect on tau, precision, or PRC	112,203,204
		No effect on frequency of splitting in LL	205
		Increased frequency of splitting in LL	206
Siberian hamster	Absent	No major effect	
Ground squirrel	?	No effect	238

aMEL rhythm indicates whether removal of the pineal gland destroys the rhythm of plasma melatonin concentrations. Pinealectomy abolishes the melatonin rhythm in most species where it has been examined, including anoles (173), ruin lizards (177), desert iguanas (180), house sparrows (140), European starlings (140), chickens (34,148, but see text), and rats (26). Pinealectomy alone does not disrupt the melatonin rhythm in pigeons (33,142) and Japanese quail (32); the melatonin rhythms in these species are obliterated by combined enucleation and pinealectomy (EX + PX). Note that enucleation alone will disrupt circadian rhythms in some species, including Japanese quail.

Table 4 Effects of Melatonin on Free-Running Circadian Rhythms in Vertebrates

	Type of treatment	Effect of melatonin treatment	Ref.
Amphibians			
Newts	Bidaily injection	Entrainment	157
Reptiles			
Texas spiny lizard	Continuous	Lengthened period (1–2 hr)	166
	Continuous	Arrhythmicity in 2 of 11 in LL	166
Western fence lizard	Single injection	Phase shift (6 hr max.)	170
	Bidaily injection	Entrainment	169
	Daily infusions	Entrainment	171
	Continuous	Lengthened period (1–2 hr)	166,167
Ruin lizard	Continuous	Lengthened period	179
Desert iguana	Continuous	Lengthened period	180
Birds			
House sparrow	Pineal transplant	Entrainment	123
	Rhythmic oral	Entrainment	118,126
	Continuous	Arrhythmicity, shortened period	121,144
House finch	Continuous	PNX birds remained arrhythmic	134
Java sparrow	Continuous	Arrhythmicity	45
European starling	Daily injection	Entrainment	141
Pigeon	Daily infusions	Entrainment	142,145
	Continuous	Arrhythmicity	143,144
Japanese quail	Rhythmic oral	Entrainment	155
	Continuous	Arrhythmicity; increased period (ca. 1 hr)	155
	Continuous	No effect	47

Mammals			
Rat	Single injection	Phase shift (60 min max.)	8,186
	Daily injections	Entrainment	6,8,49,111,113,185, 186,189,190,191
	Continuous	No effect on period	185
Siberian hamster	In vitro single tx	Phase shift (4 hr max.)	82–85
	In vitro single tx	Phase shift (4 hr max.)	86
Syrian hamster (adult)	Single injection	No effect more than vehicle	115
	Daily injections	No effect	207
	Daily injections	Entrainment	201
Syrian hamster (fetus)	Single injection	Entrainment	212
	4–9 daily injections	Entrainment	210,211
Mouse	Single injection	No effect	217
	3 daily injections	Phase shift (60 min max.)	217
	In vitro single tx	Phase shift (4 hr max.)	81
Human	Single p.o. treatment	Phase shift (60 min max.)	232,233
	Daily ingestion	Period change, entrainment	228
	Daily ingestion	No circadian effect	226
Mink	Daily injection	No effect	63
Nine-banded armadillo	Continuous	Period length increased (by 0.6 hr)	239

rhythm (34,148). Others report compensatory up-regulation of retinal melatonin content, and reappearance of detectable plasma melatonin levels weeks after pinealectomy (35). Available evidence indicates that *rhythmicity* of melatonin secretion is disrupted by pinealectomy. Thus, the persistence of behavioral rhythmicity in chickens following pinealectomy appears *not* to be due to residual rhythmicity in melatonin, but supports the existence of melatonin-independent rhythms. There has been no published report on the influence of exogenous melatonin on chicken locomotor activity rhythms.

Melatonin is nevertheless an important modulator of chicken brain function. High-affinity receptors for melatonin are widely expressed in the chicken brain and are present in the chicken vSCN, as well as in areas involved in processing of sensory information (136,137). The presence of melatonin receptors in sensory and integrative areas suggests that melatonin may have a direct modulatory role on behavior, mediated by direct actions within neuronal target sites rather than (or in addition to) binding within the SCN. The pineal is not necessary for rhythmicity in the number of melatonin receptors in optic tectum (151). Finally, circadian rhythmicity in the electroretinogram of chickens appear to be due to melatonin acting in the brain (152). The rhythmic secretion of melatonin may affect numerous aspects of chicken sensory physiology by directly modulating the activity of target areas. Nevertheless, melatonin does not appear to play a critical role in circadian organization of the chicken.

Japanese Quail (*Coturnix coturnix japonica*)

The quail has the most atypical circadian organization among avian species studied, with respect to the localization and autonomy of oscillators within the circadian system (2,31). Both the eyes and pineal contribute to blood melatonin levels, with the eyes contributing ca. two-thirds of the melatonin in blood (32). Pinealectomy alone does not abolish behavioral arrhythmicity (31,47). Combined enucleation and pinealectomy causes behavioral rhythmicity. Enucleation alone also causes arrhythmicity. The eyes are not merely functioning as a source of rhythmic melatonin, or merely as circadian photoreceptors. Instead, the eyes appear to comprise a neural circadian clock in quail (153). Optic nerve section (which does not disrupt melatonin rhythms in the retina or blood stream) results in reduced robustness of rhythmicity (31). The discrepancy between the effects of enucleation and optic nerve section suggests that the eye still has a mechanism for contributing to circadian organization even when deprived of neural connections (154,155). This mechanism appears to be melatonin, as rhythmic administration of melatonin in the drinking water entrains locomotor and temperature rhythms of normal birds (155), and constant-release implants of melatonin disrupt rhythmicity in normal birds (155). Lesions of the hypothalamus including the SCN cause arrhythmicity in quail (47). Collectively, these results demonstrate the presence of circadian oscillators in the hypothalamus and eyes, and indicate a modulatory role for melatonin in communication between these oscillators.

Amphibians

The pineal gland has been implicated as playing a role in the circadian organization of the Japanese newt (*Cynops pyrrhogaster*; 156,157). In the Japanese newt, both blinding and pinealectomy affect circadian organization, while injections of melatonin (10 μg, or ca. 2 mg/kg, every other day) resulted in entrainment in the majority of newts (156,157). Melatonin injection synchronized the end of the active period, as with other day-active species. In some animals, periods of relative coordination without entrainment occurred.

The distribution of ^{125}I-MEL binding in amphibian brain has been studied in the crested newt (*Triturus carnifex*), the green frog (*Rana esculenta*), and the leopard frog (*Rana pipiens*) (158,159). Tavolaro et al. (158) report high levels of binding in the SCN in newts and green frogs, while Weichmann and Wirsig-Weichmann (159) did not note an accumulation of ^{125}I-MEL binding over the SCN in *Rana pipiens*. However, there are significant levels of ^{125}I-MEL binding throughout the ventral hypothalamus of most nonmammalian species, and there is no apparent necessity that the level of binding in the SCN homolog be above this background level to be physiologically meaningful. These studies suggest that ^{125}I-MEL-binding sites in the hypothalamus may be involved in mediating circadian effects of melatonin in amphibians.

Fish

Pinealectomy causes changes in circadian rhythms in several species of fish (see 3 for review). Pinealectomy changes the free-running circadian period of locomotor activity rhythms in lake chub (*Couesius plumbeus*; 160), burbot (*Lota lota*; 161), and causes period changes or splitting of the activity rhythm in white suckers (*Catostomus commersoni*; 3,162). Pinealectomy disrupts the diurnal rhythm in color change in the killifish (*Fundus heteroclitus*), although these fish retain the ability to show adaptive changes in coloration in response to alterations in the background (163). Pinealectomy leads to locomotor arrhythmicity in catfish (*Heteropneustes fossilis*; 164). These effects of pinealectomy suggest that the pineal plays a role in circadian organization of these species. Little information is available on the relative contribution of the retina and pineal to circulating melatonin levels in fish, although it is known that rhythmicity persists in enucleated-plus-pinealectomized lake chub (M. Kavaliers, unpublished data cited in ref. 3, p. 288). Direct examination of the effects of melatonin treatment on circadian organization in fishes has not been reported.

Reptiles

The role of melatonin in circadian organization has not been investigated in crocodilians, turtles, or snakes. Several species of iguanid lizards have been studied, however. Comparison of the data from these species indicates that the relative contribution of melatonin to circadian organization varies dramatically. In some species, melatonin appears to play an important role in circadian organization (*Anolis* > *Sceloporus* = *Podarcis*), while in another, removal of sources of melatonin has virtually no effect on circadian organization (*Dipsosaurus*). These studies support the concept that the circadian timing system is a multioscillator system, and the relative contribution of melatonin as a coupling agent among oscillators varies considerably.

Texas Spiny Lizard (*Sceloporus olivaceus*)

Underwood has conducted an extensive series of studies on the role of melatonin in circadian organization using two species of iguanid lizards, the Texas spiny lizard (*S. olivaceus*; 165,166) and the western fence lizard (*Sceloporus occidentalis*; 166–171). Pinealectomy causes splitting of the free-running activity rhythm into two components, changes in period or, occasionally, in arrhythmicity in *S. olivaceus* (165). Continuous administration of exogenous melatonin markedly increases the period of rhythmicity in constant darkness or constant light, and in a subset of *S. olivaceus* exposed to constant illumination, melatonin treatment induces arrhythmicity (166).

Western Fence Lizard (*Sceloporus occidentalis*)

As in *S. olivaceus*, pinealectomy causes changes in period, rhythm instability, or arrhythmicity in *S. occidentalis* (167,168). The phase-response curve to light is markedly altered in pinealectomized *S. occidentalis* (168). Continuous treatment with melatonin increases free-running period (166), while single injections of melatonin induce phase-dependent phase shifts (170 see Fig. 5A). Repeated injections (every 48 hr) result in entrainment (169). Finally, infusions of melatonin to pinealectomized *S. occidentalis* (at doses that produce physiological levels of the hormone) result in entrainment (171). Collectively, these data indicate that pineal melatonin is an important component of the circadian timing system in these iguanid lizard species, and also support the idea that multiple circadian oscillators are localized to distinct tissues. It is worth noting that the activity rhythms of blinded-pinealectomized lizards can be entrained by LD cycles or 24-hr temperature cycles, which provides compelling evidence for photoreceptive capacity and a circadian oscillator within the central nervous system, independent of the retinae and pineal (172).

Anoles (*Anolis carolinensis*)

The pineal gland appears to be the primary source of circulating melatonin in anoles (*A. carolinensis*; 173). Pinealectomy abolishes the free-running circadian rhythm of locomotor activity in *A. carolinensis* (174). In animals in LD, the relative timing of activity within the day was altered by pinealectomy, and the phase of activity was unstable (animals alternated between two preferred phases of activity; 174). The anole pineal maintains rhythmicity of melatonin secretion in vitro for many days in constant darkness (129). Remarkably, individual anole pineal cells appear to possess a complete circadian system consisting of photoreceptor, oscillator, and output (melatonin secretion) (22,23). The anole pineal promises to be a useful system for determining how multiple single-cell circadian oscillators are coupled into a coherent circadian clock. Moreover, these data indicate that pineal cells contain an autonomous circadian oscillator, while the oscillator in the brain appears to be relatively more subservient.

Ruin lizard (*Podarcis sicula*)

In the ruin lizard (*P. sicula*), removal of the eyes (bilaterally), the pineal gland, or all three structures induces changes in the period of locomotor activity rhythms, but the animals are nevertheless always rhythmic (175,176). Pinealectomy completely abolishes rhythmicity of circulating melatonin (177); the retinae can secrete melatonin in low quantities, but do not contribute significantly to rhythmic levels in blood (177). Lesions of the optic chiasm also induce shortening of period, suggesting that the role of the retina in circadian organization involves neural, rather than merely humoral, interaction with other oscillatory tissues (178), and is reminiscent of the role of the eyes in quail circadian organization (see above). The ruin lizard is not completely insensitive to circadian effects of melatonin, however, as melatonin treatment does increase period (179).

Desert Iguana (*Dipsosaurus dorsalis*)

In the desert iguana, pinealectomy abolishes rhythmicity in plasma melatonin, yet has no effect on the circadian rhythm of locomotor activity (180). In contrast, electrolytic lesions of the SCN produce arrhythmicity (181). It thus appears likely that the SCN of the desert iguana are an autonomous circadian oscillator, and that melatonin is relatively unimportant for circadian control of locomotor activity. Melatonin treatment does increase period in this species, however (180).

B. Mammalian Species

In contrast to the critical role of endogenous melatonin in circadian organization in some nonmammalian vertebrates, melatonin appears to play only a modest role in the regulation of circadian rhythms in adult mammals (see Tables 3 and 4). The effects of pinealectomy on circadian function are also subtle, in general. Other reviews summarizing the effects of melatonin on mammalian circadian rhythms have appeared (6–12).

Rats

Pinealectomy Studies in Rats

Most studies indicate that pinealectomy has relatively little effect on circadian rhythmicity in rats. Pinealectomized rats continue to display free-running circadian rhythms when housed in constant darkness (111,182–188). Pinealectomy can, however, influence the rate of reentrainment following a phase shift of the LD cycle. The more rapid rate of entrainment in pinealectomized rats (183) may reflect enhanced "masking" due to an alteration in the stimulus properties of light in pinealectomized animals, rather than a direct effect of the absence of melatonin (for a full discussion, see 7,111,112). Indeed, Warren and Cassone (111) report that pinealectomized rats appear more sensitive to light, based on the finding that the circadian period length is longer at each of four light intensities, and the rate of increase in period was greater in the pinealectomized animals than in sham-operated controls. In contrast, Quay (182) reported that the dependence of period on light intensity is not altered in pinealectomized rats.

Partial destruction/isolation of the SCN (by knife cuts), in combination with pinealectomy, causes a greater disruption of circadian organization than partial SCN disruption alone in rats housed in constant light (188). Furthermore, the disruptive effects of constant white light (LL) on circadian rhythmicity are exacerbated in pinealectomized rats (111,187). These findings may also indicate that pinealectomy alters the stimulus properties of light. Alternatively, pinealectomy may alter the interactions between coupled circadian oscillators.

It is important to note that exposure to light at night suppresses melatonin production (24), and animals housed in LL are unlikely to have rhythmic melatonin production. Most studies of the effects of pinealectomy on circadian behavior in LL appear to assume that pineal melatonin production is not disrupted by LL, and that melatonin is the only relevant secretory product of the pineal gland. Neither assumption appears well founded.

Entrainment to Exogenous Melatonin in Rats

An extensive series of studies by Redman, Armstrong, Cassone, and colleagues has demonstrated that exogenous treatment with melatonin can influence circadian rhythmicity in rats (6,8,49,111,113,186,189–193). Rats placed in constant darkness and injected with melatonin at 24-hr intervals will free-run until the time of activity onset coincides with the injection time; when the injection time is coincident with activity onset, the animals will "lock on" or entrain (Fig. 8). It appears that responsiveness to melatonin is restricted to a narrow window of sensitivity late in the afternoon. Consistent with this interpretation, studies de-

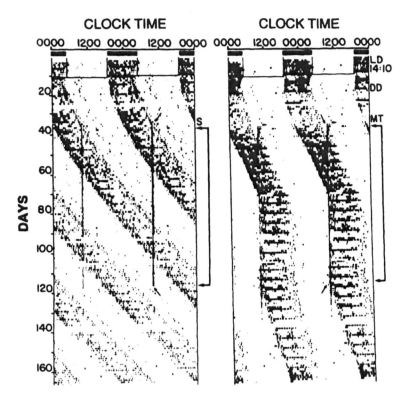

Figure 8 Entrainment of rats to daily injection of melatonin. Double-plotted actograms illustrate locomotor activity rhythms of two female rats. At the beginning of the record, the animals were housed in a light-dark cycle (14L:10D), and then were transferred to DD. The animals received daily injection of saline vehicle (S, left) or melatonin (MT, 100 μg/kg, right) for over 2 months, as indicated to the right of each record. Melatonin entrained the activity rhythm when the time of injection coincided with activity onset. (Modified from Ref. 191.)

signed to demonstrate a phase-response curve to single melatonin injections (1 mg/kg, s.c.) demonstrated significant effects of melatonin only late in the subjective afternoon (ref. 8; Fig. 6A). The finding that pinealectomized rats entrain to daily melatonin injections indicates that endogenous melatonin is not necessary for entrainment to exogenous melatonin, e.g., by entraining a window of sensitivity to melatonin (186). Furthermore, the dose-response relationships for pinealectomized and pineal-intact animals are grossly similar (113,186), with ED_{50} doses of melatonin being 332 ± 53 ng/kg (sham) and 121 ± 22 ng/kg (PNX). (A previous study reported the ED_{50} value to be 5.45 μg/kg in pineal-intact rats; 113).

Notably, injection of 1 μg/kg melatonin produces blood levels of 565 pg/ml, 10-fold in excess of normal nocturnal peak levels (ca. 1 nM), while the dose used in most studies (1 mg/kg) produces peak levels of ca. 700 ng/ml (113).

An interesting aspect of the dose response to melatonin is that the response appears to be essentially all-or-none. In response to single injections of melatonin at circadian time 10 (CT 10), rats phase-shift either 30–50 min or not at all (186). Similarly, the data of Armstrong (8) and Cassone et al. (113) indicate that the dose of melatonin influences the proportion of animals entraining to injection, rather than the magnitude of the shift induced in individuals.

Entrainment to melatonin is dependent upon the integrity of the SCN (49), but destruction of the SCN also destroys rhythmicity of the outputs examined. Entrainment to melatonin has been shown for male rats and female rats housed in constant darkness (DD) (189–191). In constant light, however, exogenous melatonin (0.1–1 mg/kg) is ineffective in entraining male or female rats (189, 191). These findings suggest that there are limits to entrainment by melatonin: melatonin administered at 24-hr intervals can synchronize free-running rhythms of animals housed in DD having a free-running period length of ca. 24.3 hr (tau-DD), but not when administered to animals housed in LL where free-running rhythms have a period length of ca. 25.3 (tau-LL) (191). Limits to entrainment are expected if melatonin exerts its influence upon a central circadian oscillator. No formal analysis of the limits of entrainment to exogenous melatonin has been reported for rats, however (e.g., through administration of melatonin by injections or infusions at non-24-hr intervals).

One curious finding from these studies is that melatonin elicits its effects on rhythmicity in rats when administered at CT 10, before the anticipated onset of activity (8). Endogenous melatonin production is also regulated by the circadian clock, and melatonin production rises *after* the onset of activity (25). Thus, exogenous melatonin is effective at a time when melatonin would never be produced in vivo. Thus, the effects of administered melatonin appear to be pharmacological, rather than physiological.

One of the most striking effects of melatonin on rhythmicity in rats was reported by Redman and Armstrong (192): following an 8-hr phase advance of the LD cycle, all uninjected and vehicle-treated rats reentrained by phase delaying, while animals treated with melatonin (1 mg/kg, s.c., administered at the dark-light transition of the old LD cycle) entrained to the shifted LD cycle by phase-advancing. S-20098, a melatonin agonist devoid of sedative effects, also influences the direction of entrainment in this 8-hr phase-advance paradigm when administered at doses of 100 μg/kg–3 mg/kg (193). The effect of melatonin and S-20098 on the direction of entrainment in this paradigm is not consistent with the narrow phase-response curve to single or repeated injections of melatonin, in that the treatment is administered at an "ineffective" time of day soon after the shift. It is only immediately after the phase shift that the phase of melatonin injection would coincide with activity onset, which is the sensitive phase determined in free-running animals. In this paradigm, then, there appears to be a clear and robust effect of melatonin, which is quite different from that observed in free-running rats. The rates of reentrainment to 5- and 8-hr phase advances of the LD cycle were also found to be more rapid in melatonin-treated rats in this study (192), although studies of the rate of reentrainment to a lighting schedule are confounded by the masking effects of light (as discussed in Section V).

Other studies have examined outputs of the circadian timing other than locomotor activity. Treatment of male rats with melatonin (1 mg/kg, s.c., daily for several days) hastened reentrainment of the pineal NAT rhythm to an 8-hr phase-advanced LD cycle (195,196). Similarly, treatment of male rats with a single injection of the melatonin agonist 6-chloromelatonin (0.5 mg/kg, single injection) hastens reentrainment of the urinary 6-sulfatoxymelatonin to the LD cycle (194). In both studies, treatment occurred at the time of lights-off of the phase-advance LD cycle. No assessment of direction of entrainment was permitted by these experimental designs. In rats housed in 10L:14D, daily injections of melatonin (1 mg/kg) phase-advance the rise of NAT activity relative to dark onset (195,196).

Few studies have examined the hypothesis that secretory products of the pineal gland exert a noncircadian, trophic effect on the circadian timing system (c.f., 197). It is possible that pinealectomy alters the rate of photic entrainment via long-term changes in the SCN (79,80). Melatonin may play a chronic role, and the circadian timing of its administration could be relatively unimportant. Existing studies have focused on the rhythmic administration of exogenous melatonin, and thus have not adequately addressed this possibility.

To summarize these findings, melatonin can influence circadian rhythms in rats. The maximum shift induced by injection of melatonin is less than 1 hr, and production of this shift requires doses of melatonin far in excess of the doses required to produce physiological levels of melatonin in blood.

Siberian Hamsters (Phodopus sungorus)

Siberian hamsters have been extensively studied as a model for reproductive effects of melatonin (12), and less intensively for effects of melatonin on circadian rhythms. Several studies indicate an effect of exogenous melatonin treatment on circadian organization. As in rats, phase advances occur in response to melatonin treatment late in subjective or actual day.

Several studies indicate that afternoon or early-evening melatonin injections can advance the evening component of the circadian oscillator. In hamsters housed in long days (16L:8D), melatonin injections phase-advance the onset of locomotor activity relative to the LD cycle (198). Similarly, Yellon (114) has shown that afternoon melatonin injections (5 μg) of animals housed in 16L:8D phase-advance the melatonin rhythm (monitored in DD). This dose of melatonin produced peak serum levels of 25–40 ng/ml, >100-fold in excess of the endogenous peak nighttime levels of 25–250 pg/ml (114).

Following transfer to short photoperiod, some Siberian hamsters fail to undergo the normal "short day responses" characterized by gonadal regression, weight loss, and molt to the winter coat. In these "photo-nonresponsive" hamsters, the nocturnal bout of activity and the nocturnal melatonin secretory episode remain compressed late in the dark phase (199). Conversely, the period of elevated SCN electrical activity remains high for a longer period, compared to photo-responsive hamsters. Repeated daily injection of melatonin (10 μg/animal, s.c., approximately 300 μg/kg) at lights-off of the 9L:15D LD cycle converts photo-nonresponsive hamsters to the responsive phenotype, by causing a phase advance

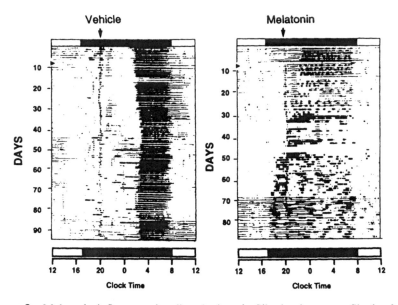

Figure 9 Melatonin influences circadian rhythms in Siberian hamsters. Single-plotted actograms illustrate locomotor activity rhythms of two representative male hamsters. Photo-nonresponsive hamsters housed in a short-day photoperiod (9L:15D) have a characteristically compressed activity pattern, beginning activity long after lights-out and remaining active for only 6–8 hr. Vehicle treatment (left) was without effect on these aspects of rhythmicity. Daily treatment with melatonin (10 μg; ca. 0.3 mg/kg) at 20:00, 3 hr after lights-off, advanced the onset of activity and increased the duration of activity. Bars at the top and bottom of each panel indicate the light/dark cycle, with darkness represented by the black portion. Injections began on day 9 of the record, as indicated by the arrow at the left margin. The time of injection (20:00) is indicated by the arrow above each panel. (Modified from Ref. 200.)

of the evening oscillator, which controls activity onset (200; Fig. 9). The daily peak of SCN electrical activity recorded in vitro was also converted to a short-day-like pattern by melatonin injection (200). Melatonin injections were effective when initially timed to occur approximately 2 hr before activity onset but 3 hr *after* dark onset (see Fig. 2 in Ref. 200). In several animals with extremely negative phase angles, the initial phase of melatonin injection was ineffective. When these animals were treated with melatonin ca. 2–3 hr before activity onset, the onset of activity gradually advanced to coincide with injection. Subsequent treatment at the initial phase (3 hr after dark onset) resulted in the photoresponder-like patterns of activity. These data suggest that the effects of melatonin are phase-dependent and, more interestingly, the effective phase of treatment was defined relative to the onset of activity rather than to the LD transition. The effects on locomotor behavior did not depend on continued melatonin treatment; animals with treatment discontinued after 2–3 weeks continued in their new phenotype. Also notable is the fact that the circadian response to melatonin required several weeks to develop (Fig. 9). Collectively, these data suggest that melatonin alters coupling among circadian oscillators. The SCN neurophysiological data suggest that the evening oscillator influenced by melatonin is localized to the SCN.

Unpublished data of Darrow and Goldman (cited in 12) indicate that infusions of melatonin (100 ng/10 hr in dim LL) on near-24-hr intervals resulted in alteration of the period of locomotor activity rhythm in some of the animals receiving infusions at $T = 23.5$, but not at $T = 24.0$ or $T = 24.6$. This low dose of melatonin produces near-physiological serum levels of the hormone, and is sufficient to induce reproductive responses. Kirsch et al. (201) infused much larger doses of melatonin (16.5 μg/hr $\times$ 6–8 hr/day) to four pinealectomized male hamsters and found that activity onset of some of the hamsters became synchronized to the onset of infusion of melatonin (but not vehicle). The findings indicate that melatonin can affect circadian rhythms in hamsters, but further work is necessary to more fully characterize its effects.

Using the hypothalamic slice preparation, Chen Liu in this laboratory has shown that the electrical activity rhythm in the Siberian hamster SCN is phase-shifted by low doses (1 nM) of melatonin (86). The presence of circadian responses to melatonin in Siberian hamsters is particularly of interest in view of our recent finding that the Mel_{1b} melatonin receptor gene does not encode a functional melatonin receptor in this species (86). The Mel_{1b} receptor is thus not necessary for circadian (or reproductive) responses to melatonin in Siberian hamsters. In contrast, the Mel_{1a} melatonin receptor does encode a functional receptor with high affinity for melatonin in this species (86), and the Mel_{1a} subtype is expressed in the Siberian hamster SCN (87). These data suggest that the Mel_{1a} receptor alone is capable of mediating circadian responses to melatonin. The presence of in vitro circadian responses to melatonin in mice lacking the Mel_{1a} melatonin receptor (81) indicates that there is likely functional redundancy between the melatonin receptor subtypes. The generation of mice lacking the Mel_{1b} melatonin receptor and of mice lacking both the Mel_{1a} and Mel_{1b} receptor subtypes will be instrumental in identifying the role of each receptor subtype in circadian responses to melatonin.

Syrian Hamsters (Mesocricetus auratus)

Most authors examining effects of pinealectomy or exogenous melatonin on rhythmicity in adult Syrian hamsters report small effects, but some investigators have reported large effects. These inconsistent findings are described below.

Pinealectomized hamsters reentrain more rapidly following a phase shift of the LD cycle (202). As in rats, the altered rate of reentrainment has been interpreted to represent altered sensitivity to light that bypasses the circadian pacemaker (masking), rather than an increase in sensitivity to the circadian effects of light (7). In support of this interpretation, the phase-response curve of pinealectomized hamsters to light pulses is not altered (112); precision of free-running rhythms in DD, and the duration of activity, were not affected by pinealectomy (112). Morin has reported that pinealectomy has little effect on circadian rhythms in Syrian hamsters (203,204).

Splitting of the locomotor activity rhythm occurs spontaneously in a proportion of Syrian hamsters upon continued exposure to LL (205). This condition results when the normally consolidated nocturnal bout of activity splits into two components that free-run, with different circadian periods, until the two components are 12 hr out of phase. Once this new phase relationship is established, the two components establish a new, stable phase relationship and free-run with the same period. Splitting has been taken as evidence for the presence of two coupled circadian oscillators (205). Splitting occurs in both normal and pinealectomized hamsters housed in LL. Once splitting has occurred in intact hamsters, its characteristics are not changed by pinealectomy (205). Finally, melatonin administration by injection (dose not stated) or by continuous release from Silastic capsules does not alter locomotor activity rhythms in hamsters with split activity rhythms (G. B. Ellis and F. W. Turek, unpublished data cited in 205). In contrast to these findings, a recent report indicates that pinealectomy facilitates splitting of the locomotor activity rhythm of hamsters housed in constant light (206). As with rats, these results are difficult to interpret in view of the suppression of melatonin rhythms by LL, and are more consistent with an effect of pinealectomy to increase the subjective intensity of light.

There are several contradictory reports regarding the effects of exogenous melatonin treatment on rhythmicity in adult Syrian hamsters. Kirsch et al. (201) report that three pinealectomized male Syrian hamsters became entrained by repeated daily infusions of melatonin (16.5 µg/hr for 6–8 hr; animals housed in dim LL). In the two animals whose data is illustrated, the activity onset during "entrainment" preceded melatonin injection by 3–5 hr, a phase relationship virtually never seen in other species whose free-running rhythms are entrained by melatonin. In both animals, the effect of melatonin became apparent only after several weeks of repeated infusions, and there is considerable instability of circadian period in the records shown. Periods of 24-hr periodicity persisted with vehicle administration for weeks after cessation of melatonin infusion. It is possible that melatonin is having irreversible "organizational effects" on the SCN, rather than inducing phase shifts or entrainment. The limited data available prevent firm conclusions.

Hastings et al. (115) demonstrated that melatonin injection (1 mg/kg) induced phase-dependent phase shifts of the locomotor activity rhythm in hamsters housed in dim red light. Injections occurring 2–4 hr before activity onset induced a small phase advance, while injections at other CTs were not effective. Notably, however, vehicle injections led to comparable phase shifts. The phase-response curves for melatonin and vehicle were virtually identical (Fig. 6B). Furthermore, melatonin delivered by a remote cannula (so that administration of melatonin did not require handling the animal) did not lead to entrainment (115). These data strongly indicate that, in the Syrian hamster at least, behavioral arousal associated with injection has a greater impact on SCN function than the administration of melatonin. Consistent with this finding, Armstrong and Redman (207) reported that melatonin injections (1 mg/kg, s.c.) do not alter free-running locomotor activity rhythms in hamsters.

Several reports indicate that melatonin injection facilitates reentrainment following a 6-hr phase advance of the LD cycle (208 and references therein), and that the benzodiazepine antagonist flumazenil blocks the effect of melatonin. It is difficult to reconcile hastened reentrainment by melatonin administration with reports that pinealectomy itself facilitate reentrainment (202).

In contrast to the subtle and inconsistent effects of melatonin on rhythmicity in adult Syrian hamsters, melatonin is extremely effective in entraining fetal hamsters (209–212) (see Section VI. C).

Mice *(Mus musculus)*

Most strains of house mice do not produce melatonin rhythmically within the pineal gland (213,214). Despite this absence, circadian rhythmicity of locomotor activity is present in all strains examined (215,216 and references therein). Pineal melatonin thus is not necessary for maintenance of rhythmicity in mice.

Exogenous melatonin can subtly influence rhythmicity in mice, however (217). Studies to date have focused on C3H/HeN mice, a strain with rhythmic pineal melatonin production. A single injection of melatonin (70 μg) does not produce a detectable phase shift in mice, but three injections given at 24-hr intervals does induce a phase-dependent phase shift (217). As in the rat, the largest phase shift in response to melatonin occurs 2–4 hr before activity onset (Fig. 6C; 217). The shape of the PRC to melatonin in mice is very similar to the PRC described in rats, except that a small phase-delay region is apparent in mice late in subjective night (217). In rats, there is suggestion of sensitivity to melatonin at this phase in neurochemical responses (71), and in phase advances (rather than delays) in vivo (8) and in vitro (83).

Melatonin also has been reported to modulate the amplitude of light-induced phase shifts in mice, and to impede the light-induced accumulation of FOS-like immunoreactivity in the SCN (11). These findings are consistent with melatonin reducing the SCN's sensitivity to light.

Using the in vitro SCN slice preparation, Chen Liu in this laboratory has shown that melatonin treatment in late afternoon (ca. CT 10) induces a 2–4-hr phase advance in mice (81). As in rats (83,84), remarkably low concentrations of melatonin and 2-iodomelatonin (10 pM) are effective in phase-shifting mouse SCN electrical activity rhythms. As noted elsewhere, the surprising persistence of a phase-shifting response to melatonin in mice lacking functional copies of the Mel_{1a} receptor (81) strongly suggest that another G-protein-coupled receptor expressed in the SCN (as the Mel_{1b} receptor is) participates in mediating the circadian responses to melatonin.

Humans

Melatonin has drawn increasing attention as a "chronobiotic" for use in humans in recent years (for reviews, see 9,10,25,217a).

Early attempts to demonstrate a circadian effect of melatonin in humans examined the effects of melatonin on jet lag. Rapid transmeridian travel is associated with difficulty in coordinating sleep to the local environmental time, resulting in sleep disturbances and fatigue. Melatonin treatment in a variety of regimens has been reported to give subjective improvement in the symptoms of jet lag (218–224). Most studies used a regimen of 5 mg p.o., and most did not adequately control the LD environment to which the subjects were exposed.

Arendt et al. reported that subjective assessment of several parameters (the severity of jet lag, sleep latency and quality) were improved by melatonin following an eastward flight crossing eight time zones (218,219). Objective measurement of circadian phase was accomplished by determining urinary 6-sulfatoxymelatonin and cortisol rhythms; these data indicate that the melatonin-treated subjects synchronized to the new environment more rapidly than the placebo-treated controls (219) Notably, however, two (of nine) subjects exhibiting no appreciable jet lag were excluded from the placebo group. It appears that the reported differences between placebo and melatonin groups may not have reached statistical significance if these subjects had remained in the placebo group (219). Petrie et al. (220) also reported that melatonin reduced the symptoms of jet lag.

While the design was improved by the use of a double-blind, crossover design, only subjective measures of jet lag were determined. Samel et al. (221) report that melatonin treatment hastens adjustment of several physiological variables and improves subjective alertness after simulated jet lag. Melatonin appears to lessen the symptoms of jet lag, but the mechanism by which it does so is a matter of conjecture. In view of other data, it is tempting to speculate that melatonin facilitates reentrainment to the phase-shifted LD cycle in the new environment. Others have also reported beneficial effects of melatonin after simulated or actual jet lag (221–224).

Rigorous examination of the effects of melatonin on circadian rhythmicity in the absence of potential interference from the LD cycle can be accomplished by assessing the effects of melatonin in blind subjects (225–228). Furthermore, periodic sleep disturbances occur in some blind subjects as they attempt to maintain the 24-hr cycle length dictated by society while exhibiting an endogenous cycle length of >24 hr (229). Melatonin treatment is thus of special interest as an agent for synchronization of blind subjects to the 24-hr solar day.

Arendt et al. (225) and Folkard et al. (226) reported administration of melatonin to a blind man. While they reported improvement of sleep consolidation during melatonin treatment (5 mg, p.o., every 24 hr), examination of urinary cortisol and core body temperature rhythms indicated that the subject had not entrained to the treatment. Even modest changes in period or small phase shifts should have been detected, considering that melatonin was given for a month. No circadian effect of melatonin was detected.

In contrast, Lewy and colleagues have reported evidence for phase shifts and entrainment of blind subjects (10,228). Melatonin (0.5 or 5 mg, p.o., for up to 21 days at 22:00) was administered at 24-hr intervals to free-running blind subjects. Treatment was initiated when the endogenous melatonin rhythm reached approximately normal phase, such that the administered melatonin would precede the endogenous melatonin secretory onset by several hours. In the five subjects, the phase shifts recorded at the end of the 21-day treatment period were: 0, 2, 3, 4, and 16 hr, based on comparison of the extrapolated and observed phases of melatonin onset and cortisol rhythms (228). Thus, the magnitude of shift is ca. 30 min/day, at most. In the five subjects studied, entrainment was not achieved, but there were changes in period in at least three of the subjects. Curiously, the period of these subjects did not revert to the pretreatment value upon cessation of treatment. Finally, an anecdotal description of a subject who self-prescribed melatonin for many years, and thus achieved stable entrainment, was included (228). Unfortunately, it is not possible to unambiguously conclude that the entrainment/period change was the result of melatonin, as it was not shown to be reversible or dependent on administered melatonin per se. Nevertheless, melatonin treatment shows promise as a treatment strategy for at least some free-running blind subjects. Importantly, small cumulative phase shifts may be sufficient to be of significant benefit for this unique population, in which the goal is to reconcile the small daily difference between the endogenous pacemaker's period and the 24-hr desired period.

Palm et al. (227) reported that daily administration of melatonin (0.5 mg, p.o., every 24 hr) to a blind, retarded boy improved the pattern of sleep consolidation and appeared to entrain the rhythm of sleep to a 24-hr cycle length. The boy's sleep-wake cycle had a period of 24.7 hr before melatonin treatment, suggesting that melatonin effected a 0.7-hr phase advance each day. It is unclear whether melatonin influenced sleep or the circadian clock regulating sleep, however.

Melatonin has also been indicated as a useful treatment in manipulating the circadian clock in subjects suffering from delayed sleep phase insomnia (230,231).

In sighted subjects, Lewy and colleagues (232) have used the onset of endogenous melatonin increase in dim light (DLMO, for dim light melatonin onset) to assess circadian phase and responses to melatonin (0.5 mg, p.o.). Melatonin shifts the DLMO phase according to a phase-response curve (232; Fig. 5F, 6D). Others also report phase-shifting effects of melatonin in sighted humans (233–235). Notably, the maximal shift attained with 0.5 mg melatonin was ca. 80 min (232). It seems unlikely that the phase-shifting effects of melatonin are due to effects on sleep (236,237). Czeisler and Wright have achieved much more robust resetting of the human circadian system with bright light exposure (237).

Melatonin receptors are present in the human SCN during adulthood and fetal development (71–73). Available evidence indicates that melatonin receptors in the human SCN are high-affinity, G-protein-coupled receptors, and likely are the product of the Mel_{1a} receptor gene. Thus, an anatomical/neurochemical substrate that could mediate circadian responses to melatonin is present in the human SCN.

Other Mammalian Species

Golden-Mantled Ground Squirrels (Spermophilus lateralis)

Martinet and Zucker (238) reported that pinealectomy was virtually without effect on circadian parameters in golden-mantled ground squirrels. Pinealectomized ground squirrels did not differ from sham-pinealectomized controls in the phase angle of entrainment in LD, in the time to stable reentrainment following a 6-hr phase advance of the LD 10:14 lighting cycle, or in the circadian period of wheel-running activity.

Mink (Mustela vison)

Bonnefond et al. (63) examined the response of mink to injections of melatonin (1 mg/kg, i.m.). Evidence for temporary entrainment to melatonin was observed in only one of eight mink free-running in DD. This is particularly of interest in view of the finding that mink and two other members of the weasel family (ferrets and skunks) lack detectable ^{125}I-MEL binding in the SCN (63–65).

Nine-Banded Armadillo (*Dasypus novemcinctus*)

Continuous-release melatonin implants increased the period length of free-running locomotor activity rhythms in the nine-banded armadillo (239).

C. Developmental Considerations

Studies in hamsters and rats have demonstrated that the mother communicates circadian phase information to the fetus during gestation (for reviews, see 209,209a,240). Entrainment of the fetal circadian clock is evident by the presence of population rhythms in AVP mRNA and metabolic activity within the fetal SCN, and by assessing postnatal endocrine or behavioral rhythms. Destruction of the maternal SCN prevents maternal communication of circadian phase information. This suggests that a rhythm whose output is regulated by the SCN is involved in synchronizing the fetal circadian clocks. The rhythm of melatonin secretion is regulated by the SCN, and melatonin crosses the placenta to reach the fetus (241–243). Thus, melatonin was a prime candidate for the maternal entraining signal.

Administration of melatonin (1.0–25 μg) to pregnant, SCN-lesioned Syrian hamsters results in entrainment of the fetuses (210). Remarkably, a single injection of melatonin on gestational day 15 is sufficient to coordinate rhythmicity of littermates to the phase of injection (212). Sensitivity to melatonin persists into the early postnatal period; melatonin injections on postnatal days 1–5 entrain the pups, while injections on days 6–10 do not (211). While these data indicate that melatonin is *sufficient* for entrainment of the fetus, melatonin does not appear *necessary* for entrainment based on studies in which the maternal pineal gland was removed from rats (see 209a,240). The most likely explanation for these apparently inconsistent results is that multiple, redundant maternal signals are responsible for entraining the fetal biological clock. Melatonin appears to be one such entraining signal, and the melatonin rhythm in the maternal circulation is reflected in the fetus (241–243). The recent finding that D1-dopamine receptor activation also entrains the fetal clock provides direct evidence for redundancy of inputs (212,244). The influence of melatonin on circadian rhythmicity in mammals appears greatest during fetal development.

There is a very marked developmental decline and anatomical redistribution of melatonin receptors within the Syrian hamster SCN (245,246). Melatonin receptor density in the adult hamster SCN is low, and is limited to dorsomedial regions (the vasopressin-expressing component of the SCN) (247). Melatonin receptor binding is excluded from the ventral, retinorecipient (input) zone of the SCN. The developmental alteration in melatonin receptor density and distribution occurs primarily in the first 2 weeks of postnatal life, which correlates well with the developmental loss in the ability of melatonin injections to entrain hamsters (211). Maywood et al. (246) have proposed that the developmental loss of responsiveness to the entraining effects of melatonin is due to loss of receptors from the "input" portion of the SCN. It seems likely that the developmental loss of ^{125}I-MEL binding in the Syrian hamster SCN is due to reduction in Mel_{1a} melatonin receptor gene expression. The Mel_{1b} melatonin receptor gene is not capable of encoding a functional receptor in this species, owing to the presence of nonsense mutations in exon 2 of the Mel_{1b} receptor gene (Weaver and Reppert, unpublished data).

VII. Conceptual Issues

As reviewed in detail above, the behavioral effects of pinealectomy include alterations in circadian period in constant conditions, splitting of the activity rhythm into two components, and frank arrhythmicity, while in some species pinealectomy has no apparent effect. Conversely, exogenous melatonin can influence the phase, period, and cohesiveness of circadian rhythmicity, or have no apparent effect. How can these diverse effects of melatonin be explained in a conceptual sense?

There is evidence for multiple circadian oscillators at each of three organizational levels within the circadian timing system. First, at an organismal level, there appear to be at least two functional circadian oscillators. Second, as clearly shown in sparrows, there can be multiple tissues that contain anatomically separate but functionally interdependent circadian oscillators. Finally, within each oscillatory tissue, there may be numerous circadian oscillatory cells. The effects of pinealectomy and melatonin on circadian rhythms likely reflect an alteration in the manner in which circadian oscillators interact at each of these three organizational levels.

A. Melatonin and Coupling of Functional Circadian Oscillators

Seminal studies by Pittendrigh and Daan (248) proposed that the mammalian circadian clock is comprised of at least two functional oscillators. These oscillators appear synchronized to the LD cycle, with one coordinated to dawn and the other to dusk [morning (M) and evening (E) oscillators, respectively]. The existence of two functional pacemakers has received broad experimental support and is perhaps most obvious in the case of splitting of the activity rhythm in hamsters (205,248,249). The free-running period of circadian rhythmicity in constant conditions is thought to reflect the interaction between two functional oscillators that are coupled in a stable phase relationship.

The circadian organization of nonmammalian vertebrates appears more complex as a result of the involvement of additional oscillatory tissues. The two-oscillator theory has not been extended to these species, but there is evidence suggesting the presence of two functional oscillators within the central nervous system, e.g., based on the presence of splitting of the activity rhythm into two free-running components after removal of the pineal gland (162,165,175). It is possible that this model will provide insights for nonmammalian species as well.

There are several possible mechanisms for the effects of pinealectomy and melatonin on circadian organization at this functional level (see refs. 1 and 7). The period of one (or both) of the functional oscillators could be influenced by pinealectomy, such that mutually coupled oscillators are no longer within each others' range of mutual entrainment. Alternatively, the range of entrainment of the individual oscillators could be restricted beyond the range necessary for mutual entrainment. In either case, splitting or arrhythmicity could result. Less severe disruption of the interaction among the functional oscillators could result in alterations in period or reduction in rhythm cohesiveness or amplitude. Changes of these types are often seen following pinealectomy in nonmammalian species. In mammals, the functional circadian oscillators may interact more efficiently, even in the absence of melatonin, and the properties of interaction may be more resilient to manipulation by melatonin. This may underlie the apparently greater "autonomy" of the mammalian circadian system, e.g., its greater independence from melatonin when compared to nonmammalian vertebrates.

B. Coupling Among Oscillatory Tissues

The most clearly understood role of melatonin in circadian organization is as a hormonal messenger coupling component tissues within the avian circadian timing system. Melatonin produced by the pineal gland appears to be a critical part of a neuroendocrine loop comprising the circadian system of sparrows (4). In the absence of feedback from the pineal in the form of melatonin, damped oscillators present within the brain are incapable of self-sustaining oscillations. Differences among species in the role of melatonin in circadian organization could be due to the relative autonomy of oscillators in different tissues, the relative (hierarchical) importance of different tissues to overall circadian organization, and the relative sensitivities of these tissues to melatonin.

C. Coupling of Single-Cell Oscillators Within Oscillatory Tissues

Several lines of evidence support the concept that multiple single-cell oscillators exist within oscillatory tissues, and that these are normally coupled to form functional pacemakers. Circadian rhythmicity has been demonstrated in individual cells isolated from several pacemaker tissues, including basal retinal neurons from the mollusk eye (250), anole photoreceptors (22,23), and recently, the rodent SCN (130). Electrical activity rhythms of dissociated SCN cells maintained in vitro free-run without apparent interactions or coordination, demonstrating that individual SCN cells are competent circadian oscillators (130,251). Rhythmicity thus can result from intracellular processes (often presumed to be intracellular molecular loops), and is not necessarily a network property. In the intact rat SCN, however, the population of cells with oscillatory potential are coupled to form a coherent circadian pacemaker. While this coupling is not reestablished in our in vitro system (130), others report that dissociated SCN neurons maintained in long-term culture secrete vasopressin rhythmically (252,253). This feat would appear to require the functional coordination of numerous SCN neurons. Dissociated SCN cells transplanted into the third ventricle of SCN-lesioned hamsters similarly form a functional unit that can restore behavioral rhythmicity (254). Furthermore, there is evidence for functional interaction between transplanted and host SCNs in "circadian chimeras," created by transplanting SCN tissue into a host whose circadian period differs from that of the donor (255,256). Thus, in some situations, dissociated SCN neurons interact and become "coupled" to re-form a circadian clock. We have virtually no information regarding the cellular and biochemical mechanisms of "coupling" among oscillatory cells within pacemaker tissues (but see 251,266). It is through this "black box" within the circadian timing system that melatonin appears most likely to exert its effects in mammals (see also Chapter 11; 257).

The rhythmic secretion of melatonin from avian pineal glands maintained in constant darkness degenerates ("dampens out to arrhythmicity") over several cycles in vitro (22,25,29). Rhythmic input, in the form of either neurohumoral or photic input, is required for the maintenance of pineal rhythmicity in vitro (20,29,30). The chicken pineal is thus a "damped oscillator," incapable of self-sustaining rhythmicity. It is likely that individual cells within the pineal maintain rhythmicity, but that the rhythmic release of melatonin from the population is lost as the cells become uncoordinated with increasing time in culture. This suggests that chicken pineal cells lack mechanisms for interaction (coupling) to form a coherent oscillator. Clearly, melatonin is not a sufficient coupling agent for synchronizing cells within the chicken pineal gland. In contrast, zebrafish and anole pineals maintain rhythmic melatonin secretion for longer periods (36,129), suggesting that these glands do possess a coupling activity/factor, or that the variability of period among clock cells within these glands is much smaller than in the chicken.

D. The Influence of Other Hormones on Circadian Organization

Another point worth considering is the extent to which the effects of melatonin on circadian organization appear to be selective and specific. Manipulation of steroid hormone levels can alter free-running period and/or lead to splitting of the activity rhythm into two components in hamsters and birds (258–262, for review, see 262). Steroid hormones are not thought to have direct access to the circadian clock; e.g., steroid receptors are not found within the SCN. Instead, the circadian effects of steroid hormones have been interpreted in terms of hormonal effects on the level of arousal. Arousal and locomotor activity are known to have a profound "feedback" effect on the circadian clock (263–265). The effect of an agent on circadian organization may have unexpected mechanisms. This emphasizes the need to utilize multiple, complementary approaches when attempting to demonstrate the role of a hormone on circadian organization. Remarkably, the only hormones shown to markedly influence circadian organization in vertebrates to date are melatonin, testosterone, and estradiol (262). Manipulation of thyroid hormone levels can also influence rhythmicity.

E. What Is Coupling?

A key area for future investigation will be to explore the coupling mechanisms that synchronize individual circadian clock cells into anatomically and functionally defined circadian oscillators, and how the two or more functional circadian oscillators are coupled to provide coherent regulation of numerous behavioral and hormonal outputs.

In 1982, Underwood wrote that "[a]lthough the exact mechanism of melatonin action is unknown, it is tempting to speculate that melatonin influences either the strength of coupling between component oscillators of a multioscillator system or the spontaneous frequency of one, or more, of the component oscillators" (Ref. 1, p. 17). This speculation remains valid, but amazingly little progress has been made in understanding of the nature of coupling, or its regulation by melatonin, in the past 15 years.

Recent studies may provide an inroad into this previously intractable question. In mice, the highly polysialated form of neuronal cell adhesion molecule (PSA-NCAM) appears necessary for maintenance of circadian rhythms (266). Mice with a targeted deletion of the NCAM-180 isoform (which bears the PSA moiety) appear rhythmic when housed under a cycling LD cycle, but become arrhythmic after several weeks in DD. Enzymatic removal of the PSA moiety within the SCN of wild-type mice also alters rhythmicity. The PSA-NCAM molecule thus appears necessary for normal circadian function in mice. The role of PSA-NCAM in cell adhesion and tissue interactions in other tissues suggests that PSA-NCAM may contribute to circadian organization by coordinating the activity of neurons within the SCN to produce a functional circadian clock. The similarity of phenotype of PSA-NCAM mutant mice with the phenotype of *Clock/Clock* mutant mice, which also show gradual decay of rhythmicity in constant darkness (267), suggests that the *Clock* mutation may also influence the ability of SCN neurons to interact and maintain coherent oscillatory function. The *Clock* gene was recently reported to encode a protein containing basic helix-loop-helix and PAS domains (268,269), suggesting it is a transcription factor. This structure suggests a role for the CLOCK protein in intracellular transcriptional loops, rather than in intercellular interactions. The mechanism by which the *Clock* mutation affects circadian rhythmicity in mice remains to be determined, however. Both the PSA-NCAM-deficient and *Clock* mutant mice provide promising models for studying the cellular and molecular mechanisms of oscillator coupling. The *Clock* gene may be essential for normal circadian behavior because it may regulate genes more directly involved in coupling or in intracellular molecular loops. Melatonin may influence intercellular adhesion molecules that can influence the functional coupling of oscillatory cells. While speculative, these are the types of questions that must be addressed to come to an understanding of the effects of melatonin on circadian organization. Simply injecting additional species with melatonin, demonstrating a phase-response curve or entrainment, will do little to advance our understanding.

VIII. Summary and Conclusions

In the sparrow, the pineal is the driving pacemaker in the circadian system, with pineal melatonin playing a critical role in the regulation of neural circadian oscillators. Pineal melatonin is necessary for coherent circadian organization in only a few other avian species. In most avian, reptilian, and fish species studied, melatonin is not necessary, but pineal melatonin does play a significant role in circadian organization. Removal of pineal melatonin alters free-running circadian period, circadian organization, or entrainment. In a few species, removal of the pineal has no discernible effect. In all mammalian species studied to date, endogenous melatonin is not necessary for the maintenance of rhythmicity. The effects of exogenous melatonin in mammals are generally subtle and are often difficult to reproduce across laboratories.

   Certain features of melatonin make it an attractive candidate for manipulating circadian function in humans. These include melatonin's low acute toxicity, rapid absorption, and short half-life, and the ability to deliver the molecule by noninvasive (oral, nasal) routes. Furthermore, melatonin appears relatively unique in having an influence on the circadian clock in humans. Whether the circadian effects of melatonin are sufficiently robust and predictable to allow manipulation of the human circadian clock remains to be seen. Effective use of melatonin as a chronobiotic will require accurate assessment of circadian phase prior to treatment, standardization of treatment regimens, and careful control of external lighting conditions. Alternatively, it may be possible to combine melatonin with other approaches or treatments to influence the circadian clock, thus maximizing the impact of the hormone. At present, it appears that environmental lighting has a much larger and more predictable impact on the human circadian timing system in sighted subjects than does melatonin (237). There is little information on the potential adverse effects of long-term melatonin administration (270). Melatonin is clearly important in vertebrate circadian systems, but its usefulness as a general-purpose "chronobiotic" remains to be demonstrated.

Acknowledgments

I thank Steve Reppert for continuing support and guidance, and Herb Underwood and Joel Levine for comments on an earlier draft of the manuscript, Work performed in this laboratory was supported by NIH Grants DK 42125, HD 14427, a grant from the Air Force Office for Scientific Research (AFOSR 92-NL-172), and a sponsored research agreement with Bristol-Myers Squibb.

References

1. Underwood H. The pineal and circadian organization in fish, amphibians and reptiles. In Reiter RJ, ed. The Pineal Gland. Vol 3. Extra-reproductive Effects. Boca Raton, FL: CRC Press, 1982:1–25.
2. Underwood H. The pineal and melatonin: regulators of circadian function in lower vertebrates. Experientia 1990; 46:120–128.
3. Kavaliers M. The pineal organ and circadian organization of teleost fish. Rev Can Biol 1979; 38:281–292.
4. Menaker M, Zimmerman N. Role of the pineal in the circadian system of birds. Am Zool 1976; 16:46–55.
5. Cassone VM, Menaker M. Is the avian circadian system a neuroendocrine loop? J Exp Zool 1984; 232:539–549.
6. Cassone VM. Effects of melatonin on vertebrate circadian systems. Trends Neurosci 1990; 13:457–464.
7. Rusak B. Circadian organization in mammals and birds: role of the pineal gland. In Reiter RJ, ed., The Pineal Gland. Vol 3. Extrareproductive Effects. Boca Raton, FL: CRC Press, 1982:27–51.
8. Armstrong SM. Melatonin and circadian control in mammals. Experientia 1989; 45:932–938.
9. Arendt J. Some effects of light and melatonin on human rhythms. In Wetterberg L, ed. Light and Biological Rhythms in Man. New York: Pergamon Press, 1993:203–216.
10. Lewy AJ, Ahmed S, Sack RL. Phase shifting the human circadian clock using melatonin. Behav Brain Res 1996; 73:131–134.
11. Dubocovich ML, Benloucif S, Masana MI. Melatonin receptors in the mammalian suprachiasmatic nucleus. Behav Brain Res 1996; 73:141–147.
12. Underwood H, Goldman BD. Vertebrate circadian and photoperiodic systems: role of pineal gland and melatonin. J Biol Rhythms 1987; 2:279–315.
13. Guerrero JM, Reiter RJ. A brief summary of pineal gland-immune system interrelationships. Endocr Res 1992; 18:91–113.
14. Iuvone PM, Alonso-Gomez AL. Melatonin in the vertebrate retina. In: Christen Y, Doly M, Droy-Lefaix M-T, eds. Les Seminaires Opthalmologiques d'IPSEN, vol. 9. Paris: Irvinn, 1998:49–62.

15. Pittendrigh CS. Circadian oscillations in cells and the circadian organization of multicellular systems. The Neurosciences: Third Study Program. Schmitt FO, Worden FG, eds. Cambridge, MA: MIT Press, 1974:437–458.

16. Klein DC, Moore RY, Reppert SM, eds. Suprachiasmatic Nucleus. The Mind's Clock. New York: Oxford University Press, 1991.

16a. Weaver DR. Suprachiasmatic nucleus: A 25-year retrospective. J Biol Rhythms 1998; 13:92–104.

17. Tosini G, Menaker M. Circadian rhythms in cultured mammalian retina. Science 1996; 272:419–421.

18. Wiechmann AF. Melatonin: parallels in pineal gland and retina. Exp Eye Res 1986; 42:507–527.

19. Cahill GM, Grace MS, Besharse JC. Rhythmic regulation of retinal melatonin: metabolic pathways, neurochemical mechanisms, and the ocular circadian clock. Cell Mol Neurobiol 1991; 11:529–560.

20. Takahashi JS, Murakami N, Nikaido SS, Pratt BL, Robertson LM. The avian pineal, a vertebrate model system of the circadian oscillator: cellular regulation of circadian rhythms by light, second messengers, and macromolecular synthesis. Recent Prog Horm Res 1989; 45:279–352.

21. Cahill GM, Besharse JC. Circadian clock functions localized in *Xenopus* retinal photoreceptors. Neuron 1993; 10:573–577.

22. Pickard GE, Tang WX. Individual pineal cells exhibit a circadian rhythm in melatonin secretion. Brain Res 1993; 627:141–146.

23. Pickard GE, Tang WX. Pineal photoreceptors rhythmically secrete melatonin. Neurosci Lett 1994; 171:109–112.

24. Bolliet V, Begay V, Taragnat C, Ravault JP, Collin JP, Falcon J. Photoreceptor cells of the pike pineal organ as cellular circadian oscillators. Eur J Neurosci 1996; 9:643–653.

25. Arendt J. Melatonin and the Mammalian Pineal Gland. London: Chapman & Hall, 1995.

26. Lewy AJ, Tetsuo M, Markey SP, Goodwin FK, Kopin IJ. Pinealectomy abolishes plasma melatonin in the rat. J Clin Endocrinol Metab 1980; 50:204–205.

27. Roseboom PH, Coon SL, Baler R, McCune SK, Weller JL, Klein DC. Melatonin synthesis: analysis of the more than 150-fold nocturnal increase in serotonin *N*-acetyltransferase messenger ribonucleic acid in the rat pineal gland. Endocrinology 1996; 137:3033–3045.

28. Coon SL, Roseboom PH, Baler R, Weller JL, Namboodiri MAA, Koonin EV, Klein DC. Pineal serotonin *N*-acetyltransferase: expression cloning and molecular analysis. Science 1995; 270:1681–1683.

29. Bolliet V, Ali MA, Lapointe FJ, Falcon J. Rhythmic melatonin secretion in different teleost species: an in vitro study. J Comp Physiol B 1996; 165:677–683.

30. Cassone VM, Menaker M. Sympathetic regulation of chicken pineal rhythms. Brain Res 1983; 272:311–318.

31. Underwood H, Siopes T. Circadian organization in Japanese quail. J Exp Zool 1984; 232:557–566.

32. Underwood H, Binkley S, Siopes T, Mosher K. Melatonin rhythms in the eyes, pineal bodies, and blood of Japanese quail (*Coturnix coturnix japonica*). Gen Comp Endocrinol 1984; 56:70–81.

33. Foa A, Menaker M. Contribution of the pineal and the retinae to the circadian rhythms of circulating melatonin in pigeons. J Comp Physiol A 1988; 164:25–30.

34. Cogburn LA, Wilson-Placentra S, Letcher LR. Influence of pinealectomy on plasma and extrapineal melatonin rhythms in young chickens (*Gallus domesticus*). Gen Comp Endocrinol 1987; 68:343–356.

35. Osol G, Schwartz B, Foss DC. Effects of time, photoperiod, and pinealectomy on ocular and plasma melatonin concentrations in the chick. Gen Comp Endocrinol 1985; 58:415–420.

36. Cahill GM, Circadian regulation of melatonin production in cultured zebrafish pineal and retina. Brain Res 1996; 708:177–181.

37. Hamm HE, Menaker M. Retinal rhythms in chicks: circadian variation in melatonin and serotonin *N*-acetyltransferase activity. Proc Natl Acad Sci USA 1980; 77:4998–5002.

38. Dubocovich ML. Melatonin is a potent modulator of dopamine release in the retina. Nature 1983; 306:782–784.

39. Heuther G. The contribution of extrapineal sites of melatonin synthesis to circulating melatonin levels in higher vertebrates. Experientia 1993; 49:665–670.

40. Ralph CL. Melatonin production by extra-pineal tissues. In: Birau N, Schloot W, eds. Melatonin—Current Status and Perspectives. New York: Pergamon Press, 1982; 35–46.

41. Waldhauser F, Waldhauser M, Lieberman HR, Deng MH, Lynch HJ, Wurtman RJ. Bioavailability of oral melatonin in humans. Neuroendocrinology 1984; 39:307–313.

42. Grace MS, Cahill GM, Besharse JC. Melatonin deacetylation: retinal vertebrate class distribution and *Xenopus laevis* tissue distribution. Brain Res 1991; 559:56–63.

43. Grace MS, Besharse JC. Melatonin deacetylase activity in the pineal gland and brain of the lizards *Anolis carolensis* and *Sceloporus jarrovi*. Neuroscience 1994; 62: 615–623.

44. Takahashi JS, Menaker M. Role of the suprachiasmatic nuclei in the circadian system of the house sparrow, *Passer domesticus*. J Neurosci 1982; 2:815–828.

45. Ebihara S, Kawamura H. The role of the pineal organ and the suprachiasmatic nucleus in the control of circadian locomotor rhythms in the Java sparrow, *Padda oryzivora*. J Comp Physiol 1981; 141:207–214.

46. Cassone VM, Forsyth AM, Woodlee GL. Hypothalamic regulation of circadian noradrenergic input to the chick pineal gland. J Comp Physiol A 1990; 167:187–192.

47. Simpson SM, Follett BK. Pineal and hypothalamic pacemakers: their role in regulating circadian rhythmicity in Japanese quail. J Comp Physiol 1981; 141:381–389.

48. Cassone VM, Brooks DS, Kelm TA. Comparative distribution of 2-[125I]-iodomelatonin binding in the brains of diurnal birds: outgroup analysis with turtles. Brain Behav Evol 1995; 45:241–256.

49. Cassone VM, Chesworth MJ, Armstrong SM. Entrainment of rat circadian rhythms by daily injection of melatonin depends upon the hypothalamic suprachiasmatic nuclei. Physiol Behav 1986; 36:1111–1121.

50. Dubocovich ML, Takahashi JS. Use of 2-125I-iodomelatonin to characterize melatonin binding sites in chicken retina. Proc Natl Acad Sci USA 1987; 64:3916–3920.

51. Vanecek J, Pavlik A, Illnerova H. Hypothalamic melatonin receptor sites revealed by autoradiography. Brain Res 1987; 435:359–362.

52. Weaver DR, Rivkees SA, Reppert SM. Localization and characterization of melatonin receptors in rodent brain by in vitro autoradiography. J Neurosci 1989; 9:2581–2590.

53. Weaver DR, Carlson LL, Reppert SM. Melatonin receptors and signal transduction in melatonin-sensitive and melatonin-insensitive populations of white-footed mice (*Peromyscus leucopus*). Brain Res 1990; 506:353–357.

54. Siuciak JA, Fang JM, Dubocovich ML. Autoradiographic localization of 2-[125I]-iodomelatonin binding sites in the brains of C3H/HeN and C37Bl/6J strains of mice. Eur J Pharmacol 1990; 180:387–390.

55. Roca AL, Godson C, Weaver DR, Reppert SM. Cloning and characterization of the mouse Mel$_{1a}$ melatonin receptor. Endocrinology 1996; 137:3469–3477.

56. Masson-Pevet M, George D, Kalsbeek A, Saboureau M, Lakhdar-Ghazal N, Pevet P. An attempt to correlate brain areas containing melatonin-binding sites with rhythmic functions: a study in five hibernator species. Cell Tissue Res 1994; 278:97–106.

57. Duncan MJ, Takahashi JS, Dubocovich ML. Characterization and autoradiographic localization of 2-[125I]-iodomelatonin binding sites in Djungarian hamster brain. Endocrinology 1989; 125:1011–1018.

58. Williams LM, Morgan PJ, Hastings MH, Lawson W, Davidson G, Howell HE. Melatonin receptor sites in the Syrian hamster brain and pituitary. Localization and characterization using [125I]iodomelatonin. J Neuroendocrinol 1989; 1:315–320.

59. Bittman EL. The sites and consequences of melatonin binding in mammals. Am Zool 1991; 33:200–211.

60. Bittman EL, Thomas EM, Zucker I. Melatonin binding sites in sciurid and hystrico-morph rodents: studies on ground squirrels and guinea pigs. Brain Res 1994; 648: 73–79.

61. Stanton TL, Siuciak JA, Dubocovich ML, Krause DN. The area of 2-[125I] iodo-melatonin binding in the pars tuberalis of the ground squirrel is decreased during hibernation. Brain Res 1991; 557:285–288.

62. Stankov B, Cozzi B, Lucini V, Capsoni S, Fauteck J, Fumagalli P, Fraschini F. Localization and characterization of melatonin binding sites in the brain of the rabbit (*Oryctolagus cuniculus*) by autoradiography and in vitro ligand-receptor binding. Neurosci Lett 1991; 133:68–72.

63. Bonnefond C, Monnerie R, Richard JP, Martinet L. Melatonin and the circadian clock in mink: effects of daily injections of melatonin on circadian rhythm of loco-motor activity and autoradiographic localization of melatonin binding sites. J Neuro-endocrinol 1993; 5:241–246.

64. Weaver DR, Reppert SM. Melatonin receptors are present in ferret pars tuberalis and pars distalis, but not in brain. Endocrinology 1990; 127:2607–2609.

65. Duncan MJ, Mead RA. Autoradiographic localization of binding sites for 2-[125I]-iodomelatonin in the pars tuberalis of the western spotted skunk (*Spirogale putorius latifrons*). Brain Res 1992; 569:152–155.

66. Stankov B, Moller M, Lucini V, Capsoni S, Fraschini F. A carnivore species (*Canis familiaris*) expresses circadian melatonin rhythm in the peripheral blood and mela-tonin receptors in the brain. Eur J Endocrinol 1994; 131:191–200.

67. Bittman EL, Weaver DR. The distribution of melatonin binding sites in neuroendo-crine tissues of the ewe. Biol Reprod 1990; 43:986–993.

68. Nonno R, Capsoni S, Lucini V, Moller M, Fraschini F, Stankov B. Distribution and characterization of the melatonin receptors in the hypothalamus and pituitary gland of three domestic ungulates. J Pineal Res 1995; 18:207–216.

69. Deveson S, Howarth JA, Arendt J, Forsyth IA. In vitro autoradiographical localization of melatonin binding sites in the caprine brain. J Pineal Res 1992; 13:6–12.

70. Stankov B, Capsoni S, Lucini V, Fauteck J, Gatti S, Gridelli B, Biella G, Cozzi B, Fraschini F. Autoradiographic localization of putative melatonin receptors in the brains of two Old World primates: *Cercopithecus aethiops* and *Papio ursinus*. Neuroscience 1993; 52:459–468.

71. Weaver DR, Stehle JH, Stopa EG, Reppert SM. Melatonin receptors in human hypothalamus and pituitary: implications for circadian and reproductive responses to melatonin. J Clin Endocrinol Metab 1993; 76:295–301.

72. Reppert SM, Weaver DR, Rivkees SA, Stopa EG. Putative melatonin receptors in a human biological clock. Science 1988; 242:78–81.

73. Weaver DR, Reppert SM. The Mel$_{1a}$ melatonin receptor is expressed in human suprachiasmatic nucleus. NeuroReport 1996; 8:109–112.

74. Cassone VM, Roberts MH, Moore RY. Effects of melatonin on 2-deoxy-[1-^{14}C]glucose uptake within rat suprachiasmatic nucleus. Am J Physiol 1988; 255:R332–337.

75. Mason R, Brooks A. The electrophysiological effects of melatonin and a putative melatonin antagonist (*N*-acetyltryptamine) on rat suprachiasmatic neurones in vitro. Neurosci Lett 1988; 95:296–301.

76. Stehle J, Vanecek J, Vollrath L. Effects of melatonin on spontaneous electrical activity of neurons in rat suprachiasmatic nuclei: an in vitro iontophoretic study. J Neural Transm 1989; 78:173–177.

77. Jiang ZG, Nelson CS, Allen CN. Melatonin activates an outward current and inhibits I$_h$ in rat suprachiasmatic nucleus neurons. Brain Res 1995; 687:125–132.

78. Shibata S, Cassone VM, Moore RY. Effects of melatonin on neuronal activity in the rat suprachiasmatic nucleus in vitro. Neurosci Lett 1989; 97:140–144.

79. Rusak B, Yu GD. Regulation of melatonin-sensitivity and firing-rate rhythms of hamster suprachiasmatic nucleus neurons: pinealectomy effects. Brain Res 1993; 602:200–204.

80. Rusak B, Yu GD. Regulation of melatonin-sensitivity and firing-rate rhythms of hamster suprachiasmatic nucleus neurons: constant light effects. Brain Res 1993; 602:191–199.

81. Liu C, Weaver DR, Jin X, Shearman LP, Pieschl RL, Gribkoff VK, Reppert SM. Molecular dissection of two distinct actions of melatonin on the suprachiasmatic circadian clock. Neuron 1997; 19:91–102.

82. McArthur AJ, Gillette MU, Prosser RA. Melatonin directly resets the rat suprachiasmatic circadian clock in vitro. Brain Res 1991; 565:158–161.

83. McArthur AJ, Hunt AE, Gillette MU. Melatonin action and signal transduction in the rat suprachiasmatic circadian clock: activation of protein kinase C at dusk and dawn. Endocrinology 1997; 138:627–634.

84. Starkey SJ, Walker MP, Beresford IJ, Hagan RM. Modulation of the rat suprachiasmatic circadian clock by melatonin in vitro. Neuroreport 1995; 6:1947–1951.

85. Starkey SJ. Melatonin and 5-hydroxytryptamine phase-advance the rat circadian clock by activation of nitric oxide synthesis. Neurosci Lett 1996; 211:199–202.

86. Weaver DR, Liu C, Reppert SM. Nature's knockout: the Mel_{1b} receptor is not necessary for circadian or reproductive responses in Siberian hamsters. Mol Endocrinol 1996; 10:1478–1487.

87. Reppert SM, Weaver DR, Ebisawa T. Cloning and characterization of a mammalian melatonin receptor that mediates reproductive and circadian responses. Neuron 1994; 13:1177–1185.

88. Reppert SM, Godson CG, Mahle CD, Weaver DR, Slaugenhaupt SA, Gusella JF. Molecular characterization of a second melatonin receptor expressed in human retina and brain: the Mel_{1b}-melatonin receptor. Proc Natl Acad Sci USA 1995; 92: 8734–8738.

89. Reppert SM, Weaver DR, Cassone VM, Godson C, Kolakowski LF Jr. Melatonin receptors are for the birds: molecular analysis of two receptor subtypes differentially expressed in chick brain. Neuron 1995; 15:1003–1015.

90. Ebisawa T, Karne S, Lerner MR, Reppert SM. Expression cloning of a high affinity melatonin receptor from *Xenopus* dermal melanophores. Proc Natl Acad Sci USA 1994; 91:6133–6137.

91. Reppert SM, Weaver DR, Ebisawa T, Mahle CD, Kolakowski LF Jr. Cloning of a melatonin-related receptor from human pituitary. FEBS Lett 1996; 386:219–224.

92. Liu F, Yuan H, Sugamori KS, Hamadanizadeh A, Lee FJ, Pang SF, Brown GM, Pristupa ZB, Niznik HB. Molecular and functional characterization of a partial cDNA encoding a novel chicken brain melatonin receptor. FEBS Lett 1995; 374: 273–278.

93. Reppert SM, Weaver DR, Godson C. Melatonin receptors step into the light: cloning and classification of subtypes. TIPS 1996; 17:100–102.

94. Dubocovich ML. Melatonin receptors: are there subtypes? TIPS 1995; 16:50–56.

95. Acuna-Castroviejo D, Reiter RJ, Menendez-Pelaez A, Pablos MI, Burgos A. Characterization of high-affinity melatonin binding sites in purified cell nuclei of rat liver. J Pineal Res 1994; 16:100–112.

96. Carlberg C, Wiesenberg I. The orphan receptor family RZR/ROR, melatonin and 5-lipoxygenase: an unexpected relationship. J Pineal Res 1995; 18:171–178.

97. Greiner EF, Kirfel J, Greschik H, Dorflinger U, Becker P, Mercep A, Schule R. Functional analysis of retinoid Z receptor beta, a brain-specific nuclear orphan receptor. Proc Natl Acad Sci USA 1996; 93:10105–10110.

98. Becker-Andre M, Schaeren-Wiemers N, Andre E. Correction. J Biol Chem 1997; 272:16707.

99. Dollins AB, Zhdanova IV, Wurtman RJ, Lynch HJ, Deng MH. Effect of inducing nocturnal serum melatonin concentrations in daytime on sleep, mood, body temperature, and performance. Proc Natl Acad Sci USA 1994; 91:1824–1828.

100. Sugden D. Psychopharmacological effects of melatonin in mouse and rat. J Pharmacol Exp Ther 1983; 227:587–591.

101. Vanecek J. Cellular mechanisms of melatonin action. Physiol Rev 1998; 78:687–721.

102. Morgan PJ, Barrett P, Hazlerigg D, Milligan G, Lawson W, MacLean A, Davidson G. Melatonin receptors couple through a cholera toxin-sensitive mechanism to inhibit cyclic AMP in the ovine pituitary. J Neuroendocrinol 1995; 7:361–369.

103. Godson C, Reppert SM. The Mel(1a) melatonin receptor is coupled to parallel signal transduction pathways. Endocrinol 1996; 138:397–404.

104. Stankov B, Biella G, Panara C, Lucini V, Capsoni S, Fauteck J, Cozzi B, Fraschini F. Melatonin signal transduction and mechanism of action in the central nervous system: using the rabbit cortex as a model. Endocrinology 1992; 130:2152–2159.

105. Lopez-Gonalez MA, Calvo JR, Osuna C, Guerrero JM. Interaction of melatonin with human lymphocytes: evidence for binding sites coupled to potentiation of cyclic AMP stimulated by vasoactive intestinal peptide and activation of cyclic GMP. J Pineal Res 1992; 12:97–104.

106. Yung LY, Tsim ST, Wong YH. Stimulation of cAMP accumulation by the cloned *Xenopus* melatonin receptor through Gi and Gz proteins. FEBS Lett 1995; 372:99–102.

107. Eison AS, Mullins UL. Melatonin binding sites are functionally coupled to phosphoinositide hydrolysis in Syrian hamster RPMI 1846 melanoma cells. Life Sci 1993; 53:393–398.

108. Popova JS, Dubocovich ML. Melatonin receptor-mediated stimulation of phosphoinositide breakdown in chick brain slices. J Neurochem 1995; 64:130–138.

109. Zisapel N, Egozi Y, Laudon M. Inhibition of dopamine release by melatonin: regional distribution in the rat brain. Brain Res 1982; 246:161–163.

110. McNulty S, Ross AW, Barrett P, Hastings MH, Morgan PJ. Melatonin regulates the phosphorylation of CREB in ovine pars tuberalis. J Neuroendocrinol 1994; 6:523–532.

111. Warren WS, Cassone VM. The pineal gland: photoreception and coupling of behavioral, metabolic, and cardiovascular circadian outputs. J Biol Rhythms 1995; 10: 64–79.

112. Aschoff J, Gerewcke V, vonGoetz C, Groos GA, Turek FW. Phase-response and characteristics of free-running activity rhythms in the golden hamster: independence of the pineal gland. In: Aschoff J, Daan S, Groos GA, eds. Vertebrate Circadian Systems. Berlin: Springer-Verlag, 1982:129–140.

113. Cassone VM, Chesworth MJ, Armstrong SM. Dose-dependent entrainment of rat circadian rhythms to daily injection of melatonin. J Biol Rhythms 1986; 1:219–229.

114. Yellon SM. Daily melatonin treatments regulate the circadian melatonin rhythm in the adult Djungarian hamster. J Biol Rhythms 1996; 11:4–13.

115. Hastings MH, Mead SS, Vindlacheruvu RR, Ebling FJP, Maywood ES, Grosse J. Non-photic phase shifting of the circadian activity rhythm of Syrian hamsters: the relative potency of arousal and melatonin. Brain Res 1992; 591:20–26.

116. Prosser RA, McArthur AJ, Gillette MU. cGMP induces phase shifts of a mammalian circadian pacemaker at night, in antiphase to cAMP effects. Proc Natl Acad Sci USA 1989; 86:6812–6815.

117. Gaston S, Menaker M. Pineal function: the biological clock in the sparrow? Science 1968; 160:1125–1127.

118. Heigl S, Gwinner E. Synchronization of circadian rhythms of house sparrows by oral melatonin: effects of changing period. J Biol Rhythms 1995; 10:225–233.

119. Gaston S. The influence of the pineal organ on the circadian activity rhythms of birds. In: Menaker M, ed. Biochronometry. Washington DC: National Academy of Sciences Press, 1971:541–548.

120. Zimmerman NH, Menaker M. Neural connections of sparrow pineal: role in circadian control of activity. Science 1975; 190:477–479.

121. Turek FW, McMillan JP, Menaker M. Melatonin: effects on the circadian locomotor rhythm of sparrows. Science 1976; 194:1441–1443.

122. In Menaker M, ed. Biochronometry. Washington DC: National Academy of Sciences Press, 1971.

123. Zimmerman NH, Menaker M. The pineal gland: a pacemaker within the circadian system of the house sparrow. Proc Natl Acad Sci USA 1979; 76:999–1003.

124. Heigl S, Gwinner E. Periodic melatonin in the drinking water synchronizes circadian rhythms in sparrows. Naturwissenschaften 1994; 81:83–85.

125. Lu J, Cassone VM. Pineal regulation of circadian rhythms of 2-deoxy[^{14}C]glucose uptake and 2[^{125}I]iodomelatonin binding in the visual system of the house sparrow, *Passer domesticus*. J Comp Physiol A 1993; 173:765–774.

126. Lu J, Cassone VM. Daily melatonin administration synchronizes circadian patterns of brain metabolism and behavior in pinealectomized house sparrows, *Passer domesticus*. J Comp Physiol A 1993; 173:775–782.

126a. Cassone VM, Brooks DS. Sites of melatonin action in the brain of the house sparrow, *Passer domesticus*. J Exp Zool 1991; 260:302–309.

127. Gwinner E. Effects of pinealectomy on circadian locomotor activity rhythms in European starlings, *Sturnus vulgaris*. J Comp Physiol 1978; 126:123–129.

128. Takahashi JS. Neural and endocrine regulation of avian circadian rhythms. PhD dissertation, University of Oregon, 1981.

129. Menaker M, Wisner S. Temperature-compensated circadian clock in the pineal of *Anolis*. Proc Natl Acad Sci USA 1983; 80:6119–6121.

130. Welsh DK, Logothetis DE, Meister M, Reppert SM. Individual neurons dissociated from rat suprachiasmatic nucleus express independently phased circadian firing rhythms. Neuron 1995; 14:697–706.

131. Barrett RK, Underwood H. The superior cervical ganglia are not necessary for entrainment or persistence of the pineal melatonin rhythm in Japanese quail. Brain Res 1992; 569:249–254.

132. Underwood H, Barrett RK, Siopes T. Melatonin does not link the eyes to the rest of the circadian system in quail: a neural pathway is involved. J Biol Rhythms 1990; 5:349–361.

133. McMillan JP. Pinealectomy abolishes the rhythm of migratory restlessness. J Comp Physiol 1972; 79:105–112.

134. Fuchs JL. Effects of pinealectomy and subsequent melatonin implants on activity rhythms in the house finch (*Carpodacus mexicanus*). J Comp Physiol 1983; 153:413–419.

135. Pant K, Chandola-Saklani A. Pinealectomy and LL abolished circadian perching rhythms but did not alter circannual reproductive or fattening rhythms in finches. Chronobiol Int 1992; 9:413–420.

136. Rivkees SA, Cassone VM, Weaver DR, Reppert SM. Melatonin receptors in chick brain: characterization and localization. Endocrinology 1989; 125:363–368.

137. Siuciak JA, Krause DN, Dubocovich ML. Quantitative pharmacological analysis of 2-^{125}I-iodomelatonin binding sites in discrete areas of the chicken brain. J Neurosci 1991; 11:2855–2864.

138. Gwinner E, Subbaraj R, Bluhm CK, Gerkema M. Differential effects of pinealectomy on circadian rhythms of feeding and perch hopping in the European starling. J Biol Rhythms 1987; 2:109–120.

139. Rutledge JT, Angle MJ. Persistence of circadian activity rhythms in pinealectomized European starlings, *Sturnus vulgaris*. J Exp Zool 1977; 202:333–338.
140. Janik D, Dittami J, Gwinner E. The effect of pinealectomy on circadian plasma melatonin levels in house sparrows and European starlings. J Biol Rhythms 1992; 7:277–286.
141. Gwinner E, Benzinger I. Synchronization of a circadian rhythm in pinealectomized European starlings by daily melatonin injections. J Comp Physiol 1978; 127: 209–213.
142. Chabot CC, Menaker M. Effects of physiological cycles of infused melatonin on circadian rhythmicity in pigeons. J Comp Physiol A 1992; 170:615–622.
143. Ebihara S, Uchiyama K, Oshima I. Circadian organization in the pigeon, *Columba livia*: the role of the pineal organ and the eye. J Comp Physiol A 1984; 154:59–69.
144. Chabot CC, Menaker M. Circadian feeding and locomotor rhythms in pigeons and house sparrows. J Biol Rhythms 1992; 7:287–299.
145. Chabot CC, Menaker M. Feeding rhythms in constant light and constant darkness: the role of the eyes and the effects of melatonin infusion. J Comp Physiol A 1994; 175:75–82.
146. MacBride S. Pineal biochemical rhythms of the chicken: light cycle and locomotor activity correlation. PhD dissertation, University of Pittsburgh, Pittsburgh, PA, 1973.
147. Nyce J, Binkley S. Extraretinal photoreception in chickens: entrainment of the locomotor activity rhythm. Photochem Photobiol 1977; 25:529–531.
148. Reppert SM, Sagar SM. Characterization of the day-night variation of retinal melatonin content in the chick. Invest Ophthalmol Vis Sci 1983; 24:294–300.
149. Pelham RW. A serum melatonin rhythm in chickens and its abolition by pinealectomy. Endocrinology 1975; 96:543–546.
150. Cassone VM, Lane RF, Menaker M. Daily rhythms of serotonin metabolism in the medial hypothalamus of the chicken: effects of pinealectomy and exogenous melatonin. Brain Res 1983; 289:129–134.
151. Siuciak JA, Dubocovich ML. Effect of pinealectomy and the light/dark cycle on 2-[^{125}I]iodomelatonin binding in the chick optic tectum. Cell Mol Neurobiol 1993; 13:193–202.
152. Lu J, Zoran MJ, Cassone VM. Daily and circadian variation in the electroretinogram of the domestic fowl: effects of melatonin. J Comp Physiol A 1995; 177:299–306.
153. Underwood H, Barrett RK, Siopes T. The quail's eye: a biological clock. J Biol Rhythms 1990; 5:257–265.
154. Underwood H. The circadian rhythm of thermoregulation in Japanese quail. I. Role of the eyes and pineal. J Comp Physiol A 1994; 175:639–653.
155. Underwood H, Edmonds K. The circadian rhythm of thermoregulation in Japanese quail. III. Effects of melatonin administration. J Biol Rhythms 1995; 10:284–298.
156. Chiba A, Kikuchi M, Aoki K. The effects of pinealectomy and blinding on the circadian activity rhythm in the Japanese newt, *Cynops pyrrhogaster*. J Comp Physiol A 1993; 172:683–691.
157. Chiba A, Kikuchi M, Aoki K. Entrainment of the circadian locomotor activity rhythm in the Japanese newt by melatonin injections. J Comp Physiol A 1995; 176: 473–477.

158. Tavolaro R, Canonaco M, Franzoni MF. Comparison of melatonin-binding sites in the brain of two amphibians: an autoradiographic study. Cell Tissue Res 1995; 279: 613–617.

159. Weichmann AF, Wirsig-Weichmann CR. Distribution of melatonin receptors in the brain of the frog Rana pipiens as revealed by autoradiography. Neuroscience 1993; 52:469–480.

160. Kavaliers M. Pineal involvement in the control of circadian rhythmicity in the lake chub, *Cousesius plumbeus*. J Exp Zool 1979; 209:33–40.

161. Kavaliers M. Circadian locomotor activity rhythms in the burbot, *Lota lota*: seasonal differences in period length and the effects of pinealectomy. J Comp Physiol 1980; 136:2165–2218.

162. Kavaliers M. Circadian organization in white suckers *Catostomus commersoni*: the role of the pineal organ. Comp Biochem Physiol 1980; 68:127–129.

163. Kavaliers M, Firth BT, Ralph CL. Pineal control of the circadian rhythm of colour change in the killifish (*Fundulus heteroclitus*). Can J Zool 1980; 58:456–460.

164. Garg SK, Sundararaj BI. Role of pineal in the regulation of some aspects of circadian rhythmicity in the catfish, *Heteropneustes fossilis* (Bloch). Chronobiologia 1986; 13:1–11.

165. Underwood H. Circadian organization in lizards: the role of the pineal organ. Science 1977; 195:587–589.

166. Underwood H. Melatonin affects circadian rhythmicity in lizards. J Comp Physiol 1979; 130:317–323.

167. Underwood H. Circadian organization in the lizard *Sceloporus occidentalis*: the effects of pinealectomy, blinding and melatonin. J Comp Physiol 1981; 141:537–547.

168. Underwood H. Circadian pacemakers in lizards: phase-response curves and effects of pinealectomy. Am J Physiol 1983; 244:R857–864.

169. Underwood H, Harless M. Entrainment of the circadian activity rhythm of a lizard to melatonin injections. Physiol Behav 1985; 35:267–270.

170. Underwood H. Circadian rhythms in lizards: phase response curve for melatonin. J Pineal Res 1986; 3:187–196.

171. Hyde LL, Underwood H. Daily melatonin infusions entrain the locomotor activity of pinealectomized lizards. Physiol Behav 1995; 58:943–951.

172. Underwood H, Menaker M. Extraretinal light perception: entrainment of the biological clock controlling lizard locomotor activity. Science 1970; 170:190–193.

173. Underwood H. Pineal melatonin rhythms in the lizard *Anolis carolinensis*: effects of light and temperature cycles. J Comp Physiol A 1985; 157:57–65.

174. Underwood H. Circadian organization in the lizard *Anolis carolinensis*: a multioscillator system. J Comp Physiol 1983; 52:265–274.

175. Foa A. The role of the pineal and the retinae in the expression of circadian locomotor rhythmicity in the ruin lizard *Podarcis sicula*. J Comp Physiol A 1991; 169:201–207.

176. Innocenti A, Minutini L, Foa A. The pineal and circadian rhythms of temperature selection and locomotion in lizards. Physiol Behav 1993; 53:911–915.

177. Foa A, Janik D, Minutini L. Circadian rhythms of plasma melatonin in the ruin lizard *Podarcis sicula*: effects of pinealectomy. J Pineal Res 1992; 12:109–113.

178. Minutini L, Innocenti A, Bertolucci C, Foa A. Electrolytic lesions to the optic chiasm affect circadian locomotor rhythms in lizards. NeuroReport 1994; 5:525–527.

179. Foa A, Minutini L, Innocenti A. Melatonin: a coupling device between oscillators in the circadian system of the ruin lizard, *Podarcis sicula*. Comp Biochem Physiol 1992; 103A:719–723.

180. Janik DS, Menaker M. Circadian locomotor rhythms in the desert iguana. I. The role of the eyes and the pineal. J Comp Physiol A 1990; 166:803–810.

181. Janik DS, Pickard GE, Menaker M. Circadian locomotor rhythms in the desert iguana. Effects of electrolytic lesions to the hypothalamus. J Comp Physiol A 1990; 166:811–816.

182. Quay WB. Individuation and lack of pineal effects in the rat's circadian locomotor rhythm. Physiol Behav 1968; 3:109–114.

183. Quay WB. Physiological significance of the pineal during adaptation to shifts in photoperiod. Physiol Behav 1970; 5:353–360.

184. Kincl FA, Chang CC, Zluzkova V. Observations on the influence of changing photoperiod on spontaneous wheel-running activity of neonatally pinealectomized rats. Endocrinology 1970; 87:38–42.

185. Cheung PW, McCormack CE. Failure of pinealectomy or melatonin to alter circadian activity rhythm of the rat. Am J Physiol 1982; 242:R261–264.

186. Warren WB, Hodges DB, Cassone VM. Pinealectomized rats entrain and phase-shift to melatonin injections in a dose-dependent manner. J Biol Rhythms 1993; 8:233–245.

187. Cassone VM. The pineal gland influences rat circadian activity rhythms in constant light. J Biol Rhythms 1992; 7:27–40.

188. Yanovski JA, Rosenwasser AM, Levine JD, Adler NT. The circadian activity rhythms of rats with mid- and parasagittal "split-SCN" knife cuts and pinealectomy. Brain Res 1990; 537:216–226.

189. Chesworth MJ, Cassone VM, Armstrong SM. Effects of daily melatonin injections on activity rhythms of rats in constant light. Am J Physiol 1987; 253:R101–107.

190. Redman J, Armstrong S, Ng KT. Free-running activity rhythms in the rat: entrainment by melatonin. Science 1983; 219:1089–1091.

191. Thomas EM, Armstrong SM. Melatonin administration entrains female rat activity rhythms in constant darkness but not in constant light. Am J Physiol 1988; 255: R237–R242.

192. Redman JR, Armstrong SM. Reentrainment of rat circadian activity rhythms: effects of melatonin. J Pineal Res 1988; 5:203–215.

193. Redman JR, Guardiola-Lemaitre B, Brown M, Delagrange P, Armstrong SM. Dose dependent effects of S-20098, a melatonin agonist, on direction of re-entrainment of rat circadian activity rhythms. Psychopharmacology (Berl) 1995; 118:385–390.

194. Kennaway DJ, Blake H, Webb HA. A melatonin agonist and *N*-acetyl-N²-formyl-5-methoxy-kynurenamine accelerate the reentrainment of the melatonin rhythm following a phase advance of the light-dark cycle. Brain Res 1989; 495:349–354.

195. Humlova M, Illnerova H. Melatonin entrains the circadian rhythm in the rat pineal *N*-acetyltransferase activity. Neuroendocrinology 1990; 52:196–199.

196. Illnerova H, Trentini GP, Masloav L. Melatonin accelerates reentrainment of the circadian rhythm of its own production after an eight-hour phase advance of the light-dark cycle. J Comp Physiol A 1989; 166:97–102.

197. Hau M, Gwinner E. Continuous melatonin administration accelerates resynchronization following phase shifts of a light-dark cycle. Physiol Behav 1995; 58:89–95.

198. Puchalski W, Lynch GR. Daily melatonin injections affect the expression of circadian rhythmicity in Djungarian hamsters kept under a long-day photoperiod. Neuroendocrinology 1988; 48:280–286.
199. Puchalski W, Lynch GR. Evidence for differences in the circadian organization of hamsters exposed to short day photoperiod. J Comp Physiol A 1986; 159:7–11.
200. Margraf RR, Lynch GR. Melatonin injections affect circadian behavior and SCN neurophysiology in Djungarian hamsters. Am J Physiol 1993; 264:R615–621.
201. Kirsch R, Belgnaoui S, Gourmelen S, Pevet P. Daily melatonin infusion entrains free-running activity in Syrian and Siberian hamsters. In: Wetterberg L, ed. Light and Biological Rhythms in Man (CIBA Foundation Symposium 183). Oxford: Pergamon Press, 1993:107–120.
202. Finkelstein J, Baum FR, Campbell CS. Entrainment of the female hamster to reversed photoperiod: role of the pineal. Physiol Behav 1978; 21:105–111.
203. Morin LP, Cummings LA. Effect of surgical or photoperiodic castration, testosterone replacement or pinealectomy on male hamster running rhythmicity. Physiol Behav 1981; 26:825–838.
204. Morin LP. Age, but not pineal status, modulates circadian periodicity of golden hamsters. J Biol Rhythms 1993; 8:189–197.
205. Turek FW, Earnest DJ, Swann J. Splitting of the circadian rhythm of activity in hamsters. In: Aschoff J, Daan S, Groos GA, eds. Vertebrate Circadian Systems. Berlin: Springer-Verlag, 1982:203–214.
206. Aguilar-Roblero R, Vega-Gonzalez A. Splitting of locomotor circadian rhythmicity in hamsters is facilitated by pinealectomy. Brain Res 1993; 605:229–236.
207. Armstrong SM, Redman J. Melatonin administration: effects on rodent circadian rhythms. In: Photoperiodism, Melatonin and the Pineal (Ciba Symposium 117). London: Pitman, 1985:188–202.
208. Golombek DA, Cardinali DP. Melatonin accelerates reentrainment after phase advance of the light-dark cycle in Syrian hamsters: antagonism by flumazenil. Chronobiol Int 1993; 10:435–441.
209. Davis FC, Frank MG, Heller HC. Ontogeny of sleep and circadian rhythms. In: Turek FW, Zee PC, eds. Regulation of Sleep and Circadian Rhythms. New York: Marcel Dekker, 1999.
209a. Weaver DR. The roles of melatonin in development. In: Oleese J, ed. Melatonin after Four Decades (Adv Exp Med Biol). New York: Plenum (in press).
210. Davis FC, Mannion J. Entrainment of hamster pup circadian rhythms by prenatal melatonin injections to the mother. Am J Physiol 1988; 255:R439–448.
211. Grosse J, Velickovic A, Davis FC. Entrainment of Syrian hamster circadian activity rhythms by neonatal melatonin injections. Am J Physiol 1996; 270:R533–R540.
212. Viswanathan N, Davis FC. Single prenatal injections of melatonin or the D1-dopamine receptor agonist SKF 38393 to pregnant hamsters sets the offsprings' circadian rhythms to phases 180 degrees apart. J Comp Physiol A 1997; 180:339–346.
213. Goto M, Oshimia I, Tomita T, Ebihara S. Melatonin content of the pineal gland in different mouse strains. J Pineal Res 1989; 7:195–204.
214. Ebihara S, Marks T, Hudson DJ, Menaker M. Genetic control of melatonin synthesis in the pineal gland of the mouse. Science 1986; 231:491–493.

215. Schwartz WJ, Zimmerman P. Circadian timekeeping in BALB/c and C57BL/6 inbred mouse strains. J Neurosci 1990; 10:3685–3694.

216. Possidente B, Hegmann JP. Gene differences modify Aschoff's rule in mice. Physiol Behav 1982; 28:199–200.

217. Benloucif S, Masana MI, Dubocovich ML. Melatonin and light induce phase shifts of circadian rhythms in the mouse. J Biol Rhythms 1996; 11:113–125.

217a. Czeisler CA, Turek FW, eds. Melatonin, Sleep and Circadian Rhythms: Current Progress and Controversies. J Biol Rhythms 1997; 12:1–708.

218. Arendt J, Aldhous M, Marks V. Alleviation of jet lag by melatonin: preliminary results of controlled double-blind trial. Br Med J 1986; 292:1170.

219. Arendt J, Aldhous M, English J, Marks V, Arendt JH, Marks M, Folkard S. Some effects of jet lag and their alleviation by melatonin. Ergonomics 1987; 30:1379–1393.

220. Petrie K, Conaglan JV, Thompson L, Chamberlain K. Effects of melatonin on jet lag after long-haul flights. Br Med J 1989; 298:705–707.

221. Samel A, Wegman HM, Vejvoda M, Maas H. Influence of melatonin treatment on human circadian rhythmicity before and after a simulated 9-hour time shift. J Biol Rhythms 1991; 6:235–248.

222. Claustrat B, Brun J, David M, Sassolas G, Chazot G. Melatonin and jet lag: confirmatory result using a simplified protocol. Biol Psychiatry 1992; 32:705–711.

223. Petrie K, Dawson AG, Thompson L, Brook R. A double-blind trial of melatonin as a treatment for jet lag in international cabin crews. Biol Psychiatry 1993; 33:526–530.

224. Lino A, Silvy S, Condorelli L, Rusconi AC. Melatonin and jet lag: treatment schedule. Biol Psychiatry 1993; 34:587.

225. Arendt J, Aldhous M, Wright J. Synchronization of a disturbed sleep-wake cycle in a blind man by melatonin treatment. Lancet 1988; 1:772–773.

226. Folkard S, Arendt J, Aldhous M, Kennett H. Melatonin stabilises sleep onset time in a blind man without entrainment of cortisol or temperature rhythms. Neurosci Lett 1990; 113:193–198.

227. Palm L, Blennow G, Wetterberg L. Correction of non-24-hour sleep-wake cycle by melatonin in a blind retarded boy. J Neural Transm 1990; 81:17–29.

228. Sack RL, Lewy AJ, Blood ML, Stevenson J, Keith LD. Melatonin administration to blind people: phase advances and entrainment. J Biol Rhythms 1991; 6:249–261.

229. Sack RL, Lewy AJ, Blood ML, Keith LD, Nakagawa H. Circadian rhythm abnormalities in totally blind people: incidence and clinical significance. J Clin Endocrinol Metab 1992; 75:127–134.

230. Dahlitz MJ, Alvarez B, Vignau J, English J, Arendt J, Parkes JD. Delayed sleep phase syndrome response to melatonin. Lancet 1991; 337:1121–1124.

231. Oldani A, Ferini-Strambini I, Zucconi M, Stankov B, Fraschini F, Smirne S. Melatonin and delayed sleep-phase syndrome: ambulatory polygraphic evaluation. NeuroReport 1994; 6:132–134.

232. Lewy AJ, Ahmed S, Jackson JM, Sack RL. Melatonin shifts human circadian rhythms according to a phase-response curve. Chronobiol Int 1992; 9:380–392.

233. Zaidan R, Geoffriau M, Brun J, Taillard J, Bureau C, Chazot G, Claustrat B. Melatonin is able to influence its secretion in humans: description of a phase response curve. Neuroendocrinology 1994; 60:105–112.

234. Deacon S, English J, Arendt J. Acute phase-shifting effects of melatonin associated with suppression of core body temperature in humans. Neurosci Lett 1994; 178:32–34.

235. Mallo C, Zaidan R, Faure A, Brun J, Chazot G, Clastraut B. Effects of a four-day melatonin treatment on the 24-hour plasma melatonin, cortisol, and prolactin profiles in humans. Acta Endocrinol 1988; 119:474–480.

236. Turek FW, Czeisler CA. Role of melatonin in the regulation of sleep. In: Turek FW, Zee PC, eds. Regulation of Sleep and Circadian Rhythms. New York: Marcel Dekker, 1999.

237. Czeisler CA, Wright KP Jr. Influence of light on circadian rhythmicity in humans. In: Turek FW, Zee PC, eds. Regulation of Sleep and Circadian Rhythms. New York: Marcel Dekker, 1999.

238. Martinet L, Zucker I. Role of the pineal gland in circadian organization of diurnal ground squirrels. Physiol Behav 1985; 34:799–803.

239. Harlow HJ, Phillips JA, Ralph CL. Circadian rhythms and the effects of exogenous melatonin in the nine-banded armadillo, *Dasypus novemcinctus*: a mammal lacking a distinct pineal gland. Physiol Behav 1982; 29:307–313.

240. Reppert SM, Weaver DR. A biological clock is oscillating in the fetal suprachias-matic nuclei. In: Klein DC, Moore RY, and Reppert SM, eds. Suprachiasmatic Nucleus. The Mind's Clock. New York: Oxford University Press, 1991:405–418.

241. Reppert SM, Chez RA, Anderson A, Klein DC. Maternal-fetal transfer of melatonin in the non-human primate. Pediatr Res 1979; 13:788–791.

242. Yellon SM, Longo LD. Effect of pinealectomy and reverse photoperiod on the circadian melatonin rhythm in the sheep and fetus during the last trimester of pregnancy. Biol Reprod 1988; 39:1093–1099.

243. McMillen IC, Nowak R. Maternal pinealectomy abolishes the diurnal rhythm in plasma melatonin concentrations in the fetal sheep and pregnant ewe during late gestation. J Endocrinol 1989; 120:459–464.

244. Viswanathan N, Weaver DR, Reppert SM, Davis FC. Entrainment of the fetal hamster circadian pacemaker by prenatal injections of the dopamine agonist SKF 38393. J Neurosci 1994; 14:5393–5398.

245. Duncan MJ, Davis FC. Developmental appearance and age related changes in specific 2-[^{125}I]iodomelatonin binding sites in the suprachiasmatic nuclei of female Syrian hamsters. Dev Brain Res 1993; 73:205–212.

246. Maywood ES, Hall SJ, Hastings MH. Developmental changes in the distribution of iodomelatonin binding sites in the SCN of the Syrian hamster and the rat. Abstracts of the 5th meeting of the Society for Research on Biological Rhythms, 1996, abstract #115.

247. Maywood ES, Bittman EL, Ebling FJ, Barrett P, Morgan P, Hastings MH. Regional distribution of iodomelatonin binding sites within the suprachiasmatic nucleus of the Syrian hamster and the Siberian hamster. J Neuroendocrinol 1995; 7:215–223.

248. Pittendrigh CS, Daan S. A functional analysis of circadian pacemakers in nocturnal rodents. V. Pacemaker structure: a clock for all seasons. J Comp Physiol 1976; 106: 333–355.

249. Illnerova H. The suprachiasmatic nucleus and rhythmic pineal melatonin production. In: Klein DC, Moore RY, and Reppert SM, eds. Suprachiasmatic Nucleus. The Mind's Clock. New York: Oxford University Press, 1991:197–216.

250. Michel S, Geusz ME, Zaritsky JJ, Block GD. Circadian rhythm in membrane conductance expressed in isolated neurons. Science 1993; 259:239–241.

251. Liu C, Weaver DR, Strogatz SH, Reppert SM. Cellular construction of a circadian clock: period determination in the suprachiasmatic circadian clock. Cell 1997; 91: 855–860.

252. Watanabe K, Koibuchi N, Ohtake H, Yamaoka S. Circadian rhythms of vasopressin release in primary cultures of rat suprachiasmatic nucleus. Brain Res 1993; 624: 115–120.

253. Murakami N, Takamure M, Takahashi K, Utunomiya K, Kuroda H, Etoh T. Long-term cultured neurons from rat suprachiasmatic nucleus retain the capacity for circadian oscillation of vasopressin release. Brain Res 1991; 545:347–350.

254. Silver R, Lehman MN, Gibson M, Gladstone WR, Bittman EL. Dispersed cell suspensions of fetal SCN restore circadian rhythmicity in SCN-lesioned adult hamsters. Brain Res 1990; 525:45–58.

255. Vogelbaum MA, Menaker M. Temporal chimeras produced by hypothalamic transplants. J Neurosci 1992; 12:3619–3627.

256. Ralph MR, Hurd MW. Circadian pacemakers in vertebrates. In: Circadian Clocks and Their Adjustment. Chichester: Wiley (Ciba Found. Symp 183), 1995:67–87.

257. Zlomanczuk P, Schwartz WJ. Cellular and molecular mechanisms of circadian rhythms in mammals. In: Turek FW, Zee PC, eds. Regulation of Sleep and Circadian Rhythms. New York: Marcel Dekker, 1999.

258. Gwinner E. Testosterone induces "splitting" of circadian locomotor activity rhythms in birds. Science 1974; 185:72–74.

259. Daan S, Damassa D, Pittendrigh CS, Smith ER. An effect of castration and testosterone replacement on a circadian pacemaker in mice (*Mus musculus*). Proc Natl Acad Sci USA 1975; 72:3744–3747.

260. Morin LP, Fitzgerald KM, Zucker I. Estradiol shortens the period of hamster circadian rhythms. Science 1977; 196:305–307.

261. Morin LP, Cummings LA. Splitting of wheelrunning rhythms by castrated or steroid treated male and female hamsters. Physiol Behav 1982; 29:665–675.

262. Turek FW, Gwinner E. Role of hormones in the circadian organization of vertebrates. In: Aschoff J, Daan S, Groos GA, eds. Vertebrate Circadian Systems. Berlin: Springer-Verlag, 1982:173–182.

263. Van Reeth O, Turek FW. Stimulated activity mediates phase shifts in the hamster circadian clock induced by dark pulses or benzodiazepines. Nature 1989; 339: 49–51.

264. Reebs SG, Mrosovsky N. Effects of induced wheel running on the circadian activity rhythms of Syrian hamsters: entrainment and phase response curve. J Biol Rhythms 1989; 4:39–48.

265. Edgar DM, Martin CE, Dement WC. Activity feedback to the mammalian circadian pacemaker: influence on observed measures of rhythm period length. J Biol Rhythms 1991; 6:185–199.

266. Shen H, Watanabe M, Tomasiewicz H, Rutishauer U, Magnuson T, Glass JD. Role of neuronal cell adhesion molecule and polysialic acid in mouse circadian clock function. J Neurosci 1997; 17:5221–5229.

267. Vitaterna MH, King DP, Chang AM, Kornhauser JM, Lowrey PL, McDonald JD,

Dove WF, Pinto LH, Turek FW, Takahashi JS. Mutagenesis and mapping of a mouse gene, *Clock*, essential for circadian behavior. Science 1994; 264:719–725.

268. King DP, Zhao Y, Sangoram AM, Wilsbacher LD, Tanaka M, Antoch MP, Steeves TDL, Vitaterna MH, Kornhauser JM, Lowrey PL, Turek FW, Takahashi JS. Positional cloning of the mouse circadian *Clock* gene. Cell 1997; 89:641–653.

269. Antoch MP, Song EJ, Chang AM, Vitaterna MH, Zhao Y, Wilsbacher LD, Sangoram AM, King DP, Pinto LH, Takahashi JS. Functional identification of the mouse circadian *Clock* gene by transgenic BAC rescue. Cell 1997; 89:655–667.

270. Weaver DR. Reproductive safety of melatonin: a "wonder drug" to wonder about. J Biol Rhythms 1997; 12:682–689.

8

The Impact of Changes In Nightlength (Scotoperiod) on Human Sleep

THOMAS A. WEHR

National Institute of Mental Health
Bethesda, Maryland

I. The Day and Night Within

Many day-active types of animals retreat to a secure refuge at night, where they rest and sleep from dusk to dawn. Moreover, as the length of the night changes during the course of the year, they adjust the duration of their nightly period of rest and sleep so that it is longer in winter and shorter in summer (1). In a natural environment, humans may have once behaved similarly; in the modern, urban environment, however, humans use artificial light to escape the confines of the solar day and extend their waking activities into the first hours of the night. This practice delays the onset of their nightly period of rest and sleep and keeps it short and unvarying year-round.

The fact that human sleep occurs primarily at night is largely a consequence of a temporal program in the circadian pacemaker that imposes a daily rhythm on one's ability to fall asleep and stay asleep, such that this ability is enhanced at night and diminished in the daytime. Thus, a nocturnal period of increased sleep propensity alternates with a diurnal period of decreased propensity (2–4) (Fig. 1). This pattern of variation in sleep propensity is a specific instance of a more general scheme in which the circadian pacemaker imposes distinct nocturnal and diurnal modes of functioning on the organism's physiology and behavior (5). In humans,

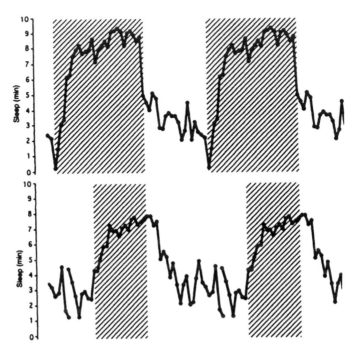

Figure 1 The duration of the nocturnal period of increased sleep propensity, measured by the amount of sleep that can be obtained in brief nap periods scheduled at regular intervals around the clock, is longer after exposure to long nights than after exposure to short nights (*N* = 7 individuals). Hatched areas indicate timing of dark periods to which individuals had been exposed during the nights preceding the 24-hr period of continuous dim (<1 lux) light during which sleep propensity was measured. Dark periods were 14 hr (6 PM–8 AM) and 8 hr (12 PM–8 AM) long, respectively. Twenty-four-hour profiles are shown in duplicate to facilitate visual inspection of the waveform. (Reprinted from Ref. 3.)

for example, melatonin is actively secreted at night and is absent in the daytime (3) (Fig. 2). Prolactin is secreted at high levels at night and at low levels in the daytime (3) (Fig. 3). And levels of cortisol rise progressively during the night and fall progressively during the daytime (3).

    This arrangement appears to be a mechanism and a consequence of a strategy whereby most organisms, during the course of their evolution, adopted specializations that make them especially fit to engage the environment during either the illuminated part of the day or the dark part of the day, but not both (5). During the part of the day to which they are less well adapted, they withdraw from their field environment into a secure refuge, where they rest and sleep. This strategy presumably maximizes the efficiency of energy-expenditure and minimizes risk-exposure.

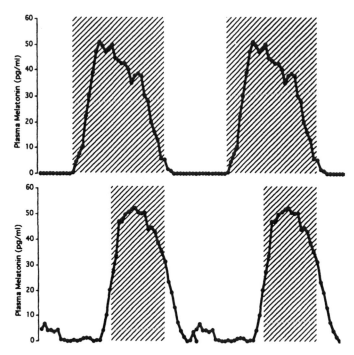

Figure 2 The duration of nocturnal melatonin secretion measured during continuous wakefulness in constant dim (<1 lux) light is longer after chronic exposure to long scotoperiods than after chronic exposure to short scotoperiods ($N = 15$ men). Hatched areas indicate timing of dark periods to which individuals had been exposed during the nights preceding the 24-hr constant-routine measurement period. Dark periods were 14 hr (6 PM–8 AM) and 8 hr (12 PM–8 AM) long, respectively. Twenty-four-hour profiles are shown in duplicate to facilitate visual inspection of the waveform. (Reprinted from Ref. 3.)

As day alternates with night, the circadian pacemaker switches the organism between alternate modes of functioning that make it suited either to seek out its field environment or to wait in its home environment. In effecting these changes, the pacemaker is not simply reacting to the rising and setting of the sun. Its endogenous temporal programs enable it to modify the organism's behavior and physiology in advance of the transitions between day and night. In this way, it creates a biological day and a biological night within the organism that mirror and anticipate the day and night outside (1). This ecological and evolutionary perspective may help to explain why circadian rhythms in sleep propensity, as well as melatonin, prolactin, and cortisol secretion, have distinct diurnal and nocturnal periods with relatively discrete transitions between them, corresponding to a biological day, biological dusk, biological night and biological dawn (3).

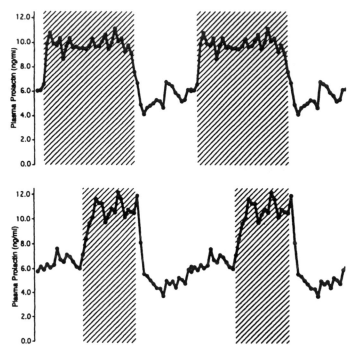

Figure 3 The duration of the nocturnal period of increased prolactin secretion is longer during long scotoperiods than during short scotoperiods ($N = 12$ men). Hatched areas indicate timing of dark periods to which individuals were being exposed during the 24-hr period when measurements were made. Dark periods were 14 hr (6 PM–8 AM) and 8 hr (12 PM–8 AM) long, respectively. Twenty-four-hour profiles are shown in duplicate to facilitate visual inspection of the waveform. (Reprinted from Ref. 3.)

Although nowadays most humans confine their sleep to a 7- or 8-hr period at night, the human biological night is actually 11 or 12 hr long (3) (Figs. 1–3). Furthermore, the use of artificial light has delayed its timing so that it begins long after sunset and ends long after sunrise in the modern urban environment, as was mentioned previously (6).

II. Human Sleep in Long Nights

In the absence of artificial light, what would human sleep be like in a natural environment, in which nights are longer (sometimes considerably longer) than 8 hr at most times of year and vary in duration over the course of the year? Clues to the answers to this question can be gleaned from the results of an experiment in which humans rested and slept in artificial long (14-hr) "nights" for periods of up to 15 weeks (7).

Human sleep has always seemed to be unique when compared with that of other animals, because it is consolidated (8). In long nights, however, human sleep occurs in bouts that are interspersed with periods of quiet wakefulness, like the sleep of other animals (9) (Fig. 5). Furthermore, the temporal distribution of these bouts is such that they are concentrated during two periods, one at the beginning and one at the end of the biological night. This bimodality of sleep has also been observed in the daily patterns of behavior of other animals (Fig. 6). Thus, the temporal organization of human sleep may not be unique, after all. Instead, the extent to which modern human sleep appears to be different from the sleep of other animals may partly be an artifact of the way we compress the nightly period of sleep when we use artificial light to extend our waking activities into the nighttime hours. The consolidation and efficiency that are hallmarks of modern human sleep may be caused by an increased propensity for sleep that results from chronic sleep restriction by the artificially shortened nights that humans impose on themselves (3).

Thus, modern humans may be more chronically sleep-deprived than they would be if they were living in a purely natural light-dark cycle. Possibly consistent with this hypothesis, we found that healthy volunteers slept almost 11 hr during their first night in experimental long (14-hr) scotoperiods, and then slept progressively less on subsequent nights until reaching a steady-state cumulative duration of 8.25 hr in the fourth week (7) (Fig. 7). This pattern might represent a process of rebound and recovery of a sleep debt that the individuals brought into the experiment from their everyday lives.

One manifestation of the efficiency of modern human sleep is that we fall asleep very quickly—10–15 minutes—after we lie down in the dark. In long nights, this is not at all the case. In experimental 14-hr nights, individuals remain awake for an average of 2 hr before they fall asleep (Fig. 7). In this situation, we interpret the onset of sleep to be governed largely by a signal that originates in the circadian pacemaker, because the sharp evening rise in sleep propensity that is characteristic of the circadian rhythm in this variable occurs at about the same time in long nights that sleep onset occurs in long nights (3) (Figs. 1 and 7). The sharp evening rise in sleep propensity is remarkably parallel to the sharp evening rise in melatonin secretion (Fig. 4). This may reflect a role for melatonin as an initiator of sleep or, more likely, the role of the circadian pacemaker as the common initiator of switches between diurnal and nocturnal modes of both melatonin secretion and sleep propensity.

Because sleep begins so rapidly after "lights-out" in modern life, it is unclear to what extent sleep onset is triggered by the nightly rise in sleep propensity that is programmed by the circadian pacemaker and to what extent it is triggered by the permissive effect on sleep of turning out the lights in someone

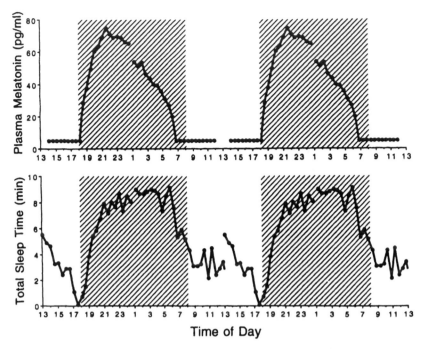

Figure 4 Circadian rhythms in melatonin secretion and sleep propensity are remarkably parallel. Profiles of the two variables were measured concurrently in constant dim (<1 lux) light in six individuals who had previously adhered to a regimen of rest and sleep in 14-hr nightly dark periods for 2 weeks. Hatched areas show the timing of the dark period to which they had been exposed on the nights preceding the period of continuous dim light in which measurements were made. Sleep propensity was measured by calculating the number of minutes individuals slept during 10-min periods that were scheduled every 30 min and during which they were encouraged to sleep. During the 20-min periods that intervened between the sleep periods, the individuals were kept awake in dim light. To preserve features of the melatonin rhythm that are characteristic of individual profiles, mean plasma levels of melatonin and mean minutes of sleep per 10-min opportunity to sleep are shown in segments that are referenced either to the time of evening onset of melatonin secretion or to the time of morning offset of melatonin secretion in the portions of the profiles that neighbored these events. Twenty-four-hour profiles are shown in duplicate to facilitate visual inspection of the waveform. An "evening wakefulness-maintenance zone" and a "sleep gate" are clearly identifiable in the profile of sleep propensity (4,13). (Previously unpublished data of the author.)

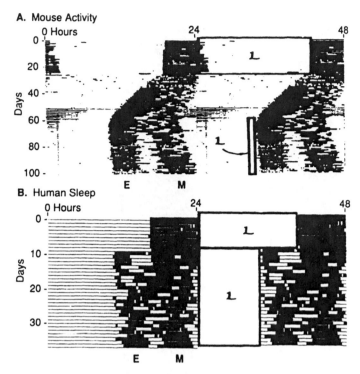

Figure 5 When schedules of daily light-exposures (L) were shifted from long days to short days, the nocturnal sleep phase of a human being (B) expanded and separated into an evening component (E) and a morning component (M). Shortening the photoperiod caused similar changes to occur in the activity phase of nocturnal rodents, as the record of mouse wheel-running activity (A) illustrates. (Mouse (*Peromyscus leukopus*) data were adapted from Ref. 38. Human data are from Ref. 20, from which the figure is reprinted.)

who has already reached a high level of sleep propensity owing to chronic sleep debt or to the circadian rise having occurred earlier, or both. Thus, the circadian onset of sleep may be masked in conventional short nights, while it appears to be unmasked in experimental long nights. This distinction may prove useful to researchers who wish to investigate the circadian regulation of sleep. The onset of sleep in long nights may be a valid phase marker of a rhythm in the circadian pacemaker, as the onset of melatonin secretion in dim light (DLMO) has been regarded to be (10).

When individuals eventually fall asleep in long nights, they usually sleep in two major bouts, one in the evening and one in the morning, with a period of quiet wakefulness between the bouts (3) (Figs. 5 and 8). Each bout lasts 2–5 hr, and the period of intervening wakefulness lasts about 1–3 hr. In long nights, transitions

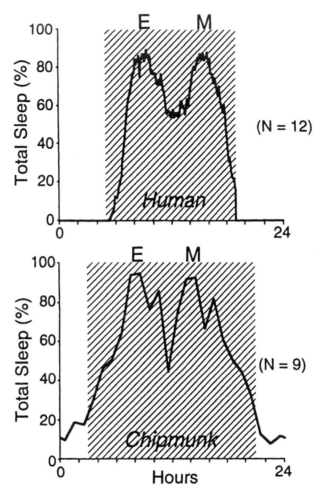

Figure 6 In long nights, Siberian chipmunks (bottom) and humans (top) exhibit homologous, symmetrically bimodal profiles of electroencephalographically monitored sleep. (Human data are from Ref. 7; chipmunk (*Eutamias sibiricus*) data are adapted from Ref. 22.)

from sleep to wakefulness almost always occur in REM sleep (3,11) (Fig. 8). From this observation, one might conclude that undisturbed sleep terminates spontaneously out of REM sleep (12,13). In this regard, it may be relevant that humans are more fully awake (have less sleep inertia) when they awaken from REM sleep than when they awaken from non-REM sleep. There is some evidence that humans may have once habitually slept in this way, prior to the Industrial Revolution. Diaries, court testimony, plays, and other documents frequently refer to two separate nightly periods of sleep, "first sleep" and "morning sleep" (A. Ekirch, personal communication).

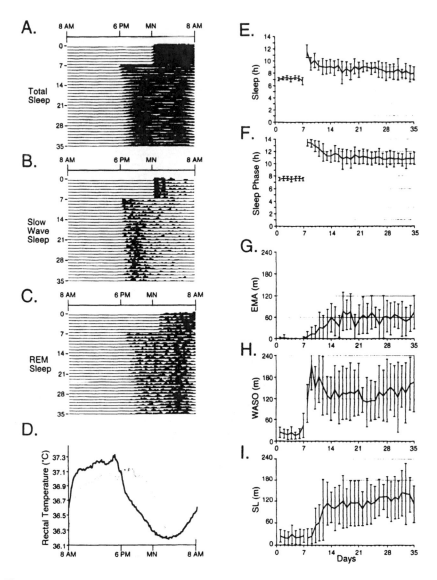

Figure 7 Average profiles of EEG-monitored sleep (A–C) and rectal temperature (D) in individuals who rested and slept in short (MN–8 AM) dark periods for 1 week and then long (6 PM–8 AM) dark periods for 4 weeks (N = 12 individuals). In long nights, the distribution of sleep (A) is bimodal, with slow-wave sleep (B) concentrated in the evening mode and REM sleep (C) concentrated in the morning mode. Total sleep time (E) is higher in long nights and rebounds after the transition from short nights to long nights. In long nights, the duration of the nightly sleep period (the interval between the first and last 30-sec epochs of sleep) (F) is about 11 hr and corresponds to the duration of the human "biological night." (G) Early morning awakening (EMA), (H) wake after sleep onset (WASO), and (I) sleep latency (SL) refer to awake time that occurs before, during, and after the sleep period, respectively. (Data are from Ref. 7.)

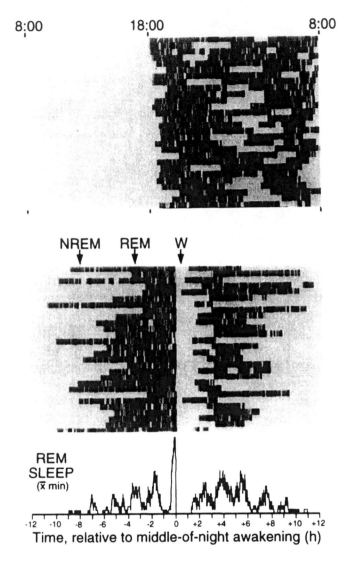

Figure 8 In long nights, transitions from sleep to quiet wakefulness are most likely to occur during REM sleep, particularly during REM periods in which phasic eye movements are especially intense. Episodes of an individual's sleep in long nights (top) have been replotted (middle) so that they are referenced to the time of onset of middle-of-the-night periods of quiet wakefulness. The average profile of REM sleep (bottom) shows that this stage of sleep is most likely to precede transitions to wakefulness. (Data adapted from Ref. 7.)

If, as some research suggests, dreams that occur during REM sleep are distinguished by their vivid and emotional nature and their narrative quality, one wonders whether dreaming associated with REM sleep might have had a greater impact on ancestors who may have slept in long nights than on their descendants who sleep in short nights today. At some time each night, they would have awakened from a vivid REM-type dream and entered an extended period of quiet wakefulness in which the effects of the dreams might reverberate in conscious awareness (Fig. 8). Perhaps these aspects of the physiology of sleep in long nights will help explain why individuals who lived in traditional societies sometimes seemed to attribute more importance to dreams than we do.

In long nights, slow-wave sleep is concentrated in the evening bout of sleep and REM sleep is concentrated in the morning bout of sleep (3) (Fig. 7). Since the occurrence of slow-wave sleep is largely governed by a homeostatic or recovery process such that slow-wave sleep propensity increases asymptotically during wakefulness and decreases exponentially during sleep (14), it is not surprising that slow-wave sleep is maximal during the evening bout of sleep and much less prevalent in the morning bout. The abundance of REM sleep in the morning bout of sleep is probably mainly a function of the circadian variation in REM sleep propensity, which has been shown to reach a maximum near the end of the biological night (15). To the extent that increased REM sleep propensity might cause increased sleep propensity, the circadian rhythm in REM sleep propensity may even help to trigger the onset and maintenance of the morning bout of sleep. Since slow-wave sleep tends to preempt REM sleep when slow-wave sleep propensity is high (16), the decline of slow-wave sleep propensity during the evening bout of sleep may also contribute to the increased prevalence of REM sleep during the morning bout of sleep.

The association of REM sleep with transitions to quiet wakefulness during the night, the increased brain activity that occurs during REM sleep, and the lower sleep inertia that is found after awakening from REM sleep compared with non-REM sleep have suggested to some that a function of REM sleep might be to prepare the organism for waking (17,18). If so, then the high prevalence of REM sleep during the morning bout of sleep might play a special role in preparing humans to enter their diurnal period of active wakefulness.

III. The Night Within Adjusts to Changes in Duration of the Night Outside

In many animals, the circadian pacemaker is able to detect seasonal changes in the length of the night outside and make proportional adjustments in the duration of the biological night that it programs within the organism (1,19). In the course of its evolution, the human circadian pacemaker has retained a similar capacity to detect such changes and make such adjustments (3). This capacity has been shown most clearly in the investigations of the effects of changes in duration of the scotoperiod on the duration of nocturnal human melatonin secretion. The duration of the nocturnal period of human melatonin secretion that is programmed by the circadian pacemaker (referred to hereafter as the "intrinsic duration") can be measured by measuring levels of melatonin in plasma that is sampled at frequent, regular intervals from individuals who remain in constant dim light. In these conditions, the endogenous pattern of melatonin secretion is protected from distortion or masking by direct secretion-suppressing effects of light. When individuals have been chronically exposed to long artificial scotoperiods, the intrinsic duration of nocturnal melatonin secretion (corresponding to the biological night) is longer than it is when they have been chronically exposed to short artificial scotoperiods (3,20) (Fig. 2). The average difference in intrinsic duration of melatonin secretion in the two scotoperiods is not more than 2 hr, even if the difference in duration of the scotoperiods was much greater.

The circadian rhythms in human sleep propensity and human melatonin secretion seem to be organized in remarkably similar ways, having a biological night characterized by increased sleep propensity and active melatonin secretion and a biological day characterized by decreased sleep propensity and absence of melatonin secretion (Fig. 4). The circadian pacemaker appears to have a capacity to adjust the intrinsic duration of the nocturnal period of increased sleep propensity in response to changes in duration of the scotoperiod as it does the intrinsic duration of nocturnal melatonin secretion.

IV. Mechanism of the Circadian Pacemaker's Response to Nightlength

How does the circadian pacemaker program the duration of the biological night and how does it modify this program in response to changes in the length of the night outside? Many years ago, in the absence of concrete knowledge of the molecular substrates of pacemaker function, Pittendrigh and Daan developed a model of the vertebrate circadian system that was based on formal analyses of the behavior of the pacemaker, as revealed in long-term recordings of rodent motor activity in a variety of experimental conditions (20). In this now-classic model, scotoperiod responses of the circadian pacemaker are effected by changes in the mutual phase relationships between two separate circadian oscillators, an evening oscillator (E), which is entrained to dusk and controls the time of onset of the biological night, and a morning oscillator (M), which is entrained to dawn and controls the time of onset of the biological day. According to the model, as the interval between dusk and dawn increases in the fall, the phase angle between the oscillations of E and M increases. Consequently, the interval between the onset and offset of the biological night increases.

The two-oscillator model of a complex circadian pacemaker was partly inspired by the observation that the behavior of many animals exhibits distinct evening and morning peaks that appear to be separately synchronized with dusk and dawn, respectively (9) (Figs. 5 and 6). In the model, these peaks represent the separate expressions of the oscillations of the dusk- and dawn-entrained circadian oscillators, E and M. The bimodal distribution of sleep that we observed in humans who slept in long nights seems comparable to the evening and morning peaks that have been observed in the behavior of other animals (Figs. 5 and 6), and they suggest that the Pittendrigh-Daan dual oscillator model could be applied to the human circadian system (3,21–23).

The strongest evidence supporting the dual-oscillator model in the rodent is "splitting" of the activity-rest cycle, a phenomenon that sometimes occurs when it free-runs in constant light (19,24). During splitting, the nocturnal activity period separates into separate evening (e) and morning (m) bouts, which oscillate independently of one another and then recouple in a new metastable phase relationship that is shifted 180° from the usual phase relationship (Fig. 9). When the dual-oscillator system shifts from its usual coupling mode to its metastable one, the intrinsic period of its free-running oscillations becomes shorter. In human sleep, we observed a phenomenon that resembles splitting in all these respects in individuals whose sleep-wake cycle was free-running in isolation from external time cues (21) (Fig. 9).

Note that in the foregoing comparisons between rodents and humans the circadian organization of sleep in humans is being compared to the circadian organization of wakefulness (locomotor activity) in rodents. This comparison can be justified by the fact that (1) the behaviors in each species are expressed during the biological night, and (2) processes that take place within the circadian pacemaker are known to be synchronized with the day-night cycle in the same way in day-active and night-active animals. With regard to this latter point, light stimuli elicit similar phase-resetting responses at different times of day in the two types of animals (25,26). In both day-active and night-active animals, light stimuli applied late in the biological night elicit phase-advance shifts in the timing of the sleep-wake cycle, while light stimuli applied early in the biological night elicit phase-delay shifts (Fig. 10). When the direction and magnitude of phase-shifts are plotted as a function of the circadian phase at which the light stimulus is applied, a phase-response curve (PRC) is generated that is roughly similar in the two types of animals (Fig. 10). Furthermore, changes in neural firing rates and rates of glucose utilization that occur in pacemaker neurons over the course of the day and night are parallel in day-active and night-active species, not reciprocal (27,28).

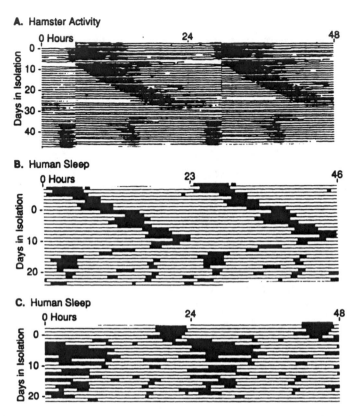

Figure 9 "Splitting" of free-running circadian rhythms of hamster running-wheel activity (A) and electroencephalographically monitored human sleep (B and C) into two components that become stably recoupled approximately 180° apart. A 23-hr plotting interval was chosen for one of the human sleep records (B) to facilitate comparison with the hamster activity record (A). Prior to splitting, the rhythms exhibit a single component with an intrinsic period greater than 24 hr. After splitting, the rhythms exhibit two components nearly 180° apart with an intrinsic period that is less than 24 hr. (Hamster activity data were adapted from Ref. 24; human sleep data were adapted from Ref. 39. Figure is reprinted from Ref. 21.)

V. Effect of Nightlength on the Pacemaker's Phase-Response to Light

The PRC for light enables one to predict the phase position that the circadian pacemaker will assume relative to the day-night cycle when it is entrained to the day-night cycle (25). Since the period of the intrinsic rhythm of the pacemaker differs slightly from 24 hr, its phase must be reset each day by an amount that is

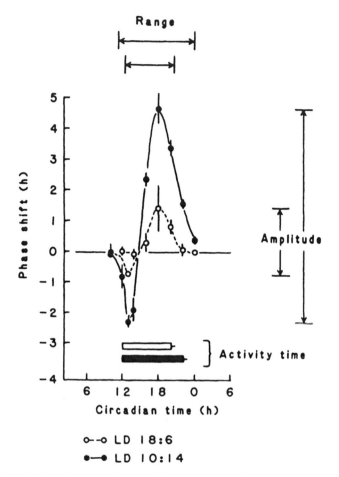

Figure 10 Photoperiod-dependence of the hamster's phase-response curve for light. Each point plots the mean phase shift for hamsters pulsed with light at the indicated circadian times on the tenth cycle in constant darkness (DD) after release from entrainment to either a long photoperiod (open symbols) or a short photoperiod (closed symbols). Following entrainment to the short photoperiod (LD 10:14), phase shift responses ($\geqslant 0.5$ hr) occur over a broader range of the circadian cycle and the phase-response curve has a larger amplitude than it does following entrainment to the long photoperiod. (Reprinted from Ref. 29.)

equal to the difference between the period of its rhythm and 24 hr if it is to maintain synchrony with the day-night cycle. Since the period of the intrinsic rhythm of the human circadian pacemaker is longer than 24 hr, it must advance the phase of its rhythm each day by exposure to morning light to maintain this synchrony.

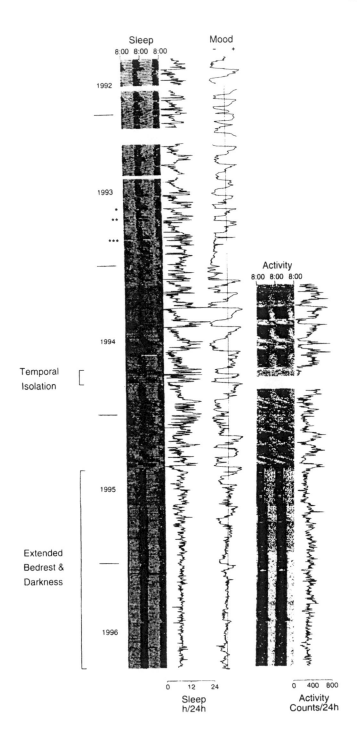

Inspection of the PRC for light (Fig. 10) shows that if the amplitude of the PRC were reduced, the phase position that the pacemaker would assume relative to the day-night cycle would be delayed in humans. Because the period of its intrinsic rhythm is longer than 24 hr, its cycles would become delayed relative to the day-night cycle until they reached a point at which morning light would elicit the magnitude of phase advance that is necessary for them to become entrained to the day-night cycle. Inspection of the PRC also shows that if the amplitude of the PRC were reduced to such an extent that the maximum possible phase advance were less than the difference between the intrinsic period of the pacemaker's rhythm and the 24-hr period of the day-night cycle, the pacemaker could no longer be entrained to the day-night cycle and would therefore free-run, following its own intrinsic rhythm.

The results of animal experiments show that the amplitude of the PRC for light is a function of the duration of the nightly scotoperiod to which it has most recently been exposed (29). As the duration of the nightly scotoperiod decreases, the amplitude of the PRC for light decreases (Fig. 10). If the PRC for light of the human pacemaker responds to changes in scotoperiod in a similar way, then one would predict that the modern practice of using artificial light to shorten the nightly scotoperiod would reduce the amplitude of the human PRC for light. As a corollary, one would predict that reduction in the amplitude of the PRC would then cause the circadian pacemaker to assume a more delayed phase position relative to the light-dark cycle. In extreme cases, the pacemaker might no longer be able to advance the phase of its rhythm to a sufficient degree to maintain its entrainment to the light-dark cycle, so that it would free-run.

These considerations may be relevant to the pathogenesis of circadian sleep disorders in humans. Reduction in the amplitude of the PRC caused by chronic exposure to short nightly scotoperiods might be a factor contributing to the

---

Figure 11   Recordings of sleep, mood, and wrist motor activity in a patient whose rapidly cycling form of manic-depressive (bipolar) illness improved after he adhered to a regimen of long, regularly scheduled nightly periods of darkness, rest, and sleep from June 1995 onward. During the regimen, the duration of the nightly dark period was gradually reduced from 14 hr (LD 10:14) to 10 hr (LD 14:10). Data for sleep (15-min sampling frequency) and activity (12-min sampling frequency) are shown in a raster format in which 24-hr (8 AM–8 AM) segments of data are plotted successively beneath one another and double-plotted to the right as a visual aid. Total daily sleep, total daily activity counts, and daily mood self-ratings as recorded on a 100-mm line are shown beside the raster plots. A vertical line indicates the center (50 mm) of the mood-rating scale. The alternating dark and light areas in the first half of the activity raster plot correspond to periods of increased and decreased activity that were associated with periods of hypomania and depression, respectively. (For details, see Ref. 34, from which figure is reprinted.)

occurrence of the most common type of circadian sleep disorder, delayed sleep phase syndrome (DSPS). The fundamental problem in DSPS is that the phase of entrainment of the sleep-wake cycle to the day-night cycle is abnormally delayed (30). Consequently, individuals with DSPS have difficulty falling asleep until late at night, and they have difficulty waking, or being fully awake, until late in the morning. They find it difficult to remedy the situation by advancing the timing of their sleep-wake cycle.

More rarely, certain individuals seem unable to entrain their sleep-wake cycle to the day-night cycle at all, so that it free-runs. In these individuals, the timing of sleep occurs progressively later each day and goes completely around the clock every few weeks (31–33) (Fig. 11).

It is not known whether the amplitude of the PRC is reduced in patients with circadian sleep disorders, and if so, whether this is a consequence of their being exposed to short nightly scotoperiods. In any case, we have shown in pilot studies that individuals with free-running sleep-wake cycles can be entrained to the 24-hr day-night cycle if they forsake the use of artificial light in the evening and adhere to a regimen of long nightly scotoperiods (33,34) (Fig. 11).

VI. Effect of Nightlength on Sleep-Related Growth Hormone Secretion in Humans

Human growth hormone is secreted in brief, vigorous bouts several times a day. The major bout of growth hormone secretion occurs about 1 hr after sleep onset and seems to be sleep-dependent (35). The results of experiments in which individuals have been chronically exposed to artificial nights of different duration indicate that the maximum level of sleep-related growth hormone secretion is markedly dependent on the duration of the nightly scotoperiod (3). During sleep in short scotoperiods, the level is twice as high as it is during sleep in long scotoperiods (Fig. 12). From this observation, one might infer that the sleep-related peak in growth hormone secretion is much higher when humans use artificial light to shorten the duration of the nightly scotoperiod than it would be at most times of year in a purely natural scotoperiod.

VII. Effect of Nightlength on Rest and Prolactin Secretion in Humans

Human prolactin is secreted at levels that are twice as high at night as they are in the daytime (36) (Fig. 3). For many years, the nocturnal rise in prolactin secretion has been interpreted as being sleep-dependent (36). This conclusion was based on the fact that the nighttime rise in prolactin secretion can be blocked with sleep deprivation, and that the timing of the rise shifts in parallel with sleep if the timing

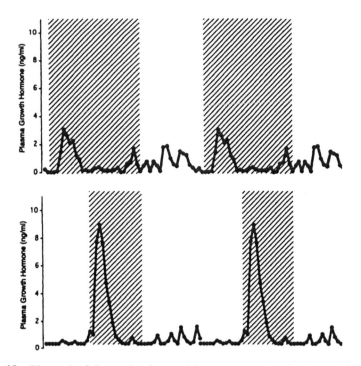

Figure 12 The peak of sleep-related growth hormone secretion is twice as high during short scotoperiods as it is during long scotoperiods ($N = 12$ individuals). Indications are the same as in Figure 1. (Reprinted from Ref. 3.)

of sleep is shifted into the daytime hours. The results of our experiments suggest, however, that the nighttime rise in prolactin secretion is not strictly sleep-dependent (3,37). The nocturnal period of increased prolactin secretion is essentially coextensive with the scotoperiod. If the duration of the scotoperiod is increased from 8 hr to 14 hr, the duration of the nocturnal period of high prolactin secretion increases from nearly 8 hr to nearly 14 hr (Fig. 3). In 8-hr scotoperiods (the conventional scotoperiod in which previous prolactin studies were carried out), the nocturnal rise in prolactin secretion and the onset of sleep occur together, giving the impression that prolactin secretion is sleep-dependent. In 14-hr scotoperiods, prolactin reaches its high nighttime levels after the first 30 min, even though sleep onset has not yet occurred (the average sleep latency in 14-hr scotoperiods is 2 hr) (Fig. 7). Moreover, there is no further increase in prolactin levels after sleep begins (37). In addition, there is no decline in prolactin secretion during the periods of quiet wakefulness that occur in the middle of the scotoperiod between the evening and morning bouts of sleep. Finally, there is no falling off of prolactin secretion after individuals awaken from the morning bout of sleep and

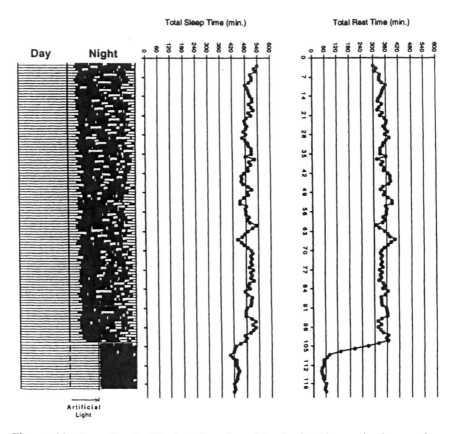

Figure 13 Extending the illuminated portion of the day into the evening hours reduces total sleep time and virtually abolishes rest. (Reprinted from Ref. 3.)

before the end of the scotoperiod. These findings seem to indicate that the nocturnal rise of prolactin secretion is dependent on a state of quiet rest that occurs either while individuals lie quietly awake waiting for sleep or while they sleep, and not on sleep itself. In 14-hr scotoperiods, individuals spend more than 5 hr/night in this state of quiet wakefulness (3) (Fig. 13). When modern humans use artificial light to shorten the duration of the scotoperiods, they spend almost no time in this state, and thus considerably less time being exposed to the high levels of prolactin that are associated with it (Fig. 3).

VIII. Conclusion

When modern humans use artificial light to extend their waking activities into the nighttime hours, they compress, reduce, and consolidate their sleep, and they delay its timing relative to the day-night cycle (Fig. 13). When human sleep is decompressed in long scotoperiods, it can appear to modern eyes to be disordered. Sleep latency is long, and sleep is fragmented and interspersed with periods of wakefulness. Because this type of sleep is similar to the sleep of many other animals, and because it occurs in healthy young individuals who sleep in long nights, we interpret it to be a natural pattern of human sleep, as it may have once occurred when humans lived from sun to sun. When modern humans find that their sleep is fragmented and interrupted by periods of wakefulness during the night, they regard it as being disordered. For some, an alternative interpretation could be that a natural pattern of human sleep is breaking through into an artificial world in which it seems unfamiliar and unwelcome.

References

1. Pittendrigh CS. The photoperiodic phenomena: seasonal modulation of the "day within." J Biol Rhythms 1988; 3:173–188.
2. Lavie P, Zvuluni A. The 24-hour sleep propensity function: experimental bases for somnotypology. Psychophysiology 1992; 29:566–575.
3. Wehr TA. A "clock for all seasons" in the human brain. In: Buijs RM, Kalsbeek A, Romihn JH, Pennartz CMA, Mirmiran M, eds. Progress in Brain Research, Vol 111. Amsterdam: Elsevier, 1996:319–340.
4. Dijk D-J, Czeisler CA. Contribution of the circadian pacemaker and the sleep homeostat to sleep propensity, sleep structure, electroencephalographic slow waves, and sleep spindle activity in humans. J Neurosci 1995; 15:3526–3538.
5. Rusak B. The mammalian circadian system: models and physiology. J Biol Rhythms 1989; 4:121–134.
6. Wehr TA. Melatonin and seasonal rhythms. J Biol Rhythms 1997; 12:517–526.
7. Wehr TA, Moul DE, Giesen HA, Seidel JA, Barker C, Bender C. Conservation of photoperiod-responsive mechanisms in humans. Am J Physiol 1993; 265:R846–R857.
8. Bert J, Kripke DF, Rhodes J. Electroencephalogram of the mature chimpanzee: twenty-four hour recordings. Electroenceph Clin Neurophysiol 1970; 28:368–373.
9. Aschoff J. Circadian activity pattern with two peaks. Ecology 1966; 47:657–662.
10. Lewy AJ, Sack RL, Singer CM. Assessment and treatment of chronobiologic disorders using plasma melatonin levels and bright light exposure: the clock-gate model and the phase response curve. Psychopharmacol Bull 1984; 20:561–565.
11. Barbato G, Barker C, Bender C, Giesen HA, Wehr TA. Extended sleep in humans in 14 hour nights (LD 10:14): relationship between REM density and spontaneous awakening. Electroenceph Clin Neurophysiol 1994; 90:291–297.
12. Langford GW, Meddis R, Pearson AJD. Spontaneous arousal from sleep in human subjects. Psychon Sci 1972; 28:228–230.
13. Lavie P, Oksenberg A, Zomer J. "It's time, you must wake up now." Percept Motor Skills 1979; 49:447–450.
14. Borbely AA. A two-process model of sleep regulation. Hum Neurobiol 1982; 1:195–204.

15. Weitzman ED, Nogeire C, Perlow M, Fukushima D, Sassin J, McGregor P, Gallagher TF, Hellman L. Effects of a prolonged 3-hour sleep-wake cycle on sleep stages, plasma cortisol, growth hormone and body temperature in man. J Clin Endocrinol Metab 1974; 38:1018–1030.

16. Feinberg I, Maloney T, March JD. Precise conservation of NREM period 1 (NREMP 1) delta across naps and nocturnal sleep: implications for REM latency and NREM/ REM alternation. Sleep 1992; 15:400–403.

17. Snyder F. Towards an evolutionary theory of dreaming. Am J Psychiatry 1966; 123: 121–136.

18. Wehr TA. A brain-warming function for REM sleep. Neurosci Biobehav Rev 1992; 16:379–397.

19. Pittendrigh CS, Daan S. A functional analysis of circadian pacemakers in nocturnal rodents. V. Pacemaker structure: a clock for all seasons. J Comp Physiol A 1976; 106: 333–355.

20. Wehr TA. The durations of human melatonin secretion and sleep respond to changes in daylength (photoperiod). J Clin Endocrinol Metab 1991; 73:1276–1280.

21. Wehr TA. In short photoperiods, human sleep is biphasic. J Sleep Res 1992; 1:103–107.

22. Dijk D-J, Daan S. Sleep EEG spectral analysis in a diurnal rodent, *Eutamias sibiricus*. J Comp Physiol A 1989; 165:205–215.

23. Tobler I, Schwierin B. Behavioural sleep in the giraffe (*Giraffa camelopardalis*) in a zoological garden. J Sleep Res 1995; 5:21–32.

24. Pickard GE, Turek FW. The suprachiasmatic nuclei: two circadian clocks? Brain Res 1983; 268:201–210.

25. Pittendrigh CS, Daan S. A functional analysis of circadian pacemakers in nocturnal rodents. IV. Entrainment. J Comp Physiol A 1976; 106:291–331.

26. Czeisler CA, Kronauer RE, Allan JS, Duffy JF, Jewett ME, Brown EN, Ronda JM. Bright light induction of strong (type 0) resetting of the human circadian pacemaker. Science 1989; 244:1328–1333.

27. Watts AG. The efferent projections of the suprachiasmatic nucleus: anatomical insights into the control of circadian rhythms. In: Klein DC, Moore RY, Reppert SM, eds. Suprachiasmatic Nucleus: The Mind's Clock. New York: Oxford University Press, 1991:77–106.

28. Schwartz WJ. SCN metabolic activity in vivo. In: Klein DC, Moore RY, Reppert SM, eds. Suprachiasmatic Nucleus: The Mind's Clock. New York: Oxford University Press, 1991:144–156.

29. Pittendrigh CS, Elliott J, Takamura T. The circadian component in photoperiodic induction. In: Photoperiodic Regulation of Insect and Molluscan Hormones (Ciba Foundation Symposium 104). London: Pitman, 1984:26–47.

30. Weitzman ED, Czeisler CA, Coleman RM, Spielman JA, Zimmer JC, Dement W. Delayed sleep phase syndrome: a chronobiological disorder with sleep-onset insomnia. Arch Gen Psychiatry 1981; 38:737–746.

31. Kokkoris CP, Weitzman ED, Pollak CP, Spielman AJ, Czeisler CA, Bradlow H. Long-term ambulatory temperature monitoring in a subject with a hypernychthemeral sleep-wake cycle disturbance. Sleep 1978; 1:177–190.

32. Kamgar-Parsi B, Wehr TA, Gillin JC. Successful treatment of human non-24-hour sleep-wake syndrome. Sleep 1983; 6:257–264.

33. Oren DA, Giesen HA, Wehr TA. Restoration of detectable melatonin secretion after entrainment to a 24-hour schedule in a "free-running" man. Psychoneuroendocrinology 1997; 22:39–52.
34. Wehr TA, Turner EH, Clark CH, Barker C, Leibenluft E. Treatment of a rapidly cycling bipolar patient by using extended bedrest and darkness to stabilize the timing and duration of sleep. Biol Psychiatry 1998; 43:822–828.
35. Sassin JF, Parker DC, Mace JW, Gotlin RW, Johjnson LC, Rossman LG. Human growth hormone release: relation to slow-wave sleep and sleep-waking cycles. Science 1969; 165:513–515.
36. Sassin JF, Frantz AG, Kapen S, Weitzman ED. The nocturnal rise of human prolactin is dependent on sleep. J Clin Endocrinol Metab 1973; 37:436–440.
37. Wehr TA. Effects of changes in nightlength on human neuroendocrine function. Horm Res 1998; 49:118–124.
38. Pittendrigh CS, Daan S. A functional analysis of circadian pacemakers in nocturnal rodents. I. The stability and lability of spontaneous frequency. J Comp Physiol A 1976; 106:223–252.
39. Wehr TA, Sack DA, Duncan WC Jr, Rosenthal NE, Mendelson WB, Gillin JC, Goodwin FK. Sleep and circadian rhythms in affective patients isolated from external time cues. Psychiatry Res 1985; 15:327–339.

9

Cellular and Molecular Mechanisms of Sleep

TARJA PORKKA-HEISKANEN*

Harvard University
Boston, Massachusetts
and Brockton VA Medical Center
Brockton, Massachusetts

DAG STENBERG

University of Helsinki
Helsinki, Finland

I. Introduction

The electrophysiological changes occurring during the sleep-wake cycle (as described in detail in Chapter 3) have intracellular correlates in terms of ionic currents across the cell membranes, changes in receptor density, energy requirements, intracellular signaling, and activation of the transcription of genes.

   The vigilance state-dependent events at the level of electroencephalographic (EEG) changes have been described in detail in the literature, but the intracellular events that correspond to those changes are largely unknown. The answer to the basic question of sleep research: "why do we have to sleep?" may lie in state-dependent changes, e.g., in the intracellular signaling pathways. Recently, there have been several reports showing changes in gene expression, not only as response to sleep deprivation (1–7) but also associated with the natural sleep-wake cycle (4,8). Such reports provide evidence that changes in vigilance state are associated with changes even at the prime level of gene expression: the transcription of genes.

Current affiliation: University of Helsinki, Helsinki, Finland.

This chapter describes the present knowledge of these events as they relate to the sleep-wake cycle and to sleep deprivation.

II. Receptors—Communication with the Surroundings

Cholinergic, monoaminergic, and histaminergic cells are more active during waking than during non–rapid eye movement sleep (NREMS). During REM sleep the monoaminergic and histamine cells further decrease their activity while cholinergic cells especially in the pons (LDT/PPT) become active (see detailed discussion in Chapter 2). Acetylcholine is released more during waking and REM sleep than during non-REM sleep (9,10), while the release of serotonin and possibly also other monoamines is highest during waking, intermediate during non-REM sleep, and lowest during REM sleep (11,12). During sleep deprivation brain tissue monoamine turnover is increased (1,13). Increase in extracellular neurotransmitter concentration can down-regulate the receptor density. It is thus possible that changes in receptor density and receptor gene expression can take place during the sleep-wake cycle and sleep deprivation.

A. Muscarinic Receptors

Acetylcholine affects sleep mainly through muscarinic receptors in the CNS. Five subtypes of muscarinic receptors (M_1–M_5) have been described (14). M_1, M_3, and M_5 act through the phospholipase C signaling pathway, while M_2 and M_4 inhibit adenylate cyclase (14). The regional distribution of different muscarinic receptor subtypes has been described using in situ hybridization (15,16) and RT-PCR combined with HPLC (17). In the basal forebrain and pons (LDT/PPT), regions that are thought to have an important role in regulation of REM sleep (18), predominantly M_2 and M_3 receptors are expressed (15–17). This is in accordance with pharmacological studies where M_2 receptor agonists induced REM sleep when administered to the pons, but M_1 agonists had no effect (19), and M_2, but not M_1 or M_3, antagonists decreased both REM sleep and slow-wave sleep (20). Further when a M_2 antagonist was injected i.c.v. it decreased REM sleep, but a M_3 antagonist had no effect (21). Thus several studies suggest that the muscarinic receptor subtype M_2 would be associated with REM sleep regulation. However, 72 hr of sleep deprivation did not affect M_2 receptor mRNA levels, while M_3 receptor mRNA was increased (22). During the normal sleep-wake cycle the number of muscarinic receptors has been shown to be higher during waking than during sleep in the pons (23).

B. Monoaminergic Receptors

Adrenergic Receptors

A major noradrenergic projection from the locus coeruleus to the forebrain serves to increase vigilance in response to sensory stimuli (24). These noradrenergic neurons show increased activity during waking (25,26). Several studies have shown that activation of α_1-adrenergic receptors increases waking and inhibits sleep (27,28), while activation of α_2-receptors with systemic drugs induces sedation and even anesthesia (29,30).

Locus coeruleus neuronal activity is markedly diminished during REM sleep, and norepinephrine has been ascribed a mostly inhibitory role in the regulation of this state (25,31). The relationship between noradrenergic neurotransmission and REM sleep is more complicated, however, which is seen in the involvement of specific receptors. Although systemic α_2-receptor agonists inhibit locus coeruleus neuronal activity and are sedative, large doses also inhibit REM sleep (29). It was also shown in the rat that blocking β_1-receptors by systemic drugs inhibited REM sleep, indication a positive involvement of β_1-receptors in REM sleep mechanisms (32). Blocking α_1-receptors, on the other hand, increases REM sleep (33), and also cataplectic attacks in narcoleptic dogs (34), whereas microinjection of α_1-agonist into the dorsal pontine tegmentum inhibits REM sleep (35). α_1-Receptors in the brainstem thus have an important role in controlling the occurrence of REM sleep.

Prolonged activation of the noradrenergic system, as during prolonged waking, might be expected to induce down-regulation of postsynaptic receptors, and as a result, decreased responsivity to noradrenergic activity. It has been hypothesized that REM sleep would serve to up-regulate or prevent down-regulation of brain noradrenergic receptors (36). After REM sleep deprivation (REMSD) of rats for 72 hr, a decrease in cortical β-receptor sites has been reported by one group (37), but another found no change (38). REMSD lasting 7 days or total sleep deprivation lasting 10 or more days had virtually no effect on adrenoceptors in rat (39,40). There is thus at present no clear experimental support for a change in noradrenergic receptor amount or sensitivity after prolonged wakefulness.

Dopaminergic Receptors

In contrast, several studies indicate hypersensitivity to dopaminergic drugs inducing stereotypic behavior after REM sleep deprivation (41,42), and D_2-receptor density is increased in several brain areas (43). Dopaminergic neurons in the substantia nigra have remarkably constant activity over a variety of behaviors and vigilance states (44,45). Nevertheless, dopaminergic transmission is apparently modulated by sleep deprivation.

Serotonergic Receptors

Serotonergic projections to the forebrain have a profound role in the regulation of sleep and wakefulness (46). Serotonergic dorsal raphe neurons are most active during waking, decrease their activity during NREMS, and are virtually silent during REM sleep (47,48). The same pattern of firing is also found in other raphe nuclei. Thus the serotonergic system is regarded as a waking-promoting and REM-sleep-inhibiting system.

Serotonergic drugs have found some use in sleep medicine. 5-HT_2-receptor blockers promote NREMS and EEG slow-wave activity in animals (49) and humans (50). Part of this effect may be subject to melatonin-mediated modulation (51). 5-HT_2-receptors mediate mainly excitatory responses through activation of phospholipase C, while 5-HT_1-receptors mediate inhibition. Systemic 5-HT_1-receptor agonists have a biphasic effect: low doses decrease waking and increase NREMS, while higher doses have the opposite effect in both rat (52) and human (53). The sleep-inducing effect of the small doses is attributed to preferential autoreceptor action, while the large doses affect also postsynaptic receptors.

Serotonergic neurons are activated during sleep deprivation, as shown by an increase in brain serotonin turnover (13,54). It might be expected that this could down-regulate postsynaptic 5-HT receptors. REM-sleep-deprived rats have shown decreased responsivity to the 5-HT_2-receptor agonist quipazine, possibly indicating receptor hyposensitivity (55). However, 4 days after a similar deprivation the responsivity was above normal (56). An interesting question is the regulation of 5-HT_1-autoreceptors. Serotonin uptake blockers (SSRIs) have found use in the treatment of severe depression. Their therapeutic effect, though, becomes evident only after several weeks. If the 5-HT_1-autoreceptors are blocked by an antagonist (and thus autoinhibiton of serotonin release is blocked), the therapeutic effect is seen within a week (57). This indicates that the therapeutic effect may be due to increased serotonergic transmission. Sleep deprivation also has therapeutic effect in depression. During sleep deprivation, desensitization of 5-HT_{1A}-autoreceptors has been observed (58). Together these findings suggest that the treatment of depression both with SSRIs and with sleep deprivation works through increased serotonergic transmission, and the therapeutic target should be looked for downstream to the postsynaptic serotonin receptors stimulated by the treatment.

Histaminergic Receptors

Microinjections of the $GABA_A$-receptor agonist muscimol to the posterior hypothalamus decrease wakefulness and increase sleep (59). From this area histaminergic neurons project to the forebrain (60). It is well known that antihistamins (H_1-receptor antagonists) induce sleepiness. Unit activity recordings have shown that there are two types of neurons in this area: one type is active during both waking and REM sleep, whereas the other type fires slowly and regularly during waking, is less active during NREMS, and is completely inhibited during REM sleep (61). The activity of the second type is typical of monoamine neurons. The fact that H_3-autoreceptor antagonists enhance wakefulness, which can be blocked by H_1-receptor antagonists, indicates that the posterior hypothalamus contains histaminergic waking neurons that promote waking through H_1-receptors, and are subject to H_3-autoreceptor regulation (62,63). Recent findings indicate that sleep-inducing and sleep-maintaining preoptic and basal forebrain mechanisms modulate the posterior hypothalamus, possibly through GABAergic connections (64,65). It is not known how the histamine receptors are influenced by prolonged waking and sleep loss.

III. Glucose and Adenosine—Energy Metabolism

Energy conservation or replenishing the energy consumed during waking might be the purpose of sleep. Neural metabolism is higher during waking than during deep slow-wave sleep (66,67), increasing the requirement for energy during waking. It is thus possible that prolonged waking consumes the energy resources available for the brain, and the lack of energy substrates would force the brain to sleep.

A. Glucose Metabolism

Several studies have examined the changes in brain energy metabolism during the sleep-wake cycle. Modern imaging methods, e.g., positron emission tomography (PET), have proved to be valuable tools, emphasizing the fact that there are state-dependent regional changes in glucose and oxygen metabolism.

Brain glucose and oxygen metabolism as well as blood flow were lower during sleep than during waking (68). The cerebral metabolic rate of oxygen decreased 24% during deep sleep, and was similar to waking values during REM sleep (69). In a PET study that enabled regional analysis of different brain areas, glucose utilization was found to decrease in six brain areas including the thalamus, especially during deeper states of sleep (66). In other PET studies, glucose metabolism has also been found to decrease during deep sleep in several brain areas (70,71). There seems to be good evidence that the energy metabolism in many brain areas is decreased during deep sleep (and no areas have shown an increase during deep sleep).

B. Adenosine

Adenosine is linked to the energy metabolism of cells through ATP/ADP/AMP (Fig. 1) and the production and release of adenosine into the extracellular space is associated with neuronal metabolic activity (72–76). Adenosine has an inhibitory effect on neurons, and it (77) and its agonists (78) increase sleep. Adenosine antagonists, such as teofylline and caffeine (79,80), induce wakefulness. These well-known properties of adenosine have made it an attractive candidate for a sleep factor.

Recent experiments using in vivo microdialysis in cats (81) have shown that adenosine levels during a natural sleep-wake cycle were about 20% higher during wakefulness than during slow-wave sleep. Further, during total sleep deprivation

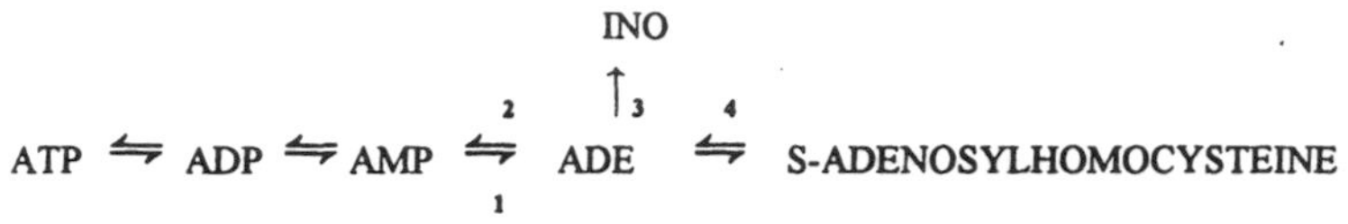

INO

$$\text{ATP} \underset{1}{\overset{2}{\rightleftharpoons}} \text{ADP} \overset{2}{\rightleftharpoons} \text{AMP} \overset{3}{\underset{1}{\rightleftharpoons}} \text{ADE} \overset{4}{\rightleftharpoons} \text{S-ADENOSYLHOMOCYSTEINE}$$

Figure 1   Main metabolic pathways of adenosine inside the cell. ADE = adenosine; INO = inosine. Enzymes: 1 = 5′-nucleotidase; 2 = adenosine kinase; 3 = adenosine deaminase; 4 = SAH hydrolase. AMP concentration is high compared to adenosine, and thus a very small increase in AMP concentration will cause a large increase in adenosine concentration.

adenosine levels in the basal forebrain increased steadily (Fig. 2) being about twofold at the end of a 6-hr total sleep deprivation period. During a 3-hr rebound sleep period the adenosine concentration declined, but did not reach the levels measured at the beginning of the deprivation. Further, the sleep-inducing effect of adenosine appeared to be site specific. The adenosine transport inhibitor *S*-(4-nitrobenzyl)-6-thioinosine (NBTI), administered through microdialysis cannula into two brain areas, increased adenosine levels to about twofold in both areas (basal forebrain and the motor nucleus of the thalamus, VA/VL), but sleep was increased only when NBTI was infused into the basal forebrain, in the vicinity of the cholinergic neurons.

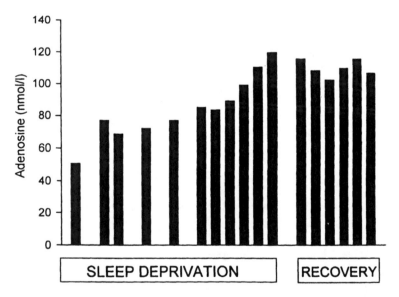

Figure 2   Adenosine in microdialysates from basal forebrain of a cat during 6 hr of sleep deprivation and 3 hr of subsequent recovery sleep. Adenosine concentration in the basal forebrain doubles during the wakefulness period, and the concentration decays only slowly during the recovery sleep.

These data are in agreement with the hypothesis that adenosine could act as a negative-feedback regulator of increased metabolic demands of wakefulness: during waking adenosine concentration in the extracellular space increases, adenosine binds to its receptors, and decreases the firing of neurons. This could take place in all active neurons, but the effects on sleep would be seen when waking-maintaining neurons such as cholinergic neurons are affected.

IV. Intracellular Signaling Pathways

A. G-Proteins

Many receptors mediate the effects of neurotransmitters and other signaling molecules through G-proteins (Fig. 3). G-proteins regulate a variety of effectors. G-proteins can either inhibit (Gi) or activate (Gs) adenylyl cyclases. In addition to enzymes, G-proteins can regulate ion channels and membrane transport proteins. Since several receptors use G-proteins as their mediators, the final outcome of receptor activation in terms of activating or inhibiting later phases of signal transduction is the result of this cross-talk, which continues at every level of intracellular signal transduction.

In one sleep-related G-protein study, carbachol was injected to the mPRF in combination with either cholera toxin (inhibits Gs), pertussis toxin (inhibits Gi), or forskolin (activates adenylyl cyclase). All these compounds inhibited carbachol-evoked REM sleep, indicating that the REM induction by carbachol acts through G-protein-mediated mechanisms (82).

B. Cyclic AMP and Cyclic GMP

Cyclic AMP (cAMP) is the classic second messenger of the cell. cAMP modifies the activity of cAMP-dependent protein kinase (pKA) through phosphorylation. Several hormones, including glucose-metabolism-regulating hormones act through cAMP.

In sleep-related studies the concentration of cAMP has been found to be lower during slow-wave sleep than during waking in several brain areas (83) including the preoptic area and anteroventromedial hypothalamus, but not the cortex (84,85).

cGMP is formed by guanylyl cyclase; the soluble form of guanylyl cyclase is activated by nitric oxide. Increase in cGMP concentration activates cGMP-dependent protein kinase. In one sleep-related study cGMP was found to be higher during paradoxical sleep than during slow-wave sleep in the midbrain, pons, and medulla (83).

Concentration changes in cAMP and cGMP can cause a multitude of changes in the later phases of the signaling cascades, and it is difficult to estimate the final outcome of these changes at the level of enzyme activation/inhibition or gene expression.

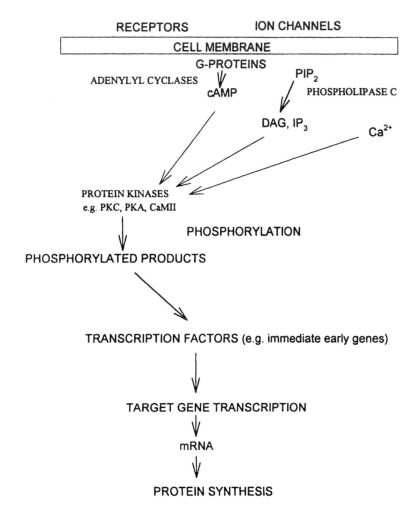

Figure 3 Signaling inside the cell. A simplified diagram of the components of intracellular signaling pathways. Emphasis on the components that have been examined in sleep-wake-cycle-related studies. PIP_2 = phosphatidylinositol; DAG = diacylglycerol; IP_3 = inositol phosphate; PKC = protein kinase C; PKA = protein kinase A; CaMII = calcium/calmodulin-dependent protein kinase.

C. Nitric Oxide

Nitric oxide (NO) is a gas that regulates neural activity. NO synthetase catalyzes the formation of NO in cells. The newly synthetized gas diffuses to interstitial space to regulate neighbouring cells. The inhibition of NO synthetase has been shown to inhibit sleep in rabbits (86) and rats (87), while NO donors increase non-REM sleep (88). In the thalamus, NO concentration was high during waking and

REM sleep and low during non-REM sleep (89), and in the cortex, NO concentration was high during waking and low during sleep (89). There is thus good evidence that NO plays a role in sleep regulation.

D. Protein Kinases

Phosphorylation and dephosphorylation of protein by kinases and phosphatases has been considered a major form of signal transduction. The main signaling pathways each have their kinases (Fig. 3): for cAMP pathway the cyclic-AMP-dependent protein kinase (pKA), for cGMP pathway the cGMP-dependent protein kinase, for PIP_2 protein kinase C (pKC), and for Ca^{2+} calcium/calmodulin-dependent (CaM) kinases. These kinases have several target proteins that they phosphorylate; in addition, they also phosphorylate each other, forming a complicated network of phosphorylation and dephosphorylation (the latter by phosphatases).

In one sleep-related study the phosphorylation of kinases at serine, threonine, and tyrosine sites (antibodies raised to the protein phosphorylated at these residues) was higher in animals that had been awake for 3 hr than in those that had been sleeping (91), indicating that protein phosphorylation is higher during waking than during sleep. A recent finding was that in CaMII-kinase knockout mice the theta band of the EEG power spectrum was shifted toward lower frequencies (92).

E. Immediate Early Genes

The activation of immediate early genes in many cascades precedes gene transcription. In the CNS immediate early genes have been used as markers of neuronal activity. Most attention in sleep research has focused on the proto-oncogene c-*fos*, though other members of this gene family as well as some other immediate early genes have also been examined.

In most brain areas of the rat c-*fos* expression is higher during the dark (= the active period of this animal) than during the light period (93,94). However, when the activity rhythm is reversed so that the activity period takes place during light period, the rhythm of c-*fos* expressions is also reversed (93), suggesting that the waking activity of the animal maintains the c-*fos* expression. This is further supported by an experiment where unilateral lesion of the locus coeruleus diminished c-*fos* activity unilaterally in the cortex and other brain areas, showing that intact noradrenergic activity, which normally is high during waking, is essential for c-*fos* expression (95). During the natural sleep-wake cycle c-*fos* expression is higher in many brain areas during waking than during sleep (4). Further, during sleep deprivation c-*fos* expression is increased in several brain areas including the cortex, thalamus, cerebellum, pons, and hypothalamus (4–7). Since c-*fos* is regarded as a marker of neuronal activity, these findings are not surprising. A more specific role for c-*fos* expression in sleep regulation is suggested by an experiment where c-*fos* antisense oligonucleotide (which binds to c-*fos* mRNA, preventing its activity) was administered to the hypothalamus of rats with the consequence that sleep in these animals was decreased (96). Recently, a c-*fos* knockout mouse has been shown to sleep less than its siblings (97).

c-*fos* has been successfully used as a marker of sleep-active cells. In the ventrolateral part of the hypothalamus, a GABAergic cell group expresses c-*fos* during sleep but not during wakefulness (65). The electrophysiologically characterized sleep-active cells lie in the ventral preoptic area (98), suggesting that at least part of these sleep-active neurons express c-*fos* during their activity period. The c-*fos*-expressing cells of this nucleus project to the histaminergic tuberomamillar nucleus in the posterior hypothalamus. Histaminergic cells are waking-promoting (99), and it is possible that the GABAergic sleep-active cells inhibit the histaminergic cells, thus promoting sleep. Further, c-*fos*-expressing cells during pharmacologically induced REM sleep have been described in the pons and brainstem (100), where the cholinergic REM-on cells become active during REM sleep.

Of the additional immediate early genes that have been examined during the sleep-wake cycle, the expression pattern of NGF1-A, NGF1-B, and *jun*-B resembles that of c-*fos* while c-*jun* expression does not show variation during the sleep-wake cycle (4,5,93).

V. Target Genes

A. Tyrosine Hydroxylase

As the release of monoamine transmitters (norepinephrine, serotonin, histamine) in the brain is higher during waking than during sleep, and at its lowest during REM sleep, it is evident that the brain must take measures to prevent the exhaustion of transmitter stores during waking, or waking time will be limited. Transmitter synthesis might be increased during waking, but on the other hand, sleep, especially REM sleep, is the ideal resting phase for transmitter replenishment. In rats, brain tissue levels of norepinephrine were diminished by REMSD lasting 24–72 hr (1). However, during continued deprivation, norepinephrine levels began to rise again. This could suggest increased production and/or activity of the synthesis-rate-limiting enzyme tyrosine hydroxylase (TH) in the locus coeruleus, the main source of noradrenergic projections to the forebrain. TH mRNA in the locus coeruleus was measured by in situ hybridization and found to be increased by the REMSD, and restored to initial levels during recovery sleep (1). This suggested that the TH gene was induced during REMSD, leading to increased de novo synthesis first of TH, then of its product L-dopa, and finally of norepinephrine. The finding has been confirmed by another group (101). Total sleep deprivation (TSD) by gentle handling, however, did not induce the formation of TH mRNA in the locus coeruleus of the rat, indicating that the effect is specific to REMSD (102).

B. Growth-Hormone-Related Neuropeptides

Human growth hormone (GH) secretion is highest during slow-wave sleep (SWS) phases (reviewed in Ref. 103), and if sleep is delayed or advanced, the GH secretion peak will shift to the new sleep time (104). During recovery from sleep deprivation, an increased magnitude of GH bursts has been observed (105). It is hypothesized that a common mechanism regulates both GH and SWS. Injections of GH increase sleep in animals; on the other hand, an increase in REM sleep but decrease in SWS has been described in humans (106). The sleep-inducing effect has been ascribed to the action of the hypothalamic neuropeptides regulating GH secretion rather than to GH itself.

GH secretion is promoted by growth-hormone-releasing hormone (GHRH), and inhibited by somatostatin (SRIH) (see 107). GHRH is feedback-inhibited by GH, whereas SRIH is feedback-stimulated by GH. GHRH and SRIH also affect each other: GHRH stimulates SRIH neurons, which in turn inhibit GHRH neurons. GHRH and SRIH have been proposed as specific sleep-regulatory factors, GHRH promoting NREM sleep and SRIH promoting REM sleep. The sleep-regulating effects, however, may be from different hypothalamic nuclei from those regulating GH secretion. While GHRH neurons regulating GH are found in the arcuate and paraventricular nuclei, other GHRH neurons are found in the ventromedial hypothalamic nucleus. SRIH neurons regulating GH are predominantly in the periventricular nucleus, but some are found also in the arcuate.

In animals, i.c.v. or systemic injections of GHRH increase both NREM and REM sleep (108), whereas application of GHRH receptor antagonists or antibody decreases sleep (109,110). Transgenic mice expressing human GH, which results in a suppression of the GHRH-GH system, expressed less NREMS than controls (111). Systemic GHRH increases NREM sleep in hypophysectomized rats, indicating a direct effect of GHRH on NREM sleep, rather than a secondary effect through GH (112). The effects on REM sleep, however, were eliminated by hypophysectomy, indicating that the REM-sleep-inducing effect was due to GH release, and possibly mediated through feedback stimulation of SRIH (112).

In humans, infusions of GHRH in the late sleep phase or in repeated i.v. pulses increase NREM sleep (113–115).

SRIH, on the contrary, increases REM sleep when administered i.c.v. to rats, whereas REM sleep is decreased by the SRIH-depleting agent cysteamine and by SRIH antibody (116). Preliminary data from our laboratory indicate that i.c.v. SRIH receptor antagonists also decrease REM sleep in rats. SRIH is thus implicated as a REM-sleep-regulating peptide.

To find out if sleep deprivation might lead to compensatory changes in the regulation of these putative sleep-regulating peptides, rats were deprived of either REM sleep using the platform method or total sleep using gentle handling, and hypothalamic messenger RNAs for GHRH and SRIH were measured using in situ hybridization. REMSD lasting 24 or 72 hr increased the number of SRIH mRNA-expressing neurons in the arcuate and paraventricular nuclei, in accordance with the hypothesis of SRIH as a REM-sleep-regulating factor, while the number of GHRH mRNA-expressing cells was decreased in the paraventricular nucleus (3). Total sleep deprivation of 6 or 12 hr duration also increased SRIH mRNA in the arcuate nucleus, but in contrast to the effect of REMSD, GHRH mRNA was now increased in the paraventricular nucleus (117,118). It is possible that GHRH neurons were stimulated by the NREM sleep loss during TSD, but that during the REMSD of longer duration that we employed, they were gradually inhibited by the strongly stimulated SRIH neurons. While a diurnal variation in total hypothalamic mRNA has been described (119), in situ hybridization measurements showed diurnal variation of GHRH mRNA only in the ventromedial hypothalamic nucleus, which, on the other hand, was not affected by TSD. Diurnal and homeostatic sleep regulations of GHRH thus seem to reside in different hypothalamic locations.

In the same series of studies it was found that hypothalamic galanin mRNA was increased by REMSD (2), but not by TSD (118). Galanin given in i.v. pulses to humans increased REM sleep (120). The closer nature of the relationship of galanin to REM sleep is not yet known. Galanin is often colocalized with other transmitters, and mostly has an inhibitory influence on target neurons.

C. VIP and Prolactin

Vasointestinal peptide (VIP) administered i.c.v. to rats restored sleep during insomnia caused by 5HT depletion by PCPA (121). In the cat, PCPA-induced insomnia could be reversed by i.c.v. infusion of CSF taken from REM sleep-deprived cats or by infusion of VIP into the fourth cerebral ventricle, and it was proposed that VIP might be the sleep-promoting factor in the transfused CSF (122). While i.c.v. VIP increases REM sleep in various species (123,124), the i.c.v. administration of VIP receptor antagonist decreases REM sleep (125). It has not been shown that VIP accumulates in the CSF during REMSD, but VIP receptors in brainstem and forebrain were found to increase during REMSD, in contrast to the situation during habituation (126). Sensitivity to VIP might thus increase during stress-inducing procedures.

In a study performed to localize the target of VIP, it was found that microinjections of either VIP or of corticotropin-like intermediate lobe peptide (CLIP) into the dorsal raphe restored sleep during PCPA insomnia in rat (127). Microinjections of VIP into the oral pontive tegmentum were able to increase REM sleep in the rat both on a short-term basis (hours) and during days (128). This indicates that VIP may have several different mechanisms of action in sleep regulation at the brainstem level, including secondary long-term effects, possibly at the genomic level. In the oral pontine tegmentum, the muscarinic agonist carbachol induces REM sleep, and it has been hypothetized that VIP and acetylcholine may act in combination to produce REM sleep. In addition, VIP has been found in abundance in the suprachiasmatic nuclei, indicating a possible role in the regulation of diurnal rhythms.

The REMS-promoting effect of systemic VIP in rats was prevented by administration of antiserum to prolactin, indicating that the REMS-promoting action of systemic VIP may be mediated by increased release of pituitary PRL (129). Prolactin has a REM-sleep-promoting effect in several species (130). Prolactin levels are highest during secretory episodes during the later half of the night in humans, when REM sleep is also more abundant (131). An interrelationship between PRL and REMS has been proposed, but whereas REMS is mostly expressed during morning sleep hours in humans, PRL is secreted during the whole sleep period, and has a correlation with EEG slow-wave activity (132,133).

D. Interleukin-1β

The relationship of IL-1 to sleep is described in detail in Chapter 15. IL-1 promotes NREM sleep, but this effect was eliminated in rats by pretreatment with GHRH antibody, suggesting that the sleep-promoting activity of IL-1 is in fact mediated by GHRH (134). Twenty-four hours of sleep deprivation resulted in an increase of IL-1β mRNA in the hypothalamus and brainstem of rats, indicating that modulation of IL-1 gene expression is also involved in the homeostatic regulation of sleep (135).

VI. Summary

Changes in receptor sensitivity during the sleep-wake cycle or sleep deprivation give some indication of which transmitters are involved in sleep regulation, but add little to our understanding of the intracellular mechanisms in sleep. The role of sleep in maintaining energy for brain work and the role of adenosine as a homeostatic sleep regulator open new views to the function of sleep. Work on intracellular signaling mechanisms during sleep and sleep loss is in its beginning stage. There is already a considerable amount of knowledge about immediate early gene expression related to sleep, and the very few studies that have been published about target gene expression after sleep loss show us that sleep and sleep loss do affect the expression of many genes, some of them directly involved in sleep regulation, and many of them probably just influenced by sleep-wake regulation, emphasizing the profound impact sleep has on body functions.

References

1. Porkka-Heiskanen T, Smith SE, Taira T, Urban JH, Levine JE, Turek FW, et al. Noradrenergic activity in the brain during REM sleep deprivation and rebound sleep. Am J Physiol 1995; 268:R1456–R1463.

2. Toppila J, Stenberg D, Asikainen M, Turek FW, Porkka-Heiskanen T. REM sleep deprivation induces galanin gene expression in the rat brain. Neurosci Lett 1995; 183:171–174.

3. Toppila J, Asikainen M, Alanko L, Turek FW, Stenberg D, Porkka-Heiskanen T. The effect of REM sleep deprivation on somatostatin and growth hormone-releasing hormone gene expression in the rat hypothalamus. J Sleep Res 1996; 5:115–122.

4. Tononi G, Pompeiano M, Cirelli C. The locus coeruleus and immediate-early genes in spontaneous and forced wakefulness. Brain Res Bull 1994; 35:589–596.

5. O'Hara BF, Young KA, Watson FL, Heller HC, Kilduff TS. Immediate early gene expression in brain during sleep deprivation: preliminary observations. Sleep 1993; 16:1–7.

6. Pompeiano M, Cirelli C, Tononi G. Effects of sleep deprivation on fos-like immuno-reactivity in the rat brain. Arch Ital Biol 1992; 130:325–335.

7. Ledoux L, Sastre JP, Buda C, Luppi PH, Jouvet M. Alterations in c-*fos* expression after different experimental procedures of sleep deprivation in the cat. Brain Res 1996; 735:108–118.

8. Basheer R, Ramanathan L, Greco MA, McCarley RW, Shiromani P. Behavioral state dependent alterations in the steady state messenger RNAs of different neuro-transmitter systems. Sleep Res 1997, 26:3.

9. Williams JA, Comisarow J, Day J, Fibiger HC, Reiner PB. State-dependent release of acetylcholine in rat thalamus measured by in vivo microdialysis. J Neurosci 1994; 14:5236–5242.

10. Marrosu F, Portas CM, Mascia MS, Casu MA, Fa M, Giagheddu M, et al. Micro-dialysis measurement of cortical and hippocampal acetylcholine release during sleep-wake cycle in freely moving rats. Brain Res 1995; 671:329–332.

11. Portas CM, McCarley RW. Behavioral state-related changes of extracellular se-rotonin concentration in the dorsal raphe nucleus: a microdialysis study in the freely moving cat. Brain Res 1994; 648:306–312.

12. Iwakiri H, Matsuyama K, Mori S. Extracellular levels of serotonin in the medial pontine reticular formation in relation to sleep-wake cycle in cats: a microdialysis study. Neurosci Res 1993; 18:157–170.

13. Asikainen M, Toppila J, Alanko L, Ward DJ, Stenberg D, Porkka-Heiskanen T. Sleep deprivation increases brain serotonin turnover in the rat. NeuroReport 1997; 8: 1577–1582.

14. Caulfield MP. Muscarinic receptors—characterization, coupling and function. Phar-macol Ther 1993; 58:319–379.

15. Buckley NJ. Localization of a family of muscarinic receptor mRNAs in rat brain. J Neurosci 1988; 8:4646–4652.

16. Vilaro MT, Palacios JM, Mengod G. Multiplicity of muscarinic autoreceptor sub-types? Comparison of the distribution of cholinergic cells and cells containing mRNA for five subtypes of muscarinic receptors in the brain. Mol Brain Res 1994; 21:30–46.

17. Wei J, Walton EA, Milici A, Buccafusco JJ. M1-M5 muscarinic receptor distribution in rat CNS by RT-PCR and HPLC. J Neurochem 1994; 63:815–821.

18. McCarley RW, Greene RW, Rainnie D, Portas CM. Brainstem neuromodulation and REM sleep. Semin Neurosci 1995; 7:341–354.

19. Velazquez-Moctezuma J, Gillin JC, Shiromani PJ. Effect of specific M_1, M_2 muscarinic receptor antagonist on REM sleep generation. Brain Res 1989; 503:128–131.
20. Imeri L, Bianchi S, Angeli P, Mancia M. Selective blockade of different brain stem muscarinic receptor subtypes: effects on the sleep-wake cycle. Brain Res 1994; 636: 68–72.
21. Imeri L, Bianchi S, Angeli P, Mancia M. Differential effects of M_2 and M_3 muscarinic antagonists on the sleep-wake cycle. NeuroReport 1991; 2:383–385.
22. Kushida CA, Zoltoski RK, Gillin JC. The expression of m1-m3 muscarinic receptor mRNAs in rat brain following REM sleep deprivation. NeuroReport 1995; 6:1705–1708.
23. Pompeiano M, Tononi G. Changes in pontine muscarinic receptor binding during sleep-waking states in the rat. Neurosci Lett 1990; 109:347–352.
24. Aston-Jones G, Chiang C, Alexinsky T. Discharge of noradrenergic locus coeruleus neurons in behaving rats and monkeys suggests a role in vigilance. Prog Brain Res 1991; 88:501–520.
25. Chu N-S, Bloom FE. Activity patterns of catecholamine-containing pontine neurons in the dorso-lateral tegmentum of unrestrained cats. J Neurobiol 1974; 5:527–544.
26. Jacobs BL, Abercrombie ED, Fornal CA, Levine ES, Morilak DA, Stafford IL. Single-unit and physiological analyses of brain norepinephrine function in behaving animals. Prog Brain Res 1991; 88:159–165.
27. Hilakivi I, Leppävuori A. Effects of methoxamine, and alpha-1 adrenoceptor agonist, and prazosin, an alpha-1 antagonist, on the stages of the sleep-waking cycle in the cat. Acta Physiol Scand 1984; 120:363–372.
28. Lin JS, Roussel B, Akaoka H, Fort P, Debilly G, Jouvet M. Role of catecholamines in the modafinil and amphetamine induced wakefulness, a comparative pharmacological study in the cat. Brain Res 1992; 591:319–326.
29. Putkonen PT, Leppävuori A, Stenberg D. Paradoxical sleep inhibition by central alpha-adrenoceptor stimulant clonidine antagonized by alpha-receptor blocker yohimbine. Life Sci 1977; 21:1059–1065.
30. Stenberg D, Porkka-Heiskanen T, Toppila J. α_2-Adrenoceptors and vigilance in cats: antagonism of medetomidine sedation by atipamezole. Eur J Pharmacol 1993; 238: 241–247.
31. Hobson JA, McCarley RW, Wyzinski PW. Sleep cycle oscillation: reciprocal discharge by two brainstem neuronal groups. Science 1975; 189:55–58.
32. Lanfumey L, Dugovic C, Adrien J. Beta-1- and beta-2 adrenergic receptors: their role in the regulation of paradoxical sleep in the rat. Electroenceph Clin Neurophysiol 1985; 60:558–567.
33. Hilakivi I, Leppävuori A, Putkonen PT. Prazosin increases paradoxical sleep. Eur J Pharmacol 1980; 65:417–420.
34. Mignot E, Guilleminault C, Bowersox S, Fruhstorfer B, Nishino S, Maddaluno J, et al. Central alpha 1 adrenoceptor subtypes in narcolepsy-cataplexy: a disorder of REM sleep. Brain Res 1989; 490:186–191.
35. Cirelli C, Tononi G, Pompeiano M, Pompeiano O, Gennari A. Modulation of desynchronized sleep through microinjection of α_1-adrenergic agonists and antagonists in the dorsal pontine tegmentum of the cat. Pflügers Arch Eur J Physiol 1992; 422:273–279.

36. Siegel JM, Rogawski MA. A function for REM sleep: regulation of noradrenergic receptor sensitivity. Brain Res Rev 1988; 13:213–233.
37. Mogilnicka E, Prezewlocka B, Van Luijtelaar EL, Klimek V, Coenen AM. Effects of REM sleep deprivation on central alpha-1- and beta-adrenoceptors in the rat brain. Pharmacol Biochem Behav 1986; 25:329–332.
38. Abel MS, Villegas F, Abreu J, Gimino F, Steiner S, Beer B, et al. The effect of rapid eye movement sleep deprivation on cortical beta-adrenergic receptors. Brain Res Bull 1983; 11:729–734.
39. Radulovacki M, Micovic N. Effects of REM sleep deprivation and desipramine on β-adrenergic binding sites in rat brain. Brain Res 1982; 235:393–396.
40. Tsai LL, Bergmann BM, Perry BD, Rechtschaffen A. Effects of chronic total sleep deprivation on central noradrenergic receptors in rat brain. Brain Res 1992; 602: 221–227.
41. Tufik S, Lindsey CJ, Carlini EA. Does REM sleep deprivation induce a supersensitivity of dopaminergic receptors in the rat brain? Pharmacology 1978; 16:98–105.
42. Demontis MG, Fadda P, Devoto P, Martellotta MC, Fratta W. Sleep deprivation increases dopamine D1 receptor antagonist ^{3}H-SCH 23390 binding and dopamine-stimulated adenylate cyclase in the rat limbic system. Neurosci Lett 1990; 117: 224–227.
43. Brock JW, Hamdi A, Ross K, Payne S. Prasad C. REM sleep deprivation alters dopamine D-2 receptor binding in the rat frontal cortex. Pharmacol Biochem Behav 1995; 52:43–48.
44. Trulson ME, Preussler DW, Howell GA. Activity of substantia nigra units across the sleep-waking cycle in freely moving cats. Neurosci Lett 1981; 26:183–188.
45. Jacobs BL. Brain monoaminergic unit activity in behaving animals. In: Epstein AN, Morrison AR, eds. Progress in Psychobiology Physiology and Psychology. New York: Academic Press, 1987:171–206.
46. Jouvet M. The role of monoamines and acetylcholine-containing neurons in the regulation of the sleep-waking cycle. Ergebn Physiol 1972; 64:166–342.
47. McGinty DJ, Harper RM. Dorsal raphe neurons: depression of firing during sleep in cats. Brain Res 1976; 101:569–575.
48. Trulson ME, Jacobs BL. Raphe unit activity in freely moving cats: correlation with level of behavioral arousal. Brain Res 1979; 163:135–150.
49. Dugovic C, Wauquier A. 5-HT2 receptors could be primarily involved in the regulation of slow-wave sleep in the rat. Eur J Pharmacol 1987; 137:145–146.
50. Idzikowski C, Mills FJ, James RJ. A dose-response study examining the effects of ritanserin on human slow wave sleep. Br J Clin Pharmacol 1991; 31:193–196.
51. Dugovic C, Leysen JE, Wauquier A. Melatonin modulates the sensitivity of 5-hydroxytryptamine-2 receptor-mediated sleep-wakefulness regulation in the rat. Neurosci Lett 1989; 104:320–325.
52. Monti JM, Jantos H, Silveira R, Reyesparada M, Scorza C, Prunell G. Depletion of brain serotonin by 5,7-DHT—effects on the 8-OH-DPAT-induced changes of sleep and waking in the rat. Psychopharmacology 1994; 115:273–277.
53. Seifritz E, Moore P, Trachsel L, Bhatti T, Stahl SM, Gillin JC. The 5-HT1A agonist ipsapirone enhances EEG slow wave activity in human sleep and produces a power spectrum similar to 5-HT2 blockade. Neurosci Lett 1996; 209:41–44.

54. Asikainen M, Deboer T, Porkka-Heiskanen T, Stenberg D, Tobler I. Sleep deprivation increases brain serotonin turnover in the Djungarian hamster. Neurosci Lett 1995; 198:21–24.

55. Santos R, Carlini EA. Serotonin receptor activation in rats previously deprived of REM sleep. Pharmacol Biochem Behav 1983; 18:149–151.

56. Mogilnicka E. REM sleep deprivation changes behavioral response to catecholaminergic and serotonergic receptor activation in rats. Pharmacol Biochem Behav 1981; 15:149–151.

57. Artigas F, Romero L, de Montigny C, Blier P. Acceleration of the effect of selected antidepressant drugs in major depression by 5-HT1A antagonists. Trends Neurosci 1996; 19:378–383.

58. Prévot E, Maudhuit C, Le Poul E, Hamon M, Adrien J. Sleep deprivation reduces the citalopram-induced inhibition of serotonergic neuronal firing in the nucleus raphe dorsalis of the rat. J Sleep Res 1996; 5:238–245.

59. Lin JS, Sakai K, Vanni-Mercier G, Jouvet M. A critical role of the posterior hypothalamus in the mechanisms of wakefulness determined by microinjection of muscimol in freely moving cats. Brain Res 1989; 479:225–240.

60. Panula P, Yang HYT, Costa E. Histamine-containing neurons in the rat hypothalamus. Proc Natl Acad Sci USA 1984; 81:2572–2576.

61. Vanni-Mercier G, Sakai K, Jouvet M. "Waking-state specific" neurons in the caudal hypothalamus. C R Acad Sci Ser III 1984; 298:195–200.

62. Lin JS, Sakai K, Vanni-Mercier G, Arrang JM, Garbarg M, Schwartz JC, et al. Involvement of histaminergic neurons in arousal mechanisms demonstrated with H3-receptor ligands in the cat. Brain Res 1990; 23:325–330.

63. Monti JM, Jantos H, Boussard M, Altier H, Orellana C, Olivera S. Effects of selective activation or blockade of the histamine-H3 receptor on sleep and wakefulness. Eur J Pharmacol 1991; 205:283–287.

64. Alam MN, Szymusiak R, McGinty DJ. Local preoptic/anterior hypothalamic warming alters spontaneous and evoked neuronal activity in the magnocellular basal forebrain. Brain Res 1995; 696:221–230.

65. Sherin JE, Shiromani PJ, McCarley RW, Saper CB. Activation of ventrolateral preoptic neurons during sleep. Science 1996; 271:216–219.

66. Maquet P, Dive D, Salmon E, Sadzot B, Franco G, Poirrier R, et al. Cerebral glucose utilization during stage-2 sleep in man. Brain Res 1992; 571:149–153.

67. Madsen PL, Holm S, Vorstrup S, Friberg L, Lassen NA, Wildschiodtz G. Human regional cerebral blood flow during rapid-eye-movement sleep. J Cereb Blood Flow Metab 1991; 11:502–507.

68. Boyle PJ, Scott JC, Krentz AJ, Nagy RJ, Comstock E, Hoffman C. Diminished brain glucose metabolism is a significant determinant for falling rates of systemic glucose utilization during sleep in normal humans. J Clin Invest 1994; 93:529–535.

69. Madsen PL, Schmidt JF, Wildschiodtz G, Friberg L, Holm S, Vorstrup S, et al. Cerebral O_2 metabolism and cerebral blood flow in humans during deep and rapid-eye-movement sleep. J Appl Physiol 1991; 70:2597–2601.

70. Buchsbaum MS, Gillin JC, Wu J, Hazlett E, Sicotte N, Dupont RM, et al. Regional glucose metabolism rate in human sleep assessed by positron emission tomography. Life Sci 1989; 45:1349–1356.

71. Heiss WD, Pawlick G, Herholz K, Wagner R, Weinhard K. Regional cerebral glucose metabolism in man during wakefulness, sleep and dreaming. Brain Res 1985; 327:362–366.

72. Pull I, McIlwain H. Metabolism of [^{14}C]adenine and derivatives by cerebral tissues, superfused and electrically stimulated. Biochem J 1972; 126:965–973.

73. Winn HR, Welsh JE, Rubio R, Berne RM. Changes in brain adenosine during bicuculline-induced seizures in rats. Effects of hypoxia and altered systemic blood pressure. Circ Res 1980; 47:568–577.

74. Schrader J, Wahl M, Kuschinsky W, Kreutzberg GW. Increase of adenosine content in cerebral cortex of the cat during bicuculline-induced seizure. Pflügers Arch Eur J Physiol 1980; 387:245–251.

75. Van Wylen DG, Park TS, Rubio R, Berne RM. Increases in cerebral interstitial fluid adenosine concentration during hypoxia, local potassium infusion, and ischemia. J Cereb Blood Flow Metab 1986; 6:522–528.

76. McIlwain H, Poll JD. Adenosine in cerebral homeostatic role: appraisal through actions of homocysteine, colchicine and dipyridamole. J Neurobiol 1986; 17:39–49.

77. Portas CM, Thakkar M, Rainnie DG, Greene RW, McCarley RW. Role of adenosine in behavioral state modulation: a microdialysis study in the freely moving cat. Neuroscience 1996; 79:225–235.

78. Radulovacki M. Role of adenosine in sleep in rats. Rev Clin Basic Pharmacol 1985; 5:327–339.

79. Nehlig A, Daval JL, Debry G. Caffeine and the central nervous system: mechanisms of action, biochemical, metabolic and psychostimulant effects. Brain Res Brain Res Rev 1992; 17:139–170.

80. Virus RM, Ticho S, Pilditch M, Radulovacki M. A comparison of the effects of caffeine, 8-cyclopentyltheophylline, and alloxazine on sleep in rats—possible roles of central nervous system adenosine receptors. Neuropsychopharmacology 1990; 3: 243–249.

81. Porkka-Heiskanen T, Strecker RE, Bjorkum AA, Thakkar M, Greene RW, McCarley RW. Adenosine: a mediator of the sleep-inducing effects of prolonged wakefulness. Science 1997; 276:1265–1268.

82. Shuman SL, Capece ML, Baghdoyan HA, Lydic R. Pertussis toxin-sensitive G proteins mediate carbachol-induced REM sleep and respiratory depression. Am J Physiol 1995; 269:R308–R317.

83. Ogasahara S, Taguchi Y, Wada H. Changes in the levels of cyclic nucleotides in rat brain during the sleep-wakefulness cycle. Brain Res 1981; 213:163–171.

84. Perez E, Zamboni G, Amici R, Jones CA, Parmeggiani PL, et al. cAMP accumulation in the hypothalamus, cerebral cortex, pineal gland and brown fat across the wake-sleep cycle of the rat exposed to different ambient temperatures. Brain Res 1995; 684:56–60.

85. Perez E, Zamboni G, Amici R, Fadiga L, Parmeggiani PL. Ultradian and circadian changes in the cAMP concentration in the preoptic region of the rat. Brain Res 1991; 551:132–135.

86. Kapas L, Shibata M, Kamura M, Krueger JM. Inhibition of nitric oxide synthesis suppresses sleep in rabbits. Am J Physiol 1994; 266:R151–R157.

87. Kapas L, Fang JD, Krueger JM. Inhibition of nitric oxide synthesis inhibits rat sleep. Brain Res 1994; 664:189–196.

88. Kapas L, Krueger JM. Nitric oxid donors SIN-1 and SNAP promote nonrapid-eye-movement sleep in rats. Brain Res Bull 1996; 41:293–298.
89. Williams JA, Vincent SR, Reiner PB. Nitric oxide production in rat thalamus changes with behavioral state, local depolarization and brain stem stimulation. J Neurosci 1997; 17:420–427.
90. Cespuglio R, Burlet R, Marinesco S, Robert U, Jouvet M. Detection voltametrique du NO cerebral chez le rat. Variations du signal a travers le cycle veille-sommeil. C R Acad Sci Ser III 1996; 19:191–200.
91. Cirelli C, Tononi G. Changes in protein phosphorylation patterns in the brain during the sleep-waking cycle. Soc Neurosci Abstr 1996; 273.2:688 (abstract).
92. Thakkar M, Rainnie DG, Hearn EF, Greene RW, McCarley RW, Shiromani PJ. Abnormal theta activity during REM sleep in calcium-calmodulin kinase II knock-out mice. Sleep Res 1997; 26:53.
93. Grassi Zucconi G, Menagazzi M, Carcereri De Prati A, Vescia S, Ranucci G, Bentivoglio M. Different programs of gene expression are associated with different phases of the 24h and sleep-wake cycles. Chronobiologia 1994; 21:93–97.
94. Yamuy J, Mancillas JR, Morales FR, Chase MH. C-*fos* expression in the pons and medulla of the cat during carbachol-induced active sleep. J Neurosci 1993; 13: 2703—2718.
95. Cirelli C, Pompeiano M, Tononi G. Neuronal gene expression in the waking state: a role for the locus coeruleus. Science 1996; 274:1211–1215.
96. Cirelli C, Pompeiano M, Arrighi P, Tononi G. Sleep-waking changes after c-*fos* antisense injections in the medial preoptic area. NeuroReport 1995; 6:801–805.
97. Shiromani P, Greco MA, Thakkar M, McCarley RW. c-*fos* knock-out mice have reduced non-REM sleep. Sleep Res 1997; 26:42.
98. Szymusiak R, McGinty DJ. Sleep-related neuronal discharge in the basal forebrain of cats. Brain Res 1986; 370:82–92.
99. Lin JS, Sakai K, Jouvet M. Hypothalamo-preoptic histaminergic projections in sleep-wake control in the cat. Eur J Neurosci 1994; 6:618–625.
100. Shiromani PJ, Kilduff TS, Bloom FE, McCarley RW. Cholinergically induced REM sleep triggers Fos-like immunoreactivity in dorsolateral pontine regions associated with REM sleep. Brain Res 1992; 580:351–357.
101. Basheer R, Magner M, Ryan K, Segall A, McCarley RW, Shiromani P. Prolonged REM sleep deprivation elevates tyrosine hydroxylase mRNA in locus coeruleus. Soc Neurosci Abstr 1996; 64.3:146 (abstract).
102. Alanko L, Toppila J, Asikainen M, Ward DJ, Tobler I, Stenberg D, et al. Tyrosine hydroxylase gene expression in the locus coeruleus is not affected by total sleep deprivation in the rat. J Sleep Res 1996; 5,suppl 1:3.
103. Van Cauter E, Plat L. Physiology of growth hormone secretion during sleep. J Pediatr 1996; 128:S32–S37.
104. Van Cauter E, Kerkhofs M, Caufriez A, Van Onderbergen A, Thorner MO, Copinschi G. A quantitative estimation of growth hormone secretion in normal man—reproducibility and relation to sleep and time of day. J Clin Endocrinol Metab 1992; 74:1441–1450.
105. Golstein J, Van Cauter E, Désir D, Noël P, Spire J, Refetoff S, et al. Effects of "jet lag" on hormonal patterns. IV. Time shifts increase growth hormone release. J Clin Endocrinol Metab 1995; 56:433–440.

106. Mendelson WB, Slater S, Gold P, Gillin JC. The effect of growth hormone adminis-
     tration on human sleep: a dose-response study. Biol Psychiatry 1980; 15:613–618.
107. Bertherat J, Bluet-Pajot MT, Epelbaum J. Neuroendocrine regulation of growth
     hormone. Eur J Endocrinol 1995; 132:12–24.
108. Obál F, Alföldi P, Cady AB, Sáry G, Krueger JM. Growth hormone releasing factor
     enhances sleep in rats and rabbits. Am J Physiol 1988; 255:R310–R316.
109. Obál F, Payne L, Kapas L, Opp M, Krueger JM. Inhibition of growth hormone-
     releasing factor suppresses both sleep and growth hormone secretion in the rat. Brain
     Res 1991; 557:149–153.
110. Obál F, Payne L, Opp M, Alföldi P, Kapas L, Krueger JM. Growth hormone-
     releasing hormone antibodies suppress sleep and prevent enhancement of sleep after
     sleep deprivation. Am J Physiol 1992; 263:R1078–R1085.
111. Zhang J, Obál F, Fang J, Collins BJ, Krueger JM. Non-rapid eye movement sleep is
     suppressed in transgenic mice with deficiency in the somatotropic system. Neurosci
     Lett 1996; 220:97–100.
112. Kalra SP, Dube MG, Sahu A, Phelps CP, Kalra PS. Neuropeptide Y secretion
     increases in the paraventricular nucleus in association with increased appetite for
     food. Proc Natl Acad Sci USA 1991; 88:10931–10935.
113. Kerkhofs M, Van Cauter E, Van Onderbergen A, Caufriez A, Thorner MO, Copin-
     schi G. Sleep-promoting effects of growth hormone-releasing hormone in normal
     men. Am J Physiol Endocrinol Metab 1993; 264:E594–E598.
114. Steiger A, Guldner J, Hemmeter U, Rothe B, Wiedemann K, Holsboer F. Effects of
     growth hormone-releasing hormone and somatostatin on sleep EEG and nocturnal
     hormone secretion in male controls. Neuroendocrinology 1992; 56:566–573.
115. Marshall L, Molle M, Boschen G, Steiger A, Fehm HL, Born J. Greater efficacy of
     episodic than continuous growth hormone-releasing hormone (GHRH) administration
     in promoting slow-wave sleep (SWS). J Clin Endocrinol Metab 1986; 367:26–30.
116. Danguir J. Intracerebroventricular infusion of somatostatin selectively increases
     paradoxical sleep in rats. Brain Res 1986; 367:26–30.
117. Toppila J, Alanko L, Asikainen M, Tobler I, Stenberg D, Porkka-Heiskanen T. Sleep
     deprivation increases somatostatin and growth hormone-releasing hormone messen-
     ger RNA in the rat hypothalamus. J Sleep Res 1997; 6:171–178.
118. Stenberg D, Toppila J, Asikainen M, Alanko L, Porkka-Heiskanen T. Differential
     effects on hypothalamic neuropeptide gene expression of total sleep deprivation and
     selective REM sleep deprivation in rats. Sleep Res 1997; 26:632.
119. Bredow S, Taishi P, Obál F Jr, Guha-Thakurta N, Krueger JM. Hypothalamic growth
     hormone-releasing hormone mRNA varies across the day in rats. NeuroReport 1996;
     7:2501–2505.
120. Steiger A, Murck H,. Frieboes RM, Maier P, Schier T, Holsboer F. Sleep endocrine
     effects of galanin in man. J Sleep Res 1996:5(Suppl 1):219.
121. Riou F, Cespuglio R, Jouvet M. Endogenous peptides and sleep in rat. III. The
     hypnogenic properties of vasoactive intestinal polypeptide. Neuropeptides 1982; 2:
     265–277.
122. Prospero-Garcìa O, Morales M, Arankowsky-Sandoval G, Drucker-Colín R. Vaso-
     active intestinal polypeptide (VIP) and cerebrospinal fluid of sleep deprived cats
     restores REM sleep in insomniac recipients. Brain Res 1986; 385:169–173.

123. Obál F, Sáry G, Alföldi P, Rubicsek G. Vasoactive intestinale polypeptide promotes sleep without effects on brain temperature in rats at night. Neurosci Lett 1986; 64: 236–240.

124. Obál F, Opp M, Cady AB, Johannesen L, Krueger JM. Prolactin, vasoactive intestinal peptide, and peptide histidine methionine elicit selective increases in REM sleep in rabbits. Brain Res 1989; 490:292–300.

125. Mirmiran M, Kruisbrink J, Bos NP, Van der Werf D, Boer GJ. Decrease of rapid-eye-movement sleep in the light by intraventricular application of a VIP-antagonist in the rat. Brain Res 1988; 458:192–194.

126. Jiménez-Anguiano A, García-García F, Mendoza-Ramirez JL, Duran-Velazquez A, Drucker-Colín R. Brain distribution of vasoactive intestinal peptide receptors following REM sleep deprivation. Brain Res 1996; 728:37–46.

127. Elkafi B, Leger L, Seguin S, Jouvet M, Cespuglio R. Sleep permissive components within the dorsal raphe nucleus in the rat. Brain Res 1995; 686:150–159.

128. Bourgin P, Lebrand C, Escourrou P, Gaultier C, Franc B, Hamon M. Vasoactive intestinal polypeptide microinjection into the oral pontine tegmentum enhance rapid eye movement sleep in the rat. Neuroscience 1997; 77:351–360.

129. Obál F, Payne L, Kacsoh B, Opp M, Kapas L, Grosvenor CE, et al. Involvement of prolactin in the REM sleep-promoting activity of systemic vasoactive intestinal peptide (VIP). Brain Res 1994; 645:143–149.

130. Roky R, Obál F, Valatx JL, Bredow S, Fang JD, Pagano LP. Prolactin and rapid eye movement sleep regulation. Sleep 1995; 18:536–542.

131. Sassin JF, Frantz AG, Weitzman ED, Kapen S. Human prolactin: 24-hour pattern with increased release during sleep. Science 1972; 177:1205–1207.

132. Spiegel K, Follenius M, Simon C, Saini J, Ehrhart J, Brandenberger G. Prolactin secretion and sleep. Sleep 1994; 17:20–27.

133. Spiegel K, Luthringer R, Follenius M, Schaltenbrand N, Macher JP, Muzet A. Temporal relationship between prolactin secretion and slow-wave electroencephalic activity during sleep. Sleep 1995; 18:543–548.

134. Obál F, Fang J, Payne LC, Krueger JM. Growth hormone-releasing hormone mediates the sleep-promoting activity of interleukin-1 in rats. Neuroendocrinology 1995; 61:559–565.

135. Mackiewicz M, Sollars PJ, Ogilvie MD, Pack AI. Modulation of IL-1 beta gene expression in the rat CNS during sleep deprivation. NeuroReport 1996; 7:529–533.

10

Cellular and Molecular Mechanisms of Circadian Rhythms in Mammals

PIOTR ZLOMANCZUK

Rydygier Medical School
Bydgoszcz, Poland
and University of Massachusetts Medical
   School
Worcester, Massachusetts

WILLIAM J. SCHWARTZ

University of Massachusetts Medical
   School
Worcester, Massachusetts

I. Introduction

Now that a circadian pacemaker in mammals has been localized to the supra-chiasmatic nucleus (SCN) of the hypothalamus, the search is underway to understand its cellular and molecular mechanisms. These investigations are in their infancy, and they owe much to insights and progress made in other organisms (beyond the scope of this review), including bacteria, algae, fungi, plants, mollusks, insects, frogs, and birds (for reviews, see 1–10). Here we survey mechanistic studies of circadian timekeeping in rodents (especially rats and hamsters) at the cellular and molecular level, highlighting areas of current research activity.

II. Assembling the Concepts and Tools

A. The Input → Oscillator → Output Paradigm

The functional role of the circadian system is to recognize local time and measure its passage. The "clock" is responsible for regulating the timing and synchronization of biochemical, physiological, and behavioral functions to external timing cues (specifically, the 24-hr light-dark cycle). To perform these tasks, the system consists of three functionally distinct elements: input (afferent) pathways for entrainment, a circadian pacemaker that actually generates a circa 24-hr oscillation, and output (efferent) pathways for translating the timing signal into rhythms in target structures. In unicellular organisms, all three of these functions are performed within a single cell, but in multicellular organisms, different structures (e.g., retina, SCN, pineal gland) can be distinguished as performing different tasks.

This notion of a linear system with serial elements has colored our views and guided experimental studies, but it may not be valid. Given the emerging complexity of cellular biochemistry—including networks of signaling and regulatory molecules with extensive cross-talk between metabolic cascades—the definition of functional borders between different elements may not be so obvious. The existence of nested feedback loops further complicates the conceptual and experimental analysis of this problem. For example (see later), the ability of photic inputs to activate the expression of certain genes in the SCN depends on time of day (i.e., the input is regulated by one or more of the pacemaker's output signals). As another example, the circadian rhythm of pineal melatonin secretion is driven by the SCN, but the hormone in turn has phase-resetting and entraining effects on overt rhythmicity (see Chapter 7). The SCN may be the site of melatonin action, as the nucleus contains high-affinity melatonin receptors (11), SCN lesions prevent melatonin entrainment (12), and SCN metabolic and electrical activities are altered by melatonin in a phase-dependent manner (13–15). Thus, melatonin secretion is a rhythmic output of the circadian pacemaker that feeds back to modulate the activity of the pacemaker itself.

B. Assays of SCN Circadian Rhythmicity

Before mechanistic experiments can begin, reliable indicators are needed as markers for the presence, phase, and period of the pacemaker's oscillation.

Neuronal Firing Rate

Circadian rhythms of single- or multiple-unit electrical activities in the SCN have been recorded extracellularly in vivo (16,17) and in vitro in hypothalamic slices (18–20), organotypic slice cultures (21), and dissociated cell cultures (22). In nocturnal rodents, the firing rate is high during the subjective light phase (subjective day) and low during the subjective dark phase (subjective night); the same is true in diurnal chipmunks (23).

Recording electrical activity has been an extremely useful assay of SCN rhythmicity, especially for in vitro studies (24), but its limitations must be noted. Na^+-dependent action potentials are not a part of the pacemaker's actual oscillatory apparatus; the internal timekeeping mechanism continues to run unperturbed in the presence of tetrodotoxin (22,25). Since electrical discharge represents one of the pacemaker's outputs, and since the nature of the coupling between the circadian pacemaker and firing rate is not known, there may be some experimental

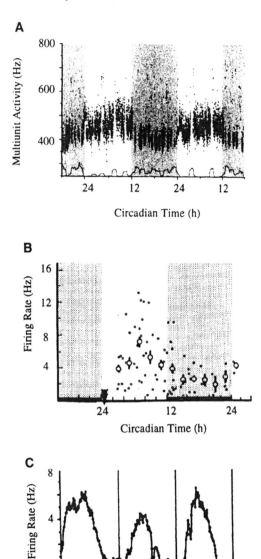

Figure 1 Circadian rhythms of SCN neuronal firing rates. Circadian rhythms of electrical activity in the SCN, recorded extracellularly as multiple units in vivo (A) and as single units in vitro in the hypothalamic slice (B) and dissociated cell culture (C). In (A), the lower trace represents the animal's movements on an arbitrary scale, and shading denotes subjective night. In (B), large circles represent mean firing rates for 2-hr circadian intervals in the SCN of a slice prepared at the ⊗. [From Refs. 17 (A), 248 (B), and 22 (C).]

conditions in which the patterns of SCN electrical activity might not reflect the state of the pacemaker's oscillation. Furthermore, most experiments record the activity of neurons sampled for short intervals across the circadian cycle, rather than attempting to monitor individual cells over 24 hr (but see 22 as an example). It is not known how this population rhythm relates to the coding of efferent signals sent by the SCN to other brain regions.

Energy Metabolism

SCN metabolic activity was the first property of the nucleus reported to exhibit circadian rhythmicity (26). These experiments utilized an autoradiographic method for in vivo determination of the rates of glucose utilization of individual structures within the brain by using tracer amounts of 2-deoxy-D-[1-^{14}C]glucose (27). This assay is useful because regional brain functional activity is closely coupled to regional brain energy utilization, and brain is dependent on the continuous provision of glucose for its energy. Like electrical activity, glucose utilization is high during the subjective day and low during the subjective night in both nocturnal and diurnal animals (28), and the rhythm persists in vitro (29).

The deoxyglucose method has been a helpful assay of SCN rhythmicity in vivo, e.g., in fetal animals (see Chapter 2), but it too has limitations. To make the measurement animals must be sacrificed. Importantly, a measured level of glucose utilization is not a unitary, indivisible quantity, but a summation of all the individual metabolic costs entailed by an assortment of cellular tasks. How these costs are apportioned among the tasks may be highly dynamic, even though the magnitude of total energy consumption remains unchanged. The corresponding autoradiographic images may appear identical and yet fail to reflect this underlying complexity. Finally, the spatial resolution of the technique ordinarily does not extend to the level of individual cells.

Neuropeptide Levels

Rhythms of neuroactive peptides synthesized in the SCN have been measured in the cerebrospinal fluid (CSF) and in tissue punches and sections at the mRNA and protein levels (30). The most intensively studied of these molecular rhythms is a circadian rhythm of the levels of the neuropeptide arginine vasopressin (AVP). CSF levels are high during the subjective day and low during the subjective night in both nocturnal and diurnal animals (31), and the rhythm persists in vitro in hypothalamic explants (32) and slices (33), organotypic slice cultures (34), and dissociated cell cultures (35,36).

Identifying such peptide rhythms is an important step in analyzing SCN rhythmicity, because they provide windows on gene expression (transcription and translation). If the biochemistry of a circadian-regulated process is known, then the molecules involved in its control should have a clear relationship to the oscillatory mechanism of the pacemaker itself. It should be possible to follow the output pathway "backward" to elucidate this mechanism. The recent demonstration (37) that two peptide rhythms can be simultaneously measured in slice cultures [AVP and vasoactive intestinal polypeptide (VIP), which express different phases in vivo] has already confirmed the power of this approach.

III. Translating the Formalism of Phase-Response Curves into the Physiology of Defined Input Pathways

One strategy for identifying elements of the circadian system is to trace the cascade of events that comprise the pacemaker's entrainment pathways; ultimately these pathways must converge and terminate on components of the oscillatory machinery to cause phase shifts of overt rhythmicity. As described previously (38,39), two categories of stimuli—photic and nonphotic—produce distinctly different patterns of phase shifts when the stimuli are applied at various time points across the free-running circadian cycle. Photic stimuli (i.e., light pulses) cause phase delays when presented during the early subjective night, phase advances during the late subjective night, and little or no phase-shifting effect during the subjective day. In contrast, nonphotic stimuli (e.g., benzodiazepines and other pharmacological agents, cage changing, social interactions) primarily lead to phase advances during the late subjective day, with small phase delays during the subjective night. These two families of phase-response curves (PRCs) are essentially opposite to each other (180° out of phase), providing a formal framework for investigating their underlying cellular and molecular substrates.

A. Photic Mechanisms

Since the 24-hr alternation of light and darkness is the most obvious indication of the earth's daily rotation about its axis, it is not surprising that evolution has selected ambient light intensity as the preeminent signal for entraining circadian rhythms to local geophysical time. Progress is being made in delineating some of the extracellular and intracellular links that comprise the SCN's photic input pathway.

Photoreception

In mammals, the retina is required for photic entrainment, but the responsible visual system is anatomically and physiologically distinct from the visual systems for oculomotor function and image formation (40). The "circadian" system includes a specialized photoreceptive mechanism (41) that may rely on green-sensitive cones (42); a subset of ganglion cells (43) that forms a monosynaptic

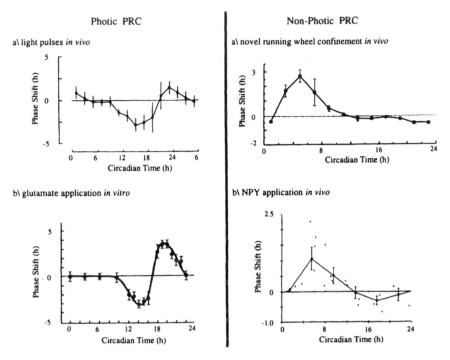

Figure 2   In vivo physiological and in vitro pharmacological PRCs. In rats (left panels), the PRC to light pulses administered to animals in vivo can be mimicked by glutamate applied to hypothalamic slices in vitro. In hamsters (right panels), the PRC to a nonphotic stimulus administered to animals in vivo can be mimicked by neuropeptide Y applied to the SCN in vivo. [From Refs. 249 (a, left), 63 (b, left), 250 (a, right), and 109 (b, right).]

pathway to the SCN [retinohypothalamic tract (RHT)]; and a population of SCN neurons that appear to function as "luminance" detectors (44). These unique features account for the preserved circadian responses of some apparently "sight-less" animals, including the blind mole rat (45), retinally degenerate mouse mutant *rd/rd* (46), and even some blind people who lack pupillary light reflexes and conscious light perception (47).

In addition to the SCN, individual ganglion cells also project to the inter-geniculate leaflet (IGL) of the thalamus (48). IGL neurons project to the SCN and also behave electrophysiologically like "luminance" detectors, and lesions of the leaflet affect the responses of circadian activity rhythms to light (49). However, the IGL's precise function in circadian timekeeping may include some "nonpho-tic" roles (see later) and remains under investigation.

Glutamate

Multiple lines of evidence suggest that the excitatory amino acid glutamate is the primary RHT neurotransmitter responsible for mediating the circadian actions of light (for review and references, see 50). Glutamate has been localized to RHT terminals innervating SCN neurons, and it is released by optic nerve stimulation of SCN slices in vitro; glutamate receptor blockers inhibit both the phase-shifting actions of light in vivo and the electrophysiological effects of optic nerve stimulation of slices in vitro; and glutamate application excites SCN neurons in vitro, with the resulting phase shifts mimicking a photic-type PRC.

NMDA, AMPA/kainate, and metabotropic receptors all seem to play some role in mediating glutamate's postsynaptic effects (50,51). In situ hybridization studies of receptor mRNA expression suggest that the SCN is relatively enriched in the NMDA-R1C and NMDA-R2C subtype variants. Interestingly, NMDA-R1C gives rise to a receptor subunit that is insensitive to phosphorylation by protein kinase C, a property that might contribute to the stability of postsynaptic responsiveness to photic stimuli.

Second Messengers and Diffusable Gases

SCN cells are electrophysiologically responsive to light during both the subjective day and subjective night, although the response is quantitatively greater during the night (52,53). Since light has no phase-shifting effect during the subjective day, recent investigations have begun to focus on SCN intracellular signal transduction pathways to account for the shape of the photic PRC (with its subjective day "dead zone").

Intracellular elevation of both Ca^{2+} and cyclic AMP leads to the phosphorylation of Ca^{2+}/cyclic AMP response element binding protein (CREB, a transcription factor). As CREB phosphorylation is thought to couple short-term extracellular stimuli to long-lasting cellular responses (53a), it is noteworthy that in the SCN CREB becomes phosphorylated after photic or glutamatergic stimulation during the subjective night (but not during the subjective day) (54,55). These data suggest the working hypothesis that CREB phosphorylation lies distal to a phase-dependent "gate" acting downstream of glutamatergic receptor activation. Recently a Ca^{2+}/calmodulin-dependent protein kinase (CaM kinase II) has been implicated, since its inhibition reduces light-induced CREB phosphorylation and attenuates (but does not abolish) light-induced behavioral phase shifts in vivo (56). In contrast, inhibition of protein kinase A has no apparent effect on light-induced phase shifts (57), and stimulation of the cyclic AMP pathway in vitro does not mimic a photic-type PRC (58).

Evidence is also accumulating that nitric oxide (NO) plays a role as an intermediate in the SCN's photic input pathway. This free radical is synthesized from L-arginine by NO synthase; the neuronal isoform of the enzyme is Ca^{2+}/calmodulin-dependent and localized to SCN neurons (59–61). Inhibition of the NO synthase blocks light-induced behavioral phase shifts in vivo (62,63) and glutamate-induced electrophysiological phase shifts (63,64) and CREB phosphorylation (55) in vitro. Conversely, generation of NO in vitro induces CREB phosphorylation during the subjective night (but not during the subjective day) (55) and mimics a photic-type PRC (63). The events that mediate NO's effects in the SCN remain uncertain and are likely to be complex. NO can diffuse across cellular membranes and activate the cyclic GMP-synthesizing enzyme guanylate cyclase (65). However, reported manipulations of the cyclic GMP pathway in vivo and in vitro seem to involve advance phase shifts only (57,66,67).

AP-1 DNA-Binding Proteins

The regulation of gene transcription is mainly accomplished by the interaction of nuclear DNA-binding proteins (like CREB, see above) with specific *cis*-acting regulatory DNA sequences on the promoters of target genes. Several laboratories have now shown that light governs the expression of such proteins in the SCN (for reviews and references, see 68–71) and IGL (72). The most intensively studied of these are proteins that make the transcription factor activity known as activator protein-1 (AP-1), which is primarily composed of structurally related DNA-binding proteins belonging to the *fos* and *jun* proto-oncogene families. Light regulates AP-1 DNA-binding activity in the SCN by altering the amount and composition of its constituent proteins (73–75). Fos B and Jun D proteins seem to be the predominant components forming AP-1-binding complexes in the SCN during darkness, whereas c-Fos and Jun B appear to compete for binding after photic stimulation. These kinds of changes in AP-1 protein composition are known to lead to alterations in the DNA-binding affinity and stability of the AP-1 complex, profoundly influencing the transcriptional regulation of target genes that have AP-1-binding sites among the response elements on their promoters.

There is compelling (although largely circumstantial) evidence that c-Fos is involved in the photic input pathway. The threshold, magnitude, and phase dependence of gene activation correlate with light-induced phase shifts of overt circadian rhythmicity (76,77); pharmacological agents that block behavioral phase shifts also block the photic stimulation of c-Fos in specific regions of the hamster SCN (69,70); and light-induced phase delays of the rat locomotor rhythm are prevented by intracerebroventricular injection of antisense oligodeoxynucleotides to both c-*fos* and *jun* B (78). It is noteworthy that mice homozygous for a c-*fos* null mutation can still entrain to a light-dark cycle and generate a photic PRC to bright light pulses (79). While these findings in knockout mice do not rule out a necessary physiological role for c-Fos in normal mice, they do emphasize the potential redundancy of the components of the photic entrainment system.

Whether or not it is necessary, c-Fos expression in the SCN does not appear to be *sufficient* for generating photic phase shifts (80 and references therein).

Modulators and Modifiers

Based on the data reviewed above, it is tempting to envision a light-induced cascade involving the release of glutamate → stimulation of ionotropic receptors → influx of Ca^{2+} → activation of NO synthase and serine/threonine protein kinases → phosphorylation of CREB → synthesis of AP-1 proteins → regulation of target gene transcription. For some of these steps, however, there is a paucity of data on their relationship to one another or to light-induced phase shifting. Moreover, since c-Fos-expressing cells in the SCN constitute a heterogenous population (81,82), it remains uncertain which of the intervening arrows are intracellular and which are intercellular.

Additional candidate neurotransmitters further complicate the simplicity of this idealized linear pathway, although their roles are not yet precisely defined. Immunoreactive substance P has been found in the SCN, both in terminals attributed to the RHT and in cell bodies that appear to contain its preprotachykinin mRNA (83,84). A substance P receptor antagonist can block light-induced c-Fos expression in the hamster SCN in vivo (but in a distribution different from glutamate receptor anatagonists) (85), and substance P application to rat SCN slices in vitro generates a photic-type PRC (86) and potentiates glutamatergic responses (87). On the other hand, glutamatergic neurotransmission may be attenuated by an adenosine A1 receptor mechanism (88).

The retino-recipient subdivision of the SCN includes neurons that synthesize VIP and gastrin-releasing peptide (GRP), colocalized in some perikarya (89). Their peptide and mRNA levels exhibit oppositely phased responses to light, with high levels of GRP during the light and of VIP during the dark (90,91), and some of the cells express c-Fos after a light pulse during the late subjective night (81). Phase shifts of the circadian locomotor rhythm have been reported after the peptides are administered in vivo, but their magnitude and direction remain controversial (92,93).

Perhaps most difficult has been elucidating a role for cholinergic neurotransmission (94,95). In vivo injection of the nonspecific agonist carbachol partially mimics a photic-type PRC, with phase delays during early subjective night and phase advances during late subjective night; but not exactly, because phase advances also can be induced during subjective day. Workers disagree on the effects of cholinergic stimulation of SCN slices in vitro (96,97), but only phase advances of the electrophysiological rhythm have been reported in these preparations. Possible factors contributing to this confusion include the presence of both nicotinic and muscarinic cholinergic receptors in the SCN (98); indirect effects of cholinergic agents in vivo that are absent in vitro (99); and the interaction of cholinergic mechanisms with glutamatergic neurotransmission and the intracellular cyclic GMP pathway (94,100).

Even more substances are being added to the list of photic modulators and modifiers (101–103). Dissecting their in vivo roles will be challenging, especially considering the likelihood of complex combinatorial and conditional interactions.

B. Nonphotic Mechanisms

The phase-resetting and entraining effects of a variety of nonphotic stimuli are now well accepted (38,39), although their precise functional significance is uncertain (nonphotic effects appear to be strongly species-specific, and the critical physiological variable responsible for their mediation remains unidentified). Elucidating the cellular and molecular substrates of nonphotic input pathway(s) should help to clarify circadian clock elements, especially since nonphotic phase shifts are affected by photic stimuli (104,105). Much of this research has focused on neuropeptide Y (NPY) and serotonin (5HT), two nonvisual afferents that innervate the retino-recipient subdivision of the SCN. The "feedback" actions of melatonin on the activity of the SCN and the expression of circadian rhythmicity are presented elsewhere (see Chapter 7).

Neuropeptide Y

About two-thirds of IGL neurons that project to the SCN contain NPY (106,107). In addition to a role for the IGL in the photic entrainment pathway (see above), there is evidence implicating it as part of the nonphotic system. In hamsters, nonphotic-type PRCs can be mimicked by electrical stimulation of the IGL (108) or injection of NPY into the SCN (109), whereas lesions of the IGL (110,111) or injections of an anti-NPY antiserum into the SCN (112) block nonphotic behavioral phase shifts [but not photically induced shifts (113)]. Some (114), but possibly not all (115,116), of the nonphotic phase-shifting stimuli lead to the expression of immunoreactive c-Fos protein in IGL cells. Over 70% of the NPY cells in the IGL express c-Fos after such nonphotic stimulation (117); this high percentage is not observed when c-Fos is induced in the IGL by photic stimulation. In rats, in vitro application of NPY to SCN slices during the subjective day causes advance phase shifts of the electrophysiological rhythm (118,119) (as would be expected with a non-photic-type PRC).

Currently, relatively little is known regarding the mechanism of action of NPY in the SCN. In vivo and in vitro phase-shifting effects appear to be mediated by the Y2 receptor subtype (120,121) and may depend on the activity of γ-aminobutyric acid (GABA) (122,123). The complex interactions recently reported between NPY and glutamate in vitro (124,125) are likely to have functional implications, since light pulses can block the phase-shifting actions of NPY in vivo (126).

Serotonin

The function of the serotonergic innervation of the SCN and IGL from the mid-brain raphe is not completely understood (127). A possible role in nonphotic phase shifting was suggested by the discovery that 5HT receptor agonists cause subjective-day phase advances of locomotor rhythmicity after drugs are injected in vivo (128,129) and electrophysiological rhythmicity after the agents are applied to SCN slices in vitro (130–132). Although these effects were mostly ascribed to stimulation of $5HT_{1A}$ receptors, their apparent dependence on the activation of the cyclic AMP pathway (133) has implicated a role for a recently characterized $5HT_7$ receptor subtype (134). Despite these pharmacological data, a physiological role for 5HT in the nonphotic system is not established; reports disagree on whether non-photic phase shifts are blocked by 5HT antagonism or depletion in vivo (135–139).

Complicating these analyses are additional data suggesting that 5HT acts to regulate SCN photic sensitivity. Manipulation of $5HT_{1A/7}$ and $5HT_{1B}$ neuro-transmission during the subjective night modulates light-induced phase shifts of locomotor rhythmicity, SCN firing rates, and SCN c-Fos expression (140–142), with heightened 5HT activity inhibiting the photic input pathway. It appears likely that the responsible interactions involve both pre- and post-synaptic mechanisms.

IV. Converting Oscillatory Outputs into Temporal Programs

The SCN governs a wide array of rhythms, from biosynthetic to behavioral, exhibiting a range of waveforms and phases different from the central oscillation that drives them. The circadian pacemaker might orchestrate this via metabolic cascades, in which the formation of one control factor regulates the formation of a second, and a third, and so on. The circadian signal would undergo even further modification as the pacemaker's outputs are coupled to effector cells via synaptic transmission, hormonal secretion, and indirectly through the rhythmic regulation of behavior. Importantly, the fidelity of the circadian signal might be altered by such multistep output pathways, so overt rhythms could differ substantially from their unidentified, underlying cellular and molecular oscillations. A rhythm in one step might not result in rhythmicity in the next or subsequent steps (see later); or, if more than one component in the pathway were clock-controlled, the resulting rhythm might have an amplitude much greater than that of any single component. Output pathways may thus provide an important substrate for plasticity in the circadian timing system.

An alternative possibility is that the multiplicity of overt rhythms reflects several pacemakers, perhaps even outside the SCN (143), each of which controls only one rhythm. In this view, the oscillator and its "hand" would be integrally related, like the observed oscillation of a feedback circuit.

A. Multiple Output Mechanisms

Clock control might be exerted in a number of ways; for example, by directly imposing rhythmicity on otherwise arrhythmic processes, by entraining secondary oscillators, or by gating the outputs of downstream systems so that their overt expression occurs only at specific times. Research is beginning to unravel multiple pathways by which the circadian timing signal from the SCN is conveyed to distal targets.

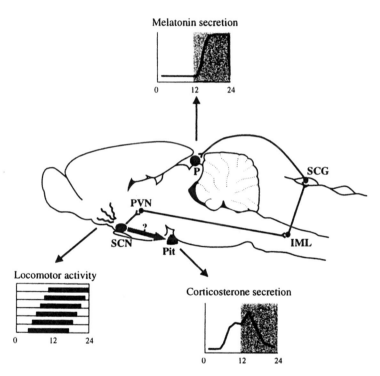

Figure 3 Multiple SCN output pathways. Clock control appears to be exerted via multiple output pathways from the SCN. The rhythm of melatonin secretion is generated via a multisynaptic neural pathway from the SCN to the pineal gland (P), involving the PVN, intermediolateral column of the spinal cord (IML), and superior cervical ganglion (SCG). The rhythm of adrenal corticosterone secretion is influenced by SCN regulation of pituitary (Pit) hormonal rhythmicity, although the precise mechanism remains unclear. The rhythm of locomotor activity may be governed by a diffusable substance, as demonstrated by its restoration in SCN-lesioned hamsters by transplants of SCN tissue isolated within semipermeable capsules.

Synaptic Connections

A multisynaptic neural pathway from the SCN drives the circadian rhythm of the activity of pineal serotonin N-acetyltransferase (NAT), the rate-limiting enzyme catalyzing the dramatic nocturnal increase in melatonin synthesis (144; also see later and Chapter 7). Light acts to suppress melatonin production and entrain its circadian rhythm, ensuring that high levels of melatonin are restricted to the dark phase in both nocturnal and diurnal animals.

SCN lesions abolish the NAT rhythm (145), and the responsible SCN → pineal neural pathway is generally understood. The paraventricular nucleus (PVN) of the hypothalamus receives extensive SCN projections (146,147) and forms a necessary link in the pathway. PVN lesions block nighttime elevation of pineal NAT activity and melatonin content (148–150) [without affecting locomotor rhythmicity (151)], whereas PVN electrical stimulation elevates normally low daytime levels of the melatonin metabolite 6-hydroxymelatonin (152). PVN projections (153) include some to preganglionic neurons of the sympathetic nervous system (in the intermediolateral column of the spinal cord); postganglionic cell bodies in the superior cervical ganglion innervate the pineal (154,155). Nighttime norepinephrine release by these terminals stimulates melatonin production (156 and see later).

While the SCN → pineal circuit is currently the best-delineated synaptic output pathway, others surely exist. For example, the timing of gonadotropin secretion is likely regulated by specific SCN efferents onto hormone-receptive neurons (157–159).

SCN Neuropeptides

AVP and VIP are found in some SCN efferent fibers, and they have been implicated in the circadian regulation of hypothalamopituitary hormonal rhythmicity. Infusions of AVP or an AVP V_1-receptor antagonist into rat dorsomedial hypothalamus suggest that endogenous AVP release by the SCN inhibits corticosterone secretion during the light phase (160,161), although the AVP rhythm cannot account for the complete corticosterone waveform (162). Studies have not been conducted in a diurnal animal, in which the corticosterone peak occurs in the early morning (a phase ~180° opposite to the peak in the nocturnal rat). The neural pathway for AVP's effect remains unclear; it is uncertain if AVP-containing SCN efferents make contact with corticotrophin-releasing hormone neurons in the PVN (163,164), nor has the possible direct role for rhythmic peptide levels in the CSF been fully investigated (165). Of note, Brattleboro rats, with a mutation in the AVP gene that prevents peptide synthesis, exhibit behavioral, pineal NAT, sleep-wake, and SCN electrophysiological circadian rhythms, although the latter two show some diminished amplitude (166,167).

Recent experiments have employed a novel technique to study the role of VIP as a circadian output signal (168,169). An antisense oligodeoxynucleotide complementary to VIP mRNA was injected into the SCN of ovariectomized female rats; control injections consisted of oligos with randomly scrambled sequences. The rhythm of corticosterone was abolished and the peak of luteinizing hormone was depressed, but prolactin rhythmicity remained unaffected.

Humoral Signals

Among the most remarkable features of the circadian system is the demonstration that grafts of fetal hypothalamic tissue containing the SCN can reestablish overt rhythmicity in arrhythmic, SCN-lesioned adult recipients (170). Restored rhythmicity expresses properties that are characteristic of the donor tissue (171,172; but see 173), indicating that the transplanted SCN cells contain an autonomous circadian pacemaker. Rhythms are not restored by transplants of other brain areas or hypothalamic tissue not containing the SCN (174). Reinstated rhythms include behavioral rhythms of locomotion, drinking, and gnawing (175) (but not melatonin). The latency for recovery after transplantation is on the order of several weeks and seems to correlate roughly with the expression of SCN-specific neuropeptides by the graft (176–179). It appears that outputs from the implanted SCN are better reconstructed than inputs to it; restored activity rhythms are not entrained by photic or nonphotic stimuli that require neural input to the SCN (180).

Although axonal fibers, some containing AVP or VIP, do grow out from the transplants (181–184), their physiological role is unclear. Importantly, rhythmicity can also be restored in SCN-lesioned hosts by the stereotaxic injection of dispersed cell suspensions of fetal hypothalamus containing the SCN, but not by suspensions prepared from superior colliculus or cortex (185). Thus, the structural integrity of the SCN is not necessary for rhythm restoration. Most recently, locomotor rhythmicity has been reestablished in SCN-lesioned hamsters by SCN tissue isolated within semipermeable polymeric capsules that prevent neural outgrowth but allow diffusion of substances with mass <500 K (186). The soluble signal released by the encapsulated tissue is not yet known.

B. Clock-Controlled Gene Expression

It is obvious that clock regulation of physiology and behavior must rely in some way on changes in cellular biochemistry. There are many ways by which enzymatic activities might be altered, ranging from dramatic changes in absolute levels to subtle changes in structure. Thus far, most of the known control mechanisms lie at the transcriptional level, stimulating research to identify the *cis*-acting elements and *trans*-acting factors that mediate circadian effects on the promoters of regulated genes.

Pineal Serotonin N-Acetyltransferase

The circadian rhythm of the enzymatic activity of pineal NAT is probably both the most robust (with a ~100-fold nocturnal increase in rat) and, with the recent cloning of the NAT gene (187,188), the best understood biochemical rhythm in mammals (189,190). Norepinephrine released at night from sympathetic terminals binds to β_1-adrenergic receptors on pinealocytes, activating membrane-bound adenylate cyclase and increasing intracellular concentrations of cyclic AMP. Concomitant activation of α_1-adrenergic receptors elevates intracellular Ca^{2+} concentrations, potentiating the β_1-adrenergic effect via the stimulation of phosphatidyl inositol turnover and the activation of protein kinase C (191). The phosphorylation of CREB likely induces NAT gene transcription (192,193). Although c-Fos and Jun B synthesis is also increased at night in the pineal (194), their roles remain uncertain; specifically, c-Fos is primarily regulated by an α_1-adrenergic (not β_1-adrenergic) mechanism (195), and NAT gene induction does not require new protein synthesis (196).

NAT mRNA levels decline during the latter part of the night, a process that does depend on new protein synthesis (196). An intriguing possibility is the β_1-adrenergic/cyclic AMP induction of inhibitory transcription factors; likely candidates include ICER (197,198), an inducible form of CREM (a member of the CREB family) that acts as a transcriptional repressor of cyclic-AMP-regulated genes, and Fos-related antigen-2 (Fra-2) (199), a *fos* family member that can inhibit AP-1-dependent gene expression. Binding sites for these factors are present on the NAT gene promoter (193,198). Since the duration of heightened NAT activity depends on photoperiod, it is noteworthy that photoperiod alters the dynamics of pineal CREM induction (200).

This research has also revealed important complications to this scheme. A key role for posttranscriptional mechanisms is demonstrated by the effects of a light pulse administered at night, which rapidly suppresses NAT enzymatic activity and melatonin synthesis while NAT mRNA levels remain unchanged (196). Posttranscriptional regulation also appears to dominate in the sheep pineal, in which a modest nocturnal increase of NAT mRNA levels (~1.5-fold) does not account for the ~10-fold increase of NAT enzymatic activity (187). These species differences may clarify differences in the phase and shape of melatonin rhythms in these animals. Investigating other enzymatic steps in the melatonin synthetic pathway [e.g., the hydroxyindole-*O*-methyltransferase gene (201)] should be instructive.

Hepatic Albumin Site-D Binding Protein (DBP)

DBP is a sequence-specific DNA-binding protein (like c-Fos) that alters the expression of target genes by regulating their transcription. In rat liver, DBP mRNA and protein levels cycle with a circadian rhythm, with a ~100-fold magnitude protein oscillation that peaks about 2 hr after lights-off in a light-dark cycle (202). The rhythm is transcriptionally generated and presumably driven by the SCN (although this latter point has not been proven), perhaps via circulating corticosterone levels. Interestingly, DBP mRNA cycles with similar phase and amplitude in other tissues, but the protein does not accumulate as in liver, thus implicating posttranscriptional mechanisms in the control of DBP tissue distribution.

Rhythmic DBP levels appear responsible for circadian transcription of the cholesterol 7α hydroxylase gene (203), which encodes the rate-limiting enzyme for the conversion of cholesterol to bile acids [maximal during the dark (feeding) phase in nocturnal rats]. Rhythmic DBP levels do not guarantee rhythms of all target genes; whether the binding of DBP generates a rhythm depends on the affinity of the binding site on the target gene's promoter (204). For example, a low-affinity site for DBP (K_d in the μM range) would be stimulated for only a short interval in the evening (when nuclear DBP concentrations are maximal), resulting in circadian transcription. But a site that binds DBP with high affinity (e.g., on the cytochrome P450 CYP2C6 gene promoter) would be occupied throughout the circadian cycle, eliminating the rhythmic signal. A second critical parameter is the stability of the target gene's transcribed product. Thus, albumin gene transcription is rhythmic but the resulting mRNA levels are not, presumably owing to a long mRNA half-life.

SCN Vasopressin

While a number of SCN neuropeptides show rhythmic properties (30), AVP is thus far the only one studied at the transcriptional level. Both AVP peptide (205) and mRNA (206–209) levels exhibit circadian rhythmicity in the rat SCN (but not in the PVN), and nuclear run-on analysis has demonstrated that mRNA abundance is regulated by transcription (210). Interestingly, the peak levels of mRNA in the rat SCN occur during the latter half of the light phase, while the peak levels of the peptide occur earlier in the morning after lights-on. This observation hints that other, posttranscriptional mechanisms contribute to the temporal pattern of SCN AVP expression. In fact, there is a circadian rhythm of AVP mRNA polyadenylate tail length, with long tails (~200 nucleotides) during the light phase and short tails (~30 nucleotides) during the dark (211,212). Since the appearance of the long-tailed AVP mRNA species coincides with the rise of peptide content, it is possible that polyadenylation state accounts for rhythmic peptide levels by altering mRNA stability and translational efficiency.

V. Constructing a Multicellular Circadian Pacemaker in the SCN

The cellular and molecular basis of the actual oscillatory mechanism of the circadian pacemaker in the SCN is unknown. Whether circadian rhythmicity is a property of individual SCN cells or instead emerges from an intercellular (network) interaction was, until recently, the focus of much debate (213). Accumulating evidence that SCN cells continue to oscillate after dissociation and culture (35,36,185) was elegantly verified using a system that simultaneously monitors the neuronal firing rates of multiple individual dispersed cells cultured on fixed microelectrode arrays (22). Single cells dissociated from rat SCN showed circadian firing rhythms with widely varying phases, in part because different cells expressed independent circadian periods. These data provide strong evidence for the existence of cellular circadian oscillators in the SCN and have also raised two key questions that remain unanswered. First, since the SCN pacemaker is not confined entirely to a single neuron, what are the mechanism(s) for intercellular communication that synchronize the disparate activities of individual cells? Second, what is the molecular nature of the intracellular circadian oscillation that actually keeps biological time?

A. Synchronization of Individual Cellular Oscillators

Local interactions within the SCN must coordinate the circadian oscillations of individual cells into a coherent, long-term rhythm with stable period and amplitude. Interest has focused on GABA, since it is believed to be the neurotransmitter common to all SCN neurons (214,215). Pharmacological manipulation of $GABA_A$ and $GABA_B$ receptor transmission leads to complex behavioral effects in vivo (216–218) and electrophysiological actions in vitro (219–221); the literature is complicated by differences in experimental preparations, and some GABAergic effects are likely to involve interactions with other neurotransmitters (e.g., glutamate and serotonin). An exciting recent report has demonstrated that GABA's actions in the SCN are phase-dependent (222). Bath application of GABA to rat SCN slices in vitro inhibits neuronal firing rate during the subjective night, while its application is excitatory during the subjective day. Both effects are mediated by a $GABA_A$ receptor; the change in sign likely follows a rhythm of the GABA equilibrium potential—positive relative to the resting membrane potential during the day, negative during the night—reflecting a circadian oscillation of intracellular chloride concentrations. Such a switching mechanism could serve a feedback function to both amplify SCN activity during the day and suppress it at night.

There is also evidence that synchronization of SCN cells can occur via Ca^{2+}-independent nonsynaptic mechanism(s) (223). Timekeeping persists when Na^+-dependent action potentials are blocked by tetrodotoxin (22,25) or the hypothermia of hibernation (224), and in the fetal SCN before most synapses form (225). Alternative intercellular coupling mechanisms have been hypothesized (226), including TTX-insensitive spikes, nonsynaptic/dendritic exocytosis (227, 228), and chemical and electrical low-resistance interneuronal junctions (229). Glial cells are also likely to be an important part of the synchronizing mechanism; intercellular Ca^{2+} waves can travel long distances across cultured SCN astrocytes

(230), presumably based on gap junctions (231). In organotypic slice cultures, synchronous rhythms of AVP and VIP release become dissociated and oscillate with independent circadian periods when glial overgrowth is prevented by treatment with an antimitotic drug (232).

B. On the Generation of Cellular Rhythmicity

Almost nothing is known of the molecular elements that produce cellular circadian oscillations in the SCN. As in other organisms, gene transcription (see above) and protein synthesis (233–236) are somehow involved. Ionic currents are being characterized in SCN neurons (219,237–241), including hints of a possibly novel K^+ channel, but these studies have not yet yielded mechanistic explanations for the generation of circadian rhythmicity.

Current ideas about the circadian oscillatory mechanism come mainly from molecular genetic studies of single-gene mutants in fruit flies (*Drosophila*) and fungi (*Neurospora*). This work has led to the identification of several molecules essential to circadian clock function, and to the general view that the pacemaker's core consists of autoregulatory feedback loops with oscillating levels of nuclear proteins negatively regulating the transcription of their own mRNAs. This literature is beyond the scope of this review and is considered elsewhere (3; and see Chapter 13). Briefly, mutations of the *period* (*per*) locus in *Drosophila* cause either fast, slow, or absent circadian rhythms. The gene codes for a cycling protein with peak levels occurring 6–8 hr after the peak of its mRNA occurs around dusk. The Per protein dimerizes with the product of a second clock gene, *timeless* (*tim*), which affects its accumulation in the cytoplasm and eventual translocation to the nucleus. Tim protein levels are rapidly reduced by light, which alters the Tim-Per interaction and affects the nuclear translocation of the proteins, leading to changes in the rhythmic transcription of their mRNAs. This dynamic can begin to account for a mechanism for entrainment to light-dark cycles and for the ability of nocturnal light pulses to cause both phase delays and advances of the circadian cycle. Still unknown, however, is how the Per and Tim proteins actually regulate transcription, since neither has known DNA-binding motifs.

In *Neurospora*, the *frequency* (*frq*) locus also leads to fast, slow, or absent circadian rhythms, and the levels of *frq* transcript and protein oscillate (although the mRNA peaks around dawn instead of dusk). Two bona fide DNA-binding proteins have also been implicated (242): transcriptional activators encoded by *white collar*-2, for making a necessary link in the loop regulating the feedback control of *frq* expression, and *white collar*-1, for permitting *frq*'s continued rhythmicity in constant darkness. An extraordinary feature of *Drosophila* and *Neurospora* clock genes is that Per, White Collar-1, and White collar-2 all share a sequence motif called a PAS domain (for protein-protein interactions), hinting that a structural element of all oscillators might have been conserved for the last

900 million years. Whether this observation speaks to the evolutionary origin of all oscillators or to common molecular mechanisms will need further work. It is already known that the likely substrate for light's phase-shifting action is different in *Drosophila* and *Neurospora* (reducing Tim levels and inducing *frq* transcription, respectively), and some unexpected differences in Per expression have now been reported in silkmoths (243).

Circadian rhythmicity in mammals is also genetically determined (244). Some workers have even claimed that *per*-like substance(s) exist in the rat SCN (245,246). A spontaneous single-gene clock mutation was discovered nearly 10 years ago in the hamster (*tau*) (247), but the lack of an adequate genetic map in this species had prevented further molecular analysis. As described elsewhere (see Chapter 12), a gene involved in mammalian circadian timekeeping (*clock*) has now been cloned in the mouse. The predicted amino acid sequence motifs suggest that

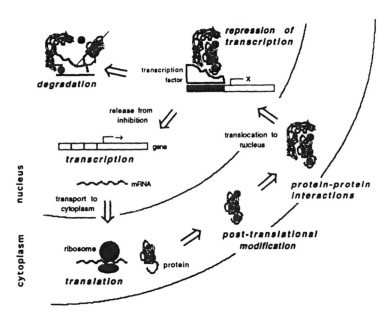

Figure 4 Hypothetical model of an autoregulatory negative feedback loop as a cellular clock mechanism. The mRNA of the clock gene is transcribed, transported to the cytoplasm, and translated into protein. There it undergoes posttranslational modification (e.g., phosphorylation) and interactions (e.g., dimerization) with other proteins, accounting for a lag period before it is translocated back to the nucleus. In concert with nuclear transcription factors, the clock protein becomes part of a DNA-binding complex that represses its own transcription. Degradation of the protein allows the cycle to resume by releasing the gene from inhibition. Light resets the rhythm by different mechanisms, depending on whether the clock product is high during day or night.

it is a transcription factor with, remarkably, a PAS domain. These findings signal that an exciting new phase has now begun in the search to dissect the molecular elements that underlie mammalian biological timekeeping.

Note Added in Proof

The molecular analysis of SCN mechanisms has now exploded with the cloning of *per* homologs (Sun et al., Cell 1997; 90:1003–1011; Tei et al., Nature 1997; 389:512–516; Shearman et al., Neuron 1997; 19:1261–1269), identification of *clock* protein binding partners (Gekakis et al., Science 1998; 280:1564–1569), and novel photopigments (Miyamoto and Sancar, Proc Natl Acad Sci USA 1998; 95:6097–6102).

Acknowledgments

W.J.S. is supported by NINDS R01 NS 24542 and AFOSR F-49620-97-1-0066.

References

1. Johnson CH, Golden SS, Ishiura M, Kondo T. Circadian rhythms in prokaryotes. Mol Microbiol 1996; 21:5–11.
2. Roenneberg T, Mittag M. The circadian program of algae. Semin Cell Dev Biol 1996; 7:753–763.
3. Dunlap JC. Genetic and molecular analysis of circadian rhythms. Annu Rev Genet 1996; 30:579–601.
4. Millar AJ, Kay SA. The genetics of phototransduction and circadian rhythms in *Arabidopsis*. BioEssays 1997; 19:209–214.
5. Block GD, Khalsa SBS, McMahon DG, Michel S, Guesz M. Biological clocks in the retina: cellular mechanisms of biological timekeeping. Int Rev Cytol 1993; 146: 83–143.
6. Rosbash M. Molecular control of circadian rhythms. Curr Opin Genet Dev 1995; 5: 662–668.
7. Hall JC. Tripping along the trail to the molecular mechanisms of biological clocks. Trends Neurosci 1995; 18:230–240.
8. Sehgal A, Ousley A, Hunter-Ensor M. Control of circadian rhythms by a two-component clock. Mol Cell Neurosci 1996; 7:165–172.
9. Cahill GM, Besharse JC. Circadian rhythmicity in vertebrate retinas: regulation by a photoreceptor oscillator. Prog Retinal Eye Res 1995; 14:267–291.
10. Zatz M. Melatonin rhythms: trekking toward the heart of darkness in the chick pineal. Semin Cell Dev Biol 1996; 7:811–820.
11. Reppert SM, Weaver DR, Ebisawa T. Cloning and characterization of a mammalian melatonin receptor that mediates reproductive and circadian responses. Neuron 1994; 13:1177–1185.

12. Cassone VM, Chesworth MJ, Armstrong SM. Entrainment of rat circadian rhythms by daily injection of melatonin depends upon the hypothalamic suprachiasmatic nuclei. Physiol Behav 1986; 36:1111–1121.

13. Cassone VM, Roberts MH, Moore RY. Melatonin inhibits metabolic activity in the rat suprachiasmatic nuclei. Neurosci Lett 1987; 81:29–34.

14. Mason R, Brooks A. The electrophysiological effects of melatonin and a putative melatonin antagonist (*N*-acetyltryptamine) on rat suprachiasmatic neurones in vitro. Neurosci Lett 1988; 95:296–301.

15. Shibata S, Cassone VM, Moore RY. Effects of melatonin on neuronal activity in the rat suprachiasmatic nucleus in vitro. Neurosci Lett 1989; 97:140–144.

16. Inouye ST, Kawamura H. Persistence of circadian rhythmicity in a mammalian hypothalamic "island" containing the suprachiasmatic nucleus. Proc Natl Acad Sci USA 1979; 76:5962–5966.

17. Meijer JH, Schaap J, Watanabe K, Albus H. Multiunit activity recordings in the suprachiasmatic nuclei: in vivo versus in vitro models. Brain Res 1997; 753:322–327.

18. Green DJ, Gillette R. Circadian rhythm of firing rate recorded from single cells in the rat suprachiasmatic brain slice. Brain Res 1982; 245:198–200.

19. Groos G, Hendriks J. Circadian rhythms in electrical discharge of rat suprachiasmatic neurones recorded in vitro. Neurosci Lett 1982; 34:283–288.

20. Shibata S, Oomura Y, Kita H, Hattori K. Circadian rhythmic changes of neuronal activity in the suprachiasmatic nucleus of the rat hypothalamic slice. Brain Res 1982; 247:154–158.

21. Belenky M, Wagner S, Yarom Y, Matzner H, Cohen S, Castel M. The suprachiasmatic nucleus in stationary organotypic culture. Neuroscience 1996; 70:127–143.

22. Welsh DK, Logothetis DE, Meister M, Reppert SM. Individual neurons dissociated from rat suprachiasmatic nucleus express independently phased circadian firing rhythms. Neuron 1995; 14:697–706.

23. Sato T, Kawamura H. Circadian rhythms in multiple unit activity inside and outside the suprachiasmatic nucleus in the diurnal chipmunk (*Eutamias sibiricus*). Neurosci Res 1984; 1:45–52.

24. Gillette MU, Medanic M, McArthur AJ, Liu C, Ding JM, Faiman LE, Weber ET, Tcheng TK, Gallman EA. Intrinsic neuronal rhythms in the suprachiasmatic nuclei and their adjustment. In: Chadwick DJ, Ackrill K, eds. Circadian Clocks and Their Adjustment. Ciba Foundation Symposium 183. Chichester: Wiley, 1995:134–153.

25. Schwartz WJ, Gross RA, Morton MT. The suprachiasmatic nuclei contain a tetrodotoxin-resistant circadian pacemaker. Proc Natl Acad Sci USA 1987; 84:1694–1698.

26. Schwartz WJ, Gainer H. Suprachiasmatic nucleus: use of ^{14}C-labeled deoxyglucose uptake as a functional marker. Science 1977; 197:1089–1091.

27. Sokoloff L, Reivich M, Kennedy C, Des Rosiers MH, Patlak CS, Pettigrew KD, Sakurada O, Shinohara M. The [^{14}C]deoxyglucose method for the measurement of local cerebral glucose utilization: theory, procedure, and normal values in the conscious and anesthetized albino rat. J Neurochem 1977; 28:897–916.

28. Schwartz WJ, Reppert SM, Eagan SM, Moore-Ede MC. In vivo metabolic activity of the suprachiasmatic nuclei: a comparative study. Brain Res 1983; 274:184–187.

29. Newman GC, Hospod FE, Patlak CS, Moore RY. Analysis of in vitro glucose utilization in a circadian pacemaker model. J Neurosci 1992; 12:2015–2021.

30. Inouye SIT, Shibata S. Neurochemical organization of circadian rhythm in the suprachiasmatic nucleus. Neurosci Res 1994; 20:109–130.

31. Reppert SM, Schwartz WJ, Uhl GR. Arginine vasopressin: a novel peptide rhythm in cerebrospinal fluid. Trends Neurosci 1987; 10:76–80.

32. Earnest DJ, Sladek CD. Circadian rhythms of vasopressin release from individual rat suprachiasmatic explants in vitro. Brain Res 1986; 382:129–133.

33. Gillette MU, Reppert SM. The hypothalamic suprachiasmatic nuclei: circadian patterns of vasopressin secretion and neuronal activity in vitro. Brain Res Bull 1987; 19:135–139.

34. Tominaga K, Inouye ST, Okamura H. Organotypic slice culture of the rat suprachiasmatic nucleus: sustenance of cellular architecture and circadian rhythm. Neuroscience 1994; 59:1025–1042.

35. Murakami N, Takamure M, Takahashi K, Utunomiya K, Kuroda H, Etoh T. Long-term cultured neurons from rat suprachiasmatic nucleus retain the capacity for circadian oscillation of vasopressin release. Brain Res 1991; 545:347–350.

36. Watanabe K, Koibuchi N, Ohtake H, Yamaoka S. Circadian rhythms of vasopressin release in primary cultures of rat suprachiasmatic nucleus. Brain Res 1993; 624: 115–120.

37. Shinohara K, Honma S, Katsuno Y, Abe H, Honma K. Circadian rhythms in the release of vasoactive intestinal polypeptide and arginine-vasopressin in organotypic slice culture of rat suprachiasmatic nucleus. Neurosci Lett 1994; 170:183–186.

38. Smith RD, Turek FW, Takahashi JS. Two families of phase-response curves characterize the resetting of the hamster circadian clock. Am J Physiol 1992; 262:R1149–R1153.

39. Mrosovsky N. Locomotor activity and non-photic influences on circadian clocks. Biol Rev 1996; 71:343–372.

40. Morin LP. The circadian visual system. Brain Res Rev 1994; 67:102–127.

41. Nelson DE, Takahashi JS. Sensitivity and integration in a visual pathway for circadian entrainment in the hamster (*Mesocricetus auratus*). J Physiol 1991; 439: 115–145.

42. Jiménez AJ, García-Fernández JM, González B, Foster RG. The spatio-temporal pattern of photoreceptor degeneration in the aged *rd/rd* mouse retina. Cell Tissue Res 1996; 284:193–202.

43. Moore RY, Speh JC, Card JP. The retinohypothalamic tract originates from a distinct subset of retinal ganglion cells. J Comp Neurol 1995; 352:351–366.

44. Meijer JH, Groos GA, Rusak B. Luminance coding in a circadian pacemaker: the suprachiasmatic nucleus of the rat and the hamster. Brain Res 1986; 382:109–118.

45. Vuillez P, Herbin M, Cooper HM, Nevo E, Pévet P. Photic induction of Fos immunoreactivity in the suprachiasmatic nuclei of the blind mole rat (*Spalax ehrenbergi*). Brain Res 1994; 654:81–84.

46. Foster RG, Provencio I, Hudson D, Fiske S, DeGrip W, Menaker M. Circadian photoreception in the retinally degenerate mouse (*rd/rd*). J Comp Physiol A 1991; 169:39–50.

47. Czeisler CA, Shanahan TL, Klerman EB, Martens H, Brotman DJ, Emens JS, Klein T, Rizzo III JF. Suppression of melatonin secretion in some blind patients by exposure to bright light. N Engl J Med 1995; 332:6–11.

48. Card JP, Whealy ME, Robbins AK, Moore RY, Enquist LW. Two α-herpesvirus strains are transported differentially in the rodent visual system. Neuron 1991; 6: 957–969.

49. Harrington ME. The ventral lateral geniculate nucleus and the intergeniculate leaflet: interrelated structures in the visual and circadian systems. Neurosci Biobehav Rev 1997; 21:705–727.

50. Ebling FJ. The role of glutamate in the photic regulation of the suprachiasmatic nucleus. Prog Neurobiol 1996; 50:109–132.

51. Scott G, Rusak B. Activation of hamster suprachiasmatic neurons in vitro via metabotropic glutamate receptors. Neuroscience 1996; 71:533–541.

52. Cui L-N, Dyball REJ. Synaptic input from the retina to the suprachiasmatic nucleus changes with the light-dark cycle in the Syrian hamster. J Physiol 1996; 497:483–493.

53. Meijer JH, Watanabe K, Détàri L, Schaap J. Circadian rhythm in light response in suprachiasmatic nucleus neurons of freely moving rats. Brain Res 1996; 741:352–355.

53a. Brindle PK, Montminy MR. The CREB family of transcription activators. Curr Opin Genet Dev 1992; 2:199–204.

54. Ginty DD, Kornhauser JM, Thompson MA, Bading H, Mayo KE, Takahashi JS, Greenberg ME. Regulation of CREB phosphorylation in the suprachiasmatic nucleus by light and a circadian clock. Science 1993; 260:238–241.

55. Ding JM, Fairman LE, Hurst WJ, Kuriashkina LR, Gillette MU. Resetting the biological clock: mediation of nocturnal CREB phosphorylation via light, glutamate, and nitric oxide. J Neurosci 1997; 17:667–675.

56. Golombek DA, Ralph MR. Circadian responses to light: the calmodulin connection. Neurosci Lett 1995; 192:101–104.

57. Mathur A, Golombek DA, Ralph MR. cGMP-dependent protein kinase inhibitors block light-induced phase advances of circadian rhythms in vivo. Am J Physiol 1996; 270:R1031–1036.

58. Prosser RA, Gillette MU. The mammalian circadian clock in the suprachiasmatic nuclei is reset in vitro by cAMP. J Neurosci 1989; 9:1073–1081.

59. Reuss S, Decker K, Robeler L, Layes E, Schollmayer A, Spessert R. Nitric oxide synthase in the hypothalamic suprachiasmatic nucleus of rat: evidence from histochemistry, immunohistochemistry and Western blot; and colocalization with VIP. Brain Res 1995; 695:257–262.

60. Wang H, Morris JF. Presence of neuronal nitric oxide synthase in the suprachiasmatic nuclei of mouse and rat. Neuroscience 1996; 74:1059–1068.

61. Chen D, Hurst WJ, Ding JM, Faiman LE, Mayer B, Gillette MU. Localization and characterization of nitric oxide synthase in the rat suprachiasmatic nucleus: evidence for a nitrergic plexus in the biological clock. J Neurochem 1997; 68:855–861.

62. Watanabe A, Ono M, Shibata S, Watanabe S. Effect of a nitric oxide synthase inhibitor, N-nitro-L-arginine methylester, on light-induced phase delay of circadian rhythm of wheel-running activity in golden hamsters. Neurosci Lett 1995; 192:25–28.

63. Ding JM, Chen D, Weber ET, Faiman LE, Rea MA, Gillette MU. Resetting the biological clock: mediation of nocturnal circadian shifts by glutamate and NO. Science 1994; 266:1713–1717.

64. Watanabe A, Hamada T, Shibata S, Watanabe S. Effects of nitric oxide synthase inhibitors on N-methyl-D-aspartate-induced phase delay of circadian rhythm of

neuronal activity in the rat suprachiasmatic nucleus in vitro. Brain Res 1994; 646: 161–164.

65. Knowles RG, Palacios M, Palmer MJ, Moncada S. Formation of nitric oxide from L-arginine in the central nervous system: a transduction mechanism for stimulation of the soluble guanylate cyclase. Proc Natl Acad Sci USA 1989; 86:5159–5162.

66. Weber ET, Gannon RL, Rea MA. cGMP-dependent protein kinase inhibitor blocks light-induced phase advances of circadian rhythms in vivo. Neurosci Lett 1995; 197:227–230.

67. Prosser RA, McArthur AJ, Gillette MU, cGMP induces phase shifts of a mammalian circadian pacemaker at night, in antiphase to cAMP effects. Proc Natl Acad Sci USA 1989; 86:6812–6815.

68. Kornhauser JM, Mayo KE, Takahashi JS. Immediate-early gene expression in a mammalian circadian pacemaker: the suprachiasmatic nucleus. In: Young MW, ed. Molecular Genetics of Biological Rhythms. New York: Marcel Dekker, 1993:271–307.

69. Abe H, Rusak B. Physiological mechanisms regulating photic induction of Fos-like protein in hamster suprachiasmatic nucleus. Neurosci Biobehav Rev 1994; 18:531–536.

70. Hastings MH, Ebling FJP, Grosse J, Herbert J, Maywood ES, Mikkelsen JD, Sumová A. Immediate-early genes and the neuronal bases of photic and nonphotic entrainment. In: Chadwick DJ, Ackrill K, eds. Circadian Clocks and Their Adjustment. Ciba Foundation Symposium 183. Chichester: Wiley, 1995:175–197.

71. Schwartz WJ, Aronin N, Takeuchi J, Bennett MR, Peters RV. Towards a molecular biology of the suprachiasmatic nucleus: photic and temporal regulation of c-*fos* gene expression. Semin Neurosci 1995; 7:53–60.

72. Peters RV, Aronin N, Schwartz WJ. c-Fos expression in the rat intergeniculate leaflet: photic regulation, co-localization with Fos-B, and cellular identification. Brain Res 1996; 728:231–241.

73. Kornhauser JM, Nelson DE, Mayo KE, Takahashi JS. Regulation of *jun*-B messenger RNA and AP-1 activity by light and a circadian clock. Science 1992; 255:1581–1584.

74. Takeuchi J, Shannon W, Aronin N, Schwartz WJ. Compositional changes of AP-1 DNA-binding proteins are regulated by light in a mammalian circadian clock. Neuron 1993; 11:825–836.

75. Peters RV, Aronin N, Schwartz WJ. Circadian regulation of Fos B is different from c-Fos in the rat suprachiasmatic nucleus. Mol Brain Res 1994; 27:243–248.

76. Kornhauser JM, Nelson DE, Mayo KE, Takahashi JS. Photic and circadian regulation of c-*fos* gene expression in the hamster suprachiasmatic nucleus. Neuron 1990; 5:127–134.

77. Trávníčková Z, Sumová A, Peters R, Schwartz WJ, Illnerová H. Photoperiod-dependent correlation between light-induced SCN c-*fos* expression and resetting of circadian phase. Am J Physiol 1996; 271:R825–R831.

78. Wollnik F, Brysch W, Uhlmann E, Gillardon F, Bravo R, Zimmermann M, Schlingensiepen KH, Herdegen T. Block of c-Fos and Jun B expression by antisense oligonucleotides inhibits light-induced phase shifts of the mammalian circadian clock. Eur J Neurosci 1995; 7:388–393.

79. Honrado GI, Johnson RS, Golombek DA, Spiegelman BM, Papaioannou VE, Ralph MR. The circadian system of c-fos deficient mice. J Comp Physiol A 1996; 178:563–570.

80. Schwartz WJ, Peters RV, Aronin N, Bennett MR. Unexpected c-*fos* gene expression in the suprachiasmatic nucleus of mice entrained to a skeleton photoperiod. J Biol Rhythms 1996; 11:35–44.

81. Romijn HJ, Sluiter AA, Pool CW, Wortel J, Buijs RM. Differences in colocalization between Fos and PHI, GRP, VIP and VP in neurons of the rat suprachiasmatic nucleus after a light stimulus during the phase delay versus the phase advance period of the night. J Comp Neurol 1996; 372:1–8.

82. Castel M, Belenky M, Cohen S, Wagner S, Schwartz WJ. Light-induced c-Fos expression in the mouse suprachiasmatic nucleus: immuno-electron microscopy reveals co-localization in multiple cell types. Eur J Neurosci 1997; 9:1950–1960.

83. Takatsuji K, Miguel-Hidalgo JJ, Tohyama M. Substance P–immunoreactive innervation from the retina to the suprachiasmatic nucleus in the rat. Brain Res 1991; 568:223–229.

84. Mikkelsen JD, Larsen PJ. Substance P in the suprachiasmatic nucleus of the rat: an immunohistochemical and in situ hybridization study. Histochemistry 1993; 100:3–16.

85. Abe H, Honma S, Shinohara K, Honma K. Substance P receptor regulates the photic induction of Fos-like protein in the suprachiasmatic nucleus of Syrian hamsters. Brain Res 1996; 708:135–142.

86. Shibata S, Tsuneyoshi A, Hamada T, Tominaga K, Watanabe S. Effect of substance P on circadian rhythms of firing activity and the 2-deoxyglucose uptake in the rat suprachiasmatic nucleus in vitro. Brain Res 1992; 597:257–263.

87. Shirakawa T, Moore RY. Responses of rat suprachiasmatic nucleus neurons to substance P and glutamate in vitro. Brain Res 1994; 642:213–220.

88. Watanabe A, Moriya T, Nisikawa Y, Araki T, Hamada T, Shibata S, Watanabe S. Adenosine A1-receptor agonist attenuates the light-induced phase shifts and fos expression in vivo and optic nerve stimulation-evoked field potentials in the suprachiasmatic nucleus in vitro. Brain Res 1996; 740:329–336.

89. Okamura H, Murakami S, Uda K, Sugano T, Takahashi Y, Yanaihara C, Ibata Y. Coexistance of vasoactive polypeptide (VIP)-, peptide histidine isoleucine (PHI)-, and gastrin-releasing peptide (GRP)-like immunoreactivity in neurons of the rat suprachiasmatic nucleus. Biomed Res 1986; 7:295–299.

90. Zoeller RT, Broyles B, Earley J, Anderson ER, Albers HE. Cellular levels of messenger ribonucleic acids encoding vasoactive intestinal peptide and gastrin-releasing peptide in neurons of the suprachiasmatic nucleus exhibit distinct 24-hour rhythms. J Neuroendocrinol 1992; 4:119–124.

91. Shinohara K, Tominaga K, Isobe Y, Inouye S-IT. Photic regulation of peptides located in the ventrolateral subdivision of the suprachiasmatic nucleus of the rat: daily variations of vasoactive intestinal polypeptide, gastrin-releasing peptide, and neuropeptide Y. J Neurosci 1993; 13:793–800.

92. Albers HE, Liou SY, Stopa EG, Zoeller RT. Interaction of colocalized neuropeptides: functional significance in the circadian timing system. J Neurosci 1991; 11:846–851.

93. Piggins HD, Antle MC, Rusak B. Neuropeptides phase shift the mammalian circadian pacemaker. J Neurosci 1995; 15:5612–5622.

94. Colwell CS, Kaufman CM, Menaker M. Phase-shifting mechanisms in the mammalian circadian system: new light on the carbachol paradox. J Neurosci 1993; 13:1454–1459.

95. Bina KG, Rusak B. Muscarinic receptors mediate carbachol-induced phase shifts of circadian activity rhythms in Syrian hamsters. Brain Res 1996; 743:202–211.

96. Trachsel L, Heller HC, Miller JD. Nicotine phase-advances the circadian neuronal activity rhythm in rat suprachiasmatic nuclei explants. Neuroscience 1995; 65: 797–803.

97. Liu C, Gillette MU. Cholinergic regulation of the suprachiasmatic nucleus circadian rhythm via a muscarinic mechanism at night. J Neurosci 1996; 16:744–751.

98. van der Zee EA, Streefland C, Strosberg AD, Schroder H, Luiten PG. Colocalization of muscarinic and nicotinic receptors in cholinoceptive neurons of the suprachiasmatic region in young and aged rats. Brain Res 1991; 542:348–352.

99. Bina KG, Rusak B, Semba K. Localization of cholinergic neurons in the forebrain and brainstem that project to the suprachiasmatic nucleus of the hypothalamus in rat. J Comp Neurol 1993; 335:295–307.

100. Liu C, Ding JM, Faiman LE, Gillette MU. Coupling of muscarinic cholinergic receptors and cGMP in nocturnal regulation of the suprachiasmatic circadian clock. J Neurosci 1997; 17:659–666.

101. Eaton SJ, Eoh S, Meyer J, Hoque S, Harrington ME. Circadian rhythm photic phase shifts are not altered by histamine receptor antagonists. Brain Res Bull 1996; 41: 227–229.

102. Bina KG, Rusak B. Nerve growth factor phase shifts circadian activity rhythms in Syrian hamsters. Neurosci Lett 1996; 206:97–100.

103. Hannibal J, Ding JM, Chen D, Fahrenkrug J, Larsen PJ, Gillette MU, Mikkelsen JD. Pituitary adenylate cyclase-activating peptide (PACAP) in the retinohypothalamic tract: a potential daytime regulator of the biological clock. J Neurosci 1997; 17: 2637–2644.

104. Joy JE, Turek FW. Combined effects on the circadian clock of agents with different phase response curves: phase shifting effects of triazolam and light. J Biol Rhythms 1992; 7:51–63.

105. Mrosovsky N. Double-pulse experiments with non-photic and photic phase-shifting stimuli. J Biol Rhythms 1991; 6:167–179.

106. Moore RY, Card JP. Intergeniculate leaflet: an anatomically and functionally distinct subdivision of the lateral geniculate complex. J Comp Neurol 1994; 344:403–430.

107. Morin LP, Blanchard J. Organization of the hamster intergeniculate leaflet: NPY and ENK projections to the suprachiasmatic nucleus, intergeniculate leaflet, and posterior limitans nucleus. Visual Neurosci 1995; 12:57–67.

108. Rusak B, Meijer JH, Harrington ME. Hamster circadian rhythms are phase-shifted by electrical stimulation of the geniculo-hypothalamic tract. Brain Res 1989; 493: 283–291.

109. Huhman KL, Albers HE. Neuropeptide Y microinjected into the suprachiasmatic region phase shifts circadian rhythms in constant darkness. Peptides 1994; 15:1475–1478.

110. Janik D, Mrosovsky N. Intergeniculate leaflet lesions and behaviorally-induced shifts of circadian rhythms. Brain Res 1994; 651:174–182.

111. Wickland C, Turek FW. Lesions of the thalamic intergeniculate leaflet block activity-induced phase shifts in the circadian activity rhythm of the golden hamster. Brain Res 1994; 660:293–300.

112. Biello SM, Janik D, Mrosovsky N. Neuropeptide Y and behaviorally induced phase shifts. Neuroscience 1994; 62:273–279.

113. Biello SM. Enhanced photic phase shifting after treatment with antiserum to neuropeptide Y. Brain Res 1995; 673:25–29.

114. Janik D, Mrosovsky N. Gene expression in the geniculate induced by a nonphotic circadian phase shifting stimulus. NeuroReport 1992; 3:575–578.

115. Cutrera RA, Kalsbeek A, Pévet P. No triazolam-induced expression of Fos protein in raphe nuclei of the male Syrian hamster. Brain Res 1993; 602:14–20.

116. Zhang Y, Van Reeth O, Zee PC, Takahashi JS, Turek FW. Fos protein expression in the circadian clock is not associated with phase shifts induced by a nonphotic stimulus, triazolam. Neurosci Lett 1993; 164:203–208.

117. Janik D, Mikkelsen JD, Mrosovsky N. Cellular colocalization of Fos and neuropeptide Y in the intergeniculate leaflet after nonphotic phase-shifting events. Brain Res 1995; 698:137–145.

118. Shibata S, Moore RY. Neuropeptide Y and optic chiasm stimulation affect suprachiasmatic nucleus circadian function in vitro. Brain Res 1993; 615:95–100.

119. Medanic M, Gillette MU. Suprachiasmatic circadian pacemaker of rat shows two windows of sensitivity to neuropeptide Y in vitro. Brain Res 1993; 620:281–286.

120. Huhman KL, Gillespie CF, Marvel CL, Albers HE. Neuropeptide Y phase shifts circadian rhythms in vivo via a Y2 receptor. NeuroReport 1996; 7:1249–1252.

121. Golombek DA, Biello SM, Rendon RA, Harrington ME. Neuropeptide Y phase shifts the circadian clock in vitro via a Y2 receptor. NeuroReport 1996; 7:1315–1319.

122. Huhman KL, Babagbemi TO, Albers HE. Bicuculline blocks neuropeptide Y–induced phase advances when microinjected in the suprachiasmatic nucleus of Syrian hamsters. Brain Res 1995; 675:333–336.

123. Chen G, van den Pol AN. Multiple NPY receptors coexist in pre- and postsynaptic sites: inhibition of GABA release in isolated self-innervating SCN neurons. J Neurosci 1996; 16:7711–7724.

124. van den Pol AN, Obrietan K, Chen G, Belousov AB. Neuropeptide Y–mediated long-term depression of excitatory activity in suprachiasmatic nucleus neurons. J Neurosci 1996; 16:5883–5895.

125. Biello SM, Golombek DA, Harrington ME. Neuropeptide Y and glutamate block each other's phase shifts in the suprachiasmatic nucleus in vitro. Neuroscience 1997; 77:1049–1057.

126. Biello SM, Mrosovsky N. Blocking the phase-shifting effect of neuropeptide Y with light. Proc Roy Soc Lond B 1995; 259:179–187.

127. Meyer-Bernstein EL, Morin LP. Differential serotonergic innervation of the suprachiasmatic nucleus and the intergeniculate leaflet and its role in circadian rhythm modulation. J Neurosci 1996; 16:2097–2111.

128. Tominaga K, Shibata S, Ueki S, Watanabe S. Effects of 5-HT1A receptor agonists on the circadian rhythm of wheel-running activity in hamsters. Eur J Pharmacol 1992; 214:79–84.

129. Edgar DM, Miller JD, Prosser RA, Dean RR, Dement WC. Serotonin and the mammalian circadian system. II. Phase-shifting rat behavioral rhythms with serotonergic agonists. J Biol Rhythms 1993; 8:17–31.

130. Medanic M, Gillette MU. Serotonin regulates the phase of the rat suprachiasmatic circadian pacemaker in vitro only during the subjective day. J Physiol 1992; 450: 629–642.

131. Shibata S, Tsuneyoshi A, Hamada T, Tominaga K, Watanabe S. Phase-resetting effect of 8-OH-DPAT, a serotonin1A receptor agonist, on the circadian rhythm of firing rate in the rat suprachiasmatic nuclei in vitro. Brain Res 1992; 582:353–356.

132. Prosser RA, Dean RR, Edgar DM, Heller HC, Miller JD. Serotonin and the mammalian circadian system. I. In vitro phase shifts by serotonergic agonists and antagonists. J Biol Rhythms 1993; 8:1–16.

133. Prosser RA, Heller HC, Miller JD. Serotonergic phase advances of the mammalian circadian clock involve protein kinase A and K⁺ channel opening. Brain Res 1994; 644:67–73.

134. Lovenberg TW, Baron BM, de Lecea L, Miller JD, Prosser RA, Rea MA, Foye PE, Racke M, Slone AL, Siegel BW, Danielson PE, Sutcliffe JG, Erlander MG. A novel adenylate cyclase-activating serotonin receptor (5-HT₇) implicated in the regulation of mammalian circadian rhythms. Neuron 1993; 11:449–458.

135. Smale L, Michels KM, Moore RY, Morin LP. Destruction of the hamster serotonergic system by 5,7-DHT: effects on circadian rhythm phase, entrainment and response to triazolam. Brain Res 1990; 515:9–19.

136. Cutrera RA, Kalsbeek A, Pévet P. Specific destruction of the serotonergic afferents to the suprachiasmatic nuclei prevents triazolam-induced phase advances of hamster activity rhythms. Behav Brain Res 1994; 62:21–28.

137. Penev PD, Turek FW, Zee PC. A serotonin neurotoxin attenuates the phase-shifting effects of triazolam on the circadian clock in hamsters. Brain Res 1995; 669: 207–216.

138. Sumová A, Maywood ES, Selvage D, Ebling FJP, Hastings MH. Serotonergic antagonists impair arousal-induced phase shifts of the circadian system of the Syrian hamster. Brain Res 1996; 709:88–96.

139. Bobrzynska KJ, Vrang N, Mrosovsky N. Persistance of nonphotic phase shifts in hamsters after serotonin depletion in the suprachiasmatic nucleus. Brain Res 1996; 741:205–214.

140. Rea MA, Glass JD, Colwell CS. Serotonin modulates photic responses in the hamster suprachiasmatic nuclei. J Neurosci 1994; 14:3635–3642.

141. Ying S-W, Rusak B. Effects of serotonergic agonists on firing rates of photically responsive cells in the hamster suprachiasmatic nucleus. Brain Res 1994; 651:37–46.

142. Pickard GE, Weber ET, Scott PA, Riberdy AF, Rea MA. 5HT1B receptor agonists inhibit light-induced phase shifts of behavioral circadian rhythms and expression of the immediate-early gene c-fos in the suprachiasmatic nucleus. J Neurosci 1996; 16:8208–8220.

143. Tosini G, Menaker M. Circadian rhythms in cultured mammalian retina. Science 1996; 272:419–421.

144. Arendt J. Melatonin and the Mammalian Pineal Gland. London: Chapman & Hall, 1995.

145. Klein DC, Moore RY. Pineal N-acetyltransferase and hydroxyindole-O-methyl-transferase: control by the retinohypothalamic tract and the suprachiasmatic nucleus. Brain Res 1979; 174:245–262.

146. Kalsbeek A, Teclemariam-Mesbah R, Pévet P. Efferent projections of the suprachiasmatic nucleus in the golden hamster (*Mesocricetus auratus*). J Comp Neurol 1993; 332:293–314.

147. Vrang N, Larsen PJ, Møller M, Mikkelsen JD. Topographical organization of the rat suprachiasmatic-paraventricular projection. J Comp Neurol 1995; 353:585–603.

148. Klein DC, Smoot R, Weller JL, Higa S, Markey SP, Creed GJ, Jacobowitz DM. Lesions of the paraventricular nucleus area of the hypothalamus disrupt the suprachiasmatic leads to spinal cord circuit in the melatonin rhythm generating system. Brain Res Bull 1983; 10:647–652.

149. Lehman MN, Bittman EL, Newman SW. Role of the hypothalamic paraventricular nucleus in neuroendocrine responses to daylength in the golden hamster. Brain Res 1984; 308:25–32.

150. Hastings MH, Herbert J. Neurotoxic lesions of the paraventriculo-spinal projection block the nocturnal rise in pineal melatonin synthesis in the Syrian hamster. Neurosci Lett 1986; 69:1–6.

151. Pickard GE, Turek FW. The hypothalamic paraventricular nucleus mediates the photoperiodic control of reproduction but not the effects of light on the circadian rhythm of activity. Neurosci Lett 1983; 43:67–72.

152. Yanovski J, Witcher J, Adler N, Markey SP, Klein DC. Stimulation of the paraventricular nucleus area of the hypothalamus elevates urinary 6-hydroxymelatonin during daytime. Brain Res Bull 1987; 19:129–133.

153. Smale L, Cassone VM, Moore RY, Morin LP. Paraventricular nucleus projections mediating pineal melatonin and gonadal responses to photoperiod in the hamster. Brain Res Bull 1989; 22:263–269.

154. Klein DC, Weller JL, Moore RY. Melatonin metabolism: neural regulation of pineal serotonin: acetyl coenzyme A *N*-acetyltransferase activity. Proc Natl Acad Sci USA 1971; 68:3107–3110.

155. Bowers CW, Zigmond RE. Electrical stimulation of the cervical sympathetic trunks mimics the effects of darkness on the activity of serotonin: *N*-acetyltransferase in the rat pineal. Brain Res 1980; 185:435–440.

156. Brownstein M, Axelrod J. Pineal gland: 24-hour rhythm in norepinephrine turnover. Science 1974; 184:163–165.

157. van der Beek EM, Wiegant VM, van der Donk HA, van den Hurk R, Buijs RM. Lesions of the suprachiasmatic nucleus indicate the presence of a direct vasoactive intestinal polypeptide-containing projection to gonadotropin-releasing hormone neurons in the female rat. J Neuroendocrinol 1993; 5:137–144.

158. de la Iglesia HO, Blaustein JD, Bittman EL. The suprachiasmatic area in the female hamster projects to neurons containing estrogen receptors and GnRH. NeuroReport 1995; 6:1715–1722.

159. Watson RE Jr, Langub MC Jr, Engle MG, Maley BE. Estrogen-receptive neurons in the anteroventral periventricular nucleus are synaptic targets of the suprachiasmatic nucleus and peri-suprachiasmatic region. Brain Res 1995; 689:254–264.

160. Kalsbeek A, Buijs RM, van Heerikhuize JJ, Arts M, van der Woude TP. Vasopressin-containing neurons of the suprachiasmatic nuclei inhibit corticosterone release. Brain Res 1992; 580:62–67.

161. Kalsbeek A, van der Vliet J, Buijs RM. Decrease of endogenous vasopressin release

necessary for expression of the circadian rise in plasma corticosterone: a reverse microdialysis study. J Neuroendocrinol 1996; 8:299–307.

162. Kalsbeek A, van Heerikhuize JJ, Wortel J, Buijs RM. A diurnal rhythm of stimulatory input to the hypothalamo-pituitary-adrenal system as revealed by timed intrahypothalamic administration of the vasopressin V1 antagonist. J Neurosci 1996; 16: 5555–5565.

163. Buijs RM, Markman M, Nunes-Cardoso B, Hou Y-X, Shinn S. Projections of the suprachiasmatic nucleus to stress-related areas in the rat hypothalamus: a light and electron microscopic study. J Comp Neurol 1993; 335:42–54.

164. Vrang N, Larsen PJ, Mikkelsen JD. Direct projection from the suprachiasmatic nucleus to hypophysiotropic corticotropin-releasing factor immunoreactive cells in the paraventricular nucleus of the hypothalamus demonstrated by means of *Phaseolus vulgaris*-leucoagglutinin tract tracing. Brain Res 1995; 684:61–69.

165. Kruisbrink J, Mirmiran M, van der Woude TP, Boer GJ. Effects of enhanced cerebrospinal fluid levels of vasopressin, vasopressin antagonist or vasoactive intestinal polypeptide on circadian sleep-wake rhythm in the rat. Brain Res 1987; 419: 76–86.

166. Brown MH, Nunez AA. Vasopressin-deficient rats show a reduced amplitude of the circadian sleep rhythm. Physiol Behav 1989; 46:759–762.

167. Ingram CD, Snowball RK, Mihai R. Circadian rhythm of neuronal activity in suprachiasmatic nucleus slices from the vasopressin-deficient Brattleboro rat. Neuroscience 1996; 75:635–641.

168. Scarbrough K, Harney JP, Rosewell KL, Wise PM. Acute effects of antisense antagonism of a single peptide neurotransmitter in the circadian clock. Am J Physiol 1996; 270:R283–R288.

169. Harney JP, Scarbrough K, Rosewell KL, Wise PM. In vivo antisense antagonism of vasoactive intestinal peptide in the suprachiasmatic nuclei causes aging-like changes in the estradiol-induced luteinizing hormone and prolactin surges. Endocrinology 1996; 137:3696–3701.

170. Ralph MR, Lehman MN. Transplantation: a new tool in the analysis of the mammalian circadian pacemaker. Trends Neurosci 1991; 14:362–366.

171. Ralph MR, Foster RG, Davis FC, Menaker M. Transplanted suprachiasmatic nucleus determines circadian period. Science 1990; 247:975–978.

172. Earnest DJ, Sladek CD, Gash DM, Wiegand SJ. Specificity of circadian function in transplants of the fetal suprachiasmatic nucleus. J Neurosci 1989; 9:2671–2677.

173. Sollars PJ, Kimble DP, Pickard GE. Restoration of circadian behavior by anterior hypothalamic heterografts. J Neurosci 1995; 15:2109–2122.

174. LeSauter J, Lehman MN, Silver R. Restoration of circadian rhythmicity by transplants of SCN "micropunches." J Biol Rhythms 1996; 11:163–171.

175. LeSauter J, Silver R. Suprachiasmatic nucleus lesions abolish and fetal grafts restore circadian gnawing rhythms in hamsters. Restor Neurol Neurosci 1994; 6:135–143.

176. Lehman MN, Silver R, Gladstone WR, Kahn RM, Gibson M, Bittman EL. Circadian rhythmicity restored by neural transplant. Immunocytochemical characterization of the graft and its integration with the host brain. J Neurosci 1987; 7:1626–1638.

177. Romero M-T, Silver R. Time course of peptidergic expression in fetal suprachiasmatic nucleus transplanted into adult hamster. Dev Brain Res 1990; 57:1–6.

178. Griffioen HA, Duindam H, van der Woude TP, Rietveld WJ, Boer GJ. Functional development of fetal suprachiasmatic nucleus grafts in suprachiasmatic nucleus-lesioned rats. Brain Res Bull 1993; 31:145–160.

179. Aguilar-Roblero R, Morin LP, Moore RY. Morphological correlates of circadian rhythm restoration induced by transplantation of the suprachiasmatic nucleus in hamsters. Exp Neurol 1994; 130:250–260.

180. Canbeyli R, Romero M-T, Silver R. Neither triazolam nor activity phase advance circadian locomotor activity in SCN-lesioned hamsters bearing fetal SCN transplants. Brain Res 1991; 566:40–45.

181. Wiegand SJ, Gash DM. Organization and efferent connections of transplanted suprachiasmatic nuclei. J Comp Neurol 1988; 267:562–579.

182. Al-Shamma HA, De Vries GJ. Fiber outgrowth from fetal vasopressin neurons of the suprachiasmatic nucleus, bed nucleus of the stria terminalis, and medial amygdaloid nucleus transplanted into adult Brattleboro rats. Dev Brain Res 1991; 64:200–204.

183. Sollars PJ, Pickard GE. Vasoactive intestinal peptide efferent projections of the suprachiasmatic nucleus in anterior hypothalamic transplants: correlation with functional restoration of circadian behavior. Exp Neurol 1995; 136:1–11.

184. Lehman MN, LeSauter J, Kim C, Berriman SJ, Tresco PA, Silver R. How do fetal grafts of the suprachiasmatic nucleus communicate with the host brain? Cell Transplant 1995; 4:75–81.

185. Silver R, Lehman MN, Gibson M, Gladstone WR, Bittman EL. Dispersed cell suspensions of fetal SCN restore circadian rhythmicity in SCN-lesioned adult hamsters. Brain Res 1990; 525:45–58.

186. Silver R, LeSauter J, Tresco PA, Lehman MN. A diffusible coupling signal from the transplanted suprachiasmatic nucleus controlling circadian locomotor rhythms. Nature 1996; 382:810–813.

187. Coon SL, Roseboom PH, Baler R, Weller JL, Namboodiri MAA, Koonin EV, Klein DC. Pineal serotonin N-acetyltransferase: expression cloning and molecular analysis. Science 1995; 270:1681–1683.

188. Borjigin J, Wang MM, Snyder SH. Diurnal variation in mRNA encoding serotonin N-acetyltransferase in pineal gland. Nature 1995; 378:783–785.

189. Klein DC, Roseboom PH, Coon SL. New light is shining on the melatonin rhythm enzyme. The first postcloning view. Trends Endocrinol Metab 1996; 7:106–112.

190. Korf HW, Schomerus C, Maronde E, Stehle JH. Signal transduction molecules in the rat pineal organ: Ca^{2+}, pCREB, and ICER. Naturwissenschaften 1996; 83:535–543.

191. Klein DC. Photoneural regulation of the mammalian pineal gland. Ciba Foundation Symposium 1985; 117:38–56.

192. Roseboom PH, Klein DC. Norepinephrine stimulation of pineal cyclic AMP response element-binding protein phosphorylation: primary role of a beta-adrenergic receptor/cyclic AMP mechanism. Mol Pharmacol 1995; 47:439–449.

193. Baler R, Covington S, Klein DC. The rat arylalkylamine N-acetyltransferase gene promoter. J Biol Chem 1997; 272:6979–6985.

194. Carter DA. A daily rhythm of activator protein-1 activity in the rat pineal is dependent upon *trans*-synaptic induction of Jun B. Neuroscience 1994; 62:1267–1278.

195. Carter DA. Neurotransmitter-stimulated immediate-early gene responses are orga-

nized through differential post-synaptic receptor mechanisms. Mol Brain Res 1992; 16:111–118.

196. Roseboom PH, Coon SL, Baler R, McCune SK, Weller JL, Klein DC. Melatonin synthesis: analysis of the more than 150-fold nocturnal increase in serotonin *N*-acetyltransferase messenger ribonucleic acid in the rat pineal gland. Endocrinology 1996; 137:3033–3045.

197. Stehle JH, Foulkes NS, Molina CA, Simonneaux V, Pévet P, Sassone-Corsi P. Adrenergic signals direct rhythmic expression of transcriptional repressor CREM in the pineal gland. Nature 1993; 365:314–320.

198. Foulkes NS, Borjigin J, Snyder SH, Sassone-Corsi P. Transcriptional control of circadian homone synthesis via the CREM feedback loop. Proc Natl Acad Sci USA 1996; 93:14140–14145.

199. Baler R, Klein DC. Circadian expression of transcription factor Fra-2 in the rat pineal gland. J Biol Chem 1995; 270:27319–27325.

200. Foulkes NS, Duval G, Sassone-Corsi P. Adaptive inducibility of CREM as transcriptional memory of circadian rhythms. Nature 1996; 381:83–85.

201. Gauer F, Craft CM. Circadian regulation of hydroxyindole-*O*-methyltransferase mRNA levels in rat pineal and retina. Brain Res 1996; 737:99–109.

202. Wuarin J, Schibler U. Expression of the liver-enriched transcriptional activator protein DBP follows a stringent circadian rhythm. Cell 1990; 63:1257–1266.

203. Lavery DJ, Schibler U. Circadian transcription of the cholesterol 7 alpha hydroxylase gene may involve the liver-enriched bZIP protein DBP. Genes Dev 1993; 7: 1871–1884.

204. Wuarin J, Falvey E, Lavery D, Talbot D, Schmidt E, Ossipow V, Fonjallaz P, Schibler U. The role of the transcriptional activator protein DBP in circadian liver gene expression. J Cell Sci 1992; 16(Suppl):123–127.

205. Tominaga K, Shinohara K, Otori Y, Fukuhara C, Inouye ST. Circadian rhythms of vasopressin content in the suprachiasmatic nucleus of the rat. NeuroReport 1992; 3: 809–812.

206. Uhl GR, Reppert SM. Suprachiasmatic nucleus vasopressin messenger RNA: circadian variation in normal and Brattleboro rats. Science 1986; 232:390–393.

207. Burbach JPH, Liu B, Voorhuis TAM, Van Tol HHM. Diurnal variation in vasopressin and oxytocin messenger RNAs in hypothalamic nuclei of the rat. Brain Res 1988; 464:157–160.

208. Young WS III, Kovacs K, Lolait SJ. The diurnal rhythm in vasopressin via receptor expression in the suprachiasmatic nucleus is not dependent on vasopressin. Endocrinology 1993; 133:585–590.

209. Cagampang FRA, Yang J, Nakayama Y, Fukuhara C, Inouye ST. Circadian variation of arginine-vasopressin messenger RNA in the rat suprachiasmatic nucleus. Mol Brain Res 1994; 24:179–184.

210. Carter DA, Murphy D. Nuclear mechanisms mediate rhythmic changes in vasopressin mRNA expression in the rat suprachiasmatic nucleus. Mol Brain Res 1991; 12:315–321.

211. Robinson BG, Frim DM, Schwartz WJ, Majzoub JA. Vasopressin mRNA in the suprachiasmatic nuclei: daily regulation of polyadenylate tail length. Science 1988; 241:342–344.

212. Carter DA, Murphy D. Diurnal rhythm of vasopressin mRNA species in the rat suprachiasmatic nucleus: independence of neuroendocrine modulation and maintenance in explant culture. Mol Brain Res 1989; 6:233–239.

213. Miller JD. On the nature of the circadian clock in mammals. Am J Physiol 1993; 264:R821–R832.

214. Moore RY, Speh JC. GABA is the principal neurotransmitter of the circadian system. Neurosci Lett 1993; 150:112–116.

215. Buijs RM, Wortel J, Hou Y-X. Colocalization of gamma-aminobutyric acid with vasopressin, vasoactive intestinal peptide, and somatostatin in the rat suprachiasmatic nucleus. J Comp Neurol 1995; 358:343–352.

216. Ralph MR, Menaker M. GABA regulation of circadian responses to light. I. Involvement of $GABA_A$-benzodiazepine and $GABA_B$ receptors. J Neurosci 1989; 9:2858–2865.

217. Smith RD, Turek FW, Slater NT. Bicuculline and picrotoxin block phase advances induced by GABA agonists in the circadian rhythm of locomotor activity in the golden hamster by a phaclofen-insensitive mechanism. Brain Res 1990; 530:275–282.

218. Golombek DA, Ralph MR. Inhibition of GABA transaminase enhances light-induced circadian phase delays but not advances. J Biol Rhythms 1994; 9:251–261.

219. Strecker GJ, Bouskila Y, Dudek FE. Neurotransmission and electrophysiological mechanisms in the suprachiasmatic nucleus. Semin Neurosci 1995; 7:43–51.

220. Gannon RL, Cato MJ, Kelley KH, Armstrong DL, Rea MA. GABAergic modulation of optic nerve-evoked field potentials in the rat suprachiasmatic nucleus. Brain Res 1995; 694:264–270.

221. Shimura M, Harata N, Tamai M, Akaike N. Allosteric modulation of $GABA_A$ receptors in acutely dissociated neurons of the suprachiasmatic nucleus. Am J Physiol 1996; 270:C1726–C1734.

222. Wagner S, Castel M, Gainer H, Yarom Y. GABA in the mammalian suprachiasmatic nucleus and its role in diurnal rhythmicity. Nature 1997; 387:598–603.

223. Bouskila Y, Dudek FE. Neuronal synchronization without calcium-dependent synaptic transmission in the hypothalamus. Proc Natl Acad Sci USA 1993; 90:3207–3210.

224. Miller JD, Cao VH, Heller C. Thermal effects on neuronal activity in suprachiasmatic nuclei of hibernators and nonhibernators. Am J Physiol 1994; 266:R1259–R1266.

225. Reppert SM, Schwartz WJ. The suprachiasmatic nuclei of the fetal rat: characterization of a functional circadian clock using ^{14}C-labeled deoxyglucose. J Neurosci 1984; 4:1677–1682.

226. van den Pol AN, Dudek FE. Cellular communication in the circadian clock, the suprachiasmatic nucleus. Neuroscience 1993; 56:793–811.

227. Castel M, Morris J, Belenky M. Non-synaptic and dendritic exocytosis from dense-cored vesicles in the suprachiasmatic nucleus. NeuroReport 1996; 7:543–547.

228. Güldner FH, Wolff JR. Complex synaptic arrangements in the rat suprachiasmatic nucleus: a possible basis for the "zeitgeber" and non-synaptic synchronization of neuronal activity. Cell Tissue Res 1996; 284:203–214.

229. Jiang Z-G, Yang Y-Q, Allen CN. Tracer and electrical coupling of rat suprachiasmatic nucleus neurons. Neuroscience 1997; 77:1059–1066.

230. van den Pol AN, Finkbeiner SM, Cornell-Bell AH. Calcium excitability and oscilla-

tions in suprachiasmatic nucleus neurons and glia in vitro. J Neurosci 1992; 12: 2648–2664.

231. Welsh DK, Reppert SM. Gap junctions couple astrocytes but not neurons in dissociated cultures of rat suprachiasmatic nucleus. Brain Res 1996; 706:30–36.

232. Shinohara K, Honma S, Katsuno Y, Abe H, Honma K. Two distinct oscillators in the rat suprachiasmatic nucleus in vitro. Proc Natl Acad Sci USA 1995; 92:7396–7400.

233. Inouye ST, Takahashi JS, Wollnik F, Turek FW. Inhibitor of protein synthesis phase shifts a circadian pacemaker in mammalian SCN. Am J Physiol 1988; 255:R1055–R1058.

234. Shinohara K, Oka T. Protein synthesis inhibitor phase shifts vasopressin rhythms in long-term suprachiasmatic cultures. NeuroReport 1994; 5:2201–2204.

235. Watanabe K, Katagai T, Ishida N, Yamaoka S. Anisomycin induces phase shifts of circadian pacemaker in primary cultures of rat suprachiasmatic nucleus. Brain Res 1995; 684:179–184.

236. Zhang Y, Takahashi JS, Turek FW. Critical period for cycloheximide blockade of light-induced phase advances of the circadian locomotor activity rhythm in golden hamsters. Brain Res 1996; 740:285–290.

237. Akasu T, Shoji S, Hasuo H. Inward rectifier and low-threshold calcium currents contribute to the spontaneous firing mechanism in neurons of the rat suprachiasmatic nucleus. Pflügers Arch 1993; 425:109–116.

238. Huang R-C. Sodium and calcium currents in acutely dissociated neurons from rat suprachiasmatic nucleus. J Neurophysiol 1993; 70:1692–1703.

239. Bouskila Y, Dudek FE. A rapidly activating type of outward rectifier K^+ current and A-current in rat suprachiasmatic nucleus neurones. J Physiol 1995; 488.2:339–350.

240. Walsh IB, van den Berg RJ, Rietveld WJ. Ionic currents in cultured rat suprachiasmatic neurons. Neuroscience 1995; 69:915–929.

241. Jiang Z-G, Yang Y, Liu Z-P, Allen CN. Membrane properties and synaptic inputs of suprachiasmatic nucleus neurons in rat brain slices. J Physiol 1997; 499.1:141–159.

242. Crosthwaite SK, Dunlap JC, Loros JJ. Neurospora wc-1 and wc-2: transcription, photoresponses, and the origins of circadian rhythmicity. Science 1997; 276:763–769.

243. Sauman I, Reppert SM. Circadian clock neurons in the silkmoth Antheraea pernyi: novel mechanisms of Period protein regulation. Neuron 1996; 17:889–900.

244. Schwartz WJ, Zimmerman P. Circadian timekeeping in BALB/c and C57BL/6 inbred mouse strains. J Neurosci 1990; 10:3685–3694.

245. Ishida N, Nishimatsu S-I, Matsui M, Mitsui Y, Nohno T, Shibata N, Noji S. Diurnal regulation of per repeat family in the suprachiasmatic nucleus of rat brain. Neurosci Biobehav Rev 1994; 18:571–577.

246. Rosewell KL, Siwicki KK, Wise PM. A period (per)-like protein exhibits daily rhythmicity in the suprachiasmatic nuclei of the rat. Brain Res 1994; 659:231–236.

247. Ralph MR, Menaker M. A mutation of the circadian system in golden hamsters. Science 1988; 241:1225–1227.

248. Gillette MU. The suprachiasmatic nuclei: circadian phase-shifts induced at the time of hypothalamic slice preparation are preserved in vitro. Brain Res 1986; 379:176–181.

249. Honma K, Honma S, Hiroshige T. Response curve, free-running period, and activity time in circadian locomotor rhythm of rats. Jpn J Physiol 1985; 35:643–658.

250. Mrosovsky N, Salmon PA, Menaker M, Ralph MR. Nonphotic phase shifting in hamster clock mutants. J Biol Rhythms 1992; 7:41–49.

11

Molecular and Genetic Aspects of Sleep

THOMAS S. KILDUFF and EMMANUEL MIGNOT

Stanford University
Stanford, California

I. Introduction

Over the past two decades, evidence has been gradually accumulating for genetic involvement in both normal sleep as well as a number of sleep disorders. In addition, determination of whether differential gene expression occurs in brain across arousal states has been the subject of increasing interest in the sleep field. The pace of this research has accelerated in the 1990s stimulated in part by the influence of the Human Genome Project, the general accessibility of molecular and genetic techniques, discoveries of specific mutations that affect the circadian system (1,2), and identification of specific genes that are induced in the supra-chiasmatic nucleus in association with the phase-shifting response (3–8). In the present chapter, we will review evidence for genetic influences on normal sleep in both humans and animal models, examine studies suggesting roles of genetic involvement in various sleep disorders, present results of recent studies that characterize gene expression in brain across arousal states, and conclude with a discussion of profitable directions for future research in this rapidly emerging field.

II. Genetic Factors in the Regulation of Normal Sleep

The ubiquitous occurrence of sleep and inactivity throughout the animal kingdom, the demonstration that sleep serves a vital function (9,10), and the conservation of slow-wave sleep (SWS) and rapid eye movement (REM) sleep in birds and mammals (11,12) indicate that phylogenetically old constitutional factors must be involved in generating sleep. On the other hand, phenotypic differences in sleep organization among species, strains, and individuals suggest the existence of polymorphic genetic factors. Taken together, these observations suggest that reverse and forward genetic techniques can be profitably applied to discover new fundamental knowledge on the physiology of sleep and its associated pathologies.

A. Twin Studies in Humans

A classic approach for estimation of the genetic component of a particular behavior is the comparison of the traits of identical versus fraternal twins. In the sleep field, research has primarily consisted of questionnaire studies comparing sleep habits (duration of sleep, schedules and quality of night sleep, frequency of napping) in monozygotic and dizygotic twin pairs (13–16). As might be expected, correlations for most of the variables analyzed are higher between monozygotic than dizygotic twins. These effects do not correlate strongly with depression or anxiety (17) and remain significant even when twins no longer live in the same environment (14). However, environmental factors also contribute significantly to the observed variance (14,15). Measures of the residual variance between monozygotic twins ($1-r_{mz}$) quantify the influence of environmental factors specific to each twin pair (18). The reported correlations barely reach 0.60, indicating that nearly half of the variance is due to environmental factors. Since twins are living in similar environments, this difference probably corresponds to short-term environmental variance.

Few authors have conducted polysomnographic sleep studies in monozygotic and dizygotic twins (19–22). The sample sizes in such studies tend to be small but generally confirm the results obtained with questionnaires. Linkowski et al. (21,22), studying 26 pairs during three consecutive nights, determined that a significant proportion of variance in stages 2, 4, and delta sleep is genetically determined in humans, but that nongenetic influences substantially determine variance in stage REM. Vogel (23) studying more particularly alpha-occipital rhythms during resting wake electroencephalogram (EEG), suggested dominant transmission for this trait, thus showing that genetic variations in the EEG are qualitative as well as quantitative. A linkage marker for low-voltage alpha EEG has now been identified on human chromosome 20q (24,25). More recently, in a study of 213 twin pairs (91 MZ and 22 DZ), the averaged heritability for the delta, theta, alpha, and beta frequencies was determined to be 76%, 89%, 89%, and 86%, respectively, indicating that brain-electrical activity is one of the most heritable characteristics in humans (26).

Most of these early studies did not take into account the fact that sleep is regulated by both circadian and homeostatic factors. Using the Horne-Ostberg questionnaire to examine morningness/eveningness in 238 twin pairs, Drennan et al. (16) found higher correlations in monozygotic pairs, suggesting the existence of genetic factors that influence the human circadian system. Linkowski et al. (27,28) addressed this issue in a laboratory study by measuring cortisol and prolactin hormonal levels in twins. The results suggested that genetic factors play a major role in the regulation of cortisol secretion but not of prolactin. Such studies could certainly be extended. To the best of our knowledge, there has yet to be a twin study in which SWS sleep/REM sleep homeostasis or circadian rhythm properties are measured under optimal experimental conditions.

B. Animal Studies

Variation Among Inbred Strains

Animal studies also support the concept of genetic influences on sleep. Major differences in SWS versus REM sleep amount and distribution can be observed within the same species, and such differences are resistant to prolonged manipulations such as forced immobilization or sleep deprivation (29–32). Significant variations in sleep/wake architecture and EEG profiles are also observed between inbred rodent strains (30–36). C57BL or C57BR strains of mice are characterized by long REM sleep episodes, short SWS episodes, and significant circadian variation under light:dark conditions (32,35). At the opposite end of the spectrum, BALB/c mice have very short REM sleep episodes and weak diurnal rhythms while DBA are intermediate for these characteristics (35). The period of the endogenous circadian period has been reported to be 50 min longer in C57BL/6J than in BALB/cByJ mice (37). Qualitative differences in EEG signals are also observed: CBA and BALB/c but not C57BR display high-amplitude spindles while REM-associated theta frequency varies significantly between strains (35). A recent study compared the distribution and amount of SWS and PS and the spectral composition of the EEG among five inbred mouse strains and supported the observation that in both PS and SWS the theta frequency peak is under genetic control (38). Intriguingly, sleep-deprivation-induced compensatory effects may also be under genetic control; in the first 3 hr of recovery sleep following 6 hr of sleep deprivation, the rebound in SWS delta power was highest in AKR and BALB, intermediate in DBA, and not present in the C57 strains (38).

The phenotypic differences described above appear to be genetically transmitted. Diallelic methods (33), simple segregation analysis in a backcross setting (35), and recombinant inbred strain studies (37,39,40) suggest that many genes are involved in the expression of each trait. The interactions observed are complex and not strictly additive, the hybrids of inbred strains occasionally presenting important deviations when compared to the average of parental strains (33).

Among rat strains, brown Norway (BN) rats have been shown to have more daily paradoxical sleep (PS) than Lewis (L) rats, while F_1 progeny had intermediate amounts, suggesting codominant or polygenic transmission (31). Since it has been known that 5 min exposures to lights-off every half-hour can trigger PS in outbred albino strains, PS triggering by dark pulse stimulation was examined in L and BN rats to determine whether PS induction has a separate genetic basis from PS amount (41). BN rats had more total daily PS than L rats but exhibited no dark pulse triggering of PS whereas L rats showed a fivefold increase in PS during dark pulse stimulation. The heritability of this trait was assessed in L × (L × BN)F_1 hybrid backcross (BC) animals. Albino BC rats increased PS % during 5-min dark pulses to three times the average PS % for the preceding 5 min of lights-on. In contrast, no significant PS triggering was observed in pigmented BC rats (42). Two other pigmented breeds of rat, dark Agouti and hooded Long-Evans, failed to show any evidence of REM sleep triggering (43). The absence of a connection between PS triggering and total daily amounts of PS suggests that the amount of PS and the induction of PS episodes may be under separate genetic control. Using anterograde tracing techniques, the distribution of retinal terminals in the ventrolateral subdivision of the hypothalamic suprachiasmatic nucleus was found to extend over a greater area in Lewis compared to brown Norway rats, which may be related (44). It has recently been found that lesion of the superior colliculus, but not the visual cortex, eliminates dark pulse triggering of PS in the albino Fischer F344 rat (45).

Pharmacogenetic Approaches

The pharmacogenetic approach is based on simple selection of animal strains relatively sensitive or resistant to pharmacological agents, for example, ethanol (46,47), benzodiazepines (48), barbiturates (49), and cholinergic compounds (49–51). These models can then be studied by pharmacological, physiological, and genetic methods. For example, rats that have been selected for their hypersensitivity to cholinergic compounds display an increase in paradoxical sleep (51), supporting the role of acetylcholine in REM sleep regulation.

LS/SS Mice

Among pharmacogenetic strains, the long-sleep (LS) and short-sleep (SS) mouse strains have been the most intensively studied (47,52). These mouse strains were created in the 1970s by selecting mice more or less sensitive to the sedative effects of ethanol, as measured by the duration of loss of the righting reflex ("sleep time") after ethanol administration. After 18 generations of selection, the resulting strains now present an average "sleep" time of 10 min (SS) or 2 hr (LS), respectively, after ingestion of a similar dose of ethanol.

LS and SS mice are of interest for several reasons. First, it is well established that there is pharmacological overlap between anesthetics, ethanol, and most benzodiazepine and barbiturate hypnotics. Partial or total cross-tolerance is observed for numerous pharmacological properties (53,54), thus suggesting that all these compounds act directly or indirectly through the GABAergic system. Studying the differential sensitivity of the LS and SS strains to various hypnotics or anesthetic agents thus allows determination of whether the genetic influence on sensitivity to these compounds overlaps with that of ethanol sensitivity (55–60). Such studies have concluded that there is some overlap for the hypnotic effect of the less liposoluble anesthetic compounds (for example, urethane and trifluorethanol) with ethanol whereas more liposoluble anesthetics, such as barbiturates, seem to produce similar effects in SS and LS strains (59), suggesting independent genetic control (49). An interaction between the effects of ethanol and cholinergic (50,58,61) and dopaminergic (47) transmission has also been suggested.

These mouse strains are also of interest for purely genetic studies. A detailed phenotypic comparison of the SS versus LS strains as well as other strains hypersensitive to ethanol suggests that the various pharmacological effects of ethanol (sedation, hypothermy, toxicity) are controlled by different genes (47,56,57). Traditional segregation studies and phenotypic analysis of SS × LS hybrids suggest that at least seven or eight genes are involved in the hypnotic effects of ethanol (62,63). Quantitative trait loci (QTL) analysis has been performed in recombinant inbred strains and multiple QTLs have been identified (52,64).

The relationships between these models and the genetic control of sleep remain uncertain. Indeed, as of today, LS and SS animals have not been studied for sleep and circadian rhythms using either polygraphic recordings or wheel-running activity. Moreover, the effects of benzodiazepines and alcohol on sleep seem to be indirect and vary depending upon previously accumulated sleep debt (65–67). A better analysis of the physiology of these models, specifically circadian rhythms and sleep during baseline condition and after deprivation, is needed.

Sleep in Transgenic and "Knockout" Strains

The creation of strains of animals in which additional copies of a specific gene have been inserted or deleted provides a powerful additional tool for behavioral studies (68). To date, there have been relatively few sleep studies on such strains. Mice homozygous for mutations disrupting the prion protein gene have been found to exhibit altered circadian activity rhythms and sleep patterns (69). Prion knockout ($PrP^{o/o}$) mice showed almost twice as many brief (<16 sec) awakenings as wild-type controls, an index of sleep fragmentation, and also a larger increase in slow-wave activity (SWA) after 6 hr of sleep deprivation. Subsequent work has shown that the increase in SWA in $PrP^{o/o}$ mice is not due to a change in sleep regulation per se, but is due to the fact that SWA reaches its maximal value after

NREM sleep onset much faster than in wild-type mice (70). These results suggest that the thalamic neurons of PrP$^{o/o}$ mice may be immediately hyperpolarized upon entering NREM sleep.

Among genes whose expression has been examined between sleep and wakefulness, the immediate early gene c-*fos* has been the most thoroughly studied (see Section IV). A preliminary report of sleep in c-*fos* knockout mice indicates an increase in wakefulness and decreased non-REM sleep without any change in REM sleep during the "lights-on" portion of the 24-hr period (71). It has yet to be determined whether this increased wakefulness during what is the normal major sleep period is due to higher metabolic demands due to the small stature of the null mutants relative to wild-type mice (72). No changes in sleep-wake percentages have been reported in mice lacking the gene for α-calcium-calmodulin kinase II (73), but further analyses revealed that the peak of the theta band was shifted to lower frequencies during REM sleep in heterozygotes relative to wild-type mice (74). However, this shift was well within the range observed in a comparison of five mice strains (38).

III. Genetic Aspects of Pathological Human Sleep

Numerous sleep pathologies, such as narcolepsy, fatal familial insomnia, sleep paralysis, hypnagogic hallucinations, sleep apnea, and restless leg syndrome, are well known to occur in certain families with a high frequency (15,75–81). These results support the existence of a group of genes whose function is related to sleep. The identification of pathological factors by genome screening in sleep disorders is therefore another possibly productive research path.

A. Molecular Genetics and Narcolepsy-Cataplexy

Narcolepsy-cataplexy is characterized by excessive daytime sleepiness and abnormal symptoms of dissociated REM sleep (cataplexy, sleep paralysis, and hypnagogic hallucinations). Narcolepsy with cataplexy affects 0.05–0.18% of the general population across various ethnic groups (82–85). Since its description in 1880 by Gélineau, familial cases have been reported by numerous authors (80,86–90), thus suggesting a genetic basis for narcolepsy. This pathology thus offers a unique opportunity to discover genes involved in the control of sleep.

More recent studies, however, suggest that narcolepsy is not a simple genetic disorder (91). The development of human narcolepsy involves environmental factors on a specific genetic background and only 25–31% of monozygotic twins reported in the literature are concordant for narcolepspy (91). One of the predisposing genetic factors is located in the major histocompatibility complex (MHC) DQ region. Ninety to one hundred percent of all narcoleptic patients with definite cataplexy share a specific human leukocyte antigen (HLA) allele, HLA DQB1*0602 (most often in combination with HLA DR2) versus 12–38% of the general population in various ethnic groups (92,93). The finding of an HLA association in narcolepsy, together with the fact that HLA DQB1*0602 is likely to be the actual HLA narcolepsy susceptibility gene, suggests that narcolepsy might be an autoimmune disorder. However, all attempts to demonstrate an immunopathology in narcolepsy to date have been negative (94–98) and the mode of HLA DQB1*0602 is still uncertain.

As indicated above, 12–38% of the general population carry HLA DQB1*0602 but only a small fraction has narcolepsy; DQB1*0602 is thus a weakly penetrant genetic factor, even if genetic association with the disorder is high. Other genetic factors, possibly more penetrant than HLA, are likely to be involved. One to two percent of the first-degree relatives of a patient with narcolepsy-cataplexy are affected by the disorder versus 0.02–0.18% in the general population in various ethnic groups, a 20–40-fold increase (80,90,91). Familial aggregation cannot be explained by the sharing of HLA haplotypes alone (91) and some families are non-HLA DQB1*0602 positive (80), thus suggesting the importance of non-HLA-susceptibility genes that could be positionally cloned using genome screening approaches in human multiplex families or isolated populations.

Studies using a canine model of narcolepsy also illustrate the importance of non-MHC genes. In this model, narcolepsy-cataplexy is transmitted as a single autosomal recessive trait with full penetrance, *canarc*-1 (99,100). This high-penetrance narcolepsy gene is unlinked to MHC class II but cosegregates with a DNA segment with high homology to the human immunoglobulin μ-switch sequence (101). This linkage marker is located very close to the actual narcolepsy gene (current LOD score 15.3 at 0% recombination) and gene isolation is ongoing both in canines and in the corresponding human syntenic region.

B. Dissociated REM Sleep Events

Sleep paralysis and hypnagogic hallucinations, two symptoms of dissociated REM sleep, occur frequently in the general population independently of narcolepsy (102). Sleep paralysis presents a high familial incidence and autosomal dominant transmission in some cases (76,87,103,104). Twin studies suggest a much higher concordance in monozygotic versus dizygotic twins for this symptom (105), which may be more frequent in the black population (104). In contrast to narcolepsy, there is no association with HLA DQB1*0602.

In REM sleep behavior disorder (RBD), automatic behavior arises during REM sleep and disturbs sleep continuity (106). RBD is frequently associated with other pathologies such as narcolepsy but may occur in isolation. The familiality of isolated RBD is not established but the disorder may be weakly associated with HLA DQ1 (107).

Cataplexy without sleepiness is exceptional (84) but some rare familial cases have been described with or without associated sleep paralysis (108,109). In the clearly isolated cases of cataplexy for which there is no other symptom from the tetrad, clinical presentation seems to differ quite significantly from narcolepsy-cataplexy and cataplexy occurred in the first months of life (109). HLA typing was not done for these families.

C. Fatal Familial Insomnia

Fatal familial insomnia is a rare neurological condition characterized by severe insomnia, neurovegetative symptoms, and intellectual deterioration and death (79,110–112). Insomnia is an early sign and sleep disruption is associated with a disappearance of stage II and SWS while brief episodes of REM sleep are usually maintained. Neuropathological lesions are mostly limited to a spongiform degeneration of the anterior ventral and mediodorsal thalamic nuclei and of the inferior olive (112). This pathology is typically associated with a mutation of the codon 178 in the prion protein gene but one recent report detected a codon 200 mutation (113). These same mutations are also found in some forms of dementia such as Creutzfeldt-Jakob, but a polymorphism on codon 129 seems to determine the phenotypic expression into FFI (111,113).

The prion protein is encoded by a gene located on human chromosome 20. The normal function of the protein is unknown but the gene is expressed in neurons. Mice homozygous for mutations disrupting the prion protein gene are behaviorally normal but may display sleep abnormalities (69). Prions are involved in a group of human and animal disorders with more or less anatomically confined spongious degeneration and neuronal atrophy (spongiform encephalopathies). A proteinase resistant form of the prion protein is probably involved in the pathology (114). These diseases can appear either in a familial context or in an infectious context, the prion protein (or an agent that cannot be distinguished from the proteic element) acting as the transmitting agent. The mechanism by which certain isoforms of the protein are infectious remains a widely discussed topic (115,116).

How a simple additional polymorphism on codon 129 alters the symptomatology from Creutzfeldt-Jakob to fatal familial insomnia is not understood but molecular studies are underway to evaluate the effect of these mutations on the metabolism of the protein (117). The differences in symptomatology are probably due to a differential anatomical localization of the lesions. In fatal familial insomnia, degeneration is mostly localized in the anterior ventral and mediodorsal thalamic nuclei while lesions are much more diffuse in Creutzfeldt-Jakob disease (115,117,118). The well-established role of the thalamus (albeit mostly of the intralaminar thalamus) and of its cortical projections in the generation of the cortical synchronization of SWS and sleep spindles (119) suggests that thalamic lesions may cause the insomnia in this disorder (118). As of today, however, no study has convincingly demonstrated that the destruction of these nuclei can produce a fatal insomnia in animal models. Bilateral lesions of these nuclei produce a persistent insomnia, which is not fatal (120). Other, more discrete anatomical lesions or a distinct pathophysiological mechanism could thus also play a role. Mouse carrying transgenes with the human prion allele specific for fatal familial insomnia have now been generated and are under study to answer these questions.

The implication of the thalamus in the pathophysiology of fatal familial insomnia suggests that this brain structure may be involved in the genesis of other, more frequent insomnias. Insomnia is a frequent symptom that affects at least 10% of the general population (121,122). Many insomnias appear to be constitutional (123) and genetic factors involving the thalamus and homeostatic abnormalities in the regulation of sleep could be involved in some cases. Other genetic factors, such as those regulating circadian rhythmicity at the level of suprachiasmatic nuclei, could be involved in other cases.

D. Restless Legs Syndrome and Periodic Limb Movements

Restless leg syndrome (RLS) is a frequent (5% of general population) (78,124) syndrome that worsens with age. RLS is almost always associated with periodic leg movements (PLMs) during sleep. RLS is best defined as painful sensations in the legs that force the patient to get up several times a night. PLMs are brief and repetitive muscular jerks of the lower limbs occurring mostly during stage II sleep. When these movements increase in strength and frequency, sleep is altered. RLS is highly familial, and in up to one-third of the reported cases the condition may be transmitted as an autosomal dominant trait (75,125–127). Unfortunately, no twin studies have been published to date and the proportion of familial cases seems to vary widely according to the geographical origin of the population studied. These differences may reflect founder effects, such as in Quebec, where one finds a high proportion of familial cases, or the influence of local environmental factors. Linkage studies using either microsatellite markers or candidate gene studies in multiplex families are ongoing to identify the gene(s) involved. Possible candidate genes are enzymes and receptors of the dopaminergic and enkephalinergic systems, two neurotransmitters involved in the pharmacological treatment of the syndrome. As of today, however, no result suggestive of linkage has been published.

E. Sleepwalking, Sleeptalking, and Night Terrors

These parasomnias generally occur during SWS (stages III and IV) (128). They are usually grouped together and considered to share a common or related pathophysiological mechanism (129), even if this notion is sometimes disputed (128). The prevalence of these symptoms is several percent among children and only rarely requires a medical consultation. Symptoms generally disappear in the adult (128,130).

The familial nature of these symptoms has been recognized by most authors (77,130,131) but the exact mode of transmission is uncertain. Twin studies have shown a high degree of concordance for sleepwalking and sleep terror (50% for monozygotes, 10–15% for dyzygotes) (105,132). The genetic predisposition of sleepwalking, sleeptalking, and, to a lesser degree, night terrors seems to overlap as the frequency of sleep terrors and enuresis might be more frequent in families with somnambulism (77,130,131). This suggests a related pathophysiological mechanism and a similar genetic control. To the best of our knowledge, there has not been any molecular study initiated on these pathologies.

F.   Obstructive Sleep Apnea Syndrome and Related Breathing Abnormalities During Sleep

Obstructive sleep apnea syndrome (OSAS) is complex syndrome in which the upper airway collapses repetitively during sleep, thus blocking breathing. Snoring is one of the cardinal symptoms. Repeated apneas prevent the patient from sleeping soundly, and the patient is frequently excessively sleepy the following day. Four to five percent of the general population suffer from OSAS (133) which, in the longer term, leads to high blood pressure and increased risk for cardiovascular accidents (134–136).

Twin studies are lacking in OSAS but two recent studies have shown higher concordance in monozygotic versus dizygotic twins for habitual snoring (105,137). Multiplex families of patients suffering from OSAS have also been reported in the literature (81,138–145) and one study found a substantial increase in HLA A2 and B39 in Japanese patients with OSAS (146). Familial aggregation is generally explained by the fact that most risk factors involved in the pathophysiology of sleep apnea are largely genetically determined. Such factors include obesity, alcoholism, facial soft tissue, and bone anatomy, all of which predispose to upper airway obstruction (81,143,145,147,148). In some cases, the genetic factor primarily involves abnormal ventilatory control by the central nervous system (81, 139). A possible genetic overlap between OSAS and sudden infant death (139, 140) and the high degree of concordance in chemoreceptor responses observed in monozygotic twins (149,150) suggests the importance of genetic factors in regulating the central control of ventilation in OSAS.

A multiplicity of genetic factors is likely to correspond to the multifactorial aspect of OSAS. A genetic linkage approach in OSAS would thus be facilitated by a careful phenotypic analysis, for example, studying sleepy or nonsleepy subjects, nonobese versus obese OSAS patients (147), or subjects with selected morphological features (151).

G.   Chromosomal and Genetic Abnormalities and Sleep Disturbances

The coincidental association of specific chromosomal breakpoints with specific pathologies can be helpful to localize susceptibility gene(s). In practice, however, karyotypes are rarely requested when a sleep disorder is the primary abnormality and few sleep studies have been performed in patients with chromosomal or genetic abnormalities. These disorders frequently produce behavioral and medical problems that have secondary effects on sleep, generally, disturbed nocturnal sleep, so it may be difficult to identify a disease-specific sleep phenotype (152). Despite these limitations, fragile X subjects have been reported to experience sleep disturbances and low melatonin levels (153,154) while subjects with Norrie disease (genetic alterations in a region encompassing the monoamine oxidase genes at Xp11.3) or Nieman Pick type C (18q11-q12) may experience cataplexy and sleep disturbances (155,156). OSAS is also frequently observed as a result of anatomical malformations, adenotonsillar enlargement, or morbid obesity (152,157,158). In a few instances, however, polygraphic studies suggest that central factors are also involved in addition to, or independently of, abnormal breathing during sleep. This may be the case for the Prader-Willi and Angelman syndromes [del(15q)] (158,159) or the Smith Magenis syndrome [del(17)(p11.2)] (158–161).

In spite of their relatively high population frequency, sleep characteristics of sex chromosomal aneuploidies have been only marginally studied, but XXY subjects may display increased 24 hr sleep time (162). These issues would be worth investigating more thoroughly because puberty is associated with established changes in sleep needs (163). Moreover, narcolepsy often starts in adolescence and, in two cases, narcolepsy started at the unusual age of 6 in a Turner syndrome patient (XO) (164) or in coincidence with a precocious puberty.

H. Other Sleep Pathologies

Insomnia, obstructive sleep apnea, narcolepsy, periodic movements and restless leg syndrome, parasomnias, and circadian disorders are the most frequent sleep pathologies. There are few or no studies on other forms of hypersomnias or parasomnias. One twin study suggests increased frequency of bruxism in monozygotic versus dizygotic twins (105) and bruxism has been reported in a multiplex context (165). Familial forms of essential hypersomnia (87), of hypersomnias associated with dystrophia myotonica (166) or sleep-responsive extrapyramidal dystonias (167,168), and of jactatio capitis nocturna (123) have also been reported. A possible association of idiopathic hypersomnias with HLA Cw2 and of hypersomnia in dystrophia myotonica with DR6 has also been found (166,169) but needs independent confirmation.

IV. Gene Expression and Arousal States

In recent years, determination of whether differential gene expression occurs in the brain across arousal states has been the subject of increasing interest in the sleep field. Such information has been stimulated for at least two reasons. First, having determined the brain regions and neurotransmitter systems likely to be involved in the regulation of sleep and wakefulness, neurobiologists have sought support at a molecular level for various hypotheses that have arisen at a systems level. A second major influence has been the expectation that the long-sought function of sleep might be explicable at a molecular level. Although molecular studies of sleep are in their infancy, at least four experimental approaches have been used to date to study the role of gene expression in association with arousal states: (1) gene expression in association with spontaneous variation in arousal state; (2) gene expression associated with sleep deprivation; (3) gene expression associated with recovery sleep after sleep deprivation; and (4) gene expression associated with drug-induced sleep.

A. Gene Expression and Spontaneous Variation in Arousal State

Grassi-Zucconi et al. (170) described a circadian rhythm of expression of the immediate early gene (IEG) c-*fos* in several brain regions using Northern analysis, with peak levels occurring in association with high levels of locomotor activity and lowest levels occurring during the lights-on (sleep) period. Pompeiano et al. (171) also used this approach and found, using in situ hybridization and immunohistochemistry, that c-*fos* and NGFI-A mRNAs and Fos and NGFI-A proteins were higher in several brain areas after periods of spontaneous wakefulness than after periods of spontaneous sleep. Furthermore, they described a strong induction of Fos-LI (Fos-like immunoreactivity) after a period of spontaneous wakefulness during the lights-on period, suggesting that arousal state influences can override any circadian variation in Fos-LI expression. Shiromani et al. (172) have suggested that the Fos protein is metabolized very rapidly during sleep, which may contribute to the higher levels observed during wakefulness, conclusions that are consistent with the observations of other investigators (173,174). On the other hand, increased Fos expression in the cortex during wakefulness (as compared to sleep) has been linked to the high levels of firing in the locus coeruleus during wake relative to sleep (174).

B. Gene Expression and Sleep Deprivation

Deprivation of sleep (SD) is a time-honored method in the sleep field to increase the homeostatic drive to sleep and this procedure has also been used in conjunction with studies of IEG expression. Pompeiano et al. (176) deprived rats of sleep for 24 hr and found that Fos-LI protein increased in some specific brain areas. O'Hara et al. (175) deprived rats of sleep beginning at light onset for 45 min, 3 hr, or 6 hr and found that c-*fos* mRNA was increased after both 45 min and 6 hr of

SD in every brain region examined. The induction of c-*fos* mRNA at 45 min was attributed to the initial sensory stimulation of deprivation whereas the induction at 6 hr was thought to more likely reflect an increase due to sleep homeostasis. Using a similar approach, Grassi-Zucconi et al. (170) found a greater increase in c-*fos* mRNA after 8 hr of SD than after 4 hr of SD. Cirelli et al. (177) deprived rats for 3 hr, 6 hr, 12 hr, and 24 hr beginning at light onset and measured c-*fos* mRNA by in situ hybridization and Fos-LI by immunohistochemistry. The increase in c-*fos* expression was not directly proportional to the time kept awake in this study; the largest increase in c-*fos* expression in many brain areas was found in the 3-hr-SD rats. Of particular interest in this study, however, was the increase in Fos-LI in the medial preoptic area in all comparisons made in this study. In contrast to these results, experiments utilizing long-term (10 day) total sleep deprivation did not reveal significant changes in the expression of the IEG Egr-1 (NGFI-A) mRNA nor immunoreactivity in most of the 25 brain regions examined, although "tendencies" for regionally specific increases in Egr-1-like immunoreactivity were reported in dorsal raphe, lateral habenula, superior colloculus, and ventral periaqueductal gray (178). Taken together, the results of the above studies suggest that there is not a simple relationship between IEG expression in brain and the amount of prior wakefulness, which may be due, at least in part, to the transient nature of IEG induction and the refractory period that follows IEG induction (179).

Rather than describing changes in expression of a known class of genes, Rhyner et al. (180) used a subtractive hybridization approach to isolate mRNA transcripts from forebrain whose expression was altered after 24 hr of SD. In this study, sleep deprivation was induced by 24 hr of forced locomotion. Four clones were isolated whose expression was lower after 24 hr of SD and six clones whose expression was greater. Two transcripts from the former class, #140 and #464, were chosen for more extensive characterization. Analysis of the primary structure of these transcripts indicated that neither of the mRNAs had been previously described. A subsequent study revealed that the deduced amino acid sequence of #140 was identical to the 17-kDa rat protein neurogranin, a phosphoprotein that contains a consensus sequence for protein kinase C phosphorylation (181). The level of neurogranin protein was subsequently found to decrease by 37% in cortex after 24 hr of SD. Subtractive hybridization has also led to the identification of cortistatin, a molecule proposed to have sleep-inducing properties (182). The results of these studies indicate that changes in gene expression may occur in association with arousal state changes in genes other than the IEGs.

Deprivation of REM sleep has been reported to increase the expression of tyrosine hydroxylase mRNA in the locus coeruleus (183), galanin mRNA in the preoptic area and periventricular nucleus (184), and to have differential effects on the expression of m1, m2, and m3 muscarinic receptor mRNAs (185). Taken together, the results of these studies indicate that changes in gene expression may occur in association with arousal state changes in genes other than the IEGs.

C. Gene Expression and Recovery Sleep

Although the homeostatic drive to sleep is thought to accrue during wakefulness, its physiological manifestation, process S, is measured only during the sleep period that ensues after prolonged wakefulness. Accordingly, determination of gene expression during recovery sleep would seem to be an attractive paradigm since the intensity of slow-wave activity can be directly measured during recovery sleep. Surprisingly, this approach has yet to be extensively used in molecular studies of sleep. Perhaps the absence of this approach has been due to the focus of the field to date on IEGs whose expression has been shown to be low in virtually all brain areas during spontaneous sleep periods. Indeed, studies that have examined recovery sleep have suggested that short periods of sleep can reverse the accumulation of Fos-LI protein, which may accrue during wakefulness (172,173). However, a cell group in the ventrolateral preoptic area (VLPO) has recently been identified in which Fos expression increased during slow-wave sleep (186). Furthermore, the VLPO was shown to project to the histaminergic group in the tuberomammillary region of the hypothalamus (186). Because of its direct correlation with a measurable physiological variable, EEG delta power, this experimental paradigm is likely to be a very strong procedure for future molecular biological studies.

D. Gene Expression and Drug-Induced Sleep

Another paradigm that has been exploited in molecular studies of sleep is microinjection of D-carbachol into the medial pons to enhance the occurrence of a state very similar to REM sleep (REMc) in conjunction with mapping of Fos-LI expression in the brainstem (187,188). Compared with vehicle control animals, carbachol-treated animals showed a significantly higher number of Fos-LI cells in pontine regions implicated in REM sleep generation, with longer REMc bouts associated with staining of more Fos-LI cells. Regions with REMc-associated Fos-LI increases included: the lateral dorsal tegmental (LDT) and pedunculopontine tegmental (PPT) nuclei, where some Fos-LI cells were immunohistochemically identified as cholinergic; the locus coeruleus, where some of the Fos-LI cells were identified to be catecholaminergic; the dorsal raphe; and the pontine reticular formation. These findings suggest IEG activation is associated with REM sleep. As in the studies of gene expression in association with spontaneous variation in arousal state described above, however, it is suspected that these changes in gene expression are a consequence rather than a cause of the arousal state change.

The literature described above indicate that changes in gene expression occur in conjunction with changes in arousal states although molecular approaches to sleep and arousal states to date have almost exclusively focused on the IEGs. Since an independent physiological variable, the intensity of slow-wave activity, can be directly measured during recovery sleep, this may be a fruitful experimental paradigm in which to examine gene expression.

V. Summary and Perspective

The complexity of sleep as a physiological phenomenon is matched by a vast number of sleep-related pathologies, Most of these pathologies are multifactorial and to a large extent genetically determined. Recent progress in molecular genetics now enables researchers to undertake a purely genetic approach to understand the pathophysiology of these disorders. Such approaches are likely to lead to the identification of genes involved in etiologically homogeneous sleep disorders such as narcolepsy. Genome screening studies in more frequent and complex sleep disorders, such as OSAS or RLS will require the inclusion of a large number of multiplex families but are now feasible. These disorders may also benefit from studies in isolated populations or even of association studies using very large numbers of single-case families; this last design will have the benefit of being readily usable for secondary candidate gene studies. Such approaches are likely to become a more viable research path as more and more genes are cloned and positioned on the human map and possible candidate genes isolated in mouse models.

Newly evolving methods for the determination of gene expression are likely to have a large impact on the study of sleep. The strategy currently employed in direct screening of candidate genes is a time-consuming and laborious process, since at least 30,000 genes are expressed in brain, and is limited in its approach to known genes. Other methods, such as subtractive hybridization (189), have been successfully used in sleep-related paradigms (180,182), but can be technically difficult and primarily suitable for applications in which there is a 10-fold or greater change in mRNA concentration between conditions (190). Newer "chip-based" approaches in which cDNAs are printed onto glass slides (191) or high-density oligonucleotide arrays are synthesized directly onto silicon chips (192) are likely to be profitably employed in sleep studies in the near future. Such methods have already been shown to enable assessment of the expression of over 1000 genes in parallel (193,194). In addition to their use in mRNA expression studies, chip-based technologies are likely to find wider use for genomic analyses (195–197) and thus are likely to contribute to both molecular and genetic studies of sleep and sleep disorders.

Acknowledgments

Research described herein has been supported in part by the National Institutes of Health (AG11084) and the Army Research Office (DAAH04-95-1-0616).

References

1. Ralph MR, Menaker M. A mutation of the circadian system in golden hamsters. Science 1988; 241:1225–1227.
2. Vitaterna MH, King DP, Chang AM, et al. Mutagenesis and mapping of a mouse gene, *clock*, essential for circadian behavior. Science 1994; 264:719–725.
3. Rea MA. Light increases Fos-related protein immunoreactivity in the rat suprachiasmatic nuclei. Brain Res Bull 1989; 23:577–581.
4. Aronin N, Sagar SM, Sharp FR, Schwartz WJ. Light regulates expression of a Fos-related protein in rat suprachiasmatic nuclei. 1990; 87:5959–5962.
5. Rusak B, Robertson HA, Wisden W, Hunt SP. Light pulses that shift rhythms induce gene expression in the suprachiasmatic nucleus. Science 1990; 248:1237–1240.
6. Kornhauser JM, Nelson DE, Mayo KE, Takahashi JS. Photic and circadian regulation of c-*fos* gene expression in the hamster suprachiasmatic nucleus. Neuron 1990; 5:127–134.
7. Kornhauser JM, Nelson DE, Mayo KE, Takahashi JS. Regulation of *jun*-B messenger RNA and AP-1 activity by light and a circadian clock. Science 1992; 255:1581–1583.
8. Sutin EL, Kilduff TS. Circadian and light-induced expression of immediate early gene mRNAs in the rat suprachiasmatic nucleus. Brain Res Mol Brain Res 1992; 15:281–290.
9. Rechtschaffen A, Gilliland MA, Bergmann BM, Winter JB. Physiological correlates of prolonged sleep deprivation in rats. Science 1983; 221:182–184.
10. Kushida CA, Bergmann BM, Rechtschaffen A. Sleep deprivation in the rat. IV. Paradoxical sleep deprivation. Sleep 1989; 12:22–30.
11. Zepelin H. Mammalian sleep. In: Kryger M, Roth T, Dement WC, eds. Principles and Practice of Sleep Medicine. Philadelphia: WB Saunders, 1994:69–81.
12. Siegel JM. Phylogeny and the function of REM sleep. Behav Brain Res 1995; 69: 29–34.
13. Gedda L, Brenci G. Twins living apart test: progress report. Acta Genet Med Gemellol 1983; 32:17–22.
14. Partinen M, Kaprio J, Koskenvuo M, Putkonen P, Langinvainio H. Genetic and environmental determination of human sleep. Sleep 1983; 6:179–185.
15. Heath AC, Kendler KS, Eaves LJ, Martin NG. Evidence for genetic influences on sleep disturbance and sleep pattern in twins. Sleep 1990; 13:318–335.
16. Drennan MD, Selby J, Kripke DF, Kelsoe J, Gillin JC. Morningness/eveningness is heritable. Soc Neurosci Abstr 1992; 18:196.
17. Kendler KS, Heath AC, Martin NG, Eaves LJ. Symptoms of anxiety and symptoms of depression. Same genes, different environments? Arch Gen Psychiatry 1987; 44: 451–457.
18. Hrubec Z, Robinette CD. The study of human twins in medical research. N Engl J Med 1984; 310:435–441.
19. Webb WB, Campbell SS. Relationships in sleep characteristics of identical and fraternal twins. Arch Gen Psychiatry 1983; 40:1093–1095.
20. Hori A. Sleep characteristics in twins. Jpn J Psychiatry Neurol 1986; 40:35–46.
21. Linkowski P, Kerkhofs M, Hauspie R, Mendlewicz J. EEG sleep patterns in man: a twin study. Electroenceph Clin Neurophysiol 1989; 73:279–284.

22. Linkowski P, Kerkhofs M, Hauspie R, Susanne C, Mendlewicz J. Genetic determinants of EEG sleep: a study in twins living apart. Electroenceph Clin Neurophysiol 1991; 79:114–118.

23. Vogel F. Brain physiology: genetics of the EEG. In: Human Genetics. New York: Springer-Verlag, 1986:590–593.

24. Anokhin A, Steinlein O, Fischer C, et al. A genetic study of the human low-voltage electroencephalogram. Hum Genet 1992; 90:99–112.

25. Steinlein O, Anokhin A, Yping M, Schalt E, Vogel F. Localization of a gene for the human low-voltage EEG on 20q and genetic heterogeneity. Genomics 1992; 12:69–73.

26. van Beijsterveldt CE, Molenaar PC, de Geus EJ, Boomsma DI. Heritability of human brain functioning as assessed by elecroencephalography. Am J Hum Genet 1996; 58:562–573.

27. Linkowski P, Kerkhofs M, Van Cauter E. Sleep and biological rhythms in man: a twin study. Clin Neuropharmacol 1992; 15:42A–43A.

28. Linkowski P, Van Onderbergen A, Kerkhofs M, Bosson D, Mendlewicz J, Van Cauter E. Twin study of the 24-h cortisol profile: evidence for genetic control of the human circadian clock. Am J Physiol 1993; 264:E173–181.

29. Webb WB, Friedman J. Attempts to modify the sleep patterns of the rat. Physiol Behav 1971; 6:459–460.

30. Kitahama K, Valatx JL. Instrumental and pharmacological paradoxical sleep deprivation in mice: strain differences. Neuropharmacology 1980; 19:529–535.

31. Rosenberg RS, Bergmann BM, Son HJ, Arnason BG, Rechtschaffen A. Strain differences in the sleep of rats. Sleep 1987; 10:537–541.

32. Valatx JL, Cespuglio R, Paut L, Bailey DW. Genetic study of paradoxical sleep in mice. Connection with coloration genes. Waking Sleeping 1980; 4:175–183.

33. Friedmann JK. A diallel analysis of the genetic underpinnings of mouse sleep. Physiol Behav 1974; 12:169–175.

34. Valatx JL, Bugat R, Jouvet M. Genetic studies of sleep in mice. Nature 1972; 238:226–227.

35. Valatx JL, Bugat R. Genetic factors as determinants of the waking-sleep cycle in the mouse. Brain Res 1974; 69:315–330.

36. Van Twyver H, Webb WB, Dube M, Zackheim M. Effects of environmental and strain differences on EEG and behavioral measurement of sleep. Behav Biol 1973; 9:105–110.

37. Schwartz WJ, Zimmerman P. Circadian timekeeping in BALB/c and C57BL/6 inbred mouse strains. J Neurosci 1990; 10:3685–3694.

38. Franken P, Malafosse A, Tafti M. Genetics of sleep EEG in mice: strain distribution pattern. Sleep Res 1997; 26:174.

39. Hofstetter JR, Mayeda AR, Possidente B, Nurnberger J Jr. Quantitative trait loci (QTL) for circadian rhythms of locomotor activity in mice. Behav Genet 1995; 25: 545–556.

40. Mayeda AR, Hofstetter JR, Belknap JK, Nurnberger J, Jr. Hypothetical quantitative trait loci (QTL) for circadian period of locomotor activity in CXB recombinant inbred strains of mice. Behav Genet 1996; 26:505–511.

41. Benca RM, Bergmann BM, Leung C, Nummy D, Rechtschaffen A. Rat strain differences in response to dark pulse triggering of paradoxical sleep. Physiol Behav 1991; 49:83–87.

42. Leung C, Bergmann BM, Rechtschaffen A. Benca RM. Heritability of dark pulse triggering of paradoxical sleep in rats. Physiol Behav 1992; 52:127–131.

43. Benca RM, Obermeyer WH, Bergmann BM, Lendvai N, Gilliland MA. Failure to induce rapid eye movement sleep by dark pulses in pigmented inbred rat strains. Physiol Behav 1993; 54:1211–1214.

44. Steininger TL, Rye DB, Gilliland MA, Wainer BH, Benca RM. Differences in the retinohypothalamic tract in albino Lewis versus brown Norway rat strains. Neuroscience 1993; 54:11–14.

45. Miller AM, Obermeyer WH, Benca RM. The superior colliculus is involved in paradoxical sleep induction by lights-off stimulation in albino rats. Sleep Res 1997; 26.

46. Morzorati S, Lamishaw B, Lumeng L, Li TK, Bemis K, Clemens J. Effect of low dose ethanol on the EEG of alcohol-preferring and -nonpreferring rats. Brain Res Bull 1988; 21:101–104.

47. Phillips TJ, Feller DJ, Crabbe JC. Selected mouse lines, alcohol and behavior. Experientia 1989; 45:850–827.

48. Korpi ER, Kleingoor C, Kettenmann H, Seeburg PH. Benzodiazepine-induced motor impairment linked to point mutation in cerebellar GABAA receptor. Nature 1993; 361:356–359.

49. Stino FK. Divergent selection for pentobarbital-induced sleeping times in mice. Pharmacology 1992; 44:257–259.

50. Overstreet DH, Rezvani AH, Janowsky DS. Increased hypothermic responses to ethanol in rats selectively bred for cholinergic supersensitivity. Alcohol Alcohol 1990; 25:59–65.

51. Shiromani PJ, Velazquez-Moctezuma J, Overstreet D, Shalauta M, Lucero S, Floyd C. Effects of sleep deprivation on sleepiness and increased REM sleep in rats selectively bred for cholinergic hyperactivity. Sleep 1991; 14:116–120.

52. Markel PD, Fulker DW, Bennett B, et al. Quantitative trait loci for ethanol sensitivity in the LS × SS recombinant inbred strains: interval mapping. Behav Genet 1996; 26:447–458.

53. Khanna JM, Kalant H, Shah G, Chau A. Tolerance to ethanol and cross-tolerance to pentobarbital and barbital in four rat strains. Pharmacol Biochem Behav 1991; 39: 705–709.

54. Le AD, Khanna JM, Kalant H. Effects of chronic treatment with ethanol on the development of cross-tolerance to other alcohols and pentobarbital. J Pharmacol Exp Ther 1992; 263:480 –485.

55. Marley RJ, Freund RK, Wehner JM. Differential response to flurazepam in long-sleep and short-sleep mice. Pharmacol Biochem Behav 1988; 31:453–458.

56. Erwin VG, Jones BC, Radcliffe R. Further characterization of LS×SS recombinant inbred strains of mice: activating and hypothermic effects of ethanol [published erratum appears in Alcohol Clin Exp Res 1990 Aug;14(4):573]. Alcohol Clin Exp Res 1990; 14:200–204.

57. Phillips TJ, Dudek BC. Locomotor activity responses to ethanol in selectively bred long- and short-sleep mice, two inbred mouse strains, and their F_1 hybrids. Alcohol Clin Exp Res 1991; 15:255–261.

58. de Fiebre CM, Collins AC. Classical genetic analyses of responses to nicotine and ethanol in crosses derived from long- and short-sleep mice. J Pharmacol Exp Ther 1992; 261:173–180.

59. de Fiebre CM, Marley RJ, Miner LL, de Fiebre NE, Wehner JM, Collins AC. Classical genetic analyses of responses to sedative-hypnotic drugs in crosses derived from long-sleep and short-sleep mice. Alcohol Clin Exp Res 1992; 16:511–521.

60. Wehner JM, Pounder JI, Parham C, Collins AC. A recombinant inbred strain analysis of sleep-time responses to several sedative-hypnotics. Alcohol Clin Exp Res 1992; 16:522–528.

61. Erwin VG, Korte A, Jones BC. Central muscarinic cholinergic influences on ethanol sensitivity in long-sleep and short-sleep mice. J Pharmacol Exp Ther 1988; 247: 857–862.

62. Dudek BC, Abbott ME. A biometrical genetic analysis of ethanol response in selectively bred long-sleep and short-sleep mice. Behav Genet 1984; 14:1–19.

63. DeFries JC, Wilson JR, Erwin VG, Petersen DR. LS × SS recombinant inbred strains of mice: initial characterization. Alcohol Clin Exp Res 1989; 13:196–200.

64. Crabbe JC, Belknap JK, Buck KJ. Genetic animal models of alcohol and drug abuse. Science 1994; 264:1715–1723.

65. Roehrs T, Zwyghuizen-Doorenbos A, Timms V, Zorick F, Roth T. Sleep extension, enhanced alertness and the sedating effects of ethanol. Pharmacol Biochem Behav 1989; 34:321–324.

66. Zwyghuizen-Doorenbos A, Roehrs T, Timms V, Roth, T. Individual differences in the sedating effects of ethanol. Alcohol Clin Exp Res 1990; 14:400–404.

67. Edgar DM, Seidel WF, Martin CE, Sayeski PP, Dement WC. Triazolam fails to induce sleep in suprachiasmatic nucleus-lesioned rats. Neurosci Lett 1991; 125: 125–128.

68. Takahashi JS, Pinto LH, Vitaterna MH. Forward and reverse genetic approaches to behavior in the mouse. Science 1994; 264:1724–1733.

69. Tobler I, Gaus SE, Deboer T, et al. Altered circadian activity rhythms and sleep in mice devoid of prion protein. Nature 1996; 380:639–642.

70. Tobler I, Deboer T. Activity in NREMS in prion protein knockout mice. Sleep Res 1997; 26:55.

71. Shiromani PJ, Greco MA, Thakkar M, McCarley RW. c-fos knockout mice have reduced non-REM sleep. Sleep Res 1997; 26:42.

72. Johnson RS, Spiegelman BM, Papaioannou V. Pleiotropic effects of a null mutation in the c-*fos* proto-oncogene. Cell 1992; 71:577–586.

73. Rainnie DG, Shiromani PJ, Thakkar M, Hearn EF, Greene RW, McCarley RW. Changes in sleep and EEG power spectra in CaM kinase II knockout mice. Sleep Res 1995; 24:38.

74. Thakkar M, Rainnie DG, Hearn EF, Greene RW, McCarley RW, Shiromani PJ. Abnormal theta activity during REM sleep in α-calcium-calmodulin kinase II knockout mice. Sleep Res 1997; 26:53.

75. Bornstein B. Restless leg syndrome. Psychiatr Neurol 1961; 141:165–201.

76. Roth B, Bruhova S, Berkova L. Familial sleep paralysis. Arch Suisses Neurol Neurochir Psychiatr 1968; 102:321–330.

77. Kales A, Soldatos CR, Bixler EO, et al. Hereditary factors in sleepwalking and night terrors. Br J Psychiatry 1980; 137:111–118.

78. Montplaisir J, Godbout R, Boghen D, DeChamplain J, Young SN, Lapierre G. Familial restless legs with periodic movements in sleep: electrophysiologic, biochemical, and pharmacologic study. Neurology 1985; 35:130–134.

79. Lugaresi E, Medori R, Montagna P, et al. Fatal familial insomnia and dysautonomia with selective degeneration of thalamic nuclei. N Engl J Med 1986; 315:997–1003.

80. Guilleminault C, Mignot E, Grumet FC. Familial patterns of narcolepsy. Lancet 1989; 2:1376–1379.

81. el Bayadi S, Millman RP, Tishler PV, et al. A family study of sleep apnea. Anatomic and physiologic interactions. Chest 1990; 98:554–559.

82. Dement WC, Carskadon M, Ley R. The prevalence of narcolepsy. II. Sleep Res 1973; 2:147.

83. Honda Y. Consensus of narcolepsy, cataplexy and sleep life among teenagers in Fujisawa city. Sleep Res 1979; 8:191.

84. Aldrich MS. Narcolepsy. Neurology 1992; 42:34–43.

85. Solomon P. Narcolepsy in Negroes. Dis Nerv Syst 1945; 6:179–183.

86. Daly D, Yoss R. A family with narcolepsy. Proc Staff Meet Mayo Clin 1959; 34: 313–320.

87. Nevsimalova-Bruhova C. On the problem of heredity in hypersomnia, narcolepsy and related disturbance. Acta Univ Carol Med 1973; 18:109–160.

88. Kessler S, Guilleminault C, Dement W. A family study of 50 REM narcoleptics. Acta Neurol Scand 1974; 50:503–512.

89. Singh SM, George CF, Kryger MH, Jung JH. Genetic heterogeneity in narcolepsy. Lancet 1990; 335:726–727.

90. Billiard M, Pasquie-Magnetto V, Heckman M, et al. Family studies in narcolepsy. Sleep 1994; 17:S54–59.

91. Mignot E. Genetic and familial aspects of narcolepsy. Neurology 1998; 50:S16–22.

92. Honda Y. Discrimination of narcolepsy by using genetic markers and the HLA. Sleep Res 1983; 12:254.

93. Matsuki K, Grumet FC, Lin X, et al. DQ (rather than DR) gene marks susceptibility to narcolepsy. Lancet 1992; 339:1052.

94. Rubin RL, Hajdukovich RM, Mitler MM. HLA-DR2 association with excessive somnolence in narcolepsy does not generalize to sleep apnea and is not accompanied by systemic autoimmune abnormalities. Clin Immunol Immunopathol 1988; 49:149–158.

95. Fredrikson S, Carlander B, Billiard M, Link H. CSF immune variables in patients with narcolepsy. Acta Neurol Scand 1990; 81:253–254.

96. Carlander B, Eliaou JF, Billiard M. Autoimmune hypothesis in narcolepsy. Neurophysiol Clin 1993; 23:15–22.

97. Mignot E, Tafti M, Dement WC, Grumet FC. Narcolepsy and immunity. Adv Neuroimmunol 1995; 5:23–37.

98. Tafti M, Nishino S, Aldrich MS, Liao W, Dement WC, Mignot E. Major histocompatibility class II molecules in the CNS: increased microglial expression at the onset of narcolepsy in canine model. J Neurosci 1996; 16:4588–4595.

99. Baker TL, Dement WC. Canine narcolepsy-cataplexy syndrome: evidence for an inherited monoaminergic-cholinergic imbalance. In: McGinty DJ, Drucker-Colin R, Morrison A, eds. Brain Mechanisms of Sleep. New York: Raven Press, 1985:199–233.

100. Mignot E, Nishino S, Sharp LH, et al. Heterozygosity at the *canarc*-1 locus can confer susceptibility for narcolepsy: induction of cataplexy in heterozygous asymptomatic dogs after administration of a combination of drugs acting on monoaminergic and cholinergic systems. J Neurosci 1993; 13:1057–1064.

101. Mignot E, Wang C, Rattazzi C, et al. Genetic linkage of autosomal recessive canine narcolepsy with a mu immunoglobulin heavy-chain switch-like segment. Proc Natl Acad Sci USA 1991; 88:3475–3478.
102. Ohayon MM, Priest RG, Caulet M, Guilleminault C. Hypnagogic and hypnopompic hallucinations: pathological phenomena? Br J Psychiatry 1996; 169:459–467.
103. Goode G. Sleep paralysis. Arch Neurol 1962; 6:228–234.
104. Bell CC, Dixie-Bell DD, Thompson B. Further studies on the prevalence of isolated sleep paralysis in black subjects. J Natl Med Assoc 1986; 78:649–659.
105. Hori A, Hirose G. Twin studies on parasomnias. Sleep Res 1995; 24A:324.
106. Mahowald M, Schenk CH. REM sleep behavior disorder. In: Kryger M, Roth T, Dement WC, eds. Principles and Practice of Sleep Medicine. Philadelphia: WB Saunders, 1994:574–588.
107. Schenck CH, Ullevig CM, Noreen H, et al. Preponderance of human leukocyte antigen (HLA) DQw1 (DQB1*05 and DQb1*06) haplotypes in human non narcoleptic REM sleep behavior disorder: a controlled study in Caucasian males. Sleep Res 1996; 25A:362.
108. Gelardi JM, Brown JW. Hereditary cataplexy. J Neurol Neurosurg Psychiatry 1967; 30:455–457.
109. Hartse KM, Zorick FJ, Sicklesteel JM, Roth T. Isolated cataplexy: a familial study. Henry Ford Hosp Med J 1988; 36:24–27.
110. Julien J, Vital C, Deleplanque B, Lagueny A, Ferrer X. Subacute familial thalamic atrophy. Memory disorders and complete insomnia Rev. Neurol 1990; 146:173–178.
111. Goldfarb LG, Petersen RB, Tabaton M, et al. Fatal familial insomnia and familial Creutzfeldt-Jakob disease: disease phenotype determined by a DNA polymorphism. Science 1992; 258:805–808.
112. Manetto V, Medori R, Cortelli P, et al. Fatal familial insomnia: clinical and pathologic study of five new cases. Neurology 1992; 42:312–319.
113. Chapman J, Arlazoroff A, Goldfarb LG, et al. Fatal insomnia in a case of familial Creutzfeldt-Jakob disease with the codon 200(Lys) mutation. Neurology 1996; 46:758–761.
114. Prusiner SB. Molecular biology of prion diseases. Science 1991; 252:1515–1522.
115. Weissmann C. A 'unified theory' of prion propagation. Nature 1991; 352:679–683.
116. Mestel R. Putting prions to the test. Science 1996; 273:184–189.
117. Petersen RB, Parchi P, Richardson SL, Urig CB, Gambetti P. Effect of the D178N mutation and the codon 129 polymorphism on the metabolism of the prion protein. J Biol Chem 1996; 271:12661–12668.
118. Lugaresi E. The thalamus and insomnia. Neurology 1992; 42:28–33.
119. Steriade M. Basic mechanisms of sleep generation. Neurology 1992; 42:9–17; discussion 18.
120. Marini G, Imeri L, Mancia M. Changes in sleep--waking cycle induced by lesions of medialis dorsalis thalamic nuclei in the cat. Neurosci Lett 1988; 85:223–227.
121. Health NIoM. Consensus conference. Drugs and insomnia. The use of medications to promote sleep. Jama 1984; 251:2410–2414.
122. Angst J, Vollrath M, Koch R, Dobler-Mikola A. The Zurich Study. VII. Insomnia: symptoms, classification and prevalence. Eur Arch Psychiatry Neurol Sci 1989; 238:285–293.

123. Thorpy MJ, Glovinsky PB. Headbanging (jacatatio capitis nocturna). In: Kryger M, Roth T, Dement WC, eds. Principles and Practice of Sleep Medicine. Philadelphia: WB Saunders, 1989:648–654.

124. Ekbom K. Restless legs syndrome. Neurology 1960; 10:863–873.

125. Jacobsen JH, Rosenberg RS, Huttenlocher PR, Spire JP. Familial nocturnal cramping. Sleep 1986; 9:54–60.

126. Walters AS, Picchietti D, Hening W, Lazzarini A. Variable expressivity in familial restless legs syndrome. Arch Neurol 1990; 47:1219–1220.

127. Walters AS, Picchietti DL, Ehrenberg BL, Wagner ML. Restless legs syndrome in childhood and adolescence. Pediatr Neurol 1994; 11:241–245.

128. Keefauver S, Guilleminault C. Sleep terrors and sleep walking. In: Kryger M, Roth T, Dement WC, eds. Principles and Practice of Sleep Medicine. Philadelphia: WB Saunders, 1994:567–573.

129. Broughton RJ. Sleep disorders: disorders of arousal? Enuresis, somnambulism, and nightmares occur in confusional states of arousal, not in "dreaming sleep." Science 1968; 159:1070–1078.

130. Abe K, Amatomi M, Oda N. Sleepwalking and recurrent sleeptalking in children of childhood sleepwalkers. Am J Psychiatry 1984; 141:800–801.

131. Debray P, Huon H. 3 cases of familial somnambulism. Ann Med Interne 1973; 124:27–29.

132. Bakwin H. Sleep-walking in twins. Lancet 1970; 2:446–447.

133. Young T, Palta M, Dempsey J, Skatrud J, Weber S, Badr S. The occurrence of sleep-disordered breathing among middle-aged adults. N Engl J Med 1993; 328:1230–1235.

134. Guilleminault C, Eldridge FL, Simmon FB, Dement WC. Sleep apnea syndrome. Can it induce hemodynamic changes? West J Med 1975; 123:7–16.

135. Koskenvuo M, Kaprio J, Partinen M, Langinvainio H, Sarna S, Heikkila K. Snoring as a risk factor for hypertension and angina pectoris. Lancet 1985; 1:893–896.

136. Hall M, Bradley T. Cardiovascular disease and sleep apnea. Curr Opin Pulm Med 1995; 1:512–518.

137. Ferini-Strambi L, Calori G, Oldani A, et al. Snoring in twins. Respir Med 1995; 89:337–340.

138. Strohl KP, Saunders NA, Feldman NT, Hallett M. Obstructive sleep apnea in family members. N Engl J Med 1978; 299:959–973.

139. Adickes ED, Buehler BA, Sanger WG. Familial lethal sleep apnea. Hum Genet 1986; 73:39–43.

140. Oren J, Kelly DH, Shannon DC. Familial occurrence of sudden infant death syndrome and apnea of infancy. Pediatrics 1987; 80:355–358.

141. Manon-Espaillat R, Gothe B, Adams N, Newman C, Ruff R. Familial "sleep apnea plus" syndrome: report of a family. Neurology 1988; 38:190–193.

142. Wittig RM, Zorick FJ, Roehrs TA, Sicklesteel JM, Roth T. Familial childhood sleep apnea. Henry Ford Hosp Med J 1988; 36:13–15.

143. Mathur R, Douglas NJ. Family studies in patients with the sleep apnea-hypopnea syndrome. Ann Intern Med 1995; 122:174–178.

144. Pillar G, Lavie P. Assessment of the role of inheritance in sleep apnea syndrome. Am J Respir Crit Care Med 1995; 151:688–691.

145. Redline S, Tishler PV, Tosteson TD, et al. The familial aggregation of obstructive sleep apnea. Am J Respir Crit Care Med 1995; 151:682–687.
146. Yoshizawa T, Akashiba T, Kurashina K, Otsuka K, Horie T. Genetics and obstructive sleep apnea syndrome: a study of human leukocyte antigen (HLA) typing. Intern Med 1993; 32:94–97.
147. Guilleminault C, Partinen M, Hollman K, Powell N, Stoohs R. Familial aggregates in obstructive sleep apnea syndrome. Chest 1995; 107:1545–1551.
148. Kronholm E, Aunola S, Hyyppa MT, et al. Sleep in monozygotic twin pairs discordant for obesity. J Appl Physiol 1996; 80:14–19.
149. Kawakami Y, Yamamoto H, Yoshikawa T, Shida A. Chemical and behavioral control of breathing in adult twins. Am Rev Respir Dis 1984; 129:703–707.
150. Thomas DA, Swaminathan S, Beardsmore CS, et al. Comparison of peripheral chemoreceptor responses in monozygotic and dizygotic twin infants. Am Rev Respir Dis 1993; 148:1605–1609.
151. Kushida CA, Guilleminault C, Mignot E. Genetics and craniofacial dysmorphism in family studies of obstructive sleep apnea. Sleep Res 1996; 25:275.
152. Carskadon MA, Pueschel SM, Millman RP. Sleep-disordered breathing and behavior in three risk groups: preliminary findings from parental reports. Child Nerv Syst 1993; 9:452–457.
153. O'Hare JP, O'Brien IA, Arendt J, et al. Does melatonin deficiency cause the enlarged genitalia of the fragile-X syndrome? Clin Endocrinol 1986; 24:327–333.
154. Staley-Gane MS, Hollway RJ, Hagerman MD. Temporal sleep characteristics of young fragile X boys. Am J Hum Genet 1996; 59:A105.
155. Challamel MJ, Mazzola ME, Nevsimalova S, Cannard C, Louis J, Revol M. Narcolepsy in children. Sleep 1994; 17:S17–20.
156. Vossler DG, Wyler AR, Wilkus RJ, Gardner-Walker G, Vlcek BW. Cataplexy and monoamine oxidase deficiency in Norrie disease. Neurology 1996; 46:1258–1261.
157. Goldberg R, Fish B, Ship A, Shprintzen RJ. Deletion of a portion of the long arm of chromosome 6. Am J Med Genet 1980; 5:73–80.
158. Kaplan J, Fredrickson PA, Richardson JW. Sleep and breathing in patients with the Prader-Willi syndrome. Mayo Clin Proc 1991; 66:1124–1126.
159. Summers JA, Lynch PS, Harris JC, Burke JC, Allison DB, Sandler L. A combined behavioral/pharmacological treatment of sleep-wake schedule disorder in Angelman syndrome. J Dev Behav Pediatr 1992; 13:284–287.
160. Greenberg F, Guzzetta V, Montes de Oca-Luna R, et al. Molecular analysis of the Smith-Magenis syndrome: a possible contiguous-gene syndrome associated with del(17)(p11.2). Am J Hum Genet 1991; 49:1207–1218.
161. Fischer H, Oswald HP, Duba HC, et al. Constitutional interstitial deletion of 17(p11.2) (Smith-Magenis syndrome): a clinically recognizable microdeletion syndrome. Report of two cases and review of the literature. Klin Padiatr 1993; 205:162–166.
162. Higurashi M, Kawai H, Segawa M, et al. Growth, psychologic characteristics, and sleep-wakefulness cycle of children with sex chromosomal abnormalities. Birth Defects 1986; 22:251–275.
163. Carskadon MA. Patterns of sleep and sleepiness in adolescents. Pediatrician 1990; 17:5–12.

164. George CF, Singh SM. Juvenile onset narcolepsy in an individual with Turner syndrome. A case report. Sleep 1991; 14:267–269.
165. Hartman E. Bruxism. In: Kryger M, Roth T, Dement WC, eds. Principles and Practice of Sleep Medicine. Philadelphia: WB Saunders, 1989:385–388.
166. Manni R, Zucca C, Martinetti M, Ottolini A, Lanzi G, Tartara A. Hypersomnia in dystrophia myotonica: a neurophysiological and immunogenetic study. Acta Neurol Scand 1991; 84:498–502.
167. Byrne E, White O, Cook M. Familial dystonic choreoathetosis with myokymia; a sleep responsive disorder. J Neurol Neurosurg Psychiatry 1991; 54:1090–1092.
168. Ishikawa A, Miyatake T. A family with hereditary juvenile dystonia-parkinsonism. Mov Disord 1995; 10:482–488.
169. Poirier G, Montplaisir J, Decary F, Momege D, Lebrun A. HLA antigens in narcolepsy and idiopathic central nervous system hypersomnolence. Sleep 1986; 9:153–158.
170. Grassi-Zucconi G, Menegazzi M, De Prati AC, et al. c-*fos* mRNA is spontaneously induced in the rat brain during the activity period of the circadian cycle. Eur J Neurosci 1993; 5:1071–1078.
171. Pompeiano M, Cirelli C, Tononi G. Immediate-early genes in spontaneous wakefulness and sleep: expression of c-*fos* and NGFI-A mRNA and protein. J Sleep Res 1994; 3:65–81.
172. Shiromani PJ, Winston S, Malik M, McCarley RW. Rapid decline in Fos-LI in association with recovery sleep that follows total sleep deprivation. Soc Neurosci Abstr 1993; 19:572.
173. Grassi-Zucconi G, Giuditta A, Mandile P, Chen S, Vescia S, Bentivoglio M. c-*fos* spontaneous expression during wakefulness is reversed during sleep in neuronal subsets of the rat cortex. J Physiol Paris 1994; 88:91–93.
174. Cirelli C, Pompeiano M, Tononi G. Neuronal gene expression in the waking state: a role for the locus coeruleus. Science 1996; 274:1211–1215.
175. Pompeiano M, Cirelli C, Tononi G. Effects of sleep deprivation on *fos*-like immunoreactivity in the rat brain. Arch Ital Biol 1992; 130:325–335.
176. O'Hara BF, Young KA, Watson FL, Heller HC, Kilduff TS. Immediate early gene expression in brain during sleep deprivation: preliminary observations. Sleep 1993; 16:1–7.
177. Cirelli C, Pompeiano M, Tononi G. Sleep deprivation and c-*fos* expression in the rat brain. J Sleep Res 1995; 4:92–106.
178. Landis CA, Collins BJ, Cribbs LL, et al. Expression of Egr-1 in the brain of sleep deprived rats. Brain Res Mol Brain Res 1993; 17:300–306.
179. Morgan JI, Curran T. Stimulus-transcription coupling in neurons: role of cellular immediate-early genes. Trends Neurosci 1989; 12:459–462.
180. Rhyner TA, Borbely AA, Mallet J. Molecular cloning of forebrain mRNAs which are modulated by sleep deprivation. Eur J Neurosci 1990; 2:1063–1073.
181. Neuner-Jehle M, Rhyner TA, Borbely AA. Sleep deprivation differentially alters the mRNA and protein levels of neurogranin in rat brain. Brain Res 1995; 685:143–153.
182. de Lecea L, Criado JR, Prospero-Garcia O, et al. A cortical neuropeptide with neuronal depressant and sleep-modulating properties. Nature 1996; 381:242–245.

183. Porkka-Heiskanen T, Smith SE, Taira T, et al. Noradrenergic activity in rat brain during rapid eye movement sleep deprivation and rebound sleep. Am J Physiol 1995; 268:R1456–1463.

184. Toppila J, Stenberg D, Alanko L, et al. REM sleep deprivation induces galanin gene expression in the rat brain. Neurosci Lett 1995; 183:171–174.

185. Kushida CA, Zoltoski RK, Gillin JC. The expression of m1-m3 muscarinic receptor mRNAs in rat brain following REM sleep deprivation. NeuroReport 1995; 6:1705–1708.

186. Sherin JE, Shiromani PJ, McCarley RW, Saper CB. Activation of ventrolateral preoptic neurons during sleep. Science 1996; 271:216–219.

187. Shiromani PJ, Kilduff TS, Bloom FE, McCarley RW. Cholinergically induced REM sleep triggers Fos-like immunoreactivity in dorsolateral pontine regions associated with REM sleep. Brain Res 1992; 580:351–357.

188. Shiromani PJ, Malik M, Winston S, McCarley RW. Time course of Fos-like immunoreactivity associated with cholinergically induced REM sleep. J Neurosci 1995; 15:3500–3508.

189. Kilduff TS, de Lecea L, Usui H, Sutcliffe JG. Isolation and identification of specific transcripts by subtractive hybridization. In: Lydic R, ed. Molecular Regulation of Conscious States. Boca Raton, FL: CRC Press, 1998:103–118.

190. Wan JS, Sharp SJ, Poirier GM-C, et al. Cloning differentially expressed mRNAs. Nature Biotechnol 1996; 14:1685–1691.

191. Schena M, Shalon D, Davis RW, Brown PO. Quantitative monitoring of gene expression patterns with a complementary DNA microarray. Science 1995; 270:467–470.

192. Lockhart D, Dong H, Byrne M, et al. Expression monitoring by hybridization to high-density oligonucleotide arrays. Nature Biotechnol 1996; 14:1675–1680.

193. Schena M, Shalon D, Heller R, Chai A, Brown PO, Davis RW. Parallel human genome analysis: microarray-based expression monitoring of 1000 genes. Proc Natl Acad Sci USA 1996; 93:10614–10619.

194. Heller RA, Schena M, Chai A, et al. Discovery and analysis of inflammatory-related genes using cDNA microarrays. Proc Natl Acad Sci USA 1997; 94:2150–2155.

195. Chee M, Yang R, Hubbell E, et al. Accessing genetic information with high-density DNA arrays. Science 1996; 274:610–614.

196. Kozal MJ, Shah N, Shen N, et al. Extensive polymorphisms observed in HIV-1 clade B protease gene using high-density oligonucleotide arrays. Nat Med 1996; 2:753–759.

197. Shoemaker DD, Lashkari DA, Morris D, Mittmann M, Davis RW. Quantitative phenotypic analysis of yeast deletion mutants using a highly parallel molecular barcoding strategy. Nat Genet 1996; 14:450–456.

12

Molecular Genetic Approaches to the Identity and Function of Circadian Clock Genes

JONATHAN P. WISOR

Stanford University School of Medicine
Palo Alto, California

JOSEPH S. TAKAHASHI

Howard Hughes Medical Institute
Northwestern University
Evanston, Illinois

I. Introduction

Revolutionary changes in mouse molecular genetics in recent years have had a profound effect on the field of mammalian circadian rhythms research. The purpose of this chapter is to describe the facts that have been learned and those that can potentially be learned with regard to mammalian rhythms from the molecular genetic approach. We will discuss the advantages and disadvantages of methods useful in identifying candidate clock genes—those genes which encode proteins necessary for circadian clock function. We will then discuss possible strategies for determining the level of clock organization at which a candidate gene's product functions.

II. Comparative Clock Genetics

As much as mammalian circadian genetics has been influenced by emerging technologies in mouse genetics, our field also owes a great debt to circadian rhythms research in nonmammalian organisms. Circadian biologists have had the extreme good fortune of being able to make comparisons across phyla and even across kingdoms. After all, where else in this volume could one find a meaningful discussion of findings in mice, fruit flies, *and* fungus? The function of circadian rhythmicity (to adapt the organism to the 24-hour cyclicity produced by the Earth's rotation) is maintained in virtually all species, and consequently similar mechanisms of circadian timing have been maintained across species. Therefore, where possible we have included information from studies on the fruit fly *Drosophila* and in the filamentous fungus *Neurospora* to augment our discussion of mouse studies. Research on the genetics of circadian rhythms in these model species can be viewed as a set of three paths, sometimes converging and sometimes diverging, but seldom far apart in a formal sense. Because we consider rhythms in *Drosophila* and *Neurospora* only as they relate to mammalian rhythms, our coverage of these two circadian systems will not be exhaustive. The interested reader should consult one of numerous recent reviews (1–6) for a more detailed discussion of the molecular basis of circadian rhythmicity in *Drosophila* and *Neurospora*. It is hoped that the reader of this chapter will come away with a sense of the immense value of cross-species comparisons in the development, both past and future, of the mammalian circadian rhythms field.

III. Macromolecular Synthesis and Cell Autonomy

In mammals, the primary circadian clock resides in the suprachiasmatic nuclei (SCN) of the hypothalamus (7,8). While the basic mechanisms of circadian rhythm generation and entrainment are only beginning to be understood, it appears that across species intracellular feedback loops involving gene expression and protein function form the basis for the circadian clock (1,5,9). Three observations indicate that electrophysiological events in the SCN are not necessary for the circadian clock to keep timing. Tetrodotoxin infusions into the SCN region block the expression of circadian rhythms at the behavioral level but do not phase shift the clock or interrupt rhythm generation (10). Also, the fetal SCN exhibit circadian rhythms of 2-deoxyglucose uptake that are entrained to maternal rhythms before synaptogenesis occurs in the SCN (11). And circadian rhythmicity is maintained in hibernating animals despite the fact that the brain temperature is well below the threshold for action potentials in the SCN (12). This observation means that in addition to being independent of action potentials, circadian rhythmicity is temperature compensated in mammals. While potential mechanisms for temperature compensation have been identified in both *Drosophila* and *Neurospora* (13), the mechanisms for mammalian clock temperature compensation will likely become an important focus of research as mammalian clock genes continue to be characterized.

The above data indicate that the circadian clock is not dependent on electrophysiological communication between cells, but do not rule out other possible intercellular interactions in generating rhythmicity. More compelling evidence for the single-celled nature of the clock in mammals comes from recent studies of electrical activity in cultured dissociated SCN neurons (14,15). In a preparation in which the activity of multiple dissociated SCN neurons was recorded simultaneously, it was found that individual cells exhibited circadian rhythms of firing differing from each other in phase and in period. Contact between the cells was not prevented entirely, but the fact that the individual firing rhythms were not synchronized strongly suggests that intercellular interactions were unnecessary for rhythmicity to occur within single cells.

The possibility that a single SCN cell is capable of circadian oscillations is consistent with the concept that the clock is an intracellular feedback loop involving transcription and translation. There is evidence for a translation-dependent circadian pacemaker in the form of biochemical data. The observation that circadian period may be altered by a protein synthesis inhibitor was first made in *Euglena gracilis* by Feldman (16). Since that time it has been shown that circadian clocks in mammals (17), *Neurospora* (18) and the marine mollusks *Bulla gouldiana* (19) and *Aplysia californica* (20) are sensitive to protein synthesis inhibitor drugs at certain circadian phases. The phase response curve measured for protein synthesis inhibitor treatment of hamsters (17) and of the SCN in vitro (21) indicates that there is a critical period of protein synthesis for the circadian clock in the early subjective day. The concept of a critical period for clock protein synthesis led naturally to the goal of identifying the critical proteins synthesized in this period. One way to address this goal is the molecular genetic approach. Finally, the existence of a spontaneous clock mutation, *tau*, in the golden hamster (22) gave credence to the genetic approach to mammalian rhythms, although the absence of suitable genetic resources in the hamster has hampered attempts to identify the *tau* gene to this day.

IV. Identifying Candidate Clock Genes in Mammals

A. Forward Genetics—From Circadian Behavior to Genes

The goal of genetics is to define causal relationships between the genetic material of an organism (genotype) and its observable anatomy, physiology, or behavior (phenotype). Genetic approaches to understanding these relationships can be classified as forward or reverse genetics (Fig. 1). In forward genetics, the experimenter identifies a phenotype of interest and utilizes the random sorting of genetic material in meiosis and its subsequent inheritance to identify the gene responsible for that phenotype. In reverse genetics, the experimenter begins with a candidate gene and determines the function of that gene in the organism. While the reverse genetic approach may be effective in determining the function of a specific gene in the system of interest, the forward approach is effective in a situation where the

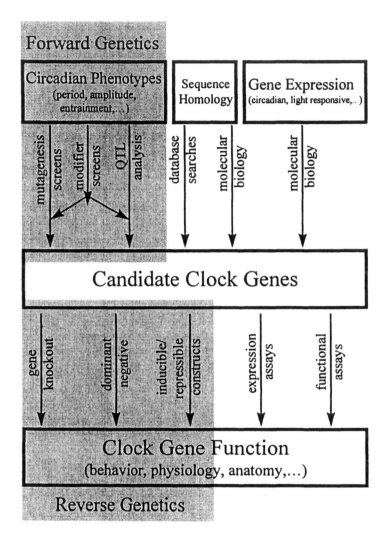

Figure 1 A schematic representation of the uses of molecular genetics in circadian biology. Forward genetic methods (upper shaded box), homology-based screens, and expression-based screens are used to identify candidate clock genes. Reverse genetics (lower shaded box), expression assays, and functional assays can be used to determine the function of a candidate gene within the circadian system.

genes underlying a particular physiological process are unknown (23). Three decades ago, the circadian rhythms field was at a stage where the molecules forming the basis of circadian rhythmicity were unknown and no obvious candidates existed. The forward genetic approach was applied in *Drosophila* (24), in *Neurospora* (25), and eventually in cyanobacteria (26), mammals (27,28), and *Arabidopsis* plants (29) to change that situation.

B. Mutagenesis Screens

The first circadian mutagenesis screen was performed in *Drosophila* by Konopka and Benzer (24). Several chemically-induced mutations which alter circadian period mapped to a single locus, called *period*, or *per*. The identity of the *per* locus was ascertained through rescue of circadian rhythmicity by insertion of a genomic region containing the candidate locus into the genome of an arrhythmic *per* mutant line (30,31). The *per* transcript was found to oscillate (32), a tentative indication that it might be a key player in the circadian clock. Subsequent biochemical and genetic experiments indicated that the PER protein is transported into the nucleus where it inhibits *per* expression, forming the basis for circadian oscillations as a negative feedback loop (4–6). While it was originally conceived that the feedback loop in *Drosophila* was dependent on the direct or indirect inhibition of *per* transcription by PER protein, it now appears that at least in certain cases, nuclear entry of PER (in the silk moth [33]) and *per* mRNA rhythmicity (in *Drosophila* [34]) may not be essential for PER protein rhythmicity. A second *Drosophila* clock gene necessary for rhythmicity, called *timeless* (*tim*), was identified by a mutagenesis screen (35,36). TIM interacts physically with PER (37), and interaction between the two proteins regulates the transport of PER into the nucleus (38).

In the case of *Neurospora*, the first circadian mutagenesis screen yielded 15 circadian mutations, nine of which mapped to a single locus, *frequency* (*frq*) (25). The isolation and sequencing of *frq* have opened the door for numerous molecular genetic experiments on *frq* function (39,40), but at this time little is known about its biochemical properties. FRQ somehow negatively feeds back on its own transcription to produce circadian oscillations (40). In addition to FRQ, the *wc-1* and *wc-2* genes were identified by mutagenesis screens in *Neurospora* (41). These genes are also essential in the production of sustained circadian oscillations (42).

The successes of the *Drosophila* and *Neurospora* mutagenesis screens in identifying circadian clock genes and the existence of the spontaneous *tau* mutation, which produces a striking circadian phenotype in golden hamsters (22), inspired mammalian circadian researchers to perform mutagenesis screens in the early 1990s. The species of choice from a genetic perspective was the laboratory mouse, *Mus musculus*, due to the power of gene mapping in this species (23). N-ethyl-N-nitrosourea (ENU) is highly effective in mutagenesis screens, as it produces point mutations at a rate of 66 to 150×10^{-5} per locus in spermatogonia. In other words, any locus within the genome will be mutated in 66 to 150 out of every 10^5 ENU-treated germ cells on average. With this mutation rate, 3× coverage of the genome could be reached on average in a screen of 2000 to 3000 gametes of ENU-treated mice. However, the circadian phenotype of the immediate offspring (G_1 generation) of mutagenized males will only be apparent in the case of semidominant or dominant mutations and not with recessive mutations. In order to identify a recessive mutation, the researcher must breed G_1 offspring with wild-type partners, and then breed the second-generation (G_2) animals with the same G_1 animal to produce G_3 offspring. A recessive mutation in any G_1 gamete can in theory be identified at 85% efficiency by testing 20 G_3 progeny. A recessive screen therefore requires about 40,000 G_3 mice (2000 gametes × 20 mice per gamete) in order to cover the genome (23,43). Such a task, though daunting, may reveal mutations such as null mutations that would go undetected in a G_1 screen.

With the goal of identifying circadian clock genes, our laboratory and at least one other (28) launched mutagenesis screens by injecting male mice with ENU and observing the circadian locomotor rhythms of their offspring. The *Clock* (circadian *locomotor output cycles kaput*) mutation, identified in our laboratory's mutagenesis screen, resulted in a lengthening of the free running circadian period to approximately 25 hours in the founder (G_1) mouse and in subsequently produced heterozygotes. Homozygous *Clock* mutants exhibit a dramatic lengthening of circadian period to 28 to 29 hours followed by arrhythmicity in constant darkness (27). *Clock* was genetically mapped to a region of mouse chromosome 5, which was then subjected to physical mapping (44).

From this point, the identification of the *Clock* gene required both the positional cloning (mapping of candidate transcriptional units in the critical region of the genome) and a functional demonstration that a transcript in the region controls circadian phenotype. A critical aspect of physical mapping was the establishment of a BAC contig, that is, the identification of a set of three bacterial artificial chromosome (BAC) vectors containing the *Clock* mutation in a 250-kb region of the mouse genome (44). The contig allowed for a three-pronged approach to identify the *Clock* gene and the *Clock* mutation at the nucleotide level. First, genetic mapping of the mutation relative to genetic markers within the contig allowed for continued refinement of the possible location of *Clock* (44). Second, the identification of expressed DNA sequences within the contig elucidated the identity of candidate genes (44). Third, the production of transgenic mice in which the BACs were integrated into the genome provided a functional assay for the location of *Clock* in the genome. Thus, when one 140-kilobase BAC but not the other two in the contig produced a wild-type circadian period in BAC-transgenic *Clock* mutant mice and a shortening of circadian period in BAC-transgenic wild type litter mates, the location of the *Clock* gene on that BAC was ascertained (45). The one candidate gene that was consistent with the data from these three approaches was sequenced in wild type and *Clock* animals and found to contain a point mutation which caused exon skipping in *Clock* mutant tissues.

The mutation was verified by reverse transcription (RT) PCR and sequencing of the mutant transcript (44). In this manner our knowledge progressed from the *Clock* phenotype to the identity of the *Clock* gene and the nature of the mutation at the molecular level.

The *Clock* gene encodes a transcriptional regulatory protein of the basic helix loop helix-PAS domain family of transcription factors. The existence of a PAS domain, a salient structural feature of PER, in CLOCK is a striking example of the parallels that are seen in diverse circadian systems. But CLOCK is unlike PER in that it's sequence is strongly suggestive of a transcriptional regulatory protein. The putative function of PER in transcriptional regulation has only been inferred from the observation that *per* transcript levels are inversely associated with PER concentration. Whether CLOCK regulates its own transcription or that of other clock-related genes remains to be determined in biochemical studies; however, with its sequence in hand a number of molecular and biochemical approaches are now tractable.

The *Wheels* mutation, which was identified by a mutagenesis screen provides a note of caution for circadian researchers in search of clock genes. This mutation results in a free-running period of 24.20 hours in heterozygous animals, compared to a period of 23.32 ± 0.02 hours in the total population of screened animals (28). The *Wheels* mutation is homozygous lethal, which is not typical of mutations in clock component genes (46). In fact, it was recognized by the authors that there are abnormalities in the inner ear development of *Wheels* mutants. The circadian changes appear to be secondary effects of those abnormalities (47). The lesson from *Wheels* is that a mutation that alters a circadian clock property, even one as basic as the circadian period, is not necessarily informative about the nature of the circadian clock. Nonetheless, the other success stories in circadian mutation screens make this approach a promising one for those who have the necessary resources and fortitude.

C. Quantitative Trait Loci Analysis

The set of genes that regulate a phenotype that is regulated quantitatively by more than one locus are collectively known as the quantitative trait loci (QTL) for that phenotype. QTL analysis is a method of mapping these loci (48–50). QTL analysis can be thought of as having three stages. The first step is to identify a trait that is regulated quantitatively by a set of genes. The quantitative trait of interest (for instance, circadian period or amplitude of locomotor rhythmicity) must be significantly different between two inbred strains of mice. The trait must then be measured in the F_1 hybrid offspring of crosses between the strains and in the offspring (either intercross or N_2 backcross offspring) of the F_1 generation. The phenotype of F_1, N_2 and F_2 mice can be used to determine the mode of inheritance and therefore the feasibility of QTL analysis for the trait.

The ideal scenario for QTL analysis would be one in which the original strains differ in the trait of interest by at least 3 standard deviations, the phenotypes of F_1 offspring are in a unimodal distribution intermediate between the original strains, and the phenotypes of F_2 offspring are widely distributed in a range of values approximating that of both originator strains combined. These distributions imply that there are a number of potentially mappable loci involved in regulating the phenotype (48). One limitation to this approach is that the only loci which are polymorphic between the two strains studies can be identified in QTL analysis. Genes that are not polymorphic between the strains may nonetheless influence the trait of interest, but their identity will not be revealed. If three loci are identified, for instance, one cannot conclude that there are only three loci influencing the trait, but that three loci that influence the trait are polymorphic between the strains. Similarly, the absence of significant QTLs between two strains does not lead to the preposterous conclusion that no genes influence the trait.

The second step in QTL analysis is to identify the genomic regions which harbor the loci regulating the quantitative trait. This step requires that the researcher survey the strain origin of regions of the genome in F_2 animals (one marker every 20 centiMorgans is a good rule of thumb) and identify those regions that are statistically significantly associated with the trait. This step can be aided by phenotypic analysis of recombinant inbred strains of mice, in which the genomic marker survey has already been performed. The third step of QTL analysis is to generate congenic strains (strains differing only at the locus of interest and identical at all other loci) for each putative locus identified in step 2 by selective breeding. The classic method of making congenic mice requires at least 10 generations of selective breeding, but recently developed methods can be used to essentially transfer a single locus from one strain to another in just a few generations (51,52). When congenic strains are produced, the role of the congenic locus in influencing the quantitative trait can be verified. If the single polymorphic locus among two otherwise identical strains has a significant effect on the trait of interest, then that locus is indeed a QTL. It can be subjected to the genetic and physical mapping techniques applied to single gene mutations and can in theory ultimately be cloned.

To date the QTL analysis approach has been taken to the stage of QTL mapping but not to the congenic stage in studies of circadian period in inbred mouse strains (53,54). In other fields, a number of putative QTLs have been lost at the stage of congenic production because the effect of the QTL is not sufficiently large for it to be statistically significant in the absence of other QTLs, or because it requires epistatic interactions with other loci (55). To address these two possibilities without actually going through the production of congenic lines, it may be possible to selectively phenotype F_2 animals in which a single putative QTL has been randomly inherited separately from all other putative QTLs (56,57). Whether any circadian QTLs will reach the point of gene isolation and sequencing remains to be seen.

D. Modifier Gene Screens

After a candidate gene has been identified through mutagenesis, QTL, or other means, it may be useful to identify other genetic loci which influence the effect of that gene on the phenotype of interest. Such loci are known as modifier loci and are said to have a genetic interaction with the previously characterized gene (58). This genetic interaction may *or may not* be associated with a physical interaction between the protein products of the two genes. For instance, a mutagenesis screen may yield a mutation that causes shortening of the circadian period to 21 hours in strain A mice. But when the mutation is bred into a mouse strain B background, the mutation may produce a circadian period of 23 hours. There is a modifier locus, or loci, in strain B which blunts the period shortening effect of the mutation. A difference in mutant phenotype due to strain background has been described as a serious concern in the interpretation of experimental results with genetically engineered mice (59). However, it can also provide the researcher with additional candidate genes to study (60).

Modifier loci may also be identified by breeding of mutagenized males with carriers of a previously identified mutation and testing of the offspring for alterations in the phenotype of the original mutation. Regardless of the source of a modifier gene, be it a mutagenesis protocol or a strain difference, a modifier gene can be mapped in the same way as any other gene and ultimately cloned. The fact that it is a modifier gene as opposed to one with a completely novel phenotype may make it easier to conceptualize and identify the function of the gene in an already characterized system.

V. Molecular Biological Approaches

The cloning of *per*, *tim*, *frq*, and *Clock* illustrates the value of the forward genetic approach in the identification of candidate clock genes. Mutations of these genes have profound effects on circadian clock function and knowledge of these genes has a similarly profound effect on the circadian *Zeitgeist*. Yet, the forward genetic approach requires significant investment. And recent advances in mammalian clock studies in particular have proven that other screening methods, particularly homology-based screens and expression-based screens, may also yield valuable candidate clock genes. Expression-based screens, which we will not discuss in detail, utilize molecular biological and biochemical techniques to identify genes expressed in clock-containing cells, genes activated by entraining stimuli such as light, and genes regulated in a circadian manner (61). The value of these screens lies in the expectation that genes with such characteristics are relevant to clock function.

A. Homology-Based Screens

The parallels between mammalian clocks and *Drosophila* clocks have led many a researcher to speculate that a mammalian *per* homolog might be found through homology screening. The existence of putative *per*-like threonine-glycine repeat-encoding DNA sequences in mouse and human DNA (62), a "mouse unusual *per* repeat" mRNA exhibiting circadian rhythmicity and light sensitivity in the SCN (63) and circadian rhythms of *per*-like immunoreactivity in the SCN (64) were all taken as evidence that a *per* homolog could be found. In 1997, two groups published the sequence of a mouse *per* homolog. One group performed PCR on mouse DNA with a series of nested, degenerate primers designed based on sequences of the PAS domain of *per* (65). One primer combination amplified a 65 base pair product, which was used to isolate a cDNA, designated *hper*, from a human cDNA library. *hper* was then used in a screen of a mouse cDNA library to isolate a mouse clone, *mper* (which subsequently has been labeled as *per1* for reasons that will soon become apparent to the reader). A second group reported the isolation of an *hper* cDNA in a search for chromosome 17-specific transcripts and the isolation of *mper* by screening of a mouse brain cDNA library with an *hper* cDNA probe (66).

DNA database searches by both groups revealed homology to the *Drosophila per* gene in certain functional domains and homology to the *Clock* gene in the PAS domain. Since the cloning of *mper* and *hper*, additional homology screens of publicly available cDNA sequences have yielded a second (*per2* [67,68]) and possibly third (68) human *per* homolog. Mouse *per1* and *per2* cDNAs were also cloned by those who cloned human *per2*. The cloning of *per* homologs in mouse and human was swiftly followed by mRNA assays showing that these genes are expressed in a manner that suggests a role in circadian clock function. *per1* and *per2* are expressed in the SCN and the retina (known to contain an autonomous clock [69]) among other tissues (65–68) and are circadian (65–68) and light (67,68,70) regulated. The identification of these mammalian *per* homologs illustrates that homology screening can be a very effective adjunct to forward genetics in identifying candidate clock genes.

VI. Determining Clock Gene Function

While the forward genetic approach has been and will continue to be a very important way of identifying genes involved in the workings of the circadian clock, forward genetics does not reveal the biochemical function of those genes. Once candidate clock genes are identified, expression assays and reverse genetics become essential in the study of these genes. In the remainder of this chapter we will present criteria for determining candidate clock gene function. We will provide illustrations of how some of these criteria have been addressed in studies of *Drosophila per* and *tim*, *Neurospora frq*, and putative mammalian clock genes. Finally, we will describe potential methods for the study of mammalian clock gene function in vivo.

A. Clock Gene Criteria

In order to discuss gene function in the circadian system, we divide clock genes into three general categories, in line with the three major components of the circadian system at the conceptual level (7–9: clock input genes, clock component genes, and clock output genes. We put forth a series of experimentally-testable criteria for each of these three types of genes (Table 1). We posit these criteria in an attempt to limit the set of genes described as "clock genes" to a manageable and meaningful size. Therefore, many genes necessary for circadian clock function (e.g., developmental genes, genes which function universally in metabolic or transcriptional machinery) will not meet criteria for any of the three categories. These criteria reflect the dynamic nature of circadian clock function, not the mere existence of circadian pacemakers in the suprachiasmatic nucleus and retina.

B. Clock Components

The idea that genes may encode clock components is rooted solidly in the conceptual and empirical background of the circadian rhythms field. Hence, the search for clock components has taken the spotlight of most molecular genetic research on clock genes. What we describe as clock components have been described variously as critical clock components (71), central clock components (2), or state variables (a term borrowed from theoretical modeling studies) in the clock mechanism (72). In each of these three cases, experimentally testable criteria were put forth for the establishment of a clock component. We have attempted a synthesis of these criteria in Table 1. We next discuss the ways in which these criteria have been tested for candidate clock genes in *Drosophila* and *Neurospora* and how they may be applied to candidate clock genes in mammals.

Oscillation

The expression level or functional activity of a gene must exhibit oscillations in the circadian range to be considered a state variable in the clock (2,9,72). After the cloning of *frq*, the mRNA was found to oscillate in a circadian fashion (40). Similarly, *per* (73) and *tim* (74) both oscillate at the mRNA level. The levels of the PER (75) and TIM (76) proteins were also found to have circadian oscillations which lag behind the mRNA oscillations by several hours. In the mouse, the

Table 1 Criteria for Determining Whether the Protein Encoded by a Candidate Clock Gene Functions as a Clock Input, a Clock Component, or a Clock Output

Clock inputs
  Necessity: expression in cells that contain a circadian oscillator or those that directly contact circadian oscillator-containing cells is necessary for at least some phase shifts to occur.
  Phase resetting: an exogenously generated change of expression or functional activity (produced by a mechanism that does not in itself cause a phase shift) must produce a phase shift of the clock at some circadian phases.
  Phase shift dependence: expression or functional activity is temporarily altered in response to at least one phase shift-inducing stimulus.
Clock components
  Oscillation: the mRNA or protein level/activity oscillates with a period in the circadian range in cells containing a circadian oscillator.
  Phase-alignment: the phase of its oscillation must be consistent with biochemical phase response data (where such data are available).
  Phase determination: an exogenously generated change of the mRNA or protein concentration (produced by a mechanism that does not in itself cause a phase shift) must produce a phase shift at some circadian phases.
  Necessity: any manipulation that abolishes expression or circadian rhythmicity of expression must abolish all other measures of circadian rhythmicity.[a]
Clock outputs
  Oscillation: the level/activity oscillates with a period of the circadian range in cells containing a circadian oscillator.
  Phase shift response: any stimulus that produces a phase shift of the clock also produces a phase shift of the same magnitude in the rhythm of clock outputs.
  Nonnecessity: any manipulation that abolishes expression or circadian rhythmicity of clock output expression may abolish some but not all other forms of circadian rhythmicity in the organism.

[a]This criterion does not apply in cases where a single cell contains two separate oscillators, as in *Gonyaulax* (92).

mRNAs of both *per1* (65,66) and *per2* (67,68) oscillate with a circadian period in the SCN and in the retina. By contrast, the mammalian *Clock* gene has not been found to oscillate at the mRNA level. While data on CLOCK protein expression are not currently available, there may be protein oscillations in the absence of mRNA oscillations. Or perhaps posttranslational modifications of CLOCK impart a circadian rhythm to its functional activity. The point in mentioning these possibilities is that a lack of circadian oscillations at the mRNA level do not disqualify a gene from encoding a clock component. The fact that quantitatively measuring gene expression at the mRNA level is technically easier than measuring protein level or protein activity should not lead the reader to conclude that mRNA is somehow more relevant to clock function.

Clock component oscillations must occur in cells that contain autonomous circadian clocks, and this is the case for *per1* and *per2*, which are rhythmic in the SCN. The identity of cells within the SCN that contain autonomous clocks remains a mystery. The identification of clock components and of clock cells will likely occur hand in hand, as discoveries in either of these two areas are likely to be of benefit to research in the other.

Necessity

While oscillation is a characteristic of clock components, it is also one of clock outputs. However, the oscillation of clock outputs is not necessary for the function of the pacemaker itself, while that of clock components is necessary. Thus, when oscillatory expression of FRQ was abolished by null mutation, constitutive overexpression, or expression of a genetically engineered nonoscillatory FRQ construct in a null mutant strain (39,40), no circadian rhythmicity at any level could be detected. Similarly, overexpression of PER protein in *Drosophila* eye cells prevented rhythmicity of *per* expression in a cell autonomous manner (77). The effects of overexpression of putative mammalian clock component proteins in the SCN on the circadian rhythmicity or RNAs should be examined in this manner. Exogenous overexpression of genes in the brain by targeted transgene expression is feasible in mice. For instance, a transgene comprised of the coding region of the dopamine transporter gene fused to the tyrosine hydroxylase promoter produces a significant increase in the level of dopamine transporter expression in the brain and at the behavioral level alters the response of the animal to dopaminergic drugs (78). This type of approach might be used to assay the necessity of candidate clock protein oscillations in mammalian rhythm production. Gene knockout technologies may be used to determine whether or not a particular gene is necessary for clock function, but it is important to note that these approaches do not indicate whether the gene encodes a clock component or a protein that plays a permissive role in the clock.

Phase Alignment

Before the oscillatory expression of specific candidate clock genes was documented, it was known that in *Neurospora* (18), in the mammalian SCN (21), and in other species (19), there is a critical period of protein synthesis for the circadian clock. This knowledge formed the basis for clock component gene searches (9) and provides a criterion for clock components in *Neurospora* and in mammals. The critical period for protein synthesis in the both the *Neurospora* and mammalian clocks, that is, the time period in which protein synthesis inhibitors shift the clock, is the late subjective night to the early subjective day. This time period closely parallels the time period during which FRQ is synthesized in *Neurospora*.

If *per1* and *per2* protein expression in the SCN is closely phased relative to mRNA synthesis, which remains to be determined, then synthesis of these proteins is also in the period for critical protein synthesis in the mammalian clock (17). There is an autonomous circadian clock in the mammalian retina (17), and the expression of *per1* and *per2* at the mRNA level in the retina is indeed rhythmic (65–68). Yet, the peak of the rhythm is delayed relative to the SCN by a period of 4 to 6 hours. This delay is difficult to reconcile with conceptual principles of entrainment to light dark cycles if in fact *per1*, *per2*, or both genes are clock components. It suggests that the light signal reaches the clock mechanism in the retina 4 hours later than it reaches the clock in the SCN, a puzzling but not physiologically impossible arrangement. An alternative but also peculiar possibility is that there is a powerful entraining stimulus from the SCN clock producing the delay in the retinal clock. Alternatively, *per* genes may be outputs of the clock in the retina and/or SCN of mammals rather than clock components.

Phase Determination

The level of expression or functional activity of clock components determines the phase of the clock's oscillation (71,72). Therefore, any exogenously generated change in the level of a clock component must reset the phase of the circadian clock. A most elegant demonstration of this characteristic of clock components comes from studies in *Neurospora* of an inducible FRQ construct (40). The construct was engineered to produce high, stable levels of FRQ expression only in the presence of quinolinic acid, a substance which does not in itself affect clock function. Cultures transfected with the inducible *frq* construct exhibited normal circadian rhythmicity in constant darkness before quinolinic acid treatment due to endogenous *frq* expression. Upon addition of quinolinic acid, the expression of *frq* became stable at high levels. When quinolinic acid was withdrawn from the medium in constant darkness, cultures began to exhibit circadian rhythmicity. The clock responded as if it were the daily onset of darkness at the time of release from quinolinic acid treatment. Regardless of the phases of the rhythms of the transgenic *Neurospora* cultures before treatment with quinolinic acid, synchronized release from quinolinic acid treatment synchronized the rhythms of all treated cultures. Thus, the timing of expression of *frq* dictates the phase of the circadian clock.

In *Drosophila*, inducible *per* constructs (though not *tim* constructs) have been employed to study the role of *per* in controlling the phase of circadian rhythms. Because the promoter used in these studies is that of a heat shock protein, the results of the studies must be interpreted with caution. The heat shock necessary to induce this construct is capable of phase shifting the clock in the absence of the construct (79,80). Thus, any influence that the heat shock induced-construct may have on the phase of the circadian clock may be due to the interaction of the induced PER protein with other molecular effects of the heat shock rather than a direct effect of the PER protein on clock function (81).

In contrast to the arrhythmicity produced by constitutive expression of *frq* under a QA-inducible promoter, constitutive *per* expression under heat shock results in rhythmic PER protein levels in otherwise arrhythmic *per*[01] flies (80). This result raises the possibility that *per* transcription is not necessary for the clock feedback mechanism and that posttranscriptional regulation of PER protein may be more central to the clock. Such a possibility gains further support from the observation that in transgenic flies PER protein can oscillate in the complete absence of mRNA oscillations. Whether the phase of the circadian rhythm after heat withdrawal is dependent on the time of heat withdrawal was not determined, as the genetically engineered *per*[01] flies used in these experiments were arrhythmic before heat treatment. In a related study, the phase response curve for 4-hour heat pulses was measured in rhythmic flies harboring a heat shock-inducible *per* construct and in wild type control flies. Phase shifts caused by *per* induction in the presence of heat shock were different from those caused by heat alone in wild type animals. This result suggests that *per* levels can influence the phase of the clock, but does not distinguish between the roles of clock input and clock component (79).

In light of the approaches taken and the relative degrees of success in *Neurospora* and *Drosophila*, mammalian clock researchers are faced with the challenge of finding a method to induce constitutive expression of candidate clock genes in the SCN and then to rapidly reverse the constitutive expression. Gene expression can be activated or inactivated at the will of the researcher by a tetracycline-dependent promoter system that has been applied in brain gene regulation (82). In this study the researchers were able to disrupt the function of a particular gene in the CA1 region of the hippocampus (calcium/calmodulin kinase II) during the postnatal period rather than during prenatal development. Furthermore, they were able to show that the disruption caused functional deficits in memory, thus implicating kinase activity in the establishment of memory. This type of approach may be useful in the circadian system for identifying genes which encode clock components. If the temporal control of expression can be engineered to sufficiently fine resolution, the hypothesis that the expression of a candidate clock gene determines circadian phase will be testable, as in *Neurospora* (2).

Phase Shift Response

The concept that any experimentally induced change in the level of a clock component results in a change in the phase of the circadian clock has a logical converse: any stimulus that produces a phase shift of the clock must change the level of one or more clock components. The canonical phase shifting agent in the environment and in experimental circadian studies is light. That the circadian rhythmicity of a clock component (or *any* measure of clock function) should phase adjust in the long term as a result of a light pulse-induced phase shift or a shift in the light-dark cycle is a banal truism. More critical to our discussion is the concept that light must adjust clock components rapidly and in *a biochemically appropriate fashion*. As Figure 2 illustrates, whether clock component oscillations peak in the day or in the night determines whether light must decrease or increase the level of those components to produce the appropriate PRC. Thus, if clock components peak in the day, light must cause an increase in those components to cause a phase shift. Conversely, if clock components peak in the night, light must cause a decrease in those components to produce the appropriate PRC.

Experimental results on the effects of light on candidate clock components in *Neurospora, Drosophila,* and mammals have borne out these conceptual predictions very nicely. *frq* in *Neurospora* (39) and *per1* in mammals (65,66) are expressed maximally in the light period or subjective day. Light exposure in the subjective night results in a dramatic and rapid increase in the mRNA levels of *frq* in *Neurospora* (42a) and of *per1* (67,68,70) in the mouse SCN. *per2* may also be light induced as a delayed response to light, as one of two articles reporting *per1* mRNA induction by light exposure also reported *per2* mRNA induction two hours after the onset of light exposure (68). Thus, in two phylogenetically diverse circadian systems, the expression of candidate clock components support the conceptual viewpoint of how a daytime peaking clock component should respond to light.

In *Drosophila*, the best candidates for clock component proteins, PER and TIM, peak in the night, approximately 180° out of phase with both the *Neurospora* and mammalian clocks. The conceptual scheme presented in Figure 2 suggests that the response of PER and TIM to light should be the opposite of that seen in *Neurospora* and mammals. Indeed, four independent studies showed that *TIM* is degraded as an immediate response to light (76,83–85). From a conceptual perspective, these results are right on the mark. But from an evolutionary perspective it is odd that the mammalian circadian clock should be more similar to the *Neurospora* clock than to the *Drosophila* clock in terms of phase alignment and the response to light, especially since the putative mammalian clock components exhibit homology to a putative *Drosophila* clock component. The resolution of this paradox will be an important aspect of future studies on clock genetics.

C. Clock Inputs

The entrainment of circadian rhythms to 24-hour sidereal cycles is an essential aspect of circadian function. Therefore, the identification of entrainment mechanisms and of genes mediating these mechanisms is a central part of rhythms research. A gene that encodes a clock input must meet three criteria. It is expressed in cells containing the clock or cells which physically contact (synaptically or otherwise) clock-containing cells. Expression of the gene or functional activity of

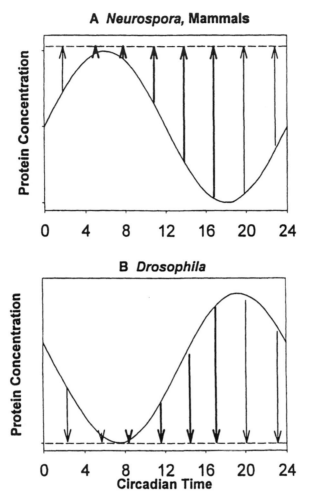

Figure 2 A schematic representation of the phase alignment of critical clock proteins and the effects of light on those proteins in (A) mammals (17) and *Neurospora* (18) and in (B) *Drosophila*. The data in (A) are based on the phase response curve for treatment of the clock with protein synthesis inhibitors. The data in (B) are based on the assumption that PER and TIM are clock component proteins, as data are not available for the effect of protein synthesis inhibitors on the *Drosophila* clock. Each graph represents the oscillation of critical clock protein concentration (y axis) across circadian times (x axis). Arrows represent the effect of light on critical clock proteins. The length of the arrow is in direct proportion to the relative magnitude of light-induced phase shifts of the clock at that circadian time. The direction of phase shifts depends on whether the light pulse occurs before or after the circadian breakpoint (corresponding to the daily minimum [A] or maximum [B] of protein concentration). Advances are indicated by thin arrows, and delays by thick arrows.

its protein product must be altered by some or all phase-shifting stimuli in a manner that is consistent with the phase shift produced by the stimulus. And rhythmicity must be maintained in the absence of the gene.

Phase Shift Dependence

In contrast to clock component genes, candidate clock input genes may be more easily and efficiently identified by expression-based screens than by forward genetic approaches. Because pharmacological data make a clear case for increased gene expression being part of the phase shifting process, any gene whose expression or functional activity is induced within the SCN or the retina may be considered a candidate clock input gene. A number of immediate early genes and post-translational signal transduction events which may be involved in immediate early gene induction have been found to be induced in the SCN by light (61). Yet, definitive tests of the necessity of these events in phase shifting are sorely needed.

Necessity

Mice in which a candidate clock input gene has been knocked out may exhibit normal entrainment (as is the case with both *vgf* and *NGFI-A* in our unpublished observations). In such a case, it is possible that related genes compensate for the one that is knocked out. Gene knockout may also produce developmental problems resulting in subtle circadian abnormalities, as in the *c-fos* knockout (86). Temporally regulated gene expression with repressible gene constructs might be helpful in addressing the issues of gene compensation and developmental defects in gene knockout mice. Constructs such as the previously described calcium/calmodulin kinase II (82) can be engineered to be expressed continuously throughout development and into adulthood, and then withdrawn at will. One would expect that when a critical clock input gene under the control of a repressible promoter is prevented from being expressed, the animal will lose the ability to entrain to a light dark cycle but will still be rhythmic. Later removal of the repressing agent will allow expression of the gene to occur and thus reentrainment of the animal to the light dark cycle.

There may be redundancy in all clock input pathways, so that preventing the function of a single gene will never prevent entrainment of the animal. In this case, it is at least theoretically possible to render several genes non-functional simultaneously with the above technologies. But another way of approaching the issue may be to work upstream from the clock component-encoding genes. Thus, if a potential binding site for leucine zipper proteins is found in the promoter region of a clock component gene, one could genetically engineer mice to express constitutively high levels of a dominant negative form of leucine zipper protein. A dominant negative is a protein that is engineered to physically interfere with the site of action of a native protein, thereby preventing the protein from performing its function. In this manner, a dominant negative form of CREB driven by an anterior pituitary-specific promoter in transgenic mice caused pituitary abnormalities and dwarfism (87). In a similar fashion, a dominant negative CREB gene could potentially be targeted to the SCN with a melatonin 1A receptor promoter (88) or other relatively SCN-specific promoter. One major limitation on the use of dominant negative constructs is that biochemical information on the functional domains of a protein must be known before this approach can be utilized.

Phase Resetting

In addition to being necessary for phase shifting, a change in the functional activity of a critical clock input gene should be sufficient to produce a phase shift of the circadian clock. Theoretically, it should be relatively simple to address this issue by measuring phase shifts produced by experimentally-induced perturbations of clock input gene expression. As in the case of a clock component gene, rapid and transient activation of a clock input gene in the SCN by a stimulus that otherwise does not phase shift the clock must result in a phase shift of the clock. The previously mentioned tetracycline-inducible promoter system that has been applied in brain gene regulation (82,89) may be effective in addressing this issue, provided that it can be targeted to the SCN or retina *in vivo* or *in vitro*. The endogenous gene of interest need not be disrupted, and consequently developmental abnormalities are not a problem. At least at some points in the circadian cycle, phase shifts must result from transient transgene induction if the gene is a clock input.

One case in which the exogenous induction approach for identifying clock inputs may not apply is when posttranslation modifications and not expression level of the protein are at the root of the clock input mechanism. Such may be case with the transcriptional regulatory protein CREB, which is constitutively expressed in the SCN, but is only transcriptionally active after phosphorylation, which occurs in response to light exposure (90). Molecular genetic methods of inducing CREB-specific phosphorylation rapidly and transiently (to determine whether CREB phosphorylation is sufficient to cause a phase shift) can only be dreamed of with the current state of technology. A similar difficulty is found in the study of *Drosophila* phase shifting responses to light. If indeed the rapid degradation of TIM is the point of entry for a light-induced phase shift in *Drosophila*, then identification of genes induced by light in *Drosophila* may provide little information about phase shift mechanisms. Indeed, the degradation of TIM is sufficiently rapid to be mediated by post-translational modification of preexistent proteins in the cell (76,83–85). Therefore, a mutagenesis approach to identify mutations resulting in an (otherwise functional) light-insensitive clock may be valuable in identifying light input genes in *Drosophila* or in mammals.

D. Clock Outputs

Oscillation

The circadian pacemaker in the SCN must confer rhythmicity to a whole spectrum of processes within the organism. The search for clock output mechanisms is consequently a major focus in the circadian field. There are a number of mRNAs and proteins that oscillate in a circadian fashion in the SCN, in *Drosophila* clock-containing cells and in *Neurospora*. The myriad of circadian rhythms in the SCN, however, make it extremely difficult to determine the mechanisms for the oscillation of any gene's expression. Whether an oscillation is regulated by direct, physical interaction with a clock component or a downstream process such as transcriptional events induced by signal transduction events activated by firing rhythms of the cell is a subtle but critical issue. This issue may most effectively be addressed in biochemical and molecular biological studies of gene function rather than by genetics.

Phase Shift Response

That the circadian rhythm of a clock output gene is phase shifted in response to a phase-shifting stimulus is simply a verification of the fact that the gene is regulated by the circadian clock, which itself is phase-shifted. Any circadian rhythm that is not shifted in parallel to the overt rhythmicity of the organism must be regulated by a separate clock with divergent light responsiveness. Such a phenomenon has never been observed in mammals or *Neurospora*, although it may exist in *Drosophila* (91) and clearly does exist in *Gonyaulax* (92).

Nonnecessity

The clock-controlled genes (*ccgs*) in *Neurospora* were identified in a screen for genes that exhibit a circadian rhythm of expression (93). In the presence of null mutations of these *ccgs*, the *frq* rhythm remains intact. These genes thus provide an example of clock outputs necessary for the expression of rhythmicity at the cellular level. Circadian studies on *Drosophila* harboring a null mutation in the *DCO* gene, which encodes a protein kinase A catalytic subunit, revealed arrhythmicity at the level of locomotor activity but an intact rhythm of PER protein (94). Similarly, the *Drosophila lark* gene, encoding an RNA binding protein, was isolated from a mutagenesis screen for mutations which produce arrhythmicity of eclosion rhythms without altering locomotor rhythmicity (95). While *lark* mRNA expression is constitutive throughout the circadian cycle, the protein is expressed rhythmically in a subset of neural cells (96). The *disco* mutation in *Drosophila* also disrupts eclosion rhythms without abolishing molecular rhythms (73). Thus, *DCO*, *lark* and *disco* are not necessary for clock function but are necessary for clock output.

In considerations of a complex organism such as the mouse, determining whether or not a specific oscillatory gene is necessary for the clock itself to function may be exceedingly facile or quite difficult. Vasopressin, for instance, oscillates at the mRNA and protein levels in the SCN (97,98). Because vasopressin-deficient Brattleboro rats have virtually no detectable circadian abnormalities, we know that vasopressin is not a clock component but rather a clock output. Our unpublished studies on behavior in *vgf* knockout mice indicate that the circadian clock continues to function in the absence of the rhythmically-expressed *vgf* gene. On the other hand, if a mutation produced by targeted mutagenesis or one identified by the forward genetic approach abolishes rhythmicity at the behavioral level, it may be exceedingly difficult (until clock components are definitively identified) to determine if the mutation disrupts the clock itself or only clock outputs. First, it is essential to show that rhythmicity is also abolished at the cellular level within the SCN, and not somewhere between the clock and its distant, systems-level outputs. This question can be addressed at the cellular level with electrophysiological recordings of SCN cells and at the intracellular level by measuring the circadian expression of genes which normally exhibit rhythmicity, such as *per1*, in the SCN.

If electrophysiological or gene expression assays indicate that cellular circadian rhythms are intact, then the mutation may affect SCN outputs or the coupling of independent oscillators. The properties of mammalian circadian rhythms at the systems and behavioral levels are consistent with the possibility that there are a number of independent but coupled circadian oscillators in the SCN. Recent studies of cultured SCN cells indicate that many individual neurons contain separate oscillators (14,15) and that the behavioral output of the clock is produced by tight coupling of those oscillators. The mechanism for this coupling is entirely unknown, but conceptually it can be thought of as a reciprocal entrainment of all the circadian clocks in the SCN. The basis for coupling may become apparent through studies of clock output genes.

VII. Conclusions

There have been rapid improvements in the technologies associated with molecular genetics in recent years, and the improvements will likely continue in the years to come. The time required for the mapping of genes is expected to decrease, resulting in more efficient identification of genes in mutagenesis screens and increased power in QTL analysis. Gene chip technologies and other technologies will be useful for identifying large numbers of genes expressed in a certain tissue under set conditions (such as in the SCN after a light pulse). This fact and the fact that the entire human genome and the mouse genome will be sequenced in the near future mean that a finite set of candidate clock genes will be in sight soon. These are exciting times for clock geneticists. They are also hopeful times for physiologists, biochemists and systems and behavioral neuroscientists. The analysis of gene function and the relevance of molecular genetic information to studies of the circadian clock will require their expertise. The book may close on clock genetics, but knowledge of clock genes will be carried onward as a powerful tool in the research arsenal of circadian biology.

Acknowledgments

Research supported by grants from the National Science Foundation Center for Biological Timing, the National Institutes of Health, Bristol-Myers Squibb Unrestricted Grant in Neuroscience and Air Force Office of Scientific Research. Jonathan Wisor is a Postdoctoral Fellow in the National Multi-site Training Program for Basic Sleep Research. J.S. Takahashi is an Investigator in the Howard Hughes Medical Institute.

References

1.   Dunlap JC. Genetics and molecular analysis of circadian rhythms. Annu Rev Genet 1996; 30:579–601.
2.   Dunlap JC, Loros JJ, Aronson BD, et al. The genetic basis of the circadian clock: identification of *frq* and FRQ as clock components in *Neurospora*. Ciba Found Symp 1995; 183:3–17.
3.   Rosbash M. Molecular control of circadian rhythms. Curr Opin Genet Dev 1995; 5:662–668.
4.   Rosbash M, Allada R, Dembinska M, et al. A *Drosophila* circadian clock. Cold Spring Harb Symp Quant Biol 1996; 61:265–278.
5.   Sehgal A. Molecular genetic analysis of circadian rhythms in vertebrates and invertebrates. Curr Opin Neurobiol 1995; 5:823–831.
6.   Young MW. The *Drosophila* genes *timeless* and *period* collaborate to promote cycles of gene expression composing a circadian pacemaker. Prog Brain Res 1996; 111: 29–39.
7.   Moore RY. Organization of the mammalian circadian system. Ciba Found Symp 1995; 183:88–99.
8.   Hastings MH. Central clocking. TINS 1997; 20:459–464.
9.   Takahashi JS. Molecular neurobiology and genetics of circadian rhythms in mammals. Annu Rev Neurosci 1995; 18:531–553.
10. Schwartz WJ, Gross RA, Morton MT. The suprachiasmatic nuclei contain a tetrodotoxin-resistant circadian pacemaker. Proc Natl Acad Sci USA 1987; 84:1694–1698.
11. Moore RY, Bernstein ME. Synaptogenesis in the rat suprachiasmatic nucleus demonstrated by electron microscopy and synapsin I immunoreactivity. J Neurosci 1989; 9:2151–2162.
12. Grahn DA, Miller JD, Houng VS, Heller HC. Persistence of circadian rhythmicity in hibernating ground squirrels. Am J Physiol 1994; 266:R1251–R1258.

13. Hall JC. Circadian pacemakers blowing hot and cold—but they're clocks, not thermometers. Cell 1997; 90:9–12.

14. Welsh DK, Logothetis DE, Meister M, Reppert SM. Individual neurons dissociated from rat suprachiasmatic nucleus express independently phased circadian firing rhythms. Neuron 1995; 14:697–706.

15. Liu C, Weaver DR, Strogatz SH, Reppert SM. Cellular construction of a circadian clock: period determination in the suprachiasmatic nuclei. Cell 1997; 91:855–860.

16. Feldman JF. Lengthening the period of a biological clock in *Euglena* by cyclohex-imide, an inhibitor of protein synthesis. Proc Natl Acad Sci USA 1967; 57:1080–1087.

17. Takahashi JS, Turek FW. Anisomycin, an inhibitor of protein synthesis, perturbs the phase of a mammalian circadian pacemaker. Brain Res 1987; 405:199–203.

18. Johnson CH, Nakashima H. Cycloheximide inhibits light-induced phase shifting of the circadian clock in *Neurospora*. J Biol Rhythms 1990; 5:159–167.

19. Khalsa SB, Whitmore D, Block GD. Stopping the circadian pacemaker with inhibi-tors of protein synthesis. Proc Natl Acad Sci USA 1992; 89:10862–10866.

20. Jacklet JW. Neuronal circadian rhythm: phase shifting by a protein synthesis inhibi-tor. Science 1977; 198:69–71.

21. Watanabe K, Katagai T, Ishida N, Yamaoka S. Anisomycin induces phase shifts of circadian pacemaker in primary cultures of rat suprachiasmatic nucleus. Brain Res 1995; 684:179–184.

22. Ralph MR, Menaker M. A mutation of the circadian system in golden hamsters. Science 1988; 241:1225–1227.

23. Takahashi JS, Pinto LH, Vitaterna MH. Forward and reverse genetic approaches to behavior in the mouse. Science 1994; 264:1724–1733.

24. Konopka RJ, Benzer S. Clock mutants of *Drosophila melanogaster*. Proc Natl Acad Sci USA 1971; 68:2112–2116.

25. Feldman JF, Hoyle MN. Isolation of circadian clock mutants of *Neurospora crassa*. Genetics 1973; 75:605–613.

26. Kondo T, Tsinoremas NF, Golden SS, Johnson CH, Kutsuna S, Ishiura M. Circadian clock mutants of cyanobacteria. Science 1994; 266:1233–1236.

27. Vitaterna MH, King DP, Chang AM, et al. Mutagenesis and mapping of a mouse gene, *Clock*, essential for circadian behavior. Science 1994; 264:719–725.

28. Pickard GE, Sollars PJ, Rinchik EM, Nolan PM, Bucan M. Mutagenesis and behavioral screening for altered circadian activity identifies the mouse mutant, *Wheels*. Brain Res 1995; 705:255–266.

29. Millar AJ, Carre IA, Strayer CA, Chua NH, Kay SA. Circadian clock mutants in *Arabidopsis* identified by luciferase imaging. Science 1995; 267:1161–1163.

30. Zehring WA, Wheeler DA, Reddy P, et al. P-element transformation with *period* locus DNA restores rhythmicity to mutant, arrhythmic *Drosophila melanogaster*. Cell 1984; 39:369–376.

31. Bargiello TA, Jackson FR, Young MW. Restoration of circadian behavioural rhythms by gene transfer in *Drosophila*. Nature 1984; 312:752–754.

32. Hardin PE, Hall JC, Rosbash M. Feedback of the *Drosophila period* gene product on circadian cycling of its messenger RNA levels. Nature 1990; 343:536–540.

33. Sauman I, Reppert SM. Circadian clock neurons in the silkmoth *Antheraea pernyi*: novel mechanisms of *Period* protein regulation. Neuron 1996; 17:889–900.

34. Cheng Y, Hardin PE. *Drosophila* photoreceptors contain an autonomous circadian oscillator that can function without *period* mRNA cycling. J Neurosci 1998; 18:741–750.

35. Sehgal A, Price JL, Man B, Young MW. Loss of circadian behavioral rhythms and *per* RNA oscillations in the *Drosophila* mutant *timeless*. Science 1994; 18:1603–1606.

36. Myers MP, Wager-Smith K, Wesley CS, Young MW, Sehgal A. Positional cloning and sequence analysis of the *Drosophila* clock gene, *timeless*. Science 1995; 270: 805–808.

37. Gekakis N, Saez L, Delahaye-Brown A, et al. Isolation of *timeless* by PER protein interaction: defective interaction between *timeless* protein and long-period mutant PERL. Science 1995; 270:811–815.

38. Saez L, Young MW. Regulation of nuclear entry of the *Drosophila* clock proteins *period* and *timeless*. Neuron 1996; 17:911–920.

39. Aronson BD, Johnson KA, Dunlap JC. Circadian clock locus *frequency*: protein encoded by a single open reading frame defines period length and temperature compensation. Proc Natl Acad Sci USA 1994; 91:7683–7687.

40. Aronson BD, Johnson KA, Loros JJ, Dunlap JC. Negative feedback defining a circadian clock: autoregulation of the clock gene *frequency*. Science 1994; 263:1578–1584.

41. Linden H, Rodriguez-Franco M, Macino G. Mutants of *Neurospora crassa* defective in regulation of blue light perception. Mol Gen Genet 1997; 254:111–118.

42. Crosthwaite SK, Dunlap JC, Loros JJ. *Neurospora wc-1* and *wc-2*: transcription, photoresponses, and the origins of circadian rhythmicity. Science 1997; 276:763–769.

42a. Crosthwaite SK, Loros JJ, Dunlap JC. Light-induced resetting of a circadian clock is mediated by a rapid increase in *frequency* transcript. Cell 1995; 81:1003–1012.

43. Shedlovsky A, McDonald JD, Symula D, Dove WF. Mouse models of human phenylketonuria. Genetics 1993; 134:1205–1210.

44. King DP, Zhao Y, Sangoram AM, et al. Positional cloning of the mouse circadian *Clock* gene. Cell 1997; 89:641–653.

45. Antoch MP, Song EJ, Chang AM, et al. Functional identification of the mouse circadian *Clock* gene by transgenic BAC rescue. Cell 1997; 89:655–667.

46. Dunlap JC. Genetic analysis of circadian clocks. Annu Rev Physiol 1993; 55:683–728.

47. Nolan PM, Sollars PJ, Bohne BA, Ewens WJ, Pickard GE, Bucan M. Heterozygosity mapping of partially congenic lines: mapping of a semidominant neurological mutation, *Wheels* (*Whl*), on mouse chromosome 4. Genetics 1995; 140:245–254.

48. Lander ES, Botstein D. Mapping mendelian factors underlying quantitative traits using RFLP linkage maps. Genetics 1989; 12:185–189.

49. Kruglyak L, Lander ES. A nonparametric approach for mapping quantitative trait loci. Genetics 1995; 139:1421–1428.

50. Zeng ZB. Precision mapping of quantitative trait loci. Genetics 1994; 136:1457–1468.

51. Markel P, Shu P, Eberling C, et al. Theoretical and empirical issues for marker-assisted breeding of congenic mouse strains. Nat Genet 1997; 17:280–284.

52. Darvasi A. Experimental strategies for the genetic dissection of complex traits in animal models. Nat Genet 1998; 18:19–24.

53. Mayeda AR, Hofstetter JR, Belknap JK, Nurnberger JI. Hypothetical quantitative trait loci (QTL) for circadian period of locomotor activity in CXB recombinant inbred strains of mice. Behav Genet 1997; 26:505–511.

54. Hofstetter JR, Mayeda AR, Possidente B, Nurnberger JI. Quantitative trait loci (QTL) for circadian rhythms of locomotor activity in mice. Behav Genet 1995; 25: 545–556.

55. Frankel WN, Johnson EW, Lutz CM. Congenic strains reveal effects of the epilepsy quantitative trait locus, *El2*, separate from other *El* loci. Mamm Genome 1995; 6: 839–843.

56. Bennett B, Beeson M, Gordon L, Johnson TE. Quick method for confirmation of quantitative trait loci. Alcohol Clin Exp Res 1997; 21:767–772.

57. Fineman RJ, de Vries SS, Jansen RC, Demant P. Complex interactions of new quantitative trait loci, *Sluc1*, *Sluc2*, *Sluc3*, and *Sluc4*, that influence the susceptibility to lung cancer in the mouse. Nat Genet 1996; 14:465–467.

58. Cormier RT, Hong KH, Halberg RB, et al. Secretory phospholipase *Pla2g2a* confers resistance to intestinal tumorigenesis. Nat Genet 1997; 17:88–91.

59. Banbury Conference on Genetic Background in Mice. Mutant mice and neuroscience: recommendations concerning genetic background. Neuron 1997; 19:755–769.

60. Frankel WN. Mouse strain backgrounds: more than black and white. Neuron 1998; 20:183.

61. Kornhauser JM, Mayo KE, Takahashi JS. Light, immediate-early genes, and circadian rhythms. Behav Genet 1996; 26:221–240.

62. Shin HS, Bargiello TA, Jackson FR, Young MW. An unusual coding sequence from a *Drosophila* clock gene is conserved in vertebrates. Nature 1985; 317:445–448.

63. Ishida N, Matsui M, Nishimatsu S, Murakami K, Mitsui Y. Molecular cloning of a gene under control of the circadian clock and light in the rodent SCN. Mol Brain Res 1994; 26:197–206.

64. Rosewell KL, Siwicki KK, Wise PM. A *period* (*per*)-like protein exhibits daily rhythmicity in the suprachiasmatic nuclei of the rat. Brain Res 1994; 659:231–236.

65. Tei H, Okamura H, Shigeyoshi Y, et al. Circadian oscillations of a mammalian homologue of the *Drosophila period* gene. Nature 1997; 389:512–516.

66. Sun ZS, Albrecht U, Zhuchenko O, Bailey J, Eichele G, Lee CC. *RIGUI*, a putative mammalian ortholog of the *Drosophila period* gene. Cell 1997; 90:1003–1011.

67. Albrecht U, Sun ZS, Eichele G, Lee CC. A differential response of two putative mammalian circadian regulators, *mper1* and *mper2*, to light. Cell 1997; 91:1055–1064.

68. Shearman LP, Zylka MJ, Weaver DR, Kolakowski LF, Reppert SM. Two *period* homologs: circadian expression and photic regulation in the suprachiasmatic nuclei. Neuron 1997; 19:1261–1269.

69. Tosini G, Menaker M. Circadian rhythms in cultured mammalian retina. Science 1996; 272:419–421.

70. Shigeyoshi Y, Taguchi K, Yamamoto S, et al. Light-induced resetting of a mammalian circadian clock is associated with rapid induction of the *mPer1* transcript. Cell 1997; 91:1043–1053.

71. Zatz M. Perturbing the pacemaker in the chick pineal cell. Disc Neurosci 1992; 8:67–72.

72. Page TL. Time is the essence: molecular analysis of the biological clock. Science 1994; 263:1570–1572.

73. Hardin PE, Hall JC, Rosbash M. Behavioral and molecular analyses suggest that

circadian output is disrupted by *disconnected* mutants in *D. melanogaster*. EMBO J 1992; 11:1–6.

74. Sehgal A, Rothenfluh-Hilfiker A, Hunter-Ensor M, Chen Y, Myers MP, Young MW. Rhythmic expression of *timeless*: a basis for promoting circadian cycles in *period* gene autoregulation. Science 1995; 270:808–810.

75. Edery I, Zwiebel LJ, Dembinska ME, Rosbash M. Temporal phosphorylation of the *Drosophila period* protein. Proc Natl Acad Sci USA 1994; 91:2260–2264.

76. Hunter-Ensor M, Ousley A, Sehgal A. Regulation of the *Drosophila* protein *timeless* suggests a mechanism for resetting the circadian clock by light. Cell 1996; 84: 677–685.

77. Zeng H, Hardin PE, Rosbash M. Constitutive overexpression of the *Drosophila period* protein inhibits *period* mRNA cycling. EMBO J 1994; 13:3590–3598.

78. Uhl GR, Gold LH, Risch N. Genetic analysis of complex behavioral disorders. Proc Natl Acad Sci USA 1997; 94:2785–2786.

79. Edery I, Rutila JE, Rosbash M. Phase shifting of the circadian clock by induction of the *Drosophila period* protein. Science 1994; 263:237–240.

80. Ewer J, Rosbash M, Hall JC. An inducible promoter fused to the *period* gene in *Drosophila* conditionally rescues adult *per*-mutant arrhythmicity. Nature 1988; 333:82–84.

81. Reppert SM, Sauman I. *period* and *timeless* tango: a dance of two clock genes. Neuron 1995; 15:983–986.

82. Mayford M, Bach ME, Huang YY, Wang L, Hawkins RD, Kandel ER. Control of memory formation through regulated expression of a CaMKII transgene. Science 1996; 274:1678–1683.

83. Zeng H, Qian Z, Myers MP, Rosbash M. A light-entrainment mechanism for the *Drosophila* circadian clock. Nature 1996; 380:129–135.

84. Lee C, Parikh V, Itsukaichi T, Bae K, Edery I. Resetting the *Drosophila* clock by photic regulation of PER and a PER-TIM complex. Science 1996; 271:1740–1744.

85. Myers MP, Wager-Smith K, Rothenfluh-Hilfiker A, Young MW. Light-induced degradation of TIMELESS and entrainment of the *Drosophila* circadian clock. Science 1996; 271:1736–1740.

86. Honrado GI, Johnson RS, Golombek DA, Spiegelman BM, Papaioannou VE, Ralph MR. The circadian system of *c-fos* deficient mice. J Comp Physiol [A] 1996; 178: 563–570.

87. Struthers RS, Vale WW, Arias C, Sawchenko PE, Montminy MR. Somatotroph hypoplasia and dwarfism in transgenic mice expressing a non-phosphorylatable CREB mutant. Nature 1991; 350:622–624.

88. Reppert SM, Weaver DR, Ebisawa T. Cloning and characterization of a mammalian melatonin receptor that mediates reproductive and circadian responses. Neuron 1994; 13:1177–1185.

89. Harding TC, Geddes BJ, Noel JD, Murphy D, Uney JB. Tetracycline-regulated transgene expression in hippocampal neurones following transfection with adenoviral vectors. J Neurochem 1997; 69:2620–2623.

90. Ginty DD, Kornhauser JM, Thompson MA, et al. Regulation of CREB phosphorylation in the suprachiasmatic nucleus by light and a circadian clock. Science 1993; 260:238–241.

91. Kaneko M, Helfrich-Forster C, Hall JC. Spatial and temporal expression of the *period* and *timeless* genes in the developing nervous system of *Drosophila*: newly identified pacemaker candidates and novel features of clock gene product cycling. J Neurosci 1997; 17:6745–6760.

92. Morse D, Hastings JW, Roenneberg T. Different phase responses of the two circadian oscillators in *Gonyaulax*. J Biol Rhythms 1994; 94:263–274.

93. Loros JJ, Denome SA, Dunlap JC. Molecular cloning of genes under control of the circadian clock in *Neurospora*. Science 1989; 243:385–388.

94. Majercak J, Kalderon D, Edery I. *Drosophila melanogaster* deficient in protein kinase A manifests behavior-specific arrhythmia but normal clock function. Mol Cell Biol 1997; 17:5915–5922.

95. Newby LM, Jackson FR. A new biological rhythm mutant of *Drosophila melanogaster* that identifies a gene with an essential embryonic function. Genetics 1993; 135: 1077–1090.

96. McNeil GP, Zhang X, Genova G, Jackson FR. A molecular rhythm mediating circadian clock output in *Drosophila*. Neuron 1998; 20:297–303.

97. Isobe Y, Nakajima K, Nishino H. Arg-vasopressin content in the suprachiasmatic nucleus of rat pups: circadian rhythm and its development. Dev Brain Res 1995; 85: 58–63.

98. Smith M, Carter DA. In situ hybridization analysis of vasopressin mRNA expression in the mouse hypothalamus: diurnal variation in the suprachiasmatic nucleus. J Chem Neuroanat 1996; 12:105–112.

13

Circadian and Sleep Control of Hormonal Secretions

EVE VAN CAUTER and KARINE SPIEGEL

University of Chicago
Chicago, Illinois

I. Introduction

A prominent feature of the endocrine system is its high degree of temporal organization. Indeed, far from obeying the concept of "constancy of the internal milieu," which was the dogma of early 20th-century endocrinology, circulating hormonal levels undergo pronounced temporal oscillations over the 24-hr cycle. The characteristics of the 24-hr patterns of hormonal secretions vary from one hormone to the other and exhibit a high degree of day-to-day reproducibility. For most hormones, the 24-hr secretory profile reflects the interaction of circadian rhythmicity (i.e., intrinsic effects of time of day, irrespective of the sleep or wake state) and sleep (i.e., intrinsic effects of the sleep state, irrespective of the time of day when it occurs). The overall waveshape of 24-hr hormonal patterns also reflects, to varying degrees, modulatory effects from rhythmic and nonrhythmic factors, such as periodic food intake, postural changes, levels of physical activity, and, within the sleep state, the alternation between non–rapid eye movement (non-REM) and REM stages.

Effects of circadian rhythmicity and sleep on endocrine function have been studied most extensively in the human. Methods of measurement, including repeated blood sampling at frequent time intervals, sensitive hormone assays, and polygraphic sleep recordings, are more readily available for humans than for laboratory animals. Human sleep is generally consolidated in a single 6–9-hr period, whereas fragmentation of the sleep period in numerous bouts is the rule in other mammals. The detection of hormonal changes during sleep is thus hindered by the short duration of the sleep cycle in laboratory species. Possibly because of the consolidation of the sleep period, the wake-sleep and sleep-wake transitions in the human are associated with hormonal and metabolic changes that are more marked than those in other mammals. Finally, humans are unique in their capacity to ignore circadian signals and to voluntarily maintain wakefulness despite an increased pressure to go to sleep. As will be described later, these behaviors are associated with alterations in endocrine function that could have long-term consequences. For these reasons, only studies in the human will be reviewed in the present chapter.

Early studies suggested that the 24-hr rhythms of certain hormones, such as growth hormone (GH) and prolactin (PRL), were entirely "sleep-related," without any input from circadian rhythmicity, and that the 24-hr profiles of other hormones, such as cortisol, were entirely "circadian-dependent," without any sleep dependence (1). Current evidence clearly indicates that there is no such dichotomy and that both circadian inputs and sleep inputs can be recognized in the 24-hr profiles of most hormones. To differentiate between effects of circadian rhythmicity and effects of sleep-wake homeostasis, experimental strategies taking advantage of the fact that the circadian pacemaker takes several days to adjust to a large abrupt shift of sleep-wake and light-dark cycles have been used. Such strategies allow for the effects of circadian modulation to be observed in the absence of sleep and for the effects of sleep to be observed at an abnormal circadian time. Figure 1 illustrates mean profiles of plasma cortisol, plasma GH, plasma prolactin (PRL), and plasma thyrotropin (TSH) observed in normal subjects who were studied before and during an abrupt 12-hr shift of the sleep-wake and dark-light cycle. The study period extended over a 53-hr span and included an 8-hr period of nocturnal sleep starting at 23:00, a 28-hr period of continuous wakefulness, and a daytime period of recovery sleep, starting 12 hr out of phase with the usual bedtime, i.e., at 11:00. To eliminate the effects of feeding, fasting, and postural changes, the subjects remained recumbent throughout the study and the normal meal schedule was replaced by intravenous glucose infusion at a constant rate. As shown in Figure 1, this drastic manipulation of sleep had only modest effects on the waveshape of the cortisol profile, in sharp contrast with the immediate shift of the GH and PRL rhythms that followed the shift of the sleep-wake cycle. As will be reviewed in subsequent sections, numerous studies have indicated that the control of diurnal rhythms of corticotropic activity is primarily dependent on circadian timing whereas sleep-wake homeostasis appears to be an important factor in the control of the 24-hr profiles of GH and PRL release (2). Nevertheless, small modulatory effects of sleep-wake homeostasis on cortisol secretion and, conversely, influences of circadian timing on somatotropic function

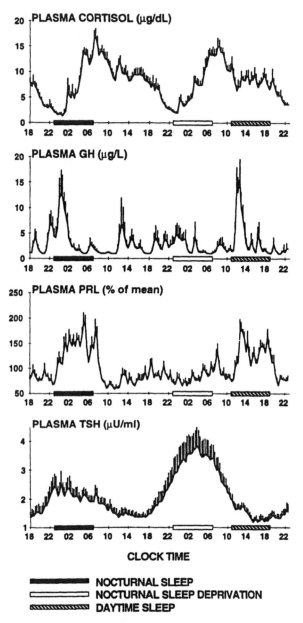

Figure 1 (From top to bottom) Mean 24-hr profiles of plasma cortisol, GH, PRL, and TSH in a group of eight normal young men (aged 20–27 years) studied during a 53-hr period including 8 hr of nocturnal sleep, 28 hr of sleep deprivation, and 8 hr of daytime sleep. The vertical bars at each time point represent the SEM. The black bars represent the sleep periods. The open bars represent the period of nocturnal sleep deprivation. The dashed bars represent the period of daytime sleep. Data were sampled at 20-min intervals.

have been clearly demonstrated (2). The diurnal variation of TSH levels includes an evening elevation thought to be under circadian control and nocturnal inhibition by sleep-dependent processes, which is clearly demonstrated during sleep deprivation, when a large increase in nocturnal TSH levels is apparent, as shown in the lowest panel of Figure 1 (2).

The pathways by which circadian rhythmicity, sleep, and their interaction (reviewed in Chapter 4) modulate hormonal release are largely unknown. At the level of the central nervous system, hormonal and/or neural signals originating from the circadian pacemaker in the hypothalamus and from brain regions involved in sleep regulation affect the activity of the hypothalamic structures responsible for the pulsatile release of neuroendocrine factors (e.g., corticotropin-releasing hormone, growth hormone-releasing hormone, gonadotropin-releasing hormone, etc.), which stimulate or inhibit intermittent secretion of pituitary hormones. As reviewed below, it appears that stimulatory or inhibitory effects of sleep on endocrine release are primarily associated with non-REM, rather than REM, sleep and therefore it appears likely that projections from the thalamus and the basal forebrain (the two primary brain structures implicated in non-REM sleep regulation) to the hypothalamus play a role in modulating the activity of the hypothalamic pulse generators. Theoretically, the modulation of neuroendocrine release by sleep and circadian rhythmicity could be achieved either by modulation of pulse amplitude, by modulation of pulse frequency, or by a combination of both. These issues have been addressed in only a handful of studies (3) examining circadian modulation in the absence of the confounding effects of sleep (i.e., during sleep deprivation) and sleep modulation at abnormal times of day (i.e., to avoid confounding circadian effects). So far, it appears that circadian rhythmicity of pituitary hormonal release is achieved primarily by modulation of pulse amplitude without changes in pulse frequency whereas sleep-wake and REM–non-REM transitions clearly affect pulse frequency. Pituitary hormones that influence endocrine systems not directly controlled by hypothalamic factors probably mediate, at least partially, the modulatory effects of sleep and circadian rhythmicity on these systems (e.g., counterregulatory effects of GH and cortisol on glucose regulation).

Finally, a number of hormones that are modulated by circadian rhythmicity and sleep appear to influence sleep quality. Prominent among these are the hormones of the somatotropic axis and PRL.

The following subsections will review the modulations of four of the hypothalamopituitary axes by circadian rhythmicity and sleep, their alterations in abnormal conditions, and the implications of disturbances of sleep and/or circadian rhythmicity for the development of endocrine abnormalities. Reciprocal effects of components of each of these axes on sleep regulation will be briefly summarized. The scope of this volume does not allow for an exhaustive review of effects of sleep and circadian rhythmicity throughout the endocrine system. For descriptions of the complex effects of sleep and/or circadian rhythmicity on the reproductive axis, on glucose regulation, and on hormones regulating water balance, the reader is referred to recent reviews focused on these subjects (4–7).

II. Temporal Organization of the Corticotropic Axis

A. Circadian and Sleep Control of Corticotropic Activity

Activity of the corticotropic axis may be measured peripherally via plasma levels of the pituitary hormone adenocorticotropin (ACTH) and of the adrenal hormone directly controlled by ACTH stimulation, cortisol. The 24-hr profiles of plasma ACTH and cortisol show an early-morning maximum, declining levels throughout the daytime, a quiescent period centered around midnight, and a rapid elevation that usually starts 2–3 hr after sleep onset. The cortisol profiles shown in the uppermost panel of Figure 1 illustrate the remarkable persistence of the wave-shape of the cortisol profile when sleep is manipulated. Indeed, the overall waveshape of the temporal profile of cortisol levels was not markedly affected by the absence of sleep or the presence of sleep at an abnormal time of day. Studies using similar experimental designs have therefore indicated that the 24-hr periodicity of corticotropic activity is primarily controlled by circadian rhythmicity. Nevertheless, modulatory effects of the sleep or wake condition have been clearly demonstrated. Indeed, a number of studies have indicated that sleep onset is reliably associated with a short-term inhibition of cortisol secretion (8–10) although this effect may not be detectable when sleep is initiated at the time of the daily maximum of corticotropic activity, i.e., in the morning (11). Under normal conditions, since cortisol secretion is already quiescent in the late evening, this inhibitory effect of sleep, which appears to be related to slow-wave stages (12), results in a prolongation of the quiescent period. Therefore, under conditions of sleep deprivation, the nadir of cortisol secretion is less pronounced and occurs earlier than under normal conditions of nocturnal sleep. Conversely, awakening at the end of the sleep period is consistently followed by a pulse of cortisol secretion (10,13,14). During sleep deprivation, these rapid effects of sleep onset and sleep offset on corticotropic activity are obviously absent, and, as may be seen in the profiles shown in Figure 1, the nadir of cortisol levels is higher than during nocturnal sleep (because of the absence of the inhibitory effects of the first hours of sleep) and the morning acrophase is lower (because of the absence of the stimulating effects of morning awakening). Overall, the amplitude of the rhythm is reduced by approximately 15% during sleep deprivation as compared to normal conditions.

B. Effects of Sleep Fragmentation and Sleep Loss

Several studies have shown that awakenings interrupting the sleep period consistently trigger pulses of cortisol secretion (3,12,13). This is illustrated in the

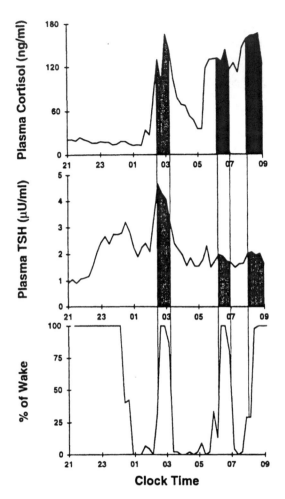

Figure 2 Nocturnal profiles of plasma cortisol (top) and TSH levels (middle) in a normal subject sampled during polygraphic sleep recording. (Bottom) The percentage of each time interval between blood samplings (i.e., 15 min) spent awake. The shaded vertical areas demonstrate the correspondence between increasing levels of cortisol and TSH during awakenings interrupting the sleep period.

uppermost panel of Figure 2, which shows the plasma cortisol levels, sampled at 15-min intervals, in a normal young man studied during polygraphically recorded sleep. The lowest panel depicts the percentage of time spent awake between two consecutive blood samplings. The subject experienced two prolonged awakenings interrupting the sleep period. The first awakening was clearly associated with a significant elevation of plasma cortisol. The relationship between the second awakening and simultaneous cortisol levels is more difficult to ascertain because the awakening occurred less abruptly, with alternating stages of wake and light sleep preceding the period of continuous wakefulness. The final morning awakening was again followed by a cortisol elevation. In an analysis of cortisol profiles during daytime sleep, it was observed that 92% of awakenings interrupting sleep coincided with or were followed within 20 min by a significant cortisol pulse (3). A study involving continuous experimentally induced arousals during sleep suggested the existence of a difference in the efficacy of feedback inhibition of corticotropic activity during sleep versus wake (13). Feedback inhibition appeared to be less effective during sleep than during wake, allowing for a more rapid increase of cortisol concentrations during the second half of the night when the subject was asleep than when the subject was awake. The mean cortisol patterns during sleep and during sleep deprivation shown in Figure 1 clearly illustrate this tendency. Sleep fragmentation, as occurs in aging (15), is thus associated with alterations of nocturnal corticotropic activity.

In addition to the immediate modulatory effects of sleep-wake transitions on cortisol levels, nocturnal sleep deprivation, even partial sleep deprivation, appears to affect corticotropic activity on the following evening (16). Figure 3 summarizes the results from a recent study involving three groups of normal young men studied before, during, and after a night of normal sleep (bedtimes: 23:00–07:00), partial sleep deprivation (bedtimes: 04:00–08:00), or total sleep deprivation. All subjects were studied over a 32-hr period, from 18:00 to 02:00. Mean levels of cortisol were similar in all three groups in the morning and in the afternoon. However, cortisol levels during the time interval 18:00–23:00 were significantly higher on the evening following partial or total sleep deprivation than on the previous day at the same clock time. Sleep loss thus appears to delay the normal return to evening quiescence of the corticotropic axis. Since the morning cortisol elevation occurs in response to increased CRH drive during the second part of the night, the daylong decrease in cortisol concentrations may be viewed as a recovery from this endogenous challenge. Therefore, a delay in the initiation of the quiescent period, as observed following sleep loss, may reflect an alteration in the rate of recovery of the HPA axis response to a challenge, often referred to as the "resiliency" of the HPA axis. Irrespective of their underlying causes, the observations illustrated in Figure 3 challenge the common belief that, on the day following a night of partial or total sleep deprivation, the primary effects of sleep loss are behavioral, rather than physiological.

C. Implications for Aging and Disease

An analysis of nearly 200 temporal profiles of plasma cortisol obtained in healthy men and women, aged 18–83 years, has demonstrated the existence of marked

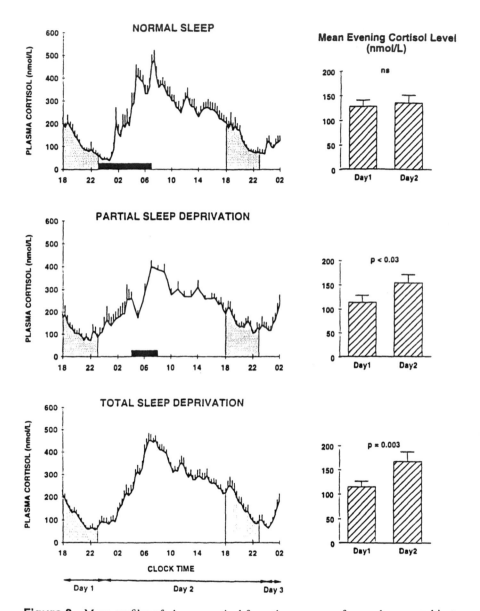

Figure 3 Mean profiles of plasma cortisol from three groups of normal young subjects studied during a 32-hr period with normal sleep (top), partial sleep deprivation (middle), or total sleep deprivation (bottom). The black bar represents the sleep period. The shaded areas highlight cortisol levels during the time interval 18:00–23:00 on day 1 and day 2. Mean cortisol levels for the same time period before and after normal sleep (top) and before and after partial or total sleep deprivation (middle and bottom) are shown for the three groups on the left panels. (Reproduced from Ref. 16.)

gender-specific effects of aging on the levels and diurnal variation of human adrenocorticotropic activity (17). In both men and women, mean cortisol levels increase by 20–50% between 20 and 80 years of age. The diurnal rhythmicity of cortisol secretion is preserved in old age, but the relative amplitude is dampened. The level of the nocturnal nadir increases progressively with aging in both sexes. This augmentation of nocturnal cortisol levels results from a dual defect: a delay in the onset of the quiescent period and an advance in the timing of the circadian elevation. The reduction in duration of the quiescent period between 25 and 65 years of age averaged 161 min in men and 272 min in women (17). An age-related elevation in morning acrophase occurs in women, but not in men. It is likely that alterations in sleep and circadian regulation with advancing age are involved in the drastic reduction of the duration of exposure to low cortisol levels. Indeed, the findings on the effects of sleep loss in young adults illustrated in Figure 3 suggest that the increased sleep fragmentation that characterizes sleep in the elderly may play a role in the failure to obtain a timely and adequate suppression of corticotropic activity in the evening. The advance of the morning elevation may represent an advance of circadian phase. There is indeed evidence that senescence is associated with neuronal loss, altered function, and decreased levels of neuro-peptides and neurotransmitters in the suprachiasmatic nucleus (18,19). These age-related alterations in the central pacemaker are consistent with a shortening of the endogenous circadian period in old age, which would be reflected, under entrained condition, in an advance of circadian phase.

Increased exposure to higher cortisol levels, particularly at a time when receptor occupancy is normally minimal, i.e., in the evening and early part of the night, could be involved in the development of cognitive and metabolic abnor-malities in aging. Indeed, animal studies have shown that cumulative exposure to glucocorticoids causes degenerative changes in the hippocampus, the brain region that normally inhibits glucocorticoid release. These alterations impair the ability to terminate glucocorticoid secretion at the end of stress, resulting in an ever-decreasing resiliency of the HPA response (20). Such hippocampal defects may underlie some of the memory deficits that occur in a majority of older adults (21). An addition to being relevant to memory impairment in aging, this loss of resiliency could result in metabolic abnormalities related to glucocorticoid excess, and in particular in age-related insulin resistance (22). Finally, age-related sleep disorders could be involved in a feed-forward cascade of negative effects because increased activity of the corticotropic axis in the evening promotes sleep fragmen-tation (23,24) and could thus further impair the ability to maintain low evening glucocorticoid concentrations for an appropriate amount of time.

Finally, it is noteworthy that depressive illness may be associated with abnormalities of the 24-hr cortisol profile that are similar, albeit more severe, than those observed in healthy older adults. In patients with major endogenous depres-sion in the acute phase of the illness, the quiescent period is drastically shortened and an advance of the morning circadian elevation is observed in the majority of subjects (25). While disorders of circadian time-keeping have often been implied in these disturbances of the cortisol profile (26), it seems likely that the profound alterations of sleep that are a hallmark of depressive illness may be partially involved in the elevation of nocturnal cortisol levels.

III. Interactions Between the Somatotropic Axis, Sleep, and Circadian Rhythmicity

A. Modulation of GH Secretion by Sleep Stages and Circadian Rhythmicity

The fact that the secretion of GH is markedly stimulated during sleep has been recognized for more than three decades. Early studies using the first available radioimmunoassays for GH demonstrated that the peripheral levels of these related hormones increased rapidly following sleep onset (27–33). In normal adult subjects, the 24-hr profile of plasma GH levels consists of stable low levels abruptly interrupted by bursts of secretion. The most reproducible pulse occurs shortly after sleep onset (29,30). This relationship between sleep onset and GH secretion appears to be most consistent in the human, and it is more difficult to uncover in other mammals. This species difference could be related to the fact that human sleep is consolidated in a single 7–9-hr period, whereas sleep is fragmented into multiple bouts in most other mammals.

In men, the sleep-onset GH pulse is generally the largest, and, after the third decade of age, often the only, pulse observed over the 24-hr span. In women, daytime GH pulses are more frequent and the sleep-associated pulse, although still present in most cases, does not generally account for the majority of the 24-hr GH release. Sleep onset will elicit a pulse in GH secretion whether sleep is advanced, delayed, or interrupted and reinitiated. The GH profiles shown in Figure 1 illustrate the maintenance of this relationship in subjects who underwent a 12-hr shift of the sleep-wake cycle. A pulse of GH secretion following sleep onset has been observed in subjects submitted to a variety of manipulations of the sleep-wake cycle, including delays by 3 hr, 5 hr, 8 hr, 12 hr, and 16 hr, daytime recovery sleep following 28 hr of continuous wakefulness, nocturnal recovery sleep following 40 hr of continuous wakefulness, a 7-hr advance associated with transmeridian travel, and a 7-hr advance following 33 hr of sleep deprivation in the laboratory (reviewed in Ref. 7). Daytime naps are more consistently associated with GH release when they occur in the afternoon, when the propensity for SW sleep is increased, than in the morning, when REM sleep predominates (34,35). A recent study of night workers indicated that the main GH secretory episode still occurred during the first half of the sleep period (36).

While sleep is clearly a major determinant of the 24-hr profile of GH secretion in humans, there is also evidence for the existence of a circadian modulation, i.e., an intrinsic effect of time of day. A careful reanalysis of an early study of normal subjects submitted to a 3-hr sleep-wake cycle (60 min of sleep every 3 hr) for a prolonged period indicated that the elevation of GH concentration following sleep onset was largest when the sleep episode began in the late evening, i.e., around the usual bedtime (37). The late evening and early part of the night appear to represent a period of increased propensity for GH secretion, as studies involving abrupt delays of the sleep period in very young adults have shown modest increases in GH pulsatility at this time of day, suggesting the existence of a weak circadian modulation of GH secretion (36,38–40). The most convincing evidence for a circadian variation has been obtained in a recent study using repeated injections of GHRH at 3-hr intervals. The mean profiles of plasma GH observed under this treatment are shown in Figure 4. A diurnal variation in peak GH response is clearly demonstrated and was interpreted as reflecting a diurnal rhythm in somatostatinergic tone (41). Importantly, elevated GH responses were already apparent in the early evening, well before sleep onset, indicating that this diurnal variation partially reflects circadian rhythmicity.

Well-documented studies published in the late sixties concurred in indicating that there is a consistent relationship between the appearance of delta waves in the EEG and GH secretion during early sleep as well as during the later part of the night (29,30,31,42). These initial findings were confirmed in a number of later reports (43–45) but not in others that suggested that the relationship with SW sleep is fortuitous (92,46,47). Indeed, nocturnal GH surges occurring independently of the presence of SW sleep were reported in one study (47), and in another, selective partial SW stage deprivation failed to suppress or delay the sleep-onset GH pulse (9). Several investigators reported marked rises in GH secretion prior to the onset of sleep (46–48). An analysis of the association of GH secretion during sleep and delta-wave activity failed to demonstrate a significant linear "dose-response" relationship between the two processes (47), and it was suggested that sleep onset per se, rather than the occurrence of SWS, is the primary determinant of sleep-related GH secretion.

Later studies of the relationship between sleep stages and GH release used deconvolution (a procedure that allows secretory rates to be derived from plasma concentrations by eliminating the effects of hormonal distribution and clearance using a mathematical model) to examine GH secretory rates, rather than plasma GH levels (38,49). The analysis of variations in GH secretory rates during the various stages of sleep is more accurate than the analysis of plasma concentration because the temporal limits of each pulse are more accurately defined and additional pulses that were masked by hormonal clearance are revealed. Using this approach, a detailed study with 30-sec sampling of plasma GH during sleep indicated that maximal GH release occurs within minutes of the onset of SW sleep (49). Furthermore, in studies examining GH secretion in normal young men of similar height and weight, it was found that approximately 70% of GH pulses

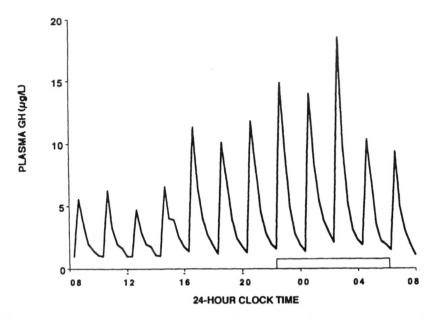

Figure 4 Mean (±SEM) profiles of plasma GH for a group of seven normal men studied on the seventh day of repeated i.v. bolus administration of 0.33 μg/kg of GHRH at 2-hr intervals. Lights were off from 23:00 to 07:00. Daytime naps were not permitted. The response to the constant GHRH stimulus increased in the evening, as compared to morning and afternoon, and increased further during the first half of the sleep period. (From Ref. 41, with permission.)

during sleep occurred during SW sleep and that there was a quantitative correlation between the amount of GH secreted during these pulses and the duration of the SW episode (38). This quantitative correlation between various markers of SW activity and amount of concomitant GH release has been confirmed in a more recent study (50).

Additional evidence for the existence of a robust relationship between SW activity and increased GH release has been obtained in studies using pharmacological stimulation of SW sleep (50,51). Reliable stimulation of SW sleep in normal subjects has been obtained with oral administration of low doses of gamma-hydroxybutyrate (GHB), a metabolite of gamma-aminobutyric acid that is normally present in the mammalian brain and is used as an investigational drug for the treatment of narcolepsy, as well as with ritanserin, a selective 5HT$_2$ receptor antagonist. The left panels of Figure 5 show mean GH profiles in a group of normal young men studied under baseline conditions and after bedtime administration of 2.5 g, 3 g, and 3.5 g of GHB. The amounts of SW sleep are shown for successive 2-hr periods from 23:00 to 07:00 in the upper panel. There was an excellent correlation (illustrated in the right panel of Fig. 5) between increases in the amount of stage IV sleep following GHB administration, and corresponding

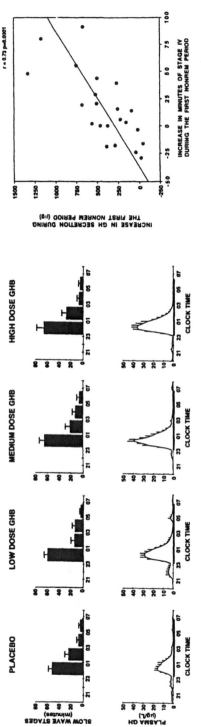

Figure 5 (Left) Mean profiles of plasma GH sampled at 15-min intervals in a group of normal young men studied under baseline conditions and after bedtime administration of 2.5 g, 3 g, and 3.5 g of gamma-hydroxybutyrate (GHB), which increases SW sleep. The black bars indicate the sleep periods. The amounts of SW sleep are shown for successive 2-hr periods from 23:00 to 07:00 in the upper panel. (Right) Correlation between increases in the amount of stage IV sleep during the first non-REM period following GHB administration, and corresponding increases in GH secretion during the same period. (Adapted from Ref. 51.)

409

increases in GH secretion during the same period. Administration of ritanserin was also found to result in parallel and highly correlated small amplitude increases of delta-wave activity and nocturnal GH release (50).

Nevertheless, the relationship between SW activity and GH secretion is not one-to-one. Indeed, nocturnal GH secretion can occur in the absence of SW sleep and approximately one-third of the SW periods are not associated with detectable GH secretion. Because GH secretion is also under inhibitory control by somatostastin, variability of somatostatinergic tone may cause a dissociation between SW sleep and nocturnal GH release. The short-term negative feedback inhibition exerted by GH on its own secretion may also explain observations of an absent GH pulse during the first SW period when a secretory pulse occurred prior to sleep onset. Such presleep GH pulses are likely to reflect the fact that the late-evening period appears to reflect a circadian rhythm in propensity to secrete GH independent of the occurrence of sleep (41).

Effects of sleep interruptions opposite to those observed for the corticotropic axis were found for GH release. Indeed, in a study where GH secretion was stimulated by the injection of the hypothalamic factor growth hormone-releasing hormone (GHRH) at the beginning of the sleep period, it was found that whenever sleep was interrupted by a spontaneous awakening, the ongoing GH secretion was abruptly suppressed (52). This inhibitory effect of awakenings on the GH response to GHRH was further demonstrated in a study where sleeping subjects who had received a GHRH injection were awakened 30 min after the injection and then allowed to reinitiate sleep 30 min later (53). A marked inhibition of the GH response was observed shortly after the awakening. These findings indicate that sleep fragmentation, as occurs in normal aging, will generally decrease nocturnal GH secretion. The inhibitory effect of nocturnal awakenings on GH secretion could be mediated by an increase in somatostatin release.

B. Alterations of GH Secretion in Aging and Sleep Disorders

Aging is associated with dramatic decreases in GH and IGF-I secretions and with pronounced alterations of sleep quality (54–59). In healthy elderly men over the age of 65, the total amount of GH secreted over the 24-hr span is generally less than one-third of the daily output of men under 30 years of age (54,56,60–63). Similarly, the amount of SW sleep in older adults is reduced in the same proportion. This decline in overall GH secretion appears to be achieved primarily by a decrease in the amplitude, rather than the frequency, of GH pulses. In a retrospective analysis involving nearly 100 simultaneous recordings of sleep and 24-hr GH secretion in adult men aged 18–82 years, we have recently shown that these dramatic effects of aging on SWS and GH secretion occur early in adulthood, in an exponential fashion, and are essentially complete by the beginning of the fifth decade (64). Similar observations in a smaller subject population studied overnight have recently been reported, although the nonlinear decrease in the amount of SW sleep failed to be detected (65). Although early studies had generally concluded that sleep-related GH pulses were absent in the elderly, the findings of more recent studies are concordant in showing persistent, but reduced, GH secretion during sleep (54,56,61,63). In our retrospective analysis, it was apparent that the proportion of daily GH output that occurs during the first few hours of sleep does not decrease with age, but remains stable or even slightly increases. A significant correlation between levels of IGF-I and delta power has been reported in older adults (66). The parallelism between decreased amount and quality of deep sleep and diminished somatotropic activity raises the interesting possibility that some of the peripheral effects of the hyposomatotropism of the elderly, such as the reduction in lean body mass, may partially reflect a central alteration in sleep control.

A few studies have examined nocturnal GH release in patients with obstructive apnea before and after treatment (67–69). Two studies that have examined the nocturnal GH profile before and after treatment with continuous positive airway pressure (CPAP) have demonstrated that treatment of the sleep disorder resulted in a clear increase in the amount of GH secreted during the first few hours of sleep (67,69). This is illustrated in the profiles shown in Figure 6. After 3 months of treatment, nasal CPAP therapy has also been shown to increase plasma IGF-1 levels in men with severe obstructive sleep apnea (70). In children, surgical correction of obstructive sleep apnea may restore GH secretion and normal growth rate (68).

C. Modulatory Effects of Components of the Somatotropic Axis on Sleep

There is good evidence that components of the somatotropic axis are involved in regulating sleep quality.

An early study involving an intramuscular injection of a pharmacological GH dosage administered 15 min before bedtime reported a marked stimulation of REM sleep and a decrease in SW sleep (71). A more recent study reported no effects on sleep quality when GH levels were elevated either by intravenous infusion or by intramuscular injection given approximately 3 hr before sleep onset (72). In GH-deficient subjects, prolonged treatment with daily injections of exogenous GH resulted in a marked increase in REM sleep (73).

A number of studies have demonstrated effects of GHRH on sleep quality and it has been suggested that GH secretion and sleep may share common regulatory mechanisms (74). No effects of GHRH on visually scored sleep stages were found when the peptide was injected during daytime or before sleep onset (75,76), or when it was given as an infusion (72,77). However, delta power during the first 100 min of sleep was significantly enhanced following bedtime injection of GHRH (76). When the intravenous injections were performed during sleep,

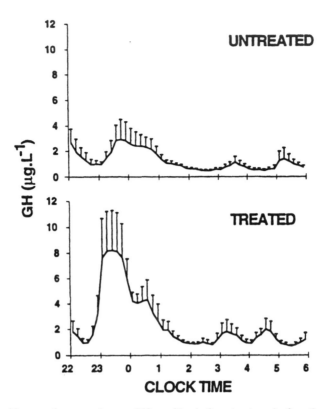

Figure 6 Nocturnal mean plasma GH profiles before (top) and after (bottom) CPAP treatment. (Data from Ref. 69.)

stimulatory effects on the duration of stages III and IV (77–79) and modest increases in REM sleep (77,79) were observed.

The findings on the effects of GHRH on human sleep are well supported by extensive studies in laboratory animals that have indicated that the stimulation of SW sleep by GHRH is likely to involve central mechanisms not mediated by GH (reviewed in Ref. 7).

IV. Circadian and Sleep Control of the Lactotropic Axis

A. Rhythms of Prolactin Secretion Under Normal Conditions

Under normal conditions, PRL levels are minimal around noon, increase modestly during the afternoon, and then undergo a major nocturnal elevation starting shortly after sleep onset and culminating around midsleep. In adults of both sexes, the nocturnal maximum corresponds to an average increase of more than 200% above the minimum level (80–82). Morning awakening is consistently associated with a brief PRL pulse (81,82). Food intake, especially at lunchtime, may also trigger a short-term PRL elevation (83).

Studies of the PRL profile during daytime naps or after shifts of the sleep period have consistently demonstrated that sleep onset, irrespective of the time of day, has a stimulatory effect on PRL release. This is well illustrated by the profiles shown in Figure 1 where elevated PRL levels occurred both during nocturnal sleep and during daytime recovery sleep whereas the nocturnal period of sleep deprivation was not associated with an increase in PRL concentrations. However, sleep is not the sole factor responsible for the nocturnal elevation of PRL concentrations. Indeed, a number of experiments involving abrupt advances or delays of sleep times have shown that the sleep-related rise of PRL may still be present, although with a reduced amplitude, when sleep does not occur at the normal nocturnal time and that maximal stimulation is observed only when sleep and circadian effects are superimposed (81,82,84). The sleep-independent circadian component of PRL secretion, sometimes referred to as the "anamnestic" peak (84,85), is expressed as a progressive increase across the late afternoon and the hours preceding the usual bedtime and is much more pronounced in women than in men (82). Decreased dopaminergic inhibition of PRL during sleep (81) and increased sensitivity of the lactotrophs to GnRH stimulation (86) have been proposed as mechanisms underlying the nocturnal PRL elevation under normal sleep conditions.

In addition to this circadian increase in PRL in the later part of the day, two studies have indicated that nocturnal PRL levels may increase in the absence of sleep (87,88). One study showed that PRL increases when the subject is awake in a dark environment, but not in the presence of light, and therefore suggested that the nocturnal elevation of PRL is partially dependent on melatonin (87). Oral administration of low doses of melatonin during the daytime indeed results in an elevation of plasma PRL levels. The other study indicated that nocturnal increases of PRL in the absence of sleep are dependent on a state of "quiet rest" and do not occur if the subjects expect to be disturbed (88). The two explanations are not mutually exclusive.

Because of the marked effect that sleep onset has on PRL release, several studies have examined the possible relationship between pulsatile PRL release during sleep and the alternation of REM and non-REM stages. An early report (89) identified an association between nadirs of PRL levels and REM stages, on the one hand, and peaks of PRL levels and non-REM stages, on the other hand. This association was not confirmed in a later, more detailed, study (90). However, a later study showed that REM sleep begins preferentially at a time of decreasing PRL secretory activity (91), suggesting that the concept of a relationship between pulsatile PRL activity and the REM–non-REM cycle should be revised rather than abandoned. Indeed, when sleep structure was characterized by power spectral analysis of the EEG, a close temporal association between increased prolactin secretion and delta-wave activity was clearly apparent (92). Conversely, awakenings inhibit nocturnal PRL release (92). Thus, fragmented sleep will generally be associated with lower nocturnal prolactin levels.

B. Effects of Sleep Disorders and Hypnotic Medications

In healthy elderly subjects, who have an increased number of awakenings and decreased amounts of non-REM stages, a nearly 50% dampening of the nocturnal PRL elevation is evident (56). This diminished nocturnal rise in aging is associated with a decrease in the amplitude of the nocturnal secretory pulses (93).

It appears that, in subjects with untreated sleep apnea, nocturnal PRL levels may not increase to the same extent as in healthy subjects with normal sleep. Treatment with CPAP does not modify the total amount of prolactin secreted during the sleep period, but the frequency of prolactin pulses is restored to values similar to those observed in normal subjects (94).

In patients with African trypanosomiasis (i.e., sleeping sickness) who are severely affected, both daytime and nighttime PRL levels are similar, reflecting the failure to consolidate sleep during the nocturnal period (85). However, the relationship between decreasing limbs of PRL pulses and REM periods is preserved (95).

Benzodiazepine as well as nonbenzodiazepine hypnotics taken at bedtime may cause an increase in the nocturnal PRL rise, resulting in concentrations in the pathological range for part of the night (96,97). The upper panels of Figure 7 illustrate the effects of bedtime oral administration of triazolam (0.5 mg) versus placebo on the 24-hr PRL profiles obtained in a group of six normal men. Sleep-related PRL release was enhanced by triazolam ingested, and in some subjects, the PRL concentrations during the early part of the night were nearly threefold higher (although still in the physiological range) than under placebo (96). The lower panels of Figure 7 show the similar effects of a nonbenzodiazepine hypnotic, zolpidem (10 mg), in eight normal women (97). Neither triazolam nor zolpidem had any effect on the 24-hr profiles of cortisol, melatonin, or GH.

C. Evidence for a Role of Prolactin in Sleep Regulation

Rapidly accumulating evidence from studies in rats and rabbits suggests that PRL is involved in the humoral regulation of sleep (98). The primary effect is a stimulation of REM sleep, which may be observed 1–2 hr posttreatment. This stimulatory effect of PRL on REM sleep depends on time of day, as it is observed only during the light (i.e., inactive) period, and may be exerted by PRL released centrally rather than by pituitary PRL. There are no direct studies of the effects of PRL on sleep in the human, although one study has used injections of vasoactive

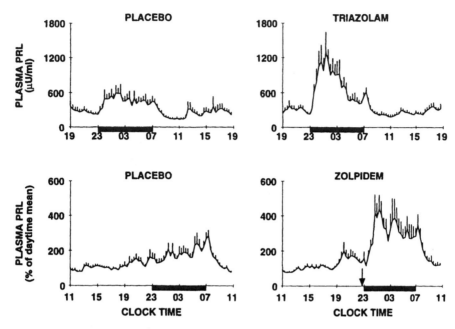

Figure 7 Effects of commonly used hypnotics on the 24-hr profile of plasma PRL in normal young subjects. Data are mean + SEM. Samples were collected at 15–20-min intervals. Sleep was polygraphically recorded. To account for interindividual variations in basal PRL levels, all data were expressed as a percentage of the daytime level (11:00–19:00). (Top) Effects of bedtime administration of triazolam (0.5 mg). (Data from Ref. 96.) (Bottom) Effects of bedtime administration of zolpidem (10 mg). (Data from Ref. 97.)

intestinal polypeptide to stimulate endogenous PRL secretion and has reported an enhancement of REM sleep. A role for PRL in the circadian regulation of REM sleep propensity has been hypothesized (98).

V. Circadian and Sleep Control of the Thyrotropic Axis

A. Twenty-Four-Hour Profiles of TSH and Thyroid Hormones

The 24-hr pattern of plasma TSH levels appears to be generated by amplitude as well as frequency modulation of secretory pulses (99). Daytime levels are low and relatively stable and are followed by a rapid elevation starting in the early evening and culminating in a nocturnal maximum occurring around the beginning of the sleep period (99,100). The later part of sleep is associated with a progressive decline in TSH levels and daytime values resume shortly after morning awakening. The first 24 hr of the study illustrated in the lowest panel of Figure 1 are typical of the diurnal TSH rhythm. Because the nocturnal rise of TSH occurs well before the time of sleep onset, it is believed to reflect a circadian effect. However, a marked effect of sleep on TSH secretion may be evidenced during sleep deprivation (clearly seen in the lowest panel of Fig. 1), when nocturnal TSH secretion is increased by as much as 200% over the levels observed during nocturnal sleep. Thus, sleep exerts an inhibitory influence on TSH secretion and this inhibition is relieved by sleep deprivation (100,101). Interestingly, when sleep occurs during daytime hours, TSH secretion is not suppressed significantly below normal daytime levels. Thus, the inhibitory effect of sleep on TSH secretion appears to be operative when the nighttime elevation has taken place, indicating once again the interaction of effects of circadian time and effects of sleep. When the depth of sleep at the habitual time is increased by prior sleep deprivation, the nocturnal TSH rise is even further decreased, suggesting that SW sleep is probably the primary determinant of the sleep-associated fall (100). Indeed, a pulse-by-pulse analysis of TSH profiles and sleep stages has revealed a consistent association between descending slopes of TSH concentrations and SW stages (102). A recent study has reevaluated this temporal relationship using spectral analysis of the sleep electroencephalogram and has demonstrated the existence of a negative cross-correlation between TSH fluctuations and relative spectral power in the delta range (103). Conversely, awakenings interrupting nocturnal sleep appear to relieve the inhibition of TSH and are consistently associated with a short-term TSH elevation. This is illustrated in the example shown in Figure 2.

Circadian and/or sleep-related variations in thyroid hormones have been difficult to demonstrate, probably because these hormones are bound to serum proteins and thus their peripheral concentrations are affected by diurnal variations in hemodilution caused by postural changes (104,105). However, under conditions of sleep deprivation, the increased amplitude of the TSH rhythm may result in a detectable increase in plasma triiodothyronine (T_3) levels, paralleling the nocturnal TSH rise (6), although negative findings have been also reported (106). If sleep deprivation is prolonged for a second night, the nocturnal rise of TSH is markedly diminished as compared to that occurring during the first night (6,106). It is likely that, following the first night of sleep deprivation, the elevated thyroid hormone levels, which persist during the daytime period because of the prolonged half-life of these hormones, limit the subsequent TSH rise at the beginning of the next nighttime period. A recent, well-documented study involving 64 hr of sleep deprivation demonstrated a more than 50% increase in the T_3 level measured at 23:00 (11 P.M.) across the study period without significant change in the concomitant thyroxin (T_4) concentration (107). During the second night of sleep deprivation, a nocturnal increase in both T_3 and T_4 levels was observed, contrasting with the decreases seen during normal sleep (107). These data suggest that prolonged sleep loss may be associated with an up-regulation of the thyroid axis.

B. Alterations During Simulated "Jet Lag" and Shift Work

The fact that the inhibitory effects of sleep on TSH secretion are time dependent may cause, under certain circumstances, elevations of plasma TSH levels that reflect the misalignment of sleep and circadian timing. An example is shown in Figure 8, which shows the mean profiles of plasma TSH and thyroid hormone levels observed in a group of normal young men in the course of adaptation to simulated "jet lag" (108). Following a 24-hr baseline period, the sleep-wake cycle and the dark period were abruptly advanced by 8 hr (i.e., from 23:00–07:00 to 15:00–23:00). In the course of adaptation to this 8-hr advance shift, TSH levels increased progressively because daytime sleep failed to inhibit TSH and nighttime wakefulness was associated with large circadian-dependent TSH elevations. As a result, mean TSH levels following awakening from the second shifted sleep period were more than twofold higher than during the same time interval following normal nocturnal sleep. This overall elevation of TSH levels was paralleled by a small increase in triiodothyronine (T_3), but not free thyroxine, concentrations (108). This study demonstrates that the subjective discomfort and fatigue that usually occur following abrupt shifts of environmental time, often referred to as the "jet lag syndrome," are associated not only with a desynchronization of bodily rhythms, but also with a prolonged elevation of a hormonal concentration in the peripheral circulation.

The 24-hr profile of plasma TSH showed a high degree of adaptation to a schedule of night work and daytime sleep in permanent shift workers, who usually reverted to daytime activities during their days off (109). Nevertheless, small, but detectable alterations were still present. In particular, the 8-hr delay in rest-activity cycle was only associated with an average 5 hr, 21 min delay in the onset of the nocturnal TSH surge. It is possible that the relative absence of disturbances of TSH secretion in the night workers included in this study is related to the delaying direction of the shift work rotation, since the phase-advance study described above suggests that adaptation to an advance rotation might be associated with more severe perturbations of the thyroid axis.

VI. Conclusions

Although this review focused on only four major neuroendocrine axes, it is clearly apparent that circadian rhythmicity and sleep are major modulators of endocrine function. Under normal conditions, the waking and sleeping periods are associated with markedly different hormonal states. Disruptions of the sleep-wake cycle, whether caused by normal aging, disease (e.g., sleep apnea), voluntary sleep curtailment, or conditions of misalignment between the imposed rest-activity cycle and endogenous circadian rhythmicity, are all associated with marked alterations in endocrine function. Recent studies have suggested that some hor-

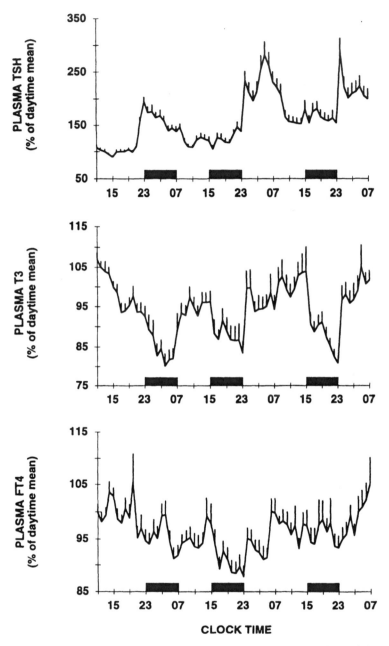

Figure 8 Mean (and SEM) profiles of plasma TSH, T_3, and free T_4 from eight normal young men who were submitted to an 8-hr advance of the sleep-wake and dark-light cycles. Black bars indicate bedtime periods. (Data from Ref. 108.)

monal and metabolic disorders may be partially caused by abnormalities in sleep, circadian rhythmicity, and their interactions.

Acknowledgments

This work was supported in part by Grants DK-41814 and AG-11412 from the National Institutes of Health, Bethesda, MD, and by Grants 94-1-0203 and 92-J0347 from the US Air Force Office of Scientific Research. The expert assistance of Rachel Leproult with data analysis and graphic presentations is gratefully acknowledged.

References

1. Moore-Ede MC, Czeisler CA, Richardson GS. Circadian time-keeping in health and disease. N Engl J Med 1983; 309:469–476, 530–536.
2. Van Cauter E. Hormones and sleep. In: Kales A, ed. The Pharmacology of Sleep. Berlin: Springer-Verlag, 1995:279–306.
3. Van Cauter E, van Coevorden A, Blackman JD. Modulation of neuroendocrine release by sleep and circadian rhythmicity. In: Yen S, Vale W, eds. Advances in Neuroendocrine Regulation of Reproduction. Norwell: Serono Symposia USA, 1990:113–122.
4. Turek FW, Van Cauter E. Rhythms in reproduction. In: Knobil E, Neill JD, eds. The Physiology of Reproduction. New York: Raven Press, 1993:1789–1830.
5. Brandenberger G, Follenius M, Goichot B, Saini J, Ehrhart J, Simon C. Twenty-four hour profiles of plasma renin activity in relation to the sleep-wake cycle. J Hypertens 1994; 12:277–283.
6. Van Cauter E, Turek FW. Endocrine and other biological rhythms. In: DeGroot LJ, ed. Endocrinology. Philadelphia: WB Saunders, 1995:2487–2548.
7. Van Cauter E, Copinschi G. Interactions between growth hormone secretion and sleep. In: Smith RG, Thorner MO, eds. Human Growth Hormone Secretion: Basic and Clinical Research. Totowa, NJ: Humana Press 1998 (in press).
8. Weitzman ED, Zimmerman JC, Czeisler CA, Ronda JM. Cortisol secretion is inhibited during sleep in normal man. J Clin Endocrinol Metab 1983; 56:352–358.
9. Born J, Muth S, Fehm HL. The significance of sleep onset and slow wave sleep for nocturnal release of growth hormone (GH) and cortisol. Psychoneuroendocrinology 1988; 13:233–243.
10. Van Cauter E, Blackman JD, Roland D, Spire JP, Refetoff S, Polonsky KS. Modulation of glucose regulation and insulin secretion by circadian rhythmicity and sleep. J Clin Invest 1991; 88:934–942.
11. Weibel L, Follenius M, Spiegel K, Ehrhart J, Brandenberger G. Comparative effect of night and daytime sleep on the 24-hour cortisol secretory profile. Sleep 1995; 18:549–556.
12. Follenius M, Brandenberger G, Bandesapt JJ, Libert J, Ehrhart J. Nocturnal cortisol release in relation to sleep structure. Sleep 1992; 15:21–27.

13. Späth-Schwalbe E, Gofferje M, Kern W, Born J, Fehm HL. Sleep disruption alters nocturnal ACTH and cortisol secretory patterns. Biol Psychiatry 1991; 29:575–584.

14. Pruessner JC, Wolf OT, Hellhammer DH, et al. Free cortisol levels after awakening: a reliable biological marker for the assessment of adrenocortical activity. Life Sci 1997; 61:2539–2549.

15. Bliwise DL. Normal aging. In: Kryger MH, Roth T, Dement WC, eds. Principles and Practice of Sleep Medicine. Philadelphia: WB Saunders, 1994; 26–39.

16. Leproult R, Copinschi G, Buxton O, Van Cauter E. Sleep loss results in an elevation of cortisol levels the next evening. Sleep 1997; 20:865–870.

17. Van Cauter E, Leproult R, Kupfer DJ. Effects of gender and age on the levels and circadian rhythmicity of plasma cortisol. J Clin Endocrinol Metab 1996; 81:2468–2473.

18. Swaab DF, Fisser B, Kamphorst W, Troost D. The human suprachiasmatic nucleus; neuropeptide changes in senium and Alzheimer's disease. Bas Appl Histochem 1988; 32:43–54.

19. Weiland NG, Wise PM. Aging progressively decreases the densities and alters the diurnal rhythms of alpha-1 adrenergic receptors in selected hypothalamic regions. Endocrinology 1990; 126:2392–2397.

20. Sapolsky RM, Krey LC, McEwen BS. The neuroendocrinology of stress and aging: the glucocorticoid cascade hypothesis. Endocr Rev 1986; 7:284–301.

21. McEwen BS, Sapolsky RM. Stress and cognitive function. Curr Opin Neurobiol 1995; 5:205–216.

22. Dallman MF, Strack AL, Akana SF, et al. Feast and famine: critical role of glucocorticoids with insulin in daily energy flow. Frontiers Neuroendocrinol 1993; 14:303–347.

23. Holsboer F, von Bardelein U, Steiger A. Effects of intravenous corticotropin-releasing hormone upon sleep-related growth hormone surge and sleep EEG in man. Neuroendocrinology 1988; 48:32–38.

24. Born J, Späth-Schwalbe E, Schwakenhofer H, Kern W, Fehm HL. Influences of corticotropin-releasing hormone, adrenocorticotropin, and cortisol on sleep in normal man. J Clin Endocrinol Metab 1989; 68:904–911.

25. Linkowski P, Mendlewicz J, Leclercq R, et al. The 24-hour profile of adrenocorticotropin and cortisol in major depressive illness. J Clin Endocrinol Metab 1985; 61:429–438.

26. Van Cauter E, Turek FW. Depression: a disorder of timekeeping? Perspect Biol Med 1986; 29:510–519.

27. Quabbe H, Schilling E, Helge H. Pattern of growth hormone secretion during a 24-hour fast in normal adults. J Clin Endocrinol Metab 1966; 26:1173–1177.

28. Hunter WM, Rigal WM. The diurnal pattern of plasma growth hormone concentration in children and adolescents. J Endocrinol 1966; 34:147–153.

29. Takahashi Y, Kipnis DM, Daughaday WH. Growth hormone secretion during sleep. J Clin Invest 1968; 47:2079–2090.

30. Sassin JF, Parker DC, Mace JW, Gotlin RW, Johnson LC, Rossman LG. Human growth hormone release: relation to slow-wave sleep and sleep-waking cycles. Science 1969; 165:513–515.

31. Honda Y, Takahashi K, Takahashi S, et al. Growth hormone secretion during nocturnal sleep in normal subjects. J Clin Endocrinol Metab 1969; 29:20–29.

32. Sassin J, Frantz A, Weitzman E, Kapen S. Human prolactin: 24-hour pattern with increased release during sleep. Science 1972; 177:1205–1207.

33. Sassin J, Frantz A, Kapen S, Weitzman E. The nocturnal rise of human prolactin is dependent on sleep. J Clin Endocrinol Metab 1973; 37:436–440.

34. Othmer E, Mendelson WB, Levine WR, Malarkey WB, Daughaday WH. Sleep-related growth hormone secretion and morning naps. Steroids Lipids Res 1974; 5:380–386.

35. Karacan I, Rosenbloom AL, Londono JH, Williams RL, Salis PJ. Growth hormone levels during morning and afternoon naps. Behav Neuropsychiatry 1974; 6:67–70.

36. Weibel L, Spiegel K, Gronfier C, Follenius M, Brandenberger G. Twenty-four-hour melatonin and core body temperature rhythms: their adaptation in night workers. Am J Physiol 1997; 272:R948–R954.

37. Aschoff J. Circadian rhythms: general features and endocrinological aspects. In: Krieger DT, ed. Endocrine Rhythms. New York: Raven Press, 1979:1–61.

38. Van Cauter E, Kerkhofs M, Caufriez A, Van Onderbergen A, Thorner MO, Co-pinschi G. A quantitative estimation of GH secretion in normal man: reproducibility and relation to sleep and time of day. J Clin Endocrinol Metab 1992; 74:1441–1450.

39. Mullington J, Hermann D, Holsboer F, Pollmacher T. Age-dependent suppression of nocturnal growth hormone levels during sleep deprivation. Neuroendocrinology 1996; 64:233–241.

40. Seifritz E, Hemmeter U, Trachsel L, et al. Effects of flumazenil on recovery sleep and hormonal secretion after sleep deprivation in male controls. Psychopharmacology (Berl) 1995; 120:449–456.

41. Jaffe C, Turgeon D, DeMott Friberg R, Watkins P, Barkan A. Nocturnal augmentation of growth hormone (GH) secretion is preserved during repetitive bolus administration of GH-releasing hormone: potential involvement of endogenous somatostatin—a clinical research center study. J Clin Endocrinol Metab 1995; 80:3321–3326.

42. Parker DC, Sassin JF, Mace JW, Gotlin RW, Rossman LG. Human growth hormone release during sleep: electroencephalographic correlations. J Clin Endocrinol Metab 1969; 29:871–874.

43. Pawel M, Sassin J, Weitzman E. The temporal relation between HGH release and sleep stage changes at nocturnal sleep onset in man. Life sci 1972; 11:587–593.

44. Moline M, Monk T, Wagner D, et al. Human growth hormone release is decreased during sleep in temporal isolation (free-running). Chronobiologia 1986; 13:13–19.

45. Golstein J, Van Cauter E, Desir D, et al. Effects of "jet lag" on hormonal patterns. IV. Time shifts increase growth hormone release. J Clin Endocrinol Metab 1983; 56:433–440.

46. Steiger A, Herth T, Holsboer F. Sleep-electroencephalography and the secretion of cortisol and growth hormone in normal controls. Acta Endocrinol 1987; 116:36–42.

47. Jarrett DB, Greenhouse JB, Miewald JM, Fedorka IB, Kupfer DJ. A reexamination of the relationship between growth hormone secretion and slow wave sleep using delta wave analysis. Biol Psychiatry 1990; 27:497–509.

48. Mendlewicz J, Linkowski P, Kerkhofs M, et al. Diurnal hypersecretion of growth hormone in depression. J Clin Endocrinol Metab 1985; 60:505–512.

49. Holl RW, Hartmann ML, Veldhuis JD, Taylor WM, Thorner MO. Thirty-second

sampling of plasma growth hormone in man: correlation with sleep stages. J Clin Endocrinol Metab 1991; 72:854–861.

50. Gronfier C, Luthringer R, Follenius M, et al. A quantitative evaluation of the relationships between growth hormone secretion and delta wave electroencephalographic activity during normal sleep and after enrichment in delta waves. Sleep 1996; 19:817–824.

51. Van Cauter E, Plat L, Scharf M, et al. Simultaneous stimulation of slow-wave sleep and growth hormone secretion by gamma-hydroxybutyrate in normal young men. J Clin Invest 1997; 100:745–753.

52. Van Cauter E, Caufriez A, Kerkhofs M, Van Onderbergen A, Thorner MO, Copinschi G. Sleep, awakenings and insulin-like growth factor I modulate the growth hormone secretory response to growth hormone-releasing hormone. J Clin Endocrinol Metab 1992; 74:1451–1459.

53. Späth-Schwalbe E, Hundenborn C, Kern W, Fehm HL, Born J. Nocturnal wakefulness inhibits growth hormone (GH)-releasing hormone-induced GH secretion. J Clin Endocrinol Metab 1995; 80:214–219.

54. Ho KY, Evans WS, Blizzard RM, et al. Effects of sex and age on the 24-hour profile of growth hormone secretion in man: importance of endogenous estradiol concentrations. J Clin Endocrinol Metab 1987; 64:51–58.

55. Prinz PN, Vitiello MV, Raskind MA, Thorpy MJ. Geriatrics: sleep disorders and aging. N Engl J Med 1990; 323:520–526.

56. van Coevorden A, Mockel J, Laurent E, et al. Neuroendocrine rhythms and sleep in aging. Am J Physiol 1991; 260:E651–E661.

57. Bliwise DL. Sleep in normal aging and dementia. Sleep 1993; 16:40–81.

58. Landin-Wilhelmsen K, Wilhelmsen L, Lappas G, et al. Serum insulin-like growth factor I in a random population sample of men and women: relation to age, sex, smoking habits, coffee consumption and physical activity, blood pressure and concentrations of plasma lipids, fibrinogen, parathyroid hormone and osteocalcin. Clin Endocrinol 1994; 41:351–357.

59. Prinz PN. Sleep and sleep disorders in older adults. J Clin Neurophysiol 1995; 12: 139–146.

60. Finkelstein JW, Roffwarg HP, Boyar RM, Kream J, Hellman L. Age-related change in the twenty-four-hour spontaneous secretion of growth hormone. J Clin Endocrinol Metab 1972; 35:665–670.

61. Vermeulen A. Nyctohemeral growth hormone profiles in young and aged men: correlation with somatomedin-C levels. J Clin Endocrinol Metab 1987; 64:884–888.

62. Iranmanesh A, Lizarralde G, Veldhuis JD. Age and relative adiposity are specific negative determinants of the frequency and amplitude of growth hormone (GH) secretory bursts and the half-life of endogenous GH in healthy men. J. Clin Endocrinol Metab 1991; 73:1081–1088.

63. Frank S, Roland DC, Sturis J, et al. Effects of aging on glucose regulation during wakefulness and sleep. Am J Physiol 1995; 269:E1006–E1016.

64. Copinschi G, Van Cauter E. Effects of ageing on modulation of hormonal secretions by sleep and circadian rhythmicity. Horm Res 1995; 43:20–24.

65. Kern W, Dodt C, Born J, Fehm HL. Changes in cortisol and growth hormone secretion during nocturnal sleep in the course of aging. J Gerontol 1996; 51A:M3–M9.

66. Prinz P, Moe K, Dulberg E, et al. Higher plasma IGF-1 levels are associated with increased delta sleep in healthy older men. J Gerontol 1995; 50A:M222–M226.

67. Cooper BG, White JE, Ashworth LA, Alberti KG, Gibson GJ. Hormonal and metabolic profiles in subjects with obstructive sleep apnea syndrome and the effects of nasal continuous positive airway pressure (CPAP) treatment. Sleep 1995; 18: 172–179.

68. Goldstein SJ, Wu RH, Thorpy MJ, Shprintzen RJ, Marion RE, Saenger P. Reversibility of deficient sleep entrained growth hormone secretion in a boy with achondroplasia and obstructive sleep apnea. Acta Endocrinol 1987; 116:95–101.

69. Saini J, Krieger J, Brandenberger G, Wittersheim G, Simon C, Follenius M. Continuous positive airway pressure treatment: effects on growth hormone, insulin and glucose profiles in obstructive sleep apnea patients. Horm Metab Res 1993; 25:375–381.

70. Grunstein RR, Handelsman DJ, Lawrence SJ, Blackwell C, Caterson ID, Sullivan CE. Neuroendocrine dysfunction in sleep apnea: reversal by continuous positive airways pressure therapy. J Clin Endocrinol Metab 1989; 68:352–358.

71. Mendelson WB, Slater S, Gold P, Gillin JC. The effect of growth hormone administration on human sleep: a dose-response study. Biol Psychiatry 1980; 15:613–618.

72. Kern W, Halder R, Al-Reda S, Späth-Schwalbe E, Fehm HL, Born J. Systemic growth hormone does not affect human sleep. J Clin Endocrinol Metab 1993; 76: 1428–1432.

73. Aström C. Interaction between sleep and growth hormone evaluated by manual polysomnography and automatic power spectral analysis. Acta Neurol Scand 1995; 92:281–296.

74. Obál FJ, Payne L, Kapás L, Opp M, Krueger JM. Inhibition of growth hormone-releasing factor suppresses both sleep and growth hormone secretion in the rat. Brain Res 1991; 557:149–153.

75. Garry P, Roussel B, Cohen R, et al. Diurnal administration of human growth hormone-releasing factor does not modify sleep and sleep-related growth hormone secretion in normal young men. Acta Endocrinol (Copenh) 1985; 110:158–163.

76. Kupfer DJ, Jarrett DB, Ehlers CL. The effect of GRF on the EEG sleep of normal males. Sleep 1991; 14:87–88.

77. Marshall L, Mölle M, Böschen G, Steiger A, Fehm HL, Born J. Greater efficacy of episodic than continuous growth hormone–releasing hormone (GHRH) administration in promoting slow-wave sleep (SWS). J Clin Endocrinol Metab 1996; 81:1009–1013.

78. Steiger A, Guldner J, Hemmeter U, Rothe B, Wiedemann K, Holsboer F. Effects of growth hormone-releasing hormone and somatostatin on sleep EEG and nocturnal hormone secretion in male controls. Neuroendocrinology 1992; 56:566–573.

79. Kerkhofs M, Van Cauter E, Van Onderbergen A, Caufriez A, Thorner MO, Copinschi G. Sleep-promoting effects of growth hormone–releasing hormone in normal men. Am J Physiol 1993; 264:E594–E598.

80. Van Cauter E, L'Hermite M, Copinschi G, Refetoff S, Desir D, Robyn C. Quantitative analysis of spontaneous variations of plasma prolactin in normal man. Am J Physiol 1981; 241:E355–E363.

81. Spiegel K, Follenius M, Simon C, Saini J, Ehrhart J, Brandenberger G. Prolactin secretion and sleep. Sleep 1994; 17:20–27.

82. Waldstreicher J, Duffy JF, Brown EN, Rogacz S, Allan JS, Czeisler CA. Gender differences in the temporal organization of prolactin (PRL) secretion: evidence for a sleep-independent circadian rhythm of circulationg PRL levels—A Clinical Research Center study. J Clin Endocrinol Metab 1996; 81:1483–1487.

83. Quigley ME, Ropert JF, Yen SS. Acute prolactin release triggered by feeding. J Clin Endocinol Metab 1981; 52:1043–1045.

84. Desir D, Van Cauter E, L'Hermite M, et al. Effects of "jet lag" on hormonal patterns. III. Demonstration of an intrinsic circadian rhythmicity in plasma prolactin. J Clin Endocrinol Metab 1982; 55:849–857.

85. Radomski MW, Buguet A, Montmayeur A, et al. Twenty-four-hour plasma cortisol and prolactin in human African trypanosomiasis patients and healthy African controls. Am J Trop Med Hyg 1995; 52:281–286.

86. Rossmanith WG, Boscher S, Ulrich U, Benz R. Chronobiology of prolactin secretion in women: diurnal and sleep-related variations in the pituitary lactotroph sensitivity. Neuroendocrinology 1993; 58:263–271.

87. Okatani Y, Sagara Y. Role of melatonin in nocturnal prolactin secretion in women with normoprolactinemia and mild hyperprolactinemia. Am J Obstet Gynecol 1993; 168:854–861.

88. Wehr T, Moul D, Barbato G, et al. Conservation of photoperiod-responsive mechanisms in humans. Am J Physiol 1993; 265:R846–R857.

89. Parker DC, Rossman LG, Vanderlaan EF. Relation of sleep-entrained human prolactin release to REM–non-REM cycles. J Clin Endocrinol Metab 1974; 38:646–651.

90. Van Cauter E, Desir D, Refetoff S, et al. The relationship between episodic variations of plasma prolactin and REM–non-REM cyclicity is an artifact. J Clin Endocrinol Metab 1982; 54:70–75.

91. Follenius M, Brandenberger G, Simon C, Schlienger JL. REM sleep in humans begins during decreased secretory activity of the anterior pituitary. Sleep 1988; 11:546–555.

92. Spiegel K, Luthringer R, Follenius M, et al. Temporal relationship between prolactin secretion and slow-wave electroencephalographic activity during sleep. Sleep 1995; 18:543–548.

93. Greenspan SL, Klibanski A, Rowe JW, Elahi D. Age alters pulsatile prolactin release: influence of dopaminergic inhibition. Am J Physiol 1990; 258:E799–E804.

94. Spiegel K, Follenius M, Krieger J, Sforza E, Brandenberger G. Prolactin secretion during sleep in obstructive sleep apnea patients. J Sleep Res 1995; 4:56–62.

95. Brandenberger G, Buguet A, Spiegel K, et al. Disruption of endocrine rhythms in sleeping sickness with preserved relationship between hormonal pulsatility and the REM-nonREM sleep cycles. J Biol Rhythms 1996; 11:258–267.

96. Copinschi G, Van Onderbergen A, L'Hermite-Balériaux M, et al. Effects of the short-acting benzodiazepine triazolam, taken at bedtime, on circadian and sleep-related hormonal profiles in normal men. Sleep 1990; 13:232–244.

97. Copinschi G, Akseki E, Moreno-Reyes R, et al. Effects of bedtime administration of zolpidem on circadian and sleep-related hormonal profiles in normal women. Sleep 1995; 18:417–424.

98. Roky R, Obál F, Valatx JL, et al. Prolactin and rapid eye movement sleep regulation. Sleep 1995; 18:536–542.

99. Veldhuis JD, Iranmanesh A, Johnson ML, Lizarralde G. Twenty-four-hour rhythms in plasma concentrations of adenohypophyseal hormones are generated by distinct amplitude and/or frequency modulation of underlying pituitary secretory bursts. J Clin Endocrinol Metab 1990; 71:1616–1623.

100. Brabant G, Prank K, Ranft U, et al. Physiological regulation of circadian and pulsatile thyrotropin secretion in normal man and woman. J Clin Endocrinol Metab 1990; 70:403–409.

101. Parker DC, Rossman LG, Pekary AE, Hershman JM. Effect of 64 hour sleep deprivation on the circadian waveform of thyrotropin (TSH): further evidence of sleep-related inhibition of TSH release. J Clin Endocrinol Metab 1987; 64:157–161.

102. Goichot B, Brandenberger G, Saini J, Wittersheim G, Follenius M. Nocturnal plasma thyrotropin variations are related to slow-wave sleep. J Sleep Res 1992; 1: 186–190.

103. Gronfier C, Luthringer R, Follenius M, et al. Temporal link between plasma thyrotropin levels and electroencephalographic activity in man. Neurosci Lett 1995; 200:97–100.

104. Greenspan SL, Klibanski A, Schoenfeld D, Ridgway EC. Pulsatile secretion of thyrotropin in man. J Clin Endocrinol Metab 1986; 63:661–668.

105. Brabant G, Brabant A, Ranft U, et al. Circadian and pulsatile thyrotropin secretion in euthyroid man under influence of thyroid hormone and glucocorticoid administration. J Clin Endocrinol Metab 1987; 65:83–88.

106. Allan J, Czeisler C. Persistence of the circadian thyrotropin rhythm under constant conditions and after light-induced shifts of circadian phase. J Clin Endocrinol Metab 1994; 79:508–512.

107. Gary KA, Winokur A, Douglas SD, Kapoor S, Zaugg L, Dinges DF. Total sleep deprivation and the thyroid axis: Efects of sleep and waking activity. Aviat Space Environ Med 1996; 67:513–519.

108. Hirschfeld U, Moreno-Reyes R, Akseki E, et al. Progressive elevation of plasma thyrotropin during adaptation to simulated jet lag: effects of treatment with bright light or zolpidem. J Clin Endocrinol Metab 1996; 81:3270–3277.

109. Weibel L, Brandenberger G, Goichot B, Spiegel K, Ehrhart J, Follenius M. The circadian thyrotropin rhythm is delayed in regular night workers. Neurosci Lett 1995; 187:83–86.

14

Relationships Between Sleep and Immune Function

JAMES M. KRUEGER and JIDONG FANG

Washington State University
Pullman, Washington

RACHAEL A. FLOYD

University of Tennessee–Memphis
Memphis, Tennessee

I. Introduction

Almost everyone is aware of the almost irresistible desire to sleep that occurs at the onset of "the flu." Indeed, most loving parents or grandparents recommend sleep as a preventive measure and as an aid for recuperation from infection. Hippocrates dispensed similar advice; nevertheless, the investigation of relationships between sleep and host defense systems had to wait 2100 years before being addressed scientifically. Within the past 10 years sleep responses to microbial challenge have been characterized as one component of the acute-phase response. These microbial-induced sleep responses may, in fact, contribute to host defense though this issue has yet to be satisfactorily resolved. Conversely, sleep loss is associated with changes in a variety of immune parameters and prolonged sleep loss with septicemia. The excess sleep associated with infectious challenge results from the amplification of physiological sleep mechanisms. Physiological sleep mechanisms include molecules that have traditionally been thought of as immune response modifiers but are now known to be constituitively expressed in brain and involved in several physiological regulations including sleep. This review will focus on some of the data that form the basis for these statements; it ends with a short discussion of the theoretical implications these data have on sleep function and brain organization as it applies to sleep.

II. Sleep Responses and Sleep Deprivation

A. Sleep Responses to Infectious Challenges

Bacterial Challenge

Sleep responses over the course of an infection were first characterized after bacterial challenge (1). After intravenous challenge with either gram-negative or gram-positive bacteria the initial response was an increase in duration of non–rapid eye movement sleep (NREMS) and amplitudes of electroencephalographic (EEG) slow waves. These enhancements were followed by a period of reduced NREMS and EEG slow-wave amplitudes (SWA). The duration of the enhancement phase depends upon the bacteria species used and the route of administration (reviewed in 2). For example, after intravenous challenge with *Staphylococcus aureus*, a gram-positive bacterium, NREMS was increased for about 20 hr. In contrast, intravenous administration of *Escherichia coli*, a gram-negative bacterium, was associated with a rapid onset in excess NREMS lasting only 4–6 hr. Throughout these biphasic NREMS responses, rapid eye movement sleep (REMS) is inhibited. The febrile response began during the enhanced NREMS period and persisted throughout the remainder of the enhanced NREMS period and the inhibitory phase.

Much work was focused at determining the mechanisms responsible for bacterial-altered sleep. Bacterial replication per se did not seem necessary since heat-killed gram-positive (1) and gram-negative (3) bacteria or isolated bacterial cell walls (4) induce sleep and fever responses similar to those observed after viable bacterial injections. Further, certain components of bacterial cell walls are capable of enhancing NREMS. These NREMS-enhancing components include endotoxin (5), its lipid A moiety (6) from gram-negative bacteria, and certain muramyl peptides from gram-positive and gram-negative bacteria (7,8). Macrophages or microglia are capable of digesting bacterial peptidoglycan and releasing muramyl peptides into the surrounding medium (9–11). Some of these muramyl peptides are also somnogenic and pyrogenic (9,11). These observations indicate that sleep responses to bacterial challenges are mediated by the bacterial products such as muramyl peptides.

An important implication from these observations is that the presence of NREMS-promoting bacterial products in the body and the brain in normal conditions suggests that they may play some role in the regulation of physiological sleep. The body is constantly exposed to bacteria under physiological conditions, bacterial product influence on sleep may thus represent an additional facet of the endosymbiotic relationships between microbes and mammals. Indeed, the somnogenic properties of muramyl peptides were first described in experiments in which sleep-promoting substances were isolated from mammalian tissues. The somnogenic muramyl peptide NAG-(1-6 anhydro)NAM-Ala-Glu-Dap-Ala was first isolated from human urine (12) and rabbit brain (13); 1 pM of this material enhances rabbit NREMS for hours (13).

Several structural requirements for somnogenic activity of muramyl peptides have been described (reviewed in 2,14). For example, conversion of certain terminal carboxyl groups to unsubstituted amides results in the complete loss of somnogenic activity (2,14). Muramyl peptides containing either a 1,6-anhydro-muramyl structure or acetylation of carbon 6 of muramaic acid have more potent sleep-promoting effects than similar substances lacking these modifications. Neither NAG nor NAM is necessary for somnogenic activity (2). The smallest muramyl peptide known to induce NREMS is muramyl dipeptide (MDP) (NAM-L-ala-D-isogln), a commercially available immune adjuvant peptide. MDP is somnogenic in rabbits (15,16), cats (15), rats (17–19), and monkeys (20). The somnogenic actions of MDP are dependent upon the steric configuration of the component amino acids (2,7,14); the LL and DD stereoisomers are inactive. Of biological importance is that some muramyl peptides are immunologically active or pyrogenic but lack sleep-promoting activity. Information about the relationships between lipid A structure and its somnogenic activity is more limited. Nevertheless, the somnogenic activity of lipid A depends on the acylation or phosphorylation patterns of the backbone structures of the molecule (6).

In summary, it seems likely that bacterial-induced sleep responses are initiated by macrophage-tailored bacterial cell wall products. These products are likely enzymatically modified as one method of regulating the host's response. These products exert their effects on sleep via induction of cytokine production (see below).

Viral Diseases

Viral diseases played an important historic role in sleep research; von Economo (21) noticed that some of his patients, who had viral-induced brain lesions, slept less while others slept more than normal people. On postmortem examination he showed that those who slept less had lesions in the anterior hypothalamus while those who slept more had posterior hypothalamic lesions. He, thus, identified areas in brain important in sleep-wake regulation; he also used his data to support the theory that sleep regulation is an active process. Despite this important historic role of viral infections, additional relationships between sleep and viral diseases remained unexplored until recently. Nevertheless, viral infections have been implicated in a number of sleep pathologies, including chronic fatigue syndrome (22), mononucleosis (23), postviral fatigue syndrome (24), and sudden infant death syndrome (25). Human immunodeficiency virus (HIV) (26) and influenza

(27,28) virus infections are also associated with sleep alterations. Recently, the effects of influenza virus and HIV infections on sleep have been experimentally studied in animal models. These animal studies provide more information about mechanisms for the sleep alterations induced by viral infection.

Intravenous injection of large doses of influenza virus induces fever, enhancement of NREMS, and increases in EEG SWA, whereas injection of heat-killed virus is ineffective in inducing these alterations (29). These alterations start within a few hours after the viral injection and last 6–8 hr. This observation provided the first experimental evidence that sleep is altered during viral infection. However, the influenza virus induces an abortive infection in rabbits since it only undergoes partial replication in this species. It was not clear to what extent the observed sleep and fever reflect the conditions during natural viral infections. More recent studies used mouse-adapted influenza virus and characterized sleep responses to full-blown viral infections in mice after intranasal viral inoculation. Total airway infection caused by intranasal inoculation of H1N1 virus induces, after a latency of 16–18 hr (about two viral replication cycles), increases in NREMS and suppression of REMS, which lasted several days in Swiss-Webster mice (Fig. 1; 30) and C57BL/6 mice (31). Body temperature and motor activity are also decreased after viral inoculation in both studies. The latency for sleep responses is longer (24 hr, or three replication cycles), if a 10-times-lower dose of virus is used (32), suggesting that the NREMS-promoting substances produced by the virus must accumulate to a certain threshold to induce sleep alterations. The magnitude of sleep responses depends on the location and severity of infection. In mice, NREMS is increased during the lethal total airway infection of H1N1 virus, but not during nonlethal upper airway infection with the same dose of the virus (32). Further, H3N2 virus, a nonlethal strain of influenza virus, causes much smaller increases in NREMS than H1N1 virus in the same total airway infection model, even if a 10-fold higher dose of H3N2 virus is used (30).

HIV infection affects human sleep. Although HIV virus affects many systems in the body, the primary targets of HIV are the immune and nervous systems. HIV is considered a neurotropic virus based on a number of findings, including the recovery of HIV from the cerebrospinal fluid (CSF) and brains of AIDS patients, increased anti-HIV antibody titers in the CSF, and high HIV DNA levels in the brain (reviewed in 33). Sleep alterations are prominent features of HIV infections. There are robust increases in slow-wave sleep during the early asymptomatic stage of HIV infection (reviewed in 34). However, sleep alterations have not been systematically characterized over the course of the HIV infection.

As early as 1974, Carter and De Clercq (35) proposed a general mechanism for viral toxicity; they posited that viral double-stranded RNA (dsRNA) is a toxic component of viruses. Such a mechanism has been extended to explain the sleep changes caused by viral infection (reviewed in 2,14). Since dsRNA is made during viral replication, this may explain why only viable, but not dead, influenza virus

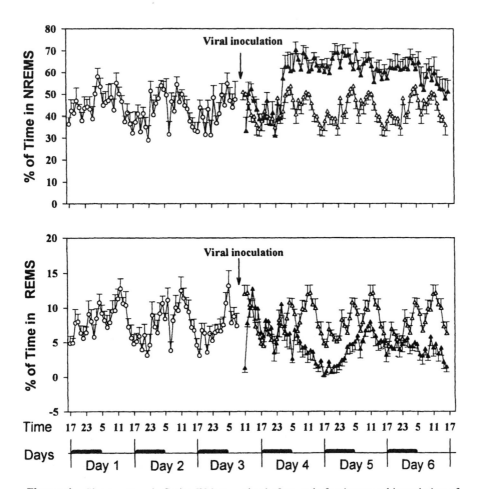

Figure 1 Sleep pattern in Swiss-Webster mice before and after intranasal inoculation of H1N1 influenza virus. (○) Baseline sleep; (△) average of baseline sleep from different days; (▲) sleep after viral inoculation. The arrows indicate the time of viral inoculation. The black bars at the bottom indicate the dark period. Viral inoculation induces pronounced and long-lasting increases in NREMS and decreases in REMS.

has the ability to induce sleep changes. Several findings indicate that dsRNA is involved in virus-induced sleep. The synthetic dsRNA poly (I:C) induces a flu-like syndrome in humans (36). In addition, both poly(I:C) (37) and isolated nuclease-resistant viral dsRNA from influenza-virus-infected mouse lungs (38) induce fever and enhance NREMS in rabbits. Pretreatment of rabbits with poly(I:C) blocks acute-phase responses, including enhanced sleep, induced by influenza virus challenge as does pretreatment with virus itself (after the initial response to these stimuli a state of tolerance is induced), suggesting that dsRNA can substitute for virus as a trigger of the acute-phase response (39). In addition, we recently observed that the synthetic dsRNA corresponding to the first 108 bases of the gene segment 3 of influenza virus dose-dependently induces sleep and fever in rabbits, whereas neither the sense nor the antisense single-stranded RNA of the same fragment is effective in inducing sleep and fever (40).

The structural requirement for the NREMS-promoting effects of dsRNA has not been determined. Nevertheless, an important factor seems to be the stability of dsRNA, which is influenced by both base composition and molecular length. Poly(I:C), a very nuclease-resistant dsRNA, is somnogenic and pyrogenic, but polyriboadenylic-polyribouridylic acid [poly(A:U)], a very nuclease-susceptible dsRNA, lacks these actions at 100-fold higher doses in an in vivo rabbit model. This suggests that base composition is likely to be very important in determining the relative stability and, therefore, toxicity of dsRNA. Molecular length also influences the stability of RNA. A longer dsRNA representing part of gene segment 3 of influenza virus (about 700 base pairs including the first 107 pairs) is more potent in inducing sleep and fever than this 107-pair fragment on a molar basis (J. Fang et al., unpublished). The attractiveness of the dsRNA hypothesis is that this mechanism is generic and may be extended to many different viruses. However, such generality has not been tested in other types of viral infections.

Other viral components, such as envelope proteins, may also be involved in the NREMS-promoting effects of virus. On a structural basis, such mechanisms must be virus-specific. In the case of HIV virus, gp120 might be one such component. Intracerebroventricular administration of gp120 induces dose-dependent enhanced NREMS and REMS without affecting body temperature. Heat inactivation of gp120 results in the loss of its sleep-promoting effects (41). gp120 also stimulates production of sleep-promoting cytokines such as interleukin-1 beta (IL-1β) and tumor necrosis factor alpha (TNFα) from human and rat glial cells in vitro and in freely moving rats.

Fungal Infections

Intravenous inoculation of rabbits with the pathogenic fungus *Candida albicans* (3) or subcutaneous inoculation of rats with nonpathogenic yeast *Saccharomyces cerevisiae* (42) induces fever and enhancement of sleep similar to those induced by bacterial infection. The somnogenic components of fungi have not been examined. Fungi have cell walls as bacteria, but these structures in fungi are more similar to those of plants and insects. Some fungal cell walls contain the same cellulose as plants but most are composed of chitin, a polymer of *N*-acetylglucosamine, or other polysaccharides such as mannan, glucan, galactosan, and chitosan (43). Mannan and glucan are the dominant polysaccharide components of pathogenic fungi such as *C. albicans* and have immunomodulatory activity (44).

Glucan, in particulate or soluble form, stimulates proinflammatory cytokines, such as TNFα and IL-1β (45,46) and modulates IL-1 receptor antagonist expression (47). Mannan is also an immunomodulatory agent via cytokines (48). Yeast cell walls comprise the classic phagocytosis activator zymosan, and mushroom cell walls comprise the active ingredients of traditional Chinese medicines. The inflammatory reaction to fungal cell walls is probably the primary source of cytokines in these insidious infections. *C. albicans* may induce TNFα (49) and interferon alpha$_2$ (IFNα$_2$) (50), which in turn stimulate neutrophils to kill *C. albicans* (51).

Protozoan Diseases

Malaria and other parasitemias are associated with fatigue and excessive sleepiness. Until recently, there were no studies of sleep over the course of a chronic parasitemia, including a disease called African sleeping sickness. Humans are infected by tsetse flies with *Trypanosoma brucei brucei*. Classically, sleeping sickness patients have been described as being sleepy in the daytime and restless at night. Recent studies indicate that these patients tend to have sleep and wake episodes equally distributed during the light/dark cycle. However, there is no hypersomnia as the name of the disease implies. In contrast, sleep is fragmented; the duration of waking and sleep episodes is inversely proportional to the severity of the disease (52,53). In rats, intraperitoneal inoculation of a related species, *Trypanosoma brucei brucei* (Tbb), induces similar sleep fragmentation and, to a lesser extent, a decrease in the amplitude of the circadian rhythm of NREMS as seen in human patients (52,54). In rabbits, Tbb induces periodic parasitemia. Toth and Krueger (55) studied the sleep alterations over a period of weeks in rabbits inoculated with Tbb. Within 4 days after inoculation, Tbb induced fever and other signs of the acute-phase response concurrent with the onset of parasitemia. NREMS was also enhanced during the periods of parasitemia, although these periodic enhancements were superimposed on a longer-term trend of decreased sleep occurring over the course of the disease. Since each episode of parasitemia presents a challenge to the host immune system, including the induction of cytokines, the enhanced sleep during parasitemia might be mediated by the immune responses to the immune stimulation.

Summary

Increased NREMS is a constitutional component of the acute-phase response. Bacterial cell wall components responsible for inducing sleep responses to infectious challenges include muramyl peptides, derived from peptidoglycan, and endotoxin and its components, LPS and lipid A, from gram-negative bacteria. The hypothesis that viral dsRNA made during replication is responsible for sleep alterations during viral infection is supported by the data from influenza virus studies. Components that are specific to particular types of viruses, such as HIV gp120 protein, may also contribute to sleep alterations during infection in a virus-specific manner. Although MDP, LPS, dsRNA, and gp120 are very different in structure, all of these microbial products share a common property, stimulation of cytokine production. Many cytokines including IL-1β, TNFα, and IFNα promote NREMS and may be the common mediators for infection-induced sleep alterations and other components of acute-phase responses, such as fever (see below).

B. Sleep Deprivation Affects Host Defenses

Long-Term Sleep Deprivation

Bacteremia

In an animal model developed by Rechtschaffen and colleagues, rats deprived of sleep for prolonged periods eventually died (56). For a number of years the cause of sleep-deprivation-induced death remained elusive until Everson showed that similarly sleep-deprived rats, but not the yoked controlled rats, developed septicemia after 16–21 days of sleep deprivation and that this was the probable cause of death (57). Her findings were subsequently confirmed by Rechtschaffen's group (58). These findings are tantalizing but leave open the question: why should host defense mechanisms breakdown as a result of sleep deprivation? To begin to answer this question requires consideration of our normal endosymbiotic relationships with microbes; this discussion will focus on bacteria.

Translocation of Bacteria

The gastrointestinal tract functions as the organ responsible for nutrient absorption, as well as a metabolic and immunological system, providing an effective barrier against endotoxin and bacteria in the intestinal lumen. Bacterial translocation is the passage of viable bacteria from the gastrointestinal tract through the epithelial mucosa. Equally important may be the passage of bacterial endotoxin through the mucosal barrier. Translocation of both endotoxin and bacteria is of clinical significance. Recent published works indicate that translocation of endotoxin in minute amounts is a physiologically important phenomenon to boost the reticuloendothelial system, especially liver Kupffer cells (reviewed in 59). However, breakdown of both the mucosal barrier and the reticuloendothelial system capacity results in systemic bacteremia and endotoxemia, which results in organ dysfunction, impairs the mucosal barrier, the clotting system, and the immune system, and depresses Kupffer cell function (reviewed in 59).

Significant translocation of viable bacteria does not normally occur from the gastrointestinal tract in healthy animals owing to: (1) the presence of an indigenous gastrointestinal microflora preventing bacterial overgrowth; (2) an intact intestinal epithelial barrier; and (3) normal host immune defenses (60). However, stressors such as sleep deprivation, starvation, protein malnutrition, and so forth can result in the disruption or impairment of any of these protective mechanisms, potentially leading to lethal systemic infections. Everson and Toth (61) found that during the early phase of sleep deprivation, before bacteremias develop, 5, 10, or 15 days of sleep deprivation was associated with the occurrence of viable bacteria in organs of filtration such as mesenteric lymph nodes. Landis et al. (62) also found viable bacteria in the mesenteric lymph nodes in female rats after sleep deprivation for 3 days in constant light. In the experiments of Landis et al., natural killer activity was slightly higher in rats with bacteria in the mesenteric lymph nodes. These investigations suggested that sleep-deprivation-induced breakdown of innate host resistance begins well before the effects of sleep deprivation are irreversible.

To test the hypothesis that the gut flora is altered during sleep deprivation, Everson measured the number of bacteria in the gut of rats subjected to sleep deprivation, yoke control, and home cage animals. She found that the number of viable bacteria in the gut of sleep-deprived rats was significantly increased after 10 days. Both Everson (57) and Bergmann et al. (58) showed that gut bacteria populations in the sleep-deprived animals are increased after less than 6 days of sleep deprivation compared to the yoked controls and home cage animals. These data suggest that the gut is the most likely source of the bacteremia seen during sleep deprivation.

The mechanisms responsible for increased gut bacterial populations and the occurrence of viable bacteria in internal organs during sleep deprivation remain unknown. Nevertheless, there is some evidence that gastrointestinal bacteria are a source of sleep-inducing bacterial cell wall muramyl peptides. In rats placed on an antibiotic regimen (neomycin and metronidazole in drinking water), there was a significant reduction in NREMS in the first 3 hr of the lights-on period as well as an increase in sleep latency (63). No other sleep parameters, including REMS sleep, were affected, suggesting that there is a specific NREMS effect due to bacterial reduction. One interpretation of these data is that bacterial products could influence everyday sleep; the relatively rapid appearance of viable bacteria in lymph nodes of sleep-deprived rats is consistent with such a hypothesis.

Short-Term Sleep Deprivation

Natural Killer Cell Activity

The immune system is composed of natural immunity and acquired immunity. Natural immunity includes phagocytes (neutrophils, eosinophils, basophils, monocytes), antibacterial proteins, natural killer (NK) cells, and the complement system. Acquired immunity occurs when foreign intruders induce production of antibodies or cytokines through clonal selection of B and T lymphocytes, respectively. Although all of these components are interdependent, the natural immune system is largely nonspecific, innately recognizing broad classes of foreign intruders, whereas the acquired immune system responds to specific antigens by preliminary sensitization and clonal expansion of B and T cells. The acquired immune system also remembers each encounter with an antigen, resulting in an increasingly effectively defensive response with repeated exposure.

NK cells are an important first line of defense against a broad range of target cells including some tumor cells and cells infected with virus, bacteria, or protozoa (reviewed in 64). NK cells also reject transplanted cells. These cells function without evidence of prior exposure to their target. They can be stimulated to attack and these activated cells are more efficient than nonstimulated cells. NK cells are an easy immunological parameter to study primarily because they require no stimulation to see their effect. The assay system is simple and the time to get the results is relatively short. It is thus easy to see why sleep researchers have chosen this cell to follow during sleep deprivation. NK cells have been shown to be low in chronic fatigue immune dysfunction syndrome (65), protein calorie malnutrition (66), repeated infections (67), vitamin and trace mineral deficiency (68), autoimmune disorders (69), chemotherapy (70), and treatment with biological modifiers such as interleukin-2 (70). Activated NK cells can secrete TNFα and kill TNFα-susceptible targets in 24-hr assays (71). Nonactivated NK cells utilize two different pathways to eliminate tumor targets: secretory (necrotic) and nonsecretory (apoptotic) cytotoxic mechanisms (72).

Human studies suggest that sleep deprivation results in immune dysfunction. One study of 20 young adults deprived of sleep for 64 hr demonstrated an increase in monocytes, granulocytes, NK cells, and NK activity (73). These authors suggested that nonspecific host defense mechanisms are enhanced after severe acute sleep loss which they viewed as distinct from a stress response. Another study of nine healthy women in a crossover design of sleeping either between 3 and 7 A.M. or between 9 P.M. and 1 A.M. showed a 25–30% decrease in NK cellular activity associated with this sleep restriction. During the recovery sleep there was a rebound to greater than baseline levels after the late (9P.M.–1 A.M.) sleep restriction but not after the early sleep restriction (74). These authors concluded that disruption of sleep imposed by a night of partial sleep deprivation reduces NK activity, and that recovery sleep restores NK activity to baseline values. A decline in cellular immunity in humans was shown in two further reports: in one study, 48 hr of sleep deprivation decreased lymphocyte proliferation responses to phytohemagglutinin (75); in another study NK activity decreased after 40 hr of wakefulness (76).

Furthermore, limited sleep loss in the form of insomnia is associated with a decline in NK activity (77). In a recent study of 23 depressed patients and 17 control individuals, total sleep time, duration of NREMS, and sleep efficiency were all positively correlated with NK activity (77). Yet another study of 11 healthy males, who were allowed to take four 2-hr naps in a 20-hr period with 4 hr of forced wakefulness between the naps, suggested that decreased NK activity is related to sleep rather than diurnal influences (78). The effects of sleep deprivation between 10 P.M. and 3 A.M. on circulating white blood cells, NK cell number and cytotoxicity, lymphokine-activated killer cell number and activity, and stimulated interleukin-2 (IL-2) production were studied in 42 medically and psychiatrically healthy male volunteers (79). A reduction of NK, lymphokine-activated killer cell activities, and suppressed concanavalin A–stimulated IL-2 production were seen after sleep deprivation. NK activity returned to normal but IL-2 production remained suppressed after recovery sleep. These data demonstrate that even a small loss of sleep (5 hr) produces a reduction of NK-cell activity and T-cell cytokine production. Thus, sleep loss and sleep deprivation clearly modulate immune functions by either increasing or decreasing various nonspecific and specific immune parameters although the nature of these effects is not consistent in the literature.

Some sleep deprivation studies showed an increase in NK-cell activity. For example, during sleep deprivation, the absolute concentration and relative fraction among peripheral blood mononuclear cells expressing characteristic NK-cell markers (CD56, CD16) were increased during sleep deprivation (80). This increase is followed by a return of NK-cell activity to baseline levels after sleep deprivation. Detectable increased NK-cell activity could be due to a change in the proportion of NK cells in the tested populations rather than to a change in the activity of individual cells or attributed to change in NK moving from the blood to organs.

Although the majority of studies in which the effects of sleep deprivation on NK activity have been performed in humans, a few studies have been performed in animals. Already mentioned are the results of Landis et al. (62), who showed increased NK activity in rats with bacteremia during long-term sleep deprivation. In contrast, after short-term sleep deprivation (6–12 hr) Floyd et al (unpublished) showed a 50% reduction in NK activity in mice (Fig. 2).

Other Measures of Immune Function

Brown et al. (81) abrogated respiratory immunity to influenza virus by subjecting orally immunized mice to 7 hr of sleep deprivation following total respiratory tract challenge with influenza virus. Virus was present in ground lung parenchyma of these mice 3 days after viral challenge. In addition, virus-specific antibody in lung homogenates was depressed. Nasobronchial protection to influenza virus in the solidly immune convalescent mouse is mediated by S-IgA (82,83) while protection of the lung parenchyma is due to serum IgG (84). Recovery from active viral infection is largely due to T cells (reviewed in 85). Thus, the reported depression of immunity within the lungs suggests the lack of both IgG-mediated and anti-influenza protection and cellular immunity to influenza virus. Since Kris et al. (84) have shown that serum influenza-specific IgG alone is capable of providing protection for the lung in nude mice lacking cellular immunity, Renegar et al.

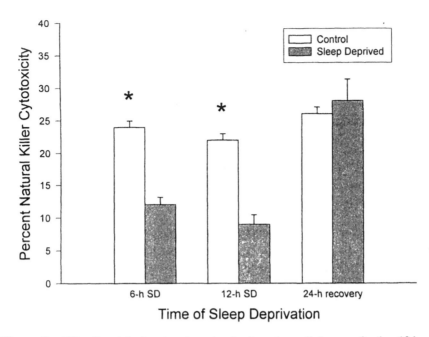

Figure 2 NK-cell activity in mice sleep-deprived starting at light onset for 6 or 12 hr or sleep-deprived for 6 hr and allowed to recover for 24 hr. Mean and SEM are shown; the values given are the percent of target cells lysed. Effector:target cell ratio was 100:1. *Indicates significant difference from control; $p < 0.05$.

investigated whether single or multiple episodes of sleep deprivation could accelerate serum IgG catabolism sufficiently to account for the reported loss of immunity (86). Mild sleep deprivation per se did not accelerate catabolism of passively administered IgG enough to account for the published results (81) and that, in the solidly immune mouse, levels of vaginal influenza-specific S-IgA, a reliable indicator of mucosal immune function (87,88) and serum anti-influenza IgG, are not depressed by sleep deprivation. Hence, Renegar's results, like those of Benca et al. (89), do not support the hypothesis that sleep deprivation depresses specific antibody formation.

In another study Renegar et al. were unable to abrogate mucosal anti-influenza viral immunity with a single postviral challenge sleep deprivation episode (90). Nor were they able to suppress this solid immunity with one pre- and two postchallenge sleep deprivation episodes in young adult or old mice. As mentioned above, sleep deprivation did not depress the level of serum influenza-specific IgG antibodies and, in fact, two episodes of sleep deprivation resulted in an increase in these antibodies compared to normal sleeping mice. These studies suggest that short-term sleep deprivation has minimal effects on solid mucosal and humoral immunity in young or old mice. Thus, sleep deprivation might be able to prevent the development of an immune response depending on the timing of sleep deprivation but has little effect on established immunity.

Brown's group also reported that sleep deprivation for 8 hr suppressed the secondary antibody response to sheep red blood cells in rats (91). This sleep-deprivation-induced suppression of the antibody response could be totally prevented by giving either IL-1β or muramyl dipeptide at the beginning of sleep deprivation in the rats; both of these substances are centrally active in the induction of NREMS (reviewed in 2). However, the role that stress played in these experiments must also be considered.

In contrast, Benca et al. (89) found no differences between immune function in rats that were totally or partially deprived of sleep and home cage controls. The parameters studied were spleen cell counts, in vitro lymphocyte proliferation responses to mitogens, and in vivo and in vitro plaque-forming cell responses to antigens. In another study, host defense against a subdermal allogenic carcinoma (as defined by reduction in tumor size) was improved by sleep deprivation (92). This improvement could be due to changes in cytokine profile. For example, TNFα is increased with sleep loss; the tumor used in this study could have been sensitive to TNFα. Further, changes in cytokine profiles could have augmented NK cellular activity and this or some other immune system cell could be responsible for the smaller tumor size in total-sleep-deprived rats. Thus, host defense in an in vivo model can be enhanced by sleep deprivation, suggesting that sleep deprivation may serve a beneficial effect in the defense against tumors.

Feng et al. studied the effects of stress on the humoral immune response during infection by influenza virus (93). Restraint stress altered the kinetics of the antibody response; seroconversion in the IgG and IgA isotypes was delayed in virus-infected C57BL/6 mice subjected to repeated cycles of physical restraint. However, the magnitude and isotype of the mature antibody response were unaffected during the plateau phase; no significant differences were observed between restrained/infected and nonrestrained/infected mice. Thus, the time during infection at which the antibody response was measured was a significant variable in the study of stress-induced alterations of the host's response to a replicating viral antigen. While restraint stress did not significantly affect the magnitude or class of the humoral response, it did alter the kinetics of response. Restraint stress is also associated with a rebound in REMS (94). Further, sleep deprivation per se can be considered a stressor; thus Feng's results may have bearing on those studies relating sleep loss to host defenses.

Collectively, these studies support the concept that sleep is a component of the homeostatic regulation of immune function, and that disturbances in sleep alter the ability of a host to defend itself against foreign invaders. We can conclude that long-term sleep loss adversely affects innate host defenses, eventually leading to death. From those studies it seems reasonable to suggest that under normal conditions at least some facets of the immune response are linked to sleep states. Regardless of whether sleep deprivation has a direct or indirect influence on immune function, it is clear that redistribution of immune cells may significantly affect the ability of immune system to respond to potential or ongoing immune challenge. Further, whether sleep-deprivation-induced effects will be beneficial or harmful will depend on the duration, chronicity, and mental state of the individual or animal.

C. Cytokine Mechanisms

Cytokines Are Involved in Physiological Sleep Mechanisms

The bacterial and viral products mentioned in the previous sections are known to induce production of an array of cytokines. Several of these cytokines are somnogenic and proinflammatory, whereas others are waking-inducing and anti-inflammatory. It seems likely, at least for the clear biphasic sleep responses occurring after bacterial inoculation, that during the initial phase of excess sleep the proinflammatory somnogenic cytokines dominate, whereas later, during the sleep inhibition phase, anti-inflammatory cytokines dominate. However, this has not yet been experimentally demonstrated. Below we discuss the limited data available suggesting the involvement of cytokines in sleep responses to bacterial products. However, in contrast, there is a wealth of data suggesting the involvement of cytokines in physiological sleep regulation and those data are reviewed first. This essay will focus on IL-1β and TNFα since much sleep data are available for these cytokines.

Cytokines in Brain and Plasma; Relationships to Sleep

Interleukin-1β.   IL-1β is a 17-kDa polypeptide with autocrine, paracrine, and endocrine roles. Although it is best known for its involvement in immunity and inflammation (reviewed in 95,96), it also has several physiological roles, including sleep regulation (see below), and regulation of appetite (97), normal development (98), gastrointestinal function (99), and several endocrine systems including the growth-hormone-releasing hormone (GHRH)–growth hormone (GH)–insulin-like growth factor (IGF) axis and the corticotropin-releasing hormone (CRH)–adrenocorticotropin hormone (ACTH)–glucocorticoids axis (100). The IL-1 family has at least nine members (see Fig. 3). The IL-1 ligands, IL-1α, IL-1β, and the interleukin-1 receptor antagonist (IL-1RA) share limited amino acid homology although their predicted 3-dimensional topologies are more similar (101); all bind to IL-1 receptors. The IL-1RA competes with IL-1α and IL-1β for binding sites, and as its name implies, it antagonizes the actions of IL-1α and IL-1β. Two IL-1 receptors are characterized, type I and type II (reviewed in 102–104); each possesses three extracellular immunoglobulin-like domains, limited

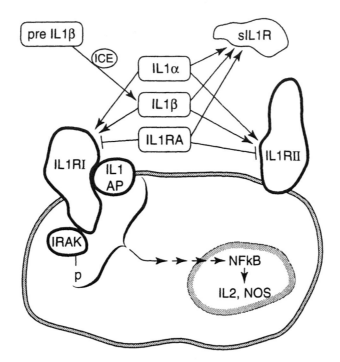

Figure 3 The interleukin-1 family of molecules. pre-IL-1β, preinterleukin-1 beta; ICE, IL-1-converting enzyme; IL-1RA, IL-1-receptor antagonist; sIL-1R, soluble IL-1 receptor; IL-1RI, IL-1 type I receptor; IL-1AP, IL-1-receptor accessory protein; IRAK, IL-1-receptor-associated kinase; IL-1RII, IL-1 type II receptor.

homology with each other, and different binding characteristics. The two receptors have distinct functions; type I is a signal-transducing receptor, whereas type II acts as a decoy receptor (105) having a truncated nonsignaling intracellular domain. There is also an IL-1 receptor accessory protein (IL-1AP) that shares limited homology with the type I and II IL-1 receptors; it also is found in brain (106,107). The IL-1AP forms a complex with the type I receptor and IL-1β or IL-1α, but not with the IL-1RA. The IL-1AP increases the binding affinity of IL-1β for the type I receptor (107) and it is important for signal transduction (108). An IL-1-receptor-associated protein kinase (IRAK) has also been identified (109); it activates nuclear factor kappa B (NFκB). Further, another IL-1 receptor protein closely related to the IL-1 receptor has also been recently cloned (110). Its function is unknown although it is in brain associated with cerebral vasculature. Both IL-1α and IL-1β are first made as precursor molecules; mature forms are released from cells. The IL-1α precursor is biologically active, whereas IL-1β requires processing. IL-1β-converting enzyme (ICE) cleaves pre-IL-1β at two sites and thus produces the 17-kDa IL-1β. Finally, an alternatively processed cDNA for rat IL-1 type II receptor that encodes only the extracellular domain of the receptor was described (111). This soluble receptor binds IL-1 and thus acts as an inhibitor of IL-1.

There are now numerous reports indicating that IL-1β, IL-1 receptors, and other members of the IL-1 family of molecules are constitutively expressed in normal brain (reviewed in 112). Important for this discussion are changes in IL-1β that correlate with changes in sleep. Lue and colleagues (113) described changes in cerebrospinal fluid levels of IL-1 that were in phase with changes in the sleep-wake cycles; highest levels occurred at sleep onset. More recently (114), IL-1β mRNA levels were shown to have a diurnal variation in rat hypothalamus, hippocampus, and cortex, but not other areas of brain. Highest levels of IL-1β mRNA occurred about 1 hr after lights were turned on; this is the period of peak NREMS in rats. In another study, IL-1β mRNA levels were shown to increase in the brainstem and hypothalamus during sleep deprivation (115).

There are also several reports describing changes in plasma levels of IL-1β correlating with sleep-wake cycles or increasing during sleep deprivation (116–119). For example, in humans, plasma levels of IL-1β peak at the onset of sleep (116). In rabbits (120) and humans (116) plasma levels of IL-1β increase during sleep deprivation. Despite these reports, the importance of circulating IL-1β in sleep regulation remains unknown. For example, inhibition of central IL-1, but not systemic IL-1, blocks sleep-deprivation-induced NREMS rebound (121).

Administration of exogenous IL-1β, whether given centrally or systemically, induces increases in NREMS in rats (122,123), rabbits (124), mice (125), cats (126), and monkeys (127). In rabbits, doses in the range of 2.5–15 ng given i.c.v. induce increases of NREMS from control values of around 45% to 65–75% for 6 or more hr depending on the dose. These very large increases in NREMS are accompanied by increases in EEG slow-wave amplitudes (SWA); this increased EEG SWA is thought to be indicative of the intensity of NREMS (128,129). During periods of peak IL-1-induced NREMS responses REMS is inhibited. In rabbits these somnogenic doses are also pyrogenic. Rats also respond to IL-1β by increasing NREMS although in this species the effects of IL-1 are more complex. After low somnogenic doses of IL-1 given i.c.v., both duration of NREMS and EEG SWA increase; these increases occur regardless of the time of day IL-1 is given. After midlevel doses of IL-1, increases in EEG SWAs occur whether IL-1 is given day or night; in contrast, this dose of IL-1β increased duration of NREMS only during nighttime hours. After high doses of IL-1, duration of NREMS is inhibited regardless of time of day of IL-1 administration. In rats, low somnogenic doses of IL-1 (i.c.v.) do not inhibit REMS nor do they induce fevers. Finally, in rats after i.p. administration of IL-1, duration of NREMS increases, but there is also a prolonged IL-1-induced decrease in EEG SWAs (130); similar responses are observed in mice after i.p. injection of IL-1 (131).

Sleep after IL-1 treatment appears to retain physiological characteristics. Thus, in all the species tested, sleep after IL-1 treatment remains episodic. Animals can be easily aroused, and after low somnogenic doses, behavior is normal. Further, the sleep after IL-1 treatment is similar to that observed after sleep deprivation in that duration of NREMS and EEG SWAs increase. After very high doses of IL-1, sleep is abnormal as are other aspects of the animal's behavior. These high doses of IL-1 induce sickness behavior, which is characterized by social withdrawal, lack of interest in the environment, and other abnormal behavior and postures (review in 132). This sickness syndrome induced by high doses of IL-1 is similar to that observed during infections or that induced by bacterial products.

In normal animals inhibition of IL-1 using either antibodies (133), the soluble IL-1 receptor or peptide fragments of the soluble receptor (134), or the IL-1RA (135) inhibits spontaneous NREMS. Such results provide evidence that IL-1β is involved in regulation of spontaneous sleep. Further, substances that inhibit IL-1 production, such as IL-10 (136), IL-4 (137), α-melanocyte-stimulating hormone (α-MSH) (138), prostaglandin E_2 (139), corticotropin-releasing hormone (CRH) (140), and glucocorticoids (141), also inhibit spontaneous sleep.

Several experimental conditions reliably result in increases in NREMS: e.g., sleep deprivation, acute mild increases in ambient temperature, and administration of bacterial products such as muramyl dipeptide (MDP). IL-1 seems to be involved in each of these experimentally induced increases in NREMS since inhibition of IL-1 blocks or attenuates NREMS induced by each of these manipulations (reviewed in 14; 120,134,142). Further, several substances that are known inducers of IL-1, e.g., dsRNA, LPS, and bacterial peptidoglycan, also promote NREMS.

It is generally believed that cytokines produced by peripheral immune cells activated by pathogens are released locally to act in an autocrine and exocrine fashion and in some cases into the circulation and transported to the brain to induce the central components of acute-phase responses such as fever. This may also be a mechanism for the sleep responses to peripheral infections. In the last few years, many central effects of LPS and cytokines were found to be inhibited by vagus nerve transections. For instance, subdiaphragmatic vagotomy blocks LPS-induced fever (143) and IL-1β mRNA in the brain (144), or attenuates LPS-induced depression of food-motivated behavior (145), adrenocorticotropin-releasing hormone responses (146), and hyperalgesia (147). Subdiaphragmatic vagotomy also blocks or attenuates IL-1-induced fever (148), taste aversion (149), and depression of food-motivated behavior (145). These observations suggest that vagal inputs play important roles in the communication of peripheral infection signals to the brain. Recently, our laboratory observed that subdiaphragmatic vagotomy also attenuates, but does not completely prevent, sleep responses to intraperitoneal (i.p.) injection of LPS in rats (L. Kapás et al, unpublished). Further, subdiaphragmatic vagotomy blocks low-dose-IL-1-induced sleep, attenuates the effects of a middle dose of IL-1 on sleep, and has no influence on high-dose-IL-1-induced sleep (150). These data indicate that the vagi play an important role in mediating peripheral LPS- and IL-1-induced sleep. The failure to completely block LPS- and IL-1-induced sleep by vagotomy suggests that other mechanisms are also involved in the communication between the peripheral immune response and the brain. For example, such communication might be mediated by other neural inputs or by transport of cytokines to brain.

The IL-1 type I receptor seems to be involved in IL-1-modulated sleep. Thus, mutant mice lacking this receptor fail to exhibit excess NREMS if given IL-1β, whereas they readily respond to other somnogens such as TNFα (151). These mutant mice also sleep significantly less than strain controls during night-time hours although this effect is small. Duration of NREMS during daylight hours is similar in IL-1 receptor knockout mice and controls.

IL-1 also induces production of several substances that also have been implicated in sleep regulation. For example, IL-1 induces GH release via a hypothalamic mechanism that involves GHRH (152). Rats pretreated with anti-GHRH fail to exhibit NREMS responses if given IL-1β (153). IL-1β also induces increases in nitric oxide synthase (NOS)-2 and IL-2; both of these substances are implicated in sleep regulation (discussed below).

Several of the biological actions of IL-1 were characterized within the context of host defense systems. Some of these actions of IL-1 associated with pathology e.g., fever, are not, in normal animals, associated with sleep. The pyrogenic actions of IL-1 are separable from its somnogenic actions (reviewed in 154). Thus, antipyretics block IL-1-induced fever but do not affect IL-1-induced NREMS responses (124). As mentioned above, in rats low doses of IL-1 induce increases of NREMS without inducing fevers (122). The pyrogenic actions of other somnogens also can be separated from their effects on sleep. Thus, IL-6 is pyrogenic but not somnogenic (155). Similarly, CRH by itself induces increases in body temperature but inhibits sleep and inhibits IL-1-induced fever (140). α-MSH is cryogenic but inhibits sleep (138). MDP is somnogenic in rabbits whether given day or night, yet is much more potent as a pyrogen during daytime than during nighttime (156). Finally, inhibition of NOS blocks IL-1-induced sleep but not IL-1-induced fevers (157). These and other data (reviewed in 154) collectively strongly suggest that fever and NREMS have separate regulatory mechanisms.

Collectively, the data briefly reviewed above clearly implicate IL-1β in physiological sleep regulation. Nevertheless, as already briefly suggested, IL-1 operates within a biochemical network involving many additional cytokines and hormones and a variety of cell types that form a sleep regulatory system. Our understanding of these molecular steps in sleep regulation is beginning to be understood; some of these steps are illustrated in Figure 4.

TNFα. TNFα is also a pleiotropic cytokine of about 17,000 Da. TNF causes cytolysis or cytostasis of several tumor cell lines; induces hemorrhagic

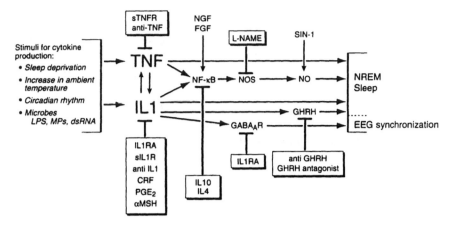

Figure 4 Sleep activational cascade indicating possible interactions between sleep-enhancing and sleep-inhibiting substances. There is substantial evidence that IL-1, TNF, and GHRH are involved in physiological NREMS regulation (see text and reviewed in 2). These substances, and associated endogenous substances involved in their regulation, affect each other's production as indicated. Substances in boxes inhibit sleep and inhibit the actions or production of NFκB, IL-1, TNF, GHRH, NO, or the GABA$_A$ receptor as indicated. Inhibition of any one step does not result in complete sleep loss. Further, it is likely the animals compensate for the loss of one step by relying on the parallel pathways shown to regulate sleep. Such redundant pathways provide stability to the sleep regulatory system as well as a mechanism whereby the variety of known sleep-promoting stimuli may affect sleep. A major challenge to sleep research is to define how and where these molecular steps interact with known regulatory mechanisms. LPS, lipopolysaccharide; MP, muramyl peptide; dsRNA, double-stranded RNA; TNF, tumor necrosis factor, sTNFR, soluble TNF receptor; IL-1, interleukin-1; IL-1RA, IL-1-receptor antagonist; sIL-1R, soluble IL-1 receptor; CRF, corticotropin-releasing factor; PGE$_2$, prostaglandin E$_2$; α-MSH, alpha-melanocyte-stimulating hormone; EGF, epidermal growth factor; NGF, nerve growth factor; FGF, fibroblast growth factor; IL-10, interleukin-10; IL-4, interleukin-4; L-NAME, an arginine analog; NOS, nitric oxide synthase; GHRH, growth-hormone-releasing hormone; SIN-1, a NO donor; GABA$_A$R, gamma-aminobutyric acid A receptor; NREMS, non–rapid-eye-movement sleep. → indicates stimulation; —— indicates inhibition.

necrosis of transplanted tumors; activates granulocyte fibroblasts and T and B lymphocytes; and induces MHC class I and intercellular adhesion proteins. TNF activates several signal transduction pathways and multiple genes (158); these actions help explain its pleiotropic nature.

The TNF family of molecules is, thus far, not as large as that of IL-1. There are two TNF ligands, TNFα and TNFβ, and two TNF receptors, 55-kDa and 75-kDa receptors. TNFα is constitutively expressed in brain; it is found in neurons (159,160) in normal brain (161,162) and astrocytes (163). It has a normal role in brain development (98,164,165). It also is up-regulated in brain by lipopolysaccharide neurotropic virus. Within the brain, TNFα seems to be regulated both transcriptionally and posttranscriptionally. Several groups demonstrated bacterial product-induced increases in brain TNFα mRNA (166–170). Posttranscriptional regulation is indicated by the observation that TNFα increases within 15 min of endotoxin given directly into brain (171). Different forms of posttranscriptional regulation also occur in brain; thus, for example, inhibition of protein kinase C results in a 10-fold decrease in TNF mRNA half-life (172). Important for sleep regulation and consistent with these findings is the occurrence of a diurnal rhythm of TNF mRNA in hypothalamus and hippocampus; TNF mRNA levels in these tissues are about twofold higher during the day than during the night (173). The amplitude of the diurnal rhythms of TNFα protein levels is greater (about 10-fold) in these tissues (174).

TNF receptors are on virtually all somatic cells including cells within the CNS (175). TNF receptors are similar to nerve growth factor receptors (176,177) and Shope fibroma virus protein (T2) (177). It appears that the 55-kDa TNF receptor, rather than the 75-kDa receptor, is important for sleep; knockout mice lacking the 55-kDa receptor do not exhibit sleep responses if given TNFα (131). Two TNF-binding proteins of 30 and 40 kDa have been isolated from tumor cell lines (178), urine (179–181), and serum; these binding proteins represent the extracellular domain of the TNF receptor. These soluble receptors inhibit TNFα (178,179,182,183) and they likely regulate, in part, the biological activity of TNF in vivo. The soluble receptors accumulate in the plasma of cancer patients (184) and in response to endotoxin challenge (185). Although very little is known about the regulation of these soluble receptors, they can affect sleep. Thus, administration of the TNF soluble receptor (186) or a peptide fragment of the soluble receptor that contains the TNF-binding site inhibits spontaneous NREMS (187).

It seems likely that brain TNFα is the pool of TNF important for sleep regulation, and indeed, as mentioned above, TNFα mRNA and TNFα protein levels in rats have their highest values during peak sleep periods. Nevertheless, there is evidence that circulating levels of TNF may also depend, in part, on sleep-wake cycles. Thus, Darko et al. (188) showed that in normal humans TNF plasma levels during NREMS covary with EEG SWA; this relationship is lost in AIDS patients. Others have described in humans a diurnal rhythm of TNF plasma levels; this rhythm is disrupted in patients with sleep apnea (189). The ability of circulating monocytes to produce TNF is coupled to the sleep-wake cycle and increases during sleep deprivation (118,119,190).

A growing body of evidence suggests that TNFα, like IL-1β, is involved in physiological sleep regulation. Thus, administration of exogenous TNFα to rabbits (191), rats (192), or mice (131) induces increases in the duration of NREMS. These effects can be large; e.g., mice receiving 3.0 μg of TNF intraperitoneally get 81 min of extra NREMS during the first 9 postinjection hr (131). TNF-induced excess NREMS appears to be normal in the sense that: (1) animals continue to cycle through waking, NREMS, and REMS episodes; (2) the clear diurnal rhythm of sleep persists; (3) asleep animals can be easily aroused; and (4) behavior after somnogenic doses appears to be relatively normal. Further, intraventricularly administered TNFα induces increases in EEG SWAs similar to those observed after sleep deprivation, during infection, or after IL-1β treatment. In rabbits, high doses of TNFα that induce very large increases in NREMS are associated with decreases in REMS duration. However, in mice, intraperitoneal injections of TNF did not inhibit REMS.

In normal animals, if TNF is inhibited using either anti-TNFα antibodies (193), the full-length soluble TNF receptor (186), or a peptide fragment of the TNF receptor (187), spontaneous NREMS is inhibited. Further, the sleep rebound occurring after sleep deprivation is attenuated if animals are pretreated with the TNF soluble receptor (186). Inhibition of TNF also attenuates bacterial-product-induced increases in NREMS and acute mild increases in ambient-temperature-induced increases in NREMS.

Substances that induce TNFα, e.g., muramyl peptides, enhance sleep. Conversely, substances that inhibit production of TNFα, such as IL-4 (137) and IL-10 (136), inhibit sleep. Finally, mice that lack the TNF 55-kDa receptor sleep less than control strains or other strains of mice recorded under the same conditions (131). Collectively, these data firmly support the hypothesis that TNFα is involved in physiological sleep regulation.

Relationships between TNFα and IL-1β in sleep regulation are not well understood although they seem to interact with each other. Thus, several somnogenic stimuli promote both IL-1β and TNFα production including sleep deprivation, mild acute increases in ambient temperature, and bacterial products (Fig. 2). Further, IL-1β and TNFα induce each other's production. If rabbits are pretreated with an inhibition of TNFα and then given IL-1, expected IL-1-induced increases in NREMS are attenuated. Conversely, inhibition of IL-1 attenuates TNFα-induced sleep responses (194). Further, both IL-1β and TNFα, as well as several other somnogens, e.g., nerve growth factor (NGF), fibroblast growth factor (FGF), and epidermal growth factor (EGF), activate NFκB. NFκB is also activated in cerebral cortex during sleep deprivation (195). Despite these close relationships between IL-1β and TNFα in sleep regulation they can, under appropriate conditions, act independently to induce sleep. Thus IL-1 type I receptor knockout mice do not mount NREMS responses if given IL-1, but exhibit robust NREMS responses if given TNF (150). Conversely, mice lacking the TNF 55-kDa receptor do not respond to TNFα, but do have large increases in NREMS if given IL-1β (131).

Other Cytokines and Growth Factors Linked to Sleep Regulation

Several additional cytokines/growth factors have been linked to sleep regulation. There is compelling evidence for the involvement of GHRH in physiological sleep regulation. Thus, administration of exogenous GHRH to humans and other animals induces increases in duration of NREMS and REMS (reviewed in 196). The enhancement of REMS seems to be via GHRH-induced GH release from the pituitary since in hypothectomized rats GHRH induces only NREMS (197) and GH by itself can induce increases in REMS (198). Within brain GHRHergic neurons are found in the arcuate nucleus and project to the median eminence; these neurons are likely involved in GH release. Another pool of GHRHergic neurons is found around the ventromedial hypothalamus; these neurons project within the hypothalamus and to the basal forebrain. It is likely these neurons are involved in NREMS regulation. There is a diurnal rhythm of hypothalamic levels of GHRH mRNA in rats with highest levels being found during peak sleep periods (199). Microinjections of GHRH into the preoptic area but not other brain areas induce excess NREMS (200).

The excess NREMS induced by GHRH is similar to that observed after sleep deprivation: (1) EEG SWAs are enhanced during periods of NREMS; (2) the duration of NREMS increases; (3) sleep remains episodic; (4) the circadian rhythm of sleep persists; and (5) sleep is readily reversible by appropriate stimuli (reviewed in 196). If GHRH is inhibited using either anti-GHRH antibodies (201) or a peptide antagonist to GHRH (202), spontaneous sleep is inhibited. Inhibition of GHRH also blocks the expected sleep rebound after sleep deprivation (201).

The biological actions of GHRH and IL-1β are linked. It has been known for some time that infection or administration of endotoxin is associated with the release of GH. More recently this was shown to be dependent upon a hypothalamic mechanism that involves IL-1 and GHRH. For example, administration of IL-1 induces GH release and this effect is blocked if animals are pretreated with anti-GHRH antibodies (203). Similarly, the NREMS-promoting activity of IL-1 is blocked in animals treated with anti-GHRH (204). Collectively, such data strongly implicate GHRH in physiological sleep regulation and in the sleep responses to infection.

Several additional cytokines, if injected into experimental animals, induce increases in the duration of NREMS. The list includes: IL-2, acidic fibroblast growth factor, nerve growth factor, epidermal growth factor, and inteferon-α (reviewed in 205). However, the involvement of these substances in sleep regulation has yet to be investigated beyond the initial experiments in which they were injected and sleep recorded; it is thus unknown whether they play any role in physiological sleep regulation. It does, however, seems likely that some are involved in the sleep responses associated with infection. Two cytokines, IL-4 and IL-10, which are known as anti-inflammatory cytokines because they inhibit production of proinflammatory cytokines such as IL-1 and TNF, inhibit spontaneous sleep. Several other cytokines seem to have no effect on sleep; the list includes basic FGF and IL-6.

In conclusion, it seems likely that within brain there is a network of cytokines and other growth factors involved in physiological sleep regulation.

This network may be similar to the cytokine networks involved in regulating the immune response although we are already aware of some differences. It also is probable that during infection the brain cytokine network is up-regulated, possibly via microbial-induced up-regulation of systemic cytokines, and this cytokine up-regulation is in turn responsible for the sleep responses observed during infection.

Mechanisms of Cytokine Somnogenic Actions

Nitric oxide (NO) is an unusual molecule in the sense that two independent literatures have extensively implicated it in what superficially appear to be very different, possibly even mutually exclusive, actions. Within the neurobiology community NO is well known as a retrograde neurotransmitter and is viewed as a normal modulator of neural activity. In contrast, within the immunology literature, the up-regulation of NO production is linked to cell killing. These differences have been somewhat dampened by the discovery of neural NO synthase (NOS-1), which is constituitively expressed in neurons, and inducible NOS (NOS-2), which, as the name implies, is induced by microbial products and cytokines such as IL-1 and TNF. These different NOSs do in fact help explain the two literatures and also help provide an explanation for the mechanisms of sleep induction during infection.

The involvement of NO in sleep regulation was first investigated because it was well known that TNF and IL-1, which were already implicated in sleep regulation, induced NO production. In the first experiments, inhibitors of NOS, arginine analogs, were found to inhibit rat (206) and rabbit (157) sleep. Further, these inhibitors blocked the somnogenic actions of IL-1β (157). In subsequent experiments, substances that spontaneously decompose to release NO were found to induce excess sleep but only after a prolonged delay (207). This delayed action was not due to a delay in NO induction of cGMP since after administration of the NO-donor cGMP levels increased within minutes. The reason for the delay remains unknown. In other experiments a diurnal rhythm of cytosolic NOS in various brain areas was demonstrated; interestingly, highest levels were found during nighttime hours in rats (208). The microinjection of either NOS inhibitors (209,210) or NO donors (211) into the pons is associated with decreases or increases in sleep, respectively. While these results are exciting, it remains to be determined exactly how NO is involved in sleep regulation and indeed which NOS and its (their) sleep-linked location in brain is (are) involved. It is also possible that the somnogenic actions of GHRH also involves NO; recently NO was implicated in GHRH-induced release of pituitary GH (212). Finally, NO may also be involved in adenosine-altered cerebral blood flow and sleep; adenosine induces NO release from astrocytes (213).

Other neurotransmitter systems are also linked to the actions or production of cytokines. For example, hypothalamic adrenergic, dopaminergic, and serotonin systems (214–220) are activated by IL-1. Several cytokines including IL-1, TNF, and FGF alter Ca^{2+} currents in neurons (221,222). Serotonin can induce IL-1 (223) or TNF (224) production. Whether any of these effects or other actions of cytokines or neurons (reviewed in 97) are involved in sleep regulation remains unknown. However, two known effects of cytokines on neurotransmitters, the effects of IL-1 on $GABA_A$ receptors (225) and of TNF on histamine neurons (226), could be of importance to EEG synchronization and wakefulness, respectively. The IL-1–IL-1 receptor complex via the $GABA_A$ receptor increases Cl^- permeability; the IL-1RA antagonizes this effect. The net effect of this action is the hyperpolarization of $GABA_A$-receptive neurons. Others (227) have shown that hyperpolarization of GABA-receptive neurons in thalamocortical circuits is responsible for EEG synchronization, which is, of course, a marker for NREMS. The association between $TNF\alpha$ and histaminergic neurons (228) is also likely important in sleep regulation. Neurotoxic lesions of histaminergic neurons decrease $TNF\alpha$ in the hypothalamus while enhancing $TNF\alpha$ production in the hippocampus (226). These data suggest that neuronal histamine is involved in the regulation of the brain $TNF\alpha$ system. Data from other cells suggest that histamine inhibits $TNF\alpha$ gene expression and TNF synthesis (229). Antihistamines are well known somnogenic agents and posterior hypothalamic histaminergic neurons are thought to be involved in the maintenance of wakefulness.

III. Conclusions

Many, beginning with Morruzzi (230), have hypothesized that sleep serves a synaptic function; the list includes Krueger and Obál (231), Marks et al. (232), Jouvet (233), and Kavenau (234). The basis for these hypotheses rests primarily on logic and the recognition that synaptic efficacy changes with synaptic use, thereby creating a mechanism for brain plasticity. At another level of organization, sleep is linked to memory and memory is thought to involve synaptic sculpturing. It is important to recognize that if one or more of these theories is even partially correct, then the functional manifestations of sleep will be at the synaptic level and thus in the association areas of cortex and other brain areas where dynamic synaptic processes are important. In this light, the changes in IL-1β mRNA and $TNF\alpha$ cortical levels with the sleep-wake cycle are of interest.

A related theoretical concern deals with how the brain is organized to produce sleep and even what level of organization within brain is capable of sleep. It has been proposed that sleep is a fundamental property of neuronal groups and that sleep begins at the neuronal group level (231). If sufficient numbers of neuronal groups are in a "sleep mode," then sleep, as defined by EEG and behavioral criteria, occurs. The coordination of neuronal groups is brought about by the neuronal networks previously tied to sleep regulation. There is experimental support for this view of sleep organization [see Pigarev (235); Krueger and

Obál (231); Kattler et al. (236) for experimental support, and Mahowald and Schenck (237) for clinical evidence]. The importance of this view in the present review is that at the neuronal group level we have developed a model that illustrates how use of a neuronal circuit results in increased production of cytokines and other growth factors that, in turn, act upon that circuit and cause a shift in the firing pattern within that circuit (neuronal group). This altered firing pattern can be considered sleep at the neuronal group level and this growth-factor-induced altered firing a mechanism of sleep. Simultaneously the cytokine/growth factors induce synthesis of synapse-related molecules, e.g., adhesion molecules via altered gene transcription, e.g., altered NFκB activation. Thus, mechanism and function cannot be separated at this level. Nothing could be more fundamental to and important for sleep, and in our opinion, neurobiology.

Acknowledgments

This work was supported, in part, by grants from the National Institutes of Health (NS25378, NS27250, NS31453, and NS01727). We thank Mrs. Maria Swayze-Nations for her secretial assistance.

References

1. Toth L, Krueger JM. Alteration of sleep in rabbits by *Staphylococcus aureus* infection. Infect Immun 1988; 56:1785–1791.
2. Krueger JM, Majde JA. Microbial products and cytokines in sleep and fever regulation. Crit Rev Immunol 1994; 14:355–379.
3. Toth L, Krueger JM. Effects of microbial challenge on sleep in rabbits. FASEB J 1989; 3:2062–2066.
4. Johannsen L, Toth LA, Rosenthal RS, Opp MR, Obál Jr F, Cady AB, Krueger JM. Somnogenic, pyrogenic and hematologic effects of bacterial peptidoglycan. Am J Physiol 1990; 259:R182–R186.
5. Krueger JM, Kubillus S, Shoham S, Davenne D. Enhancement of slow-wave sleep by endotoxin and lipid A. Am J Physiol 1986; 251:R591–R597.
6. Cady AB, Kotani S, Shiba T, Kusumoto S, Krueger JM. Somnogenic activities of synthetic lipid A. Infect Immun 1989; 57:396–403.
7. Krueger JM, Walter J, Karnovsky ML, Chedid L, Choay JP, Lefrancier P, Lederer E. Muramyl peptides: variation of somnogenic activity with structure. J Exp Med 1984; 159:68–76.
8. Krueger JM, Rosenthal RS, Martin SA, Walter J, Davenne D, Shoham S, Kubillus SL, Biemann K. Bacterial peptidoglycan as modulators in sleep. I. Anhydro forms of muramyl peptides enhance somnogenic potency. Brain Res 1987; 403:249–266.
9. Johannsen L, Wecke J, Obal Jr F, Krueger JM. Macrophages produce somnogenic and pyrogenic muramyl peptides during digestion of staphylococci. Am J Physiol 1991; 260:R126–R133.

10. Vermeulen MW, Grey GR. Processing of *Bacillus subtilis* peptidoglycan by a mouse macrophage cell line. Infect Immun 1984; 46:476–483.

11. Fincher EF, Opp MR, Johannsen L, Krueger JM. Microglia digest *Staphylococcus aureus* into low molecular weight biologically active compounds. Am J Physiol 1996; 271:R149–R156.

12. Krueger JM, Pappenheimer JR, Karnovsky ML. The composition of sleep-promoting factor isolated from human urine. J Biol Chem 1982; 257:1664–1669.

13. Krueger JM, Karnovsky ML, Martin SA, Pappenheimer JR, Walter J, Biemann K. Peptidoglycans as promoters of slow-wave sleep. II. Somnogenic and pyrogenic activities of some naturally occurring muramyl peptides; correlations with mass spectrometric structure determination. J Biol Chem 1984; 259:12659–12662.

14. Krueger JM, Majde JA. Sleep as a host defense: its regulation by microbial products and cytokines. Clin Immunol Immunopathol 1990; 57:188–199.

15. Krueger JM, Pappenheimer JR, Karnovsky ML. Sleep-promoting effects of muramyl peptides. Proc Natl Acad Sci USA 1982; 79:6102–6106.

16. Scherschlicht R, Marias J. Effects of sleep-promoting "factor S," muramyl dipeptide (MDP) and L-cycloserine on the sleep of rabbits. Experientia 1983; 39:683.

17. Meltzer LT, Serpa KA, Moss WH. Evaluation in rats of the somnogenic, pyrogenic and central nervous system depressant effects of muramyl dipeptide. Psychopharmacology 1989; 99:103–108.

18. Inoué S, Honda K, Komoda Y, Uchizona K, Ueno R, Hayaishi O. Differential sleep-promoting effects of five sleep substances nocturnally infused in unrestrained rats. Proc Natl Acad Sci USA 1984; 81:6240–6244.

19. Masek K. Immunopharmacology of muramyl peptides. Fed Proc 1986; 45:2549–2551.

20. Wexler DB, Moore-Ede MC. Effects of a muramyl dipeptide on the temperature and sleep-wake cycles of the monkey. Am J Physiol 1984; 247:R672–R680.

21. von Economo C. Sleep as a problem of localization. J Nerv Ment Dis 1930; 71:249–259.

22. Komaroff AL. Chronic fatigue syndromes: relationship to chronic viral infections. J Virol Methods 1988; 21:3–10.

23. Guillenminault C, Mondini S. Mononucleosis and chronic daytime sleepiness: a long-term follow-up study. Arch Intern Med 1986; 146:1333–1335.

24. Behan PO, Behan WM, Gow JW, Cavanagh H, Gillespie S. Enteroviruses and postviral fatigue syndrome. Ciba Found Symp 1993; 173:146–154.

25. Hoffmann HJ, Damus K, Hillman L, Krongrad E. Risk factors for SIDS: results of the National Institute of Child Health and Human Development SIDS cooperative epidemiological study. Ann NY Acad Sci 1988; 533:13.

26. Norman SE, Chediak HD, Kiel M, Cohn MA. Sleep disturbances in HIV-infected homosexual men. AIDS 1990; 4:775–781.

27. Smith A. Sleep, colds, and performance. In: Broughton RJ, Ogilvie RD, eds. Sleep, Arousal and Performance. Boston: Birkhauser, 1992:233.

28. Smith AA. Review of the effects of colds and influenza on human performance. J Soc Occup Med 1989; 39:65–68.

29. Kimura-Takeuchi M, Majde JA, Toth L, Krueger JM. Influenza virus-induced changes in rabbit sleep and acute phase responses. Am J Physiol 1992; 263:R1115–R1121.

30. Fang J, Sanborn CK, Renegar K, Majde JA, Krueger JM. Influenza viral infections enhance sleep in mice. Proc Soc Exp Biol Med 1995; 210:242–252.

31. Toth LA, Rehg JE, Webster RG. Strain differences in sleep and other pathophysiological sequelae of influenza virus infection in naïve and immunized mice. J Neuroimmunol 1995; 58:89–99.

32. Fang J, Tooley D, Gatewood C, Renegar KB, Majde JA, Krueger JM. Differential effects of total and upper airway influenza viral infection on sleep in mice. Sleep 1996; 19:337–342.

33. Johnson RT, McArthur JC, Narayan O. The neurobiology of human immunodeficiency virus infections. FASEB J 1988; 2:2970–2981.

34. Darko DF, Mitler MM, Henriksen SJ. Lentiviral infection, immune response peptides and sleep. Adv Neuroimmunol 1995; 5:57–77.

35. Carter WA, De Clercq E. Viral infection and host defense. Science 1974; 186:1172–1178.

36. Freeman AI, Al-Bussam N, O'Malley JA, Stutzman L, Bjornsson S, Carter WA. Pharmacologic effects of polyinosinic-polycytidylic acid in man. J Med Virol 1977; 1:79–93.

37. Krueger JM, Majde JA, Blatteis CM, Endsley J, Ahokas RA, Cady AB. Polyriboinosinic:polyribocytidylic acid (poly I:C) enhances rabbit slow-wave sleep. Am J Physiol 1988; 255:R748–R755.

38. Majde JA, Brown RK, Jones MW, Dieffenbach CW, Maitra N, Krueger JM, Cady AB, Smitka CW, Maassab HF. Detection of toxic viral-associated double-stranded RNA (dsRNA) in influenza-infected lung. Microb Pathogen 1991; 10:105–115.

39. Kimura-Takeuchi M, Majde JA, Toth L, Krueger JM. The role of double-stranded RNA in induction of the acute phase response in an abortive influenza virus infection model. J Infect Dis 1992; 166:1266–1275.

40. Bredow S, Fang J, Guha-Thakurta N, Majde JA, Krueger JM. Synthesis of an influenza double-stranded RNA-oligomer that induces fever and sleep in rabbits. Sleep Res 1995; 24A:101.

41. Opp MR, Hughes Jr TK, Rady P, Smith EM. Mechanisms of HIV-induced alterations in sleep: the role of cytokines in the CNS. SRS Bull 1996; 2:31–37.

42. Kent S, Price M, Satinoff B. Fever alters characteristics of sleep in rats. Physiol Behav 1988; 44:709–715.

43. Brock TD, Madigan MT. Biology of Microorganism, 5th ed. Englewood Cliffs, NJ: Prentice-Hall, 1988:72–78.

44. Domer JE. *Candida* cell wall mannan: a polysaccharide with diverse immunologic properties. Crit Rev Microbiol 1989; 17:33–51.

45. Abel G, Czop JK. Stimulation of human monocytes beta-glucan receptors by glucan particles induces production of TNF-alpha and IL-1 beta. Intl J Immunopharmacol 1992; 14:1363–1366.

46. Noble PW, Henson PM, Lucas C, Mora-Worms M, Carre PC, Riches DW. Transforming growth factor-beta primes macrophages to express inflammatory gene products in response to particular stimuli by an autocrine/paracrine mechanism. J Immunol 1993; 151:979–987.

47. Poutsiaka DD, Mengozzi M, Vannier E, Sinha B, Dinarello CA. Cross-linking of the

beta-glucan receptor antagonist but not interleukin-1 production. Blood 1993; 82:3695–3702.

48. Nelson RD, Shibata N, Podzorski RP, Herron MJ. *Candida mannan*: chemistry, suppression of cell-mediated immunity, and possible mechanisms of action. Clin Microbiol Rev 1991; 4:1–15.

49. Riipi L, Carlson E. Tumor necrosis factor (TNF) is induced in mice by *Candida albicans*: role of TNF in fibrinogen increase. Infect Immun 1990; 58:2750–2757.

50. Ausiello C, Spagnoli GC, Antonelli G, Malavasi F, Casciani CU, Dianzani F. Influence of monoclonal antibodies against HLA class I and class II antigen on interferon-gamma and -alpha induction. J Interferon Res 1987; 7:133–136.

51. Djeu JY, Blanchard DK, Halkias D, Friedman H. Growth inhibition of *Candida albicans* by human polymorphoneutrophils: activation by interferon-gamma and tumor necrosis factor. J Immunol 1986; 137:2980–2988.

52. Montmayeur A, Buguet A. Time-related changes in the sleep-wake cycle of rats infected with *Trypanosoma brucei brucei*. Neurosci Lett 1994; 168:172–174.

53. Buguet A, Bert J, Tapie P, Tabaraud F, Doua F, Londsdorfer J, Bogui P, Dumas M. Sleep-wake cycle in human African trypanosomiasis. J Clin Neurophysiol 1993; 10:190–196.

54. Bentivoglio M, Grassi-Zucconi G, Peng ZC, Kristensson K. Sleep and time keeping changes, and dysregulation of the biological clock in experimental trypanosomiasis. Bull Soc Pathol Exot 1994; 87:372–375.

55. Toth LA, Krueger JM. Somnogenic, pyrogenic, and hematologic effects of experimental pasteurellosis in rabbits. Am J Physiol 1990; 258:R536–R542.

56. Rechtschaffen A, Gilliland MA, Bergmann BM, Winter JB. Physiological correlation of prolonged sleep deprivation in rats. Science 1983; 221:182–184.

57. Everson CA. Sustained sleep deprivation impairs host defense. Am J Physiol 1993; 265:R1148–R1154.

58. Bergmann BM, Gilliland MA, Feng P-F, Russell DR, Shaw P, Wright M, Rechtschaffen A, Alverdy JC. Sleep deprivation and sleep extension: are physiological effects of sleep deprivation in the rat mediated by bacterial invasion? Sleep 1996; 19:554–562.

59. Van-Leeuwen PA, Boermeester MA, Houdijk AP, Ferwerda CC, Cuesta MA, Meyer S, Wesdorp RI. Clinical significance of translocation. Gut 1994; 35:S28–34.

60. Deitch EA, Berg R. Bacterial translocation from the gut: a mechanism of infection. J Burn Care Rehabil 1987; 8:475–482.

61. Everson CA, Toth LA. Abnormal control of viable bacteria in body tissues during sleep deprivation in rats. APSS Abstr 1997; 254.

62. Landis C, Pollack S, Helton WS. Microbial translocation and NK cell cytotoxicity in female rats sleep deprived on small platforms. APSS Abstr 1997; 188.

63. Brown R, Price RJ, King MG, Husband AJ. Are antibiotic effects on sleep behavior in the rat due to modulation of gut bacteria? Physiol Behav 1990; 48:561–565.

64. Floyd RA, Fernandes G, Good RA. The natural killer cell. Clin Bull 1979; 9:146–156.

65. Caligiuri M, Murray C, Buchwald D, Levine H, Cheney P, Peterson D, Komaroff AL, Ritz J. Phenotypic and functional deficiency of natural killer cells in patients with chronic fatigue syndrome. J Immunol 1987; 139:3306–3313.

66. Kelley DS, Daudu PA, Branch LB, Johnson HL, Taylor PC, Mackey B. Energy restriction decreases number of circulating natural killer cells and serum levels of immunoglobulins in overweight women. Eur J Clin Nutr 1994; 48:9–18.

67. Holland SM. Host defense against nontuberculous mycobacterial infections. Semin Respir Infect 1996; 11:217–230.

68. Gogos CA, Kalfarentzos FE, Zoumbos NC. Effect of different types of total parenteral nutrition on T-lymphocyte subpopulations and NK cells. Am J Clin Nutr 1990; 51:119–122.

69. Kantor TV, Whiteside TL, Friberg D, Buckingham RB, Medsger Jr TA. Lymphokine-activated killer cell and natural killer cell activities in patients with systemic sclerosis. Arthritis Rheum 1992; 35:694–699.

70. Villa ML, Ferrario E, Bergamasco E, Bozzetti F, Cozzaglio L, Clerici E. Reduced natural killer cell activity and IL-2 production in malnourished cancer patients. Br J Cancer 1991; 63:1010–1014.

71. Trinchieri G. Biology of natural killer cells. Adv Immunol 1989; 47:187–376.

72. Vujanovic NL, Nagashima S, Herberman RB, Whiteside TL. Nonsecretory apoptotic killing by human NK cells. J Immunol 1996; 157:1117–1126.

73. Dinges DF, Douglas SD, Zaugg L, Campbell DE, Whitehouse WG, McMann JM, Ieaza E, Reber D, Haupt B, Laizner A, Carlin M, Ukpah P, Orne EC, Getsy J, Orne MT. Human immune function prior to, during and following recovery from 64 hours without sleep. Sleep Res 1993; 22:329.

74. Moldofsky H, Lue FA, Davidson J, Saskin P, Gorczynski R. Comparison of sleep-wake circadian immune functions in women vs. men. Sleep Res 1989; 18:431.

75. Palmblad J, Cantell K, Strander H, Fröberg J, Karlsson C-G, Levi L, Granström M, Unger P. Stressor exposure and immunological response in man: interferon-producing capacity and phagocytosis. J Psychosom Res 1976; 20:193–199.

76. Moldofsky H, Lue FA, Saskin P, Davidson JR, Gorczynski R. The effects of sleep deprivation on immune function in humans. I. Mitogen and natural killer cell activities. Sleep Res 1987; 16:531–542.

77. Irwin M, Smith TL, Gillin JC. Electroencephalographic sleep and natural killer activity in depressed patients and central subjects. Psychosom Med 1992; 54:10–21.

78. Shahal B, Lue F, Jiang C, MacLean A, Moldofsky H. Diurnal and nap related changes in interleukins and immune functions. Sleep Res 1993; 22:634.

79. Irwin M, McClintick J, Costlow C, Fortner M, White J, Gillin JC. Partial night sleep deprivation reduces natural killer and cellular immune responses in humans. FASEB J 1996; 10:643–653.

80. Dinges DF, Douglas SD, Zaugg L, Campbell DE, McMann JM, Whitehouse WG, Orne EC, Kapoor SC, Icaza E, Orne MT. Leukocytosis and natural killer cell function parallel neurobehavioral fatigue induced by 64 hours of sleep deprivation. J Clin Invest 1994; 93:1930–1939.

81. Brown R, Pang G, Husband AJ, King MJ. Suppression of immunity to influenza virus infection in the respiratory tract following sleep disturbance. Reg Immunol 1989; 2:321–325.

82. Renegar KB, Small Jr PA. Passive transfer of local immunity to influenza virus infection by IgA antibody. J Immunol 1991; 146:1972–1978.

83. Renegar KB, Small Jr PA. Immunoglobulin A mediation of murine nasal anti-influenza virus immunity. J Virol 1991; 65:2146–2148.

84. Kris RM, Yetter RA, Cogliano R, Ramphal R, Small Jr PA. Passive serum antibody causes temporary recovery from influenza virus infection of the nose, trachea, and lung of nude mice. Immunology 1988; 63:349–353.

85. Renegar KB. Influenza virus infections and immunity: a review of human and animal models. Lab Animal Sci 1992; 42:222–232.

86. Renegar KB, Floyd RA, Krueger JM. Effect of sleep deprivation on serum influenza-specific IgG. Sleep 1998; 21:19–24.

87. de Haan A, Renegar KB, Small Jr PA, Wilschut J. Induction of a secretory IgA response in the murine female urogenital tract by immunization of the lungs with liposome-supplemented viral subunit antigen. Vaccine 1995; 13:623–616.

88. de Haan A, Geerligs HJ, Huchshorn JP, van-Scharrenburg GJ, Palache AM, Wilschut J. Mucosal immunoadjuvant activity of liposomes: induction of systemic IgG and secretory IgA responses in mice by intranasal immunization with an influenza subunit vaccine and coadministered liposomes. Vaccine 1995; 13:155–162.

89. Benca RM, Kushida CA, Everson, CA, Kalski R, Bergmann BM, Rechtschaffen A. Sleep deprivation in the rat. VII. Immune function. Sleep 1989; 12:47–52.

90. Renegar KB, Floyd RA, Krueger JM. Effects of short-term sleep deprivation on murine immunity to influenza virus in young adult and senescent mice. Sleep 1998; 21:241–248.

91. Brown R, Price RJ, King MG, Husband AJ. Interleukin-1 beta and muramyl dipeptide can prevent decreased antibody response associated with sleep deprivation. Brain Behav Immun 1989; 3:320–330.

92. Bergmann BM, Rechtschaffen A, Gilliland MA, Quintans J. Effect of extended sleep deprivation on tumor growth in rats. Am J Physiol 1996; 271:R1460–R1464.

93. Feng N, Pagniano R, Tovar CA, Bonneau RH, Glaser R, Sheridan JF. The effect of restraint stress on the kinetics, magnitude, and isotype of the humoral immune response to influenza virus infection. Brain Behav Immun 1991; 5:370–382.

94. Rampin C, Cespuglio R, Chastrette N, Jouvet M. Immobilization stress induces a paradoxical sleep rebound in rat. Neurosci Lett 1991; 126:113–118.

95. Kluger MJ. Fever: role of pyrogens and cryogens. Physiol Rev 1991; 71:93–127.

96. Dinarello CA. The biological properties of interleukin-1. Eur Cytokine Netwk 1994; 5:517–532.

97. Plata-Salaman CR. Immunoregulators in the nervous system. Neurosci Biobehav 1991; 15:185–215.

98. Merrill JE. Tumor necrosis factor alpha, interleukin 1 and related cytokines in brain development: normal and pathological. Dev Neurosci 1992; 14:1–10.

99. Uehara A, Okumura T, Kitamori S, Takasugi Y, Namiki M. Interleukin-1: a cytokine that has potent antisecretory and anti-ulcer actions via the central nervous system. Biochem Biophys Res Commun 1990; 173:585–590.

100. Besedovsky HO, del Rey A, Klusman I, Furukawa H, Monge Arditi G, Kabiersch A. Cytokines as modulators of the hypothalamus-pituitary-adrenal axis. J Steroid Biochem Mol Biol 1991; 40:613–618.

101. Auron PE, Quigley GJ, Rosenwasser CJ, Gehrke L. Multiple amino acid substitu-

tions suggest a structural basis for the separation of biological activity and receptor binding in a mutant interleukin-1 beta protein. Biochemistry 1992; 31:6632–6638.

102. Cunningham Jr ET, DeSouza EB. Interleukin-1 receptors in the brain and endocrine tissues. Immunol Today 1993; 14:171–174.

103. Liu C, Bai Y, Ganea D, Hart R. Species-specific activity of rat recombinant IL-1β. J Interferon Cytokine Res 1995; 15:985–992.

104. Ericsson A, Liu C, Hart RP, Sawchenko PE. Type I interleukin-1 receptor in the rat brain: distribution, regulation, and relationships to sites of IL-1-induced cellular activation. J Comp Neurol 1995; 361:681–698.

105. Colotta F, Dower SK, Sims JE, Mantovani A. The type II "decoy" receptor: a novel regulatory pathway for interleukin-1. Immunol Today 1994; 15:562–566.

106. Greenfeder SA, Nunes P, Knee L, Labon M, Chizzonite RA, Ju G. Molecular cloning and characterization of a second subunit of the interleukin-1 receptor complex. J Biol Chem 1995; 270:13757–13765.

107. Liu C, Chalmers D, Maki R, DeSouza EB. Rat homolog of mouse interleukin-1 receptor accessory protein: cloning, localization and modulation studies. J Neuroimmunol 1996; 66:41–48.

108. Wesche H, Neumann D, Resch K, Martin MU. Co-expression of mRNA type I and type II interleukin-1 receptors and the IL-1 receptor accessory protein correlates to IL-1 responsiveness. FEBS Lett 1996; 391:104–108.

109. Cao Z, Henzel WJ, Gao X. IRAK: a kinase associated with the interleukin-1 receptor. Science 1996; 271:1128–1131.

110. Lovenberg TW, Crowe PD, Liu C, Chalmers DT, Liu XJ, Liaw C, Clevenger W, Oltersdorf T, DeSouza EB, Maki RA. Cloning of a cDNA encoding a novel interleukin-1 receptor related protein (IL1R-rp). J Neuroimmunol 1996; 70:113–122.

111. Liu C, Hart RP, Liu XJ, Clevenger W, Maki RA, DeSouza EB. Cloning and characterization of an alternatively processed human type II interleukin-1 receptor mRNA. J Biol Chem 1996; 271:20965–20972.

112. Rothwell NT. Cytokines in the Nervous System. Austin, TX: R. G. Landes, 1996.

113. Lue FA, Bail M, Jephthah-Ocholo J, Carayanniotis K, Gorczynski R, Moldofsky H. Sleep and cerebrospinal fluid interleukin-1 like activity in the cat. Int J Neurosci 1988; 42:179–183.

114. Taishi P, Bredow S, Guha-Thakurta N, Obál Jr F, Krueger JM. Diurnal variations of interleukin-1β mRNA and β-actin mRNA in rat brain. J Neuroimmunol 1997; 75: 69–74.

115. Mackiewicz M, Sollars PJ, Ogilvie MD, Pack AI. Modulation of IL-1β gene expression in the rat CNS during sleep deprivation. NeuroReport 1996; 7:529–533.

116. Moldofsky H, Lue FA, Eisen J, Keystone E, Gorczynski RM. The relationship of interleukin-1 and immune functions to sleep in humans. Psychosom Med 1986; 48: 309–318.

117. Gudewill S, Pollmächer T, Vedder H, Schreiber W, Fassbender K, Holsboer F. Nocturnal plasma levels of cytokines in healthy men. Eur Arch Psychiatry Clin Neurosci 1992; 242:53–56.

118. Uthgenannt D, Schoolman D, Pietrowsky R, Fehm HL, Born J. Effects of sleep on the production of cytokines in humans. Psychosom Med 1995; 57:97–104.

119. Hohagen F, Timmer J, Weyerbrock A, Fritsch-Montero R, Ganter U, Krieger S, Berger M, Bauer J. Cytokine production during sleep and wakefulness and its relationship to cortisol in healthy humans. Neuropsychobiology 1993; 28:9–16.

120. Opp MR, Krueger JM. Interleukin-1 is involved in responses to sleep deprivation in the rabbit. Brain Res 1994; 639:57–65.

121. Takahashi S, Fang J, Kapás L, Wang Y, Krueger JM. Inhibition of brain interleukin-1 attenuates sleep rebound after sleep deprivation in rabbits. Am J Physiol 1997; 273:R677–R682.

122. Opp MR, Obál Jr F, Krueger JM. Interleukin-1 alters rat sleep: temporal and dose-related effects. Am J Physiol 1991; 260:R52–R58.

123. Tobler I, Borbély AA, Schwyzer M, Fontana A. Interleukin-1 derived from astrocytes enhances slow-wave activity in sleep EEG of the rat. Eur J Pharmacol 1984; 104:191–192.

124. Krueger JM, Walter J, Dinarello CA, Wolff SM, Chedid L. Sleep-promoting effects of endogenous pyrogen (interleukin-1). Am J Physiol 1984; 246:R994–R999.

125. Fang J, Kapás L, Wang Y, Krueger JM. The TNF 55kD receptor and the IL-1 type I receptor are involved in physiological sleep regulation. Soc Neurosci Abstr 1996; 22:147.

126. Susic V, Totic S. "Recovery" function of sleep: effects of purified human interleukin-1 on the sleep and febrile response of cats. Metab Brain Dis 1989; 4:73–80.

127. Friedman EM, Boinski S, Coe CL. Interleukin-1 induces sleep-like behavior and alters cell structure in juvenile rhesus macaques. Am J Primatol 1995; 35:145–153.

128. Borbély AA, Tobler I. Endogenous sleep-promoting substances and sleep regulation. Physiol Rev 1989; 69:605–670.

129. Pappenheimer JR, Koski G, Fencl V, Karnovsky ML, Krueger JM. Extraction of sleep-promoting factor S from cerebrospinal fluid and from brains of sleep-deprived animals. J Neurophysiol 1975; 38:1299–1311.

130. Hansen M, Kapás L, Fang J, Krueger JM. Cafeteria diet-induced sleep is blocked by subdiaphragmatic vagotomy in rats. Am J Physiol 1998; 274:R168–R174.

131. Fang J, Wang Y, Krueger JM. Mice lacking the TNF 55-kD receptor fail to sleep more after TNFα treatment. J Neurosci (in press).

132. Dantzer R, Bluthe RM, Aubert A, Goodall G, Bret-Dibat JL, Kent S, Goujon E, Laye S, Parnet P, Kelley KW. Cytokine actions on behavior. In: Rothwell NJ, ed. Cytokines in the Nervous System. Austin, TX: R. G. Landes Co, 1996:117–144.

133. Opp MR, Krueger JM. Anti-interleukin-1β reduces sleep and sleep rebound after sleep deprivation in rats. Am J Physiol 1994; 266:R688–R695.

134. Takahashi S, Kapás L, Fang J, Wang Y, Seyer JM, Krueger JM. An interleukin-1 receptor fragment inhibits spontaneous sleep and muramyl dipeptide-induced sleep in rabbits. Am J Physiol 1996; 271:R101–R108.

135. Opp MR, Postlethwaite AE, Seyer JM, Krueger JM. Interleukin 1 receptor antagonist blocks somnogenic and pyrogenic responses to an interleukin 1 fragment. Proc Natl Acad Sci USA 1992; 89:3726–3730.

136. Opp MR, Smith EM, Hughes Jr TK. Interleukin-10 acts in the central nervous system of rats to reduce sleep. J Neuroimmunol 1995; 60:165–168.

137. Kushikata T, Fang J, Wang Y, Krueger JM. Interleukin-4 inhibits spontaneous sleep in rabbits. Am J Physiol (in press).

138. Opp MR, Obál Jr F, Krueger JM. Effects of α-MSH on sleep, behavior, and brain temperature: interactions with IL1. Am J Physiol 1988; 255:R914–R922.

139. Krueger JM, Kapás L, Opp MR, Obál Jr F. Prostaglandins E_2 and D_2 have little effect on rabbit sleep. Physiol Behav 1992; 51:481–485.

140. Opp M, Obál Jr F, Krueger JM. Corticotropin-releasing factor attenuates interleukin 1–induced sleep and fever in rabbits. Am J Physiol 1989; 257:R528–R535.

141. Toth LA, Gardiner TW, Krueger JM. Modulation of sleep by cortisone in normal and bacterially infected rabbits. Am J Physiol 1992; 263:R1339–R1346.

142. Imeri L, Opp MR, Krueger JM. An IL-1 receptor and an IL-1 receptor antagonist attenuate muramyl dipeptide- and IL-1-induced sleep and fever. Am J Physiol 1993; 265:R907–R913.

143. Sehic E, Blatteis CM. Blockade of lipopolysaccharide-induced fever by subdiaphragmatic vagotomy in guinea pigs. Brain Res 1996; 726:160–166.

144. Laye S, Bluthe RM, Kent S, Combe C, Medina C, Parnet P, Kelley K, Dantzer R. Subdiaphragmatic vagotomy blocks induction of IL-1 beta mRNA in mice brain in response to peripheral LPS. Am J Physiol 1995; 268:R1327–R1331.

145. Bret-Dibat JL, Bluthe RM, Kent S, Kelley KW, Dantzer R. Lipopolysaccharide and interleukin-1 depress food-motivated behavior in mice by a vagal-mediated mechanism. Brain Behav Immun 1995; 9:242–246.

146. Fleshner M, Goehler LE, Hermann J, Relton JK, Maier SF, Watkins LR. Interleukin-1 beta-induced corticosterone elevation and hypothalamic NE depletion is vagally mediated. Brain Res Bull 1995; 37:605–610.

147. Watkins LR, Wiertelak EP, Goehler LE, Mooney-Heiberger K, Martinez J, Furness L, Smith KP, Maier SF. Neurocircuitry of illness-induced hyperalgesia. Brain Res 1994; 639:283–299.

148. Watkins LR, Goehler LE, Relton JK, Tartaglia N, Silbert L, Martin D, Maier SF. Blockade of interleukin-1-induced hyperthermia by subdiaphragmatic vagotomy: evidence for vagal mediation of immune-brain communication. Neurosci Lett 1995; 183:27–31.

149. Goehler LE, Busch DR, Tartaglia N, Relton J, Sisk D, Maier SF, Watkins LR. Blockade of cytokine-induced conditioned taste aversion by subdiaphragmatic vagotomy: further evidence for vagal mediation of immune-brain communication. Neurosci Lett 1995; 185:163–166.

150. Hansen MK, Krueger JM. Subdiaphragmatic vagotomy blocks the sleep and fever-promoting effects of interleukin-1β. Am J Physiol 1997; 42:R1246–R1253.

151. Fang J, Wang Y, Krueger JM. The effects of interleukin-1β on sleep are mediated by the type 1 receptor. Am J Physiol 1998; 274:R655–R660.

152. Payne L, Obál Jr F, Opp M, Krueger JM. Stimulation and inhibition of growth hormone secretion by interleukin-1β: The involvement of GHRH. Neuroendocrinology 1992; 56:118–123.

153. Obál Jr F, Fang J, Payne LC, Krueger JM. Growth hormone-releasing hormone (GHRH) mediates the sleep-promoting activity of interleukin-1 (IL-1) in rats. Neuroendocrinology 1995; 61:559–565.

154. Krueger JM, Takahashi S. Thermoregulation and sleep: closely linked but separable. Ann NY Acad Sci 1997; 813:281–286.

155. Opp M, Obál Jr F, Cady AB, Johannsen L, Krueger JM. Interleukin-6 is pyrogenic but not somnogenic. Physiol Behav 1989; 45:1069–1072.

156. Shoham S, Ahokas RA, Blatteis CM, Krueger JM. Effects of muramyl dipeptide on sleep, body temperature and plasma copper after intracerebral ventricular administration. Brain Res 1987; 419:223–228.

157. Kapás L, Shibata M, Kimura M, Krueger JM. Inhibition of nitric oxide synthesis suppresses sleep in rabbits. Am J Physiol 1994; 266:R151–R157.

158. Vilcek J, Lee TH. Tumor necrosis factor. J Biol Chem 1991; 266:7313–7316.

159. Liu T, Clark RK, McDonnell PC, Young PR, White RF, Barone FC, Feuerstein GZ. Tumor necrosis factor-alpha expression in ischemic neurons. Stroke 1994; 25:1481–1488.

160. Tchelingerian JL, Vignais L, Jacque C. TNF alpha gene expression is induced in neurones after a hippocampal lesion. NeuroReport 1994; 5:585–588.

161. Breder CD, Tsujimoto M, Terano Y, Scott DW, Saper CB. Distribution and characterization of tumor necrosis factor-alpha-like immunoreactivity in the murine central nervous system. J Comp Neurol 1993; 337:543–567.

162. Hunt JS, Chen HL, Hu XL, Chen TY, Morrison DC. Tumor necrosis factor-alpha gene expression in the tissues of normal mice. Cytokine 1992; 4:340–346.

163. Lieberman AP, Pitha PM, Shin HS, Shin ML. Production of tumor necrosis factor and other cytokines by astrocytes stimulated with lipopolysaccharide or a neutrotopic virus. Proc Natl Acad Sci USA 1989; 86:6348–6352.

164. Burns TM, Clough JA, Klein RM, Wood GW, Berman NE. Developmental regulation of cytokine expression in the mouse brain. Growth Factors 1993; 9:253–258.

165. Gendron RL, Nestel FP, Lapp WS, Baines MG. Expression of tumor necrosis factor alpha in the developing nervous system. Int J Neurosci 1991; 60:129–136.

166. Breder CD, Hazuka C, Ghayur T, Klug C, Huginin M, Yasuda K, Teng M, Saper CB. Regional induction of tumor necrosis factor alpha expression in the mouse brain after systemic lipopolysaccharide administration. Proc Natl Acad Sci USA 1994; 91:11393–11397.

167. de Kossodo S, Critico B, Grau GE. Modulation of the transcripts for tumor necrosis factor-alpha and its receptors in vivo. Eur J Immunol 1994; 24:769–772.

168. Gatti S, Bartfai T. Induction of tumor necrosis factor-alpha mRNA in the brain after peripheral endotoxin treatment: comparison with interleukin-1 family and interleukin-6. Brain Res 1993; 624:291–294.

169. Laye S, Parnet P, Goujon E, Dantzer R. Peripheral administration of lipopolysaccharide induces the expression of cytokine transcripts in the brain and pituitary of mice. Mol Brain Res 1994; 27:157–162.

170. Zhang F, zur Hausen A, Hoffman R, Grewe M, Decker K. Rat liver macrophages express the 55 kDa tumor necrosis factor receptor: modulation by interferon-γ, LPS and TNFα. Biol Chem Hoppe-Seyler 1994; 375:249–254.

171. Sanna PP, Weiss F, Samson ME, Bloom FE, Pich EM. Rapid induction of tumor necrosis factor alpha in the cerebrospinal fluid after intracerebroventricular injection of lipopolysaccharide revealed by a sensitive capture immuno-PCR assay. Proc Natl Acad Sci USA 1995; 92:272–275.

172. Lieberman AP, Pitha PM, Shin ML. Poly(A) removal is the kinase-regulated step in tumor necrosis factor mRNA decay. J Biol Chem 1992; 267:2123–2126.

173. Bredow S, Obál Jr F, Guha-Thakurta N, Taishi P, Krueger JM. Hypothalamic GHRH

mRNA and TNFα mRNA levels are higher during the day than night. Soc Neurosci Abstr 1996; 22:146.

174. Floyd RA, Krueger JM. Diurnal variations of TNFα in the rat brain. NeuroReport 1997; 8:915–918.

175. Schall TJ, Lewis M, Koller KJ, Lee A, Rice GC, Wong GHW, Gatanaga T, Granger GA, Lentz R, Raab H, Kohr WJ, Goeddel DV. Molecular cloning and expression of a receptor for human tumor necrosis factor. Cell 1990; 61:361–370.

176. Loetscher H, Pan Y-CE, Lahm H-W, Gentz R, Brockhaus M, Tabuchi H, Lesslauer W. Molecular cloning and expression of the human 55 kD tumor necrosis factor receptor. Cell 1990; 61:351–359.

177. Smith CA, Davis T, Anderson D, Solam L, Beckmann PM, Jerzy R, Dower SK, Cosman D, Goodwin RG. A receptor for tumor necrosis factor defines an unusual family of cellular and viral proteins. Science 1990; 248:1019–1023.

178. Kohno T, Brewer MT, Baker SL, Schwartz PE, King MW, Hale KK, Squires CH, Thompson RC, Vannice JL. A second tumor necrosis factor receptor gene product can shed a naturally occurring tumor necrosis factor inhibitor. Proc Natl Acad Sci USA 1990; 87:8331–8335.

179. Engelmann H, Aderka D, Rubinstein M, Rotman D, Wallach D. A tumor necrosis factor-binding protein purified to homogenity from human urine protects cells from tumor necrosis factor toxicity. J Biol Chem 1989; 264:11974–11980.

180. Olsson I, Lantz M, Nilsson E, Peetre C, Thysell H, Grubb A, Adolf G. Isolation and characterization of a tumor necrosis factor binding protein from urine. Eur J Hematol 1989; 42:270–275.

181. Seckinger P, Isaaz S, Dayer J-M. Purification and biologic characterization of a specific tumor necrosis factor α inhibitor. J Biol Chem 1989; 264:11966–11973.

182. Dayer TM, Burger D. Interleukin-1, tumor necrosis factor and their specific inhibitors. Eur Cytokine Netwk 1994; 5:563–571.

183. Peetre C, Thysell H, Grubb A, Olsson I. A tumor necrosis factor binding protein is present in human biological fluids. Eur J Hematol 1988; 41:414–419.

184. Gatanaga T, Hwang C, Kohr W, Cappuccini F, Lucci III JA, Jeffes EWB, Lentz R, Tomich J, Yamamoto RS, Granger GA. Purification and characterization of an inhibitor (soluble tumor necrosis factor receptor) for tumor necrosis factor and lymphotoxin obtained from the serum ultrafiltrates of human cancer patients. Proc Natl Acad Sci USA 1990; 87:8781–8784.

185. Spinas GA, Bloesch D, Kaufmann M-T, Keller U, Dayer J-M. Induction of plasma inhibitors of interleukin-1 and TNFα activity b endotoxin administration to normal humans. Am J Physiol 1990; 259:R993–R997.

186. Takahashi S, Kapás L, Seyer JM, Wang Y, Krueger JM. Inhibition of tumor necrosis factor attenuates physiological sleep in rabbits. NeuroReport 1996; 7:642–646.

187. Takahashi S, Tooley D, Kapás L, Fang J, Seyer JM, Krueger JM. Inhibition of tumor necrosis factor suppresses sleep in rabbits. Pflügers Arch 1995; 431:155–160.

188. Darko DF, Miller JC, Gallen C, White W, Koziol J, Brown SJ, Hayduk R, Atkinson JH, Assmus J, Munnell DT, Naitoh P, McCutchen JA, Mitler MM. Sleep electro-encephalogram delta-frequency amplitude, night plasma levels of tumor necrosis

factor α, and human immunodeficiency virus infection. Proc Natl Acad Sci USA 1995; 92:12080–12084.

189. Entzian P, Linnemann K, Schlaak M, Zabel P. Obstructive sleep apnea syndrome and circadian rhythms of hormones and cytokines. Am J Respir Crit Care Med 1996; 153:1080–1086.

190. Yamasu K, Shimada Y, Sakaizumi M, Soma G, Mizuno D. Activation of the systemic production of tumor necrosis factor after exposure to acute stress. Eur Cytokine Netwk 1992; 3:391–398.

191. Shoham S, Davenne D, Cady AB, Dinarello CA, Krueger JM. Recombinant tumor necrosis factor and interleukin-1 enhance slow-wave sleep. Am J Physiol 1987; 253: R142–R149.

192. Nistico G, DeSarro G, Rotiroti D. Behavioural and electrocortical spectrum power changes of interleukins and tumor necrosis factor after microinfusion into different areas of the brain. In: Smirne S, Franceschi M, Ferini-Strambi L, Zucconi M, eds. Sleep, Hormones and Immunological System. Milan: Masson, 1992:11–22.

193. Takahashi S, Kapás L, Fang J, Krueger JM. An anti-tumor necrosis factor antibody suppresses sleep in rats and rabbits. Brain Res 1995; 690:241–244.

194. Takahashi S, Kapás L, Fang J, Seyer JM, Krueger JM. Somnogenic relationships between interleukin-1 and tumor necrosis factor. Sleep Res 1996; 25:31, 1996 (abstract).

195. Chen Z, Fang J, Gardi J, Krueger JM. Sleep-deprivation induces activation of nuclear factor-κB. Soc Neurosci Abstr 1997; 23:792.

196. Krueger JM, Obál Jr F. Growth hormone-releasing hormone and interleukin-1 in sleep regulation. FASEB J 1993; 7:645–652.

197. Obál Jr F, Floyd R, Kapás L, Bodosi B, Krueger JM. Effects of GHRH on sleep in intact and hypophysectomized rats. Am J Physiol 1996; 270:E230–E237.

198. Obál Jr F, Bodosi B, Szilágyi A, Kacsóh B, Krueger JM. Antiserum of growth hormone suppresses sleep in the rat. Neuroendocrinology 1997; 66:9–16.

199. Bredow S, Taishi P, Obál Jr F, Guha-Thakurta N, Krueger JM. Hypothalamic growth hormone-releasing hormone (GHRH) mRNA varies across the day in rats. Neuro-Report 1996; 7:2502–2505.

200. Zhang J, Obál Jr F, Fang J, Collins BJ, Krueger JM. Sleep is suppressed in transgenic mice with a deficiency in the somatotropic system. Neurosci Lett 1996; 220:97–100.

201. Obál Jr F, Payne L, Opp M, Alföldi P, Kapás L, Krueger JM Growth hormone-releasing hormone antibodies suppress sleep and prevent enhancement of sleep after sleep deprivation. Am J Physiol 1992; 263:R1078–R1085.

202. Obál Jr F, Payne L, Kapás L, Opp M, Krueger JM. Inhibition of growth hormone-releasing hormone suppresses both sleep and growth hormone secretion in the rat. Brain Res 1991; 557:149–153.

203. Payne L, Obál Jr F, Opp M, Krueger JM. Stimulation and inhibition of growth hormone secretion by interleukin-1β: the involvement of growth hormone-releasing hormone. Neuroendocrinology 1992; 56:118–123.

204. Obál Jr F, Fang J, Payne LC, Krueger JM. Growth hormone-releasing hormone (GHRH) mediates the sleep promoting activity of interleukin-1 (IL-1) in rats. Neuroendocrinology 1995; 61:559–565.

205. Krueger JM. Cytokine involvement in sleep responses to infection and physiological sleep. In: Rothwell NJ, ed. Cytokines in the Nervous System. Austin, TX: Landes Publishing, 1996:41–71.

206. Kapás L, Fang J, Krueger JM. Inhibition of nitric oxide synthesis inhibits rat sleep. Brain Res 1994; 664:189–196.

207. Kapás L, Krueger JM. Nitric oxide donors SIN-1 and SNAP promote non-rapid eye movement sleep in rats. Brain Res Bull 1996; 41:293–298.

208. Ayers NA, Kapás L, Krueger JM. Circadian variation of nitric oxide synthase activity and cytosolic protein levels in rat brain. Brain Res 1996; 707:127–130.

209. Okabe S, Sanford LD, Veasey SC, Kubin L. Effect of nitric oxide synthase inhibitor microinjectino into rat mesopontine tegmentum on sleep. Sleep Res 1995; 24A:50.

210. Leonard TO, Lydic R. Inhibition of nitric oxide synthase (NOS) in the medial pontine reticular formation (mPRF) decreases rapid eye movement (REM) sleep. FASEB J 1996; 10:A409.

211. Datta S, Patterson E, Xie Z, Siwek DF. Role of nitric oxide in the pedunculopontine tegmental nucleus: maintenance of normal sleep. Soc Neurosci Abstr 1996; 22:1337.

212. Tena-Sempere M, Pinilla L, Gonzalez D, Aguilar E. Involvement of endogenous nitric oxide in the control of pituitary responsiveness to different elicitors of growth hormone release in prepubertal rats. Neuroendocrinology 1996; 64:146–152.

213. Janigro D, Wender R, Ransom G, Tinklepaugh DL, Winn HR. Adenosine-induced release of nitric oxide from cortical astrocytes. NeuroReport 1996; 7:1640–1644.

214. Sawada M, Hara N, Maeno T. Ionic mechanism of the outward current induced by extracellular ejection of interleukin-2 onto identified neurons of *Aplysia*. Brain Res 1991; 5454:248–256.

215. Sawada M, Hara N, Ichinose M. Interleukin-2 inhibits the GABA-induced Cl⁻ current in identified *Aplysia* neurons. J Neurosci Res 1992; 33:461–465.

216. Palazzolo DL, Quadri SK. Interleukin-1 inhibits serotonin release from the hypothalamus in vitro. Life Sci 1992; 51:1797–1802.

217. Gemma C, Ghezzi P, de Simoni MG. Activation of the hypothalamic serotonergic system by central interleukin-1. Eur J Pharmacol 1991; 209:139–140.

218. Monhankumar PS, Thyagarajan S, Quadri SK. Interleukin-1 beta increases 5-hydroxyindoleacetic acid release in the hypothalamus in vivo. Brain Res Bull 1993; 31:745–748.

219. Sawada M, Hara N, Maeno T. Reduction of the acetylcholine-induced K⁺ currents in identified *Aplysia* neurons by human interleukin-1 and interleukin-2. Cell Mol Neurobiol 1992; 12:439–445.

220. Schultzberg M, Andersson C, Unden A. Interleukin-1 in adrenal chromaffin cells. Neuroscience 1989; 30:805–810.

221. Soliven B, Albert J. Tumor necrosis factor modulates Ca^{2+} currents in cultured sympathetic neurons. J Neurosci 1992; 12:2665–2671.

222. Koike H, Saito H, Matsuki N. Effect of fibroblast growth factors on calcium currents in acutely isolated neuronal cells from rat ventromedial hypothalamus. Neurosci Lett 1993; 150:57–60.

223. Silverman DHS, Iman K, Karnovsky ML. Muramyl peptide/serotonin receptors in brain-derived preparations. Peptide Res 1989; 2:338–344.

224. Pousset F, Fournier J, Legoux P, Keane P, Shire D, Soubrie P. Effects of serotonin on cytokine mRNA expression in rat hippocampal astrocytes. Mol Brain Res 1996; 38:54–62.

225. Miller LG, Galpern WG, Lumpkin M, Chesley SF, Dinarello CA. Interleukin-1 augments γ-aminobutyric acid$_A$ receptor function in brain. Mol Pharmacol 1991; 39: 105–108.

226. Alvarez XA, Franco A, Fernandez-Novoa L, Cacabelos R. Effects of neurotoxic lesions in histaminergic neurons on brain tumor necrosis factor levels. Agents Actions 1994; 41:C70–C72.

227. Steriade M, McCarley RW. In: Brainstem Control of Wakefulness and Sleep. New York: Plenum Press, 1990.

228. Monti JM. Involvement of histamine in the control of the waking state. Life Sci 1993; 53:1331–1338.

229. Leeper-Woodford SK, Fisher BJ, Sugerman HJ, Fowler AA. Pharmacologic reduction in tumor necrosis factor activity of pulmonary alveolar macrophages. Am J Respir Cell Mol Biol 1993; 8:169–175.

230. Moruzzi G. The sleep-waking cycle. Ergeb Physiol Biol Chem Exp Pharmakol 1972; 64:1–165.

231. Krueger JM, Obál Jr F. A neuronal group theory of sleep function. J Sleep Res 1993; 2:63–69.

232. Marks GA, Shaffery JP, Oksenberg A, Speciale SG, Roffwarg HP. A functional role for REM sleep in brain maturation. Behav Brain Res 1995; 69:1–11.

233. Jouvet M. Neurophysiology of the states of sleep. Physiol Rev 1967; 47:117–177.

234. Kavanau JL. Sleep and dynamic stabilization of neural circuitry: a review and synthesis. Behav Brain Res 1994; 63:111–126.

235. Pigarev I. Partial sleep in cortical areas. World Fed Sleep Res Soc Newslett 1996; 5:7–8.

236. Kattler H, Dijk D-J, Borbély AA. Effect of unilateral somatosensory stimulation prior to sleep on the sleep EEG in humans. J Sleep Res 1994; 3:1599–1604.

237. Mahowald M, Schenck CH. Dissociated state of wakefulness and sleep. Neurology 1992; 42:44–52.

15

Intrinisic Disruption of Normal Sleep and Circadian Patterns

SCOTT S. CAMPBELL

Cornell University Medical College
and Institute for Circadian Physiology
White Plains, New York

I. Introduction

It has been recognized since the first studies of humans in time-free environments that the timing of major nighttime sleep is governed in large part by the endogenous circadian pacemaker (1,2). The dependence of sleep on the biological clock is reflected in the characteristic temporal relationship between sleep propensity and the circadian rhythm of body core temperature. When subjects are studied in environments free of time cues, the onset of major sleep episodes tends to occur in close proximity to the trough of body temperature (3,4).

Under the entrained conditions of normal daily life, this relationship is altered somewhat, such that major nocturnal sleep is typically initiated 5–6 hr prior to the temperature minimum and is terminated shortly after the minimum. For most people, this corresponds roughly to sleep onset times of between 11 P.M. and midnight, and wakeup times of between about 6 A.M. and 8 A.M. Undoubtedly, because the vast majority of humans exhibit similar timing in sleep-wake behavior, societal demands have evolved to accommodate such biological timetables.

For a small percentage of the population, however, there is a misalignment between the endogenous clock that governs the timing of sleep and the sleep/wake schedule that is desired or is regarded as the societal norm. These individuals are said to have *circadian rhythm sleep disorders* (5). This class of sleep disorders can be grouped further into two general categories: extrinsic and intrinsic types. Extrinsic types are those in which the development of the disorder is brought on by an alteration in the environment relative to sleep timing, for example, jet lag, or shift work sleep disorder.

In contrast, circadian rhythm sleep disorders of the intrinsic type are those that occur as a result of the endogenous clock being altered relative to the (social) environment. The current chapter will discuss specifically those circadian rhythm sleep disorders considered to be of the intrinsic type. There are four types of intrinsic circadian rhythm disorder, the general temporal characteristics of which are illustrated in Figure 1.

II. Delayed Sleep Phase Syndrome

Delayed sleep phase syndrome (DSPS) is a disorder in which the major sleep episode is delayed in relation to the desired clock time that results in symptoms of sleep-onset insomnia or difficulty in awakening at the desired time (5). DSPS patients typically report an inability to fall asleep before about 2 A.M. and may experience difficulties with sleep onset until as late as 6 A.M. Yet, when not required to maintain a usual schedule (i.e., awaken early for work or school) and permitted to sleep when they desire, they exhibit uninterrupted sleep of normal composition and normal duration. Therefore, as with all but one of the circadian rhythm sleep disorders (irregular sleep pattern), DSPS is not a disorder of the sleep process per se, but rather, a disorder of the *timing* of sleep relative to societal norms.

DSPS is probably the most common of the intrinsic circadian sleep disorders, or at least the most commonly diagnosed. Based on an early survey, it was estimated that approximately 7% of people diagnosed with disorders of initiating and maintaining sleep meet criteria for DSPS (6). The average DSPS patient has been reported to be younger than the average nonselected sample of insomniac patients, and many report having experienced the disorder since childhood (6–8). In a sample of 14 adult cases studied by Ito and co-workers (9), the mean age of onset was 20.9 years. (On average, another 6 years passed before patients sought treatment for the disorder.) As would be expected, DSPS patients typically score as "owls," or extreme evening types, on morningness-eveningness scales (10).

The disorder was formally proposed as a clinical entity in 1981 (6), but its existence has been known for over a century (11,12). Prior to the recognition of DSPS as a discrete disorder, sufferers were typically diagnosed with "sleep onset insomnia" since their major complaint was often an inability to fall asleep at conventional bedtimes. Even today, one criterion for differential diagnosis of DSPS is that the case "does not meet criteria for any other sleep disorder causing

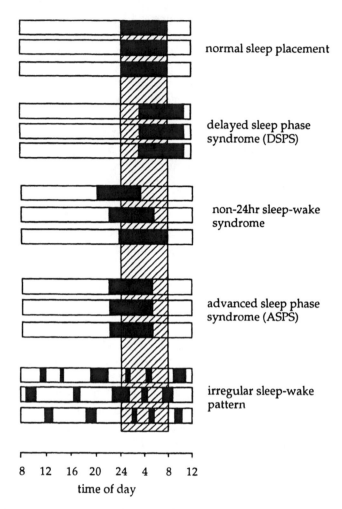

8 12 16 20 24 4 8 12

time of day

Figure 1 Schematic diagram depicting various sleep disorders associated with disturbances of the circadian timing system. The open bars represent intervals during which wakefulness typically occurs, black bars correspond to usual sleep times in normal and pathological conditions, and the hatched area represents desired sleep time. Thus, individuals with normal sleep placement obtain their sleep completely within the desired sleep interval, whereas those with advance sleep phase disturbance (ASPS), for example, sleep at a time earlier than desired.

inability to initiate sleep or excessive sleepiness." Although the disorder apparently is not gender specific in adulthood [for example, in the sample studied by Weitzman et al. (6) there were 15 men and 15 women, and in a group of DSPS patients studied by Regestein and Monk (11), there were 19 men and 14 women), there is some evidence for a predominance among males in younger samples (8).

The degree and type of psychopathology exhibited by DSPS patients, as a group, does not distinguish them from other insomniacs. Moreover, unlike individuals who suffer from non-24-hr sleep-wake syndrome (see below), case reports of DSPS sufferers reveal that while the disorder is certainly inconvenient and often disruptive, it is seldom debilitating. In many cases, individuals with DSPS are able to schedule their work, school, and social lives to accommodate the disorder (See, for example, 6). Nevertheless, there is at least one instance reported in the literature in which such accommodation was impossible. In this case, criminal proceedings were undertaken against a U.S. Marine Corps lance corporal with DSPS because he could not comply with the schedule of military duty ("failure to go") (13).

A. Pathophysiology

A number of authors have suggested that individuals with DSPS are, for one reason or another, unable to achieve the usual phase advance that is required to entrain our longer-than-24-hr endogenous clocks to the 24-hr day. Investigators have hypothesized a variety of causes that may underlie this reduced capacity to regularly advance phase. It may result from a "weak" phase-advance portion of an individual's phase-response curve (PRC) to light, or other phase-resetting stimuli. In a number of early publications describing the disorder, Weitzman and co-workers put forth the notion that sleep-wake behavior itself was responsible for initially delaying circadian phase in DSPS: "We postulate that the phase advance portion of the PRC in patients with DSPS is much less prominent than in normal subjects. When these patients retire later than usual, the phase delay mechanism is activated and cannot be fully reversed by subsequent phase advance, even when strenuous attempts are made to avoid oversleeping in the morning. The net result is a shift of sleep to a later time" (14, p. 165). This interpretation predated the now common knowledge that the environmental light-dark cycle is primarily responsible for entraining the human circadian system to the 24-hr day.

In subsequent years, several investigators have proposed a similar etiology, but with specific reference to the light PRC. For example, based on body temperature and sleep log data obtained from seven DSPS patients while they lived at home and continued their normal activities, Ozaki et al. (15) reported that these individuals not only had delayed temperature minima relative to age-matched controls, but also tended to retire at a later phase of their temperature rhythms. Moreover, once asleep, they slept longer than age-matched controls. The authors interpreted these findings as indicating that patients with DSPS may have "different circadian structures." They went on to speculate that this altered relationship between sleep timing and the circadian system, in conjunction with the tendency toward longer sleep episodes, may make it impossible for DSPS sufferers to take advantage of the phase advance portion of the light PRC, which occurs in the morning. Quite simply, they sleep through it.

Czeisler and co-workers (12) offered the alternative hypothesis that DSPS may be the expression of an endogenous period length that lies outside the range of entrainment (ROE) to the 24-hr day. As with a low amplitude, or missed phase-advance portion of the PRC, the result of an endogenous period beyond the ROE was hypothesized to result in a diminished capacity to achieve the phase advance required for execution of a conventional sleep-wake schedule. Based on this notion, other authors have suggested that the putatively greater frequency of long endogenous periods in young people may account, in part, for the tendency for DSPS to be overrepresented in this segment of the population (11). However, because no documented cases of DSPS have been studied in temporal isolation, there is little evidence that individuals with DSPS do have longer-than-normal endogenous periods.

With respect to the tendency for DSPS to be more frequent in adolescents, Carskadon and co-workers (7) have argued that other changes in the biological timing system—perhaps changes in melatonin output or gonadotropin secretion—associated with puberty may be implicated in development of the syndrome. Based on questionnaire responses from 458 sixth graders concerning sleep habits and preferences, these investigators concluded that such biological factors are more likely to be responsible for delayed sleep preference than are psychosocial factors such as peer-group influence, or the response to a "release" of parental controls on sleep timing. Such a conclusion is supported by the finding that multiple sleep latency test (MSLT) data reveal a midpubertal phase delay in daytime sleep propensity (16).

In terms of adult onset of DSPS, it has also been hypothesized that a history of recurrent phase shifts, as experienced by rotating shift workers, may predispose one to the development of DSPS (12). Likewise these authors suggested that the onset of DSPS may be triggered in some normal individuals as a consequence of a forced, acute 6–8-hr phase advance, for example that experienced during rapid transmeridian flight. In both cases, the altered schedule would "trap" the sleep period at a circadian phase ill-positioned to take full advantage of the phase advance portion of the PRC. Once trapped, those prone to development of DSPS may be unable to regain an appropriate phase position for sleep.

B. Treatment

As with most other forms of insomnia, DSPS was initially treated with sedative hypnotic drugs, but with limited success. Although such drugs were often effective in inducing sleep at earlier, more conventional bedtimes, they did nothing for—and in the case of longer-acting drugs, often exacerbated—the severe difficulties with morning alertness reported by DSPS sufferers. Self-medication with alcohol has, likewise, proven ineffective.

The first attempts to treat DSPS with strictly behavioral interventions took the logical step of advancing bedtimes across several days, using incremental steps ranging from 5 to 30 min, every 1 or 2 days (17,18). A similar attempt used acute (one night) sleep deprivation followed by a 90-min phase advance in bedtime to induce a more regular sleep-wake schedule. The procedure was repeated weekly until an acceptable bedtime was achieved (19). All such approaches, in which an advance of sleep time has been attempted, have proven largely unsuccessful.

In a similar procedure, termed "chronotherapy" by Czeisler and co-workers (12), bedtimes and rising times were systematically delayed over a period of days until the desired bedtime was achieved. The first such chronotherapy intervention was accomplished by delaying bedtimes for 3 hr each night for several consecutive nights. Incremental delays, rather than advances, were selected to take advantage of the fact that most humans exhibit endogenous periods of longer than 24 hr, and therefore, have a natural tendency to delay. Moreover, a delay in bedtime necessarily increased prior wakefulness, thereby enhancing the likelihood of relatively rapid sleep onset at the new bedtime.

Once achieved, strict adherence to the new sleep-wake schedule was thought to be critical to maintaining a positive outcome (12). Based on subjective reports and sleep logs, the author reported "significant and lasting resolution" in all five cases of DSPS studied. Two-month follow-up revealed a continued ability for subjects to remain entrained to the new schedule. Other investigators have reported less success than this original study (see for example, 18), and it should be pointed out, as well, that resolution of the delayed phase in the Czeisler et al. study did not result in improved sleep. In fact, sleep efficiencies did not change following chronotherapy when compared to pretreatment levels, remaining well below 90% on average.

Although clearly troublesome in terms of the long-term utility of the intervention, the lack of improvement in sleep quality may not be the most serious problem with the chronotherapy procedure. The fact that the investigators were able to find only five individuals willing to participate over the 2.5 years of the study is testament to the primary shortcoming of this treatment approach: The initial process of realigning the sleep-wake schedule with conventional clock time is demanding, and subsequent compliance is critical. Indeed, the authors hypothesized that a delay in scheduled bedtime for even a single night (e.g., a weekend social engagement) could result in a relapse of symptoms, thus requiring reimplementation of the round-the-clock delaying routine. Few people are willing, or able, to maintain prolonged, strict sleep-wake schedules necessary to make this treatment work.

It is this aspect of the intervention that prohibited Thorpy and co-workers from attempting the chronotherapy routine in their sample of adolescent DSPS patients (19). Additionally, there is some evidence that the treatment itself may lead to the development of a more severe form of *circadian rhythm sleep disorder* in some patients. Oren and Wehr (20) cite three cases in which the development of non-24-hr SWS (see below) appeared to be the direct consequence of treating DSPS with chronotherapy. In view of potential adverse effect of chronotherapy, these authors suggested an alternative treatment using timed exposure to light.

The demonstration in the mid-1980s that timed bright-light exposure could be used to effectively reset the human endogenous pacemaker opened the possibility that this procedure may be beneficial in the treatment of DSPS. Rosenthal et al. (21) found that a combination of morning bright-light exposure (2500 lux for 2 hr between 6 and 9 A.M.) and afternoon/evening light avoidance phase-advanced DSPS patients' temperature minima by an average of 1.45 hr. The phase advance was accompanied by enhanced morning alertness, as measured by sleep latency tests, and by subjective reports of consistently earlier bedtimes. In this study a modified (less extreme) set of criteria for the diagnosis of DSPS was employed. For example, individuals who reported usual sleep times as early as 1 A.M. and an inability to be alert in the morning between 7 and 9 A.M. were included. A substantially greater phase shift would probably be required to alleviate symptoms in more severe cases of DSPS.

Instead of light, several investigators have used timed melatonin administration in an attempt to phase-advance the circadian system, and thus sleep propensity, in DSPS. Using a randomized, double-blind, placebo-controlled design in eight subjects with DSPS, Dahlitz and co-workers found that an oral dose of melatonin (5 mg) given each night at 22:00 for 4 weeks was essentially ineffective (22). Average sleep onset time was advanced by only 23 min, and wake time by 79 min (as measured by sleep log reports). Thus, average sleep onset time following the melatonin trial continued to be quite delayed (02:07), and total sleep time was reduced by almost an hour compared to baseline. Oldani and co-workers (23) reported slightly better results in six DSPS patients studied prior to and following 1 month of treatment with daily melatonin administration (5 mg between 17:00 and 19:00). They found a mean advance in sleep onset of 115 min compared to baseline, with average bedtime after treatment still somewhat delayed (01:40), but no degradation in sleep quality.

Armstrong et al. (24) compared the effects of melatonin and a melatonin analog (S20098) on the timing of activity rhythms in an animal (rat) model of DSPS. Following daily injections for 22 consecutive days, both substances restored rhythms to a normal phase. As with light, neither of the substances was effective in permanently shifting phase: following cessation of the injections, activity rhythms showed a tendency for slow delays back toward their pretreatment phase positions.

Okawa et al. (25) reported positive therapeutic effects of daily vitamin B_{12} administration (1.5 mg/day) in a 55-year-old man suffering from DSPS, after he had failed to respond to separate interventions involving strict behavioral scheduling, timed light exposure, and hypnotic medication. Successful treatment using vitamin B_{12} (30 mg/day) was also reported in a 15-year-old girl, who showed both a phase advance in sleep onset time (from 02:00 to midnight) in response to treatment, and a reduction in average total sleep time (from 10 to 7 hr) (26). In this case, as well, previous attempts to treat the disorder with other medication and behavioral techniques had failed.

Both groups of investigators suggested that vitamin B_{12} most likely produced its beneficial effect by facilitating entrainment of the circadian clock to the 24-hr day, either by enhancing the entraining influence of light or by a direct influence on the clock. Okawa et al. also considered as a possible mechanism a simple enhancement in the capacity for sleep, i.e., a hypnotic effect. Neither group could rule out the possibility of a placebo effect associated with the therapeutic regimen.

Several different treatment approaches were used in a sample of 33 DSPS patients studied by Regestein and Monk (11), including strict behavioral scheduling, timed bright-light exposure, chronotherapy, triazolam administration, and vitamin B_{12}. Results of all of these interventions were mixed. Of all patients studied, only 27% were able to maintain controlled, socially acceptable sleep schedules as a result of treatment. The authors concluded that the outcome for any form of treatment is likely to depend largely on "individual patient characteristics," including comorbid psychopathology, particularly depression, inability to comply with treatment protocols, sedentary habits, and personal choice of the sleep-wake schedule characterizing their condition (11). A similar conclusion was reached by Thorpy and co-workers (19) in their study of adolescents with DSPS. They suggested that treatment needed to be individualized depending on patient cooperativeness, social, school, or work obligations, and the severity of the sleep disturbance.

One feature of sleep-wake patterns often observed in DSPS patients is a systematic delay in bedtime for several days, followed by an abrupt advance back to the original bedtime (which is, nevertheless, still delayed relative to normal bedtimes). This observation has led some investigators to suggest that DSPS and non-24-hr sleep-wake syndrome (SWS) are essentially the same disorder expressed with different degrees of severity. Such a notion is supported by the aforementioned observation that some individuals may develop non-24-hr SWS as a consequence of exposure to chronotherapy treatment for their DSPS symptoms (20). This putatively more severe form of DSPS is discussed next.

III. Non-24-Hr Sleep-Wake Syndrome (Hypernychthemeral Sleep-Wake Pattern)

Non-24-hr sleep-wake syndrome consists of a chronic steady pattern comprised of 1–2-hr daily delays in sleep onset and wake time in an individual living in society

(5). Depicted schematically (see Fig. 1), the characteristic sleep-wake pattern in non-24-hr SWS closely resembles that of normal subjects living in isolation without time cues. Such a free-running sleep-wake pattern occurring within the 24-hr environmental and cultural milieu means that non-24-hr patients' biologically "preferred" sleep times drift systematically in and out of phase with conventional sleep times. When in phase with societal norms, the syndrome is unrecognizable.

As sleep times become systematically delayed, the patient suffers first from symptoms common to DSPS—extended sleep onset and difficulty awakening at the desired rising time—followed by an interval in which symptoms of ASPS are reported—evening sleepiness and early-morning awakening—as the sleep propensity rhythm continues to drift through the day toward early evening. (Such a description is based on the fact that every patient with non-24-hr SWS reported in the literature has exhibited behavior consistent with an endogenous period longer than 24 hr. Indeed, the disorder was originally designated "hypernychthemeral disorder." However, in theoretical terms, there is no reason to believe that cases of "*hypo*nychthemeral" disorders may not also exist, since shorter-than-24-hr periods have been observed.)

The prevalence of the disorder is unknown, but is presumed to be quite rare. Less than two dozen cases have been reported in the literature. It is worth noting that, with only two exceptions, every adult patient described in the literature has been male [Okawa et al. (27) described non-24-hr SWS in three congenitally blind girls, aged 7, 10, and 12]. One female who exhibited non-24-hr sleep-wake patterns did so only as a consequence of "partial entrainment" to the non-24-hr cycle of her living companion (28). Based on characteristics of the documented cases, there is suggestive evidence that personality disorder may be associated with the development of non-24-hr SWS (29–31), and there is little question that blindness is a strong predisposing factor for development of the disorder.

In one of the first published descriptions of the disorder Miles et al. (32) studied a psychologically normal blind man (without light perception since birth) living and working in normal society. During a 26-day hospital stay, in which the subject was encouraged to eat, sleep, and socialize with hospital staff whenever he felt so inclined, he exhibited sleep-wake cycle of 24.9 hr. The subject's subsequent attempt to entrain to the 24-hr day while living at home failed, as he continued to show a free-running sleep-wake cycle of about 25 hr. Likewise, institution of a strict schedule of bedtimes, wake times, meals, and activity failed to induce entrainment, and the subject's sleep became progressively more disrupted and his waking performance deteriorated incrementally. Circadian rhythms of body temperature, urinary potassium, sodium, calcium, and PO_4, subjective alertness, and performance also showed periods of 24.9 hr.

A virtually identical average period length (24.8 hr) to that reported by Miles and co-workers was exhibited by a sighted, 34-year-old man during 103 days of ambulatory temperature recording while the subject continued his usual daily activities (30). Throughout the recording period, the subject also kept detailed records of his sleep-wake patterns. Although the average period was 24.8 hr, detailed examination of the sleep-wake records revealed two distinct components in his non-24-hr pattern. When the subject's sleep onset occurred within a window of between about 20:00 and noon, his period showed a mean duration of 24.7 hr. However, when sleep onset occurred between noon and 20:00, the subject routinely attempted to delay his sleep times, resulting in a transient extension of the mean period length to 25.8 hr.

This systematic alternation between two distinct periods may be a common feature of the disorder. For example, Uchiyama et al. (33) studied a sighted, 30-year-old male subject with a 12-year history of non-24-hr sleep-wake patterns. Based on sleep log data collected for 164 consecutive days, the investigators found two distinct components, the expression of which alternated with a period of 27 days. Once component was a typical free-running sleep-wake pattern with a regular, daily 30–60-min delay in sleep onset. As in the case reported by Kokkoris et al, this sleep-wake pattern predominated on days when sleep onset occurred during evening and nighttime hours. The other component, a less regular pattern characterized by clusters of "delayed-phase jumps," in which a delay in sleep onset of at least 4 hr relative to the preceding night was reported, tended to occur during intervals in which sleep was initiated during the morning hours. A remarkably similar pattern of alternating period lengths has been reported in other non-24-hr SWS subjects, with all of them exhibiting longer period lengths during intervals when sleep occurred in the daytime hours (28,31,34,35).

A. Pathophysiology

Several mechanisms underlying such alternating patterns have been proposed, and there is no reason to believe that the alternative hypotheses are mutually exclusive. Uchiyama and co-workers (33) concluded that the phase jumps that led to longer subjective days in their subject were the consequence of light-induced phase delays associated with environmental light falling intermittently on the delay portion of the subject's PRC. Based on sleep logs kept continuously for 4 years by a patient with non-24-hour SWS, Wollman and Lavie (31) hypothesized that the phase jumps that they observed were behavioral accommodations to two, distinct physiologically preferred phase positions for sleep, or "sleep gates." When the subject's physiologically preferred sleep time drifted beyond the limits of one sleep gate and into an adjacent "forbidden zone" for sleep, he was forced to delay sleep onset to reach the next sleep gate, occurring approximately 12 hr later in the circadian cycle.

Kokkoris and co-workers (30) proposed two possible causes for development of the disorder in the subject they studied. One hypothesis posited that the disorder was the result of the subject's inability to entrain to "social time cues" because his personality disorder predisposed him to social and psychological isolation. According to this hypothesis, such isolation either served to begin a psychophysiological process that interfered with the mechanism of entrainment, or from a more behavioral perspective, such a life-style resembled a temporal isolation laboratory enough to result in the subject free-running.

Alternatively, it was suggested that the subject's "hypernychthemeral" cycles were the expression of a physiologically defective biological clock, and/or the mechanisms responsible for proper entrainment—for example, low-amplitude phase response to entraining stimuli, or an endogenous period beyond the ROE to the 24-hr day. Yet, a number of studies suggest that the ROE in humans is about 23–27 hr (see, for example, 1). No patient reported in the literature has exhibited free-running cycles outside of this range. It seems unlikely, therefore, that this is the cause (29).

The predisposition for non-24-hr SWS in blind individuals clearly suggests that reduced exposure to the environmental light-dark cycle may be a major cause of the disorder. In the only study to measure light exposure levels in a sighted individual with non-24-hr SWS, average daily exposure was about half that of healthy age-matched subjects at approximately the same latitude (29). This, combined with the possibility that individuals with non-24-hr SWS may have a subsensitivity to the phase-shifting effects of light—McArthur et al. (29) found reduced sensitivity to light in their patient—may make it impossible for these patients to respond adequately to usual environmental zeitgebers.

Even if non-24-hr SWS sufferers have a normal sensitivity to light, their sleep-wake behavior may militate against appropriate use of the phase-shifting effects of natural daylight. Such an explanation, offered by Hoban and co-workers (36), is similar to that proposed by Ozaki et al. (15, see above) to explain the onset and perpetuation of DSPS. In the case of their non-24-hr SWS patient, Hoban et al. reasoned that progressively later bedtimes and wake times differentially exposed the patient to the delay portion of her PRC, while her later sleep offset caused her to sleep through the advance portion of the PRC.

B. Treatment

Citing data from a sleep-wake diary maintained over 4 years by a non-24-hr SWS subject, Weber et al. (28) concluded that one effective strategy for managing the disorder might be the imposition of stronger entraining cues, in the form of a strict work-rest schedule. This conclusion was based on the fact that, on several occasions throughout the 4-year period, the subject appeared to show periods of "relative coordination" during which he temporarily synchronized to a 24-hr schedule. One common feature of these occasions was an enhancement in the availability of social cues (e.g., trips to a farm with other students for outdoor physical work, relaxation, and socializing). However, the subject was never able to remain synchronized for prolonged intervals, and other such attempts to entrain non-24-hr sleep-wake patterns to the 24-hr day using social cues and behavioral structuring have not been successful (28,30,31). Similar failures to entrain to various self- or experimenter-imposed "social cues," including clocks, strict feeding and rest/activity schedules, and timed light exposure, have been reported by other investigators as well, in both blind and sighted subjects (27,28,30,35,37).

As with DSPS, several investigators have reported successful alleviation of symptoms associated with non-24-hr SWS with vitamin B_{12} administration (34,38,39). The patient studied by Kamgar-Parsi et al. (34) also suffered from hypothyroidism, and based on the knowledge that vitamin B_{12} deficiency is common in patients with thyroid disease, the patient decided to self-medicate with a daily dose of B_{12}. Within 3 weeks, he had advanced his sleep to a conventional bedtime and reported being able to maintain a regular 24-hr sleep-wake schedule. On two subsequent occasions when he temporarily discontinued his B_{12} regimen, the patient's sleep-wake cycle again showed a tendency to drift progressively later each day.

Okawa and co-workers (39) also reported entrainment to the 24-hr day in a non-24-hr SWS patient following daily vitamin B_{12} administration, and as in the previous case, discontinuation of the medication resulted in the reappearance of a non-24-hr sleep-wake rhythm. The mechanism by which vitamin B_{12} produced beneficial effect in non-24-hr SWS is unknown. Okawa et al. hypothesized that daily administration either changed the period of the sleep-wake rhythm or in some other way enhanced the capacity of the endogenous clock to entrain to a 24-hr day.

Hoban and co-workers (36) reported successful entrainment of the sleep-wake cycle of a non-24-hr SWS patient following a light-treatment schedule designed to advance her circadian clock. For 6 days, the patient received 2 hr of light (2500 lux) immediately on awakening at a specified time (14:00), based on her spontaneous awakening on the first day of treatment. Then the time of light exposure was advanced 30 min until she was arising at 10:00. She then continued the treatment at home, advancing rising time by 30 min/week until her desired rising time was achieved.

A similar approach, using melatonin administration rather than light exposure, was conducted by the same laboratory. McArthur et al. (29) reported successful treatment of a case of non-24-hr SWS with a daily oral dose of melatonin (0.5 g) at a time (21:00) designed to induce a phase advance of the patient's circadian clock (40). During the 4-week treatment interval sleep-wake times were reported to normalize, but sleep efficiency, as measured by actigraphy, was not improved. Moreover, interpretation of the outcome was complicated in this case by the fact that the patient had a mixed drug history during the trial.

Several investigators have reported successful treatment of non-24-hr SWS with some, but not all, benzodiazepines. For example, Kamgar-Parsi et al. (34)

found that flurazepam, but not diazepam, shortened the sleep-wake cycle length of a non-24-hr SWS patient, with accompanying relief of some symptoms. Likewise, nitrazepam administration resulted in stabilization of a 24-hr sleep-wake pattern in a 12-year-old congenitally blind patient after attempts to impose strict scheduling failed (27). Finally, there is limited evidence that the condition can remit without treatment. Kokkoris et al. (30) reported that 2 months following the end of their case study, the patient reported that his sleep-wake cycle had assumed a normal 24-hr cycle.

IV. Advanced Sleep Phase Syndrome

Advanced sleep phase syndrome (ASPS) is a disorder in which the major sleep episode is advanced in relation to the desired clock time, which results in symptoms of compelling evening sleepiness, an early sleep onset, and an awakening that is earlier than desired (5). The syndrome is considered to be far less common than DSPS. Indeed, as recently as 1981, Weitzman wrote that "there is no clear evidence as yet that there is an advanced sleep phase syndrome," (41, p. 403), and as recently as 1986, its very designation as a pathological syndrome was still uncertain (48). The fact that the disorder went unrecognized for so long is not entirely surprising. The very symptoms that define ASPS—"early to bed and early to rise"—have, for generations, constituted a philosophy to live by, rather than a disorder to treat. Simply, ASPS is more easily accommodated by societal norms, and therefore, less likely to be a source of disturbance or complaint.

In contrast to DSPS, which is often characterized by onset in childhood or adolescence (see above), ASPS is most frequently reported by middle-aged and older subjects. The late age of onset of ASPS may account, in part, for the presumed rarity of the disorder. Since its very designation as a circadian rhythm sleep disorder stipulates that ASPS is the expression of an altered circadian system, the changes in sleep associated with aging would not have met criteria for diagnosis as ASPS until they were linked to age-related changes in the circadian system. Although Weitzman speculated in 1981 that many older individuals may have ASPS (41), such an association was not clearly elucidated until quite recently (see for example, 43–46). As such, the large proportion of older individuals who exhibit sleep patterns consistent with ASPS would have been excluded from such a diagnosis.

The earliest case described as ASPS leaves questions concerning the actual nature of the disorder reported (47). Under entrained conditions, this patient reported habitual bedtimes and rising times of around 21:00 and 06:00, respectively. Particularly with regard to wake-up time, this is not drastically different from that reported by many normal sleepers with job commitments. In addition, the patient exhibited sleep episodes "fragmented by large periods of wakefulness," even during a 20-day interval in temporal isolation. If ASPS was, in fact, the cause of the patient's sleep disturbance, reasonably normal, consolidated sleep would have been expected when observed in the absence of the competing zeitgebers of usual daily life.

A second case report (48) was characterized by features that more closely conform to current criteria for diagnosis of ASPS. The 62-year-old male patient had difficulty maintaining wakefulness in the afternoon and early evening, reporting a usual bedtime of around 18:30. He would sleep soundly until around 03:00, when he would awaken alert and refreshed. Although circadian measures (e.g., round-the-clock body core temperature) were not obtained to confirm a phase advance of the biological timing system, and although the case was complicated by the presence of sleep apnea, the authors reported resolution of the problem following chronotherapy (see below).

As mentioned, a number of investigators have recently reported age-related changes in the circadian timing system that may be associated with the sleep patterns often observed in older subjects. The most consistently reported circadian change is a phase advance in various circadian rhythms, including body core temperature (43–46) and melatonin (49). Earlier and more variable bedtimes and rising times and difficulty maintaining sleep in the second half of the night are the most prominent features of age-related sleep disturbance. This combination of a phase-advanced circadian system and an apparent shift in sleep tendency to an earlier clock time clearly suggests that many middle-aged and older individuals suffer from ASPS.

A. Pathophysiology

Based on their successful treatment of an ASPS patient with phase-advance chronotherapy (see below), Moldofsky et al. (48) proposed two possible explanations for the expression of ASPS. On one hand, ASPS sufferers may have unusually short endogenous periods, outside the range of entrainment by the 24-hr day. Yet, the only report of an ASPS patient studied under conditions of temporal isolation revealed a normal endogenous rhythm in both sleep-wake and body temperature cycles of 24.5 hr (47). This result must be interpreted with caution, however, since the patient did not exhibit typical features of ASPS, as mentioned above. On the other hand, Moldofsky and co-workers raised the alternative hypothesis that in their patient, the sleep-wake cycle may have had a normal period length, but may have become dissociated from the circadian temperature rhythm in ASPS, perhaps as consequence of the patient's history of rotating shift work.

In much the same way that Wollman and Lavie (31) implicated "sleep gates" and "forbidden zones" for sleep in the manner in which non-24-hr SWS patients show occasional, systematic phase jumps in sleep onset, Lack and co-workers (50) have suggested that the early bedtimes and early-morning sleep termination characteristic of ASPS may be the consequence of phase advances of such circadian sleep gates and forbidden sleep zones. Based on body temperature and melatonin measures, these investigators found that "early-morning awakening insomniacs" were phase-advanced by 4–5 hr and 2 hr, respectively, when compared with age-matched controls. Thus, the usual timing of the nocturnal sleep gate and morning forbidden sleep zone would be advanced, leading to early-evening sleepiness and enhanced difficulty in maintaining sleep starting at around 03:00.

Certainly, the aforementioned association between changes in the circadian timing system that accompany aging and age-related sleep disturbance are consistent with the hypothesis that ASPS is the consequence of a phase advance of the endogenous circadian pacemaker. Whereas results vary concerning the effects of age on the amplitude of the circadian rhythm of body core temperature, the finding of a phase advance in the temperature minima of older subjects is quite consistent across studies. Results of a study that sought to induce "aged sleep" in younger subjects by phase advancing their circadian clocks add additional support to the notion that a phase advance of the endogenous clock may underlie ASPS (51). Following 3 days of morning bright-light exposure (07:15–11:15), healthy, young adults showed an average 97-min phase advance of the temperature minimum. This resulted in a significant decline in sleep efficiency, with the large majority of increased waking time occurring in the final 2 hr of the night.

Regardless of the pathophysiology that leads to onset of ASPS, it is likely that once the syndrome develops, ensuing behavior may have the effect of perpetuating the disorder, because the tendency for early-morning awakenings increases the likelihood that ASPS patients will be differentially exposed to the phase-advance portion of the light PRC. Thus a vicious cycle may ensue, with early-morning light exposure maintaining the aberrant, advanced phase of the circadian clock.

B. Treatment

Moldofsky et al. (48) reported the first treatment attempt in ASPS, using phase-advance chronotherapy. The intervention was implemented following several failed attempts to impose incremental delays in the subject's bedtime. Time of retiring was advanced by 3 hr every 2 days until bedtime was set at 23:00. Follow-up after 5 months indicated that the patient was able to maintain the new sleep schedule, and no longer complained of afternoon and early-evening sleepiness. As with phase-delay chronotherapy, however, stabilization of sleep time at conventional hours did not result in improved sleep quality. Indeed, wake time after initial sleep onset increased from 11% at baseline to 18% following treatment.

Several investigators have reported successful treatment of ASPS with timed (evening) exposure to bright light. In one study, Campbell and co-workers

(44) compared the effects of evening bright-light exposure (~4000 lux for 2 hr) with a dim-red-light control condition (<50 lux for 2 hr), in older subjects (mean age: 70.4) who had experienced sleep maintenance insomnia for at least 1 year prior to enrollment in the study. Following 12 consecutive days of treatment, subjects in the bright-light condition exhibited a significant increase in sleep efficiency (baseline: 77.5%; posttreatment: 90.1%), the result of an average 1-hr decline in wakefulness within the night. This reduction in waking time (stage 0) was accompanied by a significant reduction in stage 1, and by a significant *increase* in the proportion of stage 2 sleep. Nonsignificant, but perhaps clinically relevant, increases in slow-wave sleep (stage 3 and 4) and REM sleep were also observed in the bright light group. In contrast, those receiving dim-light exposure showed no significant change in sleep efficiency, wakefulness within sleep, or any other sleep parameter measured.

Similar findings were reported by Lack and Schumacher (52) using exposure to evening bright light in a group of early-morning-awakening insomniacs. In that study, subjects were exposed to either 4 hr of evening bright light (2500 lux) or dim red light (200 lux) for 2 consecutive days (20:00–24:00 on the first night, 21:00–01:00 on the second). Bright light produced improvements in sleep, as measured by wrist actigraphy and subjective assessment: Self-reported sleep duration increased significantly, whereas actigraph movement time in the first 6 hr of sleep declined significantly following treatment. The dim-light group showed no such changes in sleep measures. The same results were obtained in a subsequent study of nine subjects (mean age 53.4 years) with early-morning-awakening insomnia (49). Following 2 consecutive days of bright-light exposure (2500 lux from 20:00 to 24:00), actigraphically measured total sleep time increased by an average of 1.2 hr.

In all three studies described above, improvement in sleep was accompanied by significant delays in the circadian course of body core temperature. Campbell and co-workers (44) reported an average phase delay of 3.1 hr following treatment. Lack and Schumacher (52) observed phase delays in the temperature minimum of 3–4 hr, and Lack and Wright (49) reported an average delay of 1.85 hr. In the Lack and Wright study, enhanced sleep duration was maintained for 5 days following the acute light treatment, but in the subsequent months sleep disturbance was again reported. Likewise, our attempts to use light as a maintenance treatment for ASPS in older subjects have met with only limited success, primarily due to problems of compliance (Campbell, unpublished data).

V. Irregular Sleep-Wake Pattern

Whereas reports in the literature of ASPS are rare, studies of irregular sleep-wake pattern are virtually unheard of. Irregular sleep-wake pattern consists of temporally disorganized and variable episodes of sleep and waking behavior (5). Although total sleep per 24 hr is within a normal range, no single sleep episode continues for a normal duration. As such, unlike the other circadian rhythm sleep disorders, irregular sleep-wake pattern may be viewed not only as a circadian rhythm disorder, but also as a true sleep disorder. The disorder is thought to be most commonly observed in individuals with severe congenital, developmental, or degenerative brain dysfunction.

We could find no published report of a clear case of irregular sleep-wake pattern. However, a case that seems to closely approximate the syndrome was reported by Okawa et al. (27). They describe the sleep-wake behavior of a 4-year-old congenitally blind and mentally retarded girl who had experienced irregular sleep-wake patterns since the age of 1 year. Her sleep patterns consisted of several naps during the daytime coupled with occasions on which she remained awake throughout the night. The investigators "could not anticipate when she was likely to sleep."

Yet, a detailed record of the patient's sleep-wake behavior maintained over a period of 4 months by her mother reveals a strong tendency to obtain major sleep bouts at night, as well as weak entrainment to the 24-hr day. By the age of 6 years, the patient was able to be entrained to a 24-hr sleep-wake rhythm, with only occasional daytime napping, by means of forced awakenings and behavioral structuring during the day. Thus, the "pathological" sleep-wake profile exhibited by this patient and its subsequent resolution are similar to the developmental sleep-wake profiles of normal humans over the first 5 years of life. Indeed, one of the most impressive graphic depictions of "irregular sleep-wake pattern" is presented by Kleitman (53, p. 137) in describing sleep-wake patterns of an infant from the second to the twelfth week of life.

Because of the extreme rarity of the disorder, an extensive discussion of the pathophysiology and treatment of irregular sleep-wake pattern would be meaningless. Okawa and co-workers concluded that the sleep-wake pattern of their patient was probably due to "immaturity of the pacemaker of the circadian rhythm in the brain and/or due to weak perception of social cues in her environment" (27). The striking similarity of the patient's sleep-wake behavior to that of normal infants, in conjunction with her subsequent ability to entrain, clearly supports such a conclusion. Assuming the existence of an intact SCN in irregular sleep-wake pattern patients, strict behavioral scheduling, high-contrast photoperiods, or provision of some other form of enhanced entrainment stimuli may be effective in treating the syndrome.

VI. Summary and Conclusions

The inability to maintain a consistent sleep-wake schedule that conforms to societal norms can have a significant negative impact on one's life. Depending on the nature and severity of the disturbance, circadian rhythms sleep disorders can be little more than irritating inconveniences that interfere with and/or cause sufferers to alter social schedules, or they can be debilitating conditions that render impossible the normal interactions with society that most of us take for granted. Although the scientific study of this class of sleep disorder is less than two decades old, and despite the fact that the incidence of such disorders is relatively rare, significant steps have been taken both with respect to the elucidation of mechanisms underlying the syndromes and in terms of treatment.

Yet, two important aspects of the treatments discussed in relation to circadian rhythms sleep disorder should be emphasized. First, all such interventions may be classified as maintenance treatments; no treatment described in the literature purports to be a cure. As such, individuals suffering from circadian rhythm sleep disorders have to be willing to adhere to a course of treatment for as long as they wish to remain symptom free. And, as with any maintenance treatment, compliance is a critical factor. Issues of compliance constitute the principal drawback to long-term implementation of chronotherapy regimens, and preliminary data from our laboratory suggest that compliance is likely to be a limiting factor in the use of timed bright light as a maintenance treatment. Except in the most severe cases, patients often view such treatments as more limiting, and therefore, more troublesome than the disorder.

It should also be emphasized that the syndromes discussed in this chapter are classified as sleep disorders. The are defined by and described in the *International Classification of Sleep Disorders: Diagnostic and Coding Manual* (5), and treatment is almost always sought because patients suffer from severe sleep disturbance. As such, the primary measure of treatment efficacy and effectiveness should include improvement in sleep quality. Yet, no treatment described here has demonstrated any consistent, measurable improvement in this primary outcome variable. Can any of the interventions be considered to be truly effective then? To the extent that realignment of the internal clock with conventional social schedules leads to the alleviation of psychological stress induced by the disorders, such treatments are clearly beneficial. Yet, until a treatment can effectively combine such "social synchronization" with significant resolution of sleep disturbance, no such counter measure can be said to be efficacious.

That the biological clock can be reset to conform to social schedules, without markedly improving sleep quality, has etiological implications as well. Alleviation of the rhythm disturbance in the absence of enhanced sleep quality clearly suggests that the *sleep disorder* aspect of circadian rhythm sleep disorders may include pathophysiology beyond the confines of the circadian timing system. One current model of sleep regulation hypothesizes that the timing and composition of sleep are controlled by two interacting processes (54,55). It is quite likely that circadian rhythm sleep disorders are the result not only of alterations in process C, the model's circadian component, but also disturbances in so-called process S, the noncircadian factor proposed by the model. Future treatments that take into account the possibility of multiple etiologies, not necessarily exclusively circadian, may prove to be more effective in alleviation of both the circadian and the sleep components of the disorder.

Acknowledgment

This work was supported by PHS Grants R01 MH45067, R01 MH54617, R01 AG12112, P20 MH49762, and K02 MH01099.

References

1. Wever RA. The Circadian System of Man: Results of Experiments Under Temporal Isolation. Topics in Environmental Physiology and Medicine. K.E. Schaefer, ed., New York: Springer-Verlag, 1979.
2. Aschoff J. Circadian rhythms in man. Science 1965; 148:1427–1432.
3. Czeisler CA, Weitzman ED, Moore-Ede M, Zimmerman J, Knauer R. Human sleep: its duration and organization depend on its circadian phase. Science 1980; 210:1264–1267.
4. Zulley J, Wever R, Aschoff J. The dependence of onset and duration of sleep on the circadian rhythm of rectal temperature. Pflugers Arch 1981; 391:314–318.
5. Diagnostic Classification Steering Committee. International Classification of Sleep Disorders: Diagnostic and Coding Manual. Rochester, MN: American Sleep Disorders Associations, 1990.
6. Weitzman ED, Czeisler CA, Coleman RM, Spielman AJ, Zimmerman JC, Dement W. Delayed sleep phase syndrome. Arch Gen Psychiatry 1981; 38:737–746.
7. Carskadon MA, Vieira C, Acebo C. Association between puberty and delayed phase preference. Sleep 1993; 16(3):258–262.
8. Thorpy MJ, Korman E, Spielman AJ, Glovinsky PB. Delayed sleep phase syndrome in adolescents. J Adolesc Health Care 1988; 9:22–27.
9. Ito A, Ando K, Hayakawa T, Iwata T, Kayukawa Y, Ohta T, Kasahara Y. Long-term course of adult patients with delayed sleep phase syndrome. Jpn J Psychiatry Neurol 1993; 47(3):563–567.
10. Horne JA, Ostberg O. A self-assessment questionnaire to determine morningness-eveningness in human circadian rhythms. Int J Chronobiol 1976; 4:97–110.
11. Regestein QR, Monk TH. Delayed sleep phase syndrome: a review of its clinical aspects. Am J Psychiatry 1995; 152(4):602–608.
12. Czeisler CA, Weitzman Ed, Moore EM, Zimmerman JC, Knauer RS. Chronotherapy: resetting the circadian clocks of patients with delayed sleep phase insomnia. Sleep 1981; 4(1):1–21.
13. de Beck TW. Delayed sleep phase syndrome—criminal offense in the military? Milit Med 1990; 155:1–14.

14. Weitzman ED, Czeisler CA, Zimmerman JC, Moore-Ede MC, Ronda JM. Biological rhythms in man: organization during non-entrained (free-running) conditions and application to delayed sleep phase syndrome. In: Sleep Disorders: Basic and Clinical Research. New York: Spectrum Publications, 1983: 153–171.

15. Ozaki S, Uchiyama M, Shirakawa S, Okawa M. Prolonged interval from body temperature nadir to sleep offset in patients with delayed sleep phase syndrome. Sleep 1996; 19(1):36–40.

16. Carskadon MA. Patterns of sleep and sleepiness in adolescents. Pediatrician 1990; 17(1):5–12.

17. Regestein QR. Treating insomnia: a practical guide for managing chronic sleeplessness, circa 1975. Comprehens Psychiatry 1976; 17:517–526.

18. Alvarez B, Dahlitz MJ, Vignau J, Parkes JD. The delayed sleep phase syndrome: clinical and investigative findings in 14 subjects. J Neurol Neurosurg Psychiatry 1992; 55(8):665–670.

19. Thorpy MJ, Korman E, Spielman AJ, Glovinsky PB. Delayed sleep phase syndrome in adolescents. J Adolesc Health Care 1988; 9(1):22–27.

20. Oren DA, Wehr TA. Hypernyctohemeral syndrome after chronotherapy for delayed sleep phase syndrome. N Engl J Med 1992; 327(24):1762 (letter).

21. Rosenthal NE, Joseph-Vanderpool JR, Levendosky AA, Johnston SH, Allen R, Kelly KA, Souetre E, Schultz PM, Starz KE. Phase-shifting effects of bright morning light as treatment for delayed sleep phase syndrome. Sleep 1990 13(4):354–361.

22. Dahlitz M, Alvarez B, Vignau J, English J, Arendt J, Parkes JD. Delayed sleep phase syndrome response to melatonin. Lancet 1991; 337(8750):1121–1124.

23. Oldani A, Ferini-Strambi L, Zucconi M, Stankov B, Fraschini F, Smirne S. Melatonin and delayed sleep phase syndrome: ambulatory polygraphic evaluation. NeuroReport 1994; 6:132–134.

24. Armstrong SM, McNulty OM, Guardiola-Lemaitre B, Redman JR. Successful use of S20098 and melatonin in an animal model of delayed sleep phase syndrome (DSPS). Pharacol Biochem Behav 1993; 46:45–49.

25. Okawa M, Mishima K, Nanami T, Shimizu T, Iijima S, Hishikawa Y, Takahashi K. Vitamin B_{12} treatment for sleep-wake rhythm disorders. Sleep 1990; 13(1):15–23.

26. Ohta T, Ando K, Hayakawa T, Iwata T, Kayukawa Y, Okada T. Treatment of persistent sleep-wake schedule disorders in adolescents and vitamin B_{12}. Jpn J Psychiatry Neurol 1991; 45(1):167–168.

27. Okawa M, Nanami T, Wada S, et al. Four congenitally blind children with circadian sleep-wake rhythm disorder. Sleep 1987; 10(2):101–110.

28. Weber AL, Cary MS, Connor N, Keyes P. Human non-24-hour sleep-wake cycles in an everyday environment. Sleep 1980; 2(3):347–354.

29. McArthur AJ, Lewy AJ, Sack RL. Non-24-hour sleep-wake syndrome in a sighted man: circadian rhythm studies and efficacy of melatonin treatment. Sleep 1996; 19(7): 544–553.

30. Kokkoris CP, Weitzman ED, Pollak CP, Spielman AJ, Czeisler CA, Bradlow H. Long-term ambulatory temperature monitoring in a subject with a hypernychthemeral sleep-wake cycle disturbance. Sleep 1978; 1(2):177–190.

31. Wollman M, Lavie P. Hypernychthemeral sleep-wake cycle: some hidden regularities. Sleep 1986; 9(2):324–334.

32. Miles LE, Raynal DM, Wilson MA. Blind man living in normal society has circadian rhythms of 24.9 hours. Science 1977; 198(4315):421–423.

33. Uchiyama M, Okawa M, Ozaki S, Shirakawa S, Takahashi K. Delayed phase jumps of sleep onset in a patient with non-24-hour sleep-wake syndrome. Sleep 1996; 19(8): 637–640.

34. Kamgar-Parsi B, Wehr TA, Gillin JC. Successful treatment of human non-24-hour sleep-wake syndrome. Sleep 1983; 6(3):257–264.

35. Eliott AL, Mills NJ, Waterhouse JM. A man with too long a day. J Physiol (Lond) 1970; 212:30–31.

36. Hoban TM, Sack RL, Lewy AJ, Miller LS, Singer CM. Entrainment of a free-running human with bright light? Chronobiol Int 1989; 6(4):347–353.

37. Klein T. Martens H, Dijk D-J, Kronauer RE, Seely EW, Czeisler CA. Chronic non-24-hour circadian rhythm sleep disorder in a blind man with a regular 24-hour sleep-wake schedule. Sleep 1993; 16(4):333–343.

38. Sugita Y, A. M, Teshima Y, Egawa I, Tsutsumi T. Successful treatment with vitamin B_{12} and taking sunlight for a case of hypernychthemeral syndrome. Jpn J Psychiatry Neurol 1988; 42:177–179.

39. Okawa M, Mishima K, Hishikawa Y. Vitamin B_{12} treatment for sleep-wake rhythm disorders. Jpn J Psychiatry Neurol 1991; 45(1):165–166.

40. Lewy AJ, Ahmed S, Jackson JM, Sack RL. Melatonin shifts human circadian rhythms according to a phase-response curve. Chronobiol Int 1992; 9(5):380–392.

41. Weitzman ED. Sleep and its disorders. Annu Rev Neurosci 1981; 4:381–417.

42. Diagnostic classification of sleep and arousal disorder, 1st edition prepared by the Sleep Disorders Classification Committee, HP Roffwarg, Chairman. Sleep 1979; 2:1–130.

43. Campbell SS, Gillin JC, Kripke DF, Erikson P, Clopton P. Gender differences in the circadian temperature rhythms of healthy elderly subjects: relationships to sleep quality. Sleep 1989; 12(6):529–536.

44. Campbell S, Dawson D, Anderson M. Alleviation of sleep maintenance insomnia with timed exposure to bright light. J Am Geriatr Soc 1993; 41:829–836.

45. Czeisler CA, Dumont M, Duffy JF, et al. Association of sleep-wake habits in older people with changes in output of circadian pacemaker. Lancet 1992; 340(8825): 933–936.

46. Moe KE, Prinz PN, Vitiello MV, Marks AL, Larsen LH. Healthy elderly women and men have different entrained circadian temperature rhythms. J Am Geriatr Soc 1991; 39(4):383–387.

47. Kamei R, Hughes L, Miles L, Dement WC. Advanced sleep phase syndrome studied in a time isolation facility. Chronobiologia 1979; 6:115.

48. Moldofsky H, Musisi S, Phillipson EA. Treatment of a case of advanced sleep phase syndrome by phase advance chronotherapy. Sleep 1986; 9(1):61–65.

49. Lack L, Wright H. The effect of evening bright light in delaying the circadian rhythms and lengthening the sleep of early morning awakening insomniacs. Sleep 1993; 16(5):436–443.

50. Lack LC, Mercer JD, Wright H. Circadian rhythms of early morning awakening insomniacs. J Sleep Res 1996; 5:211–219.

51. Campbell SS, Dawson D. Aging young sleep: a test of the phase advance hypothesis of sleep disturbance in the elderly. J Sleep Res 1992 1:205–210.

52. Lack L, Schumacher K. Evening light treatment of early morning insomnia. Sleep Res 1993; 22:225.
53. Kleitman N. Sleep and Wakefulness. Chicago: University of Chicago Press, 1963.
54. Borbely AA. A two process model of sleep regulation. Hum Neurobiol 1982; 1(3): 195–204.
55. Daan S, Beersma DG, Borbely AA. Timing of human sleep: recovery process gated by a circadian pacemaker. Am J Physiol 1984; 2(1):R161–R183.

16

Sleep and Circadian Rhythm Disorders in Aging and Dementia

DONALD L. BLIWISE

Wesley Woods Geriatric Hospital
and Emory University School of Medicine
Atlanta, Georgia

I. Introduction

Studies examining the effects of aging on the circadian system can be conveniently categorized by the experimental paradigm employed in those studies. By far, the vast majority of such research, particularly in humans, has examined rhythms under entrained conditions. Far fewer studies have probed for age effects under free-running conditions, during constant routine protocols, during forced desynchrony interventions, or subsequent to experimental phase shifts. In this chapter we will review relevant studies in each of the areas; however, the reader should bear in mind the total data base on which age effects on the circadian system are inferred using protocols other than under entrainment is sparse and originates from only a handful of research groups.

Several other organizing themes and undercurrents should also be noted. First, with regard to age effects, almost every study reviewed in this chapter examines age *differences* in rhythms rather than age *changes* in those rhythms; i.e., cross-sectional studies represent the rule rather than longitudinal studies, which are the rare exception. In fact, only several animal studies (1,2) have employed such designs to examine aging. Although some human longitudinal data relevant to circadian rhythms exist (e.g., 3), the follow-ups are too short (e.g., 2 or

3 years) to derive meaningful conclusions insofar as aging and rhythms are concerned. Particularly in humans, where different cumulative disease exposures, health care experiences, and genetic predisposition may vary widely among individuals, longitudinal data would clearly represent the sine qua non of defining the effects of aging.

With regard to the examination of rhythms in dementia, only the most limited data base exists, with nearly all data collected under entrained conditions. Additionally, chronobiological studies of patients with dementing illnesses vary widely in their defining criteria, which is not surprising given many of the ambiguities in the differential diagnoses of such conditions. Examination of such "pathological" aging also pushes our limits of defining when disease begins and normality ends, this being particularly true for diseases like Parkinson's disease (PD) and Alzheimer's disease (AD), where commonplace signs and symptoms in large segments of the population often challenge notions of what such diseases represent (4–7). In this regard, geriatricians have argued for a notion of "successful" aging (8), which can be considered the optimization of the aging process. Some researchers in chronobiology have embraced this principle in defining subjects for their studies; however, a full delineation of when "successful" aging becomes "normal" or "usual" aging (i.e., aging accompanied by the customary infirmities of advancing years) remains unsettled and ill-defined.

A final theme that will be discussed in this chapter is based on exciting developments involving elucidation of the interaction between the circadian and homeostatic regulation of sleep-wakefulness. Based on the original theoretical framework developed by Feinberg (9) and Borbely (10), modeling of the interaction of these components of sleep regulation has led to a deeper and richer understanding of mammalian sleep. Since the time of our last review encompassing rhythms and aging (11), further developments involving these constructs have proven to be one of the most significant new areas of research that may shed new light on understanding mechanisms, and treatments for elderly humans with disturbances of the sleep/wake cycle. Toward the end of this chapter we will attempt to integrate some of this material and examine possible etiologies of sleep disturbance in old age from a chronobiological perspective.

II. Circadian Rhythms and Survivorship

Perhaps the most logical starting point in discussing age effects in circadian rhythms is the notion that temporal disorganization in physiology represents a key component of the aging process (12). More recently, others have characterized aging as a failure of up-regulation (13), which may represent a related perspective. Given recent developments in describing the molecular basis for the circadian clock in mammals (14) and because aging is now acknowledged to be a genomically programmed event (15), it is hardly surprising that senescence and degeneration of cellular time keeping may represent parallel processes.

Even at the systemic level it may be possible to detect manifestations of such relationships. Wax and Goodrick (2) noted, for example, that decline in light/dark wheel-running activity in several strains of mice portended death. Albers et al. (16) noted a similar result in rats subjected to constant dim-light conditions, and higher mortality associated with rapid phase shifts in adult rats has been described (17). Even in humans there is at least some evidence that temporal disorganization is associated with adverse outcomes. In infirm, largely bed-ridden nursing home patients who reside in environments where light exposure levels scarcely exceed 300 lux, "misplacement" of the body temperature nadir from its customary location in the first 5 hr after midnight predicted all-cause mortality (18,19). Those patients with such temporal disorganization lived for approximately 800 days relative to patients whose nadir occurred at 02:00 or 05:00, who survived on average approximately 2000 days.

III. Neuroanatomical Evidence for Age-Related Alterations in Rhythms

An obvious starting point in the examination of age-related alteration in circadian rhythms is neuroanatomical examination of the effects of age on the structure of the suprachiasmatic nucleus (SCN). Presumably general cell loss and decreases in number of cells with immunoreactivity in transmitter systems relevant to the SCN (e.g., vasopressin) would be expected. Evidence on this issue is actually quite equivocal and probably species and/or strain dependent. Roozendaal et al. (20) and Chee et al. (21) reported that, in brown Norway rats, the total number of SCN neurons was unchanged with age, although the number of vasopressinergic neurons did decrease, despite being larger. Earlier studies in Long-Evans and Sprague-Dawley rats showed no change in neuronal counts with age (22), whereas another study in the old mouse showed dendritic changes in SCN neurons (23). More recently Madeira et al. (24) have employed stereological techniques to provide a three-dimensional sampling of the entire SCN in six different age groups of Wistar rats. Neither neurons nor astrocytes showed age-related changes, a finding that the authors attributed primarily not to strain differences, but to biased estimates used in the previous studies. These authors were also unable to replicate previous reports suggesting that male rats had larger SCN volumes than female rats.

Dutch investigators have shown in humans that the SCN decreased in volume, total cell number and number of cells showing vasopressinergic immunoreactivity (25–27) between ages 81 and 100 relative to all other age groups, even relative to a 61–80-year-old group (see Fig. 1). Moreover these changes appeared

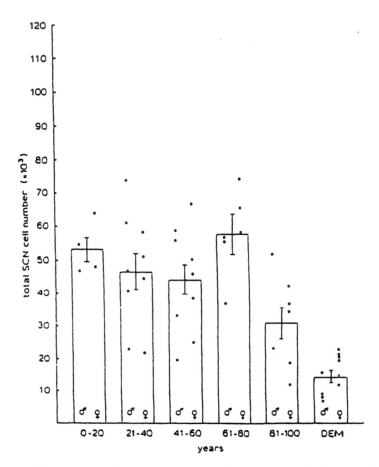

Figure 1 Neuronal counts in SCN as a function of age; Alzheimer's disease patients are shown on far right bar. (From Ref. 25.)

specific to the SCN and could be distinguished by the absence of such changes in nearby nuclei, such as the paraventricular or supraoptic nuclei. Relevant for the material to be covered in this chapter were the findings regarding AD. The total SCN number was about half of that seen in the 81–100-year-old group (Fig. 1), a finding consistent with predictions of deteriorated circadian regulation in such patients (see below). AD patients also showed decreased expression of vaso-pressinergic neurons as well. This group of investigators has also published some data suggesting gender differences in the human SCN, particularly involving vasoactive intestinal polypeptide (VIP) immunoreactivity, where women showed no age-related change and older men showed a significant drop in the number of such cells (28,29). Additionally, the SCN is elongated in women (rostral-caudal axis is 43% longer in women) whereas in men the SCN is more spherical; neither the supraoptic nor the paraventricular nuclei show such dimorphism (30,31). Relevant for aging effects, Hofman and Swaab (32) have shown diurnal variation in SCN VIP reactivity in postmortem examination of younger individuals who died at different times of day, which was not present in aged individuals.

Several observations should be made regarding these often conflicting neuroanatomical data of the SCN. First, it remains unclear to what extent species/ strain differences may contribute to the different patterns of results, even to the extent of possible differences between animals and humans. Second, the recent well-done study by Madeira et al. (24) in Wistar rats raises the possibility that most existing data (animal and human) may be artifactual due solely to differences in counting technique. Finally, regarding the Alzheimer's data in particular, it should be stressed that none of the aforementioned studies have demonstrated neuritic plaques or neurofibrillary tangles and associated Alz-50-type immunoreactivity in the SCNs of those patients. Defining vasopressinergic immunoreactive cell loss in the SCN in humans alone in the absence of such Alz-50 markers or in the absence of increased glial counts raises questions as to the precise nature of neuronal death in this region. Perhaps examining circadian rhythms subsequent to microdialysis infusion of amyloid precursor protein directly into the third ventricle of the rat (33) will provide a more critical functional test of these issues (see next section), at least in an animal model.

IV. Functional Evidence for Age-Related Alterations in Rhythms

Given the various controversies regarding the neuroanatomical results, it is not surprising that some of the most convincing evidence that the core oscillator operates differently with aging has been provided by a series of studies examining the functional properties of SCN itself as a function of age. Extracellular recordings of SCN neurons in vitro from aged rats have been reported to show lower mean firing rates as well as flattening of their unit firing rhythm (34) relative to younger animals. Some older animals showed unit firing patterns that were quite aperiodic. Wise et al. (35) reported that the circadian rhythm of glucose utilization in the SCN decreased with aging as well. Change in functional properties of the SCN per se with age is also supported by Sutin et al. (36), who reported that immediate early gene expression (c-*fos* as well as several other genes) in response to photic stimulation was decreased in old rats. Marginal changes were seen in VIP immunoreactivity in this study. However, more recently VIP changes have been reported in aged rats (37) and *jun*-B expression differences have been noted in older mice (38). In a study by Zhang et al. (39), light intensity was systematically varied to determine c-*fos* levels and the protein-regulating c-*fos* transcription (cyclic-AMP response element binding protein, CREB) in the SCN as a function of age in hamsters. In this study there were clear effects of irradiance (higher intensity) associated with higher levels of immunoreactivity. These results suggested that at the molecular level the SCN becomes less responsive to light in a roughly dose-response fashion, which is consistent with the data of Witting et al. (40) indicating that older animals required a higher light intensity to achieve entrained sleep/wake rhythms comparable to younger animals. An important aspect of the molecular data, however, was that, regardless of light intensity, the older animals expressed less protein relative to the younger animals (39), which would argue for an upper limit as to the possible effects of illumination on restoration of central pacemaker function.

In recent years, another model of restoration has been investigated extensively with regard to the SCN: fetal transplantation studies. Viswanathan and Davis (41) reported that grafts of fetal SCN into both young and old hamsters who had previously received SCN lesions successfully restored circadian rhythmicity (wheel running) in the majority of cases. The free-running period in the older recipient animals actually lengthened, suggesting a more "youthful" restoration of periodicity. Over time, however, the restored rhythm period shortened, which suggested that the ultimate source of rhythmicity changing with age resided in the SCN itself rather than its inputs. Total wheel-running activity was low in the older animals even after grafting, suggesting that, although rhythmicity in the host was restored, age-related deterioration in effector mechanisms were still operating. The data regarding periodicity essentially replicated what had been reported previously in transplants in tau mutant hamsters (42). A molecular basis for the transplantation findings has recently been suggested by Cai et al. (43), who showed that, in the fetal transplant model, c-*fos* expression was similarly enhanced.

Not only free-running period but also phase-shifting ability appears to be enhanced by fetal SCN grafts. Van Reeth et al. (44) reported that triazolam, which when injected i.p. into rats provides a powerful activity-inducing stimulus that typically results in substantial phase shifts (phase advances at circadian time 6, phase delays at circadian time 21), had far greater effects in older animals after SCN transplantation relative to a cerebellar graft control. Not only were a far greater proportion of older animals able to phase-shift, but the magnitude of the phase shift was also larger in the SCN transplantation animals.

Taken together, these recent studies suggest that fundamental molecular characteristics of the clock itself vary with age. This does not negate the importance of various neurochemically mediated photic and nonphotic inputs differentially affecting the physiological demonstration of rhythms as a function of age (see Section XI). These studies do, however, reinforce the notion that the aging process and the internal circadian clock may both represent time-keeping systems, albeit on a different scale, that are programmed at the molecular level (15).

V. Age Effects Under Entrained Conditions

The familiar observations of the elderly person going to bed early in the evening, waking up too early in the morning, and then occasionally napping during the daytime hours have, by and large, been borne out by systematic research. Monk et al. (45), for example, have shown that elderly subjects display considerably more "morningness" (i.e., "lark-like") tendencies on the owl/lark questionnaire relative to younger subjects, a finding that mirrors the earlier bedtimes and wakeup times reported in elderly subjects in a number of other surveys (46,47; for a full review see 11). This age-related change in owl/lark tendency may even be seen when younger and middle-age populations are compared (48). Increased napping and reported daytime fatigue in the elderly have been well documented as well (46,47; for a full review see 11).

In addition to these changes in sleep/wake schedule, decreases in rhythm amplitude and shifts in phase position (typically described by the nadir of the rhythm) of many physiological parameters to earlier clock times have been documented in aged humans. Body temperature, cortisol release, and, in some studies, melatonin may be phase-advanced in older persons (11). Studies suggesting decreased amplitude in older humans living under entrainment are even more numerous and include parameters such as body temperature, testosterone, thyroid-stimulating hormone, cortisol, and melatonin, but probably not prolactin and growth hormone, which show sleep-related decline with age (11,49) (see Fig. 2). Although there may be differences across species for some parameters, animal studies generally corroborate these observations. The results may be somewhat stronger for decreased amplitude than for advanced phase (50), though Zee et al. (51) reported a phase advance of older hamsters' rest/activity rhythms in a 14:10 light-dark (LD) cycle. Many reviews of circadian rhythms in humans and animals have cataloged the literally hundreds of research papers investigating such age differences cross-sectionally (11,17,50,52,53); the reader is directed to these for a listing of primary sources of this data base.

One study deserves particular mention when discussing age differences in rhythms under entrained conditions. Monk and colleagues (54) have recently challenged the conventional wisdom that the body temperature rhythm is altered in aged humans. By combining an impressive array of data from a large number of elderly men and women and younger and middle-aged men across a variety of protocols (under entrainment) and reanalyzing the temperature rhythm data, these authors were able to confirm a phase advance in the temperature rhythm but could find no convincing evidence that circadian amplitude decreased with age. Although the authors may have slightly overstated their case (i.e., a simple *t*-test between young and old men did show a higher amplitude in the former relative to the latter), their data are provocative from a number of perspectives. One possibility is that the overall health of the older group (aged 77–89) suggests a selective

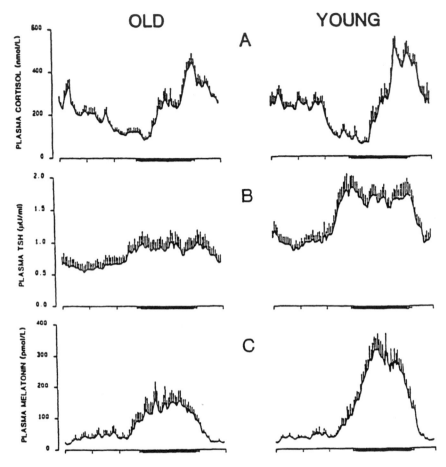

Figure 2 Mean 24-hr profiles for plasma cortisol (A), thyroid-stimulating hormone (B), and melatonin (C) in young and old men showing decreased amplitude with age. (From Ref. 49.)

survivorship-type phenomenon; i.e., these older subjects, by virtue of their optimal health and low medical burden of disease, could be atypical of older persons in general and may represent unusual "survivors." Although these older subjects are relatively healthy, an examination of medication usage suggests that, in many respects, they may well be typical of the geriatric population in general, and may thus resemble "normal" aging rather than Rowe's (8) concept of "optimal" or "successful" aging. For example, about 25% appeared to be receiving cardiovascular medication and/or nonsteroidal anti-inflammatory agents, presumably for chronic pain conditions. These data suggest the necessity that future studies in aged humans use large samples containing individuals with a wide variety of health problems.

While most, but not all, of the aforementioned data have thus generally reinforced the "advanced phase/reduced amplitude" model of age-related chronobiological changes, inference of alterations in the underlying, endogenous time-keeping system of aged mammals must be made cautiously, as the age-related alterations could be reflecting direct or indirect effects of factors that then may affect the rhythms themselves. For example, self-selected early bedtimes and wake-up times and their attendant masking effects (activity, posture, meals, self-selected illumination) may serve to synchronize all other rhythms to this timing pattern and establish stable phase angle relationships independent from rhythms generated endogenously. Although data from other mammalian species in entrained environments generally corroborate these findings in humans and thus suggest that such age-related findings do not reflect "voluntary" scheduling adjustments, human rhythms studied under entrained conditions are still more or less subject to capriciousness. Related to this issue, for example, are data suggesting that healthy older subjects (without insomnia) may actually have less intrasubject and intersubject variability in the timing of their sleep (45), which could be construed as an attempt to self-regulate rhythmicity. Using a paper-and-pencil scale called the Social Rhythm Metric, Monk et al. (55) have shown that older subjects maintain high regularity in their lives in activities such as interactions with other people and timing of meals. While the rich analysis of social zeitgebers may prove to be important in understanding how older individuals structure their days and nights in 24-hr time, there is every indication that such issues are likely to be complicated and subtle. Minors and colleagues (56) noted, for example, that age was positively correlated with napping (suggesting decreased amplitude) regardless of whether the older person lived with another person. However, an earlier bedtime (suggesting a phase advance) was apparent only if the older person lived alone. These data reiterate the fact that sleep and circadian rhythms in older persons have an exceedingly complex interaction with the individual's social milieu.

Because of all of these confounding variables, investigators in chronobiology have attempted to look beyond the simple description of rhythms in entrained conditions to use of more revealing, complex protocols to determine to what extent the clock itself or, alternatively, its response to zeitgebers or the mechanisms it affects may be altered with aging. In the following sections, we will selectively review developments using protocols attempting to circumvent such confounders.

VI. Age Effects Under Free-Running Conditions

In 1974, Pittendrigh and Daan reported the first animal study confirming that the period of the free-running period (tau) shortened with age under constant conditions (57), a finding later confirmed by others (e.g., 58). Since that time a number of animal studies have described this decrease in period using not only rest-activity but also sleep-wakefulness and temperature (59,60). Tau may not be shortened in some species; however, comparisons across species, particularly with regard to defining "old" within that species, may have contributed to those findings (17,61). Another possible confounding variable may be access to (running wheel) activity, which, as Edgar et al. have shown (62), leads to a decreased length of tau in free-running conditions. (The importance of activity is particularly relevant for studies of age effects in phase shifting. See below.) More recently, Witting et al. reported (61) that even in the absence of wheel-running access in the free-running period of the sleep-wake cycle, drinking and body temperature are significantly decreased in aged rats.

In the free-running situation, human subjects living under conditions of temporal isolation are given the opportunity to sleep and wake, as well as eat meals, at times in accordance with their own preferences. (One of the stipulations in the original studies was that subjects were also instructed not to nap, i.e., to consider that any period of intended sleep should be for an entire sleep period.) Monitoring of sleep-wake, temperature, and other parameters allows for appreciation of not only temporal patterns of any single rhythm in isolation, but also the extent of synchronization of various rhythms under such conditions.

Relatively few elderly subjects have been studied under such a protocol. Wever's book (63) described that, among 10 older (aged 40–70) subjects, spontaneous internal desynchronization (varying phase relationships between activity and temperature rhythms) occurred in 70%, relative to only 22% of his younger subjects. These high rates of internal desynchronization have not always been reported by others (64,65). Wever (63) also reported no association between the free-running period (tau) and age in his subjects; however, Monk (66) reanalyzed these data from women and showed a significant decrease in free-running period length of the body temperature cycle in older women (but not men) with one of the older women demonstrating a tau value of less than 24 hr. Weitzman et al. (64) reported results from six young and six elderly men studied in time isolation and also noted a shortening of the free-running temperature period, although the period of the sleep-wake cycle did not change. In this study the authors also noted that older subjects self-selected more time in bed than younger subjects, yet they achieved a lower total sleep time throughout the course of the time in the isolation facility. More recently Pollak et al. (67) also reanalyzed the Cornell time isolation data by examining a "fractional time" curve (68) in which successive sleep bouts of various lengths were plotted against the immediately following wake period. The results were interpreted as indicating reduced sleep "need" (lowered homeostatic drive) as a function of age.

Although technically not considered a free-run protocol, forced desynchrony protocols (see below) can, if conducted for a long-number of rest/activity (sleep/wake) cycles, afford alternative estimates of the free-running period of rhythms. Czeisler (69) recently reported that, in a 28-hr day protocol, the estimate of the intrinsic free-running period of the body temperature cycle was virtually identical in young versus elderly subjects (24.3 hr vs. 24.1 hr).

While the free-running studies (with the exception of the data from the newly presented forced desynchrony protocols) do generally corroborate the studies of rhythms made under entrained conditions, particularly in demonstrating a shorter intrinsic period of body temperature, free-running studies still are subject to masking effects. Animal studies can, of course, be conducted under conditions of constant darkness and with activity (wheel running) controlled to an extent that human free-running conditions are unable to provide. Human free-running studies, owing to uncontrolled effects of posture and activity, as well as room light, now known to be capable of shifting rhythms (70), simply do not provide an examination of endogenous rhythms or allow examination of phase angle relationships between temperature and sleep-wake rhythms.

VII. Age Effects Under Constant Routines

The constant routine protocol, originally developed by Mills and colleagues (71) and adapted more recently by the Harvard (72) and Pittsburgh groups (54,73), allows estimation of underlying endogenous circadian rhythms in a protocol in which activity, posture, meals, lighting, and even sleep are rigidly controlled for 36–40 hr. Subjects agree to a voluntary protocol of bed rest, sleep deprivation, and hourly light snacks in a dimly lit room following a series of adaptation nights on the subjects' customary sleep/wake schedule. A rectal probe and an indwelling venous catheter allow frequent sampling of core body temperature and various hormones, respectively, throughout the study. Subjects of different ages have been compared by both Czeisler et al. (72,74) and Monk et al. (54,73).

In 1992 Czeisler et al. (72) reported that circadian amplitude in the constant routine was reduced by 0.08°C in older (mean age = 70.6) relative to younger (mean age 21.9) subjects in a 40-hr constant routine whereas the nadir of the fitted body temperature curve was about 2 hr earlier in the older subjects (4:56 A.M. vs. 6:48 A.M.). Both of these differences were highly significant at the < 0.001 level. Both customary wake-up times and customary bedtimes were also earlier in the older subjects. A somewhat different pattern of results was noted by Monk et al. (54), who found no differences in amplitude or temperature minimum measures when comparing young versus older subjects; however, their elderly subjects included some individuals with circadian phase estimates that were late in the morning hours (e.g., 10:00 A.M.–11:00 A.M.), which may have contributed to the variance in the older group. It is difficult to reconcile these conflicting findings. The variability in the older subjects of Monk et al. was substantial, particularly in terms of phase estimates, suggesting that healthy older subjects may have far more variability in circadian parameters than was heretofore recognized.

The Harvard group (74) have recently presented in abstract form an update on their constant routine data expanding the sample size to 101 young men and 44 older men and women (no separate analysis by gender). While amplitude data were not yet presented, the phase data continued to show an earlier endogenous circadian phase minimum in older subjects, though that difference bordered on statistical significance and was less than an hour. Similarly habitual wake-up and bedtimes were only about an hour earlier in the older subjects in this larger, updated sample. Perhaps most interesting in these data was that older subjects had a habitual wake-up time significantly closer to their temperature minimum relative to younger subjects, thus suggesting that reduced homeostatic drive may be just as, or even more important, in explaining early-morning awakenings in the older population. In fact, when young adult "owls" and "larks" were compared in the constant routine (75), "larks," despite having a temperature minimum occurring over 2 hr earlier than the "owls," typically slept longer than their minimum by nearly an hour. Similar data were noted by Monk et al. (45) in their study of morningness/eveningness as a function of age. These data imply that age effects on sleep-wake cannot be interpreted solely on the basis of changes in circadian rhythms.

VIII.   Gender-by-Age Interactions in Rhythms

As mentioned previously, some features of the SCN suggest possible structural sexual dimorphism in this nucleus relative to other hypothalamic nuclei. There are at least several reasons to suspect that these findings could have functional significance in humans. First, a number of studies have reported gender differences in napping or daytime fatigue in old men (men napping more than women) (76–79) although some conflicting data also exist (80,81; additionally some studies have presented schedule data suggesting that older women go to bed earlier and/or wake up earlier than older men (82,83). Second, studies of temperature rhythms in both entrained conditions and under constant routines have suggested that older women may be phase-advanced relative to older men. For example, studies by Campbell et al. (83), Moe et al. (84), and Monk et al.'s (54) data on 48 elderly men and women studied for 24 hr under entrained conditions all showed a phase advance in body temperature in the older women relative to older men. With far fewer subjects under constant routines, neither Monk et al.'s data (54) nor those of Czeisler et al. (72) could show this phase advance, though the constant routine protocol did yield a larger temperature rhythm amplitude in older women in the latter data set. This higher amplitude in older women was also noted by Moe et al. (84) and Campbell et al. (83) under entrained conditions.

       An issue in those potential gender effects is whether the differences observed represent an age-by-gender interaction or whether similar gender effects might be observable at younger ages as well. Unfortunately, many of the above studies examining gender differences in temperature rhythms in the context of aging have excluded young women, perhaps owing to difficulties in controlling for phase of menstrual cycle. For example, Lee (85) has shown that the amplitude of the temperature rhythm is dampened during the luteal phase of the menstrual cycle relative to the follicular phase while the overall temperature mean is elevated, consistent with the thermogenic properties of progesterone. Phase position did not vary significantly (85). Although the Lee study did not include young males, comparison of her data to those of Vitiello et al. (86) and Winget et al. (87) suggested that, regardless of phase of menstrual cycle, younger women have *lower* amplitudes than men, in contrast to the aforementioned data suggesting *higher* amplitudes in older women. The phase advance, however, appeared similar in young and old women, arguing for a complex pattern of interaction of amplitude and phase within the circadian timing system as a function of age. A recent study comparing young (follicular) and postmenopausal (aged 54–62) women in melatonin and temperature rhythms under entrainment showed a phase advance in melatonin but not temperature and a decreased amplitude in temperature (88), findings that only underscore the complexity of gender effects. Future studies, particularly those involving protocols designed to unmask circadian rhythms in different age groups, should include younger women with menstrual cycle phase controlled (as well as excluding for oral contraceptives) and also analyze for possible gender differences at every age (c.f., 89,90), to fully appreciate the extent to which various human circadian rhythms may be manifested differently in women and men.

IX. Age Effects in Rhythms of Behavior, Mood, and Performance

In recent years the Pittsburgh group has published a convincing body of evidence suggesting that older subjects' ratings of alertness/sleepiness, as well as subjective ratings of vigor and affect and measures of performance (e.g., manual dexterity, visual search tasks, and verbal reasoning), show a different pattern over the 24-hr day relative to those patterns observed in younger subjects. It should be stressed that these data are presented in the context of minimal age differences in the circadian rhythm of body temperature amplitude reported by this research team. In the first of this series of studies, Monk et al. (73) showed that, within the constant routine paradigm, older male subjects' performance showed a pattern of decline that was better defined as a linear, monotonic decrease relative to younger male subjects (aged 20–30), who demonstrated a more pronounced tendency for a curvilinear (quadratic or cubic) pattern of decline and then recovery (see Fig. 3). In essence, the circadian influence on such measures was far more salient in younger than in older subjects, although these differential effects were somewhat

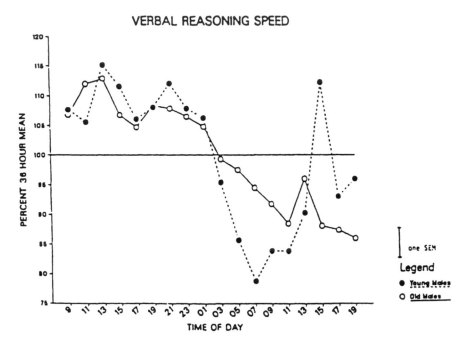

Figure 3 Performance on a verbal reasoning task in young and old men in a constant routine protocol showing linear pattern of decline in older subjects with more circadian influences apparent in the performance of the younger subjects. (From Ref. 73.)

more clear for mood ratings, manual dexterity, visual searching, and verbal reasoning than for vigilance (Mackworth Clock) or vigor ratings. Additional analyses indicated that a greater number of younger subjects showed statistically significant sinusoidal curve fitting and that the greater variance in phase measurements of the older subjects' temperature rhythms did not contribute to the observations of decreased amplitude of the performance measures. Considerable homogeneity exists among circadian variation of these various behavioral measures in younger subjects in the constant routine and the measures were relatively tightly coupled not only to temperature rhythms but also to rhythms of cortisol and melatonin (91). Buysse et al. (89) presented data on unintended sleep episodes of 3 sec or greater per 3-hr block in the constant routine as a function of age. Although in these data the linear model (i.e., more unintended sleep episodes over time) showed a more pronounced effect in younger rather than older subjects, the older subjects showed 62% fewer unintended sleep episodes than younger subjects and the older subjects appeared to be particularly vulnerable for such episodes in the final 18 hr of the protocol. Finally, Monk et al. (90) have also presented similar data on less rhythmic subjective alertness and vigor with age in comparing middle-aged (aged 37–52) and older men in time isolation instructed to go to bed and wake up at their habitual times as indicated by a prestudy sleep log. The overriding interpretation of these data is that homeostatic processes (perhaps cortical in nature) predominate in the behavioral manifestations of rhythms, at least insofar as these healthy older subjects are concerned.

X. Age Effects in Forced Desynchrony Protocols

Whereas the constant routine protocol allows for a transient unmasking of various exogenous and endogenous factors, its inherent disadvantage for modeling age differences in certain other aspects of the circadian system has led some investigators to rely on a different type of protocol to more adequately appreciate age effects. Splitting of various components of circadian rhythms, most notably the separation of the temperature versus the sleep/wake cycle, has been performed many times in many different contexts, perhaps the earliest example being Kleitman's 28-hr-day protocol (92). To be more specific, such protocols allow examination of how sleep is regulated or gated by the circadian system at different circadian times, since, because the sleep-wake schedule is made artificially independent from the temperature cycle, sleep is allowed to occur at a wide variety of circadian times. Recent attempts to examine age differences using this protocol have been made by both Czeisler et al. (69,93), using a 28-hr day, and Haimov and Lavie (94), using 20-min day.

In the Haimov and Lavie study (94), young and elderly male subjects were placed on a 7-min sleep/13-minute wake paradigm for 24 hr. In this model not only sleep propensity but also a nadir for sleep propensity (the so-called "forbidden zone") can be determined. Sleep propensity for both young and old groups was clearly circadian; however, elderly subjects showed a somewhat more evenly distributed pattern of sleep tendency with less sleep observed in the 05:00–09:00 period and some suggestion that the maximum sleep propensity was about an hour earlier in the elderly subjects (04:24 vs. 05:20) consistent with a phase advance. No age differences, however, were observed for the time period of minimum sleep propensity (19:12 vs. 19:18) across these groups.

Using a 28-hr rest/activity cycle (9 hr 20 min of scheduled sleep: 18 hr 40 min of scheduled wakefulness), Czeisler and colleagues (69,93) recently reported that, in 11 young men and 13 older men and women (9 M, 4 W) studied cumulatively for over 500 sleep episodes, total sleep time and sleep latency showed differential relationships as a function of phase position as related to age. Total sleep, for example, was significantly lower in the older subjects at all circadian phases but, perhaps more importantly, showed a different pattern of relationship to the body temperature nadir than seen in younger subjects. More specifically, after adjusting for the nadir of body temperature, elderly subjects tended to awaken at an earlier clock time relative to younger subjects. When sleep episodes were divided into quintiles, older subjects slept much less on the 4th and 5th quintile relative to the younger subjects; no differences emerged for the other 60% of the scheduled sleep period. As would be expected given these results for total sleep time, sleep latency showed no differences across age groups by circadian phase position; for both young and old subjects the shortest sleep latencies occurred 2 hr after the temperature nadir and the longest sleep latencies (akin to the forbidden zone) about 8 hr before the body temperature minimum. Among other variables, slow-wave sleep, although showing a large age difference as would be expected, did not appear to be regulated by circadian phase. On the other hand, REM sleep variables (REM latency and REM as a percentage of total sleep) appeared to be subjected to complex regulation by circadian phase, age, and total time asleep.

XI. Age Effects in Phase Shifting

It is reasonable to expect that, given at least the strong suggestion of age-related change of function at the molecular level in the SCN of the older animal, phase-shifting ability and subsequent adaptation to such shifts would be impaired in a wide variety of physiological functions. Careful and systematic evaluation of the nature of photic and nonphotic inputs in producing phase shifts, however, can provide critical clues into how zeitgebers can impact upon age differences in circadian function. Elucidation of the neurotransmitter systems affecting input systems, for example, might conceivably hold more immediate, practical implications for development of pharmacological interventions than investigation of the molecular biology of the clock per se.

   One of the first studies to investigate phase shift tolerance as a function of age in a basic science model was that of Rosenberg (95), who, in a dissertation, reported that the sleep of aged rats subjected to 180-degree phase shift showed prolonged recovery relative to the sleep of younger rats. In later work examining the phase-response curve (PRC) to light, Rosenberg et al. (96) reported that when young (<12 months) and old (>16 months) hamsters in constant darkness were pulsed with 1 hr of 500-lux illumination in the phase delay region or inactive region of the PRC the magnitude of the phase shift in activity was similar. In the phase advance region of the PRC, however, the magnitude of response was actually greater in the older animals. Zee et al. (51) reported that when young and middle-aged (13–16 month) hamsters maintained on a 14:10 LD cycle were subjected to 8-hr phase delays or 8-hr phase advances, the middle-aged hamsters shifted more quickly to the advance whereas the younger hamsters shifted more quickly to the delay (see Fig. 4).

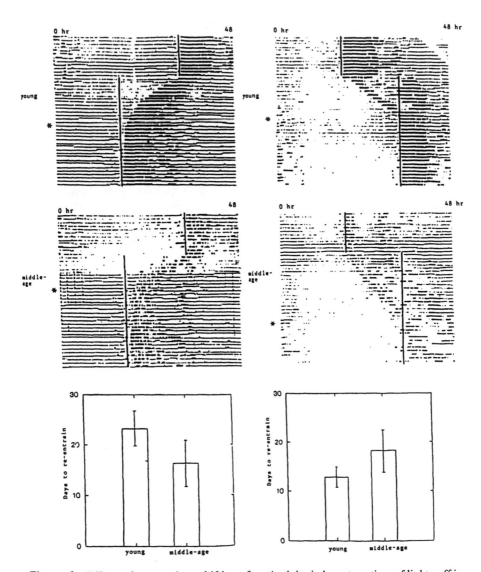

Figure 4 Effects of age on phase shifting of rest/activity in hamsters; time of lights-off is shown by vertical line; asterisk indicates day on which reentrainment criterion was met; (left) effects of 8-hr phase advance of 14:10 LD cycle; (right) effects of 8-hr phase delay of 14:10 LD cycle; (bottom) mean (± SE) of days to reentrain, suggesting that middle-aged animals have less difficulty in phase advance direction but more difficulty in phase delay direction. (From Ref. 51.)

Other studies from the Northwestern group have investigated how phase shifting in older animals may be differentially affected by photic and selected nonphotic stimulation and how various neurotransmitter systems may affect such phase shifting. A cornerstone for these later studies was the observation that a short-acting benzodiazepine hypnotic, triazolam, can induce phase shifts (advances or delays, depending on when in circadian time the medications were administered) and that those shifts were, in fact, dependent upon the animal being allowed to sustain increased motor activity (97–99). Similar effects of immobilization were noted when the stimulus was a 6-hr dark pulse. Based on these data, this group then examined how such feedback affected phase shifts in the rest/activity rhythm as a function of age. Animals were kept in constant light and at appropriate times (CT 6 for phase advance, CT 18 for phase delay) were pulsed to a 6-hr dark pulse to provide a phase-shift stimulus (100). Similar times of administration, though slightly different for the phase-delaying stimulus, were used for triazolam and ansiomycin, a protein synthesis inhibitor, the latter showing phase-shifting properties not dependent upon immobilization. Under triazolam the older animals showed greatly attenuated phase advances and phase delays with i.p. injection at the "appropriate" circadian times, respectively. However, there were equivalent increases in the amount of locomotor activity in young and old animals subsequent to triazolam injection, indicating that the drug continued to have equivalent activity induction effects. A similar dampening of response in the older animals was seen in the 6-hr dark pulse as well (both advances and delays). Only in the ansiomycin condition were phase advances and delays comparable between young and old animals (100). These results suggested that the inability of the older animals to phase-shift may have been a result of uncoupling between the clock and its sensitivity to activity feedback as a zeitgeber in aging.

Several attempts to understand the neurochemical aspects of these changes in phase shifting have been performed by selectively depleting or enhancing selected neurotransmitter systems. Depletion of monoamines (norepinephrine, dopamine, serotonin) in young animals via reserpine injection produces rest/activity rhythms far more characteristic of "older" animals, including a phase advance of activity onset under 14:10 LD conditions and a lack of responsiveness to the phase-shifting properties of triazolam (101,102). Subsequent studies have focused primarily on the serotoninergic system in this regard, largely because the SCN receives substantial 5HT input from the raphe. Studies using a selective 5HT-1A agonist administered late in the subjective day in constant darkness have demonstrated that this substance induces phase-dependent advances in a manner similar to that seen for triazolam in young animals, yet a greatly diminished response in older animals (103). Older animals pretreated with this agonist and then subjected to light pulses at CT19 (circadian time of maximal phase advance in this species) also sustained greater resistance to phase shifting than younger animals (104). These data imply not only a critical role for serotonin in the control of rhythms in aging, but also that appropriately timed use of 5HT agonists could represent an important factor in stabilizing the internal time-keeping system in human aging.

In humans, age-related voluntary increases in rigidity of scheduling of sleep (see above), the difficulty sustained by older shift workers faced with rotating shifts (105,106), and the problems older persons experience with jet lag (107,108) appear to indicate that older individuals have considerably more difficulty adjusting to phase shifts, and at least one sleep lab studied has reproduced this effect as well (109). Older shift workers, relative to younger shift workers, appear to have decreased aerobic capacity (110). Beyond age 50 there may also be some suggestion that individuals who remain in shift work may actually have fewer sleep-related problems, suggesting that middle-aged individuals often "select out" of such work (111).

Despite these observations however, it is possible, at least theoretically, to conceptualize the age-related changes in circadian timing as facilitating, rather than impeding, phase-shifting ability. Critical to this notion is the concept that any attempt to phase-shift a rhythm may also impact upon its amplitude as well. Thus, for example, the lower amplitude of the temperature rhythm seen in the elderly may render them more, rather than less, liable to phase-shift (112). One would then predict that adjustment to phase shifting would be relatively easier for the older, relative to younger, subjects.

Experimental work on phase shifting as a function of age in humans has been reported by the Pittsburgh group, who have reported results from a 6-hr phase advance performed in 15 older women and 10 older men (112,113). Following a 5-day baseline period in temporal (but not social) isolation, subjects underwent a 6-hr phase advance of their habitual bedtime and then continued to live on this schedule for 9 days postshift. Data from a previous study of similar phase shifting in eight middle-aged men (aged 37–52) served as a comparison group, although inferential statistics were not used to derive age differences, owing to some differences in experimental design.

Results from the Monk et al. phase shift studies can be summarized in terms of the measured physiological parameters, including temperature, polysomnographically defined measures of sleep and sleep architecture, subjective ratings of alertness and mood, and performance measures. Subsequent to the 6-hr phase advance, both elderly men and women demonstrated a reduction in temperature rhythm amplitude, similar to that seen in the middle-aged controls. Over the course of the 9 days postshift, the amplitude of the rhythm showed a monotonic increase over time in older women and to a lesser extent in older men as well, although the gender difference did not reach statistical significance. The findings regarding acrophase were more uniform across older men and women and, in fact, visual inspection of the graphic presentation of the data suggested that older subjects may have had a more rapid adjustment to the phase advance relative to the middle-aged comparison group.

One of the most provocative findings from this phase shift study involved age effects of recovery of sleep quality, indexed as sleep efficiency (i.e., the proportion of time spent in bed), as a function of age. Relative to the middle-aged subjects, older subjects (both men and women) appeared to have a greater decrement in sleep efficiency, which lasted the entire 9 days subsequent to the shift. In fact, by day 9, older subjects still were considerably below their baseline sleep efficiencies (approximately 70% relative to a baseline of about 83% for the women and about 68% relative to a baseline of about 73% for the men), indicating the older subjects, despite having rhythms that readjusted adequately, may have had a deficient homeostatic sleep drive, which did not allow them to achieve adequate recovery sleep even a week after the phase shift. Further analyses indicated that these changes in sleep efficiency were primarily a function of wakefulness during the first 2 hr, rather than the final 2 hr, of the sleep period (113). On the other hand, REM %, a variable previously shown to be dependent on circadian phase position (114), showed a recovery function paralleling that of acrophase. No gender differences were noted. Additionally, REM % appeared to be higher at every recovery day in the older subjects relative to the middle-aged subjects, implying less disruption of the circadian system. Aged subjects did not appear to have significant effects in terms of global mood disturbances postphase shift, though they did report lower vigor. Differences in postshift ratings and recovery of both of these variables did not appear to differ appreciably from those data seen in middle-aged subjects. Performance data also did not suggest major differences in recovery as a function of age.

Although results from the Pittsburgh 6-hr phase advance protocol implied minimal gender differences, Oginska et al. (111) and Tepas et al. (115) have shown that in the field situation women rather than men demonstrated greater intolerance for shiftwork as indicated by waking symptoms and decreased length of sleep. At the very least these data suggest that, as emphasized previously, gender differences must be considered carefully in describing how rhythms change with age.

A final approach to attempt to understand how age changes in phase-shifting ability might produce disturbance in sleep in humans was provided by Campbell and Dawson (116), who modeled the phase advance in body temperature as a cause of sleep disturbance in old age by phase-shifting younger subjects. In this study, young adults in their twenties were exposed to 4 hr of bright light in the morning, which then produced a consequent phase advance in the body temperature minimum by 97 min. Scheduled sleep time in the laboratory remained the same, however (23:00–7:00). Young subjects were thus subjected to a phase advance of body temperature at a magnitude similar to what has been reported as constituting age differences in body temperature phase under entrained conditions. Predictably, the younger subjects sustained a drop in sleep efficiency subsequent to the shift (97.6% vs. 93.6%) with the majority of the wakefulness occurring in the 05:00–07:00 interval. Although they sustained this decrement, however, the extent of the sleep disturbance was not comparable to that seen in older subjects studied under baseline conditions with similar timing of their temperature nadir, thus arguing for other, possibly noncircadian factors as additive effects insofar as sleep quality is concerned.

These results reiterate a developing theme throughout much of the research involving circadian rhythms in aging: that homeostatic factors may be far more salient in terms of sleep disturbance in old age, at least as modeled in these 4–6 hr phase advances. It remains to be seen whether still larger phase advances (e.g., 10 hr) or a phase delay of any duration would produce a similar pattern of results. [In a follow-up report to the 6-hr phase advance protocol, Monk et al. (117) reported that, in one older woman also placed in a later 6-hr phase delay protocol, sleep efficiency and temperature rhythm amplitude appeared less affected than in the 6-hr phase advance. Phase position of the temperature rhythm, however, was more severely disrupted in the phase delay condition.] Finally the possibility exists that the apparent ease of readjustment of the circadian system subsequent to this phase shift represents masking effects of other influences on these rhythms (e.g., meals, posture, or ambient "daytime" lighting) rather than a shift in the underlying core oscillator (cf. 118).

XII. Circadian Rhythms in Dementia

Although the study of the circadian timing system in human disease is of heuristic interest, there are numerous considerations that provide a compelling impetus to apply chronobiological knowledge to such conditions. Lesions of the third ventricle, the hypothalamic region, and related structures have long been known to be associated with aberrations of the sleep-wake and/or body temperature cycles in various case reports (119–122). Relative to the widespread nature of neurological disease in geriatrics, however, such conditions are relatively rare. Neurodegenerative diseases of the senium (AD, PD) and cerebrovascular disease affect huge segments of the elderly population. Perhaps as many as 30–50% of the older population suffer from at least a mild form of one of these conditions (4–6). Given this high prevalence and the fact that sleep disturbance has been shown repeatedly to precipitate institutionalization (123–125) in dementia, scientific knowledge about what mechanisms control such sleep disruption are of humanistic and economic benefit. The current annual cost of nursing home care in the United States now totals 43.1 *billion* dollars and includes a census of 1.6 million people (126). The savings of forestalling half of the yearly new nursing home admissions for even 6 months can be estimated at $6.2 million.

Scientific study of chronobiology in demented patients is by necessity difficult and limited by numerous constraints. Willingness of patients and families to participate in research probably results in a markedly biased sample with which to work. Protocols other than those performed under entrained conditions are virtually nonexistent. Limits as to the nature of patient cooperation and comprehension make usage of even mildly invasive procedures (rectal body temperature monitoring, polysomnography, indwelling venous catheterization) difficult, if not impossible, to employ. Much research in this area has therefore been behavioral, although several physiological studies have been performed. In this section we will review selected studies in this area and some recent work involving treatment.

Behaviorally, one of the major problems encountered in patients with dementia is a syndrome of agitation, seemingly temporally specific, which we and other have labeled "sundowning" (127). While there is no consensual agreement as to what constitutes this condition, we prefer the notion of agitation to describe what occurs to these patients. In short, demented patients are more likely to: remain awake and not sleep as assessed both behaviorally (128,129) and actigraphically (130,131), wander (132), vocalize (133), engage in repetitive physical behavior (134), and be physically aggressive (135) in the late-afternoon and early-evening hours, perhaps roughly equivalent to what Lavie and others have termed the "forbidden" zone (94) (see Fig. 5). The question as to whether these behaviors represent behavior manifestations of a chronobiological process remains unanswered. There is minimal evidence that measured intensity of light per se has anything to do with the increased actigraphic activity seen in this period (136). Some have speculated that the effect may be reflective of a "rebound" insomnia from previous night medication intake (137) whereas others, using PRN medication intake as proxy for sundowning, suggest the phenomenon is not real at

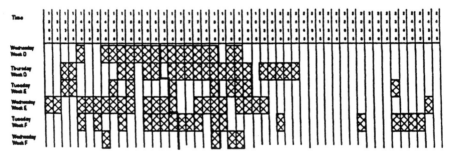

Figure 5   Occurrence of agitation (cross-hatched boxes) as a function of time of day suggestive of sundowning in a demented nursing home patient; each horizontal line refers to one (nonconsecutive) day of observation (13:00–01:00); each column refers to observations made at that time; time of sunset over period of observation is indicated by heavy dark line. Note clustering of agitation in this patient near the time of sunset (late afternoon/early evening). (From Ref. 18.)

all but merely a function of harried nursing staff loading patients on medications in the late afternoon prior to shift change (138). Perhaps another, more reasonable hypothesis is that such late-afternoon agitation might represent cumulative day-time fatigue and occurs only at the end of a prolonged period of wakefulness, thus reflecting an elevated homeostatic drive for sleep in the presence of compromised higher-order cortical control and disinhibition.

There have been a number of attempts to examine the circadian physiology of dementia patients under entrained conditions, which, in most cases, represent nursing homes or other institutional environments. Some studies report a phenomenon perhaps akin to desynchronization with different physiological parameters showing different rhythm periodicity (139,140). While not true internal desynchronization as defined by Wever's (63) studies in time-free environments, these data at least corroborate a general notion of poor coordination of time keeping in such patients. Others (141) have noted what appear to be "disorganized" temperature rhythms in demented patients as well (see Fig. 6). While these observations might well reflect masking effects rather than altered endogenous rhythmicity, it should be stressed that the typical nursing home environment is hardly an "enriched" one insofar as zeitgebers are concerned. Most patients spend many hours a day in bed rest without any physical activity and often eat in bed. Lighting during

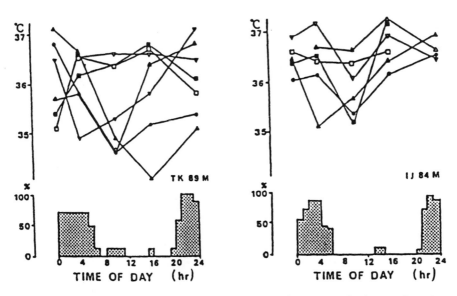

Figure 6 "Disorganized" oral temperature rhythms in two grossly demented nursing home patients; (top) 7 days of temperature data based on intermittent sampling four times a day; (bottom) percentage of observations in which sleep was observed. Note large daily variation in temperature cycles. (From Ref. 141.)

the day is exceptionally poor with minimal exposures above 300 lux (18,142), and typically, ambient light is reduced only intermittently at night (143). Optic nerve deterioration has been described in dementia as well (144). The net effect of these factors may be to minimize rather than maximize masking effects of illumination. Finally, sleep itself is widely distributed around the 24-hr day (128,131) thus probably reducing the impact of sleep-wake state as a masking influence. These factors do not make the typical nursing home environment approach the experimental rigor entailed in constant routine protocols, for example, but they do suggest that masking effects are probably less marked in this population than in other aged, fully ambulatory populations, including those with dementia (see below).

The nursing home population is quite heterogeneous in terms of the range of neurological and medical diseases that are represented. Nonetheless, given the historical studies showing that midbrain lesions displace the timing of sleep, it is reasonable to assume that circadian physiology would be more likely to be disrupted in conditions that are more likely to impact upon hypothalamic function. As reviewed earlier, some data suggest AD may involved degeneration among vasopressinergic cell bodies within the SCN. Alternatively, cerebrovascular white matter disease (often referred to as subcortical vascular dementia) might also be expected to disrupt major outputs from the SCN such as thalamus, internal capsule, and selected midbrain regions (145). On the other hand, PD, a neuro-degenerative disease characterized primarily by dopaminergic neuronal loss within the striatum and substantia nigra, may impact only minimally on hypo-thalamic structures (146). Given this rationale, it is possible to examine selected studies that have evaluated physiological parameters over the 24-hr day in such patient groups. Even under conditions of entrainment one would predict signifi-cantly less disturbance in rhythmicity in PD and, depending on the site of white matter lesions, the maximal extent of involvement in subcortical stroke, with AD perhaps falling somewhere between these two groups.

As mentioned in the Introduction, studies of circadian rhythms in demented populations are often hampered by diagnostic imprecision, so it is often difficult to know the nature and extent of disease in these cases. Nonetheless, a number of studies of institutionalized and noninstitutionalized AD patients have reported reduced amplitude in rhythms of actigraphically monitored rest/activity (147–150), polysomnographically defined sleep/wake (151–153), heart rate (154), blood pressure (155), and melatonin (156,157), but a few studies report no differences from controls in body temperature rhythms in amplitude (158–161) and phase (159,160). Data on phase relationships appear to conflict. One study of body temperature (161) and several actigraphic studies have reported apparent phase delays in AD patients (149,161,162) but another recent study did not (148). Gillin et al. (163) reported phase advances in body temperature in male AD patients but phase delays in female AD patients relative to controls. Additionally, several other studies have shown that caregivers report earlier bedtimes and wake-up times in AD patients relative to healthy elderly persons, suggesting a phase advance (164,165). Cortisol rhythms have variously been reported to be no different in AD relative to controls (166) or possibly present slightly higher amplitudes (higher nocturnal values) relative to controls (167).

Data on rhythms in other dementia subgroups are rare. Three separate studies, however, have reported comparisons between measures of rhythms in AD and subcortical vascular dementia. All reported patients with vascular dementia having greater disruption of sleep-wake or body temperature rhythms (168–170). In at least one of these studies, radiological imaging (computerized tomography, CT) suggested that cases with a greater likelihood of white matter lucencies (highest proportion in internal/outer capsule, thalamus, and paraventricular region) were related to greater 24-hr sleep/wake disruption (168). Few studies are available of rhythms in PD; however, the temperature rhythms for a small number of these patients have been reported to be similar to those of age-matched controls (171) and melatonin rhythms in de novo unmedicated patients did not differ from those of controls (172). Furthermore, PD patients did not appear sleepy on a daytime test of sleepiness even when medications were taken into account (173). In summary, although the data base relevant to disease group comparisons is small, these data, all collected from entrained conditions, are compatible with the notion that key features of the time-keeping mechanisms, including both the SCN and its outputs, may be more or less disrupted in patients with specific kinds of dementing illness.

Of note, only one patient with dementia (AD) has ever been reported in the literature to have been studied in an environment without time cues. In 1984 Wagner (174) reported on an 86-year-old man studied in the Cornell time isolation facility for 5 days. Sleep (stages 1 and 2 almost exclusively) was evenly distributed around the 24-hr day with very little REM observed. Of note were consistent episodes of agitation occurring between 17:00 and 19:00 and between 03:30 and 06:00. The former window appeared to roughly coincide with the period of minimum sleep propensity as noted in the behavioral studies mentioned above.

Traditional pharmacological approaches to the treatment of nocturnal agitation have been reviewed elsewhere (175). From a chronobiological perspective the possible use of melatonin, illumination, or enhancement of external zeitgebers represent intriguing possibilities for treating such conditions. Several recent studies have suggested utility of melatonin as a hypnotic for elderly subjects with disturbed sleep (176,177). The clinical relevance of these data for the treatment of agitation in dementia, however, remains uncertain. These studies have relied upon actigraphic measurements as their primary outcome, but perhaps a more significant limitation is that they do not specify the type or extent, if any, of dementia cases in their populations Singer et al. (178,179) have recently presented preliminary data from ongoing double-blind, crossover, placebo-controlled clinical trials examining the effects of 0.5 mg and 10 mg melatonin in a well-characterized AD population. Results have been mixed to date with only the higher dose suggesting any effect. Given the controversy over whether melatonin truly has hypnotic properties and to what extent these may be independent of its ability to induce phase shifts, use of melatonin to improve sleep/wake disturbance in dementia patients must, at this time, be considered an unproven treatment.

The unequivocal success of bright-light treatment for insomnia in older, nondemented patients with insomnia (180,181) augurs beneficial effects for such treatment in demented patients. Several such studies have been performed (141,182–186), and although differing on timing and duration of light exposure, results have been mildly encouraging. Other groups have reported some success with a structured activity intervention (187,188). Given the limitations and adverse reactions entailed by most traditional pharmacological management for sleep-wake disturbance in dementia patients (175), it is clear that more good clinical trials are needed in this area. Additionally, assessment of outcomes in these studies should employ a chronobiological perspective (i.e., examine the entire 24-hr day), rather than assuming that agitation is invariant across the day.

XIII. Circadian and Homeostatic Models in Aging and Dementia: Clinical Implications for Interacting Mechanisms

In the 25 years since the discovery of the neuroanatomical basis for circadian rhythms in mammals, the mechanisms controlling the endogenous basis for rhythmicity, and indeed the fundamental nature of the clock itself, have been well elucidated. Particularly over the last few years, substantial numbers of studies have examined how the circadian system is affected by age. In this review, we have tried to highlight points of convergence and divergence across these studies. Species differences, accuracy of neuroanatomical findings using appropriate stereological techniques, and unresolved gender differences all loom as major unresolved issues in the description of age-related alterations in the circadian timing system. To some degree, even well-accepted "facts" regarding how rhythms change with age (shortening of endogenous-phase/free-running period; decreased amplitude; decreased tolerance for phase shifting) may need further elaboration and refinement as newer data are collected under different protocols (e.g., forced desynchrony, constant routine). In light of this changing knowledge base, it may be important to evaluate what "themes" are emerging at this juncture and examine their implications for human sleep and circadian rhythm disorders.

Although for many years the homeostatic and circadian control over sleep was assumed to represent two separate, albeit overlapping, processes (9,10), present knowledge dictates that these processes interact substantially (189). This interaction may be particularly relevant for disturbed sleep/wake function in elderly humans. For example, given the widespread prevalence of insomnia in the aged population, the functional significance of such sleep loss on daytime function, in the absence of overt sleep disorders, remains uncertain (11). In this regard, the Pittsburgh group's constant routine data on behavioral measures and unintended sleep episodes clearly indicate that the omnipresent deficits in daytime alertness incurred by older persons are more likely to represent cumulative homeostatic deficits rather than truly circadian ones. A similar dissociation between polysomnographically defined sleep measures and the circadian system is evident in the studies of experimentally induced phase shifts in humans, where the former appears disrupted far more easily than the latter. On the nocturnal side, the forced desynchrony protocols indicate that the tendency for older subjects to awaken at a point closer to their body temperature minimum may be due to deficient homeostatic drive and that this factor may well be the primary cause for early-morning awakenings so common for the older population (11). If the body temperature minimum itself is also phase-advanced, this would represent an additional burden leading to inadequate sleep during the nocturnal hours. Finally, these results have considerable bearing on the typical rules of sleep hygiene, which are often given as a part of behavioral treatment for insomnia in old age (190). In such models, elderly subjects are given the general instructions to arise from bed, go to another room, and engage in some form of activity if they are unable to sleep. Given the known sensitivity of the circadian system to normal indoor levels of illumination (70) and assuming that the human PRC for light (118) does not change with age, this would imply that elderly persons receiving even minimal levels of ordinary room illumination several hours after their body temperature minimum would likely be in the phase advance portion of their PRC. Thus, a complaint of middle-of-night awakening would only be expected to become more severe over time under such commonly endorsed behavioral proscriptions.

This relative uncoupling of the homeostatic and circadian systems is evidenced not only in human studies but in animal studies as well. In this regard, the compelling series of animal studies from the Northwestern group indicating that stimuli that produce or limit motor activity are powerful factors for entrainment may be quite relevant for the human clinical condition. The fact that older animals' rest/activity (and presumably sleep/wake rhythms) appeared less affected by activity feedback than that of younger animals suggests that attempts to improve sleep in older subjects may need to: (1) focus on restoring the coupling present at younger ages and/or (2) rely upon internal zeitgebers that phase-shift in the absence of stimulating activity. Further clinical studies examining appropriate timing of 5HT agonists/reuptake inhibitors may represent a particularly fruitful approach in this regard.

Animal data suggesting interactions between circadian and homeostatic control of sleep may also be relevant for the sleep-wake disturbance in dementia. Edgar et al. (191) reported that, following SCN lesions, squirrel monkeys kept in constant light showed not only abolished circadian rhythms of body temperature but also a decreased capacity for sustained wakefulness. Subsequent to SCN lesions, animals average 4 hr more sleep per (circadian) day and had many more brief bouts of sleep. On the basis of these results, Edgar et al. (191) have argued that the circadian and homeostatic systems may, in fact, be "opponent" processes such that the circadian system effectively gates the homeostatic drive for sleep by ensuring wakefulness at maximally adaptive times of day. In dementia patients, this would lead to a prediction not of inverse day/night sleep quotas (128), but of longer amounts of total sleep (an inability to come to full alertness) associated with extent of dementia and presumed extent of neural degeneration involving the SCN and/or associated pathways. In fact, several papers have presented data showing that extent of 24-hr sleep and daytime napping were positively correlated with severity of dementia (151,152,162,165,168,192,193), though not all data concur (153). The implications of this for so-called "sundowning" and nocturnal agitation could be profound. The data may imply that the best way to help such patients achieve sleep nocturnally might be to provide appropriately timed alerting medications during the day. Although a small literature exists on use of stimulants in very old, presumably demented patients (194), full-fledged clinical trials based on chronobiological principles remain to be performed.

In summary, better understanding of the basic science of chronobiology holds huge implications for the treatment of sleep disturbance in the normal and pathological elderly population. If recent models are any indication, effective treatment will require a thorough description of the many ways the circadian timing system changes with age.

Acknowledgment

This work was supported by Grants AG-10643 and NS-35345 from the National Institutes of Health.

References

1. Richardson GS, Moore-Ede MC, Czeisler CA, Dement, WC. Circadian rhythms of sleep and wakefulness in mice: analysis using long-term automated recording of sleep. Am J Physiol 1985; 248:R320–R330.
2. Wax TM, Goodrick CL. Nearness to death and wheel running behavior in mice. Exp Gerontol 1978; 13:233–236.
3. Hoch CC, Dew MA, Reynolds CF III, Monk TH, Buysse DJ, Houck PR, Machen MA, Kupfer DJ. A longitudinal study of laboratory-and diary-based sleep measures in healthy "old old" and "young old" volunteers. Sleep 1994; 17:489–496.
4. Evans DA, Funkenstein HH, Albert MS, Scherr PA, Cook NR, Chown MJ, Hebert LE, Hennekens CH, Taylor JO. Prevalence of Alzheimer's disease in a community

population of older persons: higher than previously reported. JAMA 1989; 262: 2551–2556.

5. Skoog I, Nilsson L, Palmertz B, Andreasson LA, Svanborg A. A population-based study of dementia in 85-year-olds. N Engl J Med 1993; 328:153–158.

6. Bennett DA, Beckett LA, Murray AM, Shannon KM, Goetz CG, Pilgrim DM, Evans DA. Prevalence of parkinsonian signs and associated mortality in a community population of older people. N Engl J Med 1996; 334:71–76.

7. Crook T, Bartus RT, Ferris SH, Whitehouse P. Age-associated memory impairment: proposed diagnostic criteria and measures of clinical change: report of a National Institute of Mental Health work group. Dev Neuropsychol 1986; 2:261–276.

8. Rowe JW, Kahn RL. Human aging: usual and successful. Science 1987; 237:143–149.

9. Feinberg I. Changes in sleep cycle patterns with age. J Psychiatr Res 1974; 10: 283–306.

10. Borbely AA. A two process model of sleep regulation. Hum Neurobiol 1982; 1:195–204.

11. Bliwise DL. Sleep in normal aging and dementia. Sleep 1993; 16:40–81.

12. Samis HV. Aging: the loss of temporal organization. Perspect Biol Med 1968; 12: 95–102.

13. MacGibbon MF. Ageing as upregulation failure. Med Hypoth 1996; 46:523–527.

14. King DP, Zhao Y, Sangoram AM, Wilsbacher LD, Tanaka M, Antoch MP, Steeves TD, Vitaterna MH, Kornhauser JM, Lowrey PL, Turek FW, Takahashi JS. Positional cloning of the mouse circadian clock gene. Cell 1997; 89:641–653.

15. Finch CE. Longevity, Senescence and the Genome. Chicago: University of Chicago Press, 1990.

16. Albers HE, Gerall AA, Axelson JF. Circadian rhythm dissociation in the rat: effects of long-term constant illumination. Neurosci Lett 1981; 25:89–94.

17. Brock MA. Chronobiology and aging. J Am Geriatr Soc 1991; 39:74–91.

18. Bliwise DL, Carroll JS, Lee KA, Nekich JC, Dement WC. Sleep and sundowning in nursing home patients with dementia. Psychiatry Res 1993; 48:277–292.

19. Bliwise DL, Hughes ML, Carroll JS, Edgar DM. Mortality predicted by timing of temperature nadir in nursing home patients. Sleep Res 1995; 24:510.

20. Roozendaal B, Van Gool WA, Swaab DF, Hoogendijk JE, Mirmiran M. Changes in vasopressin cells of the rat suprachiasmatic nucleus with aging. Brain Res 1987; 409: 259–264.

21. Chee CA, Roozendaal B, Swaab DF, Goudsmit E, Mirmiran M. Vasoactive intestinal polypeptide neuron changes in the senile rat suprachiasmatic nucleus. Neurobiol Aging 1988; 9:307–312.

22. Peng MT, Jiang MJ, Hsu HK. Changes in running-wheel activity, eating and drinking and their day/night distributions throughout the life span of the rat. J Gerontol 1980; 35:339–347.

23. Machado-Salas J, Scheibel ME, Scheibel AB. Morphological changes in the hypothalamus of the old mouse. Exp Neurol 1977; 57:102–111.

24. Madeira MD, Sousa N, Santer RM, Paula-Barbosa MM, Gunderson HJG. Age and sex do not affect the volume, cell numbers, or cell size of the suprachiasmatic nucleus of the rat: an unbiased stereological study. J Comp Neurol 1995; 361:585–601.

25. Swaab DF, Fliers E, Partiman TS. The suprachiasmatic nucleus of the human brain in relation to sex, age and senile dementia. Brain Res 1985; 342:37–44.

26. Swabb DF, Fisser B, Kamphorst W, Troost D. The human suprachiasmatic nucleus; neuropeptide changes in senium and Alzheimer's disease. Bas Appl Histochem 1988; 32:43–54.

27. Hofman MA, Fliers E, Goudsmit E. Swaab DF. Morphometric analysis of the suprachiasmatic and paraventricular nuclei in the human brain: sex differences and age-dependent changes. J Anat 1988; 160:127–143.

28. Hofman MA, Zhou JN, Swaab DF. Suprachiasmatic nucleus of the human brain: an immunocytochemical and morphometric analysis. Anat Rec 1996; 244:552–562.

29. Zhou JN, Hofman MA, Swaab DF. VIP neurons in the human SCN in relation to sex, age, and Alzheimer's disease. Neurobiol Aging 1995b; 16:571–576.

30. Fliers E, Swaab DF, Pool CHRW, Verwer RWH. The vasopressin and oxytocin neurons in the human supraoptic and paraventricular nucleus; changes with aging and in senile dementia. Brain Res 1985; 342:45–53.

31. Goudsmit E, Hofman MA, Fliers E, Swaab DF. The supraoptic and paraventricular nuclei of the human hypothalamus in relation to sex, age and Alzheimer's disease. Neurobiol Aging 1990; 11:529–536.

32. Hofman MA, Swaab DF. Alterations in circadian rhythmicity of the vasopressin-producing neurons of the human suprachiasmatic nucleus (SCN) with aging. Brain Res 1994; 651:134–142.

33. Tate BA, Richardson GS. Sleep and circadian rhythm disruption in an animal model of Alzheimer's disease. Sleep Res 1997; 26:756 (abstract).

34. Satinoff E, Li H, Tcheng TK, Liu C, McArthur AJ, Medanic M, Gillette MU. Do the suprachiasmatic nuclei oscillate in old rats as they do in young ones? Am J Physiol 1993; 265:R1216–R1222.

35. Wise PM, Cohen IR, Weiland NG, London DE. Aging alters the circadian rhythm of glucose utilization in the suprachiasmatic nucleus. Proc Natl Acad Sci USA 1988; 85:5305–5309.

36. Sutin EL, Dement WC, Heller C, Kilduff TS. Light-induced gene expression in the suprachiasmatic nucleus of young and aging rats. Neurobiol Aging 1993; 14:441–446.

37. Kawakami F, Okamura H, Tamada Y, Maebayashi Y, Fukui K, Ibata Y. Loss of day-night differences in VIP mRNA levels in the suprachiasmatic nucleus of aged rats. Neurosci Lett 1997; 222:99–102.

38. Benloucif S, Masana MI, Dubocovich ML. Light-induced phase shifts of circadian activity rhythms and immediate early gene expression in the suprachiasmatic nucleus are attenuated in old C3H/HeN mice. Brain Res 1997; 747:34–42.

39. Zhang Y, Kornhauser JM, Zee PC, Mayo KE, Takahashi JS, Turek FW. Effects of aging on light-induced phase shifting of circadian behavioral rhythms, *fos* expression and creb phosphorylation in the hamster suprachiasmatic nucleus. Neuroscience 1996; 70:951–961.

40. Witting W, Mirmiran M, Bos NPA, Swaab DF. Effect of light intensity on diurnal sleep-wake distribution in young and old rats. Brain Res Bull 1993; 30:157–162.

41. Viswanathan N, Davis FC. Suprachiasmatic nucleus grafts restore circadian function in aged hamsters. Brain Res 1995; 686:10–16.

42. Ralph MR, Foster RG, Davis FC, Menaker M. Transplanted suprachiasmatic nucleus determines circadian period. Science 1990; 247:975–978.

43. Cai A, Lehman MN, Lloyd JM, Wise PM. Transplantation of fetal suprachiasmatic nuclei into middle-aged rats restores diurnal Fos expression in host. Am J Physiol 1997; 272:R422–R428.

44. Van Reeth O, Zhang Y, Zee PC, Turek FW. Grafting fetal suprachiasmatic nuclei in the hypothalamus of old hamsters restores responsiveness of the circadian clock to a phase shifting stimulus. Brain Res 1994; 643:338–342.

45. Monk TH, Reynolds CF III, Buysse DJ, Hoch CC, Jarrett DB, Jennings JR, Kupfer DJ. Circadian characteristics of healthy 80-year-olds and their relationship to objectively recorded sleep. J Gerontol Med Sci 1991; 46:M171–M175.

46. Tune GS. Sleep and wakefulness in 509 normal human adults. Br J Med Psychol 1969; 42:75–80.

47. Tune GS. The influence of age and temperament on the adult human sleep-wakefulness pattern. Br J Psychol 1969; 60:431–441.

48. Drennan MD, Klauber MR, Kripke DF, Goyette LM. The effects of depression and age on the Horne-Ostberg morningness-eveningness score. J Affect Disord 1991; 23: 93–98.

49. Van Coevorden A, Mockel J, Laurent E, Kerkhofs M, L'Hermite-Baleriaux M, Decoster C, Neve P, Van Cauter E. Neuroendocrine rhythms and sleep in aging men. Endocrinol Metab 1991; 23:E651–E661.

50. Richardson GS. Circadian rhythms and aging. In: Schneider EL, Rowe JW, eds. Handbook of the Biology of Aging, 3rd ed. San Diego: Academic Press, 1990:275–305.

51. Zee PC, Rosenberg RS, Turek FW. Effects of aging on entrainment and rate of resynchronization of circadian locomotor activity. Am J Physiol 1992; 263:R1099–R1103.

52. Myers BL, Badia P. Changes in circadian rhythms and sleep quality with aging: mechanisms and interventions. Neurosci Biobehav Rev 1995; 19:553–571.

53. Turek FW, Penev P, Zhang Y, Van Reeth O, Takahashi JS, Zee PC. Alterations in the circadian system in advanced age. In: Circadian Clocks and Their Adjustment; Number 183. Chichester: Wiley, 1995:212–234.

54. Monk TH, Buysse DJ, Reynolds CF III, Kupfer DJ, Houck PR. Circadian temperature rhythms of older people. Exp Gerontol 1995; 30:455–474.

55. Monk TH, Reynolds CF III, Machen MA, Kupfer DJ. Daily social rhythms in the elderly and their relation to objectively recorded sleep. Sleep 1992; 15:322–329.

56. Minors DS, Rabbitt PMA, Worthington H, Waterhouse JM. Variation in meals and sleep-activity patterns in aged subjects; its relevance to circadian rhythm studies. Chronobiol Int 1989; 6:139–146.

57. Pittendrigh CS, Daan S. Circadian oscillations in rodents; a systematic increase of their frequency with age. Science 1974; 186:548–550.

58. Morin LP. Age-related changes in hamster circadian period, entrainment, and rhythm splitting. J Biol Rhythms 1988; 3:237–248.

59. Mosko SS, Erickson GF, Moore RY. Dampened circadian rhythms in reproductively senescent female rats. Behav Neural Biol 1980; 28:1–14.

60. Van Gool WA, Witting W, Mirmiran M. Age-related changes in circadian sleep-wakefulness rhythms in male rats isolated from time cues. Brain Res 1987; 413: 384–387.

61. Witting W, Mirmiran M, Bos NPA, Swaab DF. The effect of old age on the free-running period of circadian rhythms in rat. Chronobiol Int 1994; 11:103–112.

62. Edgar DM, Martin CE, Dement WC. Activity feedback to the mammalian circadian pacemaker: influence on observed measures of rhythm period length. J Biol Rhythms 1991; 6:185–189.

63. Wever R. The Circadian System of Man: Results of Experiments Under Temporal Isolation. New York: Springer-Verlag, 1979.

64. Weitzman ED, Moline ML, Czeisler CA, Zimmerman JC. Chronobiology of aging: temperature, sleep-wake rhythms and entrainment. Neurobiol Aging 1982; 3: 299–309.

65. Monk TH, Moline ML. Removal of temporal constraints in the middle-aged and elderly: effects on sleep and sleepiness. Sleep 1988; 11:513–520.

66. Monk TH. Circadian rhythm. Clin Geriat Med 1989; 5:331–346.

67. Pollak CP, Wagner DR, Moline M, Monk T. Sleep need decreases with age in adults living in temporal isolation. Sleep Res 1994; 23:161 (abstract).

68. Pollak CP. Regulation of sleep rate and circadian consolidation of sleep and wakefulness in an infant. Sleep 1994; 17:567–575.

69. Czeisler CA. Sleep-wake regulation and circadian rhythms. 11th Annual Meeting, Association of Professional Sleep Societies, San Francisco, CA, June 12, 1997.

70. Boivin DB, Duffy JF, Kronauer RE, Czeisler CA. Dose-response relationships for resetting of human circadian clock by light. Nature 1996; 379:540–542.

71. Mills JN, Minors DS, Waterhouse JM. The effects of sleep upon human circadian rhythms. Chronobiologia 1978; 5:14–27.

72. Czeisler CA, Dumont M, Duffy JF, Steinberg JD, Richardson GS, Brown EN, Sanchez R, Rios CD, Ronda JM. Association of sleep-wake habits in older people with changes in output of circadian pacemaker. Lancet 1992; 340:933–936.

73. Monk TH, Buysse DJ, Reynolds CF III, Jarrett DB, Kupfer DJ. Rhythmic vs homeostatic influences on mood, activation, and performance in young and old men. J Gerontol Psychol Sci 1992; 47:P221–P227.

74. Duffy JF, Dijk DJ, Klerman EB, Czeisler CA. Altered phase relationship between body temperature cycle and habitual awakening in older subjects. Sleep Res 1997; 26:711 (abstract).

75. Hall EF, Duffy JF, Dijk DJ, Czeisler CA. Interval between waketime and circadian phase differs between morning and evening types. Sleep Res 1997; 26:716 (abstract).

76. Buysse DJ, Reynolds CF, Monk TH, Hoch CC, Yeager AL, Kupfer DJ. Quantification of subjective sleep quality in healthy elderly men and women using the Pittsburgh Sleep Quality Index (PSQI). Sleep 1991; 14:331–338.

77. Webb WB. Patterns of sleep in healthy 50-60 year old males and females. Res Communi Psychol Psychiatry Behav 1981; 6:133–140.

78. Metz ME, Bunnell DE. Napping and sleep disturbances in the elderly. Fam Prac Res J 1990; 10:47–56.

79. Middlekoop HAM, Smilde-van den Doel DA, Neven AK, Kamphuisen HAC, Springer CP. Subjective sleep characteristics of 1,485 males and females aged 50–93: effects of sex and age, and factors related to self-evaluated quality of sleep. J Gerontol Med Sci 1996; 51A:M108–M115.

80. Buysse DJ, Browman KE, Monk TH, Reynolds CF III, Fasiczka AL, Kupfer DJ. Napping and 24-hour sleep/wake patterns in healthy elderly and young adults. J Am Geriatr Soc 1992; 40:779–786.
81. Wauquier A, Van Sweden B, Lagaay AM, Kemp B, Kamphuisen HAC. Ambulatory monitoring of sleep-wakefulness patterns in healthy elderly males and females (>88 years): the "senieur" protocol. J Am Geriatr Soc 1992; 40:109–114.
82. Reyner A, Horne JA. Gender- and age-related differences in sleep determined by home-recorded sleep logs and actimetry from 400 adults. Sleep 1995; 18:127–134.
83. Campbell SS, Gillin JC, Kripke DF, Erikson P, Clopton P. Gender differences in the circadian temperature rhythms of healthy elderly subjects: relationships to sleep quality. Sleep 1989; 12:529–536.
84. Moe KE, Prinz PN, Vitiello MV, Marks AL, Larsen LH. Healthy elderly women and men have different entrained circadian temperature rhythms. J Am Geriatr Soc 1991; 39:383–387.
85. Lee KA. Circadian temperature rhythms in relation to menstrual cycle phase. J Biol Rhythms 1988; 3:255–263.
86. Vitiello MV, Smallwood RG, Avery DH, Pascualy RA, Martin DC, Prinz PN. Circadian temperature rhythms in young adult and aged men. Neurobiol Aging 1986; 7: 97–100.
87. Winget CM, DeRoshia CW, Vernikos-Danellis J, Rosenblatt WS, Hetherington NW. Comparison of circadian rhythms in male and female humans. Waking Sleeping 1977; 1:359–363.
88. Cagnacci A, Soldani R, Yen SSC. Hypothermic effect of melatonin and nocturnal core body temperature decline are reduced in aged women. J Appl Physiol 1995; 78: 314–317.
89. Buysse DJ, Monk TH, Reynolds CF III, Mesiano D, Houck PR, Kupfer DJ. Patterns of sleep episodes in young and elderly adults during a 36-hour constant routine. Sleep 1993; 16:632–637.
90. Monk TH, Buysse DJ, Reynolds CF III, Kupfer DJ, Houck PR. Subjective alertness rhythms in elderly people. J Biol Rhythms 1996; 11;268–276.
91. Monk TH, Buysse DJ, Reynolds CF III, Berga SL, Jarrett DB, Begley AE, Kupfer DJ. Circadian rhythms in human performance and mood under constant conditions. J Sleep Res 1997; 6:9–18.
92. Kleitman N. Sleep and Wakefulness. Chicago: University of Chicago Press, 1963.
93. Dijk DJ, Duffy JF, Riel E, Czeisler CA. Altered interaction of circadian and homeostatic aspects of sleep propensity results in awakening at an earlier circadian phase in older people. Sleep Res 1997; 26:710 (abstract).
94. Haimov I, Lavie P. Circadian characteristics of sleep propensity function in healthy elderly: a comparison with young adults. Sleep 1997; 20:294–300.
95. Rosenberg RS. Lifespan changes in the diurnal sleep rhythms of rats. Ph.D. dissertation, University of Chicago, Chicago, IL. 1980.
96. Rosenberg RS, Zee Pc, Turek FW. Phase response curves to light in young and old hamsters. Am J Physiol 1991; 261:R491–R495.
97. Van Reeth O, Turek FW. Stimulated activity mediates phase shifts in the hamster circadian clock induced by dark pulses or benzodiazepines. Nature 1989; 339: 49–51.

98. Van Reeth O, Turek FW. Administering triazolam on a circadian basis entrains the activity rhythm of hamsters. Am J Physiol 1989; 256:R639–R645.

99. Van Reeth O, Hinch D, Tecco JM, Turek FW. The effects of short periods of immobilization on the hamster circadian clock. Brain Res 1991; 541:208–214.

100. Van Reeth O, Zhang Y, Zee PC, Turek FW. Aging alters feedback effects of the activity-rest cycle on the circadian clock. Am J Physiol 1992; 263:R981–R986.

101. Penev PD, Turek FW, Zee PC. Monoamine depletion alters the entrainment and the response to light of the circadian activity rhythm in hamsters. Brain Res 1993; 612:156–164.

102. Penev PD, Zee PC, Turek FW. Monamine depletion blocks triazolam-induced phase advances of the circadian clock in hamsters. Brain Res 1994; 637:255–261.

103. Penev PD, Zee PC, Wallen EP, Turek FW. Aging alters the phase-resetting properties of a serotonin agonist on hamster circadian rhythmicity. Am J Physiol 1995; 268: R293–R298.

104. Penev PD, Turek FW, Wallen EP, Zee PC. Aging alters the serotonergic modulation of light-induced phase advances in golden hamsters. Am J Physiol 1997; 272:R509–513.

105. Foret J, Bensimon G, Benoit O, Vieux N. Quality of sleep as a function of age and shift work. Adv Biosci 1981; 30:149–154.

106. Harma MI, Hakola T, Akerstedt T, Laitinen JT. Age and adjustment to night work. Occup Environ Med 1994; 51:568–573.

107. Gander PH, De Nguyen BE, Rosekind MR, Connell LJ. Age, circadian rhythms, and sleep loss in flight crews. Aviat Space Environ Med 1993; 64:189–195.

108. Suvanto S, Partinen M, Harma M, Ilmarinen J. Flight attendant's desynchronosis after rapid time zone changes. Aviat Space Environ Med 1990; 61:543–547.

109. Matsumoto K, Morita Y. Effects of nighttime nap and age on sleep patterns of shift workers. Sleep 1987; 10:580–589.

110. De Zwart BCH, Bras VM, Van Dormolen M, Frings-Dresen MHW, Meijman FT. After-effects of night work on physical performance capacity and sleep quality in relation to age. Int Arch Occup Environ Health 1993; 65:259–262.

111. Oginska H, Pokorshi J, Oginski A. Gender, ageing, and shiftwork intolerance. Ergonomics 1993; 36:161–168.

112. Monk TH, Buysse DJ, Reynolds CF III, Kupfer DJ. Inducing jet lag in older people. Exp Gerontol 1993; 28:119–133.

113. Carrier J, Monk TH, Buysse DJ, Kupfer DJ. Inducing a 6-hr phase advance in the elderly: effects on sleep and temperature rhythms. J Sleep Res 1996; 5:99–105.

114. Czeisler CA, Weitzman ED, Moore-Ede MC, Zimmerman JC, Knauer RS. Human sleep: its duration and organization depend on its circadian phase. Science 1980; 210: 1264–1267.

115. Tepas DI, Duchon JC, Gersten AH. Shiftwork and the older worker. Exp Aging Res 1993; 19:295–320.

116. Campbell SS, Dawson D. Aging young sleep: a test of the phase advance hypothesis of sleep disturbance in the elderly. J Sleep Res 1992; 1:205–210.

117. Monk TH, Buysse DJ, Reynolds CF III, Kupfer DJ. Inducing jet lag in an older person: directional asymmetry. Exp Gerontol 1995; 30:137–145.

118. Czeisler CA, Kronauer RE, Allan JS, Duffy JF, Jewett ME, Brown EN, Ronda JM. Bright light induction of strong (type O) resetting of the human circadian pacemaker. Science 1989; 244:1328–1333.

119. Torch WC, Hirano A, Solomon S. Anterograde transneuronal degeneration in the limbic system: clinical anatomic correlation. Neurology 1977; 27:1157–1163.

120. Cohen RA, Albers HE. Disruption of human circadian and cognitive regulation following a discrete hypothalamic lesion: a case study. Neurology 1991; 41:726–729.

121. Schwartz WJ, Busis NA, Hedley-Whyte ET. A discrete lesion of ventral hypothalamus and optic chiasm that disturbed the daily temperature rhythm. J Neurol 1986; 233:1–4.

122. Davison C, Demuth EL. Disturbances in sleep mechanism: a clinicopathologic study. Arch Neurol Psychiatry 1946; 55:111–125.

123. Sanford JRA. Tolerance of debility in elderly dependants by supporters at home: its significance for hospital practice. Br Med J 1975; 3:471–473.

124. Pollak CP, Perlick D. Sleep problems and institutionalization of the elderly. J Geriatr Psychiatry Neurol 1991; 4:204–210.

125. Pollak CP, Perlick D, Linser JP, Wenston J, Hseih F. Sleep problems in the community elderly as predictors of death and nursing home placement. J Commun Health 1990; 15:123–135.

126. Report of the National Commission on Sleep Disorders Research. Vol 1, Executive Summary and Executive Report. Wake up America: a national sleep alert. Washington, DC: National Institutes of Health, 1993.

127. Bliwise DL. What is sundowning? J Am Geriatr Soc 1994; 42:1009–1011.

128. Bliwise DL, Bevier WC, Bliwise NG, Edgar DM, Dement WC. Systematic 24-hour behavioral observations of sleep-wakefulness in a skilled care nursing facility. Psychol Aging 1990; 5:16–24.

129. Cohen-Mansfield J, Werner P, Freedman L. Sleep and agitation in agitated nursing home residents: an observational study. Sleep 1995; 18:674–680.

130. Jacobs D, Ancoli-Israel S, Parker L, Kripke DF. Twenty-four-hour sleep-wake patterns in a nursing home population. Psychol Aging 1989; 4:352–356.

131. Ancoli-Israel S, Parker L, Sinaee R, Fell RL, Kripke DF. Sleep fragmentation in patients from a nursing home. J Gerontol Med Sci 1989; 44:M18–M21.

132. Martino-Saltzman D, Blasch BB, Morris RD, McNeal LW. Travel behavior of nursing home residents perceived as wanderers and nonwanderers. Gerontologist 1991; 31:666–672.

133. Burgio LD, Scilley K, Hardin JM, Janosky J, Bonino P, Slater SC, Engberg R. Studying disruptive vocalization and contextual factors in the nursing home using computer-assisted real-time observation. J Gerontol Psychol Sci 1994; 49:P230–P239.

134. O'Leary PA, Haley WE, Paul PB. Behavioral assessment in Alzheimer's disease: use of a 24-hr log. Psychol Aging 1993; 8:139–143.

135. Cohen-Mansfield J, Marx MS, Werner P, Freeman L. Temporal patterns of agitated nursing home residents. Int Psychogeriatr 1992; 4:197–206.

136. Hopkins RW, Rindlisbacher P, Grant NT. An investigation of the sundowning syndrome and ambient light. Am J Alzheimer's Care Relat Disord Res 1992; (March/April):22–27.

137. Little JT, Satlin A, Sunderland T, Volicer L. Sundown syndrome in severely demented patients with probable Alzheimer's disease. J Geriatr Psychiatry Neurol 1995; 8:103–106.

138. Exum ME, Phelps BJ, Nabers KE, Osborne JG. Sundown syndrome: is it reflected in

the use of PRN medications for nursing home residents? Gerontologist 1993; 33: 756–761.

139. Scheving LE, Roig C, Halberg F, Pauly JE, Hand EA. Circadian variations in residents of a "senior citizen's home". In: Scheving LE, Halberg F, Pauly JE, eds. Chronobiology. Tokyo: Igaku Shoin, 1974:353–357.

140. Tominaga M, Tsuchihashi T, Kinoshita H, Abe I, Fujishima M. Disparate circadian variations of blood pressure and body temperature in bedridden elderly patients with cerebral atrophy. Am J Hypertens 1995; 8:773–781.

141. Okawa M, Mishima K, Hishikawa Y, Hozumi S, Hori H, Takahashi K. Circadian rhythm disorders in sleep-waking and body temperature in elderly patients with dementia and their treatment. Sleep 1991; 14:478–485.

142. Ancoli-Israel S, Jones DW, Hanger MA, Parker L, Klauber MR, Kripke DF. Sleep in the nursing home. In: Kuna ST, Suratt PM, Remmers JE, eds. Sleep and Respiration in Aging Adults. New York: Elsevier Science Publishing Company, 1991:77–84.

143. Schnelle JF, Ouslander JG, Simmons SF, Alessi CA, Gravel MD. The nighttime environment, incontinence care, and sleep disruption in nursing homes. J Am Geriatr Soc 1993; 41:910–914.

144. Hinton DR, Sadun AA, Blanks JC, Miller CA. Optic-nerve degeneration in Alzheimer's disease. N Engl J Med 1986; 315:485–487.

145. Meijer JH, Rietveld WJ. Neurophysiology of the suprachiasmatic circadian pacemaker in rodents. Physiol Rev 1989; 69:671–707.

146. Matzuk MM, Saper CB. Preservation of hypothalamic dopaminergic neurons in Parkinson's disease. Ann Neurol 1985; 18:552–555.

147. Witting W, Kwa IH, Eikelenboom P, Mirmiran M, Swaab DF. Alterations in the circadian rest-activity rhythm in aging and Alzheimer's disease. Biol Psychiatry 1990; 27:563–572.

148. Pollak CP, Stokes PE. Circadian rest-activity rhythms in demented and nondemented older community residents and their caregivers. J Am Geriatr Soc 1997; 45:446–452.

149. Satlin A, Teicher MN, Lieberman HR, Baldessarini RJ, Volicer L, Rheaume Y. Circadian locomotion activity rhythms in Alzheimer's disease. Neuropsychopharmacology 1991; 5:115–126.

150. Hopkins RW, Rindlisbacher P. Fragmentation of activity periods in Alzheimer's disease. Int J Geriatr Psychiatry 1992; 7:805–812.

151. Prinz PN, Peskind ER, Vitaliano PP, Raskind MA, Eisdorfer C, Zemcuznikov N, Gerber CJ. Changes in the sleep and waking EEGs of nondemented and demented elderly subjects. J Am Geriatr Soc 1982; 30:86–93.

152. Prinz PN, Vitaliano PP, Vitiello MV, Bokan J, Raskind M, Peskind E, Gerber C. Sleep, EEG and mental function changes in senile dementia of the Alzheimer's type. Neurobiol Aging 1982; 3:361–370.

153. Allen SR, Seiler WO, Stahelin HB, Spiegel R. Seventy-two hour polygraphic and behavioral recordings of wakefulness and sleep in a hospital geriatric unit: comparison between demented and nondemented patients. Sleep 1987; 10:143–159.

154. Reynolds V, Marriott FHC, Waterhouse J, Shier P, Grant C. Heart rate variation, age, and behavior in subjects with senile dementia of Alzheimer type. Chronobiol Int 1995; 12:37–45.

155. Otsuka A, Mikami H, Katahira K, Nakamoto Y, Minamitani K, Imaoka M, Nishide

M, Ogihara T. Absence of nocturnal fall in blood pressure in elderly persons with Alzheimer-type dementia. J Am Geriatr Soc 1990; 38:973–978.

156. Uchida K, Okamoto N, Ohara K, Morita Y. Daily rhythm of serum melatonin in patients with dementia of the degenerate type. Brain Res 1996; 717:154–159.

157. Dori D, Casale G, Solerte SB, Fioravanti M, Migliorati G, Cuzzoni G, Ferrari E. Chrono-neuroendocrinological aspects of physiological aging and senile dementia. Chronobiologia 1994; 21:121–126.

158. Touitou Y, Reinberg A, Bodgan A, Auzeby A, Beck H, Touitou C. Age-related changes in both circadian and seasonal rhythms of rectal temperature with special reference to senile dementia of Alzheimer type. Gerontology 1986; 32:110–118.

159. Prinz PN, Christie C, Smallwood R, Vitaliano P, Bokan J, Vitiello MV, Martin D. Circadian temperature variation in healthy aged and in Alzheimer's disease. J Gerontol 1984; 39:30–35.

160. Prinz PN, Moe KE, Vitiello MV, Marks AL, Larsen LH. Entrained body temperature rhythms are similar in mild Alzheimer's disease, geriatric onset depression, and normal aging. J Geriatr Psychiatry Neurol 1992; 5:65–71.

161. Satlin A, Volicer L, Stopa EG, Harper D. Circadian locomotor activity and core-body temperature rhythms in Alzheimer's disease. Neurobiol Aging 1995; 16: 765–771.

162. Ancoli-Israel S, Klauber MR, Jones DW, Kripke DF, Martin J, Mason W, Pat-Horenczyk R, Fell R. Variations in circadian rhythms of activity, sleep, and light exposure related to dementia in nursing-home patients. Sleep 1997; 20:18–23.

163. Gillin JC, Kripke DF, Campbell SS. Ambulatory measures of activity, light and temperature in elderly normal controls and patients with Alzheimer disease. Bull Clin Neurosci 1989; 54:144–148.

164. Bliwise DL, Tinklenberg JR, Yesavage JA. Timing of sleep and wakefulness in Alzheimer's disease patients residing at home. Biol Psychiatry 1992; 31:1163–1165.

165. Ancoli-Israel S, Klauber MR, Gillin JC, Campbell SS, Hofstetter CR. Sleep in non-institutionalized Alzheimer's disease patients. Aging Clin Exp Res 1994; 6:451–458.

166. Christie JE, Whalley LJ, Dick H, Fink G. Plasma cortisol concentrations in the functional psychoses and Alzheimer type dementia: a neuroendocrine day approach in drug-free patients. J Steroid Biochem 1983; 19:247–250.

167. Davis KL, Davis BM, Greenwald BS, Mohs RC, Mathe AA, Johns CA, Horvath TB. Cortisol and Alzheimer's disease. I. Basal studies. Am J Psychiatry 1986; 143: 300–305.

168. Meguro K, Udea M, Kobayashi I, Yamaguchi S, Yamazaki H, Oikawa Y, Kikuchi Y, Sasaki H. Sleep disturbance in elderly patients with cognitive impairment, decreased daily activity and paraventricular white matter lesions. Sleep 1995; 18:109–114.

169. Mishima K, Okawa M, Satoh K, Shimizu T, Hozumi S, Hishikawa Y. Different manifestations of circadian rhythms in senile dementia of Alzheimer's type and multi-infarct dementia. Neurobiol Aging 1997; 18:105–109.

170. Aharon-Peretz J, Masiah A, Pillar T, Epstein R, Tzischinsky O, Lavie P. Sleep-wake cycles in multi-infarct dementia and dementia of the Alzheimer's type. Neurology 1991; 41:1616–1619.

171. Dowling GA. A comparative study of sleep and temperature rhythm in older women with and without Parkinson's disease. Sleep Res 1996; 25:409 (abstract).

172. Fertl E, Auff E, Doppelbauer A, Walkhauser F. Circadian secretion pattern of melatonin in de novo parkinsonian patients: evidence for phase-shifting properties of L-dopa. J Neural Transm (P-D Sect) 1993; 5:227–234.

173. Dihenia BH, Rye DB, Bliwise DL. Daytime alertness in Parkinson's disease. Sleep Res 1997; 26:550 (abstract).

174. Wagner DR. Sleep. Generations 1984; (Winter):31–37.

175. McGaffigan S, Bliwise DL. The treatment of sundowning: a selective review of pharmacological and nonpharmacological studies. Drugs Aging 1997; 10:10–17.

176. Haimov I, Lavie P, Laudon M, Herer P, Vigder C, Zisapel N. Melatonin replacement therapy of elderly insomniacs. Sleep 1995; 18:598–603.

177. Garfinkel D, Laudon M, Nof D, Zisapel N. Improvement of sleep quality in elderly people by controlled-release melatonin. Lancet 1995; 346:541–544.

178. Singer C, McArthur A, Hughes R, Sack R, Kaye J, Lewy A. High dose melatonin administration and sleep in the elderly. Sleep Res 1995; 24A:151 (abstract).

179. Singer CM, Moffit MT, Colling ED, Hughes RJ, Cutler NL, Sack RL, Lewy AJ. Low dose melatonin administration and nocturnal activity levels in patients with Alzheimer's disease. Sleep Res 1997; 26:752 (abstract).

180. Campbell SS, Dawson D, Anderson MW. Alleviation of sleep maintenance insomnia with timed exposure to bright light. J Am Geriatr Soc 1993; 41:829–836.

181. Murphy PJ, Campbell SS. Enhanced performance in elderly subjects following bright light treatment of sleep maintenance insomnia. J Sleep Res 1996; 5:165–172.

182. Meltzer-Brody S, Mouton A, Ge YR, Sanchez R, Zee PC. Effects of scheduled bright light exposure on subjective measurements of vigor in residents of an assisted living facility. Sleep Res 1994; 23:504 (abstract).

183. Satlin A, Volicer L, Ross V, Herz L, Campbell S. Bright light treatment of behavioral and sleep disturbances in patients with Alzheimer's disease. Am J Psychiatry 1992; 149:1028–1032.

184. Caster D, Woods D, Pigott L, Hemmes R. Effect of sunlight on sleep patterns of the elderly. J Am Acad Phys Assist 1991; 4:321–326.

185. Lovell BB, Ancoli-Israel S, Gevirtz R. Effect of bright light treatment on agitated behavior in institutionalized elderly subjects. Psychiatry Res 1995; 57:7–12.

186. Mishima K, Okawa M, Hishikawa Y, Hozumi S, Hori H, Takahashi K. Morning bright light therapy for sleep and behavior disorders in elderly patients with dementia. Acta Psychiatr Scand 1994; 89:1–7.

187. Naylor E, Janssen I, Penev PD, Orbeta L, Colecchia E, Finkel S, Zee PC. Structured physical and social activity improves sleep during the first half of the night in the elderly. Sleep Res 1997; 26:743 (abstract).

188. Penev P, Colecchia E, Finkel S, Janssen I, Keng M, Mouton A, Naylor E, Orbeta L, Ortiz R, Zee P. Effects of structured activity on performance and sleep in elderly residents of assisted living facilities: preliminary results (in press).

189. Dijk DJ, Czeisler CA. Paradoxical timing of the circadian rhythm of sleep propensity serves to consolidate sleep and wakefulness in humans. Neurosci Lett 1994; 166: 63–68.

190. Morin CM. Insomnia: Psychological Assessment and Management. New York: Guilford Press, 1993.

191. Edgar DM, Dement WC, Fuller CA. Effect of SCN lesions on sleep in squirrel monkeys: evidence for opponent processes in sleep-wake regulation. J Neurosci 1993; 13:1065–1079.

192. Bliwise DL, Carroll JS, Dement WC. Predictors of observed sleep/wakefulness in residents in long term care. Gerontol Med Sci 1990; 45:M126–M130.

193. Meguro K, Ueda M, Yamaguchi T, Sekita Y, Yamazaki H, Oikawa Y, Kikuchi Y, Matsuzawa T. Disturbance in daily sleep/wake patterns in patients with cognitive impairment and decreased daily activity. J Am Geriatr Soc 1990; 38:1176–1182.

194. Gurian B, Rosowsky E. Low-dose methylphenidate in the very old. J Geriatr Psychiatry Neurol 1990; 3:152–154.

17

Effects of Sleep and Circadian Rhythms on Performance

JULIE CARRIER* and **TIMOTHY H. MONK**

University of Pittsburgh School of Medicine
Pittsburgh, Pennsylvania

> Compared to the physical scientist, at least at levels above the atom, researchers concerned with human performance measurement should get medals of bravery. Or at least, a great deal of sympathy.
>
> —Webb (1)

I. Difficulties in Measuring Performance over Time: Methodological Issues

While the study of any circadian rhythm poses important methodological issues, this is especially true for performance rhythms. The major way in which performance efficiency differs from most physiological measures is that there is usually a large practice or "learning curve" effect. This means that performance efficiency will be better on a given trial than on the trials preceding it. The practice effect is extremely difficult to completely eliminate. To illustrate this problem, Monk (2) presented the data from a subject studied by Elliot Weitzman and Janet Zimmerman for 6 months. This subject had to sort a pack of 96 playing cards into

---

Current affiliation: University of Montreal, Montreal, Quebec, Canada.

the four suits about six times a day, 7 days/week. The subject reached his best averaged speed in about 18 weeks (after more than 750 trials). Thus even if hundreds of trials are given before an actual experiment, the researcher still has to deal with the practice effect. Failure to take practice effects into account can result in spurious conclusions regarding the trend in performance observed over the day. Performance efficiency can be worse in the morning and better in the evening, simply because morning precedes evening. One technique used to avoid this problem is to use a parallel-group design, that is, to have a separate group of unpracticed subjects perform the task at each time of day. The disadvantage of this technique is that it requires many subjects and it assumes that different subjects show similar time-of-day trends. A second way to avoid the problem is to use a within-subject counterbalanced design, that is, to have different groups of subjects experiencing the time-of-day conditions in different order. Although this design requires fewer subjects, it does assume that different subjects will show the same practice trend. In addition, a within-subject design requires a large number of different and equivalent versions of the given performance test (for example, different versions of list of words for a memory test). If a particular version of a task is repeated too soon after its initial presentation, the response obtained may reflect the subjects' memory for their previous responses rather than their true ability on the task per se. This problem is particularly serious in the case of more cognitive measures of performance, and indeed, to date, has ruled out the long-term monitoring of some types of performance. In addition, potential differences between various versions of the task have to be counterbalanced over the different times of day.

Another methodological difficulty encountered in measuring performance rhythms stems from the inherent inaccuracy of performance testing. Performance measures include errors of measurement and show reliabilities that range from only 0.6 to 0.8 at best (1). These errors of measurement can be attributable to many different chance factors. Thus, from one moment to the next, there may be differences in the motivation of the subject, differences in the subject's perception of the task, and changes in the level of distraction, mood, and attentiveness shown by the subject. To obtain reliable performance measures, each trial should thus contain a large number of measured responses. In addition, each time of day should be associated with a large number of trials coming from different subjects or from different days or from both.

From the above, it is clear that performance measurement is almost always intrusive, since subjects have to interrupt their ongoing activity for performance to be measured. Sampling can seldom be more frequent than once every 2 or 3 hr. It also means that sampling has to be reduced or eliminated during sleep hours, if one wants to study circadian fluctuations in performance without contamination by sleep deprivation. As will be discussed below, such sampling limitations have affected earlier conclusions about different time-of-day effects between performance tasks.

II. Circadian Rhythms of Performance

Until the mid-1980s most of the research into circadian fluctuations of performance examined performance over the normal working day (09:00–18:00). No attempt was made to distinguish variations in performance due to endogenous circadian factors from those linked to the amount of time since awake. In contrast, contemporary models of subjective alertness (3–7) and performance efficiency (5–7) view these variables as being determined both by a homeostatic process (amount of hours since awake) and by an input from the circadian timing system (CTS). Thus the time-of-day fluctuations observed in performance are thought to be generated by the interaction of these two processes. For example, performance efficiency on a specific task may decrease over the day because the amount of hours since awake is increasing (homeostatic drive) or because the input from the CTS is producing a less optimal "state" to perform the task, or because of both of these influences. In the same manner, performance efficiency may be stable over the day because the input from the CTS is exactly counterbalancing the effects of increasing hours awake. Thus circadian fluctuations in performance are the consequence of both the influence of the homeostatic process (time since awake) and the drive from the CTS. To dissect the individual effect of these two factors on performance efficiency is not easy. Some experimental and mathematical approaches have been proposed (e.g., forced desynchrony, mathematical removal of data trends), with each of these having underlying assumptions and limitations (see Chapter 19 for a discussion). Unless a study adopts a specific approach to separate rhythmic and homeostatic factors, it is not possible to know how they are interacting to influence the observed fluctuation in performance efficiency.

A. Early Experiments: The Arousal Theory as an Explanation for the Parallelism Between Body Temperature and Time-of-Day Effects in Performance

As reviewed by Lavie (8), the search for cycles in mental performance is not a novel interest derived from the recent development of chronobiology as an accepted field. The study of performance rhythms began in the early days of experimental and educational psychology, well before the terms "circadian" and "chronobiology" had ever been invented. This work was mainly concerned with determining the optimal time of day for the teaching of academic subjects [e.g., Gates (9), Muscio (10), Laird (11)].

It is generally accepted that Professor Nathaniel Kleitman was the investigator who made the link between the early studies and current research on the circadian fluctuation of human behavior (8,12). Kleitman (13) showed strong evidence for a parallelism between circadian rhythm in body temperature and time-of-day effects in performance for simple repetitive tasks involving muscular activity that had a small cognitive load (card sorting, mirror drawing, copying, code substitution, etc.). As with the temperature rhythm, the results showed a well-marked diurnal rhythm of performance with a maximum at midday and minima early in the morning and late at night. Kleitman and Jackson (14) went as far as to assert that fluctuations in performance could be inferred from variation in oral temperature, thus avoiding the use of "time consuming performance tests which, in themselves, interfere with, or disrupt, the scheduled activities of the persons studied." Subsequently, the work of Colquhoun (15), again concerned mainly with simple repetitive tasks, also stressed a parallelism between temperature and performance circadian fluctuations. However, whereas Kleitman's claim was based on maximum temperature and performance at about midday, the temperature rhythms measured by Colquhoun showed a sharp rise from early to late morning followed by a more gradual rise to a peak at about 9 P.M. (15). Thus, whereas Kleitman's joint peak of temperature and performance was at midday, that of Colquhoun was in the evening. This difference illustrates how cautious one should be in making generalizations about performance rhythms without first checking to see whether the reference rhythm (e.g., temperature) is the same for all studies and groups to be compared.

Colquhoun (15) studied vigilance tasks (detection of an infrequent signal), simple addition tasks (adding up six two-digit numbers), and other simple reaction time tasks. Unlike Kleitman, Colquhoun did not infer a causal relationship between the performance and body temperature. Instead, he viewed the diurnal fluctuation in performance as being mediated by a circadian rhythm in "basal arousal" (or the inverse of sleepiness). Except for a postlunch dip (see below) the arousal rhythm was considered to parallel body temperature, showing a sharp rise from early to late morning followed by a more gradual rise to a peak in the early evening (15). The arousal theory explained the increase in performance over the day by postulating that arousal is suboptimal in the morning, gradually approaching an optimal level as the day goes by.

Also during the 1970s, a research group from Germany (16) was one of the few investigative teams who were not restricting their performance evaluations to daytime hours. In their experiments, subjects were awakened a few times during the night to allow 24-hr coverage of the circadian fluctuation in performance. The results showed a good parallelism between body temperature and performance on a variety of tasks (psychomotor performance, cancellation, digit summation, and flight simulator). A good example of the parallelism held to exist between the diurnal rhythms of simple repetitive tasks and temperature is provided by a review of temporal effects in visual search (see Fig. 1).

Early in the literature, exceptions from the parallelism between body temperature and performance emerged in two areas: the postlunch dip (which will be discussed later) and performance on tasks involving short-term memory processes. As an example of the latter, Laird (11) found a decline over the waking day in short-term memory for prose. This result was later replicated by Folkard and Monk (17). Similarly, Blake (18) found a morning peak in digit span. Later

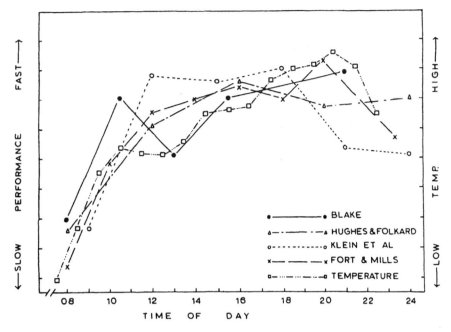

Figure 1 Time-of-day function for a series of studies involving speed scores from visual search tasks, plotted with the time-of-day function in oral temperature. (From Ref. 108.)

research by Baddeley et al. (19) and Hockey et al. (20) confirmed that for similar, relatively pure, tasks of immediate memory, performance was better in the morning than in the evening. By invoking the arousal model, these divergent results could still be explained by postulating that short-term memory tasks had a lower optimal arousal than the optimal arousal of simple repetitive tasks. The arousal model thus postulated that the arousal rhythm mediates changes in performance level through an "inverted U" relationship, whereby increases in arousal level are associated with improvements in performance up to a certain level, after which performance starts to decline (21,22). Thus, above a certain level, arousal would be superoptimal for short-term memory tasks and performance would then start to deteriorate, rather than improve, as arousal rose further. A good example of superoptimal arousal is the difficulty encountered in unlocking one's car door in an emergency situation. The superoptimal arousal interpretation is supported by studies showing that manipulations that produce high arousal (like noise) may impair short-term memory (23). A recent study on prospective memory in the elderly (simulated medication and appointment adherence) has also shown prospective memory to be better in the morning than at midday (24). No further decline was observed from midday to the evening. The authors explained their results by suggesting that a greater attentional capacity for remembering prospective events is available in the morning, since morning hours are typically less active, with people receiving less environmental stimulation.

Although immediate memory performance efficiency and prospective memory are better in the morning, studies comparing delayed retention of material presented at two different times of day have shown delayed retention to be better following afternoon or evening presentation (17,19,20,25). Since high arousal has been shown to benefit delayed retention, these results again support the arousal model. A study comparing delayed retention of material presented at six different times of day also revealed a reliable trend over the day that was clearly different from that seen in immediate retention (12). Delayed retention was particularly good following presentation at 08:00 with a second peak at 14:00–17:00. The authors suggested that the good performance following presentation at 08:00 reflected an effect of sleep per se, rather than an endogenous circadian effect (sleep allowing less potential for a buildup of proactive interference).

B. Differential Time-of-Day Variation for Different Tasks: Observations Under Normal Nycthemeral Condition

Throughout the 1960s and early 1970s, the notion was of a single performance rhythm usually parallel to the temperature rhythm, except in short-term memory tasks, whose rhythms were explained in terms of superoptimal arousal. Subsequent studies would demonstrate, however, that under nycthemeral conditions there was not one single performance rhythm, but many. This heralded a new approach to the study of circadian rhythm in performance, with more emphasis on the differences between performance rhythms than on the similarities. The new approach is epitomized by Folkard's (26) quote: "Perhaps the main conclusion to be drawn from studies on the effects of time of day on performance is that the best time to perform a particular task depends on the nature of that task." Moreover, for the first time, efforts were made to try to understand the *mechanisms* underlying circadian performance rhythms. As we shall discuss below, studies showed that both task demands, and changes of strategy adopted by the subject, could be important variables that might explain discrepancies in the time-of-day effects observed between different performance tasks.

C. Heterogeneity Between Tasks: Memory Load and Change of Strategy over the Day—Some Tentative Explanations

Working memory is involved in a wide range of tasks including the ability to understand speech or text, and to perform mental arithmetic operations. These tasks involve short-term storage and the processing of information. In general, performance on these tasks shows a maximum at about midday (11,27). This is later than the peak showed for immediate memory, but earlier than for the tasks involving simple processing, which show a parallelism with body temperature. In addition, other studies have shown different time-of-day effects depending on the requirements of the task or the particular subject population to be tested. For example, one study showed an early-morning peak of mental arithmetic performance in children (28) while another study found an evening peak for this type of performance in highly practiced young adults (18). Folkard et al. (29) have shown that the trend of performance for this type of task seems to depend on the precise size of the working memory load. These authors used a serial visual search task in which the working memory load (number of target letters to be remembered) could vary systematically. With a low working memory load, performance was positively correlated with the circadian rhythm of body temperature. However, as the memory load was increased, the relationship between performance and body temperature broke down and eventually was reversed, with peak performance occurring at the trough of temperature in a high working memory load version. These results suggest that, for a given individual, manipulations of the memory load involved in the performance of a task will affect the timing of the trend over the day. This also implies that the "effective" memory load in the performance of a given task will be influenced by a number of factors including age, intelligence, and level of practice, which may themselves give rise to different time-of-day fluctuations.

Several studies have demonstrated that a change of strategy over the day might also be an important factor when we try to understand the heterogeneity between tasks in the time-of-day literature. In a series of three experiments, Folkard (30) showed that subjects who perform a short memory task may spontaneously put more reliance on maintenance processing in the morning and elaboration processing in the evening. These experiments were based on the observation that acoustic similarity (mad, man, map) has a detrimental effect on short-term memory, and semantic similarity (large, big, huge) exerts a detrimental effect on long-term learning (31,32). Experiment 1 found the acoustic similarity effect on short-term memory to be greater at 10 A.M. than at 7 P.M. Experiment 2 showed the semantic similarity effect on long-term memory to be greater at 7:30 P.M. than at 10:30 A.M. Experiment 3 demonstrated that interposing a short-term memory task between the presentation of a list of words and its subsequent recall had a greater detrimental effect on list learning at 10:30 A.M. than at 7:30 P.M. Taken together, these results support the idea that subjects engage in different information-processing strategies at different times of day. Subjects appear to rely more on maintenance processing based on the physical characteristics of the items in the morning, but more elaboration processing based on the items' meaning in the evening.

A second example, showing how strategy can modulate the time-of-day fluctuations, comes from the field of visual search. All the tasks presented in Figure 1 are serial search tasks in which a fixed order of scanning through the material is prescribed. If no such ordering is imposed, the task is described as "free search." Monk (33) measured the time-of-day function separately for "inner" (close to the center of the display) and "outer" (close to the edge of the display) targets in a free search task. The results for inner and outer targets were in mirror image: outer targets were faster to detect in the morning and evening and slower at midday, while inner targets were faster to detect at midday and slower in the morning and evening. When inner and outer search times were averaged out, there was no evidence of an overall increase or decrease in performance over the day. These results suggested a resource reallocation mechanism, by which gains in performance for inner targets at the middle of the day were won at the expense of corresponding losses for outer targets.

Folkard (30,34) suggested that diurnal changes in performance may also reflect a morning-to-evening decrease in the degree of left-hemisphere dominance. According to this hypothesis, a lessening in left-hemisphere dominance could account for the improvement over the day typically found in the performance of many perceptual-motor tasks, and for the decrease found in short-term memory for verbal items. This hypothesis could also explain why the morning accuracy reported for verbal reasoning disappeared with an acoustically confusable version of the same task (12) since there is evidence that the use of an articulatory loop for subvocal rehearsal is exclusive to the left hemisphere (35) and that the right hemisphere directly accesses words semantically without intermediate translation into phonological code (36). Only one study has directly tested Folkard's hypothesis. Corbera et al. (37) studied 48 right-handed women who performed verbal and spatial hemifield tachistoscopic tasks at four different times a day. Supporting Folkard's hypothesis, changes in accuracy over the day showed a left-hemisphere advantage at 12:00 noon and a right-hemisphere advantage at 7:45 P.M. However, these changes only occurred when hemispheres received stimuli in the processing of which they were *not* specialized, that is, when verbal stimuli appeared at the left visual field and spatial stimuli appeared at the right visual field. Thus the explanation is not that simple.

Tasks that involve gross motor involvement also seem to constitute a further categorization that does not always show consistency across studies. Some authors, e.g., Blake (18), have found this type of performance to show a similar trend to that in serial search. However, other studies have reported motor tasks to show a midday peak with performance declining over most of the afternoon and evening (13,38–41). The reason for such heterogeneity in the timing of performance peak for these motor tasks is still not well understood. Differences in subject strategy, or in the task's requirements, might account for part of the divergence. For example, an observed linear *decline* in performance over the day on a mirror tracking was explained by suggesting that mirror-tracking proficiency might depend heavily on short-term memory for the execution of correct movements (42). Other researchers (43) have explained the specific diurnal fluctuation of a force discrimination task by its unique proprioceptive requirement and its relation to direct response-produced sensory phenomena.

There is still much work to be done before one can understand definitively which performance tasks will show different time-of-day effects and what the mechanisms are that underlie these differences. Many of the studies presented here have not yet been replicated using different populations of subjects. In addition, many of the hypotheses generated to explain the mechanisms underlying heterogeneity between the different tasks are quite stimulating but need to be more systematically tested.

D. Similar Time-of-Day Variation for Different Tasks: Recent Results from Constant Routine and Forced Desynchrony Protocols

In the previous section, we suggest that the parallelism between temperature and performance observed in the past seems to hold for only a fairly restricted range of tasks. This conclusion is based largely on studies that sampled data infrequently and/or limited data collection to normal working hours. Recent studies suggest that intertask differences observed under a normal nycthemeral condition (sleeping at night and being awake during the day) can fail to appear when data collection is extended into the night, and when non-sleep-deprived subjects are tested at all circadian phases (using the forced desynchrony protocol).

Johnson et al. (6) have replicated the decline in short-term memory over the first 10 hr of the waking day in a 40-hr wakeful bed rest protocol. Figure 2 shows the time-of-day effects observed for short-term memory reported by three studies: Laird (11), Folkard and Monk (17), and Johnson et al. (6). However, when the testing was extended to the entire 40 hr, a parallelism between short-term memory performance and temperature emerged with a coincidence in the timing of troughs of temperature and performance. These data were consistent with the results of a 72-hr sleep deprivation study in which performance on a memory-and-search task reached a trough between 2 A.M. and 6 A.M. (44). These results raise doubts about a general inversion of short-term memory and body temperature rhythms. Recently, Monk et al. (45) have studied the circadian fluctuations of performance (speed and accuracy) at serial search, verbal reasoning, and manual dexterity tasks during 36 hr of unmasking conditions (constant wakeful bed rest, temporal isolation, homogenized "meals"). Figure 3 shows the time-of-day functions for search speed, reasoning speed, vigilance hits, and dexterity speed. The linear trend of each subject's individual time series has been removed to factor out the effect of sleep deprivation. As found by Johnson et al. (6), the bathyphase of the average performance rhythms was mostly within the 5 A.M.–7 A.M. time window, broadly coincident with the timing of the trough in rectal temperature. Thus when the sleep-wake cycle is suspended and data collection is extended into the night, circadian performance rhythms appear to be generally predictable from the circadian temperature rhythm. This is true even for reasoning speed, a "working memory" task shown by Folkard (27) to exhibit a time-of-day effect (under

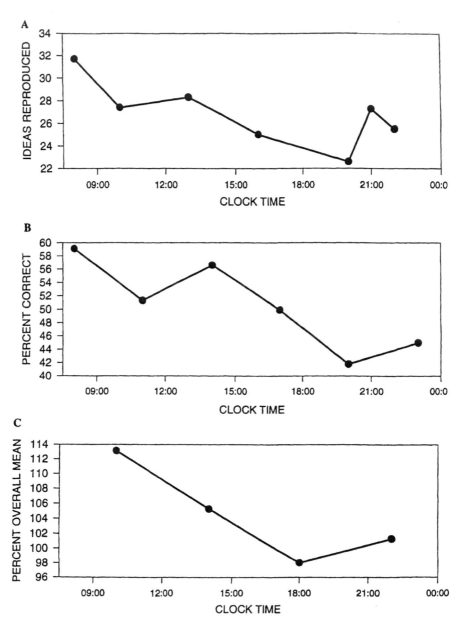

Figure 2 (A–C) Time-of-day function for three studies involving immediate memory tasks. (From Refs. 6,11,17.)

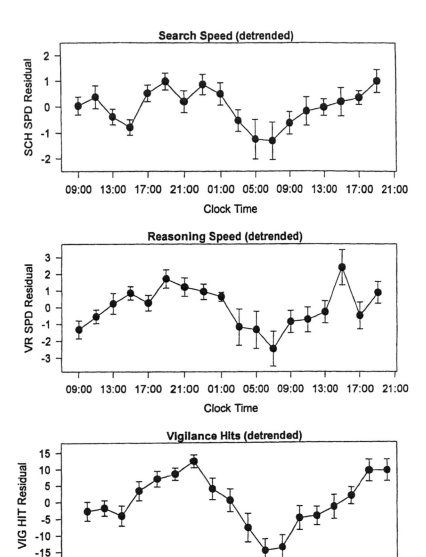

Figure 3 Detrended functions for search speed (lines/min), reasoning speed (lines/min), and vigilance hits (percentage hits). Plotted is mean ± SEM from 17 subjects. (From Ref. 45.)

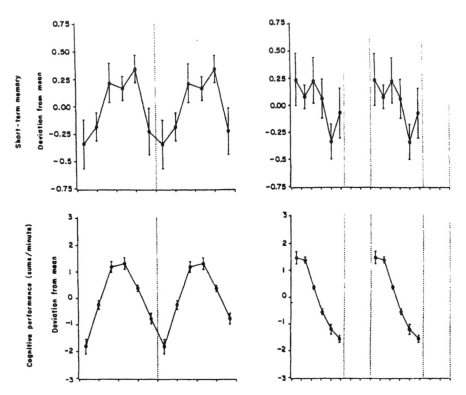

Figure 4 Circadian and sleep/wake-dependent influences on short-term memory, cognitive performance, subjective alertness, and core body temperature (°C) in nine subjects during episodes of forced desynchrony between the body temperature and sleep/wake cycles. Data are double-plotted. (Left) All data are referenced to the phase of the endogenous circadian temperature cycle educed at its intrinsic period, with 0 degrees = temperature nadir. (Right) The same data are referenced to wake time (0 min) and educed at the period of the imposed sleep/wake cycle (28 hr). (From Ref. 6.)

nycthemeral conditions) that is rather different from body temperature (a midday peak vs. an evening peak). The Johnson et al. (6) and Monk et al. (45) studies suggest that intertask differences under a normal nycthemeral condition might be driven more by the homeostatic influence of time since waking than by intertask differences in CTS influence. As mentioned earlier, it is not easy to separate the homeostatic influence from the drive of the CTS. Monk et al. (45) suggest that the parallelism observed in their study occurred because the sleep/wake cycle was suspended, and the linear buildup had been factored out by the removal of the linear trend.

The forced desynchrony protocol is one of the techniques often proposed to separate out the homeostatic influence from the drive of the CTS. Using such a protocol, Monk et al. (5) showed that intertask differences existed in the weight

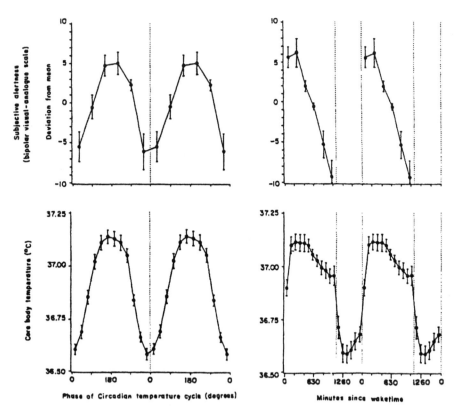

Figure 4 Continued

given to homeostatic versus CTS influences in the final circadian performance rhythm. Interestingly, however, when performance is induced at tau (the period of the CTS), a parallelism between temperature and performance occurs for all tasks (3,5–7). Figure 4 illustrates the circadian and the sleep/wake-dependent influences on short-term memory, calculation performance, subjective alertness, and core body temperature in nine subjects experiencing a forced desynchrony protocol (6). Each subject was scheduled to a 28-hr rest/activity cycle to induce a desynchrony between the body temperature rhythm and the sleep/wake cycle. Rhythms in short-term memory, subjective alertness, and calculation performance clearly varied with circadian phase and paralleled closely the educed waveform of the body temperature rhythm. Figure 4 also shows that short-term memory, subjective alertness, and performance varied with elapsed time since waking on the 28-hr day. Thus, Johnson et al. (6) confirmed that performance on these three behavioral variables is influenced by two interacting factors: an endogenous circadian process that is coupled to the temperature rhythm, and a homeostatic process related to the sleep-wake cycle.

E. Physiological Circadian Correlates of Performance: The Predictive Value of Cortisol and Melatonin Rhythms

Allowing that homeostatic ("time since waking") effects also exist, recent studies strongly suggest that endogenous circadian performance rhythms are controlled by the same pacemaker that drives the endogenous circadian rhythm of body temperature. This pacemaker also drives a number of other physiological rhythms including plasma cortisol and plasma melatonin. The body temperature rhythm has become the "gold standard" for human circadian rhythms, much as the running wheel has for hamster studies (46), at least partly because it is so easy to measure. However, there is no conceptual or mechanistic advantage to using body temperature as an index of the activity of the CTS. Only one recent study has looked at how performance rhythms are correlated with cortisol and melatonin circadian rhythms (45). Results showed that temperature and cortisol rhythms correlated with slightly more performance measures than did melatonin. While all three physiological rhythms were reasonably correlated with performance, the parallelism was far from compelling, with mean intrasubject correlations accounting for a rather small proportion of variance ($< 10\%$). Thus extreme care should be exercised in asserting, for any physiological variable, a universal parallelism between circadian rhythms in performance and physiology, resulting from some posited causal relationship between the two. Instead it would be more parsimonious to assert that performance rhythms are driven independently by the CTS (and time since waking) with a pattern that happens to yield a positive relationship with temperature and a negative one with cortisol and melatonin, without necessarily being *directly* mediated by *any* of the three physiological rhythms. It is noteworthy that in the same study global vigor (subjective alertness) correlated about as well with performance as did body temperature. Thus, although it may seem more rigorous to anchor performance rhythms to an objective index such as body temperature, in terms of predictability, a simple rating of alertness may work just as well. In conclusion, we need to be cautious in assertions regarding the mechanism by which circadian performance rhythms occur. While undoubtedly driven by the CTS and the time-since-waking effects, performance rhythms do not appear to be the simple *direct* result of circadian changes in either mood or physiology.

III. Ultradian Rhythms of Performance: The Postlunch Dip

The afternoon siesta is an integral part of many different cultures (47). A broad base of empirical evidence suggests that there is a general increase in human sleep propensity during the midafternoon hours (48–51). To account for this phenomenon, Broughton (52,53) has proposed the existence of a circasemidian rhythm of vigilance and slow-wave sleep (SWS) propensity having 12-hr and 24-hr components. Many studies of performance have also reported a short-lived decrement of performance during the midafternoon hours; as mentioned earlier, the postlunch dip was one of the first exceptions found to the parallelism between performance and temperature circadian rhythms. Blake's (18) classic studies of performance and time of day showed a clear postlunch dip in measures of simple reaction times, serial search, and signal detection. Although the postlunch dip can be exacerbated by a heavy high-carbohydrate lunch (54), it can occur even when no lunch is taken (55,56). Interestingly, postlunch dips are also apparent in "real life" studies of the frequency of "nodding off" while driving (57), missing warning signals as a train driver (58), and the traffic accident statistics of Israel (59) and the United States (60). However, some laboratory studies have failed to find evidence for a post-lunch dip, even when very similar measures of performance are considered (61). Likewise, there was little evidence of a clear postlunch dip when a meta-analysis of time of day effects in various measures of laboratory performance efficiency and subjective activation was performed by Folkard and Monk (62).

Some studies have suggested that individual characteristics may be linked to the probability of showing an afternoon dip, which might explain some of the inconsistencies found in the literature. For example, Lavie and Segal (63), using the ultrashort sleep-wake paradigm, have shown a much clearer postlunch dip in sleep length for morning types than for evening types after sleep deprivation. Along the same vein, Monk et al. (56) hypothesized that physiological characteristics of the biological clock may indicate who will, and who will not, show a postlunch performance dip. To test this, they studied rectal temperature rhythms in groups of subjects who either did, or did not, show a clear postlunch dip at a monotonous (25–30 min) vigilance task [Mackworth (64) visual vigilance task]. Performance was tested every 2 hr for the 36-hr unmasking protocol. During the protocol the subject was kept in wakeful bed rest in a temporal isolation environment. Meals were replaced by hourly food supplement comprising one-twenty-fourth of the subject's daily caloric requirement. Figure 5 shows vigilance performance (% signals correctly detected ± SEM) as a function of time of day for "dippers" and "nondippers"). Subjects showing the postlunch performance dip had a higher amplitude and later peaking 12-hr component of their rectal temperature rhythm than those not showing the performance dip. This resulted in a flat, rather than rising, function in body temperature over the 10 A.M.–3 P.M. time interval (see Fig. 6). These results suggest that the postlunch dip is linked to an endogenous phenomenon that is individually determined and that is related to the strength of the (12 hr) harmonic of the temperature circadian system.

IV. When the Biological Clock and the Sleep-Wake Cycle Are Challenged: Effects on Performance

As mentioned in the previous section, contemporary models of subjective alertness and performance efficiency (3,5,6,65) view these variables as being determined both by homeostatic processes (duration of wakefulness) and by input from

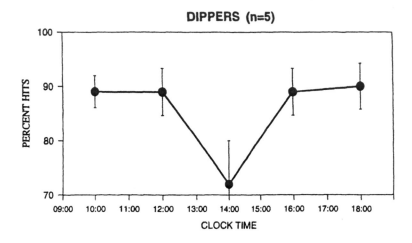

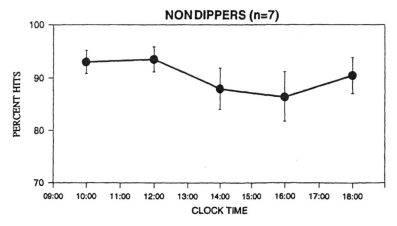

Figure 5   Vigilance performance (% signals correctly detected ± SEM) as a function of time of day for "dippers" and "nondippers." (From Ref. 56.)

the CTS. As discussed in Chapter 5, the sleep-wake cycle is itself also regulated by both of these processes (66). Many studies have reported a strong relationship between the circadian oscillator and sleep propensity. Sleep propensity increases throughout the falling limb of the circadian temperature curve and is maximal near its trough; it decreases throughout the rising limb of the circadian curve and is minimal near its peak (48,67–69).

In jet-lag and shift-work situations both the homeostatic and the circadian processes are affected, as compared to normal nycthemeral conditions. The CTS

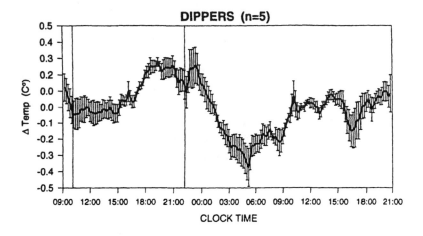

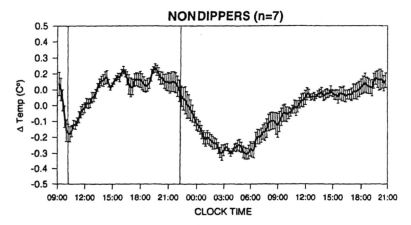

Figure 6 Mean 36-hr rectal temperature (± SEM) plotted by time of day for "dippers" and "nondippers." Each datum is expressed as deviation from that subject's 36-hr mean. The two lines indicate the time interval 10:00–22:00 on day 1. (From Ref. 56.)

usually takes time to adapt to the new time cues imposed by jet-lag or shift-work situations. As a result, there will inevitably be a period of time during which subjects will have to be active and performing at circadian phases of low performance and high sleep propensity. Moreover, these individuals will also be trying to fall asleep at circadian phases promoting high performance and low sleep propensity. Therefore, when considering performance deficits in these situations, we must take into account not only that individuals are trying to perform at a circadian phase often associated with low vigilance and poor performance, but also that their daytime sleep will be shorter and of poorer quality than their normal nocturnal sleep. This will lead to partial sleep deprivation.

In real life, one of the major differences between jet-lag and shift-work situations is that in jet lag all natural time cues are also phase-shifted, thus encouraging circadian phase adjustment in the individual. In contrast, shift workers are exposed to daylight and social time cues that appose and may prevent a complete adaptation of the circadian timing system (70). In laboratory simulation experiments the distinction between jet lag and shift work is often arbitrary, since the subjects are often not experiencing conflicting (shift work) or potent congruent (jet lag) time cues.

A. Jet Lag

During the first few days after crossing several time zones, people complain about a variety of subjective symptoms. Those reported by most reviews include sleep disturbance, daytime fatigue, gastrointestinal disturbance, headaches, reduced cognitive skills, poor psychomotor coordination, and moodiness (for reviews see 71–74). Moreover, with the rise of multinational corporations, an increasing segment of the population have occupations that require them to spend up to 2 weeks per month in time zones radically different from "home time." The term "chronic jet-lag syndrome" (CJLS) has recently been coined for this disorder (75). Because of its chronic nature, CJLS is much more similar to rotating shift work (see below) than to the usually acute, time-limited disorder of jet lag.

There are two components of jet lag: (1) stress effects that stem from the particular physical and psychological aspects of the flight itself; and (2) effects that are a product of the need to reset the biological clock. Arguably, only the latter should properly be referred to as jet lag; the effects due to the first component (malaise, nausea, headaches, aching joints) seldom last more than a few hours after the end of the flight and have nothing to do with circadian rhythms. The process of resetting the biological clock to the new time zone is, however, much longer lasting. After crossing several time zones, external time cues are phase-advanced (eastward flight) or phase-delayed (westward flight) relative to the endogenous CTS. Thus, to adjust the body's rhythms to an eastward destination, the circadian timing system needs to be phase-advanced, to adjust to a westward destination, the circadian timing system needs to be phase-delayed.

It is widely recognized that circadian rhythms do not adjust instantaneously to either real or simulated jet-lag manipulations. Complete circadian rhythm adjustment may take several days after arrival (38,76–79). The actual rate of readjustment of circadian rhythms one observes will depend not only on the size of the phase shift required, but also on its direction. Studies have usually shown more rapid adjustment following westward than following eastward flights (79–

83). A mean readjustment rate for all variables, derived from a set of different studies, was 92 min/day after a westbound flight but only 57 min/day after an eastbound flight (80). Moreover, studies have shown that the readjustment rate is not constant throughout the entire readjustment period. Readjustment rate is generally highest during the first 24 hr immediately after the shift and decreases thereafter in a nonlinear manner (38,40,84). The readjustment rate does not seem to be related to outbound or homebound flight direction, or to the time of flight departure [day or night (38,81)], but does vary with the type of activity pursued at the destination [indoor or outdoor (81)], the age of the subjects (78,85), and their personality (86). It should be noted that the *reverse* directional asymmetry effect (delay worse than advance) has also been observed in controlled laboratory conditions (87).

It is often reported that different circadian rhythms reentrain at different rates (38,88,89). According to this model, the internal relationship among the different circadian rhythms is altered, leading to a transient internal desynchronization within the individual. However, these data should be interpreted carefully. As pointed out by Boulos et al. (74), the assessment of reentrainment rate is complicated by distortions of the daily waveform following the phase shift, by the reduction in the range of oscillation, and by changes in mean daily level. Furthermore, Daan and Beersma (90) have used computer simulations to show that apparent differences in reentrainment rate can emerge. These differences may be due to waveform distortion during reentrainment, masking effects, and particular acrophase fitting procedures, even when these rhythms are generated by the same oscillator. The influence of masking on entrainment rate is also illustrated by studies using body temperature rhythms, which have shown a decrease of the reentrainment rate when the phase of the circadian temperature rhythm was estimated more precisely, either by using constant bed rest conditions or by "demasking" when the masking effect is mathematically removed (76,91).

To evaluate the effects of crossing time zones on performance efficiency is not easy. Methodological limitations related to performance evaluation, discussed at the beginning of this chapter, definitely apply to this situation. Unfortunately, many reports have used a pre/post experimental design without controlling for, or discussing limitations related to, the practice effect. Two questions are central to the study of the effects of jet lag on performance: (1) Does the mean level of performance decrease following the jet-lag manipulation? and (2) What is the reentrainment rate of the phase of circadian performance rhythms? Despite the lack of control for the practice effect, many studies have shown disruptive effects on mean levels of performance following jet-lag manipulations. Hauty and Adams (88,89) have shown detrimental effects on the first day after an eastward and a westward flight on mean reaction time, as well as mean decision time. However, this deficit was of very short duration in comparison to the rate of adjustment of the body temperature rhythm. Wright et al. (92) studied 81 healthy male soldiers for 5 days before and 5 days after an eastward flight across six time zones. Performance times for a 270-m sprint were worsened for the first 4 days following the time zone shift as were times for a 2.8-km run on the second and third days. More recently, Monk et al. (77) presented data on performance from eight middle-aged subjects submitted to a laboratory 6-hr phase advance of the sleep-wake cycle. The authors recognized that despite their efforts to minimize learning curve effects, the performance measures (verbal reasoning, serial search, manual dexterity) were still contaminated by improvements due to practice. After controlling mathematically for this effect, only the verbal reasoning task showed a significant effect, with seven of the eight subjects performing more slowly on the first 3 days after the jet-lag manipulation. Later, Monk et al. (93) evaluated the effect of the same jet-lag manipulation in older people (79–91 years). As with the middle-aged subjects, there was a practice effect in performance measures. After detrending the performance measures, manual dexterity in the 5 days following the phase shift was significantly slower than manual dexterity in the 5 baseline days. No effect was found for serial search latency.

As is the case for physiological rhythms, the direction of the flight seems to be an important factor in determining the effect of phase shift on performance. For example, Wever (87) has shown impairments in psychomotor performance following 6-hr phase advances, but not following phase delays (of the light-dark cycle). Klein et al. (38,81) presented data from eight subjects for three performance tests (psychomotor, cancellation, and digit summation tasks) after transmeridian flights involving the transition of six time zones in both directions. In contrast to many subsequent studies, measures of performance were also taken during the nighttime (by waking up the subjects) allowing the investigators to obtain performance assessments throughout the 24-hr range. Of course, this also implied that a certain level of partial sleep deprivation was induced by the experimental design. On the day of arrival after the flight, in both directions, mean performance level showed clear decrements at specific times of day. As noted by the authors, the observed impairment at certain times during the first postflight day for psychomotor performance was comparable to that resulting from the effects of a blood-alcohol concentration of 0.05% on the same performance task (94). Mean performance had returned to the preflight levels before the third postflight day. In addition to a general decrement in mean level, performance rhythms showed a phase advance compared to baseline after the westward flight, and a phase delay compared to baseline after the eastward flight. When rates of phase adjustment of different performance rhythms were compared, performance rhythms adjusted at different rates, depending on the nature of the task. The speed of reentrainment was lowest in psychomotor performance, and highest in reaction time and digit summation. Phase angle phase disruptions were more pronounced after eastbound flights than after westbound flights (81,84). By the eighth day after the flight, in either direction, the phase angles of all performance tasks were completely adjusted to the new time zone. More recently, Suvanto et al. (85) studied the effects of rapid 10-hr time zone changes on female flight attendants' circadian rhythms of oral temperature, alertness, and visual search performance. Measurements were performed before the flight from Helsinki to Los Angeles, during the second and the fourth day in the United States, and during the second and fourth day after the return flight to Finland. As found previously, the acrophase of body temperature shifted more rapidly after westward flights than after eastward flights. Circadian acrophases of alertness and visual search rhythms also shifted rapidly after arrival in the United States (after a westward flight). The acrophase of the visual search rhythm was already delayed by more than 8 hr compared to baseline by the second day in the United States. As was the case for temperature, phase adjustment of the visual search rhythm after the eastward return flight to Europe was slower than after the westward flight. The acrophase of the visual search rhythm was still delayed by more than 2 hr compared to baseline even on the fourth day after arrival in Europe.

In summary, it seems that the effects of jet lag on mean level of performance are shorter lasting than would be predicted from the duration of phase adjustment shown by physiological rhythms (e.g., body temperature). However, few of these studies were able to adequately control for the practice effect. As a consequence, it is possible that the effects of jet lag on mean level of performance have been underestimated in these studies. The phase adjustment reentrainment rate of performance have been underestimated in these studies. The phase adjustment reentrainment rate of performance measures following jet lag is also difficult to determine. Researchers in this area are left with only two choices. They can try to estimate the phase of performance rhythms using only a part of the nycthemeron, i.e., the "daytime" measures, which may lead to important errors of measurement in phase estimates, or they can wake up the subjects during the night to take performance measurements, thus worsening their subjects' sleep loss, which itself might lead to poorer postflight performance.

B. Shift Work

In discussing the issue of shirt-work performance, we have decided to concentrate upon studies that have measured the *actual* on-job performance of shift workers, rather than upon those measuring the performance of either shift workers or volunteers doing artificial laboratory tasks. Indeed, many of the latter studies have already been described earlier in this chapter. Issues related to the performance of shift workers are even more complex and multifaceted than those related to jet lag. As mentioned before, the transmeridional traveler has the advantage (in terms of phase adjustment) of an entire society, together with the natural zeitgebers of daylight and darkness, all lined up at the destination time zone. For the shift worker, the reverse is true. Daylight and darkness comprise zeitgebers that are directly opposed to a nocturnal routine. Moreover, although 20% of the working population are shift workers of one form or another (95), society is resolutely day-oriented, with none of the taboos against noise and disturbance, which protect the night sleep of day workers, there to protect the day sleep of night workers. This means that in addition to the decrements in performance stemming from performing during the down phase of the cycle, and in trying to sleep during the "up" phase, there are sleep disruptions, stresses and strains coming from society in general, and from the shift worker's own home and family in particular. Even the absence of weekends off can be a factor in this regard. Monk and Wagner (96) have shown, for example, that night workers may have an increased accident risk on Sunday nights because of the loss of day sleep time to religious and family obligations on Sunday morning.

While there are occasionally greater environmental risks associated with shift work than with day work, e.g., because of no daylight or increased chemical exposure on 12-hr shifts, these are usually outweighed by the reduction in human traffic. Indeed, when the accident rate is not expressed as a percentage of "traffic load," the number of accidents is often fewer on the night shift than on the others (97). However, when accidents are expressed in terms of rate per worker on the job, or the situation in one in which all three shifts are exactly the same in terms of tasks performed and manning levels, the night shift often comes out worst. A good example of this is Smith et al.'s (98) study of accidents in a car factory, which revealed a 23% increase in night shift accidents overall and a 42% increase in serious accidents as compared to the morning shift.

There are a number of different ways in which maladjusted shift workers can become agents of risk to themselves and others. The first way is through sleepiness at work, the etiology of which has been discussed earlier in this chapter (see also 99). Sleepiness at work can lead to the problem of missed signals (e.g., a red light, a dial going critical), or of inappropriate responses to correctly perceived signals (e.g., landing an airplane on the wrong runway) as well as that of actually dozing off to sleep on the job (100). These effects are also important in the commute to and from work, which is arguably the most dangerous activity most night workers engage in (101). Although it may seem reasonable to regard all dangerous shift workers as being overly sleepy, there are groups of maladjusted shift workers who are not sleepy at all, but who may be just as dangerous. This leads to the second means by which shift-worker performance can be impaired. Such workers may be upset and angry (102) either for biological reasons (circadian dysfunction, sleep loss) or social ones, e.g., an imminent divorce (103). Such mood changes can lead to a cavalier (or directly aggressive) attitude toward the handling of dangerous machinery, and to the absence of concern for the safety of those around them. The third means by which shift workers can become agents of risk is through simple performance decrements. Although not subjectively sleepy, workers may be suffering circadian-related performance decrements, which (e.g., in the meat packing industry, or in health care in the handling of HIV-tainted needles) might lead to personal injury because of the critical nature of the work being undertaken (104).

In considering the total "social cost" of shift work, one must realize that not only the worker, but also his or her family, and indeed the surrounding community, face hazards to their safety and well-being as a result of shift work. In evaluating these hazards one must recognize the multifaceted nature of the shift work issue. At one extreme, there may be rare, but highly significant, events such as a near meltdown at a power plant, a chemical plant explosion, or a train derailment. These highly visible events can affect thousands of lives and have a massive impact on the safety and well-being of surrounding communities (60). At the other end of the spectrum are the comparatively frequent events such as vehicular accidents on the drive home from work and minor injuries at work (105). Because of their lower profile in the news media, these events may tend to be relatively ignored. However, both in financial terms, and in an accounting of human misery, it is quite possible that the sum total of the frequent small events is as important to society as that of the occasional dramatic ones. Both are often the result of poor shift-worker performance, and both should be factored in when society evaluates the cost-effectiveness of shift work.

Moving from accidents to more continuous variables of performance, it should be noted that there are major problems in trying to determine intershift differences in "real task" performance. The most important of these is the difference in the working environment. Not only lighting levels, but also supervision levels, group morale, and distractions can all be very different indeed, between night shifts and day shifts (106). Also, poorer performance can occur on the night shift simply because there is nobody there to repair broken machines. Not only the work environment, but the work itself may be quite different between night and day. Often particular parts of the job are actually saved for the night shift, either because the process demands it (preparing things for shipment in the morning), or to make life easier for the night workers (e.g., long-running computer jobs held back to be run at night). Even in continuous-process operations, complicated development work may intrude during the day, but not at night. Despite all these problems, a number of studies [summarized by Folkard and Monk (107)] have plotted "real-life" performance as a function of time of day over the 24 hr, and these show surprisingly good agreement in the pattern of results found. Monk et al. (97) have produced a combined (Z-score) circadian rhythm from such studies, which is double-plotted in Figure 7. The major trough in performance during the night shift and subsidiary trough in the "postlunch dip" (see above) are both clearly evident. As we keep stating, though, this rhythm is a function of many different aspects of a shift worker's life, rather than just simply the status of his or her biological clock.

In conclusion, the performance rhythms of shift workers are the result of complex, multifaceted processes. As well as tapping into the "low ebb" of certain circadian performance rhythms, night work involves sleep disruption, social and

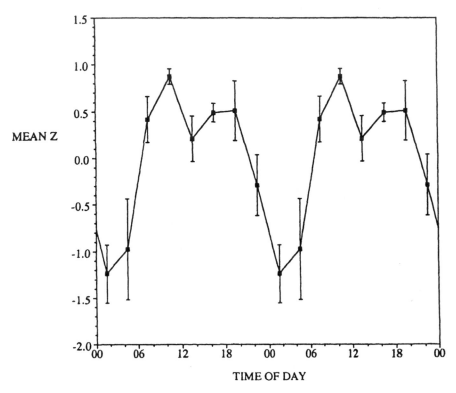

Figure 7   Meta-analysis by Monk et al. of actual "on job" performance double-plotted as a function of time of day. Increases on the ordinate represent improvements in performance. (From Ref. 97.)

domestic disruption, and the chronic equivalent of jet lag, all of which can radically affect the performance and safety of the individual concerned. Thus the amelioration of the problem must be equally multifaceted, lessening not only the stress on the shift worker (e.g., by the selection of more benign shift schedules), but also the resultant strain (e.g., by health, education, and counseling programs).

V. Conclusions

"It's as different as night and day" is an apt summary of how human performance ability fluctuates over the 24 hr. Importantly, these fluctuations are nontrivial and are predictable, given knowledge of the status of the circadian system and the amount of prior wakefulness. Because both of these factors combine to influence performance, circadian performance rhythms do not *always* parallel the body temperature rhythm, although they invariably do so when the sleep/wake cycle is suspended. Disruptions due to shift work and jet lag can have catastrophic implications for performance.

References

1. Webb WB. Problems in measuring performance: dangers in difference scores. In: Broughton RJ, Ogilvie RD, eds. Sleep, Arousal, and Performance: A Tribute to Bob Wilkinson. Boston: Birkhauser, 1992:154–163.
2. Monk TH. Circadian rhythms in subjective activation, mood, and performance efficiency. In: Kryger MH, Roth T, Dement WC, eds. Principles and Practice of Sleep Medicine, 2nd ed. Philadelphia: WB Saunders, 1994:321–330.
3. Monk TH, Moline ML, Fookson JE, Peetz SM. Circadian determinants of subjective alertness. J Biol Rhythms 1989; 4:393–404.
4. Folkard S, Akerstedt T. A three-process model of the regulation of alertness-sleepiness. In: Broughton RJ, Ogilvie RD, eds. Sleep, Arousal, and Performance: A Tribute to Bob Wilkinson. Boston: Birkhauser, 1992:11–26.
5. Monk TH, Weitzman ED, Fookson JE, Moline ML, Kronauer RE, Gander PH. Task variables determine which biological clock controls circadian rhythms in human performance. Nature 1983; 304:543–545.
6. Johnson MP, Duffy JF, Dijk DJ, Ronda JM, Dyal CM, Czeisler CA. Short-term memory, alertness and performance: a reappraisal of their relationship to body temperature. J Sleep Res 1992; 1:24–29.
7. Dijk DJ, Duffy JF, Czeisler CA. Circadian and sleep/wake dependent aspects of subjective alertness and cognitive performance. J Sleep Res 1992; 1:112–117.
8. Lavie P. The search for cycles in mental performance from Lombard to Kleitman. Chronobiologia 1980; 7:247–256.
9. Gates AI. Variations in efficiency during the day, together with practice effects, sex differences, and correlations. Univ Calif Publ Psychol 1916; 1:1–156.
10. Muscio B. Fluctuations in mental efficiency. Br J Psychol 1920, 10:327–344.
11. Laird DA. Relative performance of college students as conditioned by time of day and day of week. J Exp Psychol 1925; 8:50–63.
12. Folkard S, Monk TH. Circadian performance rhythms. In: Folkard S, Monk TH, editors. Hours of Work—Temporal Factors in Work Scheduling. New York: Wiley, 1985:37–52.
13. Kleitman N. Sleep and Wakefulness. Chicago: University of Chicago Press, 1963.
14. Kleitman N, Jackson DP. Body temperature and performance under different routines. J Appl Physiol 1950; 3:309–328.
15. Colquhoun WP. Biological Rhythms and Human Performance. London: Academic Press, 1971.
16. Klein KE, Wegmann HM, Athanassenas G, Hohlweck H, Kuklinski P. Air operations and circadian rhythms. Aviat Space Environ Med 1976; 47(3):221–230.
17. Folkard S, Monk TH. Circadian rhythms in human memory. Br J Psychol 1980; 71:295–307.

18. Blake MJF. Time of day effects on performance in a range of tasks. Psychonom Sci 1967; 9:349–350.

19. Baddeley AD, Hatter JE, Scott D, Snashall A. Memory and time of day. Q J Exp Psychol 1970; 22:605–609.

20. Hockey GRJ, Davies S, Gray MM. Forgetting as a function of sleep at different times of day. Q J Exp Psychol 1972; 24:389–393.

21. Hockey GRJ, Colquhoun WP. Diurnal variation in human performance: a review. In: Colquhoun WP, ed. Aspects of Human Efficiency: Diurnal Rhythm and Loss of Sleep. London: English Universities Press, 1972:39–107.

22. Monk TH. The arousal model of time of day effects in human performance efficiency. Chronobiologia 1982; 9:49–54.

23. Craik FIM, Blankstein KR. Psychophysiology and human memory. In: Venables PH, Christie MJ, eds. Research in Psychophysiology. London: Wiley, 1975:388–417.

24. Leirer VO, Tacke ED, Morrow D. Time of day and naturalistic prospective memory. Exp Aging Res 1994; 20:127–134.

25. Folkard S, Monk TH, Bradbury R, Rosenthall J. Time of day effects in school children's immediate and delayed recall of meaningful material. Br J Psychol 1977; 68:45–50.

26. Folkard S. Diurnal variation in human performance. In: Hockey GRJ, ed. Stress and Fatigue in Human Performance. Chichester: Wiley, 1983:245–272.

27. Folkard S. Diurnal variation in logical reasoning. Br J Psychol 1975; 66:1–8.

28. Rutenfranz J, Helbruegge T. Uber Tageschwankungen der Rechengeschwindigkeit bei 11-jahrigen kinder. Z Kinderh 1957; 80:65–82.

29. Folkard S, Knauth P, Monk TH, Rutenfranz J. The effect of memory load on the circadian variation in performance efficiency under a rapidly rotating shift system. Ergonomics 1976; 19:479–488.

30. Folkard S. Time of day and level of processing. Mem Cogn 1979; 7:247–252.

31. Baddeley AD. The influence of acoustic and semantic similarity on long-term memory for word sequences. Q J Exp Psychol 1966; 18(4):392–309.

32. Baddeley AD. Short-term memory for word sequences as a function of acoustic, semantic and formal similarity. Q J Exp Psychol 1966; 18(4):362–365.

33. Monk TH. The interaction between the edge effect and target conspicuity in visual search. Hum Factors 1981; 23:615–625.

34. Folkard S. Circadian performance rhythms: some practical and theoretical implications. Phil Trans Roy Soc Lond 1990; 327:543–553.

35. Zaidel E. Disconnection syndrome as a model for laterlety effects in the normal brain. In: Hellige JB, ed. Cerebral Hemisphere Asymmetry. New York: Praeger, 1983:95–151.

36. Zaidel E. Language in the right hemisphere. In: Benson E, Zaidel E, eds. The Dual Brain. New York: Guilford Press, 1985:205–232.

37. Corbera X, Grau C, Vendrell P. Diurnal oscillations in hemispheric performance. J Clin Exp Neuropsychol 1993; 15(2):300–310.

38. Klein KE, Wegmann HM, Hunt BI. Desynchronization of body temperature and performance circadian rhythms as a result of out-going and homegoing transmeridian flights. Aerospace Med 1972; 43(2):119–132.

39. Buck L. Psychomotor test performance and sleep patterns of aircrew flying transmeridional routes. Aviat Space Environ Med 1976; 47(9):979–985.

40. Aschoff J, Giedke H, Poppel E, Wever RA. The influence of sleep-interruption and of sleep-deprivation on circadian rhythms in human performance. In: Colquhoun WP, ed. Aspects of Human Efficiency: Diurnal Rhythm and Loss of Sleep. London: English Universities Press, 1972.

41. Monk TH, Leng VC. Time of day effects in simple repetitive tasks: some possible mechanisms. Acta Psychol 1982; 51:207–221.

42. Payne RB. Psychomotor performance as a function of time of day. Percept Mot Skills 1989; 68:455–461.

43. Miller LS, Lombardo TW. Time of day effects on a human force discrimination task. Physiol Behav 1992; 52:839–841.

44. Babkoff H, Mikulincer M, Caspy T, Kempinski D, Sing H. The topology of performance curves during 72 hours of sleep loss: a memory and search task. Q J Exp Psychol 1988; 40A:737–756.

45. Monk TH, Buysse DJ, Reynolds CF, Berga SL, Jarrett D, Begley A, et al. Circadian rhythms in human performance and mood under constant conditions. J Sleep Res 1997; 6:9–18.

46. Wever RA. The Circadian System of Man: Results of Experiments Under Temporal Isolation. New York: Springer-Verlag, 1979.

47. Dinges DF, Broughton R. Sleep and Alertness: Chronobiological, Behavioral, and Medical Aspects of Napping. New York: Raven Press, 1989.

48. Lavie P. Ultrashort sleep-waking schedule. III. Gates and "forbidden zones" for sleep. Electroenceph Clin Neurophysiol 1986; 63:414–425.

49. Campbell SS. Duration and placement of sleep in a "disentrained" environment. Psychophysiology 1984; 211:106–113.

50. Richardson GS, Carskadon MA, Orav EJ, Dement WC. Circadian variation of sleep tendency in elderly and young adult subjects. Sleep 1982; 5:S82–S94.

51. Carskadon MA, Dement WC. Multiple sleep latency tests during the constant routine. Sleep 1992; 15(6):396–399.

52. Broughton R. Biorhythmic variations in consciousness and psychological functions. Can Psychol Rev 1975; 16:217–239.

53. Broughton R. The circasemidian sleep rhythm and its relationship to the circadian and ultradian sleep-wake rhythms. In: Kalba R, Oval R, Schulz H, Vusser P, eds. Sleep '86. New York: Gustav Fisher Verlag, 1988:41–43.

54. Craig A, Baer K, Diekmann A. The effects of lunch on sensory-perceptual functioning in man. Int Arch Occup Environ Health 1981; 49:105–114.

55. Blake MJF. Temperament and time of day. In: Colquhoun WP, ed. Biological Rhythms and Human Performance. London: Academic Press, 1971:109–148.

56. Monk TH, Buysse DJ, Reynolds CF, Kupfer DJ. Circadian determinates of the post-lunch dip in performance. Chronobiol Int 1996; 13:135–145.

57. Prokop O, Prokop L. Ermunudung und Einschlafen am Steuer. Zentralbl Verkehrs-Med, Verkehrs-Psychol Angrenzende Gebiete 1955; 1:19–30.

58. Hildebrandt G, Rohmert W, Rutenfranz J. Twelve and 24 hour rhythms in error frequency of locomotive drivers and the influence of tiredness. Int J Chronobiol 1974; 2:175–180.

59. Lavie P. The 24-hour Sleep propensity function (SPF): practical and theoretical implications. In: Monk TH, ed. Sleep, Sleepiness and Performance. Chichester: Wiley, 1991:65–93.

60. Mitler MM, Hajdukovic RM, Hahn PM, Kripke DF. Circadian rhythm of death time: cause of death versus recorded death time in New York City. Sleep Res 1985; 14:306 (abstract).

61. Christie MJ, McBrearty EMT. Psychophysiological investigations of post lunch state in male and female subjects. Ergonomics 1979; 22:307–325.

62. Folkard S, Monk TH. The measurement of circadian rhythms in psychological functioning. In: Scheving LE, Halberg F, Ehret CF, eds. Chronobiotechnology and Chronobiological Engineering. Dordrecht: Martinus Nijhoff Publishers, 1987:189–201.

63. Lavie P, Segal S, Twenty-Four-Hour structure of sleepiness in morning and evening persons investigated by ultrashort sleep-wake cycle. Sleep 1989; 12(6):522–528.

64. Mackworth NH. The breakdown of vigilance during prolonged visual search. Q J Exp Psychol 1948; 1:6–21.

65. Dijk DJ, Czeisler CA. Contribution of the circadian pacemaker and the sleep homeostat to sleep propensity, sleep structure, electroencephalographic slow waves, and sleep spindle activity in humans. J Neurosci 1995; 15(5):3526–3538.

66. Borbely AA. A two-process model of sleep regulation. Hum Neurobiol 1982; 1: 195–204.

67. Czeisler CA, Zimmerman JC, Ronda JM, Moore-Ede MC, Weitzman ED. Timing of REM sleep is coupled to the circadian rhythm of body temperature in man. Sleep 1980; 2:329–346.

68. Zulley J, Wever R, Aschoff J. The dependence of onset and duration of sleep on the circadian rhythm of rectal temperature. Pflügers Arch 1981; 391(4):314–318.

69. Dijk DJ, Czeisler CA. Paradoxical timing of the circadian rhythm of sleep propensity serves to consolidate sleep and wakefulness in humans. Neurosci Lett 1994; 166(1): 63–68.

70. Eastman CI, Stewart KT, Mahoney MP, Liu L, Fogg LF. Dark goggles and bright light improve circadian rhythm adaptation to night-shift work. Sleep 1994; 17(6): 535–543.

71. Winget CM, DeRoshia CW, Markley CL, Holley DC. A review of human physiological and performance changes associated with desynchronosis of biological rhythms. Aviat Space Environ Med 1984; 55(12):1085–1096.

72. Redfern P, Minors D, Waterhouse J. Circadian rhythms, jet lag, and chronobiotics: an overview. Chronobiol Int 1994; 11(4):253–265.

73. Graeber RC. Jet lag and sleep disruption. In: Kryger MH, Roth T, Dement WC, eds. Principles and Practice of Sleep Medicine, 2nd ed. Philadelphia: WB Saunders, 1994; 463–470.

74. Boulos Z, Campbell SS, Lewy AJ, Terman M, Dijk DJ, Eastman CI. Light treatment for sleep disorders: consensus report. J Biol Rhythms 1995; 10(2):167–176.

75. Monk TH. Disorders relating to shift work and jet-lag. In: Oldham JM, ed. Sleep Disorders Section. Vol 13. Annual Review of Psychiatry 1994. Washington, DC: American Psychiatric Press, 1994; 729–756.

76. Mills JN, Minors DS, Waterhouse JM. Adaptation to abrupt time shifts of the oscillator(s) controlling human circadian rhythms. J Physiol 1978; 285:455–470.

77. Monk TH, Moline ML, Graeber RC. Inducing jet lag in the laboratory: patterns of adjustment to an acute shift in routine. Aviat Space Environ Med 1988; 59:703–710.

78. Moline ML, Pollak CP, Monk TH, Lester LS, Wagner DR, Zendell SM, et al. Age-related differences in recovery from simulated jet lag. Sleep 1992; 15(1):28–40.

79. Minors DS, Waterhouse JM. Deriving a "phase response curve" from adjustment to simulated time zone transitions. J Biol Rhythms 1994; 9(3–4):275–282.

80. Aschoff J, Hoffman K, Pohl H, Wever RA. Re-entrainment of circadian rhythms after phase-shifts of the zeitgeber. Chronobiologia 1975; 2:23–78.

81. Klein KE, Wegmann HM. The resynchronization of human circadian rhythms after transmeridian flights as a result of flight direction and mode of activity. In: Scheving LE, Halberg F, Pauly JE, eds. Chronobiology. Tokyo: Igaku Shoin 1974:564–570.

82. Graeber RC, Lauber JK, Connell JL, Gander PH. International aircrew sleep and wakefulness after multiple time zone flights: a cooperative study. Sleep Res 1986; 15:273.

83. Sasaki M, Endo S, Nakagawa S, Kitahara T, Mori A. A chronobiological study on the relation between time zone changes and sleep. Jikeikai Med J 1985; 32:83–100 (abstract).

84. Wegmann HM, Klein KE, Conrad B, Esser P. A model for prediction of resynchronization after time-zone flights. Aviat Space Environ Med 1983; 54(6):524–527.

85. Suvanto S, Harma M, Ilmarinen J, Partinen M. Effects of 10 h time zone changes on female flight attendants' circadian rhythms of body temperature, alertness, and visual search. Ergonomics 1993; 36(6):613–625.

86. Colquhoun WP. Effects of personality on body temperature and mental efficiency following transmeridian flight. Aviat Space Environ Med 1984; 55:493–496.

87. Wever RA. Phase shifts of human circadian rhythms due to shifts of artificial zeitgebers. Chronobiologia 1980; 7:303–327.

88. Hauty GT, Adams T. Phase shifts of the human circadian system and performance deficit during the periods of transition. I. East-west flight. Aerospace Med 1966; 37:668–674.

89. Hauty GT, Adams T. Phase shifts of the human circadian system and performance deficit during the periods of transition. II. West-east flight. Aerospace Med 1966; 37:1027–1033.

90. Daan S, Beersma DGM. A single pacemaker can produce different rates of reentrainment in different overt rhythms. J Sleep Res 1992; 1:80–83.

91. Minors D, Akerstedt T, Waterhouse JM. The adjustment of the circadian rhythm of body temperature to simulated time zone transitions: a comparison of the effect using raw versus unmasked data. Chronobiol Int 1954; 2:356–366.

92. Wright JE, Vogel JA, Sampson JB, Knapik JJ, Patton JF, Daniels WL. Effects of travel across time zones (jet-lag) on exercise capacity and performance. Aviat Space Environ Med 1983; 54:132–137.

93. Monk TH, Buysse DJ, Reynolds CF, Kupfer DJ. Inducing jet lag in older people: adjusting to a 6-hour phase advance in routine. Exp Gerontol 1993; 28:119–133.

94. Wegmann HM, Klein KE. Jetlag and aircrew scheduling. In: Folkard S, Monk TH, eds. Hours of Work: Temporal Factors in Work-Scheduling. Chichester: Wiley, 1985:263–276.

95. Mellor EF. Shift work and flexitime: how prevalent are they? Monthly Labor Rev 1986; 109:14–21.

96. Monk TH, Wagner JA. Social factors can outweigh biological ones in determining night shift safety. Hum Factors 1989; 31:721–724.

97. Monk TH, Folkard S, Wedderburn AI. Maintaining safety and high performance on shift work. Appl Ergonom 1996; 27:(1)17–23.

98. Smith L, Folkard S, Poole CJM. Increased injuries on night shift. Lancet 1994; 344:1137–1139.
99. Akerstedt T. Sleepiness at work: effects of irregular work hours. In: Monk TH, ed. Sleep, Sleepiness and Performance. Chichester: Wiley, 1991:129–152.
100. Akerstedt T. Sleepiness as a consequence of shift work. Sleep 1988; 11:17–34.
101. Novak RD, Auvil-Novak SE. Focus group evaluation of night nurse shiftwork difficulties and coping strategies. Chronobiol Int 1996; 13(6):457–463.
102. Lauber JK, Kayten PJ. Keynote Address: Sleepiness, circadian dysrhythmia, and fatigue in transportation system accidents. Sleep 1988; 11:503–512.
103. Colligan MJ, Rosa RR. Shiftwork effects on social and family life. Occup Med 1990; 5:315–322.
104. Monk TH. Shiftworker performance. Occup Med 1990; 5(2):183–198.
105. Richardson GS, Miner JD, Czeisler CA. Impaired driving performance in shiftworkers: the role of the circadian system in a multifactorial model. Alcohol Drugs Driving 1990; 5(4)–6(1):265–273.
106. DeVries-Griever AHG, Meijman TF. The impact of abnormal hours of work on various modes of information processing: a process model on human costs of performance. Ergonomics 1987; 30:1287–1299.
107. Folkard S, Monk TH. Shiftwork and performance. Hum Factors 1979; 21:483–492.
108. Monk TH. Temporal effects in visual search. In: Clare JN, Sinclair MA, eds. Search and the Human Observer. London: Taylor & Francis, 1979:30–39.

18

Neurological Disorders Associated with Disturbed Sleep and Circadian Rhythms

PHYLLIS C. ZEE and ZORAN M. GRUJIC

Northwestern University
Chicago, Illinois

I. Introduction

Sleep and circadian rhythms are often disrupted in neurological disorders. Increasing evidence indicates that alterations in sleep and wakefulness accompany many types of neurological disorders. In addition, disruption of the sleep-wake cycle may be responsible for some of the symptoms that are part of these neurological diseases.

In this chapter we will review the more common neurological disorders that have been associated with sleep and circadian rhythm disturbances, such as epilepsy; Alzheimer's disease and related dementias including Huntington's disease, prion disorders, and hepatic encephalopathy; cerebrovascular disease; Parkinson's disease and related movement disorders; neuromuscular disorders such as myotinic dystrophy and motor neuron disease; multiple sclerosis; and headaches.

II. Epilepsy

Epilepsy is defined as a tendency to experience recurrent seizures, which are manifestations of abnormal firing of cortical neurons. It is the third most common neurological disease behind stroke and Alzheimer's disease. Two clinical states of the brain characterize the individual with epilepsy: the ictal, or seizure, event and the interictal, or between-seizures, state. Even during the interictal state, when the patient is not having a clinical seizure, patients with epilepsy may exhibit abnormal electrical brain activity, which can manifest as difficulties in cognitive function and changes in behavior. Because seizures can arise from many areas of the brain, their clinical presentations are diverse and thus difficult to categorize into distinct syndromes. Nevertheless, the most widely accepted classification of seizures is the International Classification of Epileptic Seizures (1). A simplified version of this classification is in Table 1.

Most of the research studies in the area of epilepsy and the sleep-wake cycle have focused on the effects of sleep or sleep deprivation on both ictal and interictal neuronal discharges (IIDs) as well as on seizure threshold. Much less is known about the role of the circadian system in the temporal distribution of seizures.

A. Temporal Distribution of Seizures

It has been recognized for over one century that seizures occur preferentially at particular times of the day and are especially influenced by sleep and waking. Gowers (2), Langdon-Down and Brain (3), and Patry (4) described three major seizure patterns in adult epileptic patients. Seizures occurring predominantly during the daytime were termed diurnal while seizures occurring primarily in sleep were termed nocturnal types. Finally, patients experiencing seizures

Table 1 International Classification of Epileptic Seizures

I. Generalize seizures
    A. Tonic, clonic, or tonic-clonic
    B. Absence
    C. Lennox-Gastaut
    D. Juvenile myoclonic epilepsy
    E. Infantile spasms
    F. Atonic
    G. Myoclonic
II. Partial or focal
    A. Simple (without loss of consciousness)
        1. Motor
        2. Somatosensory
        3. Autonomic
        4. Psychic
    B. Complex (with impaired consciousness)
III. Unclassified epileptic seizures

---

throughout the sleep-wake cycle were referred to as having the diffuse, or random, type. In these studies, which sampled institutionalized epileptic patients, the diurnal type was the most common at 44% and the nocturnal type the least common at 22%. The frequency of the diffuse type was around 34%. In the diurnal pattern, seizures were typically associated with awakenings from sleep, or day-time naps. A more recent study of outpatient epileptics by Janz reaffirmed the existence of the three temporal seizure patterns: 34% diurnal, 45% nocturnal, and 21% diffuse (5). The difference in frequency of seizure patterns between Janz's study and the studies by Langdon-Down, Patry, and Gowers may be due to the types of patients (outpatient vs. institutionalized) and the types of seizures that were evaluated. For example, there is a tendency for institutionalized patients to have a greater frequency of associated neurological (mental retardation, hemiplegias, etc.) and psychiatric disturbances and therefore more extensive brain injury than the outpatients. In these types of patients with structural damage to the brain, diffuse and diurnal epilepsies are more common. A diffuse distribution of seizures has been associated with severity and duration of disease. In general, etiology and seizure type determine when a seizure occurs during the sleep wake cycle (4–7). Primary generalized seizures occur mainly during the day, while secondary generalized convulsions occur most often during sleep.

It is also well established in the clinical literature that seizures are more frequent during certain times during the sleep-wake cycle (3–7). For example, diurnal seizures have the largest peak 1–2 hr after awakening (7 A.M.–8 A.M.). A second peak occurs in the middle of the afternoon around 3 P.M. and a third peak in the early evening around 6 P.M.–7 P.M. The frequency of nocturnal seizures had a significant peak between 10 P.M. and 11 P.M. and another peak between 4 A.M. and 5 A.M. (1–2 hr before awakening). The time dependency of seizures tends to disappear with aging, so the diffuse-type seizure pattern is more prevalent in older people (5,8). The change from a diurnal or nocturnal seizure pattern to a diffuse pattern is commonly accompanied by the development of structural pathology in the course of the disease and may be related to the occurrence of multiple seizures over time (5).

B. Seizure Events and Sleep

The sleep/wake cycle has prominent influences on the expression of some types of epilepsy and on interictal phenomena. The relationship between sleep/wake state and seizure expression has been best established for the generalized epilepsies (5,23). The type of epilepsy and the location of the seizure focus may also influence the relationship between sleep and seizures. For example, seizures originating from the frontal lobes are more likely to arise during sleep than those originating from other locations.

The different stages of sleep may affect seizure threshold and neuronal excitability. For example, non-rapid-eye-movement (NREM) sleep is thought to facilitate and rapid-eye-movement (REM) sleep is thought to inhibit interictal epileptiform discharges and seizure expression (13). Both generalized and partial interictal epileptiform discharges increase during NREM sleep, whereas REM suppresses generalized epileptiform discharges (9). Stage 2 sleep in particular has been reported to facilitate generalized seizures, whereas stages 3 and 4 sleep promote both partial and generalized seizures (10). Seizures rarely occur in REM sleep when compared with arousal or NREM sleep (11–14). Focal interictal discharges may persist during REM, but rarely are of any clinical significance. One hypothesis is that NREM sleep is characterized by neuronal synchronization, which results in enhanced neuronal excitability, leading to facilitation of epileptiform discharges and seizures. On the other hand, during REM sleep, neurons discharge asynchronously, leading to localization of focal discharges and thus decreased seizure propagation (13).

C. Sleep Deprivation and Seizures

Sleep deprivation is an effective method of provoking focal and generalized seizures (15–19). In one study of 34 patients, 56% exhibited an increased number of epileptiform discharges after sleep deprivation (15). In another study of 102 patients, Degen showed that 63% of the patients had epileptiform activity in the electroencephalogram (EEG) after sleep deprivation as compared with only 19% without sleep deprivation (16). Sleep deprivation also appears to be more effective in promoting seizures in children than adults. The mechanism by which sleep deprivation promotes seizure activity is unknown. One possible explanation is that recovery sleep following sleep deprivation is associated with an increase in NREM sleep, particularly slow-wave sleep, thus enhancing neuronal excitability (6,20).

D. Polysomnographic Findings

Just as sleep and arousal states influence the expression of seizures, seizures in turn affect sleep. Sleep abnormalities appear to be more common in patients with generalized epilepsy than in those with partial epilepsy (21,23). In addition, the severity of neurological abnormalities positively correlates with the degree of sleep disturbance in patients with epilepsy (21). Sleep studies in epilepsy patients demonstrate abnormal sleep architecture and frequent microarousals (22–24). These polysomnographic findings are summarized in Table 2. In addition to the effects of epilepsy on sleep, treatment with antiepileptic medications can influence sleep. For example, the frequently used seizure medication carbamazepine increases slow-wave sleep (stages 3 and 4) and total sleep time (25). Another medication, phenobarbital, decreases REM sleep and shortens sleep latency. Therefore, abnormalities of sleep in patients with seizure disorder are likely due to the combination of the effects of the underlying neurological abnormalities and

Table 2   Effect of Epilepsy on Sleep

Increased number of awakenings after sleep onset
Increased latency to sleep onset
Increased NREM stages 1 and 2
Reduced or fragmented REM
Reduced or abnormal sleep spindles and K complexes
Reduced sleep efficiency
Frequent microarousals

Source: Refs. 21–24.

the effects of medications on sleep. Furthermore, sleep disorders such as obstructive sleep apnea or periodic leg movements in sleep may coexist with epilepsy and treatment of these other sleep disorders may result in better seizure control (26).

E.   Circadian Rhythm Disturbances in Epilepsy

In addition to the sleep/wake cycle, the circadian system may play a modulatory role in the temporal distribution of seizures. Only a few studies have attempted to distinguish the effects of endogenous circadian or ultradian rhythms from those of the sleep/wake cycle on the timing of seizures. Most studies of the circadian distribution of seizures have been limited by small numbers of patients, and often have not cited the etiologies of the seizures. Furthermore, many studies have relied on patient reports of seizure occurrence and thus may have underrepresented some seizure types. Nevertheless, there is some evidence that circadian and ultradian rhythms modulate the occurrence of clinical seizures (27,28). Based on a study of 19 epileptic patients, Kellaway et al. suggest that epileptiform activity is affected by two rhythmic processes related to sleep with periods of 24 hr and 100 min, with the latter corresponding to the REM cycle (27). Several groups have studied the EEG during a night of enforced wakefulness (29,30). Martins da Silva et al. showed that in several patients there was a marked increase in epileptiform discharges toward the end of the night paralleling these subjects' habitual patterns during nocturnal sleep, together with a fall in the amount of epileptiform activity at about the normal time of waking (29). There did not appear to be simply an increase in discharges as would be expected to occur because of the sleep deprivation. Therefore, in these epileptic patients there was evidence of an endogenous circadian rhythm of seizure activity that persisted throughout the night despite the wakefulness.

Only a few studies have looked for alterations in the circadian rhythm of core body temperature in patients with epilepsy. Although, disruption of body temperature rhythms has been reported in some epileptic patients, a consistent pattern has not been reported (31,32). These studies also failed to show a consistent change in the phase or period of circadian rhythms between the normal subjects and epileptic patients.

Evidence for circadian rhythm abnormalities in patients with epilepsy comes from studies on melatonin. Schapel et al. showed that in 30 untreated active epileptics melatonin production was increased in the 22:00–06:00 period, with a carryover into the 06:00–14:00 period (34). Thus not only was there a significant increase in melatonin levels in the epileptics when compared to the controls, but there was also a phase difference. On the other hand, treatment with anticonvulsant medications may decrease the amplitude and levels of melatonin (33). There is evidence that some anticonvulsant medications, especially carbamazapine, will suppress the production of melatonin (34).

Other neuroendocrine measures of circadian rhythms indicate that prolactin concentrations are significantly elevated during NREM sleep in patients with complex partial seizures without clinical events (35). Generalized seizures will cause elevations of prolactin levels 30–60 min after a seizure and then fall rapidly. One study also suggested that basal thyroxine and thyrotropin levels were lower in both nontreated and anticonvulsant-treated epileptic patients when compared to a normal control population (36).

III.   Dementia and Cognitive Impairment

Dementia is a syndrome of progressive loss of cognitive and/or behavioral function that interferes with activities of daily living. Alzheimer's disease (AD) is the most common dementia accounting for 70% of all dementias in most industrialized countries (37,38). The next most common type is dementia due to cerebrovascular disease. Other less common degenerative dementias include Lewy body disease (39), frontotemporal dementia, Parkinson's disease dementia, progressive supranuclear palsy (PSP), and Huntington' disease (HD). Sleep and circadian rhythm disturbances in Parkinson's disease and PSP will be discussed in Section V.

A.   Alzheimer's Disease

AD is the most common degenerative brain disorder that causes a dementia. It is characterized by progressive worsening of memory and other cognitive functioning. Although a definite diagnosis of AD can only be made by autopsy, a diagnosis of probable and possible AD (according to the NINCDS-ADRDA) is frequently used in the clinical setting (40). Therefore, most of the studies on sleep and circadian rhythms in AD are in patients with probable or possible AD.

Many studies examining the alterations in circadian rhythms and sleep architecture in AD have yielded conflicting results. These inconsistencies in the literature are due to many factors. First, investigators have not used rigorous diagnostic criteria and have thus included patients with non-Alzheimer's-type dementias in the studies. There is evidence that disturbances in sleep differ across different etiologies of dementia (41). Second, the results from several studies were confounded by lack of agreement among researchers as to the classification of the severity of AD. For example, some studies have used cognitive scales as a measure of disease severity (mild, moderate, and severe), while others emphasized activities of daily living (ADLs). Third, the influence of medications and other associated sleep disorders such as sleep apneas on sleep architecture is often difficult to separate from the effects of Alzheimer's disease. Finally, most studies have failed to screen for associated depression and thus neglected the possible role of depression in disruption of sleep and circadian rhythms (42). The lack of a clearly defined population in some of these studies may account in part for the discrepancies in the literature on sleep and circadian rhythm alteration in AD.

Sleep Disturbances in Alzheimer's Disease

Sleep disturbances appear to be more frequent in demented patients when compared to "cognitively normal," age-matched controls. Reduced sleep efficiency, long and frequent awakenings, and increased amounts of stage 1 sleep are probably the most consistent and prominent finding noted in this population (41–48,51,52). Other findings have included diminished REM, an increase in daytime sleepiness and more frequent napping, and reduced numbers of sleep spindles and K complexes (45,48,49,52). Conflicting data have been reported with regard to the extent of SWS (stage 3 and 4) reduction (42,45,46,48,50–52).

The severity of sleep disruption in AD may parallel the severity of the dementia. Vitiello and Prinz showed that reductions of SWS and REM sleep were greater in the severe Alzheimer's patients when compared to the mild-moderate patients (45,46,48). They suggested that changes in sleep architecture could be a useful marker of disease progression in AD. Other researchers, however, have found no consistent relationship between disease severity and the severity of sleep disturbance (52,53).

The decreased amounts of REM sleep noted in AD patients may be a factor contributing to poor daytime cognitive functioning. A number of studies indicate an important role of REM sleep in memory consolidation (54,55). Both demented and normal elderly individuals with more total REM sleep tend to perform better on neuropsychological tests (45,49,56).

Sleep disorders such as sleep-disordered breathing and periodic leg movements increase in frequency with age, and may be even more prevalent in patients with AD (57–62). Although the results from some studies, indicate an increase in sleep apnea in patients with AD (61), others reported no difference in the incidence of sleep apnea in AD patients and age-matched elderly subjects (62). It is therefore important to recognize that these age-associated sleep disorders may contribute to the cognitive and behavioral disturbances in AD patients and that appropriate treatment may improve the functional status of these patients. In fact, reversal of dementia after successful treatment of the sleep apnea has been reported in some patients (63,64).

Circadian Rhythm Disturbances in Alzheimer's Disease

In addition to alteration of sleep architecture, there is substantial evidence that the observed sleep/wake cycle abnormalities in patients with AD may be due to a disruption of circadian rhythmicity. Disruption of nocturnal sleep with nocturnal wandering is one of the major problems that precipitates nursing home institutionalization. We will review some of the evidence linking circadian desynchronization to AD.

Changes in the Suprachiasmatic Nucleus (SCN)

Age-related alteration in circadian pacemaker (SCN) function has been well documented in animals (65–68). However, much less is known about the age-related changes in the human SCN. Changes in neuropeptide levels in the human SCN have been associated with aging (69), and while no significant decrease in vasopressin-containing cells was associated with aging (70), an alteration of the circadian rhythm of vasopressin-producing cells has been reported in the human SCN (71). In AD patients, Swaab et al. have shown a decrease in total cell number and a decrease in vasopressin immunoreactivity in the SCN (69,72). However, studies have failed to demonstrate the typical neuritic plaques or neurofibrillary tangles in the SCN of AD patients.

Rest/Activity Cycles

Actigraphy (measurement of body movements with small wrist monitors) has been a useful tool for monitoring the sleep/wake cycles in both clinical and research patient populations. The periods of rest and activity correlate well with sleep and wakefulness. In addition to making it possible to record sleep/wake cycles over several days and weeks, actigraphy has another advantage over polysomnography in that recordings can be performed in the patient's natural environment and complete patient cooperation is not critical.

Studies of rest-activity rhythms in patients with AD have found increased fragmentation relative to controls (73–78). Van Someren et al. found considerable fragmentation in the rest-activity cycle in institutionalized AD patients but not in AD patients who were living at home (78). Satlin et al. found a decreased amplitude and an apparent phase delay in activity rhythms (delayed acrophase) (79). The finding of a phase delay in the circadian activity-rest cycle is in direct contrast with most studies suggesting earlier bedtimes (phase advance) in aging and AD (80,81). Some of these discrepancies may be explained by differences in the severity of AD in the study populations. This explanation is supported by the finding that the rest/activity pattern in patients with AD is correlated with the severity of the disease (82). Early Alzheimer's patients (defined as having the disease less than 7 years) had peak activity levels, approximately 3 hr before sunset, middle-stage Alzheimer's patients (7–10 years of disease) had the greatest activity at sunset, and finally, advanced Alzheimer's patients (greater than 10 years of disease) demonstrated the greatest activity level approximately 3 hr after sunset.

Body Temperature Rhythms

The rhythm of core body temperature is a reliable marker of circadian rhythmicity in humans. Under normal entrainment, the circadian rhythm of body temperature peaks during the afternoon and reaches a trough in the early morning. Measurements of core body temperature rhythms in patients with AD have yielded varying results. The inconsistency of the findings in due in part to the technical difficulties encountered in measuring rectal temperature in uncooperative, demented patients. In addition, factors such as the severity of the AD and gender may influence alterations in the circadian temperature rhythm. For example, a delay in the acrophase with a preserved amplitude has been reported in female patients with AD (83,85), whereas male AD patients had an advanced phase of the circadian rhythm of temperature (84,87). Disease severity appears to also play a role in whether temperature rhythms are disrupted. For example, no difference in temperature rhythms was found in patients with mild AD when compared to healthy, age-matched controls (86). In contrast, more severely impaired AD patients revealed a diminished amplitude of the temperature rhythm (83). In addition, desynchronization between the circadian rhythm of core temperature and the rest-activity cycle was seen in some patients with severe AD (83).

Melatonin

A reduction in the amplitude of the circadian rhythm of melatonin has been reported with aging (88). In the few studies that have attempted to measure melatonin levels in these patients, the results suggest a reduced amplitude to a loss of the melatonin rhythm in many AD patients (89–91). Day-night variations in melatonin levels in the elderly demented are lower in both plasma and cerebrospinal fluid (92,93). Further, melatonin release following an injection of methoxypsoralen, a compound known to increase melatonin levels, was decreased in AD patients compared to age-matched controls (94). This finding indicates that there is a reduced ability to produce melatonin. Because melatonin has both chronobiotic and hypnotic effects, replacement with melatonin has been used to improve sleep in elderly insomniacs with low 6-sulfatoxymelatonin (a metabolite of melatonin) (95,96). Employing a similar rationale, studies are underway to examine the potential effects of exogenous melatonin on sleep and behavior in patients with AD.

Other Circadian Rhythms

Further evidence that AD is associated with disruption of circadian rhythmicity comes from a limited number of studies showing a decrease in the amplitude of rhythm of growth hormone, thyroid-stimulating hormone, renin, aldosterone, estradiol, testosterone, and cortisol (97,98). In addition to hormonal rhythms, physiological rhythms such as the circadian rhythm of blood pressure is altered in some AD patients. In one study, patients with AD did not show the usual nocturnal fall in blood pressure, resulting in a blunted diurnal rhythm in blood pressure relative to elderly controls (99).

Sundowning

There is little agreement among researchers as to the definition of sundowning. Probably the best definition comes from a recent review by Bliwise, who defined sundowning as the nocturnal exacerbation of agitated and disruptive behaviors (81). Sundowning is more common in the demented elderly and in AD. Although there are different estimates of the prevalence of sundowning (100–102), Cohen-Mansfield et al. found that 14% of 408 nursing home patients had agitated behaviors in the evening (100). These types of agitated behaviors are very difficult to manage and caretaker "burnout" is high.

Although the underlying etiology of sundowning is unknown, it is well known that certain conditions promote this condition. Risk factors may include physical pain, incontinence, recent surgery, and recent room transfers (101). In addition, the loss of illumination may also play a role in sundowning. Cameron reported increased symptoms of agitation in demented patients when they were placed in darkened rooms during the day (103). Bliwise et al. observed 24 severely demented individuals and found that sleep was least likely to occur around sunset (between 3 and 7 P.M.) and that awakenings from sleep that occurred in the dark were more often associated with increased agitation than awakenings during the daytime (53,104). On the other hand, other studies have found increased agitated behaviors ("sundowning") during the morning hours, indicating that factors other than light may also be involved (105,106). Late-stage Alzheimer's patients with moderate to severe cognitive impairments were more likely to exhibit sundowning than the early-stage Alzheimer patients (83,107,108).

Disruption of circadian rhythmicity in dementia and the resultant alterations of the circadian sleep-wake cycle have been implicated in the etiology of sundowning (81,109). The circadian rhythm and sleep abnormalities have a negative impact on daytime cognition and attention, which in turn may promote confusion and agitation. Further support that the circadian system may be involved in sundowning is that these behaviors exhibit daylength (seasonal) variation. Agitated behavior was more common at sunset during the winter months compared to the fall months (110).

B. Huntington's Disease

HD is characterized by choreoathetosis (abnormal involuntary movements), dementia, and psychosis. It is a progressive, autosomal dominant neurodegenerative disorder linked to a gene on the short arm of chromosome 4. Emotional disturbances and changes in personality are often the earliest symptoms of the disease. There is no cure for HD; therefore, therapy is aimed at improvement of the manifestations of the disease. Insomnia is a common complaint in these patients (111). Polysomnographic findings include a prolonged sleep latency, an increased number of arousals, a decrease in REM and slow-wave sleep, and an increase in sleep spindles (112–114). As the disease progresses, sleep fragmentation increases and there may be total loss of slow-wave sleep and REM sleep (111,113). Choreiform movements may not be present during sleep early in the disease. However, with progression of the disease the choreiform movements will persist in sleep, in particular during NREM sleep (114). Although the pharyngeal musculature is eventually involved and swallowing may be compromised, sleep apnea is not common in HD. Medications such as haloperidol and phenothiazines are useful in the treatment of the abnormal movements and have additional sedative effects in these patients.

C. Subacute Spongiform Encephalopathies (SSEs)

The SSEs, or prion diseases, consist of a number of sporadic and inherited human diseases that have similar neuropathology and transmissibility associated with an accumulation of a protease-resistant form (PrPsc) of the normal cellular protease-sensitive amyloid or prion protein (PrPc) within the brain. There are four human prion diseases: kuru, Creutzfeldt-Jakob disease, Gerstmann-Straussler syndrome, and fatal familial insomnia (FFI). Recent studies have suggested that the PrPc may be involved in the regulation of sleep and circadian rhythms. Mice devoid of PrPc, when compared to wild-type mice, exhibited more prominent sleep fragmentation and differences in period length of circadian activity rhythm (115). Because of the involvement of the thalamus, sleep disturbance is found in many of the prion diseases in humans. However, alterations in sleep are the most prominent and salient feature in FFI.

FFI is an autosomal dominant disorder that presents with intractable, progressive insomnia, impaired motor function, and dysautonomia caused by sympathetic overactivity. Patients may also exhibit memory deficits, hallucinations and confusion. The progressive insomnia is accompanied by prominent alteration in sleep architecture, total amount of sleep, and temporal distribution of sleep. In the end-stages of FFI, the typical features of sleep stages, such as delta waves and sleep spindles, are no longer present and total sleep time is reduced to 1 hr or less (116,117). The sleep disruption that occurs in FFI patients may be secondary to the selective damage of the mediodorsal and anterior ventral thalamus. In particular, insomnia has been noted in patients with thalamic lesions caused by other disorders such as stroke (117,120).

In addition to the severe alterations in sleep, disruption of endocrine circadian rhythms has been reported in FFI. The circadian rhythms of melatonin, cortisol, prolactin, and growth hormone are abnormal or absent in FFI (116–119). For example, Portaluppi and colleagues found that mean 24-hr plasma levels of melatonin decreased as disease progressed (119). The alterations in the circadian rhythms of hormones and loss of the temporal organization of the sleep/wake cycle suggests that the central circadian pacemaker may also be affected by this disease.

D. Hepatic Encephalopathy

Hepatic encephalopathy (HE) is a clinically significant neuropsychiatric disorder characterized by alteration in mental status and behavior as a consequence of a chronic liver disease such as cirrhosis. Disturbances of the sleep/wake cycle are common in HE. A quality-of-life assessment showed that cirrhotic patients had more sleep difficulties than patients with other types of chronic illnesses (121). Patients with liver cirrhosis often complain of the inability to fall asleep and stay asleep during the night and excessive fatigue and sleepiness during the day (122). Although the mechanisms that underlie these sleep abnormalities are poorly understood, there is evidence that HE is accompanied by alterations in circadian rhythmicity.

Individuals with variable degrees of hepatic failure and portal-systemic shunting have been shown to exhibit abnormalities in the circadian rhythms of blood pressure, urinary sodium excretion, plasma amino acid, and several hormonal profiles (123–125). Disruption of the diurnal rhythm of melatonin, so that both the onset and peak levels of this hormone were delayed (126), suggests that the common complaint of sleep onset insomnia and daytime sleepiness in this population may be due to a phase delay of circadian rhythmicity (see Fig. 1).

Animal studies also indicate that liver dysfunction is associated with abnormal circadian rhythms of locomotor activity and altered entrainment patterns to the light/dark cycle (127,128). The improvement of these circadian rhythm abnormalities with a low-protein diet or neomycin (which are common therapies for HE) suggest that the observed alterations in circadian rhythmicity are due to the effects of liver dysfunction on the brain (129). The question of whether sleep and circadian rhythm abnormalities are due to a direct effect on the circadian clock or as a consequence of altered hepatic metabolism of melatonin remains an area of investigation.

IV. Cerebrovascular Disease

Cerebrovascular disease leading to stroke is one of the most common neurological problems encountered, affecting over 500,000 people in the United States annually. Sleep complaints are common in patients with stroke and may not only impair

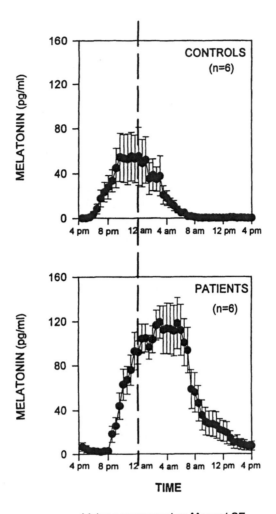

Values expressed as Mean+/-SE.

Figure 1 Mean plasma melatonin levels. Values expressed as mean ± SE. Patients with liver disease show a delay in the phase of melatonin onset when compared to controls. (Courtesy of Dr. Andres Blei.)

the quality of life, but also increase morbidity. Table 3 is a summary of stroke types and their respective frequencies.

A. Sleep Disturbances

A variety of sleep disturbances, including insomnia and hypersomnia, have been reported in stroke patients (111,130–134). Most often the type of sleep disturbance appears to be related to the location and size of the stroke. After a large

Table 3 Types of Strokes

Ischemic	85%	Hemorrhagic	15%
Cerebral thrombosis	65%	Intracerebral	10%
Large artery			
Lacunar			
Cerebral embolism	20%	Subarachnoid	5%

cerebral hemisphere infarction, a patient may have excessive daytime sleepiness and nocturnal restlessness and agitation (130). Even years after recovery from a hemispheric stroke, patients may show increased amounts of slow-wave sleep especially if the stroke was large (131). Because the brainstem and thalamus are important in regulation of sleep and wakefulness, strokes in these areas commonly affect sleep. For example, extensive lesions involving the pontine tegmentum have been associated with a decrease in total REM sleep and strokes involving the midbrain or pons may cause abnormal motor activity during REM sleep (REM sleep behavior disorder) (132). Ischemic infarction of the thalamus may be accompanied by loss of sleep spindles on the side of the lesion (133) and lesions of the paramedian thalamus are known to cause hypersomnia (134) suggesting a role of the paramedian thalamus in maintenance of wakefulness.

Recent evidence indicates that respiratory abnormalities, such as obstructive sleep apnea (OSA) may be more prevalent in patients with stroke (135–139). One study found that 72% of 47 stroke patients had evidence of OSA (144). There is an association between hypertension, a risk factor for stroke, with snoring and OSA (137–139). However, snoring and sleep apnea have also been found to be risk factors for stroke independent of other risk factors such as hypertension and obesity (136,140–143). In one study of 400 stroke patients and controls, habitual snoring was found to be a significant risk factor for stroke with a risk ratio of 3.2 (140). In another study of 177 stroke subjects that adjusted for confounding variables such as hypertension and coronary heart disease, snoring proved to be an independent risk factor for stroke (141). It has been proposed that OSA and snoring may promote cardiac arrhythmias and thus increase the risk for development of cardiac emboli to the brain (141). In addition, the hypoxemia that is associated with OSA may promote the development of atherosclerosis and thus stroke (145). Furthermore, the hemodynamic changes associated with the obstructive period in OSA may decrease cardiac index, which results in hypoperfusion to areas of the brain that are already compromised by cerebrovascular disease and thereby increases the risk of stroke (146,147). The high association between sleep apnea and cerebrovascular disease should be considered in the prevention and treatment of stroke.

B. Circadian Rhythms in Stroke

A circadian variation in time of onset of ischemic stroke has been reported in a few studies (148–153). The results from these studies suggest that ischemic stroke most often occurs in the morning after awakening (between 6 A.M. and noon). The Oxfordshire community stroke project found a peak time for stroke onset to be between 8:00 and 10:00 A.M. (153). The increased incidence of stroke in the morning hours may be related to how awakening affects the circadian rhythms of blood pressure and platelet aggregability. In the morning, when one awakens and assumes an upright posture, blood pressure and heart rate increase (154,155). Likewise, there is an increase in platelet aggregability, catecholamine levels, and plasma renin activity (154,155). All these factors interact to promote changes in vascular tone and thus increase the risk of ischemia.

In contrast, a circadian variation for other types of stroke such as cerebral hemorrhage or subarachnoid hemorrhage (SAH) has not been established. One recent study found a circadian rhythm for SAH with a peak onset between 9 and 10 A.M. in patients with hypertension (156). However, patients with normal blood pressures had a random 24-hr distribution of SAH onset. Therefore, the observed circadian variation in SAH in patients with hypertension may be explained by alterations in the circadian fluctuation in blood pressure.

In addition to a circadian rhythm in the frequency of strokes, there have been reports of a circannual rhythm in stroke and cerebral hemorrhage, with the highest number occurring in the winter months (January–March) (144,147). Again, this finding may be related to the seasonal changes in blood pressure. Mean blood pressure level is higher in the winter months increasing the risk for stroke and cerebral hemorrhage.

Only a limited number of studies have looked at the relationships between stroke and other circadian rhythms such as body temperature or hormone secretion. The circadian rhythm of cortisol does not appear to be disrupted in patients recovering from a hemispheric stroke, but prolactin levels were elevated and growth hormone levels were decreased (158,159). Probably the most interesting finding is the report of impaired nocturnal excretion of melatonin and altered cell-mediated immune function in patients with acute ischemic stroke (160). Similarly, a blunting of the melatonin rhythm has been seen after acute cerebral hemorrhage (161,162). It is interesting that melatonin-deficient rats were more susceptible to tissue damage after experimentally induced focal brain ischemia (163). Given the possible neuroprotective and immunological functions of melatonin, it is possible that alteration in circadian rhythmicity and melatonin may contribute to increased morbidity in patients with cerebrovascular disease (162).

V. Movement Disorders

The sleep and circadian manifestations of the various movement disorders have not been studied well. Most of the known information is derived from studies on case reports or small groups. However, emerging evidence indicates that abnormal motor activity during wake and sleep is often accompanied by sleep disruption. In this chapter we will discuss several movement disorders specific to sleep such as periodic leg movement in sleep, REM sleep behavior disorder, and nocturnal paroxysmal dystonia. Also the sleep abnormalities seen in several waking movement disorders will be reviewed: Parkinson's disease, progressive supranuclear palsy, and Shy-Drager syndrome.

A. Periodic Leg Movements in Sleep and Restless Syndrome

Periodic limb movements (PLMS) in sleep are repetitive, often stereotyped, movements that typically recur at intervals of 15–40 sec during NREM sleep. They are most commonly seen during the first half of the night in stages 1 and 2 of sleep. PLMS usually involve the legs with only rare involvement of the arms. The movements are often followed by changes on the EEG suggestive of an arousal (bursts of alpha activity, K complexes, and changes in sleep stages). PLMS may occur at any age, but their prevalence increases dramatically with aging, and they may affect 30–50% of individuals over 65 years of age (164). Because of the significant sleep fragmentation that occurs with PLMS, it is an important cause of insomnia and excessive daytime sleepiness. In a national cooperative study, PLMS accounted for 12.2% of the cases of insomnia (165).

Restless leg syndrome (RLS) is characterized by a sensation of discomfort in the legs (described as crawling and aching) while the person is at rest and upon retiring to bed. There is an irresistible urge to move the legs because of these uncomfortable sensations. Leg movements or walking brings some relief but delays sleep onset and disrupts sleep continuity. Although RLS is distinct from PLMS, as many as 70–90% of individuals with RLS will also have PLMS during sleep. Conversely, approximately 30% of PLMS patients will also have RLS.

The etiology of PLMS and RLS is usually unknown and therefore termed idiopathic. However, they have been associated anecdotally with a variety of medical illnesses such as anemia, uremia, iron and folate deficiency, alcohol use, vascular insufficiency, lumbar stenosis, excessive caffeine use, and Parkinson's disease. It is hypothesized that impaired central dopaminergic neurotransmission may be responsible for the abnormal motor activity in PLMS (166). Support for this hypothesis comes from a recent study that demonstrated decreased density of dopamine D2 receptors in the basal ganglia of patients with PLMS (167). Furthermore, treatment with dopamine agonist medications alleviates the symptoms of PLMS.

B. REM Behavior Sleep Disorder

REM behavior sleep disorder (RBD) is characterized by attacks of vigorous and often dangerous motor activity that occurs during REM sleep. The patient appears to be enacting a dream and often has recollection of being attacked, fighting back, and/or trying to run away. This disorder occurs more frequently in older men and has been seen in association with various brainstem abnormalities, extrapyramidal neurological disorders, and medical conditions that are summarized in Table 4.

Table 4 Medical Conditions Associated with REM Sleep Behavior Disorder

---

Idiopathic
Parkinson's disease
Progressive supranuclear palsy
Shy-Drager syndrome
Olivopontocerebellar degeneration
Multiple systems atrophy
Fatal familial insomnia
Narcolepsy
Diffuse Lewy body disease
Alzheimer's disease
Brainstem tumor
Brainstem infarct
Multiple sclerosis
Drug and alcohol withdrawal
Medication toxicity (fluoxitine, tricyclic antidepressants)

The distinguishing feature in RBD is the loss of motor inhibition during REM sleep, often referred to as REM sleep without atonia. The loss of REM atonia along with enhancement of phasic motor drive is postulated to be the mechanism behind RBD (168,169). In addition to increased motor activity during REM sleep, most patients with RBD have associated periodic and nonperiodic limb movements throughout all the stages of NREM sleep. The overall sleep architecture is usually unremarkable except for an elevated percentage of SWS (stages 3 and 4) (170,171).

C. Nocturnal Paroxysmal Dystonia

Nocturnal paroxysmal dystonia (NPD) is a syndrome of stereotyped body movements that occur during NREM sleep, particularly during slow-wave sleep (172,173). The motor attacks last less than 60 sec and may consist of ballistic, dystonic, and choreic movements. EEG and physiological monitoring reveal signs of arousal and autonomic activation at the onset of the attacks. Surface EEGs most often fail to show epileptic changes in either wakefulness or sleep. The interval between the attacks is 10–40 sec, which is in parallel to the periodic waxing and waning of EEG activation in NREM sleep (cyclic alternating pattern) (174).

The debate regarding whether NPD is the manifestation of a seizure disorder or a sleep disorder remains unresolved. However, current evidence suggests that NPD may be a variant of epilepsy rather than an unrelated paroxysmal movement disorder of sleep. The motor attacks of NPD are similar to the motor events seen with seizures that originate from the orbitomesofrontal areas (175). Surface EEGs abnormalities are often absent in seizures originating deep within the frontal lobes; thus repeated normal EEGS do not exclude an epileptic disorder. Sellal et al. have found evidence of frontal and frontotemporal hypometabolism on PET scans in four patients with NPD (176). Finally, as further support for epilepsy as a cause for NPD, several studies have found associated ictal or interictal abnormalities on EEG (175–179) and report that this disorder responds very favorably to treatment with carbamazapine, an anticonvulsant.

D. Parkinson's Disease

Parkinson's disease (PD) is a neurological disorder characterized by abnormal extrapyramidal features such as, resting tremor, rigidity, bradykinesia, and loss of postural reflexes. The underlying pathology of PD is the loss of pigmented cells in the substantia nigra and other pigmented nuclei (locus coeruleus) of the brainstem. Loss of dopaminergic neurons and impaired neurotransmission are the predominant pathology in PD. In addition, serotonergic neurons in the dorsal raphe nuclei and cholinergic neurons in the pedunculopontine nucleus are also decreased in number. These brainstem areas play an important role in the control of sleep and their damage may account for the prominent sleep abnormalities that have been reported in PD. Sleep/wake cycle disturbances are common complaints in patients with PD and severely affect their quality of life. Therefore, appropriate management of these patients must also address their sleep problems.

Sleep Disturbances in Parkinson's Disease

Sleep difficulties have been noted in greater than 70% of PD patients. The most common sleep disorder in Parkinson's patients is sleep onset and maintenance insomnia (180,181). As a result, patients often complain of multiple prolonged awakenings and excessive daytime sleepiness. It is common for patients to sleep during the day and remain awake for prolonged periods during the night, with a sleep pattern sometimes described as a reversed sleep/wake cycle.

Alterations in sleep architecture have also been reported in PD. Reduced amounts of slow-wave sleep (SWS) and REM sleep are the most frequently described changes (182–185). In addition, the number of sleep spindles during NREM sleep is reduced (185). Alterations in sleep architecture are even more prominent in patients with PD who also have nocturnal hallucinations (186).

Motor abnormalities associated with PD, such as rigidity and tremor, may interfere with sleep. Although tremor disappears after sleep onset, it may reappear in association with awakenings, arousals, and during sleep stage changes (187). Repeated eye blinking before the onset of sleep and intrusion of rapid eye movements in NREM sleep have been described and may disturb sleep (188,189). Approximately 30% of the PD patients also have periodic leg movements of sleep, which can result in multiple arousals. Finally, the treatment of PD may in itself be associated with drug-induced nocturnal dyskinesias (abnormal twisting, writing movements) that will negatively affect sleep quality (189).

In addition to the fact that motor abnormalities affect sleep, a variety of sleep-related respiratory disturbances have also been noted in PD patients (190,191). These disturbances are more frequent and severe in PD patients with autonomic dysfunction. Abnormalities of respiratory drive with resultant hypoventilation and central apneas have been described (190). Furthermore, abnormal movement in PD can involve the diaphragm or larynx and thus can impair respiration during sleep.

Parasomnias are common in PD patients and include night terrors, vivid dreams, nightmares, somnambulism, and nocturnal vocalizations. One study of 88 levodopa-treated PD patients revealed that 30.7% of the patients experienced either vivid dreaming, nocturnal vocalizations, or nightmares (192). REM RBD has also become increasingly recognized in PD patients (see Section V.B). Comella et al. showed that up to 50% of patients with moderate to severe PD had polysomnographic evidence of RBD (193).

Circadian Rhythm Disturbances in Parkinson's Disease

Motor symptoms in PD patients exhibit circadian variation. Some patients may have the mildest symptoms in the morning, while in others the symptoms are most severe in the morning and improve as the day progresses (194). Factor et al. studied 78 PD patients and found that 43.6% had better motor function in the morning, 37.2% were worse, and 19.2% were unchanged compared to the rest of the day (180). Some hereditary forms of early-onset parkinsonism (e.g., dopa-responsive hereditary dystonia and the autosomal recessive early-onset parkinsonism with diurnal fluctuation) also can exhibit a marked diurnal fluctuation in symptoms (195,196).

Abnormalities in the circadian rhythm of various hormonal and physiological variables have been reported in association with PD. Disturbed circadian rhythms in blood pressure adaptation and renal fluid handling have been described in PD patients (197,198). Cell loss in the supraoptic (SON) and paraventricular nuclei areas that are involved in the regulation of fluid balance has been reported in PD and may account for the orthostatic hypotension and loss of circadian renal water excretion that has been seen in PD patients (199). A phase advance of circadian rhythm of melatonin was seen in PD patients on dopamine replacement (200), whereas a phase advance of the melatonin rhythm was not present in an untreated group of PD patients (201). These results suggest that treatment with dopamine may in itself alter the phase of circadian rhythms. Other evidence in support of a circadian rhythm disruption in PD patients is provided in a study by Bliwise et al. (202), who compared disruptive nocturnal behaviors (sundowning) in 60 Alzheimer patients with 48 PD patients. The results showed that PD patients were more likely to exhibit sundowning than the AD patients. Sundowning may be caused by a disruption of the circadian sleep-wake cycle. Animal studies also suggest that alteration in dopamine function is associated with abnormalities in the diurnal rhythm of vasopressin levels (197). Despite some evidence that PD may be associated with circadian rhythm disturbances, other studies find no such abnormalities. For example, there was no significant difference in the circadian rhythm of body temperature in PD patients when compared to controls (203).

E. Progressive Supranuclear Palsy

Progressive supranuclear palsy (PSP) is a rare neurodegenerative disorder characterized by a subcortical dementia, axial rigidity, gait disturbance, and impaired voluntary gaze. Sleep abnormalities are present in most patients with PSP and insomnia is the most frequent complaint (189). Although a few studies indicate that alteration of sleep architecture is commonly seen in patients with PSP, information about the sleep/wake cycle and circadian rhythms is lacking.

Typical polysomnographic findings include reduced total sleep time, delayed sleep onset, fragmentation of sleep, decreased or absent sleep spindles, and reduced amounts of REM sleep (204–207). Aldrich et al. studied 10 patients with moderate to severe PSP and found that all of them had a severe insomnia, spending 2–6 hr awake per night (207). Montplaisir et al. studied the sleep architecture and quantitative EEG in six PSP patients with particular attention to quantifying REM sleep variables (206). They found a lower percentage of total REM sleep due to a reduction in the number of REM periods and mean duration of REM periods. The prominent alterations in REM sleep are probably secondary to a marked reduction of the mesopontine cholinergic nuclei, an important area in the regulation of REM sleep (208–210). Degenerative changes are also present in several other brainstem structures that play a role in sleep regulation: the pontine tegmentum, locus coeruleus, midbrain, and medical thalamus (211).

F. Multiple Systems Atrophy with Progressive Autonomic Failure (Shy-Drager Syndrome)

The Shy-Drager syndrome is characterized by progressive autonomic failure, cerebellar dysfunction, and parkinsonism (212). Pathologically, there is gliosis and degeneration in the basal ganglia, substantia nigra, locus ceruleus, cerebellar Purkinje cells, pontine nuclei, and the intermediolateral cell column. Oligodendroglial cytoplasmic inclusions are believed to be pathognomonic for the disorder (213).

Sleep abnormalities in Shy-Drager syndrome are frequently associated with the autonomic nervous system dysfunction. For example, sleep-related respiratory disturbances are common in Shy-Drager syndrome. Polysomnographic studies have demonstrated a variety of respiratory abnormalities: central, obstructive, and mixed apneas, arrhythmic breathing, and Cheyne-Stokes respirations (189,214–216). The disordered breathing is associated with nocturnal hypoxemia and frequent arousals and this in turn causes excessive daytime somnolence. Hemodynamic abnormalities may also occur during sleep, such as increased systemic arterial blood pressure during REM and slow-wave sleep (217). Also, REM sleep abnormalities have been described in this disorder, so REM SBD (see Section V.B for description) has been described in the Shy-Drager syndrome (218).

VI. Neuromuscular Disorders

Sleep complaints are common among patients with neuromuscular disease. The most frequent symptoms are daytime fatigue and sleepiness. A common problem in patients with neuromuscular disease is weakness that may affect respiration, particularly during sleep. Patients with neuromuscular disorders can have normal pulmonary function and oxygen saturations (SpO_2) while awake but have significant disordered breathing while asleep (219,220). Motor disturbances during sleep occur frequently in patients with neuromuscular disorders and may also contribute to daytime fatigue.

In this section, we will discuss the sleep abnormalities seen in some of the common neuromuscular disorders.

A. Myotonic Dystrophy

Myotonic dystrophy is the most common form of muscular dystrophy among Caucasians. It is an autosomal dominant disorder with variable penetrance. Clinical characteristics include myotonia, weakness of distal limb and facial muscles, cataracts, endocrinopathies, frontal balding, cardiomyopathy, and low intelligence or dementia. Sleep disturbances are common in patients with this disorder. Sleep studies demonstrate increased sleep fragmentation and decreased total REM sleep (221).

Excessive sleepiness is a common complaint in patients with myotonic dystrophy. The hypersomnia may be secondary to respiratory muscle involvement/weakness, resulting in abnormal breathing and oxygenation during sleep (223–226). Although sleep apnea and nocturnal hypoventilation may disturb sleep in some patients, others who complain of hypersomnia do not have significant sleep apnea and the hypersomnia is often not relieved fully by treatment of the sleep-disordered breathing (222,227,228). Patients with myotonic dystrophy have more severe respiratory disturbances during sleep than patients with respiratory muscle weakness due to other causes (227). Decreased ventilatory response to hypoxic and hypercapnic stimuli has also been reported in myotonic dystrophy (229,230). Therefore, respiratory muscle weakness alone cannot explain the level of disordered breathing that is seen in myotonic patients and raises the possibility that central mechanisms may play a role in the debilitating hypersomnia in these patients.

The neuropathology of myotonic dystrophy supports a central mechanism for the hypersomnia in this patient population. Neuronal damage in the dorso-medial nuclei of the thalamus, an area with an established role in the regulation of the sleep-wake cycle (120), has been reported in myotonic dystrophy. Both hypersomnia and insomnia have been reported after lesions of different regions of the thalamus (231).

Abnormalities in neuroendocrine function are also common in myotonic dystrophy. For example, alterations in gonadal hormones and glucose metabolism have been reported in patients with myotonic dystrophy. Therefore, infertility and diabetes mellitus are commonly present in these patients. Probably the most relevant to sleep is the finding that the normal increase in growth hormone level during slow-wave sleep was absent in three of the five patients studied by Culebras et al. (232). Given the possible role of growth hormone in aging, the loss of sleep-related growth hormone secretion may play a role in the accelerated" aging process that is characteristic in myotonic dystrophy (233).

B. Motor Neuron Disease (Amyotrophic Lateral Sclerosis and Postpolio Syndrome)

Amyotrophic lateral sclerosis (ALS) and postpolio syndrome (PPS) are disorders that primarily affect motor neurons. ALS is a progressive degenerative disorder of motor neurons in the spinal cord, brainstem, and motor cortex of the brain. It presents clinically with various combinations of muscular weakness, atrophy, and corticospinal tract abnormalities. Death is often due to respiratory failure. PPS is a disorder that affects survivors of an acute attack of poliomyelitis who may some 20 years later develop the disorder. This syndrome is characterized by impaired muscle function, fatigue, and joint pains. The weakness can involve both new muscles and muscles that were previously affected by poliomyelitis.

Fatigue is common in patients with ALS and PPS. This is particularly important in patients with PPS because fatigue and nonrefreshing sleep significantly hinder the quality of life of these patients. Despite negative consequences of poor sleep and daytime fatigue in these patients, very little work has been performed in this area. Results from a few clinical studies indicate that sleep-related respiratory disturbances, such as obstructive sleep apnea, may contribute to sleep fragmentation and daytime fatigue in patients with PPS (234–237). Ulfberg et al. studied 62 patients who had acute poliomyelitis earlier in life (on average 55 years earlier) and found that 26% of these patients had evidence of obstructive sleep apnea (238). Therefore, the fatigue that is commonly seen in PPS may be related to nocturnal respiratory dysfunction (i.e., sleep apnea) and may be at least in part reversible with treatment of the sleep apnea (233,235). Sleep-related respiratory disturbances in ALS have been even less studied (239–241). However, it is well known that respiratory muscle functioning is often impaired in patients with ALS, leading to hypoventilation and hypoxemia, which in turn will disturb sleep (239).

In addition to respiratory abnormalities during sleep, abnormal motor activity may also contribute to the sleep complaints and fatigue in patients with motor neuron disease. Nelson et al. studied a mixed population of ALS and PPS patients and found that more than 70% of the patients exhibited periodic limb movements in sleep (PLMS) (242).

C. Myasthenia Gravis

Myasthenia gravis (MG) is an autoimmune disorder affecting the neuromuscular junction, characterized by muscle fatigability. One of the most common sleep complaints includes waking up in the middle of the night with a sensation of breathlessness (243). Other complaints include daytime somnolence and morning headaches, symptoms that are often seen in patients with sleep apnea. Myasthenia gravis may involve the diaphragm and accessory respiratory muscles and thus contribute to sleep-disordered breathing (243,244). The disordered breathing may be especially severe during REM sleep where REM-related muscle atonia occurs. During REM sleep, the intercostal muscles and accessory respiratory muscles are not active leaving only the diaphragm to maintain respiration. As a result, significant oxygen desaturation and carbon dioxide retention can occur resulting in abrupt respiratory failure (243). Loss of normal metabolic ventilatory control mechanisms and central apneas will often occur in chronic MG (245). Even in patients who have been treated with medication to improve muscle strength, sleep architecture is altered. Polysomnographic recordings show decreased percentages of slow-wave sleep and REM sleep, with a concomitant increase in stage 1 sleep (233,246).

VII. Demyelinating Disease: Multiple Sclerosis

Multiple sclerosis (MS) is demyelinating disorder that affects multiple central nervous system white matter tracts. It is usually a disease of young adults and the symptoms are varied, but some, such as motor difficulties, visual problems, and sensory changes, are frequently present. Fatigue and sleep complaints are also common in patients with MS, but little is known regarding their underlying pathophysiology.

Manifestations of sleep disturbances in MS include prolonged sleep latency, frequent nocturnal awakenings, nonrestorative sleep, and early-morning awakening (247–249). Sleep abnormalities in MS are associated with a higher incidence of fatigue, excessive daytime sleepiness, and depression. Sleep disruption in MS may be explained by factors such as immobility, spasticity, urinary problems, respiratory conditions, and periodic leg movements of sleep (247,250–252). Some patients with MS have a combination of severe excessive daytime sleepiness and cataplexy, which has led to both clinical and genetic associations with narcolepsy (253–256).

Brain imaging studies suggest that the site of the demyelinating lesion may be correlated with the presence of sleep complaints. Lesions in the right and left frontal supraventricular white matter and deep white matter of the insula were found to be more frequently associated with sleep complaints (247). Based on the finding that melatonin levels are lower in patients with MS and depression, it has been speculated that abnormalities in circadian rhythmicity may be associated with alterations in the sleep cycle in MS (257,258).

VIII. Headache

Headache (of any type) affects 70–80% of the population. The most common types of headache that are associated with sleep and circadian rhythms are migraine, cluster, and chronic paroxysmal hemicrania. In this category, migraine is the most common, affecting 8% of men and 25% of women (259). Headache and sleep disorders frequently occur in the same patient either as a comorbid condition or because of factors that trigger both disorders. The same neurotransmitters that modulate sleep, such as serotonin and histamine, have also been implicated in some of these headaches (260,261). Sahota and Dexter described the potential relationship of headaches and sleep in the following ways: (1) sleep-stage related headaches, (2) sleep-relieved headaches, (3) sleep length (deprivation or excess) and headaches, (4) effects of sleep disorders on headaches, and (5) effects of headaches on sleep (262).

A. Alterations in Sleep

Migraine headache may be related to excessive amounts of slow-wave sleep and REM (263). The onset of nocturnal attacks of migraine most commonly occurs with the onset of REM sleep (264–266). Selectively depriving migraineurs of slow-wave sleep can decrease the frequency and severity of headaches. The change from one sleep phase to another can also serve as a headache trigger.

Episodic cluster headache may often be triggered by REM sleep and there is an eightfold increase in awakenings during REM sleep (267–269). On the other hand, chronic cluster appears to be triggered by non-REM sleep (267). Pfaffenrath et al. studied eight patients with chronic cluster headache and showed that majority of the attacks occurred in stage 2 (267).

Patients with chronic paroxysmal hemicrania (CPH) have reduced REM and total sleep time (270). In one case study of a patient with CPH 17 of 18 nocturnal attacks occurred during REM sleep (270).

Parasomnias are more common in migraineurs. These include night terrors, somnambulism, and enuresis (271–274). Headache is also a common feature of other sleep disorders such as sleep apnea. As many as 36–56% of sleep apnea patients report morning headaches (275). Sleep deprivation or excessive amounts of sleep can trigger headaches (276). On the other hand, a nap or a good night's sleep will often end a migraine attack. The effects of headaches on sleep have not been well documented. The increased awakenings from sleep noted in patients with cluster headache may be related to pain. Insomnia is common in patients with chronic headache disorders and may be modulated by abnormalities in serotonin metabolism (274,277).

B. Circadian Rhythms and Headache

A circadian and circannual rhythm of headache attacks is a well-known characteristic of cluster headache. Generally, cluster headache occurs most often in the spring and autumn, when there is a change in the length of light exposure. In addition to the seasonal influence, patients describe the recurrence of headaches at the same time of the day or night. This suggests that the trigger for this type of headache exhibits circadian variation. There is also evidence that there is a disruption of circadian rhythms in patients with the chronic type of cluster headache versus the episodic type (278). Findings include loss of the circadian variation of B-endorphin levels, reduced amplitude of the melatonin, cortisol, and testosterone rhythms, and altered phase of circadian rhythms during the cluster period (279–284). Other neuroendocrine abnormalities include a decrease level of prolactin during the night, in both the active and remission periods (281). The finding of circadian alterations during the remission period in cluster headache patients is important in that it suggests that these alterations are not simply due to pain, but may be part of the pathology of cluster headaches.

There is some evidence that migraine headache attacks also exhibit a circadian rhythm. One study analyzed the time of migraine onset in 15 migraine sufferers and found a peak frequency of migraine onset between 8 A.M. and 10 A.M. (285). Also noted was a decreased frequency of attacks between 8 P.M. and 4 A.M. It appears that the circadian rhythm of migraine attacks parallels those of stroke and myocardial infarction. This association suggests a possible common vascular mechanism underlying the initiation of migraine, stroke, and myocardial infarction.

References

1. Commission on Classification and Terminology and the International League Against Epilepsy. Proposal for revised clinical and electrographic classification of epileptic seizures. Epilepsia 1981; 22:489.

2. Gowers WR. Epilepsy and Other Chronic Convulsive Diseases: Their Causes, Symptoms, and Treatment. Philadelphia: P Blakiston's, 1901.

3. Langdon-Down M, Brain WR. Time of day in relation to convulsions in epilepsy. Lancet 1929; 2:1029–1032.

4. Patry FL. The relation of day, sleep and other factors to the incidence of epileptic seizures. Am J Psychiatry 1931; 87:789–813.

5. Janz D. The grand mal epilepsies and the sleep-waking cycle. Epilepsia 1962; 3:69–109.

6. Shouse MN, da Silva MA, Sammaritano M. Circadian rhythms, sleep, and epilepsy. J Clin Neurophysiol 1996; 13:32–50.

7. Hopkins H. The time appearance of epileptic seizures in relation to age, duration and type of syndrome. J Nerv Ment Disease 1933; 77:153–162.

8. Halberg F, Howard RB. 24 hour periodicity and experimental medicine; examples and interpretations. Postgrad Med 1958; 24:349–358.

9. Rossi GF, Colicchio G, Pola P, Roselli R. Sleep and epileptic activity. In: Degan R, Rodin EA, eds. Epilepsy, Sleep, and Sleep Deprivation, 2nd ed. (Epilepsy Res Suppl 2). Amsterdam: Elsevier, 1991:23–30.

10. Bazil C, Walczak T. Effects of sleep and sleep stage on epileptic and nonepileptic seizures. Epilepsia 1997; 38(1):56–62.

11. Sammaritano M, Gigli GL, Gotman J. Interictal spiking during wakefulness and sleep and the localization of foci in temporal lobe epilepsy. Neurology 1991; 41: 290–297.

12. Montplaiser J, Laverdiere M, Saint-Hilaire JM, Rouleau I. Nocturnal sleep recording in partial epilepsy: a study with depth electrodes. J Clin Neurophysiol 1987; 4:383–388.

13. Shouse MN, Siegel JM, Wu MF, et al. Mechanisms of seizure suppression during rapid-eye-movements (REM) sleep in cats. Brain Res 1989; 505:271–282.

14. Ross J, Johnson L, Walter R. Spike and wave discharge during stages of sleep. Arch Neurol 1966; 14:399–407.

15. Mattson RH, Patt KL, Calverley JR. Electroencephalograms of epileptics following sleep deprivation. Arch Neurol 1965; 13:310–315.

16. Degen R. A study of the diagnostic value of waking and sleep EEGs after sleep deprivation in epileptic patients on anticonvulsant therapy. Electroenceph Clin Neurophysiol 1980; 49:577–584.

17. Rowan AJ, Velhuisen RJ, Nagelkerke NJD. Comparative evaluation of sleep deprivation and sedated sleep EEGs as diagnostic aids in epilepsy. Electroenceph Clin Neurophysiol 1982; 54:357–364.

18. Pratt KL, Mattson RH, Weikers NJ, Williams R. EEG activation of epileptics following sleep deprivation: a prospective study of 114 cases. Electroenceph Clin Neurophysiol 1968; 24:11–15.

19. Scollo-Lavizzari G, Pralle W, Radue EW. Comparative study of afficacy of waking and sleep recordings following sleep deprivation as an activation method in the diagnosis of epilepsy. Eur Neurol 1977; 15:121–123.

20. Shouse MN, Sterman MB, Hauri PJ, Belsito O. Sleep disruption with basal forebrain lesions decreases latency to amygdala kindling in cats. Electroenceph Clin Neurophysiol 1984; 58:369–377.

21. Touchon J, Baldy-Moulinier M, Billiard M, et al. Sleep organization and epilepsy. Epilepsy Res 1991; 2:73–81.

22. Malow BA, Morton KA, Aldrich MS. Polysomnography in epilepsy patients: indications and results. Epilepsia 1995; 36:152.
23. Shouse M. Epileptic seizure manifestations during sleep. In: Kryger MH, Roth T, Dement WC, eds. Principles and Practice of Sleep Medicine, 2nd ed. Philadelphia: WB Saunders, 1994:801–814.
24. Touchon J, Baldy-Moulinier M, Billiard M, et al. Sleep organization and epilepsy. In: Degan R, Rodin EA, eds. Epilepsy, Sleep, and Sleep Deprivation, 2nd ed. (Epilepsy Res Suppl 2). New York: Elsevier, 1991:73.
25. Declerck AC, Wauquire A. Influence of antiepileptic drugs on sleep patterns. In: Degan R, Rodin EA, eds. Epilepsy, Sleep, and Sleep Deprivation, 2nd ed. (Epilepsy Res Suppl 2). New York: Elsevier, 1991:153.
26. Devinsky O, Ehrenberg B, Barthlen GM, et al. Epilepsy and sleep apnea syndrome. Neurology 1994;44:2060.
27. Kellaway P. Frost JD Jr, Crawley JW. Time modulation of spike-and-wave activity in generalized epilepsy. Ann Neurol 1980; 8:491–500.
28. Daly DD. Circadian cycles and seizures. In: Brazier MAB, ed. Epilepsy: Its Phenomena in Man. New York: Academic Press, 1973:215–233.
29. Martins da Silva A, Aarts JHP, Binnie CD, et al. Ultradian periodicity of interictal epileptiform EEG activity during sleep deprivation. Sleep Res 1983; 12:332.
30. Binnie CD. Are biological rhythms of importance in epilepsy. In Martins da Silva A, Binnie CD, Meinardi H, eds. Biorhythms and Epilepsy. New York: Raven Press; 1985:1–11.
31. Richter CF. Biological Clocks in Medicine and Psychiatry. Springfield, IL: Charles C Thomas, 1965.
32. Laasko ML, Leinonen L, Hatonen T, et al. Melatonin, cortisol and body temperature rhythms in Lennox-Gestaut patients with and without circadian rhythm sleep disorder. J Neurol 1993; 240(7):410–416.
33. Rao ML, Stefan H, Bauer J, Burr W. Hormonal changes in patients with partial epilepsy: attenuation of melatonin and prolactin circadian serum profiles. In: Dreifuss H, Meinardi H, Stefan H, eds. Chronopharmacology in Therapy of the Epilepsies. New York: Raven Press, 1990:55–70.
34. Schapel GJ, Beran RG, Kennaway DL, et al. Melatonin response in active epilepsy. Epilepsia 1995; 36(1):75–78.
35. Molai M, Culebras A, Miller M. Effect of interictal epileptiform discharges on nocturnal plasma prolactin concentrations in epileptic patients with complex partial seizures. Epilepsia 1986; 27:724–728.
36. Baust W, Irmscher K, Jorg J, Sommer T. Studies on the circadian periodicity in patients with the awakening type of idiopathic epilepsy. J Neurol 1976; 213(4):283–294.
37. Evans DA, Funkenstin HH, Albert MS, et al. Prevalence of Alzheimer's disease in a community population of older persons: higher than previously reported. JAMA 1989; 262:2551–2556.
38. American Psychiatric Association. Diagnostic and Statistical Manual of Mental Disorders, 4th ed (DSM-IV). Washington, DC: American Psychiatric Association, 1994.
39. Karla S, Bergeron C, Lang AE. Lewy body disease and dementia. A review. Arch Intern Med 1996; 156:487–493.

40. McKhann G, Drachman D, Folstein M, et al. Clinical diagnosis of Alzheimer's disease: report of the NINCDS-ADRDA Work Group. Neurology 1984; 34:939–944.

41. Aharon-Peretz J, Masiah A, Pillar T, Epstein R, et al. Sleep-wake cycles in multiinfarct dementia and dementia of the Alzheimer type. Neurology 1991; 41(10):143–159.

42. Reynolds CF III, Kupfer DJ, Taska LS, et al. EEG sleep in elderly depressed, demented, and healthy subjects. Biol Psychiatry 1985; 20:431–442.

43. Reynolds CF, Kupfer DJ, Houck PR, et al. Reliable discrimination of elderly depressed and demented patients by electroencephalographic sleep data. Arch Gen Psychiatry 1988; 45:258–264.

44. Vitiello MV, Poceta JS, Prinz PN. Sleep in Alzheimer's disease and other dementing disorders. Can J Psychol 1991; 45(2):221–239.

45. Prinz PN, Vitaliano PP, Vitiello MV, Bokan J, Raskind MA, et al. Sleep, EEG and mental function changes in mild, moderate and severe senile dementia of the Alzheimer's type. Neurobiol Aging 1982; 3:361–371.

46. Vitiello MV, Prinz PN, Williams DE, Frommlet MS, et al. Sleep disturbances in patients with mild-stage Alzheimer's disease. J Gerontol 1990; 45(4):M131–M138.

47. Prinz PN, Vitiello MV, Raskind MA. Thorpy MJ. Geriatrics: sleep disorders and aging. N Engl J Med 1990; 323(8):520–525.

48. Prinz PN, Peskind ER, Vitaliano PP, et al. Changes in the sleep and waking EEGs of nondemented and demented elderly subjects. J Am Geriatr Soc 1982; 30:86–93.

49. Vitiello M, Bokan J, Kukull W, et al. Rapid eye movement sleep measures of Alzheimer's type dementia patients and optimally healthy aged individuals. Biol Psychiatry 1984; 19:721–734.

50. Reynolds CF III, Kupfer DJ, Taska LS, et al. Slow-wave sleep in elderly depressed, demented, and healthy subjects. Sleep 1985; 8:155–159.

51. Benca RM, Obermeyer WH, Thisted RA, Gillin JC. Sleep and psychiatric disorders: a meta analysis. Arch Gen Psychiatry 1992; 49:651–668.

52. Allen SR, Seiler WO, Stahalin HB, Spiegal R. Seventy-two-hour polygraphic and behavioral recordings of wakefulness and sleep in a hospital geriatric unit: comparison between demented and nondemented patients. Sleep 1987; 10:143–159.

53. Bliwise DL, Bevier WC, Bliwise NG. Systematic 24-hr behavioral observations of sleep and wakefulness in a skilled-care nursing facility. Psychol Aging 1990; 5(1): 16–24.

54. Smith C. Sleep states, memory processes and synaptic plasticity. Behav Brain Res 1996; 78(1):49–56.

55. Stone WS, Altman HJ, Berman RF, et al. Association of sleep parameters and memory in intact old rats and young rats with lesions in the nucleus basalis magnocellularis. Behav Neurosci 1989; 103:755–764.

56. Prinz PN. Sleep patterns in the healthy aged: relationship with intellectual function. J Gerontol 1977; 32:179–186.

57. Erkinjuntti T, Partinen M, Sulkava R, et al. Sleep apnea in multinfarct dementia and Alzheimer's disease. Sleep 1987; 10:419–425.

58. Hoch CC, Reynolds CF, Nebes RD, et al. Clinical significance of sleep-disordered breathing in Alzheimer's disease: preliminary data. J Am Geriatr Soc 1989; 37:138–144.

59. Bliwise DL, Yesavage JA, Tinklenburg J, Dement WC. Sleep apnea in Alzheimer's disease. Neurobiol Aging 1989; 10:343–346.

60. Bliwise DL. Sleep apnea, dementia and Alzheimer's disease: a mini review. Bull Clin Neurosci 1989; 54:123–126.

61. Frommlet M, Prinz P, Vitiello M, et al. Sleep hypoxia and apnea are elevated in females with mild Alzheimer's disease. Sleep Res 1986; 15:189.

62. Smallwood R, Vitello M, Giblin F, Prinz P. Sleep apnea: relationship to age, sex, and Alzheimer's dementia. Sleep 1983; 6:16–22.

63. Scheltens PH, Visscher F, Keimpema ARJV, et al. Sleep apnea syndrome presenting with cognitive impairment. Neurology 1991; 41:155–156.

64. Legall D, Hubert P, Truelle JL, et al. Le syndrome d'apnee du sommeil: une cause curable de deterioration mentale. Presse Med 1986; 15:260–261.

65. Cao VH, Edgar DM, Heller HC, et al. Basal and phase-shifted neuronal rhythms in the aged SCN in vitro. Neurosci Abstr 1995; 21:1235.

66. Weiland NG, Wise PM. Aging progressively decreases the densities and alters the diurnal rhythms of alpha-1 adenergic receptors in selected hypothalamic regions. Endocrinology 1990; 126:2392–2397.

67. Wise PM, Cohen IR, Weiland NG, London DE. Aging alters the circadian rhythm of glucose utilization in the suprachiasmatic nucleus. Proc Natl Acad Sci USA 1988; 85:5305–5309.

68. Satinoff E, Li H, Tcheng TK, et al. Do the suprachiasmatic nuclei oscillate in old rats as they do in young ones? Am J Physiol 1993; 265:R1216–R1222.

69. Swaab DF, Fisser B, Kamphorst W, Troost D. The human suprachiasmatic nucleus; neuropeptide changes in senium and Alzheimer's disease. Bas Appl Histochem 1988; 32:43–54.

70. Van der Woude PF, Goudsmith E, Wierda M, et al. No vasopressin cell loss in human hypothalamus in aging and Alzheimer's disease. Neurobiol Aging 1995; 16:11–18.

71. Hoffman MA, Swaab DF. Alterations in circadian rhythmicity of vasopressin-producing neurons of the human suprachiasmatic nucleus (SCN) with aging. Brain Res 1994; 651:134–142.

72. Swaab DF, Fliers E, Partiman TS. The suprachiasmatic nucleus of the human brain in relation to sex, age, and senile dementia. Brain Res 1985; 342:37–44.

73. Anacoli-Israel S, Parker L, Sinaee R, et al. Sleep fragmentation in patients from a nursing home. J Gerontol 1989; 44(1): M18–M21.

74. Witting W, Kwa IH, Eikelenboom P, et al. Alterations in the circadian rest-activity rhythm in aging and Alzheimer's disease. Biol Psychiatry 1990; 27:563–572.

75. Regenstein QR, Morris J. Daily sleep patterns observed among institutionalized elderly residents. J Am Geriatr Soc 1987; 35:767–772.

76. Meguro K, Ueda M, Yamaguchi T, et al: Disturbance in daily sleep/wake patterns in patients with cognitive impairment and decreased daily activity. J Am Geriatr Soc 1990; 38:1176–1182.

77. Bliwise DL, Carrol JS, Dement WC. Predictors of observed sleep/wakefulness in residents in long-term care. J Gerontol Med Sci 1990; 50(A6):M303–M306.

78. Van Someren EJW, Hagebeuk EEO, Swaab DF, et al. Circadian rest-activity rhythm disturbances in Alzheimer's disease. Sleep Res 1993; 22:637.

79. Satlin A, Teicher MH, Liebermann HR, et al: Circadian locomotor activity rhythms in Alzheimer's disease. Neuropsychopharmacology 1991; 5(2):15–126.

80. Bliwise DL, Tinklenberg JR, Yesavage JA. Timing of sleep and wakefulness in Alzheimer's disease. Biol Psychiatry 1992; 31:1163–1165.

81. Bliwise DL. Dementia. In: Kryger MH, Roth T, Dement WC, eds. Principles and Practice of Sleep Medicine, 2nd ed. Philadelphia: WB Saunders, 1994:790–800.

82. Ghali LM, Hopkins RW, Rindlisbacher P. The fragmentation of rest/activity cycles in Alzheimer's disease. Int J Geriatr Psychiatry 1995; 10(4):299–304.

83. Ghali LM, Hopkins RW, Rindlisbacher P. Temporal shifts in peak daily activity in Alzheimer's disease. Int J Geriatr Psychiatry 1995; 10(6):517–521.

84. Satlin A, Volicer L, Stopa EG, Harper D. Circadian locomotor activity and core-body temperature rhythms in Alzheimer's disease. Neurobiol Aging 1995; 16(Suppl 5):765–771.

85. Campbell SS, Gillin JC, Kripke DF, Erikson P. Effects of gender and mental status in the circadian rhythm of body core temperature in elderly subjects. Sleep Res 1988; 17:362.

86. Prinz PN, Moe KE, Vitiello MV, Marks AL, et al. Entrained body temperature rhythms are similar in mild Alzheimer's disease, geriatric onset depression, and normal aging. J Geriatr Psychiatry Neurol 1992; 5(2):65–71.

87. Gillin JC, Kripke DF, Campbell SS. Ambulatory measures of activity, light and temperature in elderly normal controls and patients with Alzheimer's disease. Bull Clin Neurosci 1989; 54:144–148.

88. Van Coevorden A, Mockel J, Laurent E, et al. Neuroendocrine rhythms and sleep in aging men. Am J Physiol 1991; 260:E651–E661.

89. Uchida K, Okamoto N, Ohara K, Morita Y. Daily rhythm of serum melatonin in patients with dementia of the degenerative type. Brain Res 1996; 717:154–159.

90. Skene DJ, Vivien-Roel B, Sparks DL, Hunsaker JC, Pevet P, Ravid D, Swaab DF. Daily variation in the concentration of melatonin and 5-methoxytryptophol in the human pineal gland: effect of age and Alzheimer's disease. Brain Res 1990; 528:170–174.

91. Renfrew JW, May C, Tamarkin L, et al. Circadian rhythms in Alzheimer's disease. Soc Neurosci Abstr 1987; 3(Pt 2):1039.

92. Mishima K, Okawa M, Hishikawa Y, et al. Morning bright light therapy for sleep and behavior disorders in elderly patients with dementia. Acta Psychiatr Scand 1994; 89:1–7.

93. Tohgi H, Abe T, Takahashi S, et al. Concentrations of serotonin and its related substances in the cerebrospinal fluid in patients with Alzheimer type dementia. Neurosci Lett 1992; 141:9–12.

94. Souetre E, Salvati E, Krebs B. Abnormal melatonin response to 5-methoxypsoralen in dementia. Am J Psychiatry 1989; 146(8):1037–1040.

95. Haimov I, Laudon M, Zisapel N. Sleep disorders and melatonin rhythms in elderly people. Br Med J 1994; 309:1671.

96. Haimov I, Lavie P, Laudon M, et al. Melatonin replacement therapy in elderly insomniacs. Sleep 1995; 18(7):598–603.

97. Van Gool WA, Mirmiran M. Aging and circadian rhythms. In: Swaab DF, Fliers E, Mirmirin M, Van Gool WA, Van Haaren F, eds. Progress in Brain Research. Amsterdam: Elsevier, 1986;7:255–277.

98. Touitou Y. Some aspects of the circadian time structure in the elderly. Gerontology 1982; 28(Suppl 1):53–67.

99. Otsuka A, Mikami H, Katahira K, et al. Absence of nocturnal fall in blood pressure in elderly persons with Alzheimer type dementia. J Am Geriatr Soc 1990; 38:973–978.

100. Cohen-Mansfield J, Marx MS, Rosenthal AS. A description of agitation in a nursing home. J Gerontol 1988; 32(3):M77–M84.

101. Evans LK. Sundown syndrome in institutionalized elderly. J Am Geriatr Soc 1987; 35:101–108.

102. Jacobs D, Ancoli-Israel S, Parker L, Kripke DF. Twenty-four-hour-sleep/wake patterns in a nursing home population. Psychol Aging 1989; 4:352–356.

103. Cameron DE. Studies in senile nocturnal delirium. Psychiatr Q 1941; 15:47–53.

104. Bliwise DL. Sleep and "sundowning" in nursing home patients with dementia. Psychiatry Res 1993; 48:277–292.

105. Bliwise DL, Yesavage JA, Tinklenberg JR. Sundowning and rate of decline in mental function in Alzheimer's disease. Dementia 1992; 3:335–341.

106. Cohen-Mansfield J, Watson V, Meade W, et al. Does sundowning occur in residents of an Alzheimer's unit? Int J Geriatr Psychiatry 1989; 4:293–298.

107. Lloyd C, Hafner J, Holme G. Behavioral disturbance in dementia. J Geriatr Psychiatry Neurol 1995; 8:213–216.

108. Little J, Satlin A, Sunderland T, Volicer L. Sundown syndrome in severely demented patients with probable Alzheimer's disease. J Geriatr Psychiatry Neurol 1995; 8:103–106.

109. Vitiello MV, Bliwise DL, Prinz PN. Sleep in Alzheimer's disease and the sundown syndrome. Neurology 1992; 42(Suppl 6):83–94.

110. Bliwise DL, Carroll JS, Dement WC. Apparent seasonal variation in sundowning behavior in a skilled nursing facility. Sleep Res 1989; 18:408.

111. Wooten V. Medical causes of insomnia. In: Kryger MH, Roth T, Dement WC, eds. Principles and Practice of Sleep Medicine, 2nd ed. Philadelphia: WB Saunders, 1994:512–513.

112. Iakhno NN. Disorders of nocturnal sleep in Huntington chorea. Zh Nevropatol Psikhiatr 1985; 85:340–346.

113. Hansotia P, Wall P. Berendes J. Sleep disturbances and severity of Huntington's disease. Neurology 1985; 34:1672–1674.

114. Spire JP, Bliwise DL, Noronha ABC, et al. Sleep profiles in Huntington's disease. Neurology 1981; 31:151–152.

115. Tobler I, Gaus SE, Deboer P, et al. Altered circadian activity rhythms and sleep in mice devoid of prion protein. Nature 1996; 380:639–642.

116. Lugaresi E, Medori R, Montagna P, et al. Fatal Familial insomnia and dysautonomia with selective degeneration of the thalamic nuclei. N Engl J Med 1986; 315:997–1003.

117. Lugaresi E. The thalamus and insomnia. Neurology 1992; 42(Suppl 6):28–33.

118. Portaluppi F, Cortelli P, Avoni P, et al. Dissociated 24-hour patterns of somatotropin and prolactin in fatal familial insomnia. Neuroendocrinology 1995; 61:731–737.

119. Portaluppi F, Cortelli P, Avoni P, et al. Progressive disruption of the circadian rhythm of melatonin in fatal familial insomnia. J Clin Endocrinol Metab 1994; 78:1075–1078.

120. Tinuper P, Montagna P, Medori R, et al. The thalamus participates in the regulation of the sleep-waking cycle. A clinicopathological study in fatal familial thalamic degeneration. Electroenceph Clin Neurophysiol 1989; 73:117–123.

121. Tarter RE, Hegedus AM, Van Thiel DH, et al. Nonalcoholic cirrhosis associated with neuropsychological dysfunction in the absence of overt evidence of hepatic encephalopathy. Gastroenterology 1984; 86:1421–1427.

122. Sherlock S, Summerskill WHJ, White LP. Portal-systemic encephalopathy: neurological complication of liver disease. Lancet 1954: 2:453–455.

123. Bernardi M, De Palma R, Trevisani F, et al. Chronobiological study of factors affecting plasma aldosterone concentration in cirrhosis. Gastroenterology 1986; 91:683–691.

124. Bernardi M, Trevisani F, Ligabue A, et al. Chronobiological evaluation of sympathoadrenergic functioning in cirrhosis. Relationship with arterial pressure and heart rate. Gastroenterology 1987; 93:1178–1186.

125. Riggio O, Merli M, Pieche U, et al. Circadian rhythmicity of plasma amino acid variations in healthy subjects. Rec Prog Med 1989; 80(11):591–593.

126. Steindl PE, Finn B, Bendok B, Rothke S, Zee PC, Blei AT. Disruption of the diurnal rhythm of plasma melatonin in cirrhosis. Ann Intern Med 1995; 123:274–277.

127. Zee PC, Mehta R, Turek FW, Blei AT. Portacaval anastomosis disrupts circadian locomotor activity and pineal melatonin rhythms in rats. Brain Res 1991; 560:17–33.

128. Coy DL, Mehta R, Zee P, Salchli F, Turek FW, Blei AT. Portal-systemic shunting and the disruption of circadian locomotor activity in the rat. Gastroenterology 1992; 103:222–228.

129. Steindl PF, Coy DL, Finn B, Zee PC, Blei AT. A low protein diet ameliorates disrupted diurnal locomotor activity in rats after portacaval anastomosis. Am J Physiol 1996; 271(4 Pt 1):G555–560.

130. Culebras A, Magana R. Neurologic disorders and sleep disturbances. Semin Neurol 1987; 7(3):277–285.

131. Korner E, Flooh E, Reinhart B, et al. Sleep alterations in ischemic stroke. Eur Neurol 1986; 25(Suppl 2):104–110.

132. Culebras A, Moore JT. Magnetic resonance findings in REM sleep behavior disorder. Neurology 1989; 39:1519–1523.

133. Jurko MF, Andy OJ, Webster CL. Disordered sleep patterns following thalamotomy. Clin Electroencephal 1971; 2:213–217.

134. Bassetti C, Mathis J, Gugger M, et al. Hypersomnia following paramedian thalamic stroke: a report of 12 patients. Ann Neurol 1996; 39(4):471–480.

135. Askenasy JJ, Goldhammer I. Sleep apnea as a feature of bulbar stroke. Stroke 1988; 19:637–639.

136. Palomaki H, Partinen M, Erkinjuntti T, et al. Snoring, sleep apnea syndrome, and stroke. Neurology 1992; 42(Suppl 6):75–82.

137. Fletcher EC, DeBehnke RD, Lavoi MS, et al. Undiagnosed sleep apnea in patients with essential hypertension. Ann Intern Med 1985; 103:190–194.

138. Williams AJ, Houston D, Finberg S, et al. Sleep apnea syndrome and essential hypertension. Am J Cardiol 1985; 55:1019–1022.

139. Hoffstein V, Rubenstein I, Mateika S, et al. Determinants of blood pressure in snorers. Lancet 1988; 2:992–994.

140. Spriggs D, French JM, Murgy JM, et al. Historical risk factors for stroke: a case control study. Age Aging 1990; 19:280–287.
141. Koskenvuo M, Kaprio J, Telakivi T, et al. Snoring as a risk factor for ischemic heart disease and stroke in men. Br Med J 1987; 294:16–19.
142. Palomaki H. Snoring and the risk of ischemic brain infarction. Stroke 1991; 22: 1021–1025.
143. Partinen M, Palomaki H. Snoring and cerebral infarction. Lancet 1985; 2:1325–1326.
144. Kapen S, Park A, Goldberg J, et al. The incidence and severity of obstructive sleep apnea in ischemia. Neurology 1991; 41(Suppl 1):125 (abstract).
145. Gainer JL. Hypoxia and atherosclerosis: re-evaluation of an old hypothesis. Atherosclerosis 1987; 68:263–266.
146. Guilleminault C, Motta J, Mihm F, et al. Obstructive sleep apnea and cardiac index. Chest 1986; 89:331–334.
147. Mcginty D, Beahm E, Stern N, et al. Nocturnal hypotension in older men with sleep breathing disorders. Chest 1988; 94:305–311.
148. Cugini P, DiPalma L, Battisti P, Leone G. Ultradian, circadian, and infradian periodicity of some cardiovascular emergencies. Am J Cardiol 1990; 66:240–243.
149. Marler JR, Price TR, Clark GL, et al. Morning increase in onset of ischemic stroke. Stroke 1989; 20:473–476.
150. Argentino C, Toni D, Rasura M, et al. Circadian variation in the frequency of ischemic stroke. Stroke 1990; 21:387–389.
151. Tsementzis SA, Gill JS, Hitchcock ER, et al. Diurnal variation of and activity during the onset of stroke. Neurosurgery 1985; 17:901–904.
152. Marsh EE, Biller J, Adams HP, et al. Circadian variation in onset of acute ischemic stroke. Arch Neurol 1990; 47:1178–1180.
153. Wroe SJ, Sandercock P, Bamford J, et al. Diurnal variation in incidence of stroke: Oxfordshire community stroke project. Br Med J 1992; 304(6820):155–157.
154. Brezinski DA, Tofler GH, Muller JE, et al. Morning increase in platelet aggregability: association with upright posture. Circulation 1988; 78:35–40.
155. George CFP. Cardiovascular disease and sleep. In: Kryger MH, Roth T, Dement WC, eds. Principles and Practice of Sleep Medicine, 2nd ed. Philadelphia: WB Saunders, 1994:835–846.
156. Kleinpeter G, Schatzer R, Bock F. Is blood pressure really a trigger for circadian rhythm of subarachnoid hemorrhage? Stroke 1995; 26(10):1805–1810.
157. Pasqualetti P, Natali G, Casale R, Colantonio D. Epidemiological chronorisk of stroke. Acta Neurol Scand 1990; 81:71–74.
158. Culebras A, Miller M. Dissociated patterns of nocturnal prolactin, cortisol, and growth hormone secretion after stroke. Neurology 1984; 34(5):631–636.
159. Culebras A, Miller M. Absence of sleep-related elevation of growth hormone level in patients with stroke. Arch Neurol 1983; 40(5):283–286.
160. Fiorina P, Lattuada G, Ponari O, et al. Impaired nocturnal melatonin excretion and changes of immunological status in ischemic stroke patients. Lancet 1996; 347:692–693.
161. Pang SF, Li Y, Jiang DH, et al. Acute cerebral hemorrhage changes the nocturnal surge of plasma melatonin in humans. J Pineal Res 1990; 9:193–208.
162. Penev PD, Zee PC. Melatonin: a clinical perspective. Ann Neurol 1997; 42(4):545–553.

163. Manev H, Uz T, Kharlamov A, Joo JY. Increased brain damage after stroke or excitotoxic seizures in melatonin-deficient rats. FASEB J 1996; 10:1546–1551.

164. Ancoli-Israel S, Kripke DF, Mason W, et al. Sleep apnea and periodic movements in sleep in an aging population. J Gerontol 1985; 40:419–425.

165. Coleman R, Roffwarg HP, Kennedy SJ. Sleep-wake disorders based on a polysomnographic diagnosis: a national cooperative study. JAMA 1982; 247:997–1003.

166. Montplaisir J, Godbout R, Poirier G, Bedard MA. Restless leg syndrome and periodic movements in sleep: physiopathology and treatment with L-dopa. Clin Neuropharmacol 1986; 9:456–463.

167. Staedt J, Stoppe G, Kogler A, et al. Nocturnal myoclonus syndrome (periodic movements in sleep related to central dopamine D2-receptor alteration. Eur Arch Psychiatry Clin Neurosci 1995; 245:8–10.

168. Hishikawa Y, Sugita Y, Iijima S, et al. Mechanisms producing "stage 1-REM" and similar dissociations of REM sleep and relation to delirium. Adv Neurol Sci (Tokyo) 1981; 25:1129–1147.

169. Lapierre O, Casademont A, Montplaisir J, et al. Tonic phasic features of REM sleep behavior disorder. Sleep Res 1991; 20:276.

170. Mahowald M, Schenck CH. REM sleep behavior disorder. In: Kryger MH, Roth T, Dement WC, eds. Principles and Practice of Sleep Medicine, 2nd ed. Philadelphia: WB Saunders, 1994:574–588.

171. Schenck CH, Duncan E, Hopwood J, et al. The human REM sleep behavior disorder (RBD): quantitative polygraphic and behavioral analysis of 9 cases. Sleep Res 1988; 17:14.

172. Lugaresi E, Cirignotta F. Hypnogenic paroxysmal dystonia: epileptic seizures or a new syndrome? Sleep 1981; 4:129–138.

173. Lugaresi E, Cirignotta F, Montagna P. Nocturnal paroxysmal dystonia. J Neurol Neurosurg Psychiatry 1986; 49:375–380.

174. Sforza E, Montagna P, Rinaldi R, et al. Paroxysmal periodic motor attacks during sleep: clinical and polygraphic features. Electroenceph Clin Neurophysiol 1993; 86:161–166.

175. Tinuper P, Cerullo A, Cirignotta F, et al. Nocturnal paroxysmal dystonia with short-lasting attacks: three cases with evidence for an epileptic frontal lobe origin of seizures. Epilepsia 1990; 147:121–128.

176. Sellal F, Hirsch E, Maquet P, et al. Postures et mouvementa anormaux paroxystiques au cours du sommeil: dystonie paroxystique hypnogenique ou epilepsie partielle? Rev Neurol (Paris) 1991; 147:121–128.

177. Stoudemire A, Ninan PT, Wooten V. Hypnogenic paroxysmal dystonia with panic attacks responsive to drug therapy. Psychosomatics 1987; 28:280–281.

178. Besset A, Billiard M. Nocturnal paroxysmal dystonia. In: Koella WP, Ruther E, Schulz H, eds. Sleep '84. Stuttgart: Fischer Verlag, 1985:383–385.

179. Meierkord H, Fish DR, Smith SJM, et al. Is nocturnal paroxysmal dystonia a form of frontal lobe epilepsy? Move Disord 1992; 7:38–42.

180. Factor SA, McAlarney T, Sanchez-Ramos JR, Weiner WJ. Sleep disorders and sleep effect in Parkinson's. Move Disord 1990; 5(4):280–285.

181. Nausieda PA. Sleep disorders. In: Koller WC, ed. Handbook of Parkinson's Disease. New York: Marcel Dekker, 1987:371–380.

182. Kales A, Ansel RD, Markham CH, et al. Sleep in patients with Parkinson's disease and normal subjects prior to and following levodopa administration. Clin Pharmacol Ther 1971; 12:397–407.
183. Hardie RJ, Efthimiou J, Stern GM. Respiration and sleep in Parkinson's disease. J Neurol Neurosurg Psychiatry 1986; 49:1326.
184. Askenasy JJ. Sleep in Parkinson's disease. Acta Neurol Scand 1993; 87:167–170.
185. Aldrich MS. Parkinsonism. In: Kryger MH, Roth T, Dement WC, eds. Principles and Practice of Sleep Medicine, 2nd ed. Philadelphia: WB Saunders, 1994:835–846.
186. Comella CL, Tanner CM, Ristanovic RK. Sleep disturbances and hallucinations in Parkinson's disease: results of quantitative sleep studies. Ann Neurol 199; 30:295.
187. Fish DR, Sawyers D, Allen PJ, et al. The effects of sleep on the dykinetic movements of Parkinson's disease, Gilles de la Tourette syndrome, Huntington's disease, and torsion dystonia. Arch Neurol 1991; 48:210–214.
188. Mouret J. Differences in sleep in patients with Parkinson's disease. Electroenceph Clin Neurophysiol 1975; 38:653–657.
189. Chokroverty S. Sleep and degenerative neurological disorders. In: Aldrich M, ed. Neurologic Clinics: Sleep Disorders II. Philadelphia: WB Saunders, 1996; 791–805.
190. Feinsilver SH, Freidman JH, Rosen JM. Respiration and sleep in Parkinson's disease. J Neurol Neurosurg Psychiatry 1986; 49:964.
191. Apps MCP, Sheaff PC, Ingram DA, et al. Respiration and sleep in Parkinson's disease. J Neurol Neurosurg Psychiatry 1985; 48:1240–1245.
192. Scharf B, Moskovitz C, Lupton MD, Klawans HL. Dream phenomena induced by chronic levadopa therapy. J Neural Transm 1978; 43:143–151.
193. Comella CL, Ristanovic R, Goetz CG. Parkinson's disease patients with and without REM behavior disorder (RBD): a polysomnographic and clinical comparison. Neurology 1993; 43(Suppl 2):A301.
194. Marsden CD, Parkes JD, Quinn N. Fluctuations of disability in Parkinson's disease: clinical aspects. In: Marsden CD, Fahn S, eds. Movement Disorders. London: Butterworth Scientific 1982:96–122.
195. Sunohara N, Mano Y, Ando K, Satoyoshi E. Idiopathic dystonia-Parkinsonism with marked diurnal fluctuation of symptoms. Ann Neurol 1985; 17:39–45.
196. Segawa M, Hosaka A, Miyagawa F, et al. Hereditary progressive dystonia with marked diurnal fluctuation. Adv Neurol 1976; 14:215–233.
197. Hineno T, Mizobuchi M, Hiratani K, Inami Y, Kakimoto Y. Disappearance of circadian rhythms in Parkinson's disease model induced by 1-methyl-4-phenyl-1,2,3,6-tetrahydropyridine in dogs. Brain Res 1992; 580:92–99.
198. Miceli G, Marttignoni E, Cavallini A, Sandrini G, Nappi G. Postprandial and orthostatic hypotension in Parkinson's disease. Neurology 1987; 37:386–393.
199. Ansorge O, Daniel SE, Pearce RKB. Neuronal loss and plasticity in the supraoptic nucleus in Parkinson's disease. Neurology 1997; 49:610–613.
200. Fertl E, Auff E, Doppelbauer A, Waldhauser F. Circadian secretion pattern of melatonin Parkinsonian disease. J Neural Transm 1991; 3:41–47.
201. Fertl E, Auff E, Doppelbauer A, Waldhauser F. Circadian secretion pattern of melatonin in de novo Parkinsonian patients: evidence for phase-shifting properties of L-dopa. J Neural Transm 1993; 5:227–234.
202. Bliwise DL, Watts RL, Watts N, Rye DB, et al. Disruptive nocturnal behavior in

Parkinson's disease and Alzheimer's disease. J Geriatr Psychiatry Neurol 1995; 8(2):107–110.

203. Dowling GA. A comparative study of sleep and temperature rhythm in older women with and without Parkinson's disease. Sleep Res 1996; 25:409 (abstract).

204. Gross RA, Spehlmann R, Daniels JC. Sleep disturbances in progressive supranuclear palsy. Electroenceph Clin Neurophysiol 1978; 45:16–25.

205. Leygonie F, Thomas J, Degos JD, Bouchareine A, Barbizet J. Troubles du sommeil dans la maladie de Steele-Richardson. Rev Neurol (Paris) 1976; 132:125–136.

206. Montplaisir J, Petit D, Decary A, et al. Sleep and quantative EEG in patients with progressive supranuclear palsy. Neurology 1997; 49:999–1003.

207. Aldrich MS, Foster NL, White RF, Bluemlein L, Prokopowicz G. Sleep abnormalities in progressive supranuclear palsy. Ann Neurol 1989; 25:577–581.

208. Jellinger K. The pedunculopontine nucleus in Parkinson's disease, progressive supranuclear palsy and Alzheimer's disease. J Neurol Neurosurg Psychiatry 1988; 51:540–543.

209. Zweig RM, Whitehouse PJ, Casanova MF, Walker LC, Jankel WR, Price DL. Loss of pedunculopontine neurons in progressive supranuclear palsy. Ann Neurol 1987; 22:18–25.

210. McCarley RW, Greene RW, Rainnie D, Portas CM. Brainstem neuromodulation and REM sleep. Semin Neurosci 1995; 7:341–354.

211. Steele JC, Richardson JC, Olszeweski J. Progressive supranuclear palsy. Arch Neurol 1964; 10:333–359.

212. Quinn N. Multiple systems atrophy: the nature of the beast. J Neurol Neurosurg Psychiatry 1989; 52(Suppl):S78–S89.

213. Papp MI, Lantos PL. The distribution of oligodendroglial inclusions in multiple systems atrophy and its relevance to clinical symptomatology. Brain 1994; 117:235–243.

214. Chokroverty S, Sharp JT, Barron KD. Periodic respiration in erect posture in Shy-Drager syndrome. J Neurol Neurosurg Psychiatry 1978; 41:980–986.

215. Guilleminault C, Tilkian A, Lehrman K, et al. Sleep apnea syndromes: states of sleep and autonomic dysfunction. J Neurol Neurosurg Psychiatry 1997; 40:718–725.

216. Guilleminault C, Briskin J, Greenfield M, et al. The impact of autonomic nervous system dysfunction on breathing during sleep. Sleep 1981; 4:263–278.

217. Martinelli P, Coccagna G, Rizzuto N, Lugarasi E. Changes in systemic arterial pressure during sleep in Shy-Drager syndrome. Sleep 1981; 4:139–146.

218. Shimizu T, Sugita Y, Iijima S, et al. Sleep study in Shy-Drager syndrome. Clin Neurol (Jpn) 1981; 21:218–227.

219. Labanowski M, Schmidt-Nowara W, Guilleminault C. Sleep and neuromuscular disease: frequency of sleep disordered breathing in a neuromuscular disease clinic population. Neurology 1996; 47:1173–1180.

220. Smith PE, Calverly PM, Edwards RH. Hypoxemia during sleep in Duchenne muscular dystrophy. Am Rev Respir Dis 1988; 137:884–888.

221. Broughton R, Stuss D, Kates M, et al. Neuropsychological deficits and sleep in myotonic dystrophy. Can J Neurol Sci 1990; 17:410.

222. Hansotia P, Frens D. Hypersomnia associated with alveolar hypoventilation in myotonic dystrophy. Neurology 1981; 31:1336.

223. Park YD, Radtke RA. Hypersomnolence in myotonic dystrophy. Neurology 1992; 42(Suppl 3):352.

224. Coccagna G, Martinelli P, Lugaresi E. Sleep and alveolar hypoventilation and hypodynamics in myotonic dystrophy. Acta Neurol Belg 1982; 82:185–194.

225. Guilleminault C, Cummiskey J, Motta J, Lynne-Davies P. Respiratory and hypodynamics study during wakefulness and sleep in myotonic dystrophy. Sleep 1978; 1:19–31.

226. Cirignotta F, Mondini S, Zucconi M, et al. Sleep related breathing impairments in myotonic dystrophy. J Neurol 1987; 235:80–85.

227. Gilmartin JJ, Cooper BG, Griffiths CJ, et al. Breathing during sleep in patients with myotonic respiratory muscle weakness. Q J Med 1991; 78:21–31.

228. Meche FGA, Boogaard JM, Sluys JCM, et al. Daytime sleep in myotonic dystrophy is not caused by sleep apnea. J Neurol Neurosurg Psychiatry 1994; 57:626.

229. Serisier DE, Mastaglia FL, Gibson GJ. Respiratory muscle function and ventilatory control. I. In patients with motor neurone disease. II. In patients with myotonic dystrophy. Q J Med 1982; 202:205–226.

230. Begin R, Bureau MA, Lupien L, et al. Pathogenesis of respiratory insufficiency in myotonic dystrophy. Am Rev Respir Dis 1982; 125:312–318.

231. Guilleminault C, Quera-Salva M-A, Goldberg MP. Pseudoexcessive sleepiness and bilateral paramedian thalamic lesions. In: Guilleminault C, Lugaresi E, Montagna P, Gambetti P, eds. Fatal Familial Insomnia: Inherited Prion Diseases, Sleep, and the Thalamus. New York: Raven Press, 1994:15–24.

232. Culebras A, Podolsky S, Leopold NA. Absence of sleep-related growth hormone elevations in myotonic dystrophy. Neurology 1977; 27:165.

233. Culebras A. Sleep and neuromuscular disorders. In: Aldrich M, ed. Neurologic Clinics: Sleep Disorders II. Philadelphia: WB Saunders, 1996:791–805.

234. Guilleminault C, Motta J. Sleep apnea syndrome as long term sequela of poliomyelitis. In: Guilleminault C, Dement WC, eds. Sleep Apnea Syndromes. New York: Alan R Liss, 1978:309–315.

235. Steljes DG, Kryger MH, Kirk BW, et al. Sleep in post-polio syndrome. Chest 1969; 98:133.

236. Hill R, Robbins AW, Messing R, Aurora NS. Sleep apnea syndrome after poliomyelitis. Am Rev Respir Dis 1983; 127:129–131.

237. Bach JR, Alba AS, Shin D. Management alternatives for post-polio respiratory insufficiency. Assisted ventilation by nasal or oral-nasal interface. Assisted Ventilation 1989; 68:264–271.

238. Ulfberg J, Jonsson R, Ekeroth G. Sleep apnea syndrome among poliomyelitis survivors. Neurology 1997; 49:1189.

239. Gay PC, Westbrook PR, Daube JR, et al. Effects of alterations in pulmonary function and sleep variables on survival in patients with amyotrophic lateral sclerosis. Mayo Clin Proc 1991; 66:686.

240. Minz M, Autret A, Laffont F, et al. A study on sleep in amyotrophic lateral sclerosis. Biomedicine 1979; 30:40.

241. Parhad IM, Clark AW, Barron KD, et al. Diaphragmatic paralysis in motor neuron disease. Neurology 1978; 28:18.

242. Nelson J, Sufit R, Janssen I, Zee PC. Fatigue in motor neuron disorders is associated with pathological motor activity during sleep. Ann Neurol 1995; 38(2):350 (abstract).

243. Quera-Salva MA, Guilleminault C, Chevret S, et al. Breathing disorders during sleep in myasthenia gravis. Ann Neurol 1992; 31:86–92.

244. Mier-Jedrejowicz A, Brophy C, Green M. Respiratory muscle function in myasthenia gravis. Am Rev Respir Dis 1988; 138:867.

245. Plum F, Leigh RJ. Abnormalities of central mechanisms. In: Hornbein TF. Regulation of Breathing: Part II. New York: Marcel Dekker, 1981; 989–1067.

246. Culebras A. Neuroanatomic and neurologic correlates of sleep disturbances. Neurology 1992; 42(Suppl 6):19–27.

247. Clark CM, Fleming JA, Oger J, Klonnoff H, Paty D. Sleep disturbance, depression and lesion site in patients with multiple sclerosis. Arch Neurol 1992; 49:641–643.

248. Saunders J, Whitham R, Schaumann B. Sleep disturbance, fatigue and depression in multiple sclerosis. Neurology 1991; 41:320.

249. Leo GJ, Rao SM, Bernardin L. Sleep disturbance in multiple sclerosis. Neurology 1991; 41:320.

250. Smith CR, Aisen ML, Scheinberg L. Symptomatic management in multiple sclerosis. In: McDonald WI, Silberg DA, eds. Multiple Sclerosis. London: Butterworth, 1986:1–11.

251. Valiquette G, Herbert J, Maede-D'Alisera P. Desmopressin in the management of nocturia in patients with multiple sclerosis. A double-blind, crossover study. Arch Neurol 1996; 53(12):1270–1275.

252. Compston DAS: The management of multiple sclerosis. Q J Med 1989; 70:93–101.

253. Berg O, Hanley J. Narcolepsy in two cases of multiple sclerosis. Acta Neurol Scand 1963; 39:252–257.

254. Schrader H, Gotibesen OB, Skomedal GM. Multiple sclerosis and narcolepsy/cataplexy in a monozygotic twin. Neurology 1980; 30:105–108.

255. Younger DS, Pedley TA, Thorpy MJ. Multiple sclerosis and narcolepsy: possible similar genetic susceptibility. Neurology 1991; 41:447–448.

256. Poirier G, Montplaisir J, Dumont M, et al. Clinical and sleep laboratory study of narcoleptic symptoms in multiple sclerosis. Neurology 1987; 37:693–695.

257. Sandyk R. The pineal gland, cataplexy, and multiple sclerosis. Int J Neurosci 1995; 83(3–4):153–163.

258. Sandyk R. Multiple sclerosis: the role of puberty and the pineal gland in its pathogenesis. Int J Neurosci 1993; 68(3–4):209–225.

259. Lipton R, Rapoport A. Headache syndromes and their treatment. In: Samuels M, Feske S, eds. Office Practice of Neurology. New York: Churchill Livingstone, 1996: 1105–1111.

260. Blau JN. Resolution of migraine attacks: sleep and the recovery phase. J Neurol Neurosurg Psychiatry 1982; 45:223–226.

261. Lance JW, Lambert GA, Goadsby PJ, et al. Brainstem influences on the cephalic circulation: experimental data from cat and monkey of relevance to the mechanism of migraine. Headache 1983; 23:258–265.

262. Sahota PK, Dexter JD. Sleep and headache syndrome: a clinical review. Headache 1990; 30:80–84.

263. Dexter JD. The relationship between stage III+IV+REM sleep and arousals with migraine. Headache 1979; 19:364–369.

264. Dexter JD, Riley TL. Studies in nocturnal migraine. Headache 1975; 15:51–62.

265. Paiva T, Batista A, Martins P, Martins A. The relationship between headaches and sleep disturbances. Headache 1995; 35:590–596.

266. Hsu LK, Crisp AH, Kalucy RS, et al. Early morning migraine. Lancet 1977; 1:447–451.

267. Pfaffenrath V, Pollmann W, Ruther E, Lund R, Hajak G. Onset of nocturnal attacks of chronic cluster headache in relation to sleep stages. Acta Neurol Scand 1986; 73:403–407.

268. Kudrow L, McGinty DJ, Phillips ER, Stevenson M. Sleep apnea in cluster headache. Cephalalgia 1984; 4:33–38.

269. Dexter J, Weitzman E. The relationship of nocturnal headaches to sleep stage patterns. Neurology 1970; 20:513–518.

270. Kayed K, Godtlibsen QB, Sjaastad O. Chronic paroxysmal hemicrania IV. "REM sleep locked" nocturnal headache attacks. Sleep 1978; 1:91–95.

271. Barabas G, Ferrari M, Matthews WS. Childhood migraine and somnambulism. Neurology 1983; 6:95–100.

272. Pradalier A, Guroud M, Dry J. Somnambulism, migraine, and propanolol. Headache 1987; 27:143–145.

273. Dexter JD. The relationship between disorders of arousal from sleep and migraine. Headache 1986; 26:322.

274. Paiva T, Martins P, Batista A, et al. Sleep disturbances in chronic headache patients: a comparison with healthy controls. Headache Q 1994; 5:135–141.

275. Guilleminault C, Hold J, Mitler MM. Clinical overview of the sleep apnea syndromes. In: Guilleminault C, Dement WC, eds. Sleep Apnea Syndromes. New York: Alan R Liss, 1978:1–12.

276. Blau JN. Sleep deprivation headache. Cephalalgia 1990; 10:157–160.

277. Paiva T, Esperanca P, Martins A, Batista A, Martins P. Sleep disorders in headache patients. Headache Q 1992; 3:438–442.

278. Ferrari E, Canepari C, Bossolo PA, Vailati A, Martignoni E, Micieli G, et al. Changes of biological rhyth.ms in primary headache syndromes. Cephalalgia 1983; 3(Suppl 1):56–68.

279. Facchinetti F, Nappi G, Cicoli C, Micieli G, Ruspa M, Bono G, Genazzani AR. Reduced testosterone levels in cluster headache: a stress-related phenomenon? Cephalalgia 1986; 6:29–34.

280. Waldenlind E, Ekborn K, Friberg Y, Saaf J, Watterberg L. Decreased nocturnal serum melatonin levels during active cluster headache periods. Opusc Med 1984; 29:109–112.

281. Waldenlind E, Gustafsson SA, Ekborn K, Wetterberg L. Circadian secretion of cortisol and melatonin in cluster headache during active cluster periods and remission. J Neurol Neurosurg Psychiatry 1987; 50:207–213.

282. Nappi G, Micieli G, Facchinetti F, et al. Changes in rhythmic temporal structure in cluster headache. In: Sicuteri F, et al, eds. Trends in Cluster Headache. Amsterdam: Elsevier 1987:351–359.

283. Chezot M, Claustrat B, Brun J, Jordan D, Sassolas G, Schott B. A chronobiological

study of melatonin, cortisol, growth hormone and prolactin secretion in cluster headache. Cephalalgia 1984; 4:213–220.

284. Leone M, Bussone G. A review of hormonal findings in cluster headache. Evidence for hypothalamic involvement. Cephalalgia 1993; 13:309–317.

285. Solomon GD. Circadian rhythms and migraine. Cleveland Clin J Med 1992; 59: 326–329.

19

Psychiatric Disorders Associated with Disturbed Sleep and Circadian Rhythms

DANIEL J. BUYSSE, ERIC A. NOFZINGER, MATCHERI S. KESHAVAN, CHARLES F. REYNOLDS III, and DAVID J. KUPFER

University of Pittsburgh School of Medicine
Pittsburgh, Pennsylvania

I. Introduction

Psychiatric disorders are linked to disturbances of sleep and circadian rhythms on many different levels:

> Epidemiological studies indicate that insomnia symptoms are significant risk factors for new-onset depression and anxiety disorders.

> Sleep disturbances are among the most common symptoms of psychiatric disorders. Conversely, many patients with sleep disorders have symptoms of depression and anxiety.

> Neurobiological mechanisms regulating sleep, mood, and circadian rhythms share common components.

> Manipulations of sleep and circadian rhythms are useful treatments for mood disorders.

> Subjective and objective measures of sleep are associated with acute and long-term clinical outcome in patients with psychiatric disorders.

Examining these bidirectional associations has led to an improved understanding of the neurobiology of psychiatric disorders, as well as to insights regarding the functions of sleep and circadian rhythms.

II.   Associations Between Mood, Sleep, and Circadian Rhythms in Healthy Adults

Mood is typically defined as the pervasive and sustained emotional "climate" that colors one's perception of the world (1) and is usually defined in terms of the subject's own perception. Nevertheless, mood is clearly affected by both sleep (or sleep loss) and circadian fluctuations. In fact, the most reliable effect of sleep deprivation on performance and self-reports in healthy subjects is a worsening of mood (2), often characterized by fatigue, loss of vigor, irritability, and depression. The effects of sleep loss on mood are typically larger in magnitude than those seen for cognitive or motor functioning. A recent study of sleep restriction over the course of a week in healthy young adults showed evidence of progressive increases in fatigue, stress, and exhaustion, but not in actual sadness (3). Healthy subjects also report diurnal variation in mood, with worsening mood in the evening hours (4).

Studies in healthy adults under conditions of constant wakeful bed rest ("constant routine" studies) demonstrate a linear trend toward mood deterioration with progressive wakefulness. Superimposed upon this linear trend is a circadian rhythm of mood, in phase with the circadian rhythm of core body temperature (5). Recent data from "forced desynchrony" studies focused on mood ratings when subjects were placed on a 28- or 30-hr sleep-wake schedule. In these studies, the long sleep-wake cycle dissociates from the rhythm of core body temperature, which continues to have a period of approximately 24 hr. These studies not only confirmed the presence of wake-dependent and circadian fluctuations in mood (Fig. 1), but have further demonstrated that these two processes interact in a nonadditive fashion (6). The wake-dependent variation in mood is greatest during the first half of the circadian cycle.

The mechanisms by which sleep, sleep deprivation, and circadian variation affect mood in healthy adults are not yet well defined. Gerner and colleagues (7) reported that healthy controls with the most negative mood change following sleep deprivation had smaller core body temperature rhythm amplitude than those with less mood change; the implication is that reduced circadian rhythm amplitude may predispose to negative mood. Total sleep deprivation also affects brain metabolism, as measured by ^{18}F-FDG positron emission tomography (PET). Specifically, PET studies conducted during wakefulness before and immediately after one night of sleep deprivation in healthy controls showed no overall decrease in whole brain absolute metabolic rate, although regional decreases were noted in thalamus, basal ganglia, and white matter (8). However, areas of the cortex now thought to be implicated in mood regulation, such as the dorsolateral prefrontal cortex, medial/ventral frontal cortex, and regions of the cingulate, were not specifically examined, nor were measures of subjective mood reported. Thus, the brain mechanisms by which sleep deprivation produces mood deterioration in healthy subjects have yet to be described.

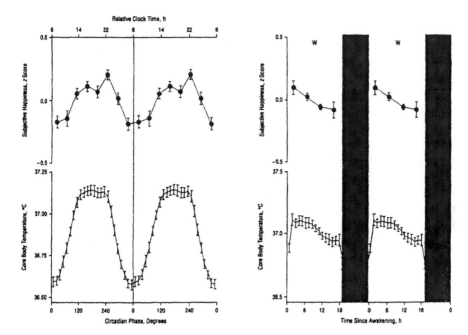

Figure 1 Circadian and wake-dependent variation in self-rated mood from 10 healthy young adult subjects during a forced desynchrony protocol. Subjects lived in conditions of temporal isolation with a scheduled rest/activity cycle of 28 hr. In this way, sleep-wake and body temperature rhythms were dissociated. A statistically significant circadian variation of self-rated happiness can be seen, as well as an evident (but statistically nonsignificant) effect of wake duration. W = awakening; black bars = sleep. (Reprinted by permission from Ref. 6.)

III. Sleep and Circadian Rhythms in Mood Disorders

Sleep and circadian rhythms have been most extensively studied in patients with mood disorders—major depressive disorder (MDD), bipolar mood disorder (manic-depressive illness), and dysthymia (chronic minor depression). Several lines of evidence suggest that the associations between sleep and circadian disturbances and mood disorders are not merely epiphenomena, but are fundamental to the pathophysiology of these disorders.

A. Subjective Sleep Complaints

Patients with mood disorders almost universally complain of disturbed sleep. Approximately two-thirds of patients with depression have some type of insomnia, with about 40% complaining of the specific symptoms of sleep onset difficulty, frequent awakenings, and early-morning insomnia (9,10). The insomnia complaints of patients with depression can be more severe than those of patients with primary insomnia (i.e., insomnia related to behavioral and conditioning factors) (11). Approximately 15% of depressed patients complain of hypersomnia.

Conversely, a high percentage of individuals with sleep complaints also have a mood disorder. Epidemiological and cohort studies indicate that 35–50% of individuals with insomnia or hypersomnia also met criteria for a current mood or anxiety disorder, compared to 15–20% of individuals with no sleep complaint (12–14). Among patients referred to sleep disorders centers for chronic insomnia, psychiatric disorders (and, in particular, mood and anxiety disorders) are identified as the major cause in 35–46% of cases (15,16).

Further evidence suggests that insomnia is actually a risk factor for the new onset of depression. Ford and Kamerow reported that persistent insomnia complaints were a significant risk factor for incident mood disorders during a 1-year follow-up, with an odds ratio of 39.8 (12). A subsequent study found that individuals with a prior history of insomnia had a fourfold increase in the relative risk for incident depression over a 3½-year follow-up interval, even after controlling for the presence of other depressive symptoms at baseline (17).

B. Polysomnographic Sleep Findings

Numerous laboratory investigations conducted over the past 25 years have documented the polysomnographic sleep characteristics of patients with mood disorders (18,19). These characteristics fall into four major categories, as illustrated in Figure 2:

> Decreased sleep continuity (prolonged sleep latency, increased number of awakenings, increased early-morning awakening, decreased sleep efficiency).
>
> Decreased slow-wave sleep (decreased percentage of stage 3/4 sleep, decreased delta activity by period-amplitude or power spectral analysis).
>
> Enhanced rapid-eye-movement (REM) sleep (increased percentage of REM sleep, increased phasic eye movements during REM sleep).
>
> Alterations in temporal characteristics of sleep. These include reduced REM sleep latency (i.e., the duration of the first NREM period), reduced delta electroencephalographic (EEG) activity in the first NREM period relative to the second (reduced "delta sleep ratio"), and increased duration and phasic eye movement activity during the first REM period.

The above findings have been described using both visually defined sleep stage scoring and quantitative measures such as period-amplitude analysis and power spectral analysis. Using the latter techniques, additional differences between depressed and control subjects, such as differences in interhemispheric coherence and differences in coherence between different frequency bins, have also been described (20). Depressed patients show consistently more power in beta-, theta-, and delta-frequency bands in the right hemisphere, particularly during REM sleep, whereas healthy control subjects have less consistent right-left differences (21).

Although each of these findings is characteristic of major depression, most patients do not have each feature, and no single finding can be considered a sensitive or specific marker of depressive disorders. Studies comparing patients with major depressive disorder (MDD) to patients with other psychiatric disorders show that individual sleep variables are associated with sensitivity and specificity in the range of 60–75% (22). In a meta-analysis of polysomnographic findings in psychiatric disorders, Benca et al. (23) found that REM sleep measures were consistently more abnormal in patients with MDD compared to patients with other diagnoses (Table 1). Multivariate statistical techniques such as discriminant function analysis have accuracy of approximately 80% for distinguishing depressed patients, nondepressed psychiatric patients, and controls (24–27).

A number of clinical and historical features may influence EEG sleep findings in depressed patients:

Age affects polysomnographic sleep variables in healthy subjects (see Chapter 17) and in patients with depression. Depressed-control differences in sleep are magnified with increasing age. Sleep studies conducted in adolescents with depression have inconsistently reported any abnormalities relative to healthy control samples, particularly in terms of reduced REM latency (28–31). Within midlife and late-life depressed patients, depressed-control differences in sleep continuity, slow-wave sleep, and REM latency increase with age (32–34). Sleep-onset REM latencies are noted primarily among older subjects (35).

Many sleep measures are also affected by *sex*, both in healthy and in depressed samples. For instance, elderly depressed men and women have consistent differences in slow-wave sleep (36), which are similar to those described in healthy subjects. Other quantitative EEG analyses have indicated that sex differences are more pronounced in depressed patients than in healthy control subjects, with depressed women showing more high-frequency beta activity than men, particularly in the right hemisphere (37).

Severity of depression, usually measured by scores on composite rating scales such as the Hamilton Rating Scale for Depression (HRSD; 38), has been associated with more abnormal sleep findings such as reduced REM latency in some studies (e.g., 39–41), but not in others (e.g., 42–44). In a similar vein, specific complaints of sleep continuity disturbance have been inconsistently related to their corollary polysomnographic measures. Using canonical correlation to address the issue of which depressive symptoms relate to EEG sleep variables, Perlis and colleagues found that depressive symptoms and sleep variables were each expressed along a single dimension of severity, and that symptoms and sleep measures were significantly correlated (10). In other words, more severe depres-

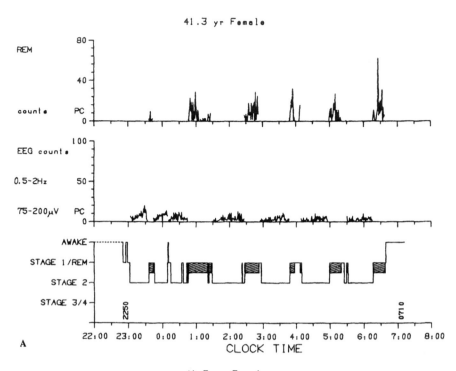

A

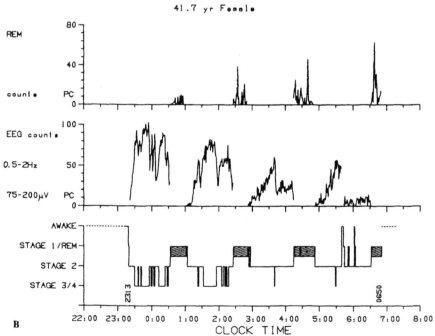

B

sive symptoms (including neurovegetative, affective, and cognitive symptoms) were related to more abnormal sleep findings (including slow-wave sleep, REM sleep, sleep continuity, and REM latency).

Depression history and clinical characteristics have been explored in relation to sleep features. Shorter duration of the index depressive episode is associated with increased phasic REM activity and sleep continuity disturbance (45,46). Other depression history variables (such as earlier age of onset, larger number of past episodes, past hospitalizations) and psychosocial factors (such as increased life stressors, low levels of social support, and low levels of education) are also associated with more "abnormal" polysomnographic sleep (46). Elderly individuals with depression who have not sought treatment do not have the sleep abnormalities typically described in treatment samples (47). Conversely, patients hospitalized for treatment of depression typically have more severe EEG sleep abnormalities than those studied as outpatients (48,49).

Finally, *subtype of depression* may also affect subjective and polysomnographic sleep characteristics. Endogenous depression, a subtype characterized by prominent neurovegetative symptoms and diurnal variation in mood, is generally associated with the more severe EEG sleep abnormalities than the nonendogenous subtype (50–52). Major depression with psychotic features, which refers to depression accompanied by hallucinations or delusions, is more commonly associated with very short REM latencies in both adolescents and adults (53,54). The reduced REM sleep percentage and phasic REM activity in some patients with psychotic depression may simply reflect the gross sleep continuity disruption associated with this subtype (55). In keeping with their subjective complaints, nocturnal sleep continuity disturbances and reduced REM latency are not as severe among the minority of depressed patients complaining of hypersomnia (56), but their complaints of daytime sleepiness are not verified with objective testing (57). Bipolar mood disorder, or manic-depressive disorder, is characterized by episodes of mania (elevated, grandiose, and irritable mood) as well as depressive episodes. Perhaps because of the frequent occurrence of hypersomnia in patients with bipolar depression, their nocturnal sleep is longer than in patients with unipolar depression, but the two groups are otherwise similar (58).

Seasonal affective disorder (SAD) deserves special mention because of the hypothesized links between this form of depression and circadian/sleep dysregula-

---

Figure 2 Representative all-night sleep results from a patient with major depressive disorder (A) and an age-matched healthy control subject (B). (Top panel) The time course of rapid eye movements detected with an automated algorithm; (middle panel) delta EEG activity derived from period-amplitude analysis; (bottom panel) visually scored sleep stages. The patient has reduced REM sleep latency, increased percentage of REM sleep and phasic REM activity, and reduced delta EEG activity and stage 3/4 sleep relative to the healthy control. This particular patient did not have severe sleep continuity difficulty.

Table 1 Polysomnographic Sleep Findings in Depression and Other Disorders

Disorder	Sleep continuity	REM %	REM density	REM latency	Slow-wave sleep %
Depression	↓	↑	↑	↓	↓
Schizophrenia	↓	↓→	→	→↓	↓
Anxiety disorders	↓	→	→↑	→	→
Chronic (primary) insomnia	↓	→	→	→	↓

Arrows indicate consensus findings from a majority of studies and indicate direction of change relative to healthy control subjects. ↑ = increase; → = no change; ↓ = decrease.

tion. Typically, patients with SAD are described as having hypersomnolence during their winter depressions. Studies using self-report questionnaires and prospective sleep diaries have found a significant increase in sleep duration and difficulty awakening during winter months, at times when patients are symptomatic (59–61). However, sleep duration correlates poorly with overall symptom severity (59,61). Polysomnographic studies of symptomatic SAD patients show reduced sleep efficiency, reduced stage 3/4 percentage, and increased REM density (as in other forms of depression), but less evidence of reduced REM sleep latency (61). Successful treatment with bright light does not reliably alter EEG sleep measures, even though patients report less hypersomnia following treatment (62,63).

C. Sleep Deprivation Studies in Depression

Relationships between sleep, circadian rhythms, and depression are further substantiated by numerous observations of the therapeutic effects of sleep deprivation in depressed patients. Beginning with Schulte (64) and Pflug and Tölle (65), investigators have repeatedly confirmed that approximately 60% of depressed patients have a transient antidepressant response to one night of total sleep deprivation (66,67). Subsequent studies have suggested that other forms of sleep deprivation, such as partial sleep deprivation in the second half of the usual sleep period (68–70) and selective deprivation of REM sleep (71,72), have equally robust effects. The mood improvement during sleep deprivation typically begins between 04:00 and 08:00 (67). The magnitude of the antidepressant response is moderate, with responders typically showing a reduction of > 30% in depressive symptoms. Sleep deprivation in bipolar depressed patients often has the undesired effect of precipitating mania, and manic patients who recover quickly are characterized by longer sleep compared to those who do not recover quickly (73).

     One of the vexing features about sleep deprivation as a therapeutic tool in depression is the fact that recovery sleep, whether taken as a nap or as the following night's sleep, leads to a complete relapse of symptoms. The sleep stage composition of nap sleep does not appear to influence this relapse, although afternoon naps are more often associated with relapse compared to morning naps (74,75). The effects of selective REM sleep deprivation appear to persist over time, but no replication of Vogel's pioneering studies has ever been conducted. Consequently, various investigators have tried to maintain sleep deprivation's therapeutic effects by conducting serial partial sleep deprivations (76–79), combining sleep deprivation with antidepressant drugs (80–82), and combining sleep deprivation with a subsequent phase advance of the sleep cycle (83).

Numerous physiological and clinical measures have been investigated as possible correlates of a positive response to sleep deprivation. In general, patients with the endogenous subtype of depression (67) and more abnormal baseline EEG sleep (84) fare better with this intervention, and there is some evidence that prominent diurnal mood variation is also associated with a positive response (85–87). Patients who show a more robust increase in sleep continuity and slow-wave sleep during "recovery sleep" following deprivation also appear to have more favorable responses (88). As discussed below, neuroimaging studies have also identified correlates of positive responses to sleep deprivation.

D. Circadian Rhythm Disturbances in Mood Disorders

One possible explanation for the subjective and polysomnographic sleep findings in depression is that they are part of a larger disturbance in the regulation of circadian rhythms. Unfortunately, most circadian rhythm studies in depression have not been as rigorous methodologically as studies in healthy adults (89). For instance, no constant routine studies have been reported in symptomatic depressed patients.

Core Body Temperature

Studies of body temperature rhythms in depressed patients have typically been conducted with subjects on a normal diurnal sleep-wake cycle. As a result, measurement of temperature rhythms in depression is influenced by the masking effects of sleep (or the sleep disruption characteristic of depression). Early studies suggested a phase advance of core body temperature in depression (90,91), but this finding has not been consistently replicated. Rather, most studies have found increased variability of temperature phase in depressed patients. The most consistent finding from temperature studies is reduced amplitude of core body temperature amplitude, which is primarily related to a reduced fall in temperature during sleep (92–94) (Fig. 3). Temperature rhythm amplitude is lower among patients with reduced REM sleep latency, compared to patients with longer REM latency (95,96). Patients with bipolar mood disorder may have more disruption of temperature rhythms than unipolar patients. Wehr and Goodwin demonstrated a phase advance of temperature before switches from mania to depression, and phase

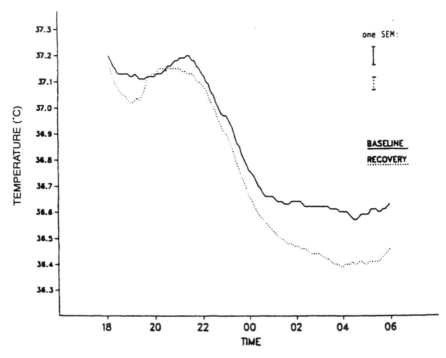

Figure 3 Core body temperature during evening and sleep hours in depressed out-patients. The solid line indicates core body temperature taken every 1 min when patients were symptomatic; the dotted line indicates temperature data from the same patients taken during remission, while taking either imipramine or fluoxetine. (Reprinted with permission from Ref. 99.)

delays preceding switches in the opposite direction (97). Only one study has examined the free-running temperature rhythms of three depressed patients (two of whom had bipolar disorder) under temporal isolation conditions (98). Two of the three patients continued to have entrained temperature rhythms with period length in the range observed in healthy volunteers; the remaining patient showed internal desynchronization with a sleep-wake period of less than 24 hr and temperature rhythm period of greater than 24 hr. In our laboratory, depressed outpatients were studied during a constant routine procedure after they had recovered from depression. The period and amplitude of their core body temperature rhythms were indistinguishable from those in a healthy contrast group (99), suggesting that any changes in the temperature rhythm that may have been present during depression did not persist into the asymptomatic state.

Constant routine studies have been conducted in a small sample of individuals with SAD (100). This study showed a significant phase delay of core body temperature in patients relative to controls, and an advance in the nadir of core body temperature of patients following successful bright-light treatment. Thus, patients with bipolar disorder and SAD may represent more chronobiologically disturbed groups of depressed patients.

Melatonin

Unlike temperature rhythms, melatonin concentrations are not affected by sleep or wakefulness and may therefore serve as a more reliable marker of circadian phase when unmasking and free-running studies are difficult to conduct. A number of studies have reported reduced melatonin concentrations in depressed patients (e.g., 101–103), but most of these studies have included small numbers of subjects, poorly matched controls, and infrequent melatonin sampling. The best-controlled studies of melatonin in depressed patients found neither reduced levels nor phase alterations in patients with endogenous depression (104,105).

A phase delay in melatonin onset has been described in SAD with winter depression (106,107). The dim-light melatonin onset (DLMO) of patients with SAD advances with morning phototherapy, and the timing of DLMO is correlated with reduced depression ratings (108). However, other investigators have not consistently confirmed the finding of delayed melatonin rhythms in patients with SAD, nor the finding that morning bright light is superior to bright light at other times of day (109,110).

Other Neuroendocrine Measures of Circadian Rhythms

A variety of other endocrine rhythms, including those of cortisol, growth hormone, thyroid hormone, and prolactin, have been measured in patients with major depression. Interest in cortisol rhythms was sparked not only by interest in circadian physiology in depression, but also by the observation that approximately 50% of patients fail to suppress cortisol in response to dexamethasone administration (111). Dexamethasone nonsuppression indicates faulty feedback mechanisms in the hypothalamic-pituitary-adrenal (HPA) axis, probably related to tonically elevated circulating cortisol levels and hypersecretion of corticotropin-releasing factor (CRF) in the hypothalamus and ACTH in the pituitary (112,113). Several reports have indicated that patients with depression have reduced cortisol rhythm amplitude, which is specifically related to higher nadir values during the first half of the sleep period (114,115). A phase advance in cortisol is suggested by reducing latency between sleep onset and the major circadian rise in cortisol (116). Findings of reduced amplitude and phase advance have been substantiated by an unpublished meta-analysis of cortisol rhythms in depression that combined data from several laboratories (E. Van Cauter, personal communication). Unlike most depressed patients, those with SAD have been reported to have a phase delay of cortisol in a constant routine study (100). Abnormal cortisol circadian rhythms may be part of a cascade of dysregulation in the HPA axis: Excessive cortisol production may damage corticoid receptors in the hippocampus, which impairs feedback inhibition of CRH, leading to further excessive secretion of ACTH and cortisol (117,118).

Growth hormone (GH) secretion normally occurs in pulsatile fashion, with secretory peaks stimulated by food and exercise (see Chapter 14). The secretion of GH is also strongly masked by sleep, with the largest secretory episode typically occurring in conjunction with the first NREM sleep period. Thus, the endogenous circadian rhythm of GH is weak. In patients with major depression, the relative balance between GH secretion during sleep and wakefulness is altered, so relatively more GH is secreted during wakefulness, and relatively less during sleep (e.g., 115,119–121). In addition, patients with depression show impaired responses to stimuli that typically increase GH secretion, such as clonidine and GH-releasing hormone (122,123). Therefore, abnormal GH secretory patterns in depression appear to indicate a more general dysfunction of the somatotropic axis and its integration with sleep-wake regulatory mechanisms.

Sleep-Wake Cycles and Social Rhythms

The circadian pattern of sleep-wake propensity may itself be affected in patients with mood disorders, but very few studies have examined this possibility directly. Reduced amplitude of the circadian pattern of sleep propensity was reported in depressed patients with continuous EEG monitoring (124), but not in a study of remitted depressed patients monitored during a constant routine study (125). Thus, if unipolar depression affects the circadian rhythm of sleep propensity, this effect does not appear to persist beyond the symptomatic phase.

A series of observations reported by Wehr, Leibenluft, and colleagues suggest that sleep-wake rhythms are more seriously affected in patients with the bipolar form of depression, particularly those with rapid cycling between depression and mania. Among a group of 11 rapid-cycling patients followed longitudinally for 18 months, decreased nocturnal sleep time and wake onset time were significant predictors of mania or hypomania the following day (126). This suggests that reduced sleep time and a phase advance of sleep are associated with elevated mood, and that decreased sleep may serve as a "final common pathway" to mania (127). Based on observations that an extended nocturnal dark period can significantly affect sleep duration and circadian rhythm profiles (128), Wehr and colleagues have also observed the effect of an extended dark period on a patient with rapid-cycling bipolar disorder (129). On a 10:14 light-dark cycle, the rapidity and intensity of mood cycling markedly diminished. Thus, the pattern of light-dark exposure can have a substantial impact on mood in bipolar patients, which may be mediated at least in part by changes in sleep-wake cycles.

The pattern of daily activities and social contact that we experience can also be expressed as a type of rhythm (130). This "social rhythm" may be disrupted as either a cause or consequence of depression, with its attendant social isolation and anhedonia. In a preliminary test of this theory, Szuba and colleagues reported that depressed inpatients did in fact have lower scores for social rhythms than nondepressed control subjects (131).

E. Sleep and the Functional Neuroanatomy of Depression

Another avenue for understanding the sleep and circadian features of depression comes from functional neuroimaging techniques, particularly PET. A number of studies have examined patterns of brain glucose metabolism during wakefulness in patients with MDD. Most evidence has pointed to resting hypometabolism in dorsolateral prefrontal cortex and anterior cingulate regions, as well as relative hypermetabolism in mesial temporal regions (amygdala, hippocampus) and in the ventral and medial orbitofrontal cortex (132–134). These findings indicate dysfunction in circuits linking frontal lobe function, emotional regulation, and neurovegetative function.

A more limited number of studies have examined brain metabolism during sleep in patients with depression. Ho and colleagues reported that patients with MDD had globally higher glucose metabolic rates during the first NREM period relative to healthy control subject, with no regionally specific findings (135). The usual finding in healthy subjects is a significant global decrease in glucose metabolism during NREM (136). Thus, in contrast to relative hypometabolism in several regions during wakefulness, patients with MDD appear to be globally "activated" during early NREM sleep. This abnormality correlated with the loss of slow-wave sleep.

In healthy humans and in animal studies, REM sleep is associated with a cortical metabolic rate comparable to that during wakefulness, as well as more specific increases in brain metabolism in limbic and paralimbic structures, including the amygdala, cingulate cortex, habenula, substantia innominata, and anteroventral thalamus (136–138) (Fig. 4). Nofzinger and colleagues examined the difference between glucose metabolism during wakefulness and REM sleep in women with depression and in healthy controls (139). In several areas such as the left inferior orbitofrontal cortex, patients with MDD showed waking hypermetabolism relative to controls, and a failure to further activate these structures during REM (Fig. 5). In other words, patients with MDD showed waking overactivation and a blunted response to the naturalistic limbic system probe of REM sleep. This pattern is in some ways consistent with that seen on several neuroendocrine challenge paradigms, where depressed patients having resting overactivity and blunted responses to physiological challenge.

Other studies have examined waking glucose metabolism or single photon emission computed tomography (SPECT) before and after sleep deprivation in patients with depression. Patients who showed an antidepressant response to sleep deprivation were characterized by elevated anterior cingulate metabolism prior to

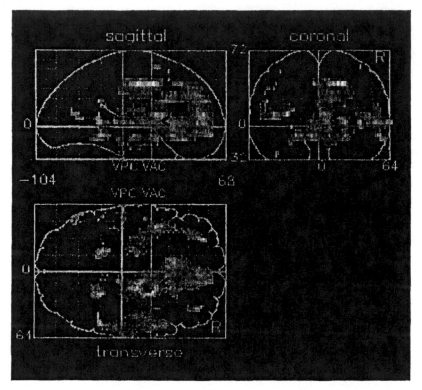

Figure 4 All brain regions that have statistically greater ($p < 0.01$) relative cerebral glucose metabolism during REM sleep than during wakefulness are shown in this statistical parametric map. Data are from six healthy subjects who had ^{18}F-FDG PET studies during wakefulness and during REM sleep. (Reprinted with permission from Ref. 138.)

sleep deprivation and a decrease to normal levels following sleep deprivation; patients who showed no response to sleep deprivation showed no change in anterior cingulate metabolism (140,141).

F. Sleep and Circadian Rhythms: Longitudinal Studies in Mood Disorders

The preceding discussion has focused on changes in sleep, circadian rhythms, and brain metabolism that characterize depressed patients during the symptomatic episode. A number of other studies have examined sleep in the context of the clinical course of depression, primarily to address two questions: (1) Are the abnormalities related to the symptoms of depression themselves, or to some more persistent trait? and (2) Are sleep and circadian features related to the clinical course of the disorder?

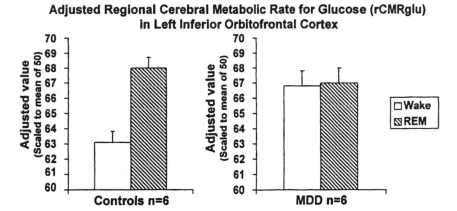

Figure 5 Example of an interaction change in normalized glucose metabolism from waking to REM sleep between depressed and control subjects. Depressed subjects show an apparent waking hypermetabolism in the paralimbic orbitofrontal cortex with no further activation during REM sleep. In contrast, control subjects show activation of this region, similar to other paralimbic regions, during REM sleep.

State-Trait

Patients who are treated successfully for MDD with medications and psychotherapy report improvement in subjective sleep quality (142). Findings with regard to EEG sleep measures are more complicated. Sleep-onset REM periods are less likely to be observed in recovered patients than in those who are still symptomatic (35). Phasic and tonic REM activity tend to decrease over time during the symptomatic episode (143–145), and several comparisons of sleep during symptomatic and recovered periods have shown reductions in REM sleep measures after successful pharmacotherapy or psychotherapy (45,146,147). In addition, measures of sleep continuity such as early-morning awakening and sleep efficiency show improvement during recovery (146,148). The changes reported in REM sleep and sleep continuity related to current clinical state are not large, and several other studies have shown stability in mean polysomnographic sleep variables, as well as strong correlations between baseline and recovery (149–151). Measures of slow-wave sleep do not change with clinical phase in most investigations, although a few reports have shown decreased slow-wave sleep or delta activity after recovery (151,152). Categorically defined REM latency (153) is also stable between symptomatic and recovered periods. Thus, the bulk of evidence of evidence suggests small changes toward "normal" values in REM and sleep continuity domains, with stability across clinical state in most NREM sleep measures.

The trait-like nature of EEG sleep measures can also be examined in the context of family studies. Sleep quality, sleep-wake patterns, and NREM sleep characteristics including slow-wave sleep show strong correlations among family members and in twin studies, suggesting genetic transmission; by contrast, the genetic component of REM sleep is much weaker (154–156). Giles and colleagues identified depressed probands with categorically reduced or nonreduced REM latency and studied the family members of each group. Reduced and nonreduced REM latency showed strong familial aggregation (157). Moreover, family members with reduced REM latency had a threefold greater risk for depression relative to family members with normal REM latency. Family members with reduced REM latency also had other EEG sleep features consistent with depression (158). Lauer and colleagues found that 18% of individuals with a strong family history of depression, but no personal history, had EEG sleep findings typical for depression (159). Thus, the strong familial aggregation of the EEG sleep characteristics of depression suggests that they are a genetic vulnerability trait.

Circadian rhythms have also been examined as a function of phase of illness. The blunted amplitude of core body temperature observed in some depressed patients increases toward normal values during recovery (160,161) (Fig. 3). Similarly, the hypercortisolemia and altered feedback inhibition of cortisol tend to assume more "normal" profiles after the episode (161). These findings are in contrast to those for the somatotropic axis. GH secretion and responses to various stimuli for GH release continue to show altered regulation even after patients have recovered from an episode of depression (119). As previously noted, the abnormal phase of melatonin onset observed in some patients with SAD advances toward a more normal phase with morning bright-light treatment (106).

Sleep and Clinical Course

Sleep measures are associated with treatment response and clinical outcome in depression. As noted above, insomnia complaints are an independent risk factor for incident depression. Persistent complaints of insomnia are also associated with nonresponse to antidepressant medication (162), chronic depression (163), and suicide (164). Poor subjective sleep quality is also a risk factor for recurrence of depression among elderly patients who are not treated with maintenance medication (165).

EEG sleep characteristics are also associated with clinical outcomes. Among elderly depressed patients, elevated measures of tonic and phasic REM are associated with a slower response, or nonresponse, to acute treatment with combined medication and psychotherapy (166). Younger depressed patients with reduced REM latency at baseline and those who show prolongation of REM latency and improved sleep efficiency during initial treatment are more likely to respond to pharmacotherapy with tricyclic antidepressants (167,168). Although these features do not predict a favorable response to psychotherapy (150,169,170), other

EEG sleep measures do. As mentioned earlier, Thase and colleagues used a multivariate classification based on REM latency, REM density, and sleep efficiency to distinguish outpatient depressives with a "normal" (i.e., like healthy controls) or "abnormal" (i.e., like depressed inpatients) sleep profile. In two independent samples, the "abnormal" sleep profile was associated with reduced likelihood of response to psychotherapy, but not to subsequent pharmacotherapy (171,172). Finally, a series of investigations have demonstrated that certain aspects of slow-wave sleep, such as reduced total delta activity by period-amplitude analysis and the temporal distribution of delta activity, are associated with the likelihood of recurrence of depression during maintenance treatment (152,173,174). Reduced REM latency has also been associated with recurrences of depression (153).

G. Models of Sleep and Circadian Dysregulation in Mood Disorders

Several theoretical and mechanistic models have been developed to explain the sleep and circadian rhythm changes associated with depressive disorders. Although each model is supported by a certain amount of empirical data, none of them explains all of the key observations. On the other hand, most of these models are not mutually incompatible, and, because mood disorders are certainly heterogeneous with respect to etiology, it seems likely that several mechanisms of sleep and circadian dysregulation may also be at play.

Abnormal Phase Relationships Among Circadian Rhythms

Papousek (175) and later Wehr and Goodwin (97) and Kripke (176) proposed that mood disorders may result from a phase advance of the circadian rhythm governing temperature, cortisol, and REM sleep relative to other circadian rhythms such as sleep-wakefulness. This theory was later refined as the "internal phase coincidence" model (177), which posits that depressed mood occurs when awakening occurs at an abnormally early or "sensitive" phase of circadian rhythms. Phase advance theories may explain findings such as short REM latency and increased REM sleep early in the night, phase advance of cortisol, and temperature rhythms. A limited amount of experimental data in nondepressed subjects and from computer simulations also supports these theories: Abruptly delaying the major sleep period under these circumstances produces reduced REM latency and increased amounts of REM sleep early in the sleep period, as well as (in some cases) depressed mood (178). Furthermore, a small number of studies have demonstrated that advancing the phase of sleep (thereby "normalizing" phase relationships between sleep-wake and other rhythms) can improve symptoms of depression (83,179,180). Longitudinal studies in patients with rapid-cycling bipolar disorder show lengthening of sleep and relative advances of REM sleep prior to depressive episodes, and short sleep with relative delays of REM sleep in mania, further supporting the concept of abnormal phase relationships in mood disorders (97).

The "social zeitgeber" hypothesis of depression proposes a cascade of rhythm disruptions that lead to depression in vulnerable individuals (181). Initially, a stressful life event precipitates a change in social prompts for daily activities (social zeitgebers), and day-to-day instability in those social and activity rhythms results. This instability in turn destabilizes biological rhythms, which, in vulnerable individuals, can lead to pathological entrainment (or nonentrainment) or biological rhythms. This pathological entrainment produces depressed mood.

However, several lines of evidence also suggest the limitations of abnormal phase models in depression. Significant phase advances are not present in the majority of depressed patients. Moreover, the magnitude of phase shifts required to make normal sleep look "depressed" is very large—approximately 6 hr. Finally, there have been very few studies of circadian rhythms in depression that account for the masking effects of sleep and activity with paradigms such as constant routine studies or forced desynchrony. It seems likely that phase advances or delays are present only in a subset of patients with depression, including those with bipolar mood disorder.

Reduced Amplitude of Circadian Rhythms

Although reduced amplitude of rhythms is the most consistent observation in depressed patients, this observation alone has limited explanation power. Reduced amplitude could be due to a number of factors, including masking effects of sleep disturbance, intrasubject dyssynchrony of various circadian rhythm phases, intersubject differences in phase, or impaired expression of circadian rhythmicity in organs that express circadian rhythms "downstream" from the central pacemaker(s).

The "S-Deficiency Hypothesis"

This theory, based on the "two process" model of sleep regulation (182,183), proposes that depressed patients have deficient build-up of process S, the homeostatic sleep regulatory process, which could result in decreased slow-wave sleep, exaggerated expression of REM sleep, and (through undefined mechanisms) depressed mood (184). According to this hypothesis, interventions such as sleep deprivation permit greater buildup of process S, more slow-wave sleep (during recovery sleep), and antidepressant effect. However, the antidepressant effects of sleep deprivation occur *before* recovery sleep and increased slow-wave sleep expression. An alternative explanation of the effects of total and REM sleep deprivation is that they *decrease* slow-wave sleep expression during the therapeutic interval. Many antidepressant drugs decrease the amount of slow-wave sleep, which is also apparently inconsistent with predictions of the S-deficiency hypothesis. These observations led Beersma and van den Hoofdakker (185) and Wu and Bunney (66) to propose that NREM sleep, or some process associated with sleep, is actually depressogenic. The two types of theories relating to slow-wave sleep may be reconciled by the possibility that increased "drive" or "pressure" for slow-wave sleep, rather than its actual expression, may be associated with a therapeutic response.

"Depressogenic REM Sleep"

Based on the observation that selective REM sleep deprivation has antidepressant effects, and that most efficacious antidepressant medications suppress REM sleep, Vogel has proposed and modified the hypothesis that REM sleep (or its neurobiological substrate) is depressogenic (71,186). More specifically, effective antidepressant interventions are generally characterized by "arousal-type REM suppression," i.e., the reduced expression of REM sleep during the intervention, followed by a REM sleep rebound upon discontinuation of the intervention. Indeed, REM suppression followed by REM rebound is characteristic of total sleep deprivation, REM sleep deprivation, and antidepressant drugs in the tricyclic, monoamine oxidase inhibitor, and selective serotonin reuptake inhibitor classes. However, newer data indicate that some atypical antidepressants, such as bupropion, trazodone, and nefazodone, are associated with little REM sleep suppression, or even increased REM. The exact mechanism underlying the "REM depressogenic" theory is not entirely clear, but may relate to alterations in sensitivity of serotonergic and noradrenergic receptors that may normally occur during REM sleep.

Neurochemical Theories

These theories often focus on a functional cholinergic "overdrive" relative to monoaminergic neurotransmission (serotonin and norepinephrine) (187). Evidence in support of this theory comes from studies showing more rapid induction of REM sleep when cholinomimetic agents are infused during NREM sleep in depressed patients, as compared to healthy control subjects (188–190). Although the acute dietary depletion of the serotonin precursor L-tryptophan leads to acute worsening of depression in patients treated with serotonergic antidepressants (113), depressed patients do not have differential EEG sleep responses to this intervention compared to healthy control subjects (191). The main difficulty with neurochemical theories of sleep and depression is that they must remain rather general. While it is true that most efficacious antidepressants enhance serotonergic and/or noradrenergic neurotransmission, they do not all have anticholinergic effects. Furthermore, the specific mechanisms by which different antidepressants act, and their effects on sleep, are quite different. For instance, effective antidepressants include those that inhibit serotonin reuptake into presynaptic neurons, those that antagonize postsynaptic $5HT_2$ receptors, and those that are $5HT_{1A}$ receptor agonists.

Functional Neuroanatomical Models

Such models are currently being developed from observations using PET and other neuroimaging modalities, as discussed above. One model suggests a globally increased metabolic rate during NREM sleep in depressed patients compared to controls (135). This hypermetabolism could relate to other findings such as reduced slow-wave sleep, increased nocturnal core body temperature, and reduced sleep-related GH secretion.

Neuroimaging studies of REM sleep suggest a different pattern of control-depressed differences, with greater regional specificity. As reviewed above, REM sleep is associated with robust limbic and paralimbic system activation relative to wakefulness in healthy control subjects (137,138). Depressed patients show waking hypermetabolism relative to controls in these areas, with a lesser degree of activation during REM sleep (139). Thus, heightened measures of REM sleep in depressed patients may be an indicator of limbic and paralimbic system "overdrive" during both wakefulness and REM sleep.

Taken together with studies of sleep and treatment outcome, functional neuroimaging studies suggest the possibility of two different types of arousal that may affect depressed patients. One type of arousal may be indicated by reduced slow-wave sleep, global hypermetabolism during NREM, and greater likelihood of recurrences during long-term outcome. A second type of arousal, indicated by increased REM sleep and reduced activation of limbic structures during REM sleep, may be associated with worse short-term response to treatment. These two types of arousal can also be understood within the context of the "type 1/type 2" schema of reduced latency proposed by Kupfer and Ehlers (192), in which physiological abnormalities may be related to either a persistent, genetic dysregulation (type 1, exemplified by reduced slow-wave sleep), an acute symptom-related dysregulation (type 2, exemplified by increased REM sleep), or some combination. Future neuroimaging studies can add to this model by examining specific receptor distributions and binding capacities, which will lend neurochemical specificity to the regional metabolic data currently being generated.

H. Summary

Patients with mood disorders have characteristic, if not pathognomonic, changes in subjective sleep patterns, polysomnographic sleep patterns, and circadian rhythms. These alterations are related to cross-sectional clinical features and to longitudinal treatment outcome. Recent studies have begun to clarify the functional neuroanatomical basis of sleep and circadian dysregulation in depression. Future studies combining polysomnography, brain imaging using specific receptor radioligands, and controlled treatment or neurochemical challenge paradigms promise to further elucidate this pathophysiology.

IV. Sleep and Circadian Rhythms in Schizophrenia

Like depressed patients, patients with schizophrenia consistently describe disturbances in sleep and dreaming. This observation, and the phenomenological similarity between dreaming and psychosis, has led to investigations of the association between sleep and schizophrenia. In this section, we review empirical data on sleep in this illness and discuss their clinical and pathophysiological significance.

A. Subjective Sleep Complaints

Patients with schizophrenia complain of sleep onset problems and frequent awakenings, particularly leading up to exacerbations of psychosis. In our sleep disorders clinic, we have seen several patients who have complained of total or neartotal insomnia, even in the presence of polysomnographically defined sleep. Thus, complaints of severe sleeplessness may actually become part of the patient's delusional beliefs. Finally, patients with schizophrenia often have social withdrawal and isolation, which in extreme cases can lead to a non-24-hr ("freerunning") sleep-wake pattern (193).

B. Polysomnographic Sleep Findings

Patients with schizophrenia may have several types of EEG sleep abnormalities:

Decreased sleep continuity (increased sleep latency, increased awakenings, decreased total sleep time)
Decreased REM sleep, particularly during psychotic exacerbations
Decreased slow-wave sleep
Reduced REM sleep latency

None of these findings have been replicated in all studies, although reduced slow-wave sleep and impaired sleep continuity are more consistent than REM sleep and REM latency results.

Early sleep EEG studies sought to test the intriguing hypothesis that the symptoms of schizophrenia represent an intrusion of the dream state (REM sleep) into wakefulness. No evidence has accrued to support this prediction. Furthermore, studies of REM sleep in schizophrenia have been conflicting, with increases, decreases, as well as no change being found (23,194). Studies examining treatment-naïve schizophrenia patients show no increases in REM sleep (195,196); the increases in REM sleep observed in previously treated subjects may reflect effects of medication withdrawal and/or changes related to the acute psychotic state (196). Thus the decreased REM latency reported in seven of 10 studies in a recent review (194) is unlikely to result from primary abnormalities in REM sleep or increased "REM pressure."

Studies of slow-wave sleep (SWS) abnormalities in schizophrenia have been somewhat more consistent. Several studies have shown a reduction of SWS in acute, chronic, as well as remitted states, and in never-medicated, neuroleptic-treated, as well as unmedicated patients (194). However, not all studies show these deficits (196–198). Studies that have failed to find differences in SWS have generally used conventional visual scoring, whereas all three studies that have used quantitative sleep EEG analysis have consistently shown reductions in SWS (199–201). Thus, quantitative EEG measures may be more sensitive indicators of SWS deficiency in schizophrenia.

Sleep continuity disturbances, shortening of REM latency, and reduced SWS are also seen in affective disorders (23) raising the question of diagnostic specificity. However, while affective disorders are associated with significant increases in REM density, REM amount, and REM sleep percentage in controlled studies, no consistent changes have been seen in the amounts of REM sleep in schizophrenia (23,196). Thus, a combination of EEG sleep abnormalities (decreased REM latency and SWS deficits without appreciable increases in REM sleep amounts) might be seen in schizophrenia (202).

As with depression, a number of clinical features have been associated with specific sleep findings in schizophrenia. Research during the past decade has focused increasingly on the *positive and negative symptoms* in schizophrenia. Positive symptoms refer to more flagrant psychotic symptoms such as hallucinations and delusions, whereas negative symptoms refer to the apathy, amotivation, and withdrawal that also accompany this disorder. Tandon et al. (196) reported an inverse association between REM latency and negative symptoms. Alterations of SWS appear to be stable over the course of 1 year in patients with schizophrenia, suggesting that SWS deficits are trait-related (203). Given the association between negative symptoms and other trait-related neurobiological abnormalities such as ventriculomegaly, one might predict an association between SWS deficits and negative symptoms. Ganguli et al. (195) originally observed that negative symptoms were inversely correlated with delta EEG counts. This finding was subsequently replicated by van Kammen et al. (204), Kajimura et al. (199), and Keshavan et al. (205). However, there have been negative reports as well (196,198, 206). Reduced SWS at baseline has also been found to correlate with impaired outcome at 1 and at 2 years (207).

An association between *attentional impairment* and SWS deficits has also been reported in early studies (208). As the major relay station receiving input from the reticular activating system, limbic, and cortical association areas, the thalamus plays a crucial role in attention and gating of information. A defect in this structure could therefore explain much of the psychopathology of schizophrenia and alterations in SWS (see below).

No association has been seen between sleep abnormalities and *depressive symptoms* in schizophrenia (196). However, two studies have showed that increased REM sleep may correlate with *suicidal behavior* in schizophrenia (209,210).

Studies in animals and humans have suggested a role for REM sleep in learning and memory consolidation. To test the hypothesis that REM sleep is associated with *cognitive function* in schizophrenia, Taylor et al. (211) examined the relationship between REM parameters and neuropsychological test performance. An inverse relationship was seen between REM amounts in the first REM period and neuropsychological performance. The authors argued that phasic REM sleep regulation at the beginning of the night may serve a compensatory function for cognitive dysfunction in schizophrenia.

C. Sleep Deprivation Studies in Schizophrenia

Sleep deprivation in healthy subjects and depressed patients is usually followed by an immediate increase in SWS and a delayed rebound in REM sleep. An intriguing reduction in REM rebound following REM sleep deprivation has been described in several studies in patients with acute psychotic symptoms, but a normal or exaggerated REM rebound in remitted schizophrenic patients (194). In keeping with the "REM intrusion" hypothesis mentioned above, the absence of rebound in acute schizophrenia was initially attributed to a possible "leakage" of phasic REM events from REM sleep into NREM sleep. However, noninvasive examination of phasic REM events such as middle-ear muscle activity has failed to show any significant change in their distribution between REM and NREM sleep in schizophrenia (212).

Sleep deprivation studies can also help clarify the mechanisms of SWS abnormalities in schizophrenia. Recovery of stage 4 sleep is diminished in schizophrenia following total sleep deprivation (206,213). SWS deprivation in healthy subjects causes impaired attention and vigilance similar to that attributed to frontal lobe dysfunction in schizophrenia (214). Thus, impaired SWS recovery in schizophrenia may be a further indicator of impaired frontal lobe function, consistent with impaired cognitive processes such as psychomotor vigilance. SWS deficits may occur in only a subgroup of schizophrenic patients, consistent with the probable pathophysiological heterogeneity of this disorder.

D. Circadian Rhythms in Schizophrenia

A relatively small number of studies have examined circadian rhythms in patients with schizophrenia, compared to studies in depressed patients. This undoubtedly relates to the challenges posed by schizophrenic patients' clinical symptoms, but is also related to the absence of theoretical frameworks for suspecting circadian abnormalities in schizophrenia. No studies have been reported using techniques such as constant routines or temporal isolation in these patients; thus, the criticism that any observed abnormalities represent masking effects rather than endogenous abnormalities applies to studies in schizophrenia, as well as studies in depression. Nevertheless, few circadian abnormalities have been described under these relatively uncontrolled conditions in patients with schizophrenia.

Oral temperature rhythms appear to be normal in patients with schizophrenia (215). Cortisol and corticotropin levels are also normal with respect to phase and amplitude (215,216); one study found higher cortisol levels during early sleep within overall normal rhythm parameters (217). Studies regarding sleep-related GH have been inconsistent in unmedicated patients with schizophrenia, with both decreased (218) and normal (217) patterns reported. Melatonin secretion is consistently reduced (216,219,220), and one study indicated a phase advance in the melatonin rhythm as well (221). Finally, prolactin secretion is normally suppressed by dopamine and shows a circadian peak in early-morning hours. One well-designed study found a significant increase in sleep-related prolactin secretion in patients with schizophrenia compared to healthy controls (217). In summary, sleep-related and circadian patterns of neuroendocrine rhythms in schizophrenia differ from those in depression, and in general indicate normal phase measures but abnormal mean values in some hormones. This suggests dysregulation at the level of secretory organs and their neurochemical control, rather than abnormalities in the biological clock itself.

E. Models of Sleep and Circadian Dysregulation in Schizophrenia

Clues to the pathophysiological significance of SWS deficits in schizophrenia have been found in the context of a neurodevelopmental framework. Converging evidence suggests a substantial reorganization of human brain function during adolescence, which is also the period when schizophrenia usually manifests. Adolescence is marked by a large decline in synaptic density, cortical gray matter volume, and regional cerebral metabolism in the prefrontal cortex. A substantial decrease in SWS (222) also occurs during this period. The time courses for maturational changes in SWS, cortical metabolic rate, and synaptic density during the postnatal phase of human development are strikingly similar. Feinberg (223) suggested that cortical neurons go through a phase of regression (pruning) of neural elements during adolescence. Thus, the maturational processes during adolescence in sleep EEG, cortical synaptic density, and regional cerebral metabolism might reflect a common underlying biological change, i.e., a large-scale synaptic elimination. Polysomnographic abnormalities in schizophrenia may be viewed in relation to these brain maturational parameters. In addition to SWS deficits, consistent alterations in structure and function of cortical and subcortical brain regions have been observed in schizophrenia. Thus, studies of the associations between structural and functional brain lesions and sleep may elucidate the pathophysiological substrate of schizophrenia.

Altered Neuroanatomy

There is some evidence for reductions in cortical gray matter, perhaps more prominently in the frontal and temporal cortex (224), as well as in thalamic volume in schizophrenia (225). SWS generation appears to be regulated by a complex neural system involving the frontal cortex and thalamus. The relationship between alterations in these brain structures and SWS would therefore be of interest. One computed tomography scan study showed a trend for an inverse association between SWS percent and anterior horn ratio, a measure of frontal lobe size (226). Van Kammen et al. (204) found that SWS deficits were associated with cerebral ventricular enlargement. This association may perhaps be related to reductions in subcortical structures such as the thalamus, which forms a substantial part of the ventricular boundaries. A correlation between thalamic volume and SWS deficits would be predicted and would be consistent with the role of the thalamus in SWS generation.

Altered Neurochemistry

SWS results from several processes that produce neural inhibition, functional deafferentation, and EEG synchrony. Activation of the cholinergic system facilitates arousal and hastens REM sleep. Therefore, cholinergic hyperfunction, postulated to contribute to schizophrenia, could account for SWS and REM latency reductions (202). Like depressed patients, patients with schizophrenia show supersensitive REM sleep induction with the cholinergic agonist RS 86 (227), suggesting increased cholinergic sensitivity. Alternatively, serotonergic abnormalities may also be involved, as indicated by an inverse correlation between serotonin metabolites in the CSF and SWS in schizophrenia (228). Disturbances in monoaminergic mechanisms may also underlie SWS deficits in schizophrenia; norepinephrine and dopamine, which are presumed to be hyperfunctional in schizophrenia, inhibit REM. Thus, cholinergic and monoaminergic abnormalities may mediate the constellation of reduced REM latency and SWS deficit without increases in REM sleep amounts in schizophrenia (202).

Finally, the possible relationship between hormonal substances and SWS in schizophrenia has also received some attention. Van Kammen et al. (229) showed an association between reduction in delta sleep induction peptide, a putative endogenous sleep modulator, and SWS decrements, but this observation needs to be replicated.

Altered Functional Neuroanatomy

Some evidence, albeit modest, for decreased frontal lobe metabolism has been documented in schizophrenia using a variety of techniques, including PET, SPECT, ^{31}P magnetic resonance spectroscopy (MRS), and ^{131}Xe inhalation technique (230). SWS deficits can be considered in the context of such physiological alterations. An association has been demonstrated between SWS deficits and reduced frontal lobe membrane phospholipid metabolism as examined by ^{31}P MRS (231).

The decreased synaptic density postulated to underlie schizophrenia could conceivably result in reduced SWS by decreased membrane surface (fewer dendrites/neuron), causing a smaller voltage response to the synchronizing stimulus (223). Studies in cats using single cell recordings (232) have shown that slower ($<$ 1 Hz) synchronized oscillations originate mainly in the neocortex, while delta EEG activity (1–4 Hz) arises primarily from activity of thalamocortical circuits. A finer analysis of these oscillations may therefore clarify the nature of pathophysiology in schizophrenia. Preliminary analysis of this question (201) using period amplitude analyses suggested more prominent deficits in the $<$ 1-Hz range in schizophrenia, pointing to a corticothalamic dysfunction. This finding deserves further study and replication.

F. Summary

Considerable evidence confirms abnormalities in sleep architecture in schizophrenia, though several methodological issues remain to be addressed in this area of inquiry. The constellation of reduced SWS and shortened REM latency without alterations in REM sleep characterizes schizophrenia. Further, this constellation appears to be a trait feature of the disorder. The precise nature of the link between sleep and the pathophysiology of schizophrenia needs to be further delineated by using quantitative EEG methods, physiological probes of sleep-wake regulation, and functional neuroanatomical studies.

V. Sleep and Circadian Rhythms in Anxiety Disorders

Anxiety disorders include a heterogeneous group of conditions that share the emotional features of fear, worry, or apprehension about future danger. Specific disorders in this category include: generalized anxiety disorder (GAD), characterized by continuous fearful ruminations about many different concerns; panic disorder, characterized by sudden, overwhelming fear and physical symptoms of arousal; posttraumatic stress disorders (PTSD), characterized by intrusive recollections of a traumatic event and heightened arousal; and obsessive compulsive disorder, characterized by repetitive, intrusive, and unwanted thoughts or actions (1). Not surprisingly, many patients with anxiety disorders also have disordered sleep. In fact, the epidemiological survey of Ford and Kamerow (12) found that anxiety disorders were the psychiatric conditions most strongly associated with insomnia complaints. Patients with any of the specific anxiety disorders may complain of sleep onset or sleep maintenance insomnia due to excessive anxiety and apprehensive expectation about one or more life circumstances (18). In this section we will briefly review sleep and circadian rhythm findings in patients with the most clinically significant anxiety disorders: GAD, panic disorder, and PTSD.

A. Generalized Anxiety Disorder

Subjectively, patients with GAD may complain of virtually any type of insomnia, although sleep onset difficulties are often most prominent because of anxious ruminations at bedtime. Polysomnographic studies have been performed in these patients to clarify their objective sleep disturbances and to distinguish these patients from patients with other psychiatric disorders (18,233–238). A consensus EEG sleep profile of these patients includes a general lightening of sleep (prolonged sleep latencies, increased awakenings throughout the night; increased stages 1 and 2 sleep) with little reliable change in REM sleep timing or duration. These EEG sleep changes have been shown to reliably discriminate these patients both from healthy subjects (lighter sleep in the patients) and from patients with mood disorders (shorter REM latencies and increased phasic REM in mood disorders patients). Patients with GAD and panic disorder do not show the cholinergic supersensitivity (short latency to REM sleep following administration of the muscarinic cholinergic agonist arecoline) that typifies patients with mood disorders (239,240). They also do not exhibit a generalized hypersensitivity of brain $5HT_2$ receptors (induction of SWS following administration of the $5HT_2$ antagonist ritanserin) in comparison with healthy subjects (241). A study that compared brain morphology and EEG sleep measures in patients with anxiety disorders versus those with schizophrenia did not reveal clear associations between EEG sleep and cerebral ventricle size (242). Pathophysiological models involving cholinergic sensitivity, $5HT_2$ receptor dysfunction, or brain dysmorphology, therefore, would not be supported in describing the sleep disturbances in patients with GAD.

No laboratory studies of circadian rhythms have been reported in patients with GAD. In a mixed sample of anxiety disorder patients including those with GAD, Shear and colleagues reported lower levels of regularity in daily routines, as measured by the Social Rhythm Metric (243). Whether this lower level of regularity is associated with endogenous circadian rhythms is unknown.

B. Panic Disorder

A distinctive sleep complaint of patients with panic disorder is the occurrence of spontaneous sleep-related panic attacks (244–247). Nearly 70% of these patients will exhibit sleep-related panic episodes at some point in their life, and 33% have sleep-related panic attacks on a recurrent basis. These episodes are behaviorally similar to their daytime panic attacks and occur in the transitional period between stages 2 and delta sleep in the first third of the night (247,248). Panic disorder patients who also have sleep panic attacks are characterized by early-onset anxiety and comorbid mood and anxiety disorders as adults (244). Clinically, sleep-related panic attacks can be distinguished from NREM parasomnias (sleepwalking, sleep terrors) by the presence of a full awakening, complete recall and symptoms consistent with daytime attacks. Sleep-related panic attacks respond to treatment with antidepressants and/or benzodiazepine hypnotics. Sleep EEG profiles of patients with panic disorder are also characterized by reduced sleep continuity, normal amounts of REM sleep and SWS and variable REM latency findings (normal to reduced) (249–254).

Although cortisol and catecholamine levels have been measured in patients with panic disorder, few 24-hr studies have been reported. Circadian rhythms of cortisol and ACTH may be abnormal in patients with panic disorder. Specifically, patients have been observed to have elevated nocturnal cortisol secretion, larger ultradian secretory episodes, and variable phase and mean level changes in corticotropin relative to control subjects (255). Moreover, patients with elevated cortisol were less likely to achieve drug-free remission (256).

C. Posttraumatic Stress Disorder

Subjectively, the distinctive sleep complaint of patients with posttraumatic stress disorder (PTSD) is the occurrence of stereotypic repetitive traumatic nightmares related to a traumatic life event (257–261). Controversy exists as to whether these disturbing events are variants of nightmares (REM-sleep-related) or of night terrors (NREM-sleep-related). During polysomnographic monitoring, these events have been observed to occur from both REM and NREM sleep, further obscuring their origin. In addition to demonstrating reduced sleep efficiency and reduced slow-wave sleep, patients with PTSD have been reported to demonstrate a variety of alterations related to REM sleep. Several studies have reported REM sleep findings characteristic of mood disorders, such as a reduction in REM latency (262–266). Other studies have found PTSD patients to have significant reductions in measures of REM sleep including prolongation of REM latency and diminished durations of REM sleep (259). However, many of these studies are confounded by the high comorbidity of mood disorders and substance abuse with PTSD (267).

Pathophysiologcally, the distinction between disorders of NREM or REM sleep may clarify whether PTSD is more related to a disturbance in general arousal or a disturbance in emotional processing. NREM parasomnias implicate brain structures modulating global cortical arousal such as the brainstem and components of the ascending reticular activating system. REM sleep parasomnias involve brain structures that modulate emotion, motivation, memory, and emotionally mediated attention. Alternatively, an interaction between the two systems may also be hypothesized. For example, arousal thresholds from slow-wave sleep have been reported to be higher in PTSD patients than in healthy controls (268,269). This contrasts with the findings of reduced sleep efficiency, increased overall awakenings, and an increased incidence of motor dysfunction (periodic limb movements) during both NREM and REM sleep, which suggest hyper-arousal (270–273). One way of explaining these discrepant findings is to hypothesize that the sleep of patients with PTSD is characterized by an internally focused locus of attention related to traumatic memories that often result in awakenings. However, this internally directed focus of attention may minimize arousals related to external stimuli.

The sleep disturbance of PTSD patients may also be reflected in neuroendocrine measures. For instance, patients with PTSD have increased 24-hr catecholamine excretion (274), the ratio of nocturnal to daytime urinary catecholamines is increased relative to control subjects, and this ratio correlates with total sleep time (275). Unlike depressed patients, individuals with PTSD do not show nocturnal hypersecretion of cortisol (176).

VI. Conclusions

Psychiatric disorders are frequently associated with disturbed sleep and circadian rhythms. The patterns of sleep and circadian rhythms disturbances in specific disorders suggest that these findings are not merely epiphenomena, but that they are intimately related to specific disease pathophysiologies. A new generation of studies promises an even richer understanding of the connections between sleep, circadian rhythms, and mental function through the combination of quantitative electroencephalographic measures, clinical neuroscience tools such as functional neuroimaging, specific neurochemical and behavioral probes, and data from controlled treatment trials.

Acknowledgment

This work was supported by NIH Grants MH 48891, MH 30915, MH 37869, MH 00295, MH 24652, and NIA 15138.

References

1. American Psychiatric Association. Diagnostic and Statistical Manual of Mental Disorders (DSM-IV), 4th ed. Washington, DC: American Psychiatric Association, 1994.
2. Pilcher JJ, Huffcutt AI. Effects of sleep deprivation on performance: a meta-analysis. Sleep 1996; 19(4):318–326.

3. Dinges DF, Pack F, Williams K, Gillen KA, Powell JW, Ott GE, et al. Cumulative sleepiness, mood disturbance, and psychomotor vigilance performance decrements during a week of sleep restricted to 4–5 hours per night. Sleep 1997; 20(4): 267–277.

4. Wood C, Magnello ME. Diurnal changes in perceptions of energy and mood. J R Soc Med 1992; 85:191–194.

5. Monk TH, Buysse DJ, Reynolds CF, Jarrett DB, Kupfer DJ. Rhythmic versus homeostatic influences on mood, activation and performance in the elderly. J Gerontol Psych Sci 1992; 47:P221–P227.

6. Boivin DB, Czeisler CA, Dijk D, Duffy JF, Folkard S, Minors DS, et al. Complex interaction of the sleep-wake cycle and circadian phase modulates mood in healthy subjects. Arch Gen Psychiatry 1997; 54:145–152.

7. Gerner RH, Post RM, Gillin JC, Bunney WE. Biological and behavioral effects of one night's sleep deprivation in depressed patients and normals. J Psychiatr Res 1979; 15:21–40.

8. Wu JC, Gillin JC, Buchsbaum MS, Hershey T, Hazlett E, Sicotte N, et al. The effect of sleep deprivation on cerebral glucose metabolic rate in normal humans assessed with positron emission tomography. Sleep 1991; 14:155–162.

9. Hamilton M. Frequency of symptoms in melancholia (depressive illness). Br J Psychiatry 1989; 154:201–206.

10. Perlis ML, Giles DE, Buysse DJ, Thase ME, Tu X, Kupfer DJ. Which depressive symptoms are related to which sleep EEG variables? Biol Psychiatry 1997; 42: 904–913.

11. Buysse DJ, Reynolds CF, Monk TH, Berman SR, Kupfer DJ. The Pittsburgh Sleep Quality Index (PSQI): a new instrument for psychiatric research and practice. Psychiatry Res 1989; 28:193–213.

12. Ford DE, Kamerow DB. Epidemiologic study of sleep disturbances and psychiatric disorders. JAMA 1989; 262:1479–1484.

13. Vollrath M, Wicki W, Angst J. The Zurich Study. VIII. Insomnia: association with depression, anxiety, somatic syndromes, and course of insomnia. Eur Arch Psychiatry Clin Neurosci 1989; 239(2):113–124.

14. Foley DJ, Monjan AA, Brown SL, Simonsick EM, Wallace RB, Blazer DG. Sleep complaints among elderly persons: an epidemiologic study of three communities. Sleep 1995; 18(6):425–432.

15. Coleman RM, Roffwarg HP, Kennedy SJ, Guilleminault C, Cinque J, Cohn MA, et al. Sleep-wake disorders based on a polysomnographic diagnosis; a national cooperative study. JAMA 1982; 247:997–1033.

16. Buysse DJ, Reynolds CF, Hauri PJ, Roth T, Stepanski EJ, Thorpy MJ, et al. Diagnostic concordance for sleep disorders using proposed DSM-IV categories: a report from the APA/NIMH DSM-IV field trial. Am J Psychiatry 1994; 151(9): 1351–1360.

17. Breslau N, Roth T, Rosenthal L, Andreski P. Sleep disturbance and psychiatric disorders: a longitudinal epidemiological study of young adults. Biol Psychiatry 1996; 39(6):411–418.

18. Nofzinger EA, Buysse DJ, Reynolds CF, Kupfer DJ. Sleep disorders related to another mental disorder (nonsubstance/primary): a DSM-IV literature review. J Clin Psychiatry 1993; 54(7):244–255; discussion 256–259.

19. Benca RM. Mood disorders. In: Kryger MH, Roth T, Dement WC, eds. Principles and Practice of Sleep Medicine, 2nd ed. Philadelphia: WB Saunders, 1994:899–913.
20. Armitage R. Microarchitectural findings in sleep EEG in depression: diagnostic implications. Biol Psychiatry 1995; 37:72–84 (editorial).
21. Armitage R, Roffwarg HP, Rush AJ. Digital period analysis of EEG in depression: periodicity, coherence, and interhemispheric relationships during sleep. Prog Neuropsychopharmacol Biol Psychiatry 1993; 17:363–372.
22. Buysse DJ, Kupfer DJ. Diagnostic and research applications of electroencephalographic sleep studies in depression: conceptual and methodological issues. J Nerv Ment Dis 1990; 178:405–414.
23. Benca RM, Obermeyer WH, Thisted RA, Gillin JC. Sleep and psychiatric disorders: a meta-analysis. Arch Gen Psychiatry 1992; 49:651–668.
24. Gillin JC, Duncan W, Pettigrew KD, Frankel BL, Snyder F. Successful separation of depressed, normal, and insomniac subjects by EEG sleep data. Arch Gen Psychiatry 1979; 36:85–90.
25. Kerkhofs M, Hoffmann G, DeMartelaere V, Linkowski P, Mendlewicz J. Sleep EEG recordings in depressive disorders. J Affect Disord 1985; 9:47–53.
26. Reynolds CF, Kupfer DJ, Houck PR, Hoch CC, Stack JA, Berman SR, et al. Reliable discrimination of elderly depressed and demented patients by EEG sleep data. Arch Gen Psychiatry 1988; 45:258–264.
27. Thase ME, Simons AD, Reynolds CF. Abnormal EEG sleep profiles in major depression. Association with response to cognitive behavior therapy. Arch Gen Psychiatry 1996; 53:99–108.
28. Goetz RR, Puig-Antich J, Ryan N, Rabinovich H, Ambrosini PJ, Nelson B, et al. Electroencephalographic sleep of adolescents with major depression and normal controls. Arch Gen Psychiatry 1987; 44:61–68.
29. Emslie GJ, Rush AJ, Weinberg WA, Rintelmann JW, Roffwarg HP. Children with major depression show reduced rapid eye movement latencies. Arch Gen Psychiatry 1990; 47:119–124.
30. Kutcher S, Williamson P, Marton P, Szalai J. REM latency in endogenously depressed adolescents. Br J Psychiatry 1992; 161:399–402.
31. Dahl RE, Puig-Antich J, Ryan ND, Nelson B, Dachille S, Cunningham SL, et al. EEG sleep in adolescents with major depression: the role of suicidality and inpatient status. J Affect Disord 1990; 19:63–75.
32. Gillin JC, Duncan WC, Murphy DL, Post RM, Wehr TA, Goodwin FK, et al. Age-related changes in sleep in depressed and normal subjects. Psychiatry Res 1981; 4:73–78.
33. Knowles JB, MacLean AW. Age-related changes in sleep in depressed and healthy subjects: a meta-analysis. Neuropsychopharmacology 1990; 3:251–259.
34. Lauer CJ, Riemann D, Wiegand M, Berger M. From early to late adulthood changes in EEG sleep of depressed patients and healthy volunteers. Biol Psychiatry 1991; 29:979–993.
35. Schulz H, Lund R, Cording C, Dirlich G. Bimodal distribution of REM sleep latencies in depression. Biol Psychiatry 1979; 14:595–600.
36. Reynolds CF, Kupfer DJ, Thase ME, Frank E, Jarrett DB, Coble PA, et al. Sleep, gender, and depression: an analysis of gender effects on the electroencephalographic sleep of 302 depressed outpatients. Biol Psychiatry 1990; 28:673–684.

37. Armitage R, Hudson A, Trivedi M, Rush AJ. Sex differences in the distribution of EEG frequencies during sleep: unipolar depressed outpatients. J Affect Disord 1995; 34(2):121–129.

38. Hamilton M. Development of a rating scale for primary depressive illness. Br J Soc Clin Psychol 1967; 6:278–296.

39. Spiker DG, Coble P, Cofsky J, Foster FG, Kupfer DJ. EEG sleep and severity of depression. Biol Psychiatry 1978; 13:485–488.

40. Reynolds CF, Taska LS, Jarrett DB, Coble PA, Kupfer DJ. REM latency in depression: is there one best definition? Biol Psychiatry 1983; 18:849–863.

41. Giles DE, Roffwarg HP, Schlesser MA, Rush AJ. Which endogenous depressive symptoms relate to REM latency reduction? Biol Psychiatry 1986; 21:473–482.

42. Giles DE, Schlesser MA, Rush AJ, Orsulak PJ, Fulton CL, Roffwarg HP. Polysomnographic findings and dexamethasone nonsuppression in unipolar depression: a replication and extension. Biol Psychiatry 1987; 22:872–882.

43. Ansseau M, Kupfer DJ, Reynolds CF, McEachran AB. REM latency distribution in depression: clinical characteristics associated with sleep onset REM. Biol Psychiatry 1985; 19(12):1651–1666.

44. Kumar A, Shipley JE, Eiser AS, Feinberg M, Flegel P, Grunhaus L, et al. Clinical correlates of sleep onset REM periods in depression. Biol Psychiatry 1987; 22: 1473–1477.

45. Kupfer DJ, Ehlers CL, Frank E, Grochocinski VJ, McEachran AB. EEG sleep profiles and recurrent depression. Biol Psychiatry 1991; 30:641–655.

46. Dew MA, Reynolds CF, Buysse DJ, Houck PR, Hoch CC, Monk TH, et al. EEG sleep profiles during depression: effects of episode duration and other clinical and psychosocial factors in older adults. Arch Gen Psychiatry 1996; 53:148–156.

47. Vitiello MV, Prinz PN, Avery DH, Williams DE, Ries RK, Bokan JA, et al. Sleep is undisturbed in elderly, depressed individuals who have not sought health care. Biol Psychiatry 1990; 27(4):431–440.

48. Reynolds CF, Newton TF, Shaw DH, Coble PA, Kupfer DJ. Electroencephalographic sleep findings in depressed outpatients. Psychiatry Res 1982; 6:65–75.

49. Buysse DJ, Jarrett DB, Miewald JM, Kupfer DJ, Greenhouse JB. Minute-by-minute analysis of REM sleep timing in major depression. Biol Psychiatry 1990; 28:911–925.

50. Feinberg M, Carroll BJ. Biological "markers" for endogenous depression: effect of age, severity of illness, weight loss, and polarity. Arch Gen Psychiatry 1984; 41: 1080–1085.

51. Mendlewicz J, Kerkhofs M, Hoffmann G, Linkowski P. Dexamethasone suppression test and REM sleep in patients with major depressive disorder. Br J Psychiatry 1984; 145:383–388.

52. Kupfer DJ, Targ E, Stack J. Electroencephalographic sleep in unipolar depressive subtypes: support for a biological and familial classification, J Nerv Ment Dis 1982; 170:494–498.

53. Kupfer DJ, Broudy D, Coble PA, Spiker DG. EEG sleep and affective psychosis. J Affect Disord 1980; 2:17–25.

54. Naylor MW, Shain BN, Shipley JE. REM latency in psychotically depressed adolescents. Biol Psychiatry 1990; 28:161–164.

55. Thase ME, Kupfer DJ, Ulrich RF. Electroencephalographic sleep in psychotic depression: a valid subtype? Arch Gen Psychiatry 1986; 43:886–893.

56. Thase ME, Himmelhoch JM, Mallinger AG, Jarrett DB, Kupfer DJ. Sleep EEG and DST findings in anergic bipolar depression. Am J Psychiatry 1989; 146:329–333.
57. Nofzinger EA, Thase ME, Reynolds CF, Himmelhoch JM, Mallinger A, Houck P, et al. Hypersomnia in bipolar depression: a comparison with narcolepsy using the multiple sleep latency test. Am J Psychiatry 1991; 148:1177–1181.
58. Giles DE, Rush AJ, Roffwarg HP. Sleep parameters in bipolar I, bipolar II, and unipolar depressions. Biol Psychiatry 1986; 21:1340–1343.
59. Shapiro CM, Devins GM, Feldman B, Levitt AJ. Is hypersomnolence a feature of seasonal affective disorder? J Psychosom Res 1994; 38(1):49–54.
60. Putilov AA, Booker JM, Danilenko KV, Zolotarev DY. The relation of sleep-wake patterns to seasonal depressive behavior. Arctic Med Res 1994; 53(3):130–136.
61. Anderson JL, Rosen LN, Mendelson WB, Jacobsen FM, Skwerer RG, Joseph-Vanderpool JR, et al. Sleep in fall/winter seasonal affective disorder: effects of light and changing seasons. J Psychosom Res 1994; 38:323–337.
62. Partonen T, Appelberg B, Partinen M. Effects of light treatment on sleep structure in seasonal affective disorder. Eur Arch Psychiatry Clin Neurosci 1993; 242(5): 310–313.
63. Brunner DP, Krauchi K, Dijk D, Leonhardt G, Haugh H, Wirz-Justice A. The sleep EEG in seasonal affective disorder and in control women: effects of midday light treatment and sleep deprivation. Biol Psychiatry 1996; 40(6):485–496.
64. Schulte W. Kombinierte psycho-und pharmakotherapie bei melancholikern. In: Kranz HN, Petrilowitsch, eds. Probleme der pharmakopsychiatrischen Kombinations- und Langzeitbehandlung. Basel: Rothenburger Gesprach, 1966:150–169.
65. Pflug B, Tölle R. Disturbance of the 24-hour-rhythm in endogenous depression and the treatment of endogenous depression by sleep deprivation. Int Pharmacopsychiatry 1971; 6:187–196.
66. Wu JC, Bunney WE. The biological basis of an antidepressant response to sleep deprivation and relapse: review and hypothesis. Am J Psychiatry 1990; 147:14–21.
67. Kuhs H, Tölle R. Sleep deprivation therapy. Biol Psychiatry 1991; 29:1129–1148.
68. Schilgen B, Tölle R. Partial sleep deprivation as therapy for depression. Arch Gen Psychiatry 1980; 37:267–271.
69. Sack DA, Duncan W, Rosenthal NE, Mendelson WE, Wehr TA. The timing and duration of sleep in partial sleep deprivation therapy of depression. Acta Psychiatr Scand 1988; 77:219–224.
70. Giedke H, Geilenkirchen R, Hauser M. The timing of partial sleep deprivation in depression. J Affect Disord 1992; 25:117–128.
71. Vogel GW, Thurmond A, Gibbons P, Sloan K, Boyd M, Walker M. REM sleep reduction effects on depression syndromes. Arch Gen Psychiatry 1975; 32:765–777.
72. Vogel GW, Vogel F, McAbee RS, Thurmond AJ. Improvement of depression by REM sleep deprivation. Arch Gen Psychiatry 1980; 37:247–253.
73. Nowlin-Finch NL, Altshuler LL, Szuba MP, Mintz J. Rapid resolution of first episodes of mania: sleep related? J Clin Psychiatry 1994; 55(1):26–29.
74. Gillin JC, Kripke DF, Janowsky DS, Risch SC. Effects of brief naps on mood and sleep in sleep-deprived depressed patients. Psychiatry Res 1989; 27:253–265.
75. Weigand M, Riemann D, Schreiber W, Lauer CJ, Berger M. Effect of morning and afternoon naps on mood after total sleep deprivation in patients with major depression. Biol Psychiatry 1993; 33:467–476.

76. van Bemmel AL, Van den Hoofdakker RH. Maintenance of therapeutic effects of total sleep deprivation by limitation of subsequent sleep: a pilot study. Acta Psychiatr Scand 1981; 63:453–462.

77. Dessauer M, Goetze U, Tölle R. Period sleep deprivation in drug-refractory depression. Neuropsychobiology 1985; 13:113–116.

78. Holsboer-Trachsler E, Wiedermann K, Holsboer F. Serial partial sleep deprivation in depression—clinical effects and dexamethasone suppression test results. Neuropsychobiology 1988; 19:73–78.

79. Papadimitriou GN, Christodoulou GN, Katsouyanni K, Stefanis CN. Therapy and prevention of affective illness by total sleep deprivation. J Affect Disord 1993; 27: 107–116.

80. Shelton RC, Loosen PT. Sleep deprivation accelerates the response to nortriptyline. Prog Neuropsychopharmacol Biol Psychiatry 1993; 17:113–123.

81. Baxter LR, Jr., Liston EH, Schwartz JM, Altshuler LL, Wilkins JN, Richeimer S, et al. Prolongation of the antidepressant response to partial sleep deprivation by lithium. Psychiatry Res 1986; 19(1):17–23.

82. Leibenluft E, Moul DE, Schwartz PJ, Madden PA, Wehr TA. A clinical trial of sleep deprivation in combination with antidepressant medication. Psychiatry Res 1993; 46:213–227.

83. Riemann D, Hohagen F, Konig A, Schwarz B, Gomille J, Voderholzer U, et al. Advanced vs. normal sleep timing: effects on depressed mood after response to sleep deprivation in patients with a major depressive disorder. J Affect Disord 1996; 37: 121–128.

84. Duncan WC, Post RM, Wehr TA. Relationship between EEG sleep patterns and clinical improvement in depressed patients treated with sleep deprivation. Biol Psychiatry 1980; 15:879–889.

85. Elsenga S, Van den Hoofdakker RH. Response to total sleep deprivation and clomipramine in endogenous depression. J Psychiatr Res 1987; 21(2):157–161.

86. Reinink E, Bouhuys N, Wirz-Justice A, van den Hoofdakker R. Prediction of the antidepressant response to total sleep deprivation by diurnal variation of mood. Psychiatry Res 1990; 32:113–124.

87. Bouhuys AL. Towards a model of mood responses to sleep deprivation in depressed patients. Biol Psychiatry 1991; 29:600–612.

88. Reynolds CF, Kupfer DJ, Hoch CC, Stack JA, Houck PR, Berman SR. Sleep deprivation effects in older endogenous depressed patients. Psychiatry Res 1987; 21: 95–109.

89. Wirz-Justice A. Biological rhythms in mood disorders. In: Bloom FE, Kupfer DJ, eds. Psychopharmacology: The Third Generation of Progress. New York: Raven Press, 1995:999–1017.

90. Wehr TA, Goodwin FK. Desynchronization of circadian rhythms as a possible source of manic-depressive cycles. Psychopharmacol Bull 1980; 16:19.

91. Lund R, Schulz H. The relationship of disturbed sleep in depression to an early minimum of the circadian temperature rhythm. In: Collegium Internationale Neuro-Psychopharmacologicum Program, Goteborg, Sweden, 1980:233.

92. von Zerssen D, Barthelmes H, Dirlich G, Doerr P, Emrich HM, von Lindern L, et al. Circadian rhythms in endogenous depression. Psychiatry Res 1985; 16:51–63.

93. Tsujimoto T, Yamada N, Shimoda K, Hanada K, Takahashi S. Circadian rhythms in depression. Part II. Circadian rhythms in inpatients with various mental disorders. J Affect Disord 1990; 18:199–210.

94. Avery DH, Wildschiodtz G, Rafaelsen OJ. Nocturnal temperature in affective disorder. J Affect Disord 1982; 4:61–71.

95. Schulz H, Lund R. On the origin of early REM episodes in the sleep of depressed patients: a comparison of three hypotheses. Psychiatry Res 1985; 16:65–77.

96. Avery DH, Wildschiodtz G, Smallwood G, Martin D, Rafaelsen OJ. REM latency and core temperature relationships in primary depression. Acta Psychiatr Scand 1986; 74:269–280.

97. Wehr TA, Goodwin FK. Biological rhythms in manic-depressive illness. In: Wehr TA, Goodwin FK, eds. Circadian Rhythms in Psychiatry. Pacific Grove, CA: The Boxwood Press 1983:129–184.

98. Wehr TA, Sack DA, Duncan WC. Sleep and circadian rhythms in affective patients isolated from external time cues. Psychiatry Res 1985; 15:327–339.

99. Monk TH, Buysse DJ, Frank E, Kupfer DJ, Dettling J, Ritenour A. Nocturnal and circadian body temperatures of depressed outpatients during symptomatic and recovered states. Psychiatry Res 1994; 51:297–311.

100. Avery DH, Dahl K, Savage MV, Brengelmann GL, Larsen LH, Kenny MA, et al. Circadian temperature and cortisol rhythms during a constant routine are phase-delayed in hypersomnic winter depression. Biol Psychiatry 1997; 41(11):1109–1123.

101. Arendt J. Melatonin: a new probe in psychiatric investigation? Br J Psychiatry 1989; 155:585–590.

102. Claustrat B, Chazot G, Brun J, Jordan D, Sassolas G. A chronobiological study of melatonin and cortisol secretion in depressed subjects: plasma melatonin, a biochemical marker in major depression. Biol Psychiatry 1984; 19:1215–1228.

103. Beck-Friis J, Ljunggren J, Thoren M, von Rosen D, Kjellman BF, Wetterberg L. Melatonin, cortisol and ACTH in patients with major depressive disorder and healthy humans with special reference to the outcome of the dexamethasone suppression test. Psychoneuroendocrinology 1985; 10:173–186.

104. Rubin RT, Heist EK, McGeoy SS, Hanada K, Lesser IM. Neuroendocrine aspects of primary endogenous depression. XI. Serum melatonin measures in patients and matched control subjects. Arch Gen Psychiatry 1992; 49:558–567.

105. Thompson C, Franey C, Arendt J, Checkley SA. A comparison of melatonin secretion in depressed patients and normal subjects. Br J Psychiatry 1988; 152:260–265.

106. Lewy AJ, Sack R, Miller L, Hoban T. Antidepressant and circadian phase-shifting effects of light. Science 1987; 235:352–354.

107. Sack RL, Lewy AJ, White DM, Singer CM, Fireman MJ, Vandiver R. Morning vs evening light treatment for winter depression. Arch Gen Psychiatry 1990; 47:343–351.

108. Lewy AJ, Sack RL, Singer CM, White DM, Hoban TM. Winter depression and the phase-shift hypothesis for bright light's therapeutic effects: history, theory, and experimental evidence. J Biol Rhythms 1988; 3:121–134.

109. Jacobsen FM, Wehr TA, Skwerer RA, Sack DA, Rosenthal NE. Morning versus midday phototherapy of seasonal affective disorder. Am J Psychiatry 1987; 144:1301–1305.

110. Wirz-Justice A, Graw P, Krauchi K, Gisin B, Jochum A, Arendt J, et al. Light therapy in seasonal affective disorder is independent of time of day or circadian phase. Arch Gen Psychiatry 1993; 50(12):929–937.

111. Carrol BJ. The dexamethasone suppression test for melancholia. Br J Psychiatry 1982; 140:292–304.

112. Nemeroff CB, Widerlov E, Bissette G, Walleus H, Karlsson I, Eklund K, et al. Elevated concentrations of CSF corticotropin-releasing factor-like immunoreactivity in depressed patients. Science 1984; 226:1342–1344.

113. De Souza EB. Corticotropin-releasing factor receptors: physiology, pharmacology, biochemistry and role in central nervous system and immune disorders. Psychoneuroendocrinology 1995; 20(8):789–819.

114. Dahl RE, Ryan ND, Puig-Antich J, Nguyen NA, Al-Shabbout M, Meyer VA, et al. 24-hour cortisol measures in adolescents with major depression: a controlled study. Biol Psychiatry 1991; 30:25–36.

115. Linkowski P, Mendlewicz J, Kerkhofs M, Leclercq R, Golstein J, Brasseur M, et al. 24-hour profiles of adrenocorticotropin, cortisol, and growth hormone in major depressive illness: effect of antidepressant treatment. J Clin Endocrinol Metab 1987; 65:141–152.

116. Jarrett DB, Coble PA, Kupfer DJ. Reduced cortisol latency in depressive illness. Arch Gen Psychiatry 1983; 40:506–511.

117. Sapolsky RM, Krey LC, McEwen BS. Glucocorticoid-sensitive hippocampal neurons are involved in terminating the adrenocortical stress response. Proc Natl Acad Sci USA 1984; 81:6174–6177.

118. Stokes PE. The potential role of excessive cortisol induced by HPA hyperfunction in the pathogenesis of depression. Eur Neuropsychopharmacol 1995; 5(Suppl):77–82.

119. Jarrett DB, Miewald JM, Kupfer DJ. Recurrent depression is associated with a persistent reduction in sleep-related growth hormone secretion. Arch Gen Psychiatry 1990; 47:113–118.

120. Dahl RE, Ryan ND, Williamson DE, Ambrosini PJ, Rabinovich H, Novacenko H, et al. Regulation of sleep and growth hormone in adolescent depression. J Am Acad Child Adolesc Psychiatry 1992; 31(4):615–621.

121. Franz B, Kupfer DJ, Miewald JM, Jarrett DB, Grochocinski VJ. Growth hormone secretion timing in depression: clinical outcome comparisons. Biol Psychiatry 1995; 38:720–729.

122. Lesch K, Laux G, Pfuller H, Erb A, Beckmann H. Growth hormone (GH) response to GH-releasing hormone in depression. J Clin Endocrinol Metab 1987; 65:1278–1281.

123. Matussek N, Ackenheil M, Hippius H. Effect of clonidine on growth hormone release in psychiatric patients and controls. Psychiatry Res 1980; 2:25–36.

124. Kerkhofs M, Linkowski P, Lucas F, Mendelwicz J. Twenty-four-hour patterns of sleep in depression. Sleep 1991; 14(6):501–506.

125. Buysse DJ, Monk TH, Kupfer DJ, Frank E, Stapf D. Circadian patterns of unintended sleep episodes during a constant routine in remitted depressed patients. J Psychiatr Res 1995; 29(5):407–416.

126. Leibenluft E, Albert PS, Rosenthal NE, Wehr TA. Relationship between sleep and mood in patients with rapid-cycling bipolar disorder. Psychiatry Res 1996; 63(2–3): 161–168.

127. Wehr TA, Sack DA, Rosenthal NE. Sleep reduction as a final common pathway in the genesis of mania. Am J Psychiatry 1987; 144:201–204.

128. Wehr TA, Moul DE, Barbato G, Giesen HA, Seidel JA, Barker C. Conservation of photoperiod-responsive mechanisms in humans. Am J Physiol 1993; 265(4 Pt 2): R846–R857.

129. Wehr TA, Turner EH, Shimada JM, Clark CH, Barker C, Leibenluft E. Treatment of a rapidly cycling bipolar patient by using extended bedrest and darkness to stabilize the timing and duration of sleep. Biol Psychiatry 1998; 43:822–828.

130. Monk TH, Flaherty JF, Frank E, Hoskinson K, Kupfer DJ. The Social Rhythm Metric (SRM): an instrument to quantify the daily rhythms of life. J Nerv Ment Dis 1990; 178:120–126.

131. Szuba MP, Yager A, Guze BH, Allen EM, Baxter JR. Disruption of social circadian rhythms in major depression: a preliminary report. Psychiatry Res 1992; 42:221–230.

132. Drevets WC, Price JL, Simpson JR, Jr., Todd RD, Reich T, Vannier M, et al. Subgenual prefrontal cortex abnormalities in mood disorders. Nature 1997; 386: 824–827.

133. George MS, Ketter TA, Post RM. Prefrontal cortex dysfunction in clinical depression. Depression 1994; 2:59–72.

134. Mayberg HS. Limbic-cortical dysregulation: a proposed model of depression. J Neuropsychiatry Clin Sci 1997; 9(3):471–481.

135. Ho AP, Gillin JC, Buchsbaum MS, Wu JC, Abel L, Bunney WE. Brain glucose metabolism during non-rapid eye movement sleep in major depression: a positron emission tomography study. Arch Gen Psychiatry 1996; 53:645–652.

136. Buchsbaum MS, Gillin JC, Wu J, Hazlett E, Sicotte N, DuPont RM, et al. Regional cerebral glucose metabolic rate in human sleep assessed by positron emission tomography. Life Sci 1989; 45:1349–1356.

137. Maquet P, Peters JM, Aerts J, Delfiore G, Degueldre C, Luxen A, et al. Functional neuroanatomy of human rapid-eye-movement sleep and dreaming. Nature 1996; 383:163–166.

138. Nofzinger EA, Mintun MA, Wiseman MB, Kupfer DJ, Moore RY. Forebrain activation in REM sleep: an FDG PET study. Brain Res 1997; 770:192–201.

139. Nofzinger EA, Mintun MA, Kupfer DJ. Limbic glucose metabolism is enhanced during REM sleep in mood disorders patients. Sleep Res 1996; 25:170 (abstract).

140. Wu JC, Gillin JC, Buchsbaum MS, Hershey T, Johnson JC, Bunney WE. Effect of sleep deprivation on brain metabolism of depressed patients. Am J Psychiatry 1992; 149:538–543.

141. Ebert D, Feistel H, Barocka A. Effects of sleep deprivation on the limbic system and the frontal lobes in affective disorders: a study with Tc-99m-HMPAO SPECT. Psychiatr Res Neuroimag 1991; 40:247–251.

142. Buysse DJ, Monahan JP, Cherry CR, Kupfer DJ, Frank E. Persistent effects on sleep EEG following fluoxetine discontinuation. Sleep Res 1997; 26:285 (abstr).

143. Coble PA, Kupfer DJ, Spiker DG, Neil JF, McPartland RJ. EEG sleep in primary depression: a longitudinal placebo study. J Affect Disord 1979; 1:131–138.

144. Kupfer DJ, Frank E, Grochocinski VJ, Gregor M, McEachran AB. Electroencephalographic sleep profiles in recurrent depression: a longitudinal investigation. Arch Gen Psychiatry 1988; 45:678–681.

145. Dew MA, Reynolds CF, Buysse DJ, Houck PR, Hoch CC, Monk TH, et al. Electro-

encephalographic sleep profiles during depression. Effects of episode duration and other clinical and psychosocial factors in older adults. Arch Gen Psychiatry 1996; 53(2):148–156.

146. Riemann D, Berger M. EEG sleep in depression and in remission and the REM sleep response to the cholinergic agonist RS 86. Neuropsychopharmacology 1989; 2: 145–152.

147. Buysse DJ, Kupfer DJ, Frank E, Monk TH, Ritenour A, Ehlers CL. Electroencephalographic sleep studies in depressed patients treated with psychotherapy. II. Longitudinal studies at baseline and recovery. Psychiatry Res 1992; 40:27–40.

148. Lee JH, Reynolds CF, Hoch CC, Buysse DJ, Mazumdar S, George CJ, et al. EEG sleep in recently remitted, elderly depressed patients in double-blind placebo-maintenance therapy. Neuropsychopharmacology 1993; 8(2):143–150.

149. Rush AJ, Erman MK, Giles DE, Schlesser MA, Carpenter G, Vasavada N, et al. Polysomnographic findings in recently drug-free and clinically remitted depressed patients. Arch Gen Psychiatry 1986; 43:878–884.

150. Thase ME, Simons AD. The applied use of psychotherapy in the study of the psychobiology of depression. J Psychother Prac Res 1992; 1:72–80.

151. Steiger A, von Bardeleben U, Guldner J, Lauer C, Rothe B, Holsboer F. The sleep EEG and nocturnal hormonal secretion studies on changes during the course of depression and on effects of CNS-active drugs. Prog Neuropsychopharmacol Biol Psychiatry 1993; 17:125–137.

152. Buysse DJ, Frank E, Lowe KK, Cherry CR, Kupfer DJ. Electroencephalographic sleep correlates of episodes and vulnerability to recurrence in depression. Biol Psychiatry 1997; 41:406–418.

153. Giles DE, Jarrett RB, Roffwarg HP, Rush AJ. Reduced rapid eye movement latency: a predictor of recurrence in depression. Neuropsychopharmacology 1987; 1:33–39.

154. Linkowski P, Kerkhofs M, Hauspie R, Susanne C, Mendlewicz J. EEG sleep patterns in man: a twin study. Electroencephalogr Clin Neurophysiol 1989; 73:279–284.

155. Linkowski P, Kerkhofs M, Hauspie R, Mendlewicz J. Genetic determinants of EEG sleep: a study in twins living apart. Electroencephalogr Clin Neurophysiol 1991; 79(2):114–118.

156. Heath AC, Kendler KS, Eaves LJ, Martin NG. Evidence for genetic influences on sleep disturbance and sleep pattern in twins. Sleep 1990; 13(4):318–335.

157. Giles DE, Biggs MM, Rush AJ, Roffwarg HP. Risk factors in families of unipolar depression. I. Psychiatric illness and reduced REM latency. J Affect Disord 1988; 14:51–59.

158. Giles DE, Kupfer DJ, Roffwarg HP, Rush AJ, Biggs MM, Etzel BA. Polysomnographic parameters in first-degree relatives of unipolar probands. Psychiatry Res 1989; 27:127–136.

159. Lauer CJ, Schreiber W, Holsboer F, Krieg JC. In quest of identifying vulnerability markers for psychiatric disorders by all-night polysomnography. Arch Gen Psychiatry 1995; 52(2):145–153.

160. Avery D, Gildschiodtz G, Rafaelsen O. REM latency and temperature in affective disorder before and after treatment. Biol Psychiatry 1982; 17(4):463–470.

161. Souetre E, Salvata E, Belugou J, Pringuey D, Candito M, Krebs B, et al. Circadian rhythms in depression and recovery: evidence for blunted amplitude as the main chronobiological abnormality. Psychiatry Res 1989; 28:263–278.

162. Casper RC, Katz MM, Bowden CL, Davis JM, Koslow SH, Hanin I. The pattern of physical symptom changes in major depressive disorder following treatment with amitryptyline or imipramine. J Affect Disord 1994; 31(3):151–164.

163. Kennedy GJ, Kelman HR, Thomas C. Persistence and remission of depressive symptoms in late life. Am J Psychiatry 1991; 148(2):174–178.

164. Fawcett J, Scheftner WA, Fogg L, Clark DC, Young MA, Hedeker D, et al. Time-related predictors of suicide in major affective disorder. Am J Psychiatry 1990; 147: 1189–1194.

165. Reynolds CF, Frank E, Houck PR, Mazumdar S, Dew MA, Cornes C, et al. Which remitted elderly depressed patients benefit from continued interpersonal psychotherapy after discontinuation of antidepressant medication? Am J Psychiatry 1997; 154(7):958–962.

166. Dew MA, Reynolds CF, Houck PR, Hall MH, Buysse DJ, Frank E, et al. Temporal profiles of the course of depression during treatment: predictors of pathways toward recovery in the elderly. Arch Gen Psychiatry 1997; 54:1016–1024.

167. Rush AJ, Giles DE, Jarrett RB, Feldman-Koffler F, Debus JR, Weissenburger J, et al. Reduced REM latency predicts response to tricyclic medication in depressed outpatients. Biol Psychiatry 1989; 26:61–72.

168. Kupfer DJ, Spiker DG, Coble PA, Neil JF, Ulrich R, Shaw DH. Sleep and treatment prediction in endogenous depression. Am J Psychiatry 1981; 138:429–434.

169. Jarrett RB. Does the pretreatment polysomnogram predict response to cognitive therapy in depressed outpatients? A preliminary report. Psychiatry Res 1990; 33: 285–299.

170. Buysse DJ, Kupfer DJ, Frank E, Monk TH, Ritenour A, Ehlers CL. Electroencephalographic sleep studies in depressed patients treated with psychotherapy. I. Baseline studies in responders and nonresponders. Psychiatry Res 1992; 40:13–26.

171. Thase ME, Simons AD, Reynolds CF. Abnormal electroencephalographic sleep profiles in major depression. Arch Gen Psychiatry 1996; 53:99–108.

172. Thase ME, Buysse DJ, Frank E, Cherry CR, Cornes CL, Mallinger AG, et al. Which depressed patients will respond to interpersonal psychotherapy? The role of abnormal electroencephalographic sleep profiles. Am J Psychiatry 1997; 154(4):502–509.

173. Kupfer DJ, Frank E, McEachran AB, Grochocinski VJ. Delta sleep ratio: a biological correlate of early recurrence in unipolar affective disorder. Arch Gen Psychiatry 1990; 47:1100–1105.

174. Kupfer DJ, Ehlers CL, Frank E, Grochocinski VJ, McEachran AB, Buhari A. Electroencephalographic sleep studies in depressed patients during long-term recovery. Psychiatry Res 1993; 49:121–138.

175. Papousek M. Chronobiological aspects of cyclothymia. Fortschr Neurol Psychiatr Ihrer Grenzgeb 1975; 43:381.

176. Kripke DF. Phase-advance theories for affective illness. In: Wehr TA, Goodwin FK, eds. Circadian Rhythms in Psychiatry. Pacific Grove, CA: Boxwood Press, 1983.

177. Wehr TA, Wirz-Justice A. Internal coincidence model for sleep deprivation and depression. In: Sleep 1980, 5th European Congress on Sleep Research. Basel: Karger, 1981:26–33.

178. MacLean AW, Knowles JB, Vetere C. REM sleep and depression: further use of computer simulation to test the phase-advance hypothesis. Psychiatry Res 1986; 19: 25–36.

179. Wehr TA, Wirz-Justice A, Goodwin FK, Duncan WC, Gillin JC. Phase advance of the circadian sleep-wake cycle as an antidepressant. Science 1979; 206:710–713.

180. Sack DA, Nurnberger J, Rosenthal NE, Ashburn E, Wehr TA. Potentiation of antidepressant medications by phase advance of the sleep-wake cycle. Am J Psychiatry 1985; 142:606–608.

181. Ehlers C, Frank E, Kupfer DJ. Social zeitgebers and biological rhythms; a unified approach to understanding the etiology of depression. Arch Gen Psychiatry 1988; 45:948–952.

182. Borbely AA. A two-process model of sleep regulation. Hum Neurobiol 1982; 1: 195–204.

183. Feinberg I, Floyd TC. The regulation of human sleep: clues from its phenomenology. Hum Neurobiol 1982; 1:185–195.

184. Borbely AA, Wirz-Justice A. A two-process model of sleep regulation. II. Implications for depression. Hum Neurobiol 1982; 1:205–210.

185. Beersma DGM, van den Hoofdakker RH. Can non-REM sleep be depressogenic? J Affect Disord 1992; 24:101–108.

186. Vogel GW, Buffenstein A, Minter K, Hennessey A. Drug effects on REM sleep and on endogenous depression. Neurosci Biobehav Rev 1990; 14:49–63.

187. Janowsky DS, El-Yousef MK, Davis JM, Sekerke HJ. A cholinergic-adrenergic hypothesis of mania and depression. Lancet 1972; 2:632–635.

188. Gillin JC, Sutton L, Ruiz C, Kelsoe J, DuPont RM, Darko D, et al. The cholinergic rapid eye movement test with arecoline in depression. Arch Gen Psychiatry 1991; 48:264–270.

189. Riemann D, Hohagen F, Bahro M, Berger M. Sleep in depression: the influence of age, gender and diagnostic subtype on baseline sleep and the cholinergic REM induction test with RS 86. Eur Arch Psychiatry Clin Neurosci 1994; 243:279–290.

190. Sitaram N, Nurnberger JI, Gershon ES, Gillin JC. Cholinergic regulation of mood and REM sleep: potential model and marker of vulnerability to affective disorders. Am J Psychiatry 1982; 139:571–576.

191. Bhatti T, Gillin JC, Golshan S, Clark C, Demodena A, Schlossr A, et al. The effect of a tryptophan-free amino acid drink on sleep and mood in normal controls. Sleep Res 1995; 24:153.

192. Kupfer DJ, Ehlers CL. Two roads to rapid eye movement latency. Arch Gen Psychiatry 1989; 46:945–948.

193. Tagaya H, Matsuno Y, Atsumi Y. A schizophrenic with non-24-hour sleep-wake syndrome. Jpn J Psychiatry Neurol 1993; 47(2):441–442.

194. Zarcone VP, Benson KL. Sleep and schizophrenia. In: Principles and Practice of Sleep Medicine. Philadelphia: WB Sanders, 1994:105–214.

195. Ganguli R, Reynolds CF, Kupfer DJ. Electroencephalographic sleep in young, never medicated schizophrenics. Arch Gen Psychiatry 1987; 44:36–44.

196. Tandon R, Shipley JE, Taylor S, Greden JF, Eiser A, DeQuardo J, et al. Electroencephalographic sleep abnormalities in schizophrenia: relationship to positive/negative symptoms and prior neuroleptic treatment. Arch Gen Psychiatry 1992; 49: 185–194.

197. Kempenaers C, Kerkhofs M, Linkowski P, Mendlewicz J. Sleep EEG variables in young schizophrenic and depressive patients. Biol Psychiatry 1988; 24:833–838.

198. Lauer CJ, Schreiber W, Pollmacher T, Holsboer F, Krieg JC. Sleep in schizophrenia: a polysomnographic study on drug-naive patients. Neuropsychopharmacology 1997; 16(1):51–60.

199. Kajimura N, Kato M, Okuma T, Sekimoto M, Watanabe T, Takahashi K. Relationship between delta activity during all-night sleep and negative symptoms in schizophrenia: a preliminary study. Biol Psychiatry 1996; 39:451–454.

200. Hiatt JF, Floyd TC, Katz PH, Feinberg I. Further evidence of abnormal non-rapid-eye-movement sleep in schizophrenia. Arch Gen Psychiatry 1985; 42:797–802.

201. Keshavan MS, Reynolds CF, Miewald, MJ, Montrose DM, Sweeney JA, Vasko RC, Kupfer DJ. Delta sleep deficits in schizophrenia: evidence from automated analyses of sleep data. Arch Gen Psychiatry 1998; 55:443–448.

202. Keshavan MS, Tandon R. Sleep abnormalities in schizophrenia: pathophysiological significance. Psychol Med 1993; 23:831–835 (editorial).

203. Keshavan MS, Reynolds CF, Miewald JM, Montrose DM. A longitudinal study of EEG sleep in schizophrenia. Psychiatry Res 1996; 59:203–211.

204. Van Kammen DP, van Kammen WB, Peters J, Goetz K, Neylan T. Decreased slow-wave sleep and enlarged lateral ventricles in schizophrenia. Neuropsychopharmacology 1988; 1:265–271.

205. Keshavan MS, Reynolds CF, Ganguli R, Haas GL, Sweeney J, Miewald J, et al. Slow wave and symptomatology in schizophrenia and related psychotic disorder. J Psychiatr Res 1995; ;29:303–314.

206. Benson KL, Sullivan EV, Lim KO, Zarcone VP. The effect of total sleep deprivation on slow wave recovery in schizophrenia. Sleep Res 1993; 22:143.

207. Keshavan MS, Reynolds CF, Miewald J, Montrose D. Slow-wave sleep deficits and outcome in schizophrenia and schizoaffective disorder. Acta Psychiatr Scand 1995; 91:289–292.

208. Orzack MH, Hartman EL, Kornetsky C. The relationship between attention and slow wave sleep in schizophrenia. Psychopharmacol Bull 1977; 13:59–61.

209. Keshavan MS, Reynolds CF, Montrose D, Miewald J, Downs C, Sabo EM. Sleep and suicidality in psychotic patients. Acta Psychiatr Scand 1994; 89(2):122–125.

210. Lewis C, Tandon R, Shipley JR, Dequardo JR, Gibson M, Taylor SF, et al. Biological predictors of suicidality in schizophrenia. Acta Psychiatr Scand 1996; 94:416–420.

211. Taylor S, Goldman RS, Tandon R, Shipley JE. Neuropsychological function and REM sleep in schizophrenic patients. Biol Psychiatry 1992; 32(6):529–538.

212. Benson KL, Zarcone VP. Testing the REM sleep phase event intrusion hypothesis of schizophrenia. Psychiatry Res 1985; 15:163–173.

213. Luby ED, Caldwell DF. Sleep deprivation and EEG slow wave activity in chronic schizophrenia. Arch Gen Psychiatry 1967; 17:361–364.

214. Horne JA. Human sleep, sleep loss and behavior: implications for the prefrontal cortex and psychiatric disorder. Br J Psychiatry 1993; 162:413–419.

215. Rao ML, Strebel B, Halaris A, Gross G, Braunig P, Huber G, et al. Circadian rhythm of vital signs, norepinephrine, epinephrine, thyroid hormones, and cortisol in schizophrenia. Psychiatry Res 1995; 57(1):21–39.

216. Monteleone P, Maj M, Fusco M, Kemali D, Reiter RJ. Depressed nocturnal plasma melatonin levels in drug-free paranoid schizophrenia. Schizophren Res 1992; 7(1):77–84.

217. VanCauter E, Linkowski P, Kerkhofs M, Hubain P, L'Hermite-Baleriaux M, Leclercq R, et al. Circadian and sleep-related endocrine rhythms in schizophrenia. Arch Gen Psychiatry 1991; 48(4):348–356.

218. Kahn RS, Davidson M, Hirschowitz J, Stern RG, Davis BM, Gabriel S, et al. Nocturnal growth hormone secretion in schizophrenic patients and healthy subjects. Psychiatry Res 1992; 41(2):155–161.

219. Robinson S, Rosca P, Durst R, Shai U, Ghinea C, Schmidt U, et al. Serum melatonin levels in schizophrenic and schizoaffective hospitalized patients. Acta Psychiatr Scand 1991; 84(3):221–224.

220. Fanget F, Claustrat B, Dalery J, Brun J, Terra JL, Marie-Cardine M, et al. Melatonin and schizophrenia. Encephale 1989; 15(6):505–510.

221. Rao ML, Gross G, Strebel B, Halaris A, Huber G, Braunig P, et al. Circadian rhythm of tryptophan, serotonin, melatonin, and pituitary hormones in schizophrenia. Biol Psychiatry 1994; 35(3):151–163.

222. Smith JR, Karacan I, Yang M. Ontongeny of delta activity during human sleep. Electroencephalogr Clin Neurophysiol 1977; 43:229.

223. Feinberg I. Schizophrenia: caused by a default in programmed synaptic elimination during adolescence. J Psychiatr Res 1982; 17:319–330.

224. Pearlson GD, Petty RG, Ross CA, Tien AY. Schizophrenia: a disease of heteromodal association cortex? Neuropsychopharmacology 1996; 14(1):1–17.

225. Andreasen NC, Arndt S, Swayze V, Cizadlo T, Flaum M, O'Leary D, et al. Thalamic abnormalities in schizophrenia visualized through magnetic resonance imaging averaging. Science 1994; 266:294–298.

226. Keshavan MS, Reynolds CF, Ganguli R, Brar J, Houck P. Electroencephalographic sleep and cerebral morphology in functional psychoses: a preliminary study with computed tomography. Psychiatry Res 1991; 39(3):293–301.

227. Riemann D. Cholinergic REM induction test: muscarinic supersensitivity underlies polysomnographic findings in both depression and schizophrenia. J Psychiatr Res 1994; 28:195–210.

228. Benson KL, Faull KF, Zarcone VPJ. Evidence for the role of serotonin in the regulation of slow wave sleep in schizophrenia. Sleep 1991; 14:133–139.

229. Van Kammen DP, Widerlov E, Neylan TC, Ekman R, Kelley ME, Mouton A, et al. Delta sleep-inducing-peptide-like immunoreactivity (DSIP-LI) and sleep in schizophrenic volunteers. Sleep 1992; 15(6):519–525.

230. Gur RC, Gur RE. Hypofrontality in schizophrenia: RIP. Lancet 1995; 345(8962): 1383–1384.

231. Keshavan MS, Pettegrew JW, Reynolds CF, Panchalingam KS, Montrose D, Miewald J, et al. Slow wave sleep deficits in schizophrenia: pathophysiological significance. Psychiatry Res 1995; 571:91–100.

232. Steruade M. Brain electrical activity and sensory processing during waking and sleep states. In: Principles and Practice of Sleep Medicine. Philadelphia: WB Sanders, 1994; 105–214.

233. Reynolds CF, Shaw DF, Newton TF, Coble PA, Kupfer DJ. EEG sleep in outpatients with generalized anxiety: a preliminary comparison with depressed outpatients. Psychiatry Res 1983; 8:81–89.

234. Papadimitriou GN, Linkowski P, Kerkhofs M, Kempenaers C Mendlewicz J. Sleep

EEG recordings in generalized anxiety disorder with significant depression. J Affect Disord 1988; 15:113–118.

235. Papadimitriou GN, Kerkhofs M, Kempenaers C, Mendlewicz J. EEG sleep studies in patients with generalized anxiety disorder. Psychiatry Res 1988; 26:183–190.

236. Saletu B, Anderer P, Brandstatter N, Frey R, Grunberger J, Klosch G, et al. Insomnia in generalized anxiety disorder: polysomnographic, psychometric and clinical investigations before, during and after therapy with a long- versus a short-half-life benzodiazepine (quazepam versus triazolam). Neuropsychobiology 1994; 29(2):69–90.

237. Arriaga F, Paiva T. Clinical and EEG sleep changes in primary dysthymia and generalized anxiety: a comparison with normal controls. Neuropsychobiology 1990; 24(3):109–114.

238. Saletu B, Klosch G, Gruber G, Anderer P, Udomratn P, Frey R. First-night-effects on generalized anxiety disorder (GAD)-based insomnia: laboratory versus home sleep recordings. Sleep 1996; 19(9):691–697.

239. Dube S, Kumar N, Ettedgui E, Pohl R, Jones D, Sitaram N. Cholinergic REM induction response: separation of anxiety and depression. Biol Psychiatry 1985; 20: 408–418.

240. Sitaram N, Dube S, Jones D, Pohl R, Gershon S. Acetylcholine and alpha 1-adrenergic sensitivity in the separation of depression and anxiety. Psychopathology 1984; 17(3):24–39.

241. da Roza Davis JM, Sharpley AL, Cowen PJ. Slow wave sleep and 5-HT2 receptor sensitivity in generalised anxiety disorder: a pilot study with ritanserin. Psychopharmacology 1992; 108(3):387–389.

242. Lauer CJ, Krieg JC. Sleep electroencephalographic patterns and cranial computed tomography in anxiety disorders. Compr Psychiatry 1992; 33(3):213–219.

243. Shear MK, Randall J, Monk TH, Ritenour A, Tu X, Frank E, et al. Social rhythm in anxiety disorder patients. Anxiety 1994; 1:90–95.

244. Labbate LA, Pollack MH, Otto MW, Langenauer S, Rosenbaum JF. Sleep panic attacks: an association with childhood anxiety and adult psychopathology. Biol Psychiatry 1994; 36(1):57–60.

245. Stokes PE, Fraser A, Casper R. Unexpected neuroendocrine relationships. Psychopharmacol Bull 1981; 17:72–75.

246. Mellman TA, David D, Kulick-Bell R, Hebding J, Nolan B. Sleep disturbance and its relationship to psychiatric morbidity after Hurricane Andrew. Am J Psychiatry 1995; 152(11):1659–1663.

247. Mellman TA, Uhde TW. Electroencephalographic sleep in panic disorder: a focus on sleep-related panic attacks. Arch Gen Psychiatry 1989; 46:178–184.

248. Hauri PJ, Friedman M, Ravaris CL. Sleep in patients with spontaneous panic attacks. Sleep 1989; 12(4):323–337.

249. Dube S, Jones DA, Bell J, Davies A, Ross E, Sitaram N. Interface of panic and depression: clinical and sleep EEG correlates. Psychiatry Res 1986; 19:119–133.

250. Uhde TW, Boulenger JP, Roy-Byrne PP. Longitudinal course of panic disorder: clinical and biological considerations. Prog Neuropsychopharmacol Biol Psychiatry 1985; 9:39–51.

251. Uhde TW, Tancer ME, Black B, Brown TM. Phenomenology and neurobiology of social phobia: comparison with panic disorder. J Clin Psychiatry 1991; 52:31–40.

252. Lydiard RB, Zealberg J, Laraia MT, Fossey M, Prockow V, Gross J, et al. Electro-encephalography during sleep of patients with panic disorder. J Neuropsychiatry Clin Sci 1989; 1(4):372–376.

253. Lauer CJ, Krieg JC, Garcia-Borreguero D, Ozdaglar A, Holsboer F. Panic disorder and major depression: a comparative electroencephalographic sleep study. Psychiatry Res 1992; 44(1):41–54.

254. Stein MB, Enns MW, Kryger MH. Sleep in nondepressed patients with panic disorder. II. Polysomnographic assessment of sleep architecture and sleep continuity. J Affect Disord 1993; 28(1):1–6.

255. Abelson JL, Curtis GC. Hypothalamic-pituitary-adrenal axis activity in panic disorder. 24-hour secretion of corticotropin and cortisol. Arch Gen Psychiatry 1996; 53(4):323–331.

256. Abelson JL, Curtis GC. Hypothalamic-pituitary-adrenal axis activity in panic disorder: prediction of long-term outcome by pretreatment cortisol levels. Am J Psychiatry 1996; 153(1):69–73.

257. Ross RJ, Ball WA, Sullivan KA, Caroff SN. Sleep disturbance as the hallmark of post-traumatic stress disorder. Am J Psychiatry 1989; 146(6):697–707.

258. Glaubman H, Mulincer M, Povet A, Wasserman O, Birger M. Sleep of chronic post-traumatic patients. J Traum Stress 1990; 3:255–263.

259. Kramer M, Kinney L. Sleep patterns in trauma victims with disturbed dreaming. Psychiatr J Univ Ottawa 1988; 13(1):12–16.

260. Fuller KH, Waters WF, Scott O. An investigation of slow-wave sleep processes in chronic PTSD patients. J Anxiety Disord 1994; 8(3):227–236.

261. Lavie P, Hefez A, Halperin G, Enoch D. Long-term effects of traumatic war-related events on sleep. Am J Psychiatry 1979; 136:175–178.

262. Ross RJ, Ball WA, Dinges DF, Kribbs NB, Morrison AR, Silver SM, et al. Rapid eye movement sleep disturbance in post-traumatic stress disorder. Biol Psychiatry 1994; 35:195–202.

263. Lund HG, Bech P, Eplov L, Jennum P, Wildschiodtz G. An epidemiological study of REM latency and psychiatric disorders. J Affect Disord 1991; 23(3):107–112.

264. Mellman TA, Nolan B, Hebding J, Kulick-Bell R, Dominguez R. A polysomnographic comparison of veterans with combat-related PTSD, depressed men, and non-ill controls. Sleep 1997; 20(1):46–51.

265. Reist C, Kauffmann CD, Chicz-Demet A, Chen CC, Demet EM. REM latency, dexamethasone suppression test, and thyroid releasing hormone stimulation test in post-traumatic stress disorder. Prog Neuropsychopharmacol Biol Psychiatry 1995; 19(3):433–443.

266. Southwick SM, Bremner D, Krystal JH, Charney DS. Psychobiologic research in post-traumatic stress disorder. Psychiatr Clin North Am 1994; 17(2):251–264.

267. Woodward SH, Friedman MJ, Bliwise DL. Sleep and depression in combat-related PTSD inpatients. Biol Psychiatry 1996; 39(3):182–192.

268. Dagan Y, Lavie P, Bleich A. Elevated awakening thresholds in sleep stage 3–4 in war-related post-traumatic stress disorder. Biol Psychiatry 1991; 30:618–622.

269. Schoen L, Kramer M, Kinney L. Auditory thresholds in the dream disturbed. Sleep Res 1984; 13:102.

270. Ross RJ, Ball WA, Dinges DF, Kribbs NB, Morrison AR, Silver SM, et al. Motor

dysfunction during sleep in posttraumatic stress disorder. Sleep 1994; 17(8): 723–732.
271. Brown TM, Boudewyns PA. Periodic limb movements of sleep in combat veterans with post-traumatic stress disorder. J Traum Stress 1996; 9(1):129–136.
272. Mellman TA, Kulick-Bell R, Ashlock LE, Nolan B. Sleep events among veterans with combat-related post-traumatic stress disorder. Am J Psychiatry 1995; 152: 110–115.
273. Lavie P, Hertz G. Increased sleep motility and respiration rates in combat neurotic patients. Biol Psychiatry 1979; 14:983–987.
274. Yehuda R, Southwick S, Giller EL, Ma X, Mason JW. Urinary catecholamine excretion and severity of PTSD symptoms in Vietnam combat veterans. J Nerv Ment Dis 1992; 180(5):321–325.
275. Mellman TA, Kumar A, Kulick-Bell R, Kumar M, Nolan B. Nocturnal/daytime urine noradrenergic measures and sleep in combat-related PTSD. Biol Psychiatry 1995; 38(3):174–179.
276. Yehuda R, Teicher MH, Trestman RL, Levengood RA, Siever LJ. Cortisol regulation in posttraumatic stress disorder and major depression: a chronobiological analysis. Biol Psychiatry 1996; 40(2):79–88.

AUTHOR INDEX

Italic numbers give the page on which the complete reference is listed.

SUBJECT INDEX

A

Adenosine, 35, 118, 119, 291,292
Advanced sleep phase syndrome
  (ASPS), 477–480
  age-related changes, 478
  pathophysiology, 478
  phase advance in various
    circadian rhythms, 478
  treatment, 479
  use light, 480
African trypanosomiasis (sleeping
  sickness), 414
Age-related alterations in rhythms,
  489–498
  advanced phase, 495
  Alzheimer's disease (AD), 490
  behavior, mood, and performance,
    499–501
  c-*fos*, 492
  changes in SCN, 489
  constant routines, 497–498
  entrained conditions, 493–495
  free-running conditions, 495
  glucose utilization, 491

[Age-related alterations in rhythms]
  molecular characteristics, 492
  owl-lark questionnaire, 493
  sex difference, 491
  transplantation studies, 492
Aging, 487–514
  GH and IGF-I secretion, 410
  plasma cortisol, 403
  SW sleep, 410
α-calcium calmodulin kinase II, 348
Altricial species, 20, 25, 50
Alzheimer's disease (AD), 562–567
  circadian rhythm disturbances,
    564
  cortisol rhythms, 510
  melatonin, 510, 565
  phase delays, 510
  sleep disturbances, 563
  temperature rhythms, 510, 565
Angelman syndrome, 353
Antisense oligodeoxynucleotide,
  321
Anxiety disorders, 622-625
  circadian rhythms, 622
  sleep, 622

Printed in the United States
201971BV00001B/1/A